The
CHEMICAL ANALYSIS
OF
FOODS AND FOOD PRODUCTS

By

MORRIS B. JACOBS, Ph.D.

Associate Professor of Occupational Medicine,
School of Public Health and Administrative Medicine,
Columbia University

Facsimile of

THIRD EDITION

CBS PUBLISHERS & DISTRIBUTORS PVT. LTD.
New Delhi • Bengaluru • Chennai • Kochi • Kolkata • Mumbai • Pune

ISBN: 81-239-0643-9

First Indian Edition: 1999

Published by:
Satish Kumar Jain for CBS Publishers & Distributors Pvt. Ltd.,
4819/XI Prahlad Street, 24 Ansari Road, Daryaganj, New Delhi - 110002
delhi@cbspd.com, cbspubs@airtelmail.in • www.cbspd.com
Ph.: 23289259, 23266861, 23266867 • Fax: 011-23243014

Corporate Office: 204 FIE, Industrial Area, Patparganj, Delhi - 110 092
Ph: 49344934 • Fax: 011-49344935
E-mail: publishing@cbspd.com • publicity@cbspd.com

Branches:
• *Bengaluru:* 2975, 17th Cross, K.R. Road, Bansankari 2nd Stage,
  Bengaluru - 70 • Ph: +91-80-26771678/79 • Fax: +91-80-26771680
  E-mail: cbsbng@gmail.com, bangalore@cbspd.com
• *Chennai:* No. 7, Subbaraya Street, Shenoy Nagar, Chennai - 600030
  Ph: +91-44-26681266, 26680620 • Fax: +91-44-42032115
  E-mail: chennai@cbspd.com
• *Kochi:* Ashana House, 39/1904, A.M. Thomas Road, Valanjambalam,
  Ernakulum, Kochi • Ph: +91-484-4059061-65
  Fax: +91-484-4059065 • E-mail: cochin@cbspd.com
• *Kolkata:* 6-B, Ground Floor, Rameshwar Shaw Road, Kolkata - 700014
  Ph: +91-33-22891126/7/8 • E-mail: kolkata@cbspd.com
• *Mumbai:* 83-C, Dr. E. Moses Road, Worli, Mumbai - 400018
  Ph: +91-9833017933, 022-24902340/41 • E-mail: mumbai@cbspd.com
• *Pune:* Bhuruk Prestige, Sr. No. 52/12/2+1+3/2,
  Narhe, Haveli (Near Katraj-Dehu Road Bypass), Pune - 411041 .
  Ph: +91-20-64704058/59, 32342277 • E-mail: pune@cbspd.com

Representatives:

• Hyderabad: 0-9885175004   • Nagpur: 0-9021734563
• Patna: 0-9334159340   • Vijayawada: 0-9000660880

Printed at Neekunj Print Process, Haryana, India

DEDICATED

to

my wife, MARGARET
and my children, ROBERTA and CHARLES, BERENICE and ROBERT,
and ROGER
in the hope that it will insure not only them but also all sons
and daughters a surer, better, and safer food supply.

PREFACE TO THIRD EDITION

The analytical methods of the food chemist are applied: in the development and enforcement of standards of identity, purity, or value; in problems of decomposition under either normal or abnormal storage conditions; in studies designed to improve or control the quality of natural or processed foods; in the determination of the nutritive value of foods for scientific, dietary, or labeling purposes; in the technical control of foods being processed; in the technical control or supervision of raw materials being purchased; or in problems of a toxicological or forensic nature. The scope of this book has been enlarged in order to serve these purposes.

The fundamental matter of the first and second editions has been retained but several new ones on radiochemical determinations, pesticide residues, and artificial sweetening agents have been added in order to obtain the objectives noted above.

The arrangement of the first edition has been largely retained because this arrangement proved to be successful when the book was employed as a class text. Although it has been enlarged, by a judicious selection of the methods detailed, students can be taught procedures which are actually in use in industry for control and in governmental agencies for regulatory supervision.

I wish to thank Mr. Charles Gellis, my son-in-law, for his assistance in providing illustrations for this edition.

Acknowledgment is made for the abstraction of methods, procedures, tables, etc., of other well-known texts. These are noted, at times, throughout the actual text and at other times in the selected references at the end of each chapter. If any reference is omitted, such omission is entirely unintentional. The author will be only too happy to correct such omissions if brought to his attention.

MORRIS B. JACOBS

Bayside, New York
August 1958

PREFACE TO FIRST EDITION

This book was written with a number of objectives. First and foremost was the desire of the author to give systematic coverage to the salient facts of the chemical analysis of foods and food products and to include certain of the newer aspects of food analysis such as the chemical assay of vitamins, the detection of improper pasteurization of milk, the homogenization of milk, the detection of gums and methods for the detection of newer types of sophistication of foods.

The author also has the desire to present a number of topics such as gums, jams and jellies, milk products in other foods, soy bean flour in meats, pumped smoked meats and other timely topics which properly belong in the modern literature of the subject and on which information has heretofore been difficult to obtain.

The author has attempted to present short practical methods which are usable and efficient and are, of course, of vast importance in routine analysis and in general control work. Throughout the book stress is placed on the fact that in all probability the analysis of a food product should fall within known normal limits and that the tendency to adulterate is closely tied to economic factors.

The book is designed for use as an educational text and as a manual for manufacturers for control work. It is hoped that it will prove useful in regulatory laboratories, both governmental and commercial and as a reference particularly in those subjects treated herein which are not treated in other texts in the subject.

The author wishes to thank Mr. A. A. Singer for his persistent encouragement, Mr. H. J. Kothe for his aid in the chapter on vitamins and for reviewing the manuscript, Mr. F. E. Nussberger for aid in experimentation, Mr. E. Buxbaum for photography and Dr. C. V. King, Dr. L. Sattler, and Mr. N. I. Goldstone for reviewing various portions of the text.

The author also wishes to thank for their cooperation in extending the loan of electrotypes for illustrative purposes:

Bausch and Lomb	Rochester, New York
Eimer and Amend	New York, New York
Leeds and Northrup	Philadelphia, Pa.
Arthur H. Thomas	Philadelphia, Pa.
Central Scientific	Chicago, Illinois
Fisher Scientific	Pittsburgh, Pa.
Wilkens-Anderson	Chicago, Illinois
Mojonnier Bros.	Chicago, Illinois
Kimball Glass	Vineland, New Jersey
Spencer Lens	Buffalo, New York

Acknowledgment is made for the abstraction of methods, procedures, tables, etc., of other well known texts. These are noted at times throughout the actual text and at other times in the selected references at the end of each chapter.

MORRIS B. JACOBS

Brooklyn, New York
March 1938

CONTENTS

CONTENTS

CONTENTS

CONTENTS

CONTENTS

DEFINITIONS OF TERMS AND EXPLANATORY NOTES

1. The term *water* used in the methods means distilled water.

2. The terms *alcohol* and *ether* refer respectively to 95 per cent ethyl alcohol and ethyl ether. The terms *high-grade gasoline* and *petroleum ether* are often used interchangeably.

3. The foods and food products discussed in the text have, if defined, the definitions and standards given in United States Federal Security Agency, Food and Drug Administration, Service and Regulatory Announcements, Food, Drug, and Cosmetic No. 2, Rev. 1, 1949, as amended and extended.

4. The following reagents, unless otherwise specified or qualified in the text, have the approximate strength stated and conform in purity with the requirements of the United States Pharmacopoeia.

Sulfuric acid .... specific gravity 1.84

Hydrochloric acid .... specific gravity 1.184

Nitric acid .... specific gravity 1.42

Glacial acetic acid .... specific gravity 1.048 (25° C.)

Hydrobromic acid .... specific gravity 1.38

Phosphoric acid .... 85 per cent strength by weight

Ammonium hydroxide ...... specific gravity 0.90

5. All other reagents and test solutions, unless otherwise described in the text, conform to the specifications of the American Chemical Society or of the United States Pharmacopoeia. When the anhydrous salt is intended to be used, it is so stated; otherwise the salt referred to is the crystallized product.

6. In the expressions (1:2), (5:4), etc., used in connection with the name of a reagent, the first numeral indicates the volume of the reagent used, and the second numeral indicates the volume of water. For example, hydrochloric acid (1:2) means a reagent prepared by mixing one volume of hydrochloric acid with two volumes of water. When one of the reagents is a solid the expression means parts by weight, the first numeral representing the solid reagent and the second numeral the water.

7. In making up solutions of definite percentage it is understood that x grams of substance are dissolved in water and made up to 100

ml. Although not theoretically correct, this procedure will not result in any appreciable error in any of the methods given in this book.

8. All calculations are based on the table of international atomic weights, given in the frontispiece.

9. The following abbreviations are used and have the indicated meaning:

g.	gram
ml.	milliliter
C.	centigrade degrees
F.	Fahrenheit degrees
p.p.m.	parts per million
N	normal, with reference to solutions
mm.	millimeters
mg.	milligrams
e.m.f.	electric motive force
r.p.m.	revolutions per minute
lb.	pound
sp. gr.	specific gravity
A.O.A.C.	Methods of Analysis of the Association of Official Agricultural Chemists.

The abbreviations of periodicals, bulletins, circulars, etc., referred to in the footnotes follow the system of *Chemical Abstracts.*

10. As a general rule, the author has given the preparation of the reagent immediately following the naming of the reagent in the method rather than giving a numbered or lettered list, in the beginning of the method. The author has found that in reading methods it is a great disadvantage to read, "and then add 5 ml. of reagent A ...." Very often the analyst must look back to the beginning of the method to find out what reagent A is. Only in the case of very complex reagents is the method of preparation detailed in advance.

The proper way to perform a procedure is to read the method at least twice to completion. On the third reading the reagents are prepared. Then if the determination has never been performed by the analyst before, known samples should be used and only after some experience with known samples has been obtained should unknown samples be analyzed.

CHAPTER I

GENERAL METHODS

GENERAL INTRODUCTION

THE student learns, early in his educational career, that chemistry is a mathematical science, and that there is no sharp line of demarcation separating the major branches, namely, inorganic, physical, analytical, or organic. Thus the fundamental laws are studied and restudied in each branch. The constitution of the atom is treated fully in both inorganic and physical chemistry texts. Identity tests for elements form as important a part of inorganic chemistry as they do in analytical chemistry. Organometallic compounds are treated as exhaustively in some organic texts as in some more advanced books on inorganic chemistry. In a similar manner, there is no sharp differentiation of chemical analysis of food and food products from chemical analysis of other materials. The object in both cases is to recognize the elements composing or compounds comprising the material and to ascertain the percentage composition. Necessarily, methods peculiar to the field of food analysis have been perfected in order to expedite results.

The chemical analysis of food enables us to know the composition of such materials and, with the aid of nutritional and biochemical knowledge, to know what we should eat and what we should avoid eating. Accepting the above as a base, food analysis had its beginnings in times immemorial, for we might say, by organoleptic analysis—that is, analysis by use of the senses of smell, taste, sight, and touch—man learned that certain materials were not fit for food, either because they made one ill or had little food value. In later times, the diétary rules of Christians, Hebrews, Mohammedans, and others codified the trial-and-error analyses of their predecessors. In more recent times, the fact that man ate relatively enormous quantities of food over a period of time without appreciable gain in weight was clearly recognized and became an object of study.

Possibly the earliest analytical work recorded in a related field was that of Andreas Libavius (1546-1616), the eminent iatrochemist, who published about 1606 his work *De Judicio Aquarum Mineralium* containing methods for the analysis of mineral waters. Libavius was also one of the first to use the term, alcohol, instead of spirit

1

of wine and published a very important work, *Opera medica-chymica*, which contained data on the composition of wine. Francesco Redi (1626-1697) was one of the earliest chemists to write about the adulteration of foods, about 1660. In 1673 Leeuwenhoek (1632-1723) described graphically his original investigations with the aid of his microscopes on milk, vinegar, coffee, and tea. Louis Lemery's *Traite des Aliments, ca.* 1702, was the classical work on foods of the 18th century. Andreas Marggraf (1709-1782) as early as 1747 demonstrated the sugar was present in beet juice, although it was not until after the Napoleonic era that methods for its extraction on a commercial scale were available.

Frederick Accum (1769-1838), a British chemist, published a book on the adulteration of food and culinary poisons in 1820 under the interesting title, *Death in the Pot*.

Lavoisier (1743-1794) was one of the first to recognize that life is a chemical function and that food was the fuel of the body. With the demonstration by Wöhler of the conversion of ammonium cyanate, NH_4CNO, an inorganic substance, to urea, NH_2CONH_2, an organic substance, the development by Dumas (1800-1884) in 1830 of a method for the exact determination of nitrogen in organic compounds, and the modern methods of organic analysis devised by Liebig, as a consequence of the growth of organic chemistry, the tools needed for the complete development of the science were at hand.

Among the earliest quantitative analyses of food materials recorded were those of Pearson,[1] in England, in 1795. In these, Pearson estimated the proportions of water, starch, fibrous matter, extractive matters and ash in kidney potatoes. He recognized the presence of fat, acids, and sugar. The earliest European analyses made comparable with those of recent times are perhaps those of milk by Peligot[2] in 1836, those of feeding stuffs by Boussingault[3] in 1836 and 1838, and those of milk by Boussingault and Le Bel[4] in 1839. The earliest American analyses were made by Shephard[5] in 1845 on the ash of rice and those of Salisbury[6] on maize in 1848.

After these earliest investigations the stress was laid on the proportion of carbon and nitrogen in various food materials. Then Liebig and his followers, Playfair, Boeckman and others, during the period from 1840 to 1865, made the first systematic investigations

[1] Atwater and Bryant, *U. S. Dept. Agr., Bull.* 28, rev. (1906).
[2] Peligot, *Ann. chim. phys.* [2] 62, 432 (1836).
[3] Boussingault, *Ann. chim. phys.* [2] 67, 225 (1838).
[4] Boussingault and Le Bel, *Ann. chim. phys.* [2] 71, 65 (1839).
[5] Shephard, *Am. Quart. J. Agr. Sci.* 1, 122 (1845).
[6] Salisbury, *Trans. New York State Agr. Soc.* p. 678 (1848).

of foods and feeding stuffs by methods more or less similar to those of today. A great advance was made when Henneberg[7] and his associates elaborated the methods for the proximate analysis of foods. A "proximate" analysis is distinguished from an "ultimate" analysis in that it is not a determination for a particular element or compound but is rather an estimation of a certain type of component as "volatile matter," "moisture," "fat," "carbohydrate," "ash," "nitrogenous matter," etc. Proximate analyses are more easily made and generally give more useful information.

From these earlier analyses and nutrition investigations grew the belief that a proper diet must consist of a correct amount of protein, fat, carbohydrate, water, and ash. These were the substances whose percentage was required for the necessary nutritional interpretation; hence, these were the substances which were determined by analysis. Methods for the analysis of these components form a preponderant portion of food analytical literature.

In still more recent times, the realization that the diet problem was not solved merely by adequate utilization of a proper portion of protein, fat, and carbohydrate, the three great foodstuffs, lead to the successful search for other substances necessary for the maintenance of health and life. We now know that vitamins, vitagens, minute amounts of certain metals, iodine, and other substances and elements are necessary. With this knowledge has grown the development of methods and procedures for the determination and estimation of minute quantities of these substances, so that the science of food analysis is as replete with inorganic, as with organic, qualitative and quantitative methods.

The growth of food analysis as a science is based on five main factors:

1. The desire to obtain nutritional and biochemical knowledge, that is, the knowledge necessary to provide for the well-being of the living organism from the food point of view.

2. The standardization of production and manufacture of food products by means of control analyses as a commercial development.

3. The use of food assays as the means of regulating the purchase of foods on commercial, industrial, and governmental levels, for food purchased by large commercial and industrial establishments and by governmental agencies must conform to certain specifications generally set by the purchaser.

4. The need for governmental regulation in order to protect its citizens from deleterious foods and from being defrauded by adulterated and sophisticated foods.

---

[7] Atwater and Bryant, *U. S. Dept. Agr., Bull.* 28, rev. (1906).

5. The use by the government of analysis as a base for revenue, by taxation of foods and beverages.

The latter two factors have given a governmental and therefore regulatory cast to the science of food analysis from earliest times in contradistinction to other analytical sciences. In the United States, the Federal Government passed the Wiley food act on June 30, 1906. The function of this law was to control the interstate commerce of adulterated, deleterious, and misbranded foods and drugs. It was termed an act preventing the manufacture, sale, or transportation of adulterated or misbranded or poisonous or deleterious foods, drugs, medicines, and liquors.

Standards, both mandatory and discretionary, for food products, have been set by governmental agencies for the protection of food manufacturers, dealers, and consumers. However, minimum standards have one drawback. Some manufacturers and producers have a tendency to keep their products just at, or slightly above, the minimum standard, thus actually causing a lowering of the average product. For example, vinegar produced under normal fermentation conditions has an acidity varying between 4.5 to 5.5 per cent. A minimum was set by the Department of Agriculture of the United States of 4 per cent. This minimum was an outside lower limit. The result has been that most of the vinegar sold is now diluted down to the quoted minimum acidity.

Food analysis is then, in the main, a branch of analytical chemistry. It has need of both qualitative and quantitative analysis. Its interest lies in determining not only what, but also how much, of a component may be present in a food. As Oser [8] has pointed out the analytical methods of the food chemist are applied in the development and enforcement of standards of identity, purity, or value; in problems of decomposition under either normal or abnormal storage conditions; in studies designed to improve or control the quality of natural or processed foods; or in the determination of the nutritive value of foods for scientific, dietary, or labeling purposes.

The degree of accuracy desired in food analysis is conditioned by a number of factors not ordinarily considered in other analytical fields. Thus analytical results for nutritional and biochemical investigational work need be more accurate than those of control work, where the primary purpose is to determine whether or not the food product falls within required limits. A further factor concerns the legal aspect of much food analytical work. Here, too, great care must be taken.

---

[8] Oser, *Anal. Chem.* 21, 216 (1949).

Nevertheless, in the measurement of these analytical quantities, the beginner and often the regular analyst is prone to attempt being too accurate. Every measurement entails some error, and to work far outside the limits of that error is to involve useless labor with no gain in accuracy. An error, in the scientific sense, is defined as a deviation from the truth, and the accuracy of a determination depends on the accuracy of the data on which it is based. Errors should, however, be distinguished from mistakes or blunders. Thus if, in reading a burette, the correct reading is 17.68 ml. and this is read and recorded as 17.66 ml., the difference of 0.02 ml. is considered an error in reading. If, however, the burette reading was recorded as 16.66 ml., then a mistake will have been made.

The accuracy of the final result is quantitatively governed by the accuracy of the least accurate measurement. Thus if we are performing an analysis in which an instrument is used that can be read to only 1 part per thousand, there is waste of time and effort in weighing the original material to be analyzed with an accuracy much greater than 1 part per thousand.

Let us assume that we are analyzing butter for moisture. We have the use of an analytical balance that is accurate to 0.1 mg.

```
The dish weighs ..........................15.6028 g.
Butter weighs ............................ 3.0006
Combined weight .........................18.6034
Weight after heating ....................18.1553
Moisture ................................. 0.4481
Per cent moisture .......................14.94%
```

Suppose instead of weighing to 0.1 mg., we weigh to the nearest milligram.

```
The dish weighs ..........................15.603 g.
Butter weighs ............................ 3.001
Combined weight .........................18.604
Weight after heating ....................18.155
Moisture ................................. 0.449
Per cent moisture .......................14.9%
```

It is clear that the accuracy desired is obtained by weighing to the nearest milligram for whether the butter contains 14.94 per cent or 14.9 per cent is not of great moment in food analytical chemistry. Indeed, it has been demonstrated that the error involved in the sampling of the butter is greater.

Of course, if we had to make an analysis that required accuracy to 1 part in a thousand, we would have to weigh to at least 0.1 mg.

for less than a 1-g. sample, in order to obtain that precision. On the other hand, if we were weighing a substance that lost or gained moisture rapidly, the additional time needed to give an exact weight would be greatly overbalanced by the error involved in the respective gain or loss of moisture.

If, instead of using arithmetical means for calculating, we were to use a 10-in. slide rule, our accuracy would be reduced to 1 part in 500 or 1 part in 800, for we could not read the rule closer than that. Very often accuracy with 1 per cent, 10 per cent, and even 100 per cent error is not at all flagrant. Thus, whether a foodstuff contains 0.1 p.p.m. of lead or 0.05 p.p.m. is not of material difference in most cases. If 0.05 p.p.m. is the correct result, then 0.1 p.p.m. is 100 per cent in error.

We should be as accurate as possible but not so much so that we become impractical. The point involved and to be stressed is that accuracy is indissolubly linked with practicality in food analysis.

Throughout the text more than one procedure is often detailed for any particular determination. This is done intentionally, for the analyst is wont to find one method preferable to another. Furthermore, in the opinion of the author, a check result obtained by different methods is more indicative of the true estimation than a check result obtained by the same method.

SAMPLING

A most important matter to be considered by the food analyst, although not directly his province, is the proper sampling of the food or food product to be analyzed. There are probably as many incorrect determinations made because a sample was improperly taken as because of the combined errors of preparation of sample, manipulation, calculation of result, etc. The failure to obtain a proper sample makes a subsequent analysis worthless.

Factors to Consider in the Field.—To take a sample truly representative of a batch is often difficult. As a general rule, the material to be sampled is mixed thoroughly and a sample is taken after mixing or a composite sample is obtained by taking specimens according to some definite rule from a large stock of material. These methods are detailed below.

In order to be able to identify a sample, certain data must be obtained in its collection. The *label* should be taken or as complete a copy as possible should be made. The *amount* of the original material should be noted as well as the *quantity* of the sample itself. The *lot number,* the *date,* the *owner,* the *place* of sampling are all pertinent data.

Certain precautions must be observed during and after the collection of sample. It is necessary to note the physical conditions which might influence the sample such as heat and light. Sometimes cold weather may affect the sample. The sample should be taken in such a manner that it will not (1) undergo bacterial decomposition, (2) spoil because of autolytic action of enzymes or become rancid as a result of light or heat, and (3) become contaminated. It is also necessary that it should be placed in a container so that the contents cannot be tampered with.

By keeping certain samples cold, decomposition and increase in bacterial population will generally be prevented. This may be done by refrigeration. However, it is unwise to use solid carbon dioxide, except with samples like ice cream, because the dry ice may freeze the product so that its texture is damaged.

Drawing a proper sample from small containers, packages or batches, such as are customarily sold at retail, is relatively easy for, if small enough, the entire batch or package may be taken as the sample. Since these are usually manufactured or processed foodstuffs, they may be taken as representative.

Moderately large batches present greater difficulties. If the product is a liquid or a powder, it should be mixed thoroughly by stirring or by pouring from container to container, if possible, and then an adequate portion withdrawn for analysis. If the product is a solid, such as butter or margarine, the withdrawal of sections by means of a trier is probably the best procedure. If the product is frozen, such as frozen eggs, sections may be withdrawn by use of a *drill*.

Large batches, lots or loads of food, such as grain, nuts, fruits, etc., are very difficult to sample properly. It is impracticable, for example, to mix carload lots or shiploads. Hence, when large lots are being sampled, portions of the material are taken from various sections. These portions are mixed thoroughly and from the mixture a suitable sample is drawn. In the case of foods stored in boxes, crates, bags or large cans, portions taken from a representative number of containers are mixed and from the composite a sample is withdrawn for analysis. At times, especially if it is difficult to obtain a representative composite, it is preferable to obtain a number of samples from a large lot rather than to attempt to get one composite representative sample.

Sample Containers.—The container in which the sample is placed is an important item. When sampling a material for *filth*, it is clear that the sample container must be clean both physically and bacteriologically. This can only be the case with specially prepared containers. When sampling for *moisture*, dry screw cap glass jars preferably should be used. If the analysis is to be performed shortly after receipt of the sample waxed cartons are sometimes used. For determinations of *fat* and *foreign fat,* glass jars are preferable. The original container is preferable for *canned goods,* consequently cans from the same lot number should be used as duplicates. In sampling *ice cream* for total solids per unit volume, it is necessary to take the original container with the ice cream intact. This may be done by use of solid carbon dioxide.

All food materials sampled and not in original sealed containers that are likely to lose moisture or otherwise undergo change should be placed in hermetically sealed containers, if practicable, by the person taking the sample. The sample, so bottled or jarred, should be sealed as a precaution against tampering before receipt by the analyst.

Sampling Instruments.—Various implements have been devised to assist in proper sampling. The more important of these are the "thief," sampling tube, and trier. The *thief,* which is used for sampling liquids, is a long tube, about 2 to 3 ft. in length, which has holes in a cap at the bottom end. The tube is inserted in the product to be sampled and, when the liquid has risen to the same level as the surrounding liquid, the tube is closed by pushing the cap against the bottom of the container. The thief is then withdrawn and the sample transferred to the sample bottle. An alternative type of thief is the oil thief, Fig. 1, used for sampling oil in drums. It is a copper tube, about 3 ft. long and $1\frac{1}{4}$ in. in diameter, with cone shaped ends having an opening, $\frac{3}{8}$ in. in diameter. Three legs are placed on the lower end to hold the opening $\frac{1}{8}$ in. from the bottom of the drum. Two rings, soldered to opposite sides, at the upper end, permit holding the thief with two fingers leaving the thumb free to close the upper opening and thus withdraw the sample.

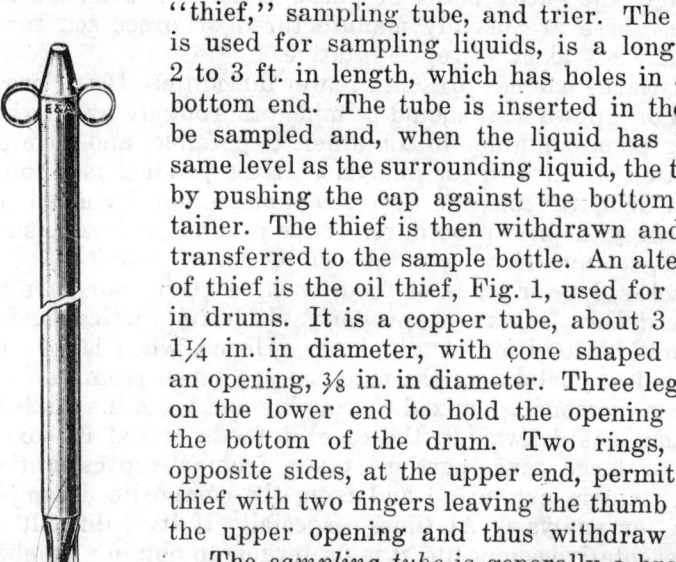

Fig. 1. Oil Thief

(Courtesy of Eimer & Amend)

The *sampling tube* is generally a brass tube 2 to 3 ft. long and $\frac{1}{2}$ in. to 1 in. wide, with a conical sharp tip at one end, and a handle at the other. A slot extends the length of the tube almost from the tip to the handle. The tube is used mainly for sampling powders, such as milk powder, and coarser materials such as grains. The tube is introduced into the container, which may be a bag, with the slot on the under side. It is then turned so that the tube is filled and a core of the material can be withdrawn. Another type, Fig. 2,

FIG. 2. Sampling Tube (Courtesy of Eimer & Amend)

for securing samples representative of different levels in a container, consists of two brass telescopic tubes with registering slots opened or closed by rotation of the inner tube, the outer tube being provided with a sharp point to facilitate penetration into the sample to the full depth of the tube.

A *trier* is in reality a very long gouge. It is about 2 to 3 ft. long and ¾ to 1 in. wide. The tip and

FIG. 3. Sampling Trier

sides are sharpened so that after insertion into the food material to be sampled, say butter or margarine, rotation of the trier will cut a core of foodstuff that can be removed and transferred to the sample jar or bottle. A special type of corkscrew-type of trier, made of ⅜-in. steel, bent to form an open cylinder about 4 in. in diameter, with the pitch of the twist being 2 in. and the screw portion about 34 in. long, has been designed for the sampling of cottonseed. It can withdraw about 5 lbs. of sample at one stroke.

To use the trier or sampling tube properly, the instrument should be inserted practically its full length from a point near a top edge or corner through the center to a point diagonally opposite the point of entry. Usually 2 more triersful are taken from points equidistant from the first.

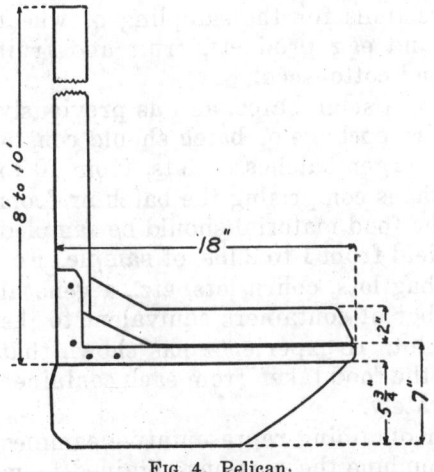

FIG. 4. Pelican·

(From Triebold, *Quantitative Analysis.*
Van Nostrand)

The *drill* or auger, as mentioned above, are the samplers of choice when sampling hard solid materials such as frozen whole eggs, egg yolks, or egg whites and the like. The shavings are placed as rapidly as possible into sample containers. It is also common practice for the inspector or other person performing the sampling to smell at the hole produced and thus note any decomposition. This type of organoleptic analysis requires training.

The *pelican* is another sampling instrument employed for obtaining samples from a continuous flow of grain or analogous particulate material. It consists of a handle from 8 to 10 ft. long to which a scoop about 18 in. wide with an overall depth of 9 in. is attached. The pelican is commonly used for securing samples of grain being transferred to holds of ships and for similar transfers. This is done by pushing the pelican, with the aid of its long handle, through the stream of falling material at known intervals. The samples withdrawn are then made into a composite from which a laboratory sample is taken as described below.

In addition to the sampling instruments described above, it should be noted that scoops, shovels, dippers, and other devices are also often used for sampling.

Procedure of Sampling.—The number of cases, cans, tubs, bags, boxes, and other containers of a lot to sample to obtain a representative sample is not a simple matter. For some foodstuffs the proper procedure has been studied, for others no set rule has been established. The A.O.A.C.[9] details directions for the sampling of wheat flour, butter, milk, cheese, eggs and egg products, fruit and fruit products, grain and stock feeds, and cottonseed.

In the case of single packages or a small batch, as was previously stated, if it be convenient, the entire package or batch should constitute the sample. For moderately larger batches or lots, from 10 to 20 per cent of the number of packages comprising the batch or from 5 to 10 per cent of the weight of the food material should be sampled and sufficient should be taken to yield from 1 to 2 lbs. of sample. For very large lots such as case lots, bag lots, churn lots, etc., a general rule to follow is to sample a number of containers equivalent to the square root of the number in the lot. If experience has shown that no composite sample can be made, the food taken from each container sampled should be analyzed separately.

A method in common usage for obtaining representative samples is the procedure of quartering. Combine the portions obtained from various sections of the lot and, after mixing as thoroughly as possible by rolling in a sheet or blanket, if the sample is large, or paper, if the sample is of moderate size, form the material into a cone. Flatten the cone into a circular shape and divide into quarters. Take two opposite quarters, that is, quadrants 1 and 3, and repeat the above process. However, after dividing into quarters this time, the

⁹ *Methods of Analysis*, A.O.A.C., Washington, 1945.

opposite quarters to those used before, namely, quadrants 2 and 4, are taken. This process is continued until a sample small enough for submission for analysis is secured. If permissible, the material is ground and then reground to a finer mesh before each quartering. From 1 to 5 lbs., depending on the amount of material comprising the lot and the amount needed for analysis, should constitute the size of the sample.

The foregoing are minimum sampling requirements and, in case of doubt, the food material should be resampled, especially if analysis shows that commercial agreements or sanitary regulations are being violated.

There are further legal factors regarding the question of proper sampling. In some communities the regulatory codes governing foods require that the food material sold to the purchaser be within the standards set by the code. In these instances the purchased article constitutes the representative sample. Another factor concerns the matter of leaving a portion of the sample taken with the owner of the foodstuff for check analysis by his own chemist. This is a fair precaution, not only for regulatory agencies but also in commercial relationships. If the material being sampled is small canned or bottled goods, other cans or bottles of the same lot number are usually sufficient for the duplicate sample.

NET CONTENTS

In general, we are interested not only in the food and the food product itself, but also in its container and label. This interest arises because some packaged articles are either short-weight or short-volume and often are mislabeled and misbranded. Consequently, the food analyst almost invariably determines the net contents, the net weight, and net volume. This may be done in a number of ways. If a large number of empty and filled containers are available, weigh the empty containers and determine the average weight. Repeat this process with the filled containers and obtain the average weight of a filled container; then the difference between the weights of the average filled container and the average empty container equals the average net weight of the contents. To determine the volume of the container, fill the empty containers with water at a definite temperature and weigh. The average weight of the water divided by the weight of 1 ml. of water at that temperature is equal to the volume of the container. It is assumed that all the weights are in grams. If the material is a liquid, the net volume is determined by dividing the net weight in grams by the specific gravity of the material.

The general procedure for a container of glass or tinned or other material that may be washed and dried is the following. Weigh the container and the contents as received. Remove the contents, wash, and dry the container. Weigh. The difference in the weights is the net weight of the contents. Fill the container with water at a defi-ι.ite temperature and weigh. The weight of the water divided by the weight of 1 ml. of water at that temperature is equivalent to the volume of the container. If the contents of the container is a liquid, the net weight divided by the specific gravity yields the net volume.

If the food product is a paper packaged frozen or solid article, such as ice cream or cheese, weigh the contents in the frozen or solid state. Measure the dimensions of the wrapper with an ordinary or micrometer rule. Remove the wrapper. Clean, dry, and weigh. The gross weight less the weight of the wrapper equals the net weight of the contents. In the case of a frozen product, the difference in the temperature of the material being weighed and the surroundings introduces an error. The net weight divided by the volume calculated from the measured dimensions of the container yields the approximate density of the material. This method can be used for any size or shape container such as a cone or frustrum of a pyramid. Thus if the package has the shape of a frustrum of a pyramid

$$V = \frac{h}{3}\,[a + a_1 + \sqrt{aa_1}]$$

where

V = volume
h = height
a = area of top
a_1 = area of bottom.

Similar formulas may be obtained for other shapes in many reference texts.

In calculating net volume from net weight, weights being in avoirdupois ounces and volume in fluid ounces, the following formula should be used.

$$V = \frac{W}{d_{20}^{20}} \times 0.961$$

where

V = net volume in fluid ounces at 20° C.
W = net weight in avoirdupois ounces
d_{20}^{20} = specific gravity of the liquid at 20° C. referred to water at 20° C.

General Method for Water Capacity of Containers.[10]—(1) In the case of a container with lid attached by double seam, cut out the lid without removing or altering the height of the double seam.

(2) Wash, dry, and weigh the empty container.

(3) Fill the container with distilled water at 68° F. to $\frac{3}{16}$-in. vertical distance below the top level of the container, and weigh the filled container.

(4) Subtract the weight found in (2) from the weight found in (3). The difference is considered to be the weight of water required to fill the container.

In the case of a container with lid attached otherwise than by double seam, remove the lid and proceed as directed in subparagraphs(2)-(4), except that under (3) fill the container to the level of the top.

General Method for Fill of Containers.[10]—(1) In the case of a container with lid attached by double seam, cut out the lid without removing or altering the height of the double seam.

(2) Measure the vertical distance from the top level of the container to the top level of the food.

(3) Remove the food from the container; wash, dry, and weigh the container.

(4) Fill the container with water to $\frac{3}{16}$-in. vertical distance below the top level of the container. Record the temperature of the water, weigh the filled container, and determine the weight of the water by subtracting the weight of the container found in (3).

(5) Maintaining the water at the temperature recorded in (4), draw off water from the container as filled in (4) to the level of the food found in (2), weigh the container with remaining water, and determine the weight of the remaining water by subtracting the weight of the container found in (3).

(6) Divide the weight of water found in (5) by the weight of water found in (4), and multiply by 100. The result is considered to be the per cent of the total capacity of the container occupied by the food.

In the case of a container with lid attached otherwise than by double seam, remove the lid and proceed as directed in (2)-(6), except that under (4), fill the container to the level of the top.

Drained Weight.—For some types of packaged products, it is necessary to obtain the drained net weight as in the case of canned oysters, or the net weight free of brine as in the case of anchovies or other fish. To obtain the drained net weight, open the container and transfer the contents to a skimmer or sieve. After allowing to drain for a stipulated time, weigh the contents. This weight is the drained net weight. To obtain the net weight of salted fish, remove the fish

[10] *U. S. Food Drug Admin., S.R.A., F.D.&C.* 2, Parts 51, 52, 53, and Appendix, March, 1954.

from the container and free the fish as well as possible from the salt and weigh the fish. This weight is the net weight of the fish.

The methods for drained weight for canned peas, canned beans, and oysters are empirical and must be followed precisely in order to obtain adequate results. They will be detailed in the sections of the text devoted to these items.

Particular attention should be paid to the label of packaged articles for misleading and misbranded statements.

PREPARATION OF SAMPLE

In order to be reasonably certain that the results obtained at the end of an analysis are correct, the sample to be analyzed must be properly prepared. This preparation consists in insuring homogeneity of the sample so that any portion subsequently taken for analysis will represent the whole foodstuff. In general, such homogeneity is obtained in samples that are liquid, pasty, or emulsified by shaking the sample and stirring it thoroughly or by transfer from vessel to vessel immediately prior to removing a portion for analysis. Care should be taken not to incorporate air in case specific gravity determinations are to be made. Solid samples should be ground and reground at least three times or comminuted by suitable means. For this purpose, a meat grinder, coffee mill, power grinder, colloid mill, mortar and pestle, and knives may be used, in accordance with the particular food product being prepared. The use of high-speed mixing machines or mechanical beaters such as blendors is particularly valuable in preparing homogeneous samples of vegetables, fruits, and animal tissues.

Samples that are likely to lose moisture should be stored in mason jars or other rubber sealed or hermetically sealed containers. Samples that deteriorate easily and that must be kept for a long time should have a preservative added, after analysis for preservatives shows their absence. Samples that are likely to spoil should be kept in a refrigerator. Other samples should be dried, ground, and stored in a suitable container.

DETERMINATION OF SPECIFIC GRAVITY

The term used to express the relative masses of equal volumes of a material being measured and water at a stated temperature is *specific gravity*. Thus in the expression d_{20}^{20} referring to specific gravity, the term means the specific gravity of a material at 20° C.

referred to water at 20° C. as unity. Sometimes specific gravity is referred to as *relative density*.

Density is mass per unit volume and is expressed in grams per milliliter. One ml. of water at 4°·C., at which temperature water has its maximum density, weighs 1 g. *in vacuo*. It is customary to refer densities to the density of water at 4° C. In the expression d_4^{20}, referring to density, the term means the density of a material at 20° C. referred to water at 4° C.

Westphal Balance.—In food analysis, specific gravity determinations are generally performed only on liquid foodstuffs. Specific

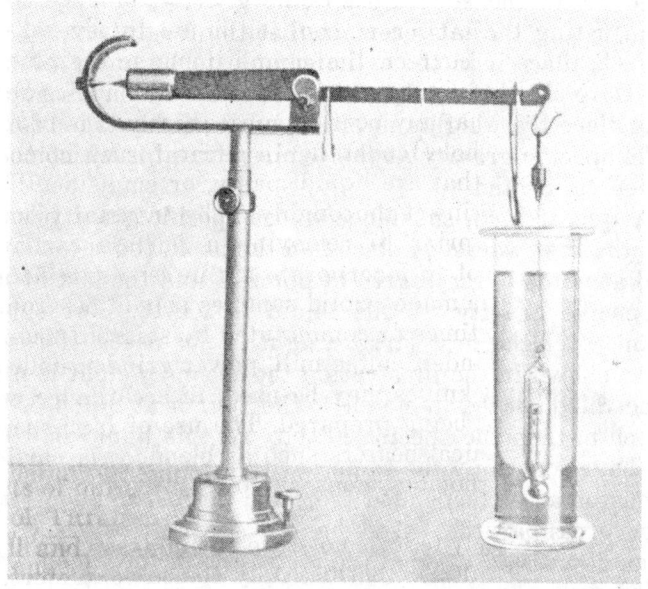

FIG. 5. Westphal Balance

gravity may be measured in a number of ways. A simple and convenient method, where sufficient sample is obtainable, is the use of the Westphal balance. This instrument is illustrated in Fig. 5. It is based upon the principle that a body immersed in a fluid is buoyed up by a force equal to the weight of the liquid displaced. The balance consists of a stand with a leveling screw supporting a beam balanced on a knife edge. The beam has a plummet suspended from one end and counterpoised by a fixed weight with a pointer at the other end. The plummet is made of glass, contains a thermometer, and is weighted at the bottom. It is constructed so that it will displace exactly 5 g. of water at 15.5° C. The distance on the beam between the knife edge and the point of support for the sinker is divided into ten equal parts. The balance is equipped with 5 horseshoe weights or riders. Two are large and of equal weight, namely 5 g., so that one of them placed at the end from which the plummet is suspended, will just counteract a buoyancy of 5 g. The other three weights are respectively one-tenth the weight of the next greater weight. These weights when placed on the notches; that is, the points of equal division of the scale, read respectively, unit, tenths, hundredths, thousandths, and ten thousandths.

Procedure.—To measure the specific gravity, balance the beam in air by adjusting the set-screw, cool the liquid to several degrees below 15.5° C., place it in the cylinder and immerse the plummet in the liquid. Care should be taken to see that no air bubbles adhere to the sinker. Place the weights in order, until the beam is balanced.

Read the specific gravity at the temperature for which the plummet has been calibrated.

For example, if for a substance whose specific gravity is greater than water the first weight is placed on the hook, the second at notch 1, the third at notch 5, the fourth at notch 1, and the last at notch 6, then the specific gravity of this liquid is 1.1516. For substances lighter than water the first large weight is not used. If for such a substance as shown in the illustration the second weight is placed at notch 9, the third placed at notch 1, the fourth at notch 5, and the fifth at notch 5, then the specific gravity of this liquid is 0.9155. If two riders have to occupy the same position on the beam, the lighter weight is customarily suspended from the larger weight. There are other types of apparatus based on the same principle. Readings should be made at 15.5° C. or at the temperature for which the plummet is calibrated.

Specific Gravity Balance.—The Westphal balance is a special type of specific gravity balance. In these devices use is made of the Archimedes' principle that a body immersed in a fluid is buoyed up by a force equal to the weight of the fluid displaced. Such devices are available commercially.

Procedure.—Suspend a plummet from the arm of the balance with the aid of a platinum wire. Obtain its weight in air, after immersion in distilled water, and after immersion in the liquid to be tested. Calculate the specific gravity by means of the formula:

$$d = \frac{P - W_1}{P - W}$$

where

d = specific gravity of liquid
P = weight of plummet in air
W_1 = weight of plummet in liquid
W = weight of plummet in water.

Reduce the weights in air to weights in vacuum, taking into consideration the density of air, the barometric pressure, and the temperature, to calculate the true density.

The specific gravity of vegetables can be obtained by means of a balance. This property has been used as a means of determining the maturity of frozen vegtables and the details will be found in Chapter XIII.

Pyknometer.—A pyknometer is a small, light flask, usually made of glass, of definite volume. It can be used, therefore, to weigh equal volumes of liquids at a given fixed tempera-

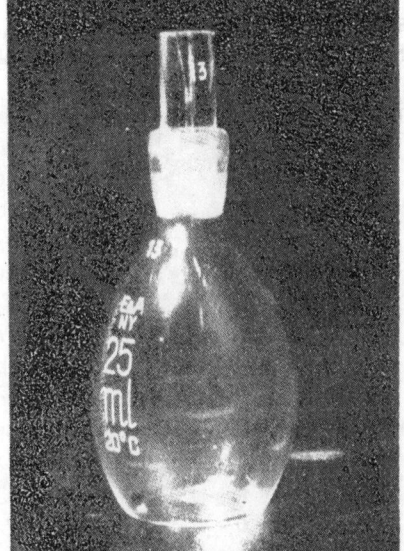

FIG. 6. Specific Gravity Bottle

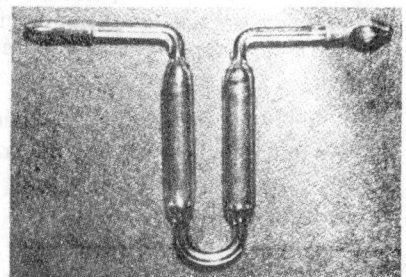

FIG. 7. Sprengel-Ostwald Tube

ture. The ratio of the weights of these equal volumes yields the specific gravity or if using the metric system, the density. The common forms are the specific gravity bottle, Fig. 6, and the Sprengel-Ostwald tube, Fig. 7. The pyknometer method is a very accurate means of determining specific gravity. There are various forms of specific gravity bottles. Some are equipped with a capillary side arm and thermometer. Others consist of a small flask fitted with a ground stopper having a capillary, as in the illustration.

Procedure.—Obtain the weight of the pyknometer, when it is clean, empty, and dry. Fill it carefully so that no air is incorporated with the liquid cooled to several degrees below the temperature at which it will be weighed. Allow the pyknometer and its contents to come to temperature either by allowing to stand or by placing in a bath. Wipe it dry. Remove the excess liquid by means of blotting or filter paper. In the case of the Sprengel-Ostwald tube, adjust the volume to the mark by this means. In the case of the weighing bottle, carefully wipe the capillary tube and place the cap on. Weigh the pyknometer. This is generally done at 20° C.

The weight of the liquid, divided by the weight of the same volume of water, determined by calibrating the pyknometer by weighing it dry and filled with boiled, cooled water, is the specific gravity. If the specific gravity of the liquid is to be referred to water at the same temperature as that at which the measurements were made, then the formula

$$d_{t^\circ}^{t^\circ} = \frac{W_1}{W}$$

where

d = specific gravity
t° = temperature designated, usually 20° C.
W_1 = weight of a volume of liquid
W = weight of an equal volume of water

is used.

In order to refer this specific gravity to water at its maximum density 4° C., the following formula is used.

$$d_{4^\circ}^{t^\circ} = \frac{W_1}{W} \times d_{W t^\circ}$$

where

$d_{4^\circ}^{t^\circ}$ = density of liquid referred to water at 4° C.
W_1 = weight of volume of liquid
W = weight of an equal volume of water
$d_{W t^\circ}$ = density of water at temperature t°.

Hydrometer.—Another but less accurate means of determining specific gravity is the use of a specific gravity spindle or hydrometer, Fig. 8. The spindle is a float usually cylindrical in shape with a bulb in the middle, a slender upper stem and weighted bottom. The stem is graduated with a scale to read specific gravities directly or some arbitrary scale such as Brix, Twaddell, Baumé, Quevenne or New

York Board of Health. The lactometer, alcoholometer, saccharometer, and Baumé spindle are types of specific gravity spindles.

Procedure.—Place the liquid in a suitable cylinder at the definite temperature for which the spindle is calibrated and immerse the hydrometer in the liquid. Push the hydrometer slightly below the equilibrium point and allow it to rise to its proper level. This is done to overcome surface tension and viscosity effects. When the hydrometer comes to rest and is free of air bubbles, read the specific gravity directly or read the degrees of the arbitrary scale and then obtain the density from tables. To read the hydrometer correctly, place the eye slightly below the surface level of the test liquid. Raise it slowly until the surface, first seen as an ellipse, becomes a straight line. Take the reading at the position where this line cuts the hydrometer scale.

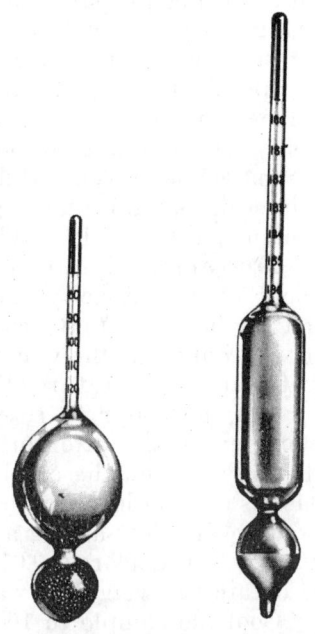

A convenient method of using hydrometers is to use a set consisting of two range finding hydrometers, one for liquids lighter than water and the other for liquids heavier than water, and nineteen others more finely graduated. These nineteen are graduated from specific gravity 0.700 to specific gravity 1.850, each hydrometer measuring an interval of 0.060 except the last which measures an interval of 0.070. One of the two range finding hydrometers is used first and from the reading the appropriate finely graduated hydrometer is selected and subsequently used.

Balling Scale.—The Balling hydrometer is a saccharometer which is used to indicate directly the percentage of sucrose (sugar) by weight at 17.5° C. or at 60° F.

Fig. 8. Left—Lactometer,
N. Y. Board of Health
Right—Hydrometer

present in a solution. The latter temperature is preferred by the National Bureau of Standards. It is mainly used for the estimation of the per cent of extract in wort in the brewing industry.

Brix Scale.—The Brix hydrometer is similar to the Balling hydrometer for the Balling scale was recalculated by Brix, however, most Brix hydrometers [11] refer to the per cent by weight of sucrose in a solution at 20° C.

---

[11] Bates and Bearce, *Nat. Bur. Standards (U. S.) Technol. Paper* **T115** (1918).

Baumé Scale.—Based on a modulus of 145, which has been most generally adopted, the relation between specific gravity at 20°/20° C. and degrees Baumé for liquids heavier than water is

$$°Bé. = 145 - \frac{145}{sp. \ gr. \ 20°/20° \ C.}$$

Flotation Method.—This method is based on the principle that a solid will just float in a liquid of the same specific gravity. The method is often used for waxes and hard fats. A series of suitable liquids of graded specific gravities must be made—e.g., water diluted with varying proportions of alcohol. That mixture of definite specific gravity in which a small smooth piece of substance free from air just floats is the specific gravity of the substance.

Specific Gravity Centrifuge Bottle.[12]—This method is of great value in determining the specific gravity of materials of high viscosity which retain air bubbles. This may be done by centrifuging the materials in weighed flasks. The specific gravity of tomato products such as pulp, purée, paste, catsup, and chili sauce may be conveniently estimated by this method.

Procedure.—Obtain the weight of an empty, clean, dry flask of about 125-ml. capacity that is of such shape as to fit a Babcock centrifuge holder. A Babcock flask, Fig. 50, with its neck removed, though only half the volume, serves well. Fill it to overflowing with water at a temperature slightly below 20° C. and allow the temperature of the water to rise slowly to exactly 20° C., having a thermometer in the water in the flask. Remove the thermometer, wipe the flask dry on the outside, and add water drop by drop until the flask is exactly level full, or slightly overfill the flask and remove the excess water by passing a microscope slide over the opening. Weigh at once to the nearest 0.01 g. Subtract the weight of the empty flask to obtain the weight of water this flask holds at 20° C.

Cool the sample to 16-18° C. Fill the flask with the sample and place it in the centrifuge with a suitable counterpoise. Whirl the centrifuge cup for from one-half to one minute at a speed of about 1000 r.p.m. The air bubbles are removed by the whirling and the surface of the pulp may now be considerably below the top of the flask. Fill the flask to the top and whirl in the centrifuge again. Remove the flask from the centrifuge, wash the outside, wipe dry, then insert a thermometer. When the temperature is just 20° C., remove it and add a few more drops of pulp or sample so that the level of the product comes exactly even with the top of the flask or slightly overfill and strike off with a straight edge. Weigh the flask and con-

---

[12] Bigelow, Smith, and Greenleaf, *Nat. Canners' Assoc. Bull.* 27-L (1934).

tents, at once, to the nearest 0.01 g. Calculate the apparent specific gravity by dividing the weight of sample in the flask by the weight of water at 20° C. that the flask was found to hold.

DETERMINATION OF MOISTURE

Moisture is the material lost by a foodstuff on heating not much higher than the temperature of boiling water, or by allowing to stand over a dehydrating agent, or by heating in a vacuum. It is generally considered to be water but actually is the total volatile matter lost or driven off at this temperature. When water content alone is to be determined, direct heating methods are not applicable. The residue remaining after using the direct heating, vacuum oven, or over sulfuric acid methods, is termed total solids.

Direct Heating Method.—The estimation of moisture is one of the most often performed determinations in food analysis. Usually, this is done as follows: Weigh a definite amount of material into a low, flat-bottomed, dried and tared dish. Add, if necessary because of the tackiness of the material, some dried sand or asbestos and reweigh. Place the dish and its contents in a water oven, or air oven thermostatically controlled at 100° C. to 105° C., or some similar oven and heat for a stipulated time, or until successive weighings show no further loss. At the end of this time, remove the dish from the oven, place it in a desiccator, allow to cool, and weigh.

The loss in weight equals the moisture. The loss in weight divided by the weight of the original sample is the per cent moisture. The sand and asbestos aid the drying process probably by providing focal points for evaporation and by decreasing the possibility of superheating. Aluminum milk bottle caps, which are inexpensive enough to be used only once, make suitable dishes.

Vacuum Oven Method.—Some foodstuffs contain substances, for example D-fructose, levulose, that are decomposed by heating at 100° C. under atmospheric pressure. Such materials should be dried in a vacuum oven under reduced pressure at 70° C. by the procedure detailed above until successive weighings show no further loss in weight.

Sulfuric Acid Method.—Drying without heat over sulfuric acid is slow and tedious, but nevertheless should be used where very exact determinations are necessary. Weigh into a flat-bottomed, tared, metal dish, fitted with a tight cover, 2-4 g. of material. Place the dish, uncovered, in a vacuum desiccator, Fig. 9, containing 200 ml. of fresh sulfuric acid. Exhaust the desiccator by means of a vacuum pump. Turn the desiccator around 4 or 5 times during the first 12

hours. After 24 hours open the desiccator, causing the incoming air
to bubble through the sulfuric acid. The cover is then replaced and
the dish weighed. Repeat, using fresh sulfuric acid until the weight
is practically constant. The percentage moisture is calculated as
directed above.

Immiscible Solvent Distillation Method.—A rapid and fairly accurate method for determining relatively small quantities of moisture is the immiscible solvent dis-
tillation method. This method is
of especial importance, when it is
desired to distinguish between
water and volatile matter, as for
example in spices, which contain
volatile oils.

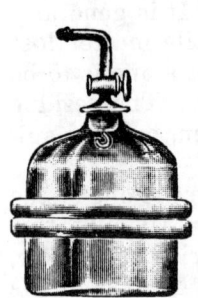

FIG. 9. Vacuum
Desiccator
(Courtesy of Eimer
& Amend)

Procedure.—Weigh into a
250-ml. flask or transfer to this
flask sufficient sample to yield
2-5 ml. of water. Cover the material with about 75 ml. of toluene.
In case of syrups add 10 g. Filter-
Cel before addition of syrup and
toluene to the flask and another
10 g. Filter-Cel just prior to distillation. Other
solvents immiscible with water may be used in
place of toluene, as for example xylol. Tate and
Warren [13] recommend heptane. Attach the flask,
Fig. 10, with a Liebig condenser in a reflux position
by means of a Bidwell and Stirling receiver, Fig.
11. Fill the receiving tube with toluene by pouring
the liquid down the condenser. Distill slowly at
the rate of about 2 drops per second and then more
rapidly when practically all of the water is over.
Just prior to ending the distillation, wash down
the condenser with toluene by pouring some through
the top. Continue distilling and if no more water
comes over, the determination is finished. If any

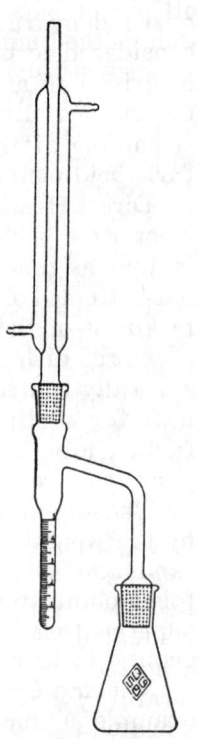

FIG. 10. Immiscible
Solvent Apparatus
(Courtesy of
Scientific Glass
Apparatus)

water adheres to the side of the condenser or tube, rub it down by
means of a rubber band attached to copper wire, while washing down
with toluene. Now distill for a short time longer. No water should
adhere if the apparatus has been thoroughly and properly cleaned
with cleaning solution and then washed and dried.

Read the volume of the water in the distillation tube and calcu-

[13] Tate and Warren, *Analyst* 61, 367 (1936).

late the percentage moisture assuming that the water volume is equivalent to its weight in grams, by dividing by the weight of the original sample taken and multiplying by 100.

As was previously stated, the direct heating method does not determine the true water content. It estimates the total volatile matter at 100° C. When it is desired to distinguish between volatile matter or apparent moisture and water content, some other method such as the immiscible solvent distillation method must be used.

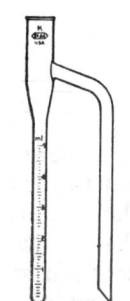

Brown-Duvel Method. —The Brown-Duvel method,[14] for which a special apparatus has been designed, may be considered a variation of the immiscible solvent method. It was originally designed for the determination of moisture in grain but has been adopted for the determination of moisture in vegetables, spices, and the like.

FIG. 11. Bidwell-Stirling Receiver (Courtesy of Kimball Glass)

Procedure.—A known weight of sample, sufficient to yield about 20 ml. of water, is placed in a specially designed liter, round-bottom, distillation flask, having a short neck and a side tube with two bends so that it can fit into a vertical condenser. About 150 ml. of mineral oil with a flash point of 200-205° C. (392-401° F.) and a viscosity of 10-15 (Engler) is added. The flask is stoppered with a rubber stopper carrying a special thermometer graduated from 0-200° C. so adjusted that the bulb of the thermometer just dips into the oil layer. The distillation flask is connected to an air condenser about 34 cm. long and 6-7 cm. in diameter and the condenser is surrounded by a cold water tank. Under the tip of the condenser is placed a 25-ml. graduated cylinder which acts as the receiver. The distillation flask is heated with a burner so that the oil reaches the appropriate temperature for the material being tested in 20 minutes. Usually 100 g. is heated to 190° C. but for wheat 180° C. is used. The heat is turned off and the flask is permitted to stand until water no longer drops from the condenser into the graduate. This generally requires an additional 4 to 6 minutes. The volume of water in the cylinder is read and the moisture is calculated as detailed above.

Fischer Method.—The Karl Fischer [15, 16] method for the estimation of water in organic solvents is based on the reduction of iodine by sulfur dioxide in the presence of water and a base like pyridine. The stoichiometry of this reaction has been studied by Smith,

---

[14] Duvel, *U. S. Dept. Agr., Bur. Plant Ind. Circ.* 72 (1910).

[15] Fischer, *Angew. Chem.* 48, 394 (1935).

[16] Mitchell, Jr., and Smith, *Aquametry.* Interscience, New York, 1948.

Bryant, and Mitchell [17] who concluded that a stepwise reaction is involved, the water taking part in only one step:

$$I_2 + SO_2 + 3\;\langle\!\!\!\rangle N + H_2O \longrightarrow 2\;\langle\!\!\!\rangle N \overset{H}{\underset{I}{}} + \langle\!\!\!\rangle N \overset{SO_2}{\underset{O}{}}$$

$$\langle\!\!\!\rangle N \overset{SO_2}{\underset{O}{}} + CH_3OH \longrightarrow \langle\!\!\!\rangle N \overset{H}{\underset{SO_4CH_3}{}}$$

It appears that 1 mole of water is involved with 1 mole of iodine, but in practice this is seldom achieved for, because of side reactions, the freshly prepared reagent is usually equivalent to 70 per cent of the theoretical water content. In the actual method this is of no concern since the reagent is standardized in known equivalents of water.[18]

In this method, the Fischer reagent, which is a solution of iodine, sulfur dioxide, and pyridine in anhydrous methyl alcohol or other anhydrous solvent like methyl Cellosolve, is added to a solution of the sample in anhydrous methyl alcohol until all the water reacts. The excess iodine acts as its own indicator and the end point of the reaction is evidenced by the persistence of the red-brown color. In a number of the variations used in food analysis, it is common to add an excess of the Karl Fischer reagent and backtitrate with a standard methanol-water solution.

This method is finding wide use in food analysis. For instance, Brobst [19] used it to determine moisture in lecithin and crude soybean oil; Schroeder and Nair [20] and also Johnson [21] as a general method for water in dehydrated foods; and Zerban and Sattler [22] as a method for moisture in sugar solutions. It very likely will prove to be the method of choice for a number of food materials such as oils, fats, shortenings, dehydrated foods, certain candies for which other types of moisture determination are difficult. Marked advantages of the method are ease of duplication of results and speed. The disadvantages are the need to restandardize the reagent daily and the cost of equipment. Commercial titrimeters designed for the Karl Fischer method are available.

[17] Smith, Bryant, and Mitchell, Jr., *J. Am. Chem. Soc.* 61, 2407 (1939):
[18] Almy, Griffin, and Wilcox, *Ind. Eng. Chem., Anal. Ed.* 12, 392 (1940).
[19] Brobst, *Anal. Chem.* 20, 939 (1948).
[20] Schroeder and Nair, *Anal. Chem.* 20, 452 (1948).
[21] Johnson, *Ind. Eng. Chem., Anal. Ed.* 17, 312 (1945).
[22] Zerban and Sattler, *Ind. Eng. Chem., Anal. Ed.* 18, 138 (1946).

Brobst Variation.—*Reagents.*—Fischer Reagent.—Dissolve 169 g. of resublimed iodine in 511 ml. of dry methyl alcohol and 425 ml. of reagent-grade pyridine, and add 128 g. of sulfur dioxide gas at a rate of about 40 g. per hour. The molar ratios of iodine, sulfur dioxide, and pyridine in this reagent are 1:3:8. Allow to age 2 days before using. It deteriorates at a rate of about 1 per cent per day becoming more stable as it ages.

Standard Water-Methyl Alcohol Solution.—Add 4 ml. of water to 1 liter of commercial absolute methyl alcohol. This solution contains about 5 mg. per ml.

Solvent.—Add 3 parts of dry chloroform to 1 part of dry methyl alcohol. Dry technical chloroform over 6- to 16-mesh silica gel for 3 to 4 days. Dehydrate commercial absolute methanol by refluxing over magnesium ribbon. Both solvents can be prepared with a water content of less than 0.01 per cent.

Procedure.—Place 50 ml. of solvent in a clean, dry sample container, add 1 to 2 ml. of Fischer reagent, and backtitrate with the standard water-methyl alcohol solution. Allow the titrated solvent to remain in operating position on the container support of the titrimeter until ready for use.

Weigh accurately about 5 g. of the sample into a second dry sample container. Run in about 15 ml. of Fischer reagent and immediately add the titrated solvent. Place in operating position and stir for 2 to 3 minutes to disperse the sample completely. Slowly add water-methyl alcohol solution until the first slight shift of the galvanometer needle is noted. Close the burette, allow to stir for 30 seconds, and proceed with the titration dropwise until the galvanometer comes to rest. This end titration must be done with care allowing at least a 30-second stirring time between drops because the characteristic lag is especially evident in this titration. Generally only 3 to 5 drops are needed after the first slight shift of the galvanometer.

Determine the exact water value of the standard water-methyl alcohol solution by titration of weighed amounts of distilled water, checking at least weekly. Set up a value for the equivalence of the number of milligrams of water equal to 1 ml. of Karl Fischer reagent. From these values and the weight of the sample the moisture content can be calculated.

Schroeder and Nair Variation.—This variation of the Fischer method is particularly designed for the determination of moisture in dehydrated materials. It is similar to the Johnson method. The reagents detailed above may be used.

Procedure.—Grind solid samples to pass a 40-mesh screen. Transfer an accurately weighed portion, 2-g. portion for materials containing up to 10 per cent moisture and correspondingly smaller portions for materials with higher moisture contents, to a glass-stoppered titration flask, previously dried 1 hour at 100° C. Cool with stopper in place, add 25 ml. of anhydrous methyl alcohol, and immediately reflux 5 minutes or longer to extract the moisture, by heating in a bath at 90° C. containing Carbowax 1000.[23] Cool the flask to room temperature while still attached to the reflux condenser, wipe dry, detach, and connect immediately to the titration assembly. Titrate directly with the Fischer reagent to an end point lasting 30 seconds. Subtract the methyl alcohol blank, and calculate the percentage moisture from the titer of the reagent.

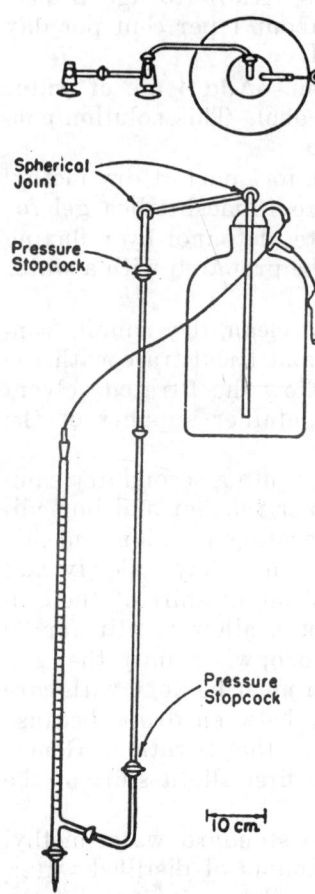

Spherical Joint

Pressure Stopcock

Pressure Stopcock

10 cm

FIG. 12. All-Glass Burette Assembly (Courtesy of Shell Chemical Corp.)

Shell Variation.—*Apparatus.*—It is necessary to use an all-glass titration assembly similar to that shown in Fig. 12. All openings should be protected by use of adapters containing indicating Drierite or equivalent material. All stopcocks should be adequately lubricated. Some procedures recommend Silicone stopcock grease. Such apparatus is commercially available as is electrometric type apparatus.

Reagents. — Fischer Reagent. — Mix 425 ml. of anhydrous pyridine and 425 ml. of anhydrous methyl Cellosolve[24] in a glass-stoppered bottle and cool in an ice bath. Add slowly while swirling constantly 70 ml. of anhydrous liquid sulfur dioxide with the aid of a graduated cylinder. Remove the bottle from the bath, add 133 g. of finely ground iodine crystals, and allow the mixture to stand with occasional mixing until the iodine is completely dissolved. Mix thoroughly and transfer to the all-glass burette assembly.

---

[23] Carbide and Carbon Chemicals Corp., New York, N. Y.
[24] *Allyl Alcohol.* Shell Chemical Corp., New York, 1946.

In the preparation of this reagent anhydrous methyl Cellosolve has been substituted for anhydrous methyl alcohol to improve the stability of the reagent.

Standardize the reagent each day it is used against a weighed portion of water, 0.15 to 0.20 g., using a Lunge weighing bottle, according to the procedure detailed below. Express the standardization as an equivalency factor F, in mg. of water equivalent to one ml. of the Fischer reagent.

Ethylene Glycol-Pyridine Mixture.—Mix one volume of anhydrous pyridine with four volumes of anhydrous ethylene glycol.

Procedure (in the Absence of Ketones and Amines).—Place 10 ml. of anhydrous methyl alcohol into each of two 100-ml. glass-stoppered flasks. Titrate both to the red-brown end point. Stopper one of these flasks and retain it for use as an end point color standard. Transfer to the other flask an accurately weighed or measured quantity of sample containing between 0.15 and 0.20 g. of water. Limit the sample size to 10 ml. Mix well and titrate with the Fischer reagent until a red-brown color is obtained which persists for at least 0.5 minute after vigorous shaking of the flask and which is equivalent to the color of the standard. Avoid wetting the stopper and upper end of the titration flask with either the reagent or the titration mixture. If such wetting occurs, dry thoroughly with a clean cloth to minimize absorption of moisture from the atmosphere. Each time the titration is interrupted, touch the tip of the burette with the neck of the flask to remove the adhering droplet, for this, if not removed, would absorb moisture from the air. Dry the burette stem and tip frequently for the same reason, using a downward stroke.

Procedure for Ketones.—Place 10 ml. of ethylene glycol-pyridine mixture into each of two glass-stoppered flasks and titrate to the red-brown end point. Stopper one flask and retain it as a standard. Chill the other flask in an ice bath, add a measured or weighed quantity of the sample to this flask, and proceed with the method as detailed above.

Procedure for Amines.—Place 10 ml. of anhydrous methyl alcohol or of ethylene glycol-pyridine mixture and 10 to 15 ml. of glacial acetic acid into each of two glass-stoppered flasks. Titrate to the red-brown end point with the Fischer reagent. Stopper one flask and retain it as an end point control. Chill the other flask in an ice bath, add a weighed or measured quantity of the test sample, and proceed as detailed above.

Calculation.—Calculate the water content of the test sample by use of the following equation:

$$\text{Water (weight per cent)} = \frac{VF}{10W}$$

in which

V = volume of Fischer reagent used for the sample in ml.
F = equivalency factor, mg. water equivalent to 1 ml. of the Fischer reagent
W = weight of the sample in grams.

Smith and Bryant Method.—A rapid quantitative procedure [25] for determining varying amounts of water in organic liquids utilizes the fact that acetyl chloride in the presence of pyridine produces two moles of titratable acid with water, whereas reacting with an alcohol only 1 mole of available acid is produced. Fatty acids can be analyzed for water by this scheme.

Procedure.—Pipette 10 ml. of 1.5 molal acetyl chloride in toluene into a dry 250-ml. glass-stoppered volumetric flask, using an automatic pipette. Place the flask in a beaker containing a slurry of finely chopped ice and, after allowing it to stand for a minute or so, add 2 ml. of pyridine from a pipette. Stopper the mixture and shake. A thin paste of acetyl pyridinium chloride in toluene is formed. Add a known weight or volume of sample to be analyzed for water content in such proportions that an excess of 0.5 mole of acetyl chloride remains for each mole reacted. Shake the mixture vigorously to insure intimate contact between the reagent and the sample. After standing for at least 2 minutes at room temperature, in which time all of the water present should have reacted, decompose the excess of reagent by adding absolute ethanol in two installments. Add the first 1 ml. of absolute ethanol with a pipette and follow by vigorous shaking to decompose the major portion of the reagent. Then after at least 5 minutes, add 25 ml. more of absolute ethanol to complete

[25] Smith and Bryant, *J. Am. Chem. Soc.* **57**, 841 (1935).

the decomposition and produce a homogeneous solution suitable for titration. Allow the solution to stand an additional 10 minutes at room temperature before titration. The addition of absolute alcohol in two portions is necessitated by its small but nevertheless significant water content which would otherwise introduce errors of varying magnitude.

Titrate the mixture with 0.5 N sodium hydroxide solution to a phenolphthalein end point. At least one blank determination should be made along with each group of samples. The increase in acidity of the sample over the blank is a direct measure of the water present in the sample, one mole of water liberating an extra mole of acid.

Refractive Index Method.—Other methods for the determination of water and moisture in foods are available. Some of these have the advantage of being very rapid and in certain instances, as in the use of the refractive index method for the determination of total solids in jellies, are equal to and even better than other more tedious methods. The refractive index method for the determination of water in jellies and like products, see Chapter XII, is very rapid and requires only a drop or two of material. This method is described in Chapter II and in Chapter XII.

Density Method.—Total solids in sugar solutions and in products consisting principally of sugar syrups and conversely moisture content may readily be determined by measuring the density. These methods have been detailed in prior sections of this chapter. See also the analysis of maple syrup in Chapter X.

Conductivity Method.—This method is based on the principle that, as the moisture content of a food product varies, its resistance to the passage of an electric current will vary proportionately. Thus the electrical resistance of wheat containing 13 per cent moisture is approximately 7 times as high as that of wheat containing 14 per cent moisture and 50 times as high as wheat with 15 per cent moisture. The Tag-Happenstahl meter, described by Coleman,[26] requires only about 1 minute for the entire determination. The method is described in Chapter XI.

Dielectric Method.—Since starches, proteins, and similar materials have a dielectric constant of about 10, whereas water has a dielectric constant of about 80, a small change in water content will make a rather large change in the dielectric constant of a given food. This is the basis of a rapid method [27] for the measurement of moisture in flour.

A graph is constructed with samples of flour containing known

[26] Coleman, *Cereal Chem.* 8, 328 (1931).
[27] Coleman, *Cereal Chem.* 8. 317 (1931).

moisture concentrations plotting moisture against dielectric constant. The dielectric constant of test samples is measured in standard metal condensers and the moisture content can be obtained from the graph.

This method as applied to grain is not considered as accurate as other methods.

Calcium Carbide Method.—Another rapid method [28,29] which has had some use in food analysis is based upon the reaction of calcium carbide with water to yield acetylene.

$$CaC_2 + 2H_2O \rightarrow C_2H_2 + Ca(OH)_2$$

The method has been used for determination of moisture in flour, butter, and fruit juices.

The amount of acetylene, and hence the moisture, may be estimated by determining the loss in weight of a test sample after the reaction and the escape of the acetylene, by determining the pressure produced in a closed system, or by trapping the acetylene and measuring its volume.

DETERMINATION OF ASH

Ash is the residue remaining after a foodstuff is ignited until it is carbon free, usually at a temperature not exceeding red heat.

Procedure.—Weigh into a tared platinum or porcelain or other suitable dish, a quantity of substance representing 2 g. of the dry material. Dry as usual at 100° C., burn at a low red heat, ash in a muffle oven at a dull red heat until free from carbon, remove from muffle oven, allow to cool for a moment, place in a desiccator until cool, and weigh.

If a carbon free ash cannot be obtained, leach the ash with water, filter on ashless filter paper. Return the filter and residue to the dish and burn to a white or nearly white ash. Cool, add the filtrate, evaporate to dryness, heat at dull red heat until a white or nearly white ash is obtained, remove from muffle oven, allow to cool for a moment, place in a desiccator until cool, and weigh.

Do not ash materials high in phosphates or containing lead, arsenic, or antimony in a platinum dish, for such ashing may make the platinum brittle. A small heat resistant glass beaker [30] may be used advantageously for ashing in preference to both porcelain or platinum, especially where the temperature of the ashing is low and where an ash that might fuse into the porcelain is likely to be obtained.

---

[28] McNeil, *U. S. Dept. Agr., Circ.* **97** (1912).

[29] Blish and Hites, *Cereal Chem.* **7**, 99 (1930).

[30] Pyrex beakers made by Corning Glass Works are suitable.

An alternative procedure is the following. Dry a weighed portion of the sample in a tared dish at 100° C. Add a few drops of olive oil. Heat slowly over a free flame until swelling stops. Place the dish and charred material in a muffle oven and treat as above.

If a carbon-free ash cannot be obtained, some analysts add a small volume of ammonium carbonate solution to the residue, evaporate to dryness, and then treat as above.

The addition of olive oil serves to minimize frothing and swelling. The addition of ammonium carbonate solution assists in giving a white ash and since it is completely volatile, it does not enter into the final weight.

It is advantageous, sometimes, to combine the direct moisture determination and ash determination by first drying the material at 100-105° C., weighing the loss in moisture, and then performing the ash determination.

Water Soluble and Insoluble Ash.—Add water to the ash obtained as directed above and heat almost to boiling. Filter through an ashless filter paper and wash with hot water until the combined filtrate and washings measure about 60 ml. Reserve the filtrate and washings for the determination of alkalinity of soluble ash. Return the filter paper and its contents to the original dish in which the ashing was performed, ignite carefully, cool, and weigh. From the respective weights found, the percentage of water soluble and insoluble ash may be calculated, for the total ash minus insoluble ash equals the soluble ash.

Alkalinity of Ash.—Cool the filtrate and washings reserved as directed immediately above and titrate with 0.1 N hydrochloric acid, using methyl orange indicator solution, made by dissolving 1 g. of the dye in a liter of water. Alkalinity of ash is generally expressed as the number of milliliters of normal acid per 100 g. of sample. This determination gives the alkalinity of the water soluble ash.

To estimate the alkalinity of the total ash or water insoluble ash, add an excess of 0.1 N hydrochloric acid to the dish containing the ash and heat to boiling cautiously on a hot plate. Cool, and titrate the excess hydrochloric acid with 0.1 N sodium hydroxide solution, using methyl orange as indicator. The alkalinity may be expressed in terms of the number of milliliters of normal acid per 100 g. of sample or as the alkalinity number, which is defined as the number of milliliters of normal acid required to neutralize 1 g. of ash.

DETERMINATION OF NITROGEN

Another usual analysis is the determination of nitrogen, for from the nitrogen content, the protein content of materials may be calculated. Proteins are complex organic substances consisting of chains of amino acids. They are a major constituent of all living cells, both plant and animal. As was previously discussed in the general introduction, proteins are necessary components of animal foods. The nitrogen content of different proteins is nearly alike and is approximately 16 per cent; hence multiplying the nitrogen estimated by the factor 6.25 yields the amount of protein. In certain cases as, for example, casein, a higher factor, namely, 6.38 is used for this conversion and more nearly represents the true proportion of nitrogen in those cases.

The estimation of nitrogen is generally done by a modified Kjeldahl digestion method. This digestion should be done only in a hood with a good draught. This method depends upon the decomposition of organic nitrogen compounds by boiling with sulfuric acid. The carbon and hydrogen of the organic material are oxidized to carbon dioxide and water. A part of the sulfuric acid is simultaneously reduced to sulfur dioxide which in turn reduces the nitrogenous material to ammonia. The ammonia combines with the sulfuric acid and remains as ammonium sulfate, a substance with a high boiling point. The ammonia is subsequently liberated by the addition of sodium hydroxide; is distilled into a known amount of standard acid and the excess acid is estimated by titration with standard alkali. In the method detailed, that is, the Kjeldahl-Gunning-Arnold Method, copper sulfate or mercury is added to act as a catalyst. Potassium sulfate or sodium sulfate is added in order to raise the temperature of the reaction mixture and thus hasten the digestion.

Gerritz and St. John [31] recommend the addition of 10 g. of anhydrous dipotassium phosphate or 12 g. of dipotassium phosphate trihydrate to be substituted for $1\%_{16}$ths of the potassium or sodium sulfate to obtain more rapid digestion.

Kjeldahl-Gunning-Arnold Method.—Transfer a weighed portion of about 0.7-3.5 g. of the material according to its nitrogen content into a digestion flask, Fig. 13. This may be done by weighing by difference or by weighing the material directly on filter paper or ungummed cigarette paper and transferring the paper and its contents to the flask. If the material is moist, a convenient method is to support a weighed piece of filter paper on the balance pan by means of a watch-glass and rubber washer or gasket of a mason jar. Then weigh the material directly and rapidly on the filter paper and, after recording the weight, transfer the paper and contents to the flask. The rubber washer prevents the filter paper from touching the

---

[31] Gerritz and St. John, *Ind. Eng. Chem., Anal. Ed.* **7**, 380 (1935).

watch-glass and thus prevents any loss of moisture by wetting the watch-glass.

Weigh out and add 18 g. of anhydrous potassium sulfate or anhydrous sodium sulfate or an equivalent amount of the crystallized hydrated salts, 1 g. of crystallized copper sulfate, $CuSO_4 \cdot 5H_2O$, or approximately 0.7 g. of mercuric oxide, HgO, or its equivalent of metallic mercury and 25 ml. of sulfuric acid. Do not add mercuric oxide or mercury if it is possible to carry out the digestion easily without these materials, for then the subsequent addition of potassium or sodium sulfide solution may be avoided. Heat the mixture gently until frothing ceases, then boil briskly and continue the digestion for a time after the mixture is colorless or nearly so, or until the oxidation is complete. The digestion usually requires at least 2 hours and the flask should be rotated at intervals during the digestion. Cool, add about 200 ml. of water, and, if mercuric oxide or metallic mercury has been used, add also 50 ml. of potassium sulfide solution, 40 g. K_2S per liter, or sodium sulfide solution, 40 g. Na_2S per liter, or sodium thiosulfate solution, 80 g. $Na_2S_2O_3 \cdot 5H_2O$ per liter. Then make strongly alkaline by pouring 70-75 ml. concentrated sodium hydroxide solution, 454 g. NaOH plus 1 liter of water, down the side of the flask so that it does not mix at once with the acid

FIG. 13. Kjeldahl Nitrogen Apparatus

solution. Add a pinch of zinc dust to prevent bumping and reduce frothing. Connect the flask to the condenser by means of a Kjeldahl connecting bulb, taking care that the tip of the condenser extends below the surface of the standard acid in the receiver and that the contents of the flask are mixed completely by shaking the flask at first carefully and cautiously and then vigorously. Distill until all of the ammonia has passed over into a measured quantity of standard acid. The first 150 ml. of the distillate will generally contain all the ammonia. Titrate with standard alkali using methyl red indicator, 1 g. of the dye in a mixture of 50 ml. 95 per cent alcohol plus 50 ml. of water.

The Winkler [32] modification of the Kjeldahl method is very useful. In this method the ammonia is distilled as usual but is fixed in 50 ml. of a saturated solution of pure recrystallized boric acid with the formation of ammonium borate. The ammonia may then be titrated directly with standard acid, because the boric acid is too weak an acid to affect the hydrogen-ion concentration to an appreciable extent during the titration. The advantages of this method are that it needs only one standard solution, namely, acid; it saves time and the boric acid need be measured only approximately. Care must be taken, however, that the receiver of the distillate be kept cool during the distillation for ammonium borate is somewhat volatile.

Pregl-Parnas-Wagner Micro Method.—In this method which is a modification [33] of the Pregl-Parnas-Wagner method,[34] as in the macro method detailed above, protein and other forms of nitrogen are converted to ammonia and fixed as ammonium sulfate by digestion with sulfuric acid. The ammonia is liberated by the addition of sodium hydroxide solution, is distilled, trapped in standard hydrochloric acid, and the excess hydrochloric acid is estimated titrimetrically with standard sodium hydroxide solution. The apparatus used is shown in Fig. 14.

Procedure.—Dilute an aliquot portion of the material being analyzed, if necessary, or use a weighing variation as detailed in the macro method, so that the amount of protein nitrogen or other form of nitrogen will be 1 or 2 mg. per ml. Transfer 1 ml. to a micro Kjeldahl digestion flask. Add 1 ml. of concentrated sulfuric acid, 1 ml. of a 4 per cent copper sulfate solution to act as the catalyst, and 0.8 g. of potassium sulfate, and digest on a digestion oven. Raise the heat slowly, boil vigorously, and after the material has been digested, as evidenced by a clear, straw yellow or light green color, reduce and cut off the heat. This process generally takes about 20 minutes. If the mixture does not clear in this time, reduce the heat, carefully add 2 to 3 drops of 30 per cent hydrogen peroxide, and then continue heating for 5-10 minutes. Allow to cool, add 4 ml. of water, and stir to dissolve the salts.

Add 7.0 ml. of 0.01 N hydrochloric acid, accurately measured (generally transferred by use of a reservoir type micro or semi-micro burette) to a 25-ml. flask and add a trace of methyl red indicator solution. Allow the water in the steam generator to boil gently

[32] Winkler. *Z. angew. Chem.* **26**, 231 (1913).
[33] Jacobs, and Jacobs and Shepard, *J. Am. Pharm. Assoc., Sci. Ed.* **40**, 151, 154 (1951).
[34] Parnas and Wagner, *Biochem. Z.* **125**, 253 (1931).

and open the pinch clamp or stopcock at the bottom of the steam trap so that the steam can escape. Transfer the digest from the micro Kjeldahl digestion flask to the distillation tube through the small funnel. Wash out the micro Kjeldahl digestion flask with two

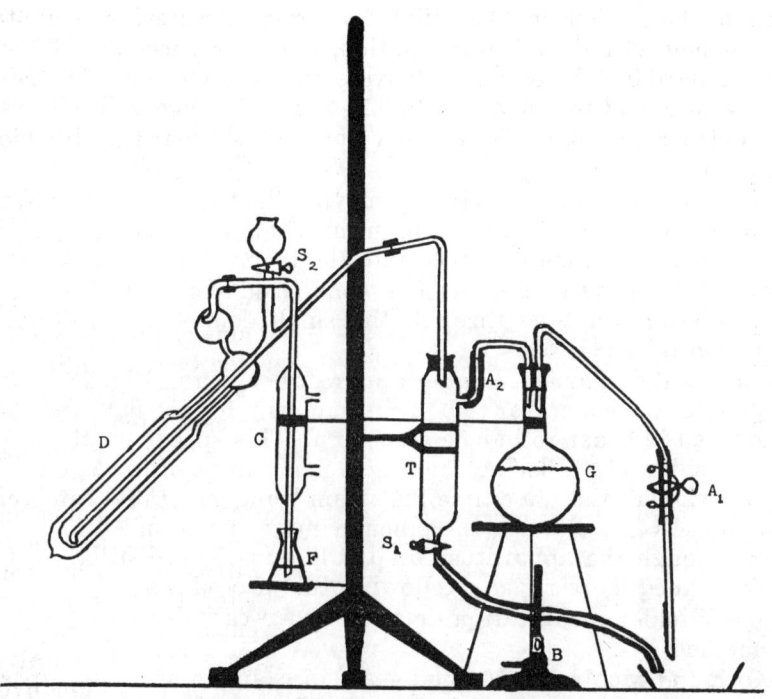

FIG. 14. Modified Parnas-Wagner Apparatus

A₁— Position 1 of pinch clamp
A₂— Position 2 of pinch clamp
B— Burner
C— Condenser
D— Distillation tube
F— Receiving flask
G— Steam generator
S₁— Stopcock attached to steam trap
S₂— Stopcock connecting funnel to distillation tube
T — Steam trap

2-ml. portions of water and add these washings to the distillation tube. Place the receiving flask under the condenser so that the tip of the silver tube condenser is below the standard acid. Add with the aid of a pipette 7 ml. of 30 per cent sodium hydroxide solution to the mixture in the distillation tube through the small funnel.

Close the stopcock or pinch clamp of the small funnel and on the steam trap thus compelling the steam to pass through the distillation tube. Distill for exactly 3 minutes. Lower the receiving flask so that the tip of the condenser is about 1 cm. above the surface of the distillate. Continue the distillation for another minute. Rinse the tip of the condenser tube with a few drops of water. Add another trace of methyl red indicator solution, if this is necessary. Titrate with standard 0.01 N sodium hydroxide solution. One ml. of standard hydrochloric acid is equivalent to 0.14 mg. of nitrogen. Run a blank and subtract the blank from the volume of standard hydrochloric acid used.

Clean the distillation flask by removing the flame under the steam generator. This creates a vacuum in the steam trap and the material in the distillation tube is sucked into the trap. Open the steam trap to reject this mixture. Replace the flame. Add about 10 ml. of water to the distillation tube through the small funnel and repeat the cleaning process.

The modified Parnas-Wagner apparatus shown in Fig. 14 permits the water in the generator to boil continuously for the flame need not be removed in order to induce a vacuum. This speeds up the operations considerably. Placing the pinch clamp in position A_2 causes a vacuum while at the same time the steam being generated can escape through the vent. Placing the pinch clamp in position A_1 forces the steam through the apparatus for distillation. The distillation tube can be cleaned more rapidly following this procedure.

The Winkler modification detailed above can also be adapted to this method.

Koch and McMeekin Method.—As an example of many types of micro nitrogen determinations the Koch and McMeekin[35] method is detailed. It is a direct nesslerization method in which the organic material is destroyed by digestion with sulfuric acid and 30 per cent hydrogen peroxide and a portion of the resulting solution is nesslerized and compared with prepared standards in a colorimeter.

Procedure.—Transfer a weighed portion of material according to the nitrogen content, from 0.25 g. of a meat product to 5 ml. of a low nitrogen content material, such as orange juice, to a 250 mm. x 25 mm. lipped Pyrex test tube. Cigarette paper is most conveniently used for this type of transfer, especially where practically dry materials are being used. Care must be taken to cut away the gummed portion of the cigarette paper.

Add 2 ml. of sulfuric acid (1:1) and 6 drops of copper sulfate

[35] Koch and McMeekin, *J. Am. Chem. Soc.* **46**, 2066 (1924).

solution, 1 g. $CuSO_4 \cdot 5H_2O$ in 100 ml. water and, if a dry material is being used, 2 ml. of water. Place the tube in a clamp at an angle, add two glass beads, boil off the water with the aid of a micro burner, and continue heating until fumes of sulfur trioxide fill the bottom of the tube. Adjust the tube to a vertical position and add 2-3 drops of 30 per cent hydrogen peroxide solution to the hot mixture. Heat again with the micro burner until the fumes are produced, remove the flame for 30 seconds and again add 2-3 drops of hydrogen peroxide. Continue the alternate heating and addition of hydrogen peroxide until the digestion is complete. During the alternate heating ·and addition of the peroxide, the solution will become clear almost immediately after the addition of the peroxide and the further heating will cause the mixture to darken again. Digestion is complete when no further darkening occurs on continued heating, and the solution is perfectly clear. The process is usually complete in about 20 minutes. As much as 30 drops of the peroxide are needed at times. The tube is cooled. Transfer the contents to a 100-ml. volumetric flask. Make to volume. Transfer from 1 to 5 ml., according to the nitrogen content, to a 50-ml. volumetric flask, dilute to 35 ml., add 6 ml. of Folin's Nessler reagent, prepared as directed immediately below, and make to volume. Read in a colorimeter against a standard prepared by diluting 10 ml. of ammonium sulfate solution containing 471.6 mg. per liter to 100 ml. Pipette 20 ml. of this solution into a 50-ml. volumetric flask, dilute to 35 ml., add the same. quantity of Folin's Nessler reagent as above and make to volume. Each milliliter of the standard solution now contains 0.004 mg. of nitrogen. If the nitrogen content of the unknown is much higher than the standard, less than 4 ml. of the unknown is nesslerized and, conversely, if the nitrogen content is much lower than the standard, more than 4 ml. is nesslerized.

Care must be taken not to heat the tube too strongly when sulfur trioxide fumes are produced because some ammonium sulfate may be lost mechanically by bumping or by other means.

Preparation of Folin's Nessler Reagent.—Nessler's solution is an alkaline solution of the double iodide of mercury and potassium [$HgI_2 \cdot 2KI$]. Transfer to a 200-ml. flask 30 g. of potassium iodide and 22.5 g. of iodine; add 20 ml. of water and after solution is complete, an excess of metallic mercury, that is, approximately 30 g. Shake the flask continuously and vigorously until the dissolved iodine has nearly all disappeared which takes about 7 to 15 minutes. The solution becomes hot. When the red iodine solution has begun to become visibly pale, although still red, cool in running water and continue shaking until the reddish color of the iodine has been re-

placed by the greenish color of the double iodide. The whole operation generally takes 15 minutes. Test a portion of the solution with
starch solution. Unless the starch test is positive, the solution may
contain mercurous compounds. Decant the solution, washing the
mercury and flask with water. Dilute the solution and washings to
200 ml. and mix well. If the cooling was begun in time, the resulting
reagent is clear enough for immediate dilution with 10 per cent
alkali and water and the finished solution can be used at once for
nesslerization. From this stock solution of potassium mercuric
iodide, prepare the final Nessler's solution as follows: To 975 ml.
of an accurately prepared 10 per cent sodium hydroxide solution,
add the 200 ml. of the double iodide solution. Mix thoroughly and
allow to clear by standing.

The 10 per cent sodium hydroxide solution should be made from
a 1:1 solution of sodium hydroxide and water which has been allowed
to stand until the carbonate has settled, the clear solution being
decanted and used. This solution should be standardized to an
accuracy of at least 5 per cent by titration and subsequent adjustment by the addition of more water or alkali as the case may be.
The alkalinity of Nessler's reagent is important and should be
checked against N hydrochloric acid. Twenty ml. of N hydrochloric
acid should require 11 to 11.5 ml. of Nessler's solution.

Determination of Nitrogen Including Nitrates.—Any nitrogen
present in a sample in the form of nitrate would be lost as nitric
acid by volatilization during the Kjeldahl digestion. When it is
desired to include this form of nitrogen, the Kjeldahl-Gunning-
Arnold Method must be modified. This may be done by adding some
substance such as salicylic acid or phenol, which is readily nitrated
and holds the nitric acid formed as the nitro-derivative. The nitro
group is then subsequently reduced to an amino group and finally to
ammonia.

Procedure.—To the weighed sample transferred to a digestion
flask as directed above, add 30 ml. of sulfuric acid containing 1 g.
of salicylic acid, shake until thoroughly mixed, and allow to stand
for at least 30 minutes with frequent shaking or until complete solution results. Add 5 g. of crystallized sodium thiosulfate and digest
as directed below. As an alternate procedure, add to the substance
30 ml. of sulfuric acid containing 2 g. of salicylic acid, allow to
stand at least 30 minutes with frequent shaking or until complete
solution results. Add gradually 2 g. of zinc dust, shaking the contents of the flask at the same time and digest as follows: Heat over
a low flame until all danger from frothing has passed. Then increase
the heat until the acid boils briskly and continue the boiling for 5-10

minutes or until white fumes no longer escape from the flask. Cool, add 10 g. of potassium sulfate or anhydrous sodium sulfate, and proceed as directed in the Kjeldahl-Gunning-Arnold Method.

EXTRACTIONS AND SEPARATIONS

One operation performed very often in food analysis·is that of extraction and subsequent separation. Thus in nearly every chapter in this book a procedure is detailed in which extractions either by immiscible solvents or by continuous methods are made.

The fundamental law governing the process of extraction, more particularly known as the law of partition, is a special case of Henry's law which states: The concentration of any single molecular species in two phases at equilibrium bear a constant ratio to each other, the temperature remaining constant. As applied to the special case of its use in food analysis,·the law is more simply stated: A solute will distribute itself, at constant temperature, between immiscible solvents in a definite manner which depends on the solubility of the substance in each of the solvents separately. This is conditioned by the fact that the distribution of the solute is also governed by its molecular weight and species in each solvent. A distribution in which the solute has the same molecular weight and species in both solvents is stated mathematically:

$$\frac{c_1}{c_2} = \text{coefficient of partition} = \text{constant,}$$

in which

c_1 = concentration in moles per liter in the first solvent
c_2 = concentration in moles per liter in the second solvent.

In many cases in food analysis, the solute is not in the same molecular state in both solvents. Thus for example, in the distribution of benzoic acid between water and benzene, the benzoic acid is in the form of dimeric molecules in the benzene. Then for such a system the concentration of the single molecules in the second solvent $nA \rightleftarrows A_n$ would be according to the law of mass action proportional to the nth root of the total concentration, and therefore if c_1 is the concentration of the solute in the 1st solvent and c_2 is the concentration in the 2nd solvent, the mathematical statement becomes

$$\frac{c_1}{\sqrt{c_2}} = \text{coefficient of partition} = \text{a constant}$$

It can be seen, then, that the solute will merely distribute itself between the two solvents in proportion to its solubility as explained above and that therefore it will, in general, be necessary to make repeated extractions with the second solvent before the concentration of the solute in the first solvent is reduced to a negligible quantity.

The most common form of apparatus used for extractions and separations is the separatory funnel, which is well adapted for multiple extractions by one solvent which has a higher specific gravity than another solvent. Thus, for example, in the extraction of lead from an aqueous ammoniacal cyanide solution by a solution of dithizone in chloroform, repeated extractions may be made by chloroform portions without removing the aqueous layer from the separatory funnel.

Jacobs-Singer Separatory Flask.—The separatory funnel is not well adapted for multiple extractions in which the supernatant

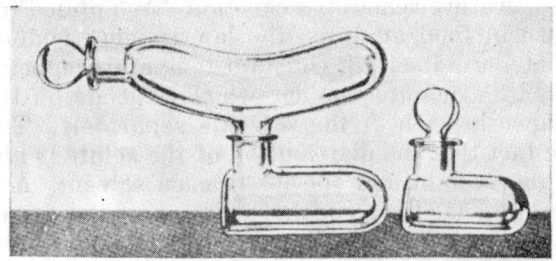

FIG. 15. Jacobs-Singer Separatory Flask

layer—that is, the lighter specific gravity liquid—is the extracting medium because the higher specific gravity layer must first be drawn off through the stopcock and then the lighter layer is generally poured through the upper orifice in order to reduce contamination.

There are several types of apparatus designed to achieve this purpose. One of the most useful is the Jacobs-Singer separatory flask. This apparatus, Fig. 15, consists of two sections either separable or integral substantially horizontal to one another. If separable, the lower section is an independent flask and is designed so as to permit direct weighing without the use of any auxiliary attachments and to permit its use as a chemical vessel in ordinary operations. The lower section, when separable, is fitted with an offset orifice or opening ground with standard taper to fit not only the upper section but also, because of the interchangeable joint, the stopper or any other piece of apparatus, such as a condenser, having the same taper. The lower section is also fitted with a flange or shoulder to aid in the handling of the apparatus. Furthermore, it has a flat bottom so that this section of the flask can, as the illustration shows, support the upper section.

The upper section of the flask is a curved tube having two orifices. One, at the lower portion of the middle, is externally ground to fit into the lower section as if it were a stopper. The upper section of the flask is designed so that there is a downward slope from both ends to the middle orifice. One of the ends has a pouring orifice which is internally ground at the same standard taper.

The use of standard taper permits the interchange of upper and, more often, lower sections of divers capacities as illustrated in Fig. 15. Thus, as shown in the illustration, the glass stoppers, the upper section and the two lower sections are all interchangeable. The more convenient capacities for general analytical work consist of an upper section of 100 to 120 ml. and lower sections of 25 to 30 ml. and 55 to 60 ml. capacities, respectively.

The general method of use of this separatory flask is the following. The material to be extracted is either weighed directly into the lower section, or a weighed portion is transferred to this section by means of a weight pipette or by weighing by difference, or an aliquot is placed in the flask by a suitable means such as a pipette, burette, or cylinder, depending on the accuracy desired. The sample in the flask is then subjected to whatever processing is required in the method—for example, see the Roese-Gottlieb method for fat in cheese, or the ethyl ether method for unsaponifiable matter in oils, or the separation of colors by the immiscible solvent method. Then the upper section is placed in position and the remaining agents added, or else the appropriate solvent is added, until the lower lever of the connecting joint is reached. The amount of additional solvent to add depends on whether, after shaking with the extracting solvent, the volume of the lower liquid—that is, the heavier specific gravity solvent—is increased or decreased because of the respecti e solubilities of the one solvent in the other. Then the requisite amount of extracting solvent is added.

The stopper is inserted in the outlet or orifice of the upper section. Now holding the upper section in the palm of the hand and holding the lower section firmly against the upper section by means of the fourth and fifth fingers, while at the same time holding the stopper firmly with the thumb, the liquids are separated so that some of each solvent is in each section of the flask, by a few deft movements, and then the contents of the flask are thoroughly shaken. If a volatile extractive solvent is being used, after shaking, hold the flask with the outlet upward and relieve the pressure by removing the stopper. Restopper, set the apparatus on its base, and allow the liquids to separate.

The higher specific gravity liquids will of necessity flow to the bottom flask due to the taper of the upper section. After the separation is complete, the supernatant liquid may be poured out by simply removing the stopper and slightly tilting the flask. The extraction may now be repeated in the same manner the required number of times. At the end of the extraction, any of the extracting solvent left in the lower section of the flask, or in the lower part of the connecting joint, may be removed by adding more of the lower solvent, cautiously to avoid mixing and over-shooting the mark, from a wash bottle, thus raising the level of the extracting liquid high enough in the connecting joint to be poured off.

If very stubborn emulsions form, pass all of the liquid into the upper section after stoppering the orifice. Invert this section, remove the lower section, and, after stoppering the middle orifice by means

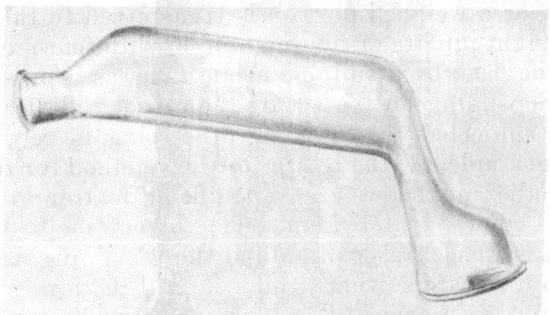

FIG. 16. Mojonnier Extraction Tube.

of a rubber nipple, centrifuge by placing the upper section in a suitable centrifuge cup.

If series of extractions are being made, a box designed to hold a number of these separatory flasks so that they may all be shaken at once is very useful.

Mojonnier Extraction Tube.—Another device [36] used for multiple extractions with supernatant layers is the Mojonnier extraction tube, Fig. 16. This device is of limited use, being mainly devised for use as the extraction apparatus needed in the Roese-Gottlieb method for the estimation of fat in milk products. This extraction apparatus consists of a tube having two chambers, a mixing chamber and a settling chamber at an angle to one another and connected by a constriction in the tube. The material to be analyzed is weighed into the flask which

---

[36] Mojonnier and Troy, *Technical Control of Dairy Products*, Mojonnier Bros., Chicago.

must be suspended from hooks in the balance or else a definite volume of the sample is transferred to the tube from some auxiliary apparatus. The various reagents required by the method are added, the tube is stoppered with a cork stopper and then shaken. The extracting solvent is then added and again the tube is stoppered and shaken and the solvents are permitted to separate. When separation is complete, the supernatant layer can be poured off through the

FIG. 17. Soxhlet Extraction Apparatus

mouth of the tube without disturbing the lower layer. Exact details are to be found in other sections of the book, see particularly the Roese-Gottlieb method for milk.

The flask or rather tube must be equipped with some box holder for it cannot stand independently. The same box can be used to hold an entire series of tubes and thus aid in routine analyses.

Soxhlet Continuous Extraction Apparatus.—In contradistinction to the extraction process in which multiple portions of an extracting immiscible solvent are used, there is the continuous extraction process. In this procedure, the apparatus is designed so that a fresh portion of solvent comes in contact with the material to be extracted over a relatively long period of time. There are many forms of this type of apparatus. One of the most commonly used is the Soxhlet type of continuous extractor, Fig. 17.

The apparatus consists of a tared flask, containing a volatile solvent and resting on some type of heating device, generally and preferably a multi-regulated electric hot plate. The tared flask is connected preferably by means of interchangeably ground joints with a tube having a siphon arrangement and side arm. The extraction tube is connected, again preferably by interchangeably ground joints with a condenser. The apparatus is used in the following manner. The dry material is weighed into or a weighed amount of dry material is transferred to a thimble which may be of alundum, cotton, or other porous material. The thimble is placed in the extracting tube and this tube is then connected with the weighed flask and also the condenser.

The heat vaporizes the volatile solvent which passes up the side arm and is condensed in the condenser. The condensed solvent falls drop by drop into the thimble. When sufficient solvent has been thus transferred to the extracting tube to fill the siphon arm, it siphons back over into the weighed flask. This process is continued until the extraction is complete, varying at times from 8 to 24 hours. Then the tared flask is removed, the volatile solvent is evaporated and the residue is the extracted material.

For other types of continuous extractors see Blasdale [37] or some similar text.

Continuous Extractors.—*Liquid-Liquid Type.*—The continuous extractors presently used are generally of the Palkin [38] or Widmark [39, 40, 41] type. In the Palkin model of continuous extractor, the extracted substance is collected in a distilling flask from which the solvent is continuously distilled back into the vessel holding the sample. The extracted substance is subjected to a temperature equivalent to the boiling point of the solvent, hence mixed solvents of greatly different volatility cannot be used.

The Widmark type continuous extractor is based on the principle that the extracting solvent comes in contact with the sample solution and also with an absorbing solution which reacts with and thus

37 Blasdale, *Fundamentals of Quantitative Analysis*. Van Nostrand, New York, 1928.
38 Palkin, Murray, and Watkins, *Ind. Eng. Chem.* 17, 612 (1925).
39 Widmark, *Skand. Arch. Physiol.* 48, 61 (1926).
40 Pucher and Vickery, *Ind. Eng. Chem., Anal. Ed.* 11, 656 (1939).
41 Yakowitz and Meuron, *J. Assoc. Official Agr. Chem.* 31, 127 (1948).

changes the composition of the extracted substance so that it is no longer soluble in the solvent. The extracted substance is collected in the absorbing solution; hence the extraction can be carried out at any desired temperature, even room temperature. This type has been modified by Yakowitz and Meuron so that the solvent is circulated rapidly from a compartment containing the sample solution into a compartment containing the absorbing solution by means of a stirrer-pump.

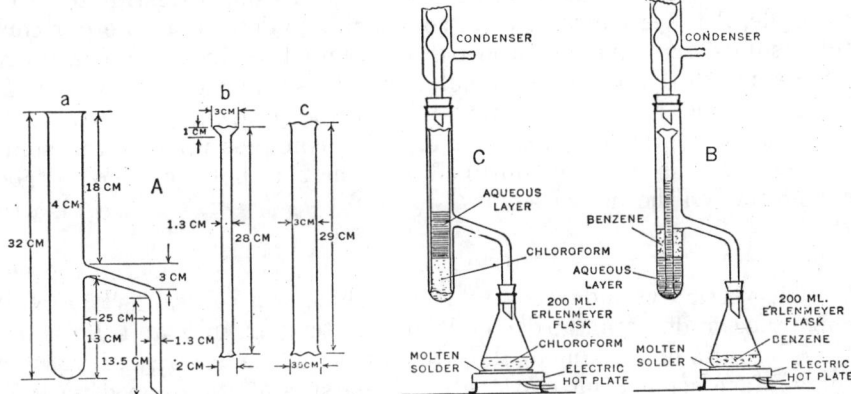

FIG. 18. Continuous Extraction Apparatus
(After *Methods of Analysis*, A.O.A.C.)

SELECTED REFERENCES

Blasdale, *Fundamentals of Quantitative Analysis.* Van Nostrand, New York, 1928.
Brooks, *Critical Studies in the Legal Chemistry of Foods.* Reinhold, New York, 1927.
Findlay, *Practical Methods of Physical Chemistry.* Longmans, London, 1933.
Fryer and Weston, *Oils, Fats, and Waxes.* Cambridge, 1920.
Jacobs, Ed., *Chemistry and Technology of Food and Food Products*, 2nd ed. Interscience, New York, 1951.
Leach, *Food Inspection and Analysis.* Wiley, New York, 1920.
Methods of Analysis, A.O.A.C., 8th ed. Washington, 1955.
Moore, *History of Chemistry.* McGraw-Hill, New York, 1918.
Neto, *Subsidio à Historia da Bromatologia.* Naval, Bahia, Brazil, 1946.
Oser, *Anal. Chem.*, 21, 216 (1949).
Sherman, *Chemistry of Food and Nutrition.* Macmillan, New York, 1941.
Triebold, *Quantitative Analysis.* Van Nostrand, New York, 1946.
Winton and Winton, *Analysis of Foods.* Wiley, New York, 1945.
Woodman, *Food Analysis.* McGraw-Hill, New York, 1941.

CHAPTER II

PHYSICAL CHEMICAL METHODS

PHYSICAL CHEMICAL methods are important in food analysis. They lend themselves to rapid, simple means of determining with a great degree of accuracy many factors we are anxious to estimate. For example, it takes hours or days sometimes to determine the per cent total solids or conversely the per cent moisture in sugar solutions, whereas with the use of a refractometer it is a matter of minutes. It is beyond the scope of this text to describe in great detail any and all of these instruments. However, the principles upon which some of them are based will be outlined, and the use that they find in food chemistry will be indicated.

REFRACTOMETRY

The refractive index is a quantity which is a constant for a pure substance under standard conditions of temperature and pressure. It is the ratio of the sine of the angle of incidence of a ray of light on the surface separating two media to the sine of its angle of refraction. The ray passing from a dense to a denser medium is bent toward the normal. Expressed mathematically,

$$\frac{\sin i}{\sin r} = n = \text{index of refraction}$$

where

$i =$ angle of incidence
$r =$ angle of refraction.

Abbé Refractometer.—With the Abbé refractometer, Fig. 19a, the refractive index can be read directly, only a few drops of the liquid are needed, and either white or monochromatic light can be used. This refractometer consists, mainly, of a fixed telescope and two matched right angle prisms. The liquid is placed in contact with the prisms. A ray of light passing through the prism and liquid and entering the prism ABC at grazing incidence, Fig. 19b, will emerge from AC at less than a right angle for any value of n other than

$$n = \sin' A.$$

Hence, the prism must be rotated through an angle in order that the rays may be parallel to the telescope. This angle of emergence determines the refractive index. Rays striking the surface at an angle greater than the angle of emergence are totally reflected. By adjusting the light and dark portions of the field so that the line of demarcation is sharp and coincides with the cross hairs, the refractive index can be read directly on the scale. The prism box has a jacket so that the temperature can be controlled and the telescope has a set of Amici prisms, Fig. 25, for compensating light aberration.

Since the index of refraction of a pure substance is constant at constant temperature and pressure, it can be used as a means of

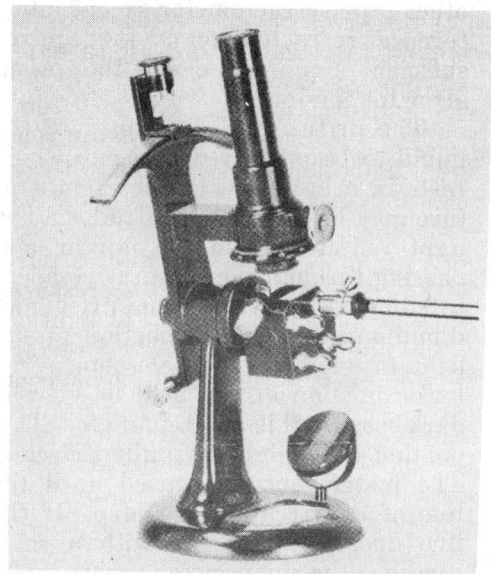

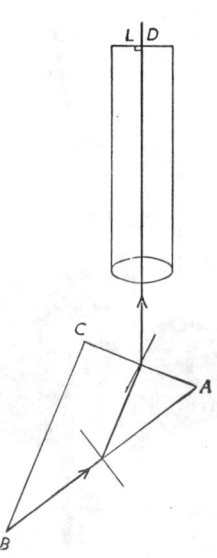

Fig. 19a. Abbé Refractometer
(Courtesy of Bausch & Lomb)

Fig. 19b. Diagram of
Refractometer Prism

identification. It is used to determine the purity of oils, fats, and waxes, which, although not pure substances in the strict sense of the word, have indices which vary over a slight range. It is used to determine the amount of sugar in sugar solutions and, in general, for determining total solids where sugar is a main component, as in fruit juices, tomato products, honey, syrups, and soda water. Methods have been developed for the determination of the percentage fat in cacao products, the percentage sugar in chocolate products, and the estimation of total solids in eggs, all of which will be referred to in their appropriate section of the book.

Immersion Refractometer.—The theoretical considerations of the immersion or dipping refractometer are the same as those of the

Abbé refractometer. Hence in this instrument, also, determination of refractive index depends on the observation of the line of total reflection. It is an instrument designed to cover a small index range with a higher degree of accuracy than is possible with the Abbé type. It consists of three essential units, a telescope, a compensator, and an immersion prism. Originally the immersion instrument was designed to read indices only from 1.32 to 1.36. However, by using interchangeable prisms the range of the instrument may be extended up to 1.54. This type of instrument gives greater accuracy than any

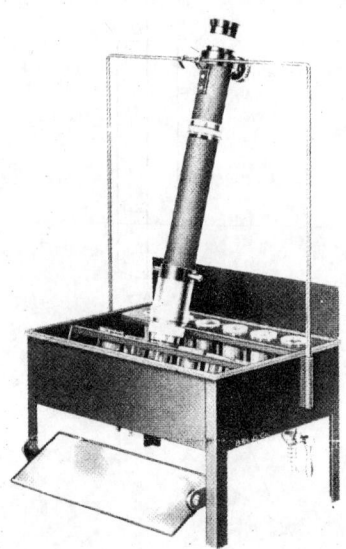

FIG. 20. Immersion Refractometer
(Courtesy of Bausch & Lomb)

other type refractometer except interference refractometers. It requires sufficient liquid to cover the prism after immersion.

The prism, Fig. 20, dips into the liquid to be measured in a cup which rests in a bath, so that the temperature may be carefully controlled. The light reflected from the mirror and passing through the prism is compensated by the internal Amici prism and illuminates the upper portion of the field of view of the telescope. The lower portion of the field of view is dark because it is not illuminated. This portion of the light is totally reflected. The instrument is focused until the line of demarcation is sharp. If the line does not coincide with a scale reading, the micrometer screw is set at 0, the next lower scale division is read, and then the micrometer screw is turned until the line coincides with a scale division. The vernier reading is added to the scale reading and the refractive index is obtained from Table 1 or Table 4 appendix. This instrument is used to determine the refractive index of milk serum, the per cent methyl alcohol present in ethyl alcohol, and has many other uses.

COLORIMETRY

The term colorimetry [1] as applied in chemistry includes both the measurement of color as color and the measurement of light absorptive capacity. Colorimetric analysis has received such extensive

---

[1] Mellon, *Colorimetry for Chemists*, Smith, Columbus, 1945.

COLORIMETRY 49

investigation that it can be considered an entire science[2, 3, 4] in its own right.

TABLE 1. INDICES OF REFRACTION, CORRESPONDING TO SCALE READINGS OF IMMERSION REFRACTOMETER

Scale Reading	n_D	Scale Reading	n_D	Scale Reading	n_D
0	1.32736	35	1.34086	70	1.35388
1	1.32775	36	1.34124	71	1.35425
2	1.32814	37	1.34162	72	1.35461
3	1.32854	38	1.34199	73	1.35497
4	1.32893	39	1.34237	74	1.35533
5	1.32932	40	1.34275	75	1.35569
6	1.32971	41	1.34313	76	1.35606
7	1.33010	42	1.34350	77	1.35642
8	1.33049	43	1.34388	78	1.35678
9	1.33087	44	1.34426	79	1.35714
10	1.33126	45	1.34463	80	1.35750
11	1.33165	46	1.34500	81	1.35786
12	1.33204	47	1.34537	82	1.35822
13	1.33242	48	1.34575	83	1.35858
14	1.33281	49	1.34612	84	1.35894
15	1.33320	50	1.34650	85	1.35930
16	1.33358	51	1.34687	86	1.35966
17	1.33397	52	1.34724	87	1.36002
18	1.33435	53	1.34761	88	1.36038
19	1.33474	54	1.34798	89	1.36074
20	1.33513	55	1.34836	90	1.36109
21	1.33551	56	1.34873	91	1.36145
22	1.33590	57	1.34910	92	1.36181
23	1.33628	58	1.34947	93	1.36217
24	1.33667	59	1.34984	94	1.36252
25	1.33705	60	1.35021	95	1.36287
26	1.33743	61	1.35058	96	1.36323
27	1.33781	62	1.35095	97	1.36359
28	1.33820	63	1.35132	98	1.36394
29	1.33858	64	1.35169	99	1.36429
30	1.33896	65	1.35205	100	1.36464
31	1.33934	66	1.35242		
32	1.33972	67	1.35279		
33	1.34010	68	1.35316		
34	1.34048	69	1.35352		

In the years 1925-1945, during which there was great development in this field, differences in nomenclature arose because of the different meanings assigned in analytical chemistry and in optics.

2 Snell and Snell, *Colorimetric Methods of Analysis.* Van Nostrand, New York, 1936-7.
3 Yoe, *Colorimetry.* Wiley, New York, 1928.
4 Sandell, *Colorimetric Determination of Traces of Metals.* Interscience, New York, 1945.

Color may be considered a combination of three factors—namely, the radiator or light source, the object transmitting or reflecting the radiated light, and the observer. The instruments used to measure the color of an object or a material can be placed into three major groups:

(1) *Stimulimeters* are instruments which are used to match, by a given system of known colors, the color of the solvent or material being measured. Mellon assigned the term stimulimeters to these instruments since they use a combination of stimuli to match a given stimulus. Examples of these are the Munsell System, the Donaldson instrument and the Lovibond Tintometer.

(2) *Color comparators* are devices by which the color of an unknown solution or solvent is compared with a standard or a series of known solutions or standards. In the former instance the matching may be made by variation of the depth of the liquid color, whereas in the latter instance the known concentration of standards of equal volume is varied systematically. Examples of the former type devices are the Duboscq colorimeter, and Nessler tubes are generally used for comparison of the latter type.

(3) *Color absorptometers* are of two types—filter photometers and spectrophotometers. The former devices provide for the measurement of light transmitted or, at times, reflected from an object or system in comparison with the light incident upon the object. The latter also are used to estimate the proportion of light transmitted as related to the incident light but they have a monochromator by means of which measurements can be made at any average wave length, and with any width of spectral band within the limit of proper use. The latter are briefly discussed in the section concerning spectrometers.

It is beyond the scope of this text to describe these instruments in detail. The reader is referred to Mellan and other texts noted at the end of the chapter.

Hue or *tone* is dependent entirely upon the wave length of the light. Red, yellow, green, blue, and violet are different hues.

Brightness (*brilliance* or *luminosity*) or Munsell *value* is determined by the intensity of the rays, whatever their wave length, falling upon the retina. There can be many degrees of brightness within a single hue. Of two objects of the same hue and illuminated to the same degree, the one that absorbs the smaller proportion of the rays of the specific wave length and reflects the greater proportion appears the brighter. The different brightness of a given hue may be compared to a number of mixtures of black and white made up in various proportions to produce a graded series of shades of gray. For example, just as any shade of gray may be produced by adding black to white, so a paint of a given hue can be reduced to any degree of brightness by mixing with it a suitable quantity of black. It is very difficult for one to match the brightness of two colors (heterochromatic photometry).

As in the case of brightness, color *saturation* (or *purity* or Munsell *chroma*) may show infinite variations in degree within a given hue. Red is a more saturated color than pink. The difference is due to the relative quantities of white light with which the red light is mixed; the saturation of a paint of a given hue can be reduced by mixing white with it. The mixture reflects white light as well as light of the wave length by which its hue is perceived. The saturation of a colored light can be determined by means of the spectroscope; a fully saturated red or green of homogeneous or monochromatic light gives a red or a green color at the corresponding part of the spectrum, whereas, if either color contained an admixture of white light, all the spectral colors would appear, their intensity depending upon the quantity of white in the mixture.

Lovibond Tintometer.—The Lovibond tintometer may be considered a stimulimeter. The depth of color of unknown materials or solutions is matched against the color of standard glasses. These standards consist of a series of slips of colored glass in the three primary colors—red, yellow, and blue—accurately graded in very small steps from nearly colorless to "deep" colors so that virtually all colors can be matched and measured either by use of a single glass standard or by comparison against an appropriate combination of glass standards.

For example, an orange color can be matched by a suitable combination of red and yellow; green by combinations of blue and yellow; violet by combinations of blue and red; and neutral tints by combinations of all three primary standards. The scales are additive. Thus a 1.0 red glass plus a 3.0 red glass gives the same color as a 4.0 red glass. Equal amounts of all three colors give an equal unit of neutral tint. Thus a 1.0 red, plus a 1.0 yellow, plus a 1.0 blue glass combine together to give 1 unit of neutral tint.

Liquid samples are placed in specially constructed glass vessels and are viewed by transmitted light' against a white background. Solid samples are viewed by reflected light.

Various models of this instrument are available. Thus there is a model for general laboratory work which is of particular interest in food analysis; models for vegetable oils generally, cottonseed and other oils specially; for sugar refining; and for wines and liquors. The Lovibond-Schofield tintometer is an elaboration of the regular tintometer that employs a special mechanical device instead of neutral tint glasses for measuring the dullness or brightness of a sample.

The Munsell System.—The standards for color in canned tomatoes under the *Definitions and Standards for Food* of the Food and Drug Administration [5] and at one time for tomato juice, tomato pulp, and tomato catsup formulated by the Bureau of Agricultural Economics [6] are all expressed in the Munsell System.[7, 8] Color measurement and its application to the grading of agricultural products is discussed in detail by Nickerson.[9] It is beyond the scope of this text to give a detailed consideration of the apparatus and technique necessary for the successful measurement of color by this system. However, an explanation of the notation used in the standards follows.

The definite valuation of the color of a particular product according to the Munsell System consists of two parts: The first is the percentage of the different specific colors which when blended together give a composite color that exactly matches the sample. The percentage notations of a particular color must always add up to 100 per cent. The second essential part of each color notation is the exact description of each of the color cards used to match the sample in terms of accurately determined values for "hue," "brilliance" and "chroma." Each class of agricultural product requires a particular set of color cards. For example, the color of canned tomatoes and tomato products may be matched by varying the proportions of four particular colors, namely, definite shades of red, yellow, neutral black and neutral gray. These colors are designated by specific formulas in each of which there is an empirical designation of first, hue; second, value; and third, chroma.

The formulas for the red color used in matching tomato products is 5R2·6/13. The portion 5R indicates the *hue*, which is the attribute of color that permits colors to be classed as reddish, yellowish, greenish, or bluish. Five principal hues are used in the Munsell System—namely, red, yellow, green, blue, and purple designated R.Y.G.B. and P. Midway between these are five similar intermediate hues: yellow-red, green-yellow, etc., designated YR, GY, etc. The designation 5R is a pure réd free from purple or yellow.

The second notation, "2·6" indicates the *value* or intensity of the black constituent of the color. This is expressed in an arbitrary scale from 0 to 10 where zero is absolute black and 10 is absolute white. A "value" of 2·6 indicates a considerable proportion of black constituent.

[5] *U. S. Food Drug Admin., S.R.A., F.D.&C.* 2, Parts 51, 52, and 53, March, 1934.

[6] *U. S. Dept. Agr., Bur. Agr. Econ.,* "U. S. Standard for Grades of Canned Tomato Pulp and Tomato Catsup," January, 1934.

[7] Cooper, *Munsell Manual of Color.* Munsell Color Co., Baltimore, 1929.

[8] Nickerson, *U. S. Dept. Agr., Bur. Agr. Econ., Tech. Bull.* 154 (1928).

[9] Nickerson, *U. S. Dept. Agr., Misc. Pub.* 580 (1946).

The final designation shows the *chroma,* which expresses the saturation or purity of the color. The notation of chroma is in an arbitrary scale from 0 to 10 or even further, which in this case is 13, which indicates a very strong, intense red color.

The required color cards, which can be obtained only from the Munsell Color Company, are cut in the form of "Maxwell discs," which are uniform circles of each of the cards with a small hole at the center and a single radial slit from the center to the edge so that the discs may be placed one on the other and slipped together by means of this radial slit so that any desired proportion of one or more of the color discs may be exposed as a segment of the single circle that is visible. When these discs are held together with a suitable binding post at the center and spun at a speed great enough to eliminate flicker, the resulting color seen by the eye is the sum of the different segments exposed. The discs should be slipped together in such a direction that the spinning will cause them to lie flat rather than to fly apart. The amount of each color card exposed is changed until the combined effect exactly matches the color of the sample. The amount of each color exposed is then measured by the per cent of the circumference occupied by each segment. The discs are mounted on a stiff card containing a permanent circular chart for measuring these percentages. So it is that the expression in the standard for color in tomato pulp "YR21(2·5YR5/12)" means that 21 per cent of the yellow-red disc is exposed and this yellow-red disc has the formula Hue 2·5YR, Value 5, Chroma 12.

It will be noted that the requirement for color in canned tomatoes according to the Food and Drug Administration *Definitions and Standards for Food* is that it shall be equal or better than that produced by spinning a combination of the following Munsell color discs. Red (5R 2·6/13)-(glossy finish); Yellow (2·5YR 5/12)-(glossy finish); Black N1 (glossy finish); Gray N4 (mat finish). This means that when the color of the whirling color discs matches the color of the sample there must be at least one third the area of the red disc exposed and not more than one third the area of the yellow-red disc exposed and the remaining percentage may be any combination of the black N1 and the gray N4. All of the color discs come in either the glossy finish or the mat finish so that it is necessary to specify which finish is desired when such color cards are being ordered.

Color Comparator.—Duboscq Type.—There are many types of color comparator used. One of the oldest and most frequently used types is the Duboscq colorimeter, Fig. 21. The instrument consists of two movable cups set on stages attached to rack and pinion, two plungers, a set of prisms, an eyepiece, and 2 millimeter scales with verniers. Diffuse light is reflected by a mirror through the cups and the plungers and is observed in the eyepiece. One cup is partially filled with the unknown solution. The other cup is partially filled with the standard solution. The cups are raised so that the plungers dip into the liquid, care being taken that no air bubbles adhere to the plungers. Setting the standard at a definite position on the scale, usually at 15 or 20 mm., the unknown is moved up or down until the two halves in the eyepiece appear equally colored. It is good practice to approach the point of equality a number of times from each direction and then average the readings. The light transmitted is now equal. Since the transmitted light varies inversely as the depth and the concentration, we have mathematically for one side of the field:

$$\text{light transmitted for known} = \frac{1}{C_1 R_1}$$

and for the other side of the field

$$\text{light transmitted for unknown} = \frac{1}{C_2 R_2}$$

and since the light transmitted when the fields are equally colored is equal

$$\frac{1}{C_1 R_1} = \frac{1}{C_2 R_2}.$$

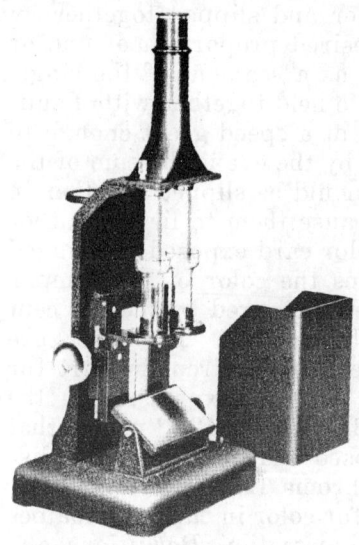

Fig. 21. Duboscq Colorimeter
(Courtesy of Bausch & Lomb)

Hence,

$$C_2 R_2 = C_1 R_1$$

and

$$C_2 = \frac{C_1 R_1}{R_2}$$

where

C_1 = concentration of known
C_2 = concentration of unknown
R_1 = scale reading of known
R_2 = scale reading of unknown.

There are other types of colorimeters based on very similar principles.

Photoelectric Colorimeter.[10, 11, 12]—A photoelectric cell consists of a photo-responsive cathode surface of an alkali or alkaline earth metal or compound enclosed in a glass envelope which is either evacuated or filled with an inert gas under low pressure. The cell contains one or more anodes according to the design. Light falling on the active surface of this cell through its window will cause electrons to be hurled out of the active surface with a velocity depending on the frequency of the incident light and in number proportional to the intensity of illumination. The conditions are given by the well-known Einstein expression:

$$\frac{Ee}{300} = \frac{1}{2} mv^2 = h\nu - h\nu_0$$

where

E = potential in volts required to hold electron, e, at surface of emitter
m = mass of electron
v = velocity of emerging electron
h = Planck constant
ν = frequency of incident illumination
ν_0 = threshold frequency or lowest frequency at which electrons will be ejected.

If now an accelerating potential is applied between the active surface and the anode through a sensitive galvanometer, the electrons emitted from the active surface will travel across the interior of the cell to the anode and manifest themselves as a feeble current by causing a deflection of the galvanometer. Under proper conditions the reading on the galvanometer will be proportional to the intensity of the illumination reaching the active surface.

The photoelectric colorimeter, Fig. 22, consists essentially of an optical system, an absorption cell, and an electrical measuring system. The optical system consists of a source of light in a well-ventilated housing, a precision iris diaphragm, and a light filter with suitable transmission characteristics. The electrical measuring system consists of a photoelectric cell and a sensitive low resistance microammeter or a thermionic amplifier to amplify the photoelectric current derived as described above which may then be read on a milliammeter. The use of proper filters is important, or preferentially monochromatic light should be used.

Fundamentally, this instrument measures the difference in light falling on a photoelectric cell because of different absorptive powers

[10] Partridge, *Ind. Eng. Chem.*, *Anal. Ed.* 2, 207 (1930).
[11] Partridge and Muller, *Ind. Eng. Chem.*, *Anal. Ed.* 3, 169 (1931).
[12] Mellon, *Colorimetry for Chemists*. Smith, Columbus, 1945.

of liquids. Curves are plotted for known values of each particular absorptive substance, as, for example, hydrogen-ion indicators on semilogarithmic graph paper.

According to the law of Bouguer, each layer of equal thickness traversed absorbs an equal fraction of the passing light. Since the intensity of the emitted light decreases exponentially as the thickness of the absorbing medium increases arithmetically, the absorption varies directly as the logarithm of the thickness.

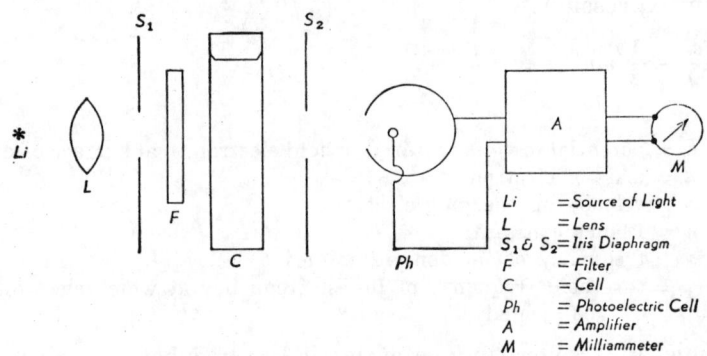

FIG. 22. Photoelectric Colorimeter

This may be expressed mathematically

$$I = I_0 t^l$$

where

I_0 = intensity of incident light
I = intensity of emergent light

from a solution of concentration c and thickness l. A thickness l will transmit the fraction t^l if a layer of unit thickness transmits the fraction t.

According to the law of Beer, the absorptance of a solution is directly proportional to the concentration of the solute, actually to the number of molecules of the absorbing substance present. That is the transmittancy T for a given thickness of a solution of concentration c is

$$T = t^c$$

in which t equals the transmittancy of the same thickness of a solution having unit concentration.

The laws of Bouguer and Beer may be combined so that

$$T = t^{cl}$$

where

$T =$ the transmittancy of concentration c in a thickness l

and

$t =$ the transmittancy for a solution of unit concentration and unit thickness

and more readily from a more useful form of Beer's law

$$c = -\frac{k}{1} \log_{10}[I/I_0]$$

where

$c =$ concentration of the substance in solution
$I_0 =$ intensity of the incident light
$I =$ the intensity of the light transmitted through a 1-cm. thickness of solution

Hence a logarithmic relationship exists between the concentration of the absorbing medium and the intensity as measured by the electrical measuring device. These curves act as calibration curves for each particular substance and are especially useful where Beer's law does not hold.

The following terms are often used in connection with absorption phenomena. The symbols and the defining equations are given. The defining equations are expressions in the typical form for the transmitted light according to the Beer law.

Extinction Coefficient $= K$	$I = I_0 10^{-lK}$
Molecular Extinction Coefficient $= \epsilon$	$I = I_0 10^{-\epsilon cl}$
Absorption Coefficient $= \mu$	$I = I_0 e^{-\mu l}$

The photoelectric colorimeter has virtually displaced the visual colorimeter in the food laboratory, for it is timesaving and diminishes the personal factor in colorimeter readings. It is widely used in biochemical determinations and some instruments are equipped with scales reading percentage directly, for a particular substance, provided a fixed method is followed. It is also used for hydrogen-ion determinations and for titrations. The change in color occurring during a titration can be followed by the photocell, and the apparatus can be arranged so that the amplified current or output of the cell can be made to interrupt the addition of the titrating agent at any desired end point.

This type of apparatus may also be used as a nephelometer or turbidometer because different turbidities will also cause variations in the photoelectric current. This factor is important, and it is necessary to have no turbidity in all measurements depending on color, because the photoelectric cell cannot distinguish the cause of reduction or increase in intensity of light, it merely measures that intensity.

Colorimeters of all types have many uses. In food chemistry determinations of nitrogen, phosphorus, lead, aluminum, citral, vanilla, and sugar can be conveniently made colorimetrically. These methods will be detailed in other sections of the text. The selection of the proper colorimetric procedure depends on the method and the experience of the analyst. Thus where colors and tints are to be matched as in honey, beer, or vanilla extract, a Lovibond comparator or some similar device must be used. In lead spray determinations and in other determinations in the chapter on metals, methods using Nessler tubes advantageously are described. Where routine analyses are being made, as, for example, nitrogen determinations, it is convenient to calibrate a curve on a photoelectric colorimeter and estimate the amount of nitrogen by that means. Where it is possible to obtain fairly uniform suspensions, a nephelometer may be used as in silver chloride determinations. In short, the instrument to use is the one that will give the best results for the determination and method involved.

Nephelometer.—A nephelometer [13] is an optical device for determining the amount of suspended matter in a solution by comparison with standard suspensions. The method is based on the measurement of the brightness of the light reflected by the cloud—that is, by the particles in suspension. The intensity of the light reflected is a function of the quantity of suspended particles when other conditions are constant.

Some colorimeters can be changed easily into nephelometers by changing cups and the source of light so that the light shines at the cups instead of through the cups. Hence, the amount of light reflected from the suspended particles in the liquids instead of the light transmitted may be measured by exposing varying amounts of the cup to the light. In a similar way then to colorimetry, nephelometry implies that the smaller the concentration, the greater the depth traversed for an equal reflection of light.

Fluorescence Analysis.—All substances absorb electromagnetic vibrations or light, usually over a characteristic range of wave lengths, and many emit or re-emit such radiations. This emission phenomenon is known as luminescence. Luminescent light may be in

---

[13] Kober, *Ind. Eng. Chem.* **10**, 556 (1918).

the visible and invisible regions. Luminescence may be exhibited by solids, liquids, or gases and is classified under two sub-heads:

(1) Fluorescence, if the luminescence lasts only during the period of excitation,

(2) Phosphorescence, if the luminescence persists after the exciting source is removed.

When the luminescence produced is characteristic of the substance irradiated, it may be used as a means of analysis. For most cases it is sufficient to note the intensity and color; and, where greater specificity is desired, the light emitted may be examined spectroscopically or photometrically. In general, the intensity is proportional to the amount of active substance present consequently quantitative measurements may be made.

Radley and Grant [14] group fluorescent analysis methods as follows:

1. *Qualitative.*—a. Direct Irradiation.—The substance is placed in ultraviolet light, and the nature, color, and intensity of the luminescence are noted and compared with the characteristics of that from genuine samples of known origin. This may be carried out in acid, alkaline, or neutral solutions, at various concentrations, and in a number of solvents.

b. A refinement of the above is to observe the luminescence under a luminescence-microscope or to determine its spectral characteristics.

c. Chemical Reactions.—The suspected substance is treated with a chemical reagent which should produce a luminiscent compound, and the appearance of such a compound observed. Conversely, if the substance is itself luminiscent, a reagent may be chosen which destroys the luminescence of the substance concerned.

d. Capillary Analysis.—If a filter paper is held vertically with one edge in the solution, it draws up the liquid by capillary attraction; and when the wet portion is examined under the lamp, characteristic zones are obtained from many substances.

2. *Quantitative.*—a. Trial and Error.—A number of mixtures containing known quantities of the luminescent substance are compared with the sample in ultraviolet light, and an approximate match is obtained with one of them.

b. Photometry.—The intensity of the luminescence, which, other conditions being equal, is proportional to the amount of luminescent

---

[14] Radley and Grant, *Fluorescence Analysis.* Van Nostrand, New York, 1933.

substance present, may be determined photometrically. Photometers and spectrophotometers equipped for fluorescence analysis are commercially available. See Chapter XVII.

c. Luminescent Indicators.—Determinations of pH may be made by means of indicators where luminescence in ultraviolet light changes with change in hydrogen-ion concentration. These, and similar compounds, may be used to indicate end points of neutralization or oxidation-reduction titrations, and are usually sensitive in very small quantities and in very great dilutions of reagent, as well as being free from many of the usual errors inherent in the use of indicators. Thus Grant [15] points out the use of quinine sulfate as a fluorescent indicator for precipitation reactions. Its action appears to depend on the fact that quinine sulfate fluoresces in the presence of an excess of some ions but not of others, so that if the end point of the reaction involves a change from one to another of these types of ions it will be indicated by the production or disappearance of fluorescence.

The titration should take place in a thin-walled 200-ml. conical flask, which is supported on a black base in such a position that the filtered ultraviolet light falls on it from the side; it is preferable to arrange the lamp so that it is slightly above the level of the liquid in the flask. The solution in the flask should be as concentrated as is conveniently possible, and it should be shaken well after each addition of reagent to stimulate coagulation, as this leaves a supernatent liquid in which the fluorescence can be observed. Only a small pinch of solid quinine sulfate need be used, and a darkened room is not essential if colorless 0.1 N solutions are being titrated. Titrating sodium chloride against silver nitrate, for example, a change from bright pale blue to dull purple is obtained at the end point.

If solvents are used in this type of analysis, they should be optically inert and nonfluorescent. Ether, petroleum ether, amyl alcohol, and chloroform are often used. The materials should be examined in dishes of nonfluorescent glass or porcelain. Containers made from black filter paper are effective for solids which give a white fluorescence. Petri dishes covered with Cellophane are useful in keeping out contamination.

Fluorescent analysis is a relatively newer form of analytical procedure, nevertheless, a large amount of work has been done on foods. These will be referred to in the appropriate sections of the text.

---

15 Grant, *Analyst* 62, 285 (1937).

SPECTROMETRY

When a parallel beam of light passes through a prism, it is refracted and dispersed because the index of refraction of glass is greater for violet than for red light. Such dispersed bands of light are called spectra. If the light coming from an incandescent substance is passed through a prism, a spectrum characteristic of that substance is formed. Hence by observing the emission spectrum— that is, a flame, arc, spark, or other spectrum of an unknown material —its nature may be ascertained from the definite lines and bands its spectrum forms.

Emission spectra can be produced in a number of ways. When a more or less volatile metallic salt is placed in contact with a flame and is subsequently volatilized by the flame, flame spectra are produced. A convenient method is to place the salt in solution and then feed it into the flame by means of a platinum wick. Arc spectra may be produced from substances that are not easily fusible. Rods of the metal may serve as the poles of the arc or the metal may serve as the negative pole and carbon as the positive pole. Alternative means of producing arc spectra are to use rods of carbon having a central core of the metal or placing small pieces of the substance in a crater of the positive pole of the carbon rod. Spark spectra are produced by means of sparks from an induction coil made to pass between small poles of the substance or discharged through gases or passed between a platinum wire and a solution of a salt of a metal. Another form is the use of cathode streams for the production of phosphorescence in solid substances.

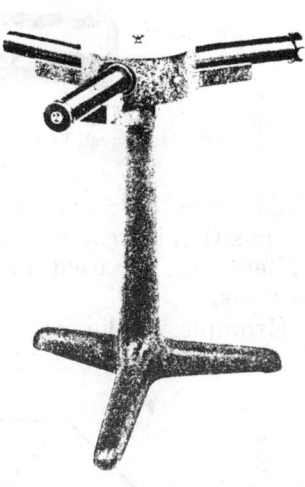

FIG. 23. Spectrometer
(Courtesy of Bausch & Lomb)

If white or other light is passed through an absorbing medium before being passed through the prism, a dark band will appear in the spectrum and will be characteristic of the absorbing medium. Such a spectrum is called an absorption spectrum.

The spectrometer, Fig. 23, consists of a collimator which is a tube having a narrow slit and lens to provide a parallel ray of light, a prism to refract the light and form the spectrum, a telescope to view the spectrum, and a third tube which has a scale that may be reflected into the ocular and so be used for calibrating the spectrum. This instrument is called an angular vision spectrometer.

The principle of the direct vision spectrometer, Fig. 24, is to make the dispersed light emerge from the same direction from which it came by means of a suitable system of prisms, as, for example, the

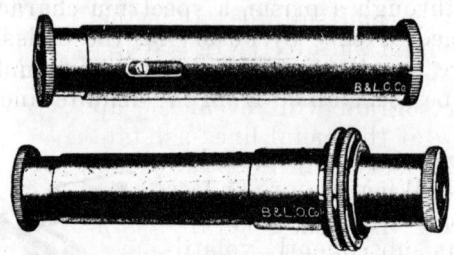

FIG. 24. Direct Vision Spectrometers (Courtesy of Bausch & Lomb)

Amici prism, Fig. 25. A small direct vision spectrometer screwed into position instead of the eyepiece of a colorimeter, as for example, a Klett, can be used to compare the absorption spectra of two solutions.

Examples of the use of spectroscopy in food analysis are given by Harrison.[16]

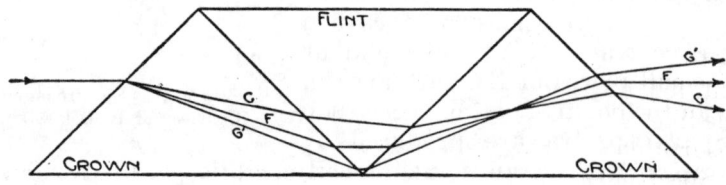

FIG. 25. Amici Prism (Courtesy of Bausch & Lomb)

All methods of quantitative spectrum analysis are based on the fact that, the greater the percentage of a metal in a sample, the stronger will be the spectrum lines of that metal, that is, the concentration of the metal varies directly with the intensity. The density of such lines may be measured photometrically. A useful method is that of Lewis[17] and is based on the principle of synthesizing a specimen so that its spectrum exactly matches that of the sample.

[16] Harrison, *Food Research* **3**, 121 (1938).
[17] Lewis, *Analyst* **60**, 11 (1935).

It is necessary to bring the whole sample of ash or other material into some constant homogeneous condition and to arc or spark under well-defined uniform control. This may be done by treating the ash including any phosphate or silicate or other insoluble matter with a small excess of sulfuric acid. The mixture is evaporated and ignited to drive off excess sulfuric acid. The mass may now be assumed to consist mainly of normal sulfates. It is then mixed with an equal amount of ammonium sulfate, $(NH_4)_2SO_4$, to regularize its action while burnt in the arc and is ready for spectrographing.

The first spectrogram will reveal the approximate composition from the intensity of the metal lines as stated above. If precise results are desired a special ratio powder may now be prepared in accordance with the findings and the spectrographing repeated.

The spectrometer is used, generally, as a qualitative instrument in contradistinction to the colorimeter and polarimeter. There are, however, newer spectrographs and photospectrometers (spectrometers equipped with photographic apparatus) which are quantitative instruments. Essentially a spectrophotometer is a device for determining wave length by wave length the proportion of radiant energy incident upon a body that is reflected or transmitted by it. Whereas a colorimeter is limited to colored solutions, the spectrophotometer is not limited to colored solutions, but can be used for ultraviolet or infrared spectra and for both opaque and transparent substances.

A spectrophotometer generally consists of (1) a spectrometer for dispersing the light into its component wave lengths, (2) a photometer for obtaining a two-part photometric field equipped with means for varying the brightness of one or both parts, (3) a source of light, and (4) auxiliary equipment such as cuvettes, light filters, and the like.

Absorption spectrograms are discussed in some detail in the section "Spectrographic Method" for vitamin A in the Chapter XVII on Vitamins.

These instruments are used in food analysis to detect metallic contamination, to distinguish dyes both natural and artificial, to detect the presence of foreign oils in olive oil and for general absorption determinations.

POLARIMETRY

When light is propagated, the wave front of the light is in motion in all directions. If a ray of this ordinary light is passed through certain crystals such as those of Iceland spar, calcite, two rays are produced which now vibrate in planes perpendicular to each other and the rays are said to be plane polarized. If such a ray

of plane polarized light is passed through a solution of a substance
or through a crystal of a substance which is molecularly asymmetric,
the polarized ray will be rotated to the right or left. Organic substances in whose molecules there are four different groups or radicals attached to a given carbon atom are asymmetric and exhibit the
phenomenon of rotating plane polarized light or optical rotation
activity. Many other substances both inorganic and organic having
different radicals or groups arranged around an atom or in which
the molecule as a whole is asymmetric will also turn a polarized ray.
Substances which rotate the ray of polarized light to the right are
termed dextrorotatory, and those which rotate it to the left are
termed levorotatory.

In general, in apparatus employed to measure optical rotation,
crystals of Iceland spar are used because the velocity of the two rays
formed when ordinary light is passed
through them is markedly different
and they diverge to a greater degree.
If two such crystals are joined together by means of a thin layer of
Canada balsam or analogous bonding
material into a prism, known as a
Nicol prism as shown in Fig. 26, the
so-called ordinary ray o, which has a
refractive index greater than that of the Canada balsam, is totally
reflected and absorbed because it reaches the balsam at more than the
critical angle, whereas the extraordinary or emergent ray e is transmitted by the principal section of the prism. In most Nicol prisms
the line of transmission of the emergent ray is through the faces
having the short diagonal.

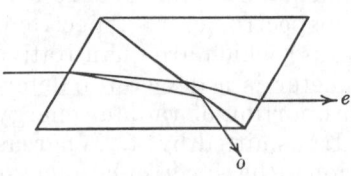

Fig. 26. Nicol Prism

If light is passed through two parallel Nicol prisms, the first
prism will permit the emergent ray to pass and the second prism
will do the same. This is known as maximum transmission. If, however, the principal section of the second prism is placed or rotated
so that it is perpendicular to that of the first prism, a position
commonly termed crossed Nicols, the emergent ray from the first
prism will now be in the ordinary ray vibration plane of the second
prism and consequently will be totally absorbed or reflected. This
is termed total extinction. Intermediate positions of the second or
rotatable Nicol will permit some light to be transmitted varying
between total extinction and maximum transmission.

If a substance or a solution capable of rotating the plane of vibration of polarized light is placed between crossed Nicols at total
extinction, some light will be transmitted because of the rotation

produced by the optically active material. Compensatory rotation of the second prism can be used to restore total extinction and this rotation can be measured by a suitable arrangement.

The polarimeter is an instrument which measures the angle through which the polarized ray is turned to the right or left. The instrument consists of a slit and lens to make the rays from a source of monochromatic light parallel, a polarizer or fixed Nicol prism to make the light vibrate in one plane, a tube to contain the rotating solution, an analyzer prism which can be turned, a system of lenses to observe the rotation and a scale by which the angle of rotation is measured. The tube, Figs. 27a and 27b, which is of definite length

 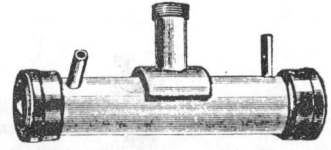

Fig. 27a. Polarimeter Tube   Fig. 27b. Polarimeter Tube, Jacketed

(From Thurston, *Pharmaceutical and Food Analysis*. D. Van Nostrand Co., Inc.)

is filled with the unknown solution. It is placed in position and then the analyzer is turned to maximum transmission of light or complete extinction. The rotation is given in degrees.

Specific Rotation.—The specific rotation $[a]$ is given by

$$[a] = \frac{100a}{lc}$$

where

a = the rotation observed in degrees
l = the length of the tube in decimeters
c = the concentration of the active substance in grams per 100 ml.

The molecular (or equivalent) rotation is usually given as:

$$[a_M] = \frac{M[a]}{100} = \frac{Ma}{lc}$$

where

M = the molecular (or equivalent) weight of the active substance.

$[a_M]$ is thus the rotation per mole (or per equivalent) per 1 ml. of solution for a column 1 mm. long, or per mole (or per equivalent) per liter of solution for a column 1 meter long. The rotation is usually given for monochromatic light, generally the sodium line or one of the mercury lines. If the polarimeter scale is calibrated to read the arbitrary units of the sugar scales of Ventzke, or Bureau of

Standards, etc., it is called a saccharimeter, Fig. 28. The tube generally used is termed a normal tube and is exactly 2.0 dm. long. At times depending on high rotation, on depth of color and other factors, a half-normal or 1.0 dm. tube is used. For low rotation, a double normal or 4.0 dm. tube is used. For readings at higher temperatures, for example, 87° C., a jacketed tube is used. The polarimeter in food analysis is used mainly for the estimation of sugar solutions and essential oils.

Polarimeters and saccharimeters are discussed in detail by Bates.[18] The theory and application of polarimetric measurements is discussed by Browne and Zerban.[19] The use of the polarimeter in food analysis is also considered by Woodman.[20]

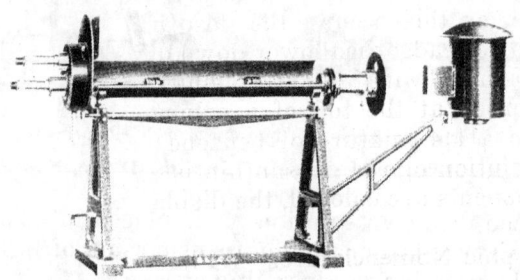

Fig. 28. Saccharimeter (Courtesy of Bausch & Lomb)

CHROMATOGRAPHY

Chromatography is a procedure by which different adsorbed materials may be spatially separated on a single adsorbent. A preferable term might be *adsorption analysis*. It is an analytical method based on the phenomenon of adsorption, that is, the concentration at a surface of dissolved particles from a liquid phase surrounding the adsorbing surface. This process is generally reversible, consequently; adsorbed substances may be removed from the adsorbent by washing. The history, scope, and methods of chromatography are discussed by Zechmeister.[21]

Tswett, a Russian botanist considered to be the originator of chromatographic methods of analysis, recognized that an adsorbent

[18] Bates, *Polarimetry, Saccharimetry, and the Sugars, Nat. Bur. Standards, (U. S.) Circ.* C440, 1942.

[19] Browne and Zerban, *Physical and Chemical Methods of Sugar Analysis.* Wiley, New York, 1941.

[20] Woodman, *Food Analysis.* McGraw-Hill, New York, 1941.

[21] Zechmeister, *Ann. N. Y. Acad. Sci.* 49, 145 (1948).

which is saturated with respect to one substance may still be able to take up a small quantity of a second substance, thus substitution of the first by the second may take place. A substance B may thus be displaced by a substance A, but the reverse substitution cannot take place. Consequently substances can displace each other according to an adsorption series, the precise order of which is dependent on the solvent employed.

As a solution containing different components percolates through an adsorbing column, those with a high affinity for the adsorbent displace those with a weaker affinity for it. The final result is that, as the percolation continues, the components become arranged on the column in the order of their adsorption affinities. Since the process is a reversible one, the bands can be desorbed and moved downward by washing and, as this occurs, the desorbed substance goes into solution again to be readsorbed lower down on the column displacing, at times, a substance with a lower affinity. For instance, yellow zeaxanthin adsorbed at the top of a column of calcium carbonate from solution in a 1:4 mixture of benzene and petroleum ether is displaced if a solution of red capsanthin is poured into the column. Since both components are colored, the displacement can be observed visually.

Chromatographic Nomenclature.—Williams [22] has defined a number of the terms commonly used in chromatography. The vertical tube containing the adsorbent is said to be packed with a *column of adsorbent*. The column with its adsorbed components is termed a *chromatogram*. The process of *development* is the procedure of washing the column with a pure solvent to separate the *zones* or *bands* of adsorbed components. After development is complete, the column of adsorbent may be pushed out of the tube or *extruded*. The extruded column may be mechanically separated into the differently colored zones, and these may be eluted or washed off by means of an appropriate solvent or *eluent*. This term is also used to designate the solvent employed for development. The liquid which is washed through the column is called the *filtrate*, whereas the solution obtained by eluting adsorbed components is known as the *eluate*. The ability of an adsorbent to adsorb dissolved substances is designated its *activity*. More active adsorbents hold adsorbed substances more strongly.

Location of Adsorbed Zones.—The location of adsorbed zones on a column of adsorbent, particularly when colorless or relatively colorless materials are being analyzed by adsorption analysis, can be performed by use of several different techniques. In the case of

<hr>

[22] Williams, *Introduction to Chromatography*. Chemical Publishing, Brooklyn, 1947.

adsorbed substances which can react with another substance to form a color, a reagent may be used. Thus, for example, vitamin A may be located on a chromatogram by use of the Carr-Price reagent, by painting a very thin vertical strip of the column with the reagent. A variation of this method is to convert the entire amount of adsorbed material into a colored compound and, after elution, regenerate the original compound. This can be done with ketones and aldehydes by converting them to the 2,4-dinitrophenylhydrazones. Substances which are colorless in visible light but which fluoresce in the ultraviolet may be located by noting the fluorescent portions of the column.

Separation of Components.—As noted above, the adsorbed components may be separated mechanically by cutting the colored zones of the extruded column. This is the classical method of separation.

It is relatively common to employ the process of obtaining a *liquid chromatogram*. In this procedure, development or washing is continued until the lowest zone on the column is washed through to be collected in a separate container. The development is continued in this manner, substituting a different receiver for each zone that is washed through. The completeness of removal of successive zones can often be checked by means of chemical tests.

Recording Chromatograms.—There are three principal methods of recording chromatograms. In the first method an approximate scale diagram is made. Second, chromatograms can be photographed and thus recorded. Third, the narrowest zone is taken as the unit zone, and the depths of the other zones of the chromatogram are recorded in terms of the narrowest one, for instance, as $3\times$, $5\times$, $8\times$, etc.

Apparatus.—Chromatographic apparatus generally consists of a reservoir containing the solution to be analyzed connected to a tube containing the adsorbent which is connected in turn to a receiving flask. Pressure can be applied to the reservoir or vacuum to the receiving tube. It is to be noted, however, that when low boiling solvents like petroleum ether and ethyl ether are used, it is inadvisable to use vacuum because of boiling away of the solvent.

Small devices are available in which a micro tube can be centrifuged so that centrifugal force can be utilized to assist in the separation.

Adsorbents.—The principal adsorbents used in chromatographic analysis may be listed following Strain [23] as in Table 2.

---

[23] Strain, *Chromatographic Adsorption Analysis.* Interscience, New York, 1945.

TABLE 2. ADSORBENTS

Increasing	Weak	Medium	Strong
	Sucrose	Calcium carbonate	Activated magnesium silicate
	Starch	Calcium phosphate	Activated alumina
	Inulin	Magnesium carbonate	Activated charcoal
	Talc	Magnesia	Activated magnesia
Activity	Sodium carbonate	Slaked lime	Fuller's earth

Eluents.—Among the solvents commonly used in chromatographic analysis are petroleum ether, carbon tetrachloride, cyclohexane, carbon disulfide, anhydrous and alcohol-free ether, acetone, benzene, toluene, esters, chloroform, dichloromethane, 1,2-dichloroethane, alcohols, water and solutions of different pH and salt concentration, pyridine, organic acids, and mixtures of the aforementioned.

Partition Chromatography.—In this method of chromatography equilibrium is established between two liquid phases, one of which is held in the form of a gel. Thus in one system [24] a chloroform solution is one phase and water adsorbed on silica gel, approximately 70 per cent water being present, is the other phase. This system has been used for the analysis of protein hydrolyzates, the individual amino acids being identified by use of ninhydrin and other reagents.

Other procedures of partition chromatography using paper and starch are described by Martin.[25]

Several methods in which chromatography is employed particularly for the isolation of squalene from olive oil and other oils and in a method for the estimation of vitamin B_1 will be detailed in this text.

ION EXCHANGE

Methods of analysis utilizing ion exchangers have not in the past been exploited but an increasing number of papers have appeared in which such methods are detailed.

Ion exchange has been defined as the reversible exchange of ions between a liquid phase and a solid phase which is not accompanied by any radical change in the solid structure. Ion exchangers are

[24] Martin and Synge, *Biochem. J.* 35, 1358 (1941).
[25] Martin, *Ann. N. Y. Acad. Sci.* 49, 249 (1948).

insoluble acids or bases and consequently they are capable of reacting as ordinary acids and bases with the particular exception that one of the reaction products is removed from solution as an insoluble salt.

Ion-Exchange Materials.—Mindler [26] has classified ion-exchange materials according to the following scheme

A. Cation-Exchange Materials
  1. Inorganic cation exchangers
    a. Processed naturally occurring minerals such as the glauconites or greensands
    b. Synthetic inorganic products such as precipitated sodium aluminosilicates
  2. Organic cation exchangers
    a. Processed naturally occurring materials such as sulfonated coal
    b. Synthetic resin cation exchangers
      (1) Modified phenol formaldehyde resins
      (2) Sulfonated hydrocarbons
      (3) Carboxylic acid resins
B. Anion-Exchange Materials
  1. Aromatic amine resins
  2. Aliphatic amine resins

Ion-Exchange Reactions.—The principal reaction of the ion-exchange type used industrially for more than 50 years was that of base exchange. With the development of the organic cation, and anion-exchange materials, other reactions—namely, acid formation, or hydrogen exchange, acid removal, and anion exchange—were made possible. These reactions may be illustrated by the following equations:

Base Exchange:

$$Na_2Z + Ca^{++} \underset{\text{regeneration}}{\overset{\text{reaction}}{\rightleftarrows}} CaZ + 2Na^+$$

in which Na_2Z represents the cation exchanger in a sodium exchange cycle.

Acid Formation:

$$H_2Z + Ca^{++} \underset{\text{regeneration}}{\overset{\text{reaction}}{\rightleftarrows}} CaZ + 2H^+$$

---

[26] Mindler, *Food Technol.* **3**, 43 (1949).

in which H_2Z represents the cation exchanger in a hydrogen exchange cycle. The exchangeable hydrogen ion is generally provided by one of the following radicals: $—SO_3H$, $—CH_2SO_3H$, $—COOH$, $—OH$, $—CH_2OH$, or $—CH_2SH$.

Acid Removal:

$$2R_3N + 2HCl \xrightarrow{\text{reaction}} 2(R_3NH)Cl$$

$$2(R_3NH)Cl + Na_2CO_3 \xrightarrow[\text{regeneration}]{} 2R_3N + 2NaCl + H_2CO_3$$

in which R_3N represents an alkali regenerated anion exchanger.

Anion Exchange:

$$2(R_3NH)Cl + SO_4^= \underset{\text{regeneration}}{\overset{\text{reaction}}{\rightleftarrows}} (R_3NH)_2SO_4 + 2Cl^-$$

Aromatic and aliphatic groups such as $NH_2—$, $NH=$, or $N\equiv$ endow the anion exchangers with their anion binding properties.

One of the first uses of ion exchange in analysis was the determination of ammonia by Folin and Bell[27] employing a sodium permutit. Other examples of such employment for analysis of foods are the estimation of vitamin B_1 and the separation and estimation of mixtures of fruit acids.

ELECTROMETRIC DETERMINATIONS

Hydrogen-ion Concentration and pH.—Electrometric determinations are so often concerned with the estimation of hydrogen-ion concentration that a brief discussion is appropriate.

Quantity and Intensity Acidity Factors.—Clark[28] suggested that it is convenient to consider the term normality as the quantity factor of acidity or alkalinity and the hydrogen-ion concentration as the intensity factor. By quantity or capacity we mean the total available amount of hydrogen ion in a solution or mixture as shown by estimation of the total titratable acidity, a factor generally expressed as normality. By the intensity factor we mean the apparent hydrogen-ion concentration, hence its activity, generally expressed in terms of pH.

[27] Folin and Bell, *J. Biol. Chem.* 29, 329 (1917).
[28] Clark, *Determination of Hydrogen Ions*. Williams & Wilkins, Baltimore, 1928.

Method of Expression.—Hydrogen-ion concentrations vary ordinarily from 1 g. of dissociated ions per liter to less than 1 million-millionth. These concentrations are expressed as:

$$1 = 1 \times 10^0$$
$$0.1 = 1 \times 10^{-1}$$
$$0.01 = 1 \times 10^{-2}$$
$$0.000001 = 1 \times 10^{-6} \qquad \text{etc.}$$

Rarely, however, is the concentration an "even" decimal fraction. Other concentrations may also be written in the same notation, namely,

$$0.000472 \text{ g. per liter} = 4.72 \times 10^{-4} \text{ g. per liter.}$$

Sørensen and Clark introduced a simpler means of notation. Since a logarithm is an exponent of the power of 10, the concentration can be stated in terms of the logarithm, thus:

$[H^+] = 1 \times 10^{-4}$ can be written log $[H^+] = -4$ or $-\log [H^+] = 4$

and since a negative logarithm indicates that the quantity is a fraction, the concentration can be expressed as the reciprocal of $[H^+]$, that is, $\log \dfrac{1}{[H^+]} = 4$. Clark assumed that a symbol

$$pH = \log \frac{1}{[H^+]} \text{ or } [H^+] = 10^{-pH},$$

hence, in the above example, pH = 4. The pH value of a solution may be considered the logarithm of the reciprocal of the hydrogen-ion concentration.

Actually, however, as MacInnes [29] points out, it is not necessary to consider the meaning of pH in terms of theory of solution. The numbers need only be accepted as a practical scale of acidity and alkalinity.

In general, pH values are obtained by measurements of the potentials, E, of cells of the type

$(Pt)H_2$ | Solution X | Saturated KCl | Reference electrode

by use of the equation

$$pH = \frac{E - E_0}{2.306 \, RT/F}$$

where
Solution X = the solution whose pH is being measured
E_0 = a constant depending on the reference electrode
R = gas constant
T = absolute temperature
F = the faraday.

---

[29] MacInnes, *Science* 108, 693 (1948).

Exponential expressions of hydrogen-ion concentration can be converted to pH expressions and vice versa. For example, a solution of acetic acid has a hydrogen-ion concentration equal to 1.36×10^{-3} g. per liter. The corresponding pH may be found as follows:

$$pH = \log \frac{1}{[H^+]}$$
$$= \log \frac{1}{1.36 \times 10^{-3}} = \log \frac{10^3}{1.36}$$
$$= \log 10^3 - \log 1.36$$
$$= 3.00000 - 0.13354$$
$$= 2.86646$$

or as more usually expressed, to the nearest hundredth, pH = 2.87.

For the converse example, let us assume a solution of ammonium hydroxide has a pH = 10.77. The hydrogen-ion concentration may be estimated as follows:

$$pH = 10.77 \text{ is equivalent to } [H^+] = 1 \times 10^{-10.77}$$

The exponent $-10.77 = \overline{11}.000 + 0.230$. These numbers are logarithms and looking up the antilogarithms—that is, numbers whose logarithms have the given value—we find the antilogarithm of 0.230 is 1.7, therefore:

$$[H^+] = 1.7 \times 10^{-11}.$$

On the pH scale, the hydrogen-ion concentration of pure water is 7 at 22° C., whereas a normal solution of hydrochloric acid approaches 0 pH and a normal solution of sodium hydroxide approaches 14 pH.

Buffer Solutions.—Measurement of pH in solution in common use shows results widely different from those with pure solutions, because of the effect of certain substances in the solution on the dissociation of acid and base molecules. When a neutralizing reagent is added to an acid solution containing substances which repress dissociation, the change in pH is less than that occurring when the same quantity of reagent is added to the pure acid solution. This effect is called the buffer action of the substances other than the acid. The power of certain solutions to resist change in hydrogen-ion or hydroxyl-ion concentration on the addition of acids or alkalis, is known as buffer action and such solutions are known as buffer solutions.

Potentiometer.—A potentiometer, Fig. 29, is a device for measuring potential differences by either totally or partially balancing the unknown e. m. f. against a variable potential difference, the value of which is known by reference to a standard of electromotive force. If the two potential differences are exactly balanced, we have the

TABLE 3. COMPARISON OF pH SCALE WITH H-ION CONCENTRATION

	Strong acid						Pure water							Strong base	
pH scale	0	1	2	3	4	5	6	7	8	9	10	11	12	13	14
$[H^+]$ in g/liter	$\frac{1}{1}$	$\frac{1}{10}$	$\frac{1}{100}$	$\frac{1}{1000}$			$\frac{1}{1000000}$			$\frac{1}{1000000000}$			$\frac{1}{1000000000000}$		Fractional notation
	10^0	10^{-1}		10^{-3}			10^{-6}			10^{-9}			10^{-12}		Exponential notation

usual "null" potentiometer, if partially balanced, the galvanometer gives a measure of their difference and we have a deflection potentiometer. This instrument, Fig. 29, consists essentially of a source of

FIG. 29. Potentiometer (Courtesy of Leeds and Northrup)

current generally a 2-volt storage battery or two 1-volt dry cells, a slide wire arrangement such as a Leeds and Northrup potentiometer, a slide rheostat to control the fall of potential along the slide wire and act as a reference potential, a galvanometer to show the direction

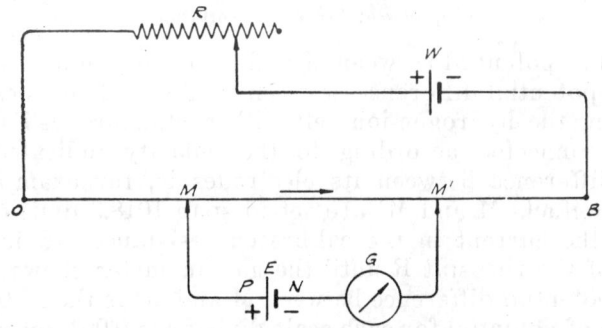

FIG. 30. Potentiometer Diagram
(Courtesy of Leeds and Northrup)

of current flow, and an unknown solution which is turned into a cell with the aid of electrodes. If E in Fig. 30 represents the cell for hydrogen-ion measurements the calomel electrode is the positive pole P, which is connected to M, and the hydrogen electrode is the

negative pole N, which is connected to M^1 through the galvanometer
G. This puts like polarities in opposition. By adjusting the position
of M and M^1 the potential difference between them can be made
exactly equal to that between P and N. When the voltages are balanced the deflection of the galvanometer is zero, and since this is a
current-indicating instrument, it proves that no current is flowing
between the electrodes. Then if the fall of the potential between is
known, the voltage of D is measured without any flow of current
from E.

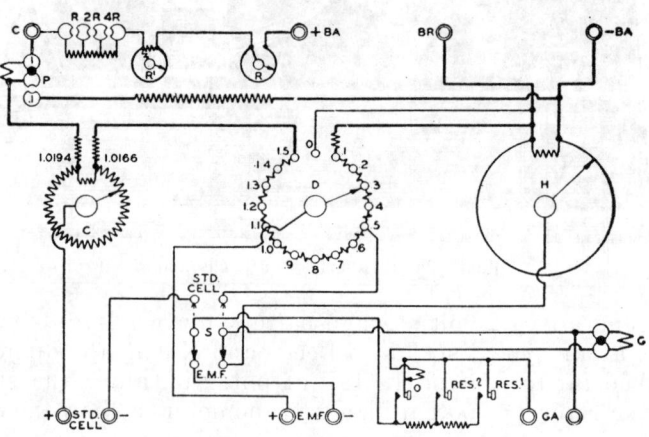

Fig. 31. Potentiometer Diagram, Detailed
(Courtesy of Leeds and Northrup)

The fall of potential between M and M^1 is determined when there
is a known potential difference between O and B. This is established
by replacing the hydrogen-ion cell with a standard cell having its
electrodes connected according to the polarity indicated. If the
potential difference between its electrodes is, for example, 1.0183
volts, the contacts M and M^1 are set to span 1018.3 divisions of the
scale, and the current in the calibrated resistance OB is adjusted
by means of the rheostat R until the galvanometer shows no deflection. The potential difference between M and M^1 is then 1.0183 volts,
and the fall of potential for each scale division is 0.001 volt or 1 milli-
volt. If the hydrogen-ion cell is now substituted for the standard
cell, the potential difference between its electrodes will be shown
directly in millivolts by the number of scale divisions between M and
M^1 when they are now adjusted for voltage balance.

Hydrogen Electrode.—The unknown solution may be set up as a
cell by the use of a saturated calomel half-cell. Fig. 32a, as the
reference electrode and a hydrogen electrode, Fig. 32b, as in Fig. 33.

The electromotive force of this cell is measured with the aid of the potentiometer and the hydrogen-ion concentration or pH of the solution is ascertained by the formula:

$$pH = \frac{E - 0.245}{0.0591} \qquad \text{at } 25° \text{ C.}$$

where E = the observed potential.

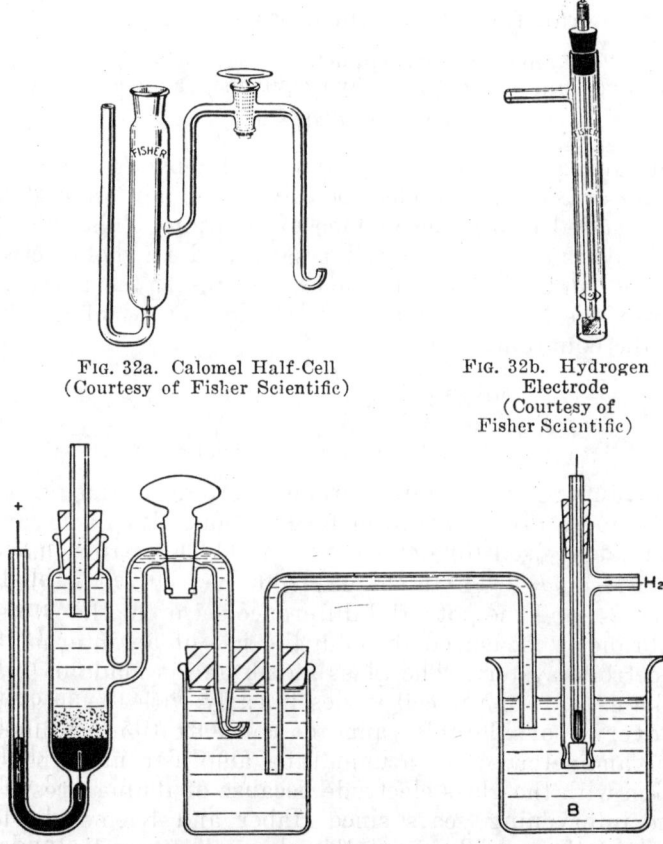

FIG. 32a. Calomel Half-Cell
(Courtesy of Fisher Scientific)

FIG. 32b. Hydrogen
Electrode
(Courtesy of
Fisher Scientific)

FIG. 33. Cell (Courtesy of Leeds and Northrup)

Although the hydrogen electrode is of high accuracy and is applicable to the entire pH range, it is not applicable in the presence of dissolved gases, foaming solutions, oxidizing or reducing solutions, and is easily poisoned by hydrogen sulfide, alkaloids, and hydrocarbons.

Quinhydrone Electrode.—Some of the difficulties encountered in the use of the hydrogen electrode are avoided by using a quinhydrone electrode. Quinhydrone is an equimolecular compound of quinone and hydroquinone. It is only slightly soluble in water, as only 3.94 g. are present in a liter of solution at 25° C., but it is 93 per cent dissociated to quinone and hydroquinone.

$$(C_6H_4)_2O_2(OH)_2 \rightleftarrows C_6H_4O_2 + C_6H_4(OH)_2$$

The probable transformations which occur are:

quinone $+ 2\epsilon \rightleftarrows$ anion of hydroquinone, and
anion of hydroquinone $+ 2H^+ \rightleftarrows$ hydroquinone, or alternatively expressed

$$QH_2 \rightleftarrows Q + 2H^+ + 2\epsilon$$

The quinhydrone half-cell is very simple. It consists of a metal such as platinum or gold as the electrode and a certain amount of quinhydrone is added to a given volume of solution. This half-cell may then be placed in combination with a saturated calomel electrode and the difference in e. m. f. determined by means of the potentiometer. In this case, the hydrogen-ion concentration or pH of a solution is given by the equation:

$$pH = \frac{0.453 - E}{0.0591} \qquad \text{at } 25° \text{ C.}$$

where E = the observed potential.

The advantage of the quinhydrone electrode is that it is simple and needs no catalytic metal surface or gas. Its disadvantage is that it cannot be used for determination of pH greater than 9.

Glass Electrode.—The glass electrode consists of a glass membrane, Fig. 34, and the potential difference set up at the two sides of the membrane is measured through the use of an internal and external electrode system. The glass membrane is made of soda lime glass. The e. m. f. of the cell varies linearly with the hydrogen-ion activity over a considerable range of concentration. A high sensitivity galvanometer or a vacuum tube amplifier must be used in conjunction with the glass electrode because of its high resistance.

In the intervening years since Haber and Klemenziewicz first demonstrated the usefulness of the glass electrode there has been considerable development in their manufacture. Commercial devices, in which the vacuum tube amplifier is inserted into the potentiometric circuit powered by batteries and by direct-line connection, "pH meters," are available in which the glass electrode is used.

In general, in the operation of the direct-line connected type, the instrument is turned on for a specified time to permit it to come to equilibrium. The scale reading is checked on the potentiometric portion of the instrument by measuring the pH of two standard buffer solutions, usually pH 4 and pH 7: If the scale reading does not coincide with the standard buffer solutions, the various rheostats are adjusted to give the proper reading. The temperature control knob is also regulated, if the instrument is equipped with such a device. The cuvette or cup or cell of the device is washed and filled with the liquid to be measured and the pH is then read. Some devices have been devised for the continuous measurement of pH.

The pH, using a glass electrode, may be calculated, as with the quinhydrone electrode by the expression:

$$pH = \frac{0.453 - E}{0.0591} \quad \text{at } 25° \text{ C.}$$

The glass electrode may be used to measure oxidation-reduction systems and is the only pH electrode adaptable to non-buffer solutions. Its disadvantages are that it is easily broken and that it has a high resistance. Special glass electrodes with high resistance to shock and breakage are available.

The potentiometer is not limited to the measurement of a single pH but can be used too for making routine titrations and for many pro-

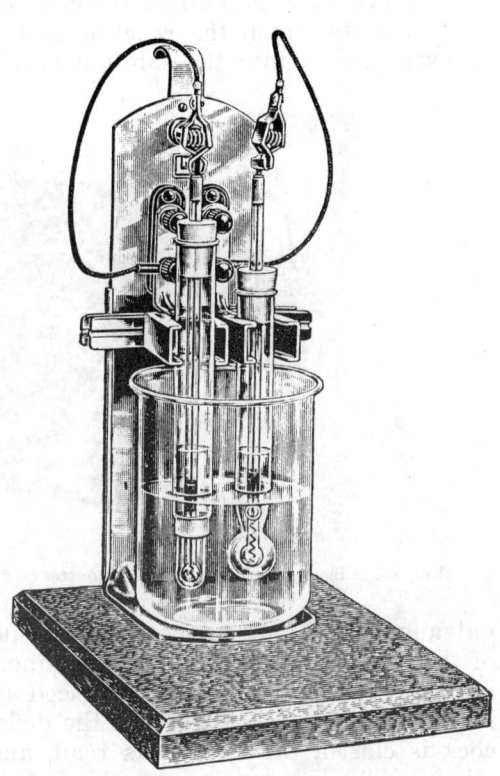

Fig. 34. Glass Electrode Cell
(Courtesy of Leeds and Northrup)

cedures in which a potentiometric end point can be obtained. Thus in following a titration, the e. m. f. may be measured upon successive small additions of acid or alkali in a titration. The pH values derived are then plotted as ordinates and the ml. of reagent as abscissae. First, there is a more or less gradual slope due to the slowly increasing effect of the base upon the acid, then a decided bend as the quantity of base becomes more nearly equivalent to that of the acid, and finally a precipitous drop in the curve with the addition of a very slight amount of alkali, indicating that the reaction has passed the neutral point. With further addition of base, the curve bends again but in the opposite direction, and returns to a gradual slope as the solution becomes increasingly alkaline.

In making routine titrations it is not necessary to develop a complete curve. The potentiometer can be set at the hydrogen-ion concentration to which the reaction is to be carried, and the reagent allowed to flow into the solution quite rapidly, the key closing the

FIG. 35. Line-operated pH Meter (Courtesy of American Instrument Company)

galvanometer circuit being tapped frequently to observe the progress of the neutralization. At the beginning, the deflections will be lively, but they will become less pronounced as the solution approaches the neutral condition. Just before the deflection becomes zero the stopcock is closed, the burette is read, and the solution is thoroughly stirred. The reagent is now added slowly until the equivalent point is reached, at which point the addition of a single drop of reagent will cause a large change in the concentration of hydrogen ions, as will be observed on resetting the potentiometer.

The potentiometer can be used for many reactions in which a difference of potential can be obtained. Many of these methods are applicable to the analysis of foodstuffs. The simple instrument is being replaced in order to do away with reference electrodes by the use of mono-metallic polarized electrodes, the bi-metallic platinum-tungsten system and electronic methods using a continuous reading micro-ammeter. One of the newer potentiometric devices uses a 6E5 cathode-ray tube, the "magic eye" visual tuning indicator of radio, and merely requires plugging into the usual 110-volt alternating-current line. The titrating solution is added as usual from a burette and the end point is noted by the nonfluorescent sector of the screen attached to the apparatus opening a full 90°.

In the food laboratory, the potentiometer and pH meter are used for the determination of the hydrogen-ion concentration of materials such as milk, other dairy products, plant and fruit juices, for following titrations and in general when accurate measurements of acidity and alkalinity need be made. For example, Mitchell [30] made a study of the pH of tomato juice, using the glass, quinhydrone, and hydrogen electrodes for comparison.

Conductometric Measurements.—The conductivity of an electrolyte is not measured directly, but is estimated from a measurement of the resistance of the solution between two electrodes immersed in it. The resistance offered by a 1 cm. cube of a conductor is termed the *specific resistance* or *resistivity*. The reciprocal of this is called the *specific conductance* or *conductivity* and is denoted κ. *Equivalent conductivity* is the conductance of a solution containing 1 gram-equivalent of a solute placed between 2 electrodes of indefinite size and 1 cm. apart and is denoted Λ. It is equal to the conductivity κ multiplied by the volume in ml. [ϕ] containing 1 gram-equivalent of solute, that is

$$\Lambda = \kappa\phi.$$

The resistivity or conversely the conductivity of a solution is generally measured by means of a Wheatstone bridge. The four resistances of this system are arranged so that if three of them are known, the fourth may be calculated. A conductivity apparatus, Fig. 66, has a cell to contain the unknown solution which acts as the unknown resistance. This cell is connected in series with a known but variable resistance. The other two arms of the bridge consist of a slide wire of uniform resistivity with a sliding contact to provide a known ratio. An alternating current and a telephone or an equivalent device to tell the point of minimum sound or current and thus the equality of the ratios completes the set-up. To make a measurement, the sliding contact is moved until the sound in the telephone is a minimum. Since the variable resistance is known and the slide wire ratio may be read, the unknown resistance is given by:

$$\frac{X}{R} = \frac{A}{1000 - A}$$

[30] Mitchell, *J. Assoc. Official Agr. Chem.* **18**, 128 (1935).

and

$$X = \frac{A}{1000 - A} R$$

in which

X = the resistance furnished by the unknown in its cell

R = the variable known resistance

$\dfrac{A}{1000 - A}$ = the variable ratio obtained from the slide wire which is divided into 1000 parts.

Zerban and Sattler [31] and other investigators have used the conductometric method to determine the ash content in raw cane sugar, refined cane sugar and allied sugar products. The principle of the method is that sugar is a nonconductor and that only the electrolytes, which are present in the sugar will conduct. Hence a relationship exists between the conductivity value of a sugar and its ash. The conductivity value of maple products is an official A. O. A. C. method.[32] Conlin [33] describes a method which is given in detail in the chapter on sugars. Bates [34] discusses the electrical conductance of sugar solutions and the determination of ash by this method.

Conductivity measurements may be used to follow usual titrations, neutralizations, precipitation reactions, and in general, any reaction where a minimum in conductivity may be obtained. In fact, whenever a reaction between substances dissolved in two liquids produces such changes in the number or mobility of the ions in the first liquid as to cause its conductivity to reach a maximum or minimum at the completion of the reaction, the end point can be conveniently determined by following the conductivity changes occurring in the solution. Thus whether the solutions are very highly colored or very dilute, reactions taking place within them are still susceptible to this type of measurement.

If then conductance values or more simply bridge values $\left[\dfrac{A}{1000 - A} \right]$ are plotted as ordinates against milliliters of reagent added as abscissae, a titration graph is obtained which consists of two intersecting straight lines. Because of the withdrawal of hydrogen ions and hydroxyl ions or other reacting ions as the neutralization or precipitation proceeds, the conductivity will reach a minimum and then will increase on the further addition of the titrating agent.

[31] Zerban and Sattler, *Ind. Eng. Chem., Anal. Ed.* 3, 41 (1931).

[32] *Methods of Analysis, A.O.A.C.,* 8th ed. Washington, 1955.

[33] Conlin, *Ind. Eng. Chem., Anal. Ed.* 7, 426 (1935).

[34] Bates, *Polarimetry, Saccharimetry, and the Sugars; Nat. Bur. Standards (U. S.) Circ.* C440. Government Printing Office, Washington, 1942.

The point of minimum conductivity is the end point and is obtained graphically.

Conductivity methods for the determination of moisture in foods, principally cereal foods, have been discussed in Chapter I.

FOOD RHEOLOGY

The customary definition of rheology is that it is the science of the deformation and flow of matter. This definition as Blair [35] points out is too broad and thus hydro- and aerodynamics and parts of the theory of elasticity are usually excluded. In the food industry rheological methods are of importance particularly in the milling and baking industries and in the production of certain fat products. Thus it is not possible to mill wheat or other grain adequately if the grains are too hard or too soft for breaking; hence the wheat is tempered in order to adjust its moisture content and distribution so that a proper balance between brittleness and plasticity is attained. Evaluation of these factors can be obtained as noted in Chapter I by rapid moisture determinations such as by measurement of the dielectric constant or conductivity, but often a rheological method is of greater value. Rheological methods are discussed in various sections of the text, particularly in Chapter IX on fats, in Chapter VIII in connection with the testing of the curd of milk, and in Chapter XI in the testing of dough. Other examples of rheological methods applied to foods are assessing when the proper amount of whey has drained out syneretically from cheese curd in the scalding process in making hard-pressed cheese by measuring the superficial density and in noting the correlation of certain fats with their iodine numbers.[36]

VISCOSITY

The viscosity of a fluid is the internal friction which tends to bring to rest portions of the fluid which are moving relative to one another. It is the resistance which particles of a fluid offer to free motion one over the other. It is measured in relation to some standard viscosity generally water at 25° C. or sugar solutions. Liquids which do not have a high viscosity may be measured by determining the time rate of flow through a capillary of a definite volume of the fluid and then applying Poiseuille's formula:

$$\eta = \frac{\pi \Delta p r^4 t}{8Vl} = Ct\Delta p$$

---

[35] Scott Blair, *Analyst* 74, 387 (1949).
[36] Scott Blair, and Coppen, *J. Dairy Research* 11, 187 (1940).

in which

η = coefficient of viscosity
Δp = the driving force = difference in pressure between the two
ends of the capillary tube
r = the radius of the capillary tube
t = the time required for
V = the volume of liquid to flow through the tube
l = length of the capillary tube

and

C = a constant for a given capillary viscosimeter equal to
$$\pi r^4/8Vl$$

Capillary Tube Viscosimeters.—The *Ostwald viscosimeter* is a U-shaped tube of glass, Fig. 36, one arm of which contains a bulb

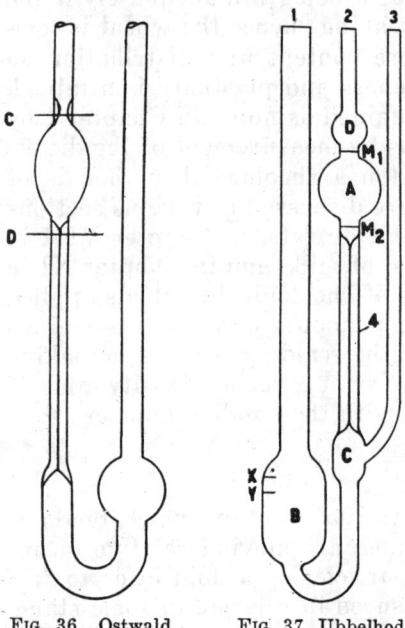

connected to a capillary and the other arm contains a wider tube and a bulb. Because the hydrostatic head of these tubes cannot be varied, a series of tubes of different capillary size are required to cover a range of different viscosities.

Procedure.—Transfer an adequate volume of the liquid to be tested to the wide arm of the Ostwald tube. Force the liquid up through the capillary and the adjoining bulb to the mark C. Measure the time the liquid requires to fall to the mark D.

The Ostwald-Cannon-Fenske modification [37] is designed after the original Ostwald viscosity tube but the capillary and the lower portion of the receiving limb are bent at an angle to minimize errors in standardization.

FIG. 36. Ostwald
Viscosimeter

FIG. 37. Ubbelhode
Viscosimeter

Ubbelhode Viscosimeter.—One of the forms of the Ubbelhode viscosimeter, which is a suspended level device, is shown in Fig. 37.

Procedure.—Fill bulb B through tube 1 until the meniscus lies between marks X and Y. Close tube 3 at the top and apply suction to tube 2 until the liquid being tested is drawn above mark M_1. Open

[37] Cannon and Fenske, *Oil Gas J.* 34, No. 47, 45 (1936).

tubes 2 and 3 to the air. This divides the liquid in C into two parts producing the suspended level at the bottom end of the capillary. With the formation of the suspended level the liquid flows in a thin layer on the vertical walls of C into B. Determine the time interval required for the meniscus of the liquid being tested to fall from M_1 to M_2.

Short-tube Efflux Viscosimeters.—Fluids which have a high viscosity cannot be measured by such means, and other devices are used. The Engler and Saybolt viscosimeters use the principle of measuring the time rate of flow of a given quantity of material, as, for example, oil through a standard orifice.

Rotation Viscosimeters. — The MacMichael, Fig. 38, and the Doolittle viscosimeters operate on the torsion principle. A plunger of standard dimensions is suspended by a torsion wire of exact length from the top of the instrument. The material is placed in a cup, which is revolved at a constant rate of speed on a motor driven platform mounted on ball bearings until the drag of the liquid is balanced by the resistance of the suspended wire to twisting. The amount of twist imparted to the wire, depending upon the viscosity of the material,

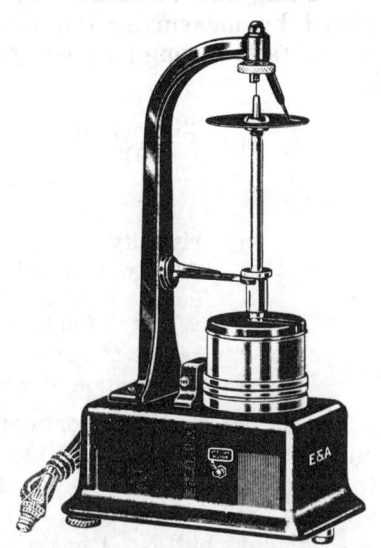

FIG. 38. MacMichael Viscosimeter
(Courtesy of Eimer & Amend)

is read on a graduated disc attached to the spindle. The readings are in arbitrary degrees M ($\frac{1}{300}$ circle). However, by standardizing against solutions of known viscosity in poises, the results can be interpreted in poises, $i.e.$, C. G. S. absolute units of viscosity.

Viscosity in poises may be calculated by use of the equation

$$\eta = C \, \frac{M^\circ}{HN}$$

where
η = viscosity
C = a constant for the instrument
M° = deflection in arbitrary degrees MacMichael
H = depth to which the pendulum plunger is suspended in the liquid
N = number of revolutions per minute of the cup.

To provide for the measurement of a range of viscosities, the instrument is equipped with several wires of different gage.

In the Stormer viscosimeter the viscosity is measured by determining the speed achieved by a central rotor in a stationary concentric cylinder containing the liquid whose viscosity is to be measured by the action of a weight acting through a spindle, a gear, and a pinion. The number of revolutions made by the cylinder in distilled water is compared with the number made in the sample liquid.

Falling Ball Viscosimeters.—The viscosity of a liquid can be determined by measuring the velocity with which a sphere of known radius falls through a medium. This relationship is given by Stokes law

$$\eta = \frac{2(\rho_b - \rho)gr^2}{9V}$$

where

$\eta = $ viscosity
$\rho = $ density of liquid
$\rho_b = $ density of ball
$V = $ velocity of sphere
$r = $ radius of sphere
$g = $ acceleration of gravity.

In the Hoeppler viscosimeter, the absolute viscosity can be determined by measuring the time of fall of a glass or steel ball through the liquid held in an inclined glass tube of precise internal diameter. A set of steel balls is required to cover a wide range of viscosity. Thus ball No. 1 has a time interval of 68.62 seconds in benzene equivalent to a viscosity of 0.65. With water the time interval of fall is 96.51 equivalent to a viscosity of 1.0046. Ball No. 2 has a fall of 43.71 seconds in petroleum D equivalent to a viscosity of 1.885. With such a set of balls a range of viscosity from 0.6 to 100,000 centipoises can be measured.

For a given tube and a given ball, the ratio of the viscosities of two liquids is

$$\frac{\eta_1}{\eta_2} = \frac{(\rho_b - \rho_1)}{(\rho_b - \rho_2)} \frac{t_1}{t_2}$$

in which t_1 and t_2 are the times of fall in two different liquids and ρ_1 and ρ_2 are their densities.

The viscosities of oils, gum solutions, gelatin solutions, ice cream mixes, sugar solutions, acidulated flour, serums, and like materials and prepared solutions and mixtures can be obtained by the use of this instrument and similar instruments and are valuable indices of whether a food material is standard or sub-standard.[38]

SURFACE TENSION APPARATUS

The surface particles on any given liquid have greater molecular forces tending to pull these particles toward the interior of the liquid than outward. Hence, a fluid surface acts like a stretched elastic membrane. This effect is called surface tension. It measures the work bringing molecules from the interior to the surface of the liquid and numerically is the force necessary to balance the tendency to contract in a strip of surface film one centimeter in width. The surface tension of a substance is characteristic of that substance. Surface tension is measured in dynes per centimeter.

There are two classes of methods of measurements for surface tension, the static and dynamic. Among the static methods are the capillary rise, weight of hanging drops, adhesion of a disc to a surface and direct measurement of curvature of the surface at contact with a plane. Among the dynamic methods are oscillating jet, capillary undulations, etc., in which new surfaces are being formed. An instrument designed to measure surface tension easily, rapidly and accurately is that of Du Noüy. This apparatus is a torsion balance which, instead of measuring the tension by means of weights, uses the torsion of the wire to counteract the tension of the liquid film and break it. A single reading on a dial indicating the degree of torsion of the wire gives a figure which, if the apparatus has been previously standardized with water, gives the surface tension of the liquid by a simple proportion.

The instrument, Fig. 39, consists essentially of a stand provided at the top with a fine steel wire stretched between end supports. One end of the wire is tightly clamped, the other being attached to a worm wheel controlled by a thumb-screw. To the worm wheel is also attached a pointer which moves over a metal scale graduated in degrees. To the middle of the wire is clamped a hollow, light steel lever with a small hook in the outer end. A stirrup is attached to this hook carrying a carefully made loop of platinum-iridium wire with a periphery exactly 4 cm. long.

The liquid in a watch glass is placed on the platform and carefully raised by means of the adjusting screw until the platinum loop has made contact with the liquid. The pointer having been previously set at zero, the torsion of the wire is gradually increased by means of the thumbscrew controlling the worm gear, until the loop of wire

[38] Harbel, *J. Assoc. Official Agr. Chem.* **18**, 577 (1935).

tears loose from the liquid. The number of degrees is then read from the scale and by a simple calculation is converted directly into dynes per centimeter.

Another method of measuring surface tension is to determine the drop-number by means of a Traube stalagmometer, to ascertain the density and then calculate the surface tension.

Surface tension methods are valuable for indicating the purity of oils and alcohols. Instruments of this type have been used to determine constants of serums such as those made from milk.

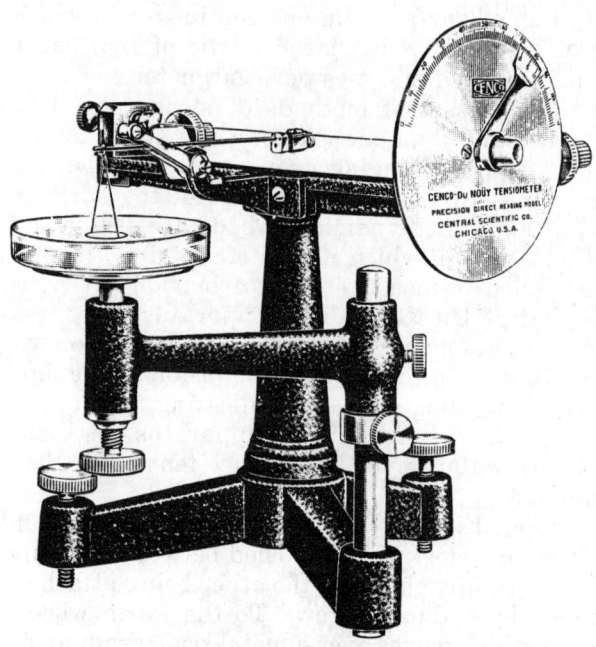

FIG. 39. Du Noüy Surface Tension Apparatus
(Courtesy of Central Scientific Company)

POLAROGRAPHY

Polarographic analysis is based on the principle that the kind and amount of electro-reducible and electro-oxidizable components present in a solution can be determined from the current-voltage relationships revealed by use of the dropping mercury electrode. This is an electrolytic device which has one electrode from which mercury can emerge in small drops from a capillary. While such electrodes have been in use for more than 50 years, it was not until the work of Heyrovsky [39] was interpreted by Kolthoff and his co-workers,[40, 41] that wider use was made of this technique.

There are a number of advantages in using the polarographic method. These have been stated by Feicht, Schrenk, and Brown [42] as (1) organic as well as inorganic substances can be determined, (2) simultaneous qualitative and quantitative determinations may be made, (3) samples containing but a few micrograms of the substance being determined are adequate, (4) many samples require little preparation, (5) several components may be estimated in the same solution, (6) determinations may be made very rapidly, (7) the sample solution is not altered and hence may be used for other determinations, and (8) equipment is commercially available.

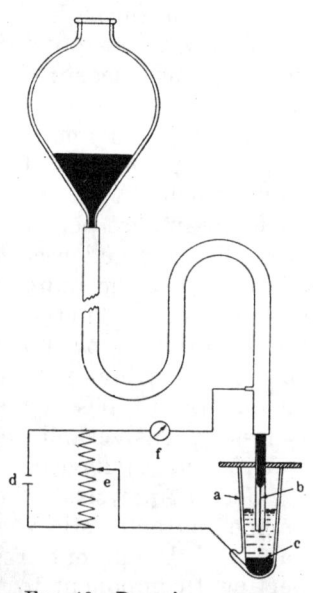

As representative of this method, the determination of lead will be described following the procedure of Feicht, Schrenk, and Brown.

Apparatus.—The essential parts of the dropping-mercury electrode are shown in Fig. 40. These are: Cell *a* for holding the solution that contains the component to be estimated; dropping-mercury electrode *b* which is ordinarily used as a cathode but may be used as an anode in determining electro-oxidizable components; electrode *c* consisting of a

FIG. 40. Dropping-mercury Electrode [41]

pool of mercury on the bottom of the cell; battery *d* for supplying electromotive force; potentiometer *e* for varying the potential across the cell; and galvanometer *f* for measuring current that results from the diffusion of ions of the substance to be estimated to the dropping electrode.

Various types of cells are available to accommodate different volumes of solution and for special purposes.

The dropping electrode consists of a reservoir of mercury usually connected by pressure rubber tubing to a glass capillary, 0.03-0.04

[39] Heyrovsky, *Phil. Mag.* [6] **45**, 303 (1923).
[40] Kolthoff and Lingane, *Polarography.* Interscience, New York, 1939.
[41] Kolthoff and Laitinen, *pH and Electro Titrations.* Wiley, New York, 1941.
[42] Feicht, Schrenk, and Brown, *U. S. Bur. Mines, Rept. Invest.* 3639 (1942).

mm. in diameter, from which small drops of mercury, about 0.5 mm. in diameter, issue at intervals of about 3 to 4 seconds or 15 to 20 drops per minute.

The second electrode may be a quiet mercury electrode, greater than 1 cm.^{2}, in the bottom of the cell, or an external calomel, Hg_2Cl_2, electrode connected to the cell by a low-resistance salt bridge.

Any battery such as a lead storage battery or dry cells capable of providing potentials up to 3 volts is suitable. Ordinarily the current passing through the solution is less than 50 microamperes.

Slide-wire or potentiometer-type radio rheostats are used for varying and measuring the potentials applied to the electrolytic cell.

Sensitive galvanometers of the D'Arsonval or reflecting type are employed to measure the current. The maximum sensitivity of the galvanometer, with suitable shunts in the circuit, should be at least 0.1 microampere per mm. and preferably 0.01 per mm.

Indicating and recording type dropping-mercury electrodes are available. In the indicating type, the voltage regulator is operated by hand and the current is indicated by the galvanometer, whereas in the recording type, the voltage is regulated automatically and the current at various voltages is recorded photographically or by other means. Both types give equally reliable results. The indicating type is less expensive and is generally as satisfactory as the recording type for most purposes.

Use in Analysis.—The qualitative features of the method depend on the characteristic values of the half-wave potentials at the half values of the currents resulting from the diffusion of the ions of the pertinent component to the dropping electrode and the quantitative features upon the final or limiting value of the diffusion current. Consideration of a particular determination is helpful in understanding the principles involved. A current-voltage curve for lead is shown in Fig. 41.

Air must be removed from the solution to be tested, as oxygen is reduced readily at the dropping-mercury cathode and consequently interferes with the determination of other components. It can be removed by passing some inert gas, such as nitrogen or hydrogen, through the solution for several minutes before the determination but not during the determination because the stirring produced would disturb the dropping of the mercury and cause abnormal current fluctuations. Commercial nitrogen or hydrogen may be used, but these gases should not contain more than a fraction of a per cent of oxygen. Excess oxygen may be removed from the nitrogen or 'hydrogen by passing the gas through a Pyrex or quartz tube containing copper turnings heated to about 450° C.

As the applied potential is increased to a certain value, a small, steadily increasing residual current flows. This current results from the building up of a charge on the constantly changing dropping-mercury surface corresponding to the applied potential and by acquisition of electrons by any substance reducible in this range of potentials.

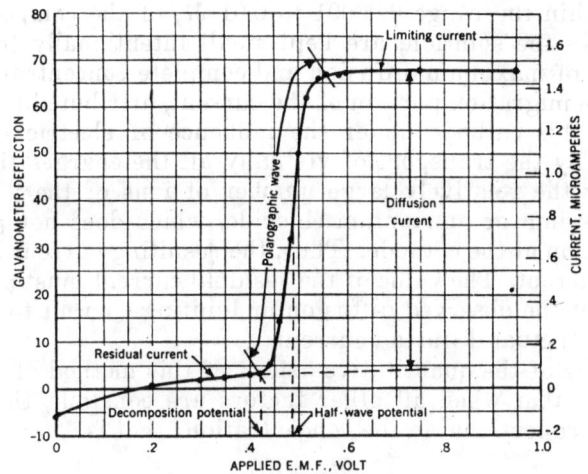

FIG. 41. Current-voltage Characteristics [41]

(Air-free lead nitrate solution 0.0003 molar with respect to lead and about 0.1 molar with respect to potassium chloride)

Continuous electrolysis, that is, passage of current other than the residual current through the solution, begins when the applied potential reaches the decomposition value. The reduction of lead and its combination with the dropping mercury to form a very dilute amalgam begin at the dropping cathode, and the loss of electrons or dissolution of a corresponding amount of mercury from the anode and the combination of these mercury ions with the chloride ions to form insoluble mercurous chloride, Hg_2Cl_2, begin at the anode. These reactions may be illustrated by the following equations:

$$Pb^{++} + 2e \rightarrow Pb(Hg) \qquad \text{at the dropping cathode}$$

$$2Hg - 2e \rightarrow 2Hg^+ + 2Cl^- \rightarrow Hg_2Cl_2 \qquad \text{at the anode}$$

As the applied potential is increased above the decomposition value, reduction at the dropping cathode increases and causes the concentration of the lead ions in the immediate vicinity to decrease; this decrease in turn causes the diffusion of the lead ions from the solution to the cathode and the consequent concentration polarization and diffusion current to increase. Increase in diffusion current continues with increase in the applied potential until the lead ions are reduced as fast as they reach the cathode. Thus the concentration of these ions at the cathode reaches a minimum value, and the diffusion and concentration polarization a maximum value. The size of the mercury drops at the dropping electrode and the concentration, usually within the range 0.00001 to 0.01 M, of the component to be estimated in the solution, are kept small intentionally to facilitate attainment of maximum diffusion and complete concentration polarization. The migration portion of the current, attributed to the travel of ions to this cathode under the influence of electrical forces, is eliminated by the transport of virtually all the charges through the solution by the relatively large number of ions of the "indifferent salt" potassium or ammonium chloride, which does not participate in the reaction at the cathode. Thus, the resulting current is purely a diffusion current. The value of the residual current must be deducted from that of the observed diffusion or limiting current to obtain the value of the actual diffusion current.

The basis of the quantitative aspect of this method of analysis is the relation that when all other factors are constant, the diffusion current is proportional to the concentration; that is:

$$i_d = KC$$

However, this relationship holds only when the dropping time is equal to or more than 3 seconds per drop.

The diffusion coefficient of the component being determined and the amount of mercury issuing from the cathode per second are affected by the temperature; consequently the temperature should be controlled to at least 0.5° C. to keep variations attributed to this factor within 1 per cent in the measurement of the diffusion current. The temperature can be readily controlled within this range by means of a large beaker of water.

Instead of rising only to the limiting value, as shown in Fig. 41, the current may increase almost directly with the applied potential to a so-called maximum before falling more or less suddenly to the limiting value. These maxima usually can be prevented by the use of a solution of gelatin, tylose, or a mixture of methyl red and bromocresol green.

The relatively large, quiet pool of mercury ordinarily used as the second electrode, having an area larger than 1 cm.^{2}, is depolarized completely and has a constant potential when present in solutions containing halide or other ions that form insoluble salts with mercury. In solutions containing chloride ions the potential of the electrode is about the same as that of a calomel electrode at the same

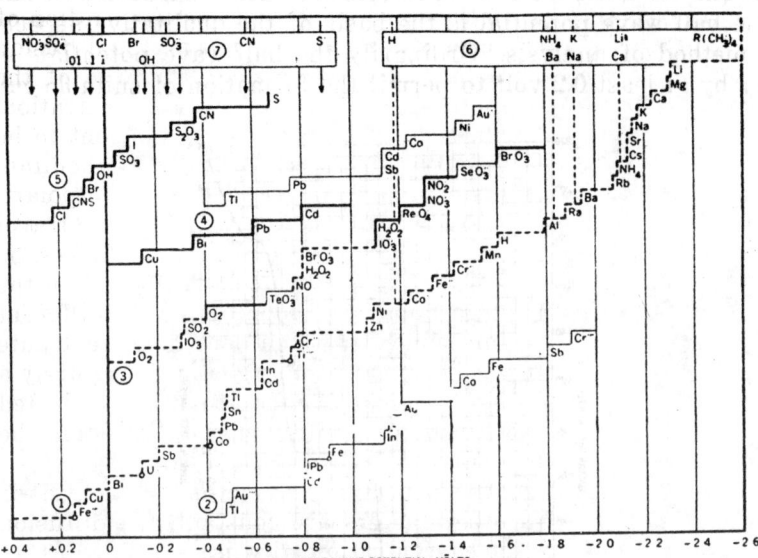

FIG. 42. Vladimir Majer's ''Polarographic Spectrum'' of Half-wave Potentials of Inorganic Ions in Various Media [41]

1. Reduction and deposition potentials of cations in neutral or acidic solution.
2. Reduction and deposition potentials of ions in alkaline solution.
3. Reduction potentials of anions and molecules in neutral (————), or in acidic (- - - -) solution and in NH₃-buffer (———). NO₃' and NO₂' in 0.1 N LaCl₃.
4. Reduction potentials of complexes.———: in 10 per cent sodium potassium tartrate,———: in I N KCN.
5. Depolarizing potentials of anions. (Anodic polarization)
6. Deposition potentials of cations of the most used indifferent electrolytes present in excess (1000 X) in the solution.
7. Potentials of the large mercury reference electrode in solutions containing the usual anions in about I N concentration. (Chloride in concentrations: 1.0, 0.1, 0.01 N).

All values give the potentials at the half-diffusion current except those which are marked by a little circle signifying the contact point of a 45° tangent at a 2-cm. wave. The values of the potentials plotted refer to the I N calomel electrode as zero and to room temperature. (By Vladimir Majer)

chloride-ion activity. The potential of the dropping electrode is the difference between the observed potential and the potential of this quiet electrode.

Although characteristic of the electro-reducible or electro-oxidizable component present, the decomposition potential of a solution depends partly upon the concentration of the pertinent component.

The half-wave potential (see Fig. 41) is constant and independent of the concentration of the reducible metal ions, the particular capillary used, and the dropping time of the capillary, provided the temperature and composition of the solution are constant with respect to substances other than the reducible one. This characteristic nature of the half-wave potential is the basis of the qualitative aspects of this method of analysis. Ordinarily the half-wave potentials must differ by at least 0.2 volt to permit the formation of suitable curves

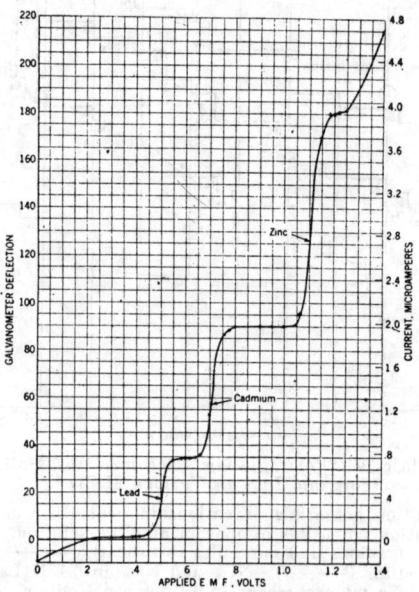

FIG. 43. Current-voltage characteristics of Lead, Cadmium, and Zinc in the Same Solution [41]

for the determination of more than one component in the same solution.

At least three methods of using the dropping-mercury electrode for quantitative analysis are available.

General Method.—The dropping-mercury electrode is calibrated against known concentrations of the substance to be estimated under the same conditions planned for the actual analysis. These resulting data are plotted to give a calibration curve.

Standard-addition Method.—A current-voltage curve for the unknown amount of a standard solution of the pertinent substance is determined in the regular manner, and then a known amount is added and a second curve is obtained. The original concentration of the substance to be estimated is then calculated from the values of the diffusion currents indicated by the two curves after a correction has been applied for the change in volume.

Pilot-ion or Internal-standard Method.—The relationship between the diffusion current of the substance to be estimated and that of another substance used as the internal standard in the solution, having the same kind and concentration of indifferent salt or supporting electrolyte and the same temperature as the solution to be used in the actual analysis, is obtained from actual determination. The ratio of the diffusion current of the pertinent substance to that of the internal standard may then be plotted against the concentration of the substance being determined to give a calibration curve; the concentration of the pertinent substance may be determined from the diffusion current of this and of the internal standard in the actual solution. Results obtained by this procedure are independent of the particular capillary and, within certain small limits, temperature changes in the solution. The need for only the internal-standard solution rather than for a separate one for each substance being determined, as in the standard-addition method, is one of the important advantages of this procedure.

After the complete current-voltage curve has been determined for a particular substance the concentration of the same substance in similar solutions under the same conditions may be obtained from the difference between the current found at a potential slightly below the point at which the current increase begins and that found at a potential slightly above the point at which the current increase ends.

Procedure.—Ash biological material at 500° C. as described in Chapter V. Dissolve the ash by the addition of nitric acid and water and rinse into glass-stoppered Pyrex cylinders or volumetric flasks, adjusting the volumes with water as dictated by experience. Extract an aliquot or the entire sample, as for instance $\frac{1}{20}$ of a day's mixed food, with dithizone solution, 0.03 g. in 1 liter of chloroform, preferably in the presence of citrate and cyanide (see Chapter V). Wash the dithizone extract with 50 ml. of water. Shake out the lead with 20 to 40 ml. of nitric acid (1:100). If the lead content is sufficiently high, the acid washing of the extract may be polarized as described below. If the lead content is low, evaporate to a small volume, 3 to 5 ml., and proceed with the method.

Place 1 or 5 ml. of the diluted or reconcentrated sample solution, as the case may be, in the electrolytic cell of the dropping mercury electrode. Add 1 ml. of a solution 0.7 N with respect to potassium chloride containing 2.5 ml. per liter of a "maxima suppressor" solution, 3 parts of a 0.2 per cent alcoholic solution of methyl red, and 2 parts of a 0.2 per cent alcoholic solution of bromocresol green. If zinc or cadmium is known to be absent, add 0.5 mg. of the absent metal from a nitrate solution containing 0.5 mg. per ml. to serve as the internal standard. Dilute the solution to about 7 ml. with water and acidify to about pH 3, as indicated by an indicator paper and the methyl red and bromocresol green indicator solution with dilute nitric acid and potassium hydroxide or sodium hydroxide solution. Bubble nitrogen, shown by analysis to contain less than 0.25 per cent oxygen, through the solution for at least 10 minutes. Seal the electrolytic cell to the dropping-mercury electrode with a rubber connection to prevent the re-entry of oxygen. Adjust the dropping time to maintain a rate of 6 to 7 seconds per drop. Use of the internal-standard procedure at room temperature eliminates the necessity for exact temperature control.

FREEZING POINT DETERMINATIONS

A property depending entirely on the number of molecules present is termed a colligative property. Such colligative properties are the volume of a gas and for substances in solution, vapor pressure, osmotic pressure, elevation of boiling point, depression of freezing point and others. In food analysis, we seldom use methods that estimate any of these colligative properties other than that of freezing point depressions. Van't Hoff showed that these properties were all interrelated.

Since the freezing point depression is dependent on the number of dissolved particles, then quantities proportional to the molecular weights—that is, quantities containing the same number of molecules, dissolved in identical weights of a solvent—will cause equal depressions of the freezing point of that solvent.

It is known that the osmotic pressure of body fluids, such as blood and milk are almost constant. Hence the freezing points of these fluids are also almost constant. This constant freezing point in milk is due to the combined effects of three of the soluble components of milk, namely, lactose, the salts of potassium and sodium, and the salts of calcium and magnesium. The smallest osmotic pressure of genuine milk ever observed is represented by a freezing point depression of about 0.53° C., as determined by the Hortvet method. The average depression obtained by different observers varies from 0.54° to 0.55° C. If, therefore, a fresh milk gives a freezing point depression of less than 0.53° C., it is definite proof of the presence of added water.[43] Thus from an estimation of the freezing point of a milk the amount of added water may be ascertained.

Hortvet Cryoscope Method.—The method in general use, an official method of the A. O. A. C., is that of Hortvet.[44] The apparatus, Fig. 44, consists of an inner glass tube for the sample; a larger metal tube to hold the glass tube; a vacuum flask to insure constant tem-

FIG. 44. Freezing Point Apparatus
(Courtesy of Eimer & Amend)

perature conditions; a metal stirrer; a metal tube for vaporizing the solvent producing the low temperatures; a metal "T" shaped tube; a large cork stopper; an inner and outer air drying tube; a glass tube for holding thermometers when not in use; a freezing starter; glass funnel for pouring in the solvent; a vapor disposal tube; a special thermometer graduated from 1° above zero C. to 2° below

[43] Evans, *Analyst* **61**, 666 (1936).
[44] Hortvet. *Ind. Eng. Chem.* **13**, 198 (1921).

zero, subdivided into 0.01° C.; a control thermometer graduated
from 30° above zero C. to 30° below zero, subdivided in 1/1° C.; a
thermometer scale magnifier for observing accurately and estimating
to 0.001° C.; a small cork mallet for tapping thermometer; and a
supporting stand including a milk glass scale.

Procedure.—Insert a small caliber funnel tube into the vertical
portion of the "T" tube at one side of the apparatus and pour in
400 ml. of ether previously cooled to 10° C. or lower. Close the vertical tube by means of a small cork and connect the pressure pump to
the inlet tube of the air drying attachment. Adjust the pump so as
to pass air through the apparatus at a moderate rate, as may be
judged by the agitation of the sulfuric acid in the drying tube. Continuous vaporization of the ether will cause a lowering of the temperature in the flask, from ordinary room temperature to 0° C. in
about 8 minutes. Continue the temperature lowering until the control thermometer registers near −3° C. At this stage, by lowering a
narrow gauge graduated glass tube into the ether bath, then closing
the top by means of the forefinger and raising to a suitable height, an
estimate can be made as to the amount of ether necessary to pour in
for the purpose of restoring the 400-ml. volume. When the apparatus
has once been cooled down to the proper temperature, an additional
10 to 15 ml. of ether is sufficient on an average for each succeeding
determination. Measure into the freezing test tube 30 to 35 ml. of
boiled distilled water, cooled to 10° C. or lower. Enough water
should be measured in to fairly submerge the thermometer bulb.
Insert the thermometer together with the stirrer and lower the test
tube into the larger tube. A small quantity of alcohol, sufficient to
fill the space between the two test tubes, will serve to complete the
conducting medium between the interior of the apparatus and the
liquid to be tested. A sufficiently tight connection between the inner
and outer tubes is afforded by means of a narrow section of thin
walled rubber tubing. Keep the stirrer in steady up and down motion
at a rate of approximately one stroke each 2 or 3 seconds, or even at
a slower rate providing the cooling proceeds satisfactorily. Maintain passage of air through the apparatus until the temperature of
the cooling bath reaches −2.5° C., at which time the top of the mercury thread in the standard thermometer usually recedes to a position in the neighborhood of the probable freezing point of water.
Maintain the temperature of the cooling bath at −2.5° C. and continue the manipulation of the stirrer until a super-cooling of the
sample of 1.2° is observed. As a rule, by this time the liquid will

begin to freeze, as may be noted by the rapid rise of the mercury thread. Manipulate the stirrer slowly and carefully three or four times as the mercury column approaches its highest point. By means of a suitable light-weight cork mallet tap the upper end of the thermometer cautiously a number of times until the top of the mercury column remains stationary a couple of minutes. Taking the necessary precautions to avoid parallax, observe the exact reading on the thermometer scale and estimate to 0.001° C. When the observation has been satisfactorily completed make a duplicate determination, then remove the thermometer and stirrer and empty the water from the freezing tube.

Rinse out the test tube with about 25 ml. of the sample of milk, previously cooled to 10° C. or lower, measure into the tube 35 ml. of the milk, or enough to fairly submerge the thermometer bulb, and insert the tube into the apparatus. Maintain the temperature of the cooling bath at 2.5° C. below the probable freezing point of the sample. Make the determination on the milk, following the same technique as that employed in determining the freezing point of water. As a rule, however, it is necessary to start the freezing action in the sample of milk by inserting the freezing starter, carrying a fragment of ice, at the time when the mercury column has receded to 1.2° below the probable freezing point. A rapid rise of the mercury column results almost immediately. Manipulate the stirring device slowly and carefully two or three times while the mercury column approaches its highest point. Complete the adjustment of the mercury column in the same manner as in the preceding determination; then, avoiding parallax, observe the exact reading on the thermometer scale and estimate to 0.001° C. The algebraic difference between the reading obtained on the sample of water and the reading obtained on the sample of milk represents the freezing point depression of the milk.

The percentage of added water may be calculated as follows:

$$W = \frac{100(t - t^1)}{t}$$

in which

t = the freezing point of normal milk which averages −0.55° C.

t^1 = the observed freezing point of the sample

W = added water, per cent by weight.

A more accurate formula which corrects for the fact that the added water should be calculated by using weights of solvent and not of solution, that is, one agreeing with usual custom and hence referring always to 100 g. of water, is:

$$W = \frac{t - t^1}{t} (100 - T.S.)$$

in which

T.S. = the percentage of total solids in the milk and the other symbols have the defined meaning.[45]

It is customary to use 7 and 10 per cent (W/V) solutions of pure sucrose in the standardization of the thermometers used in this method. A 7 per cent sucrose solution should give by the Hortvet method a freezing-point depression of 0.422° C. and a 10 per cent sucrose solution should give a depression of 0.621° C. Assuming that the spacing of the graduations between the two freezing point depressions is equidistant, corrections can be applied in the standardization of the thermometers.

Sutton [46] and co-workers suggest the use of an auxiliary 8.75 per cent sucrose solution with a freezing point depression of 0.536-7° C. as an additional check point. They discuss methods of eliminating errors in the Hortvet method.

It is essential that the freezing point determination be made only on fairly fresh milk, owing to the fact that the development of acidity to the extent of 0.10 per cent beyond normal for fresh milk, approximately, 0.15 per cent, lowers the freezing point about 0.025° to 0.030° C. Therefore, milk that shows an acidity greater than 0.18 per cent as determined by the method detailed in the chapter on milk and cream, Chapter VII, should not be subjected to the test.

The percentage of added water in cream may be estimated in a similar manner. However, the following formula is used for the calculation:

$$W = \frac{t - t^1}{t} \% \text{ serum in cream}$$

% serum in cream = 100% − (% fat + % protein) in which, if per cent protein is not estimated, it may be assumed to be 38 per cent of the solids-not-fat. The other symbols have the defined meanings.

While as mentioned on pages 96 and 97, a freezing point depression of less than 0.53° C. is definite proof of added water, it is the convention to use 0.55° C., the average freezing point depression of milk, in the formulas noted above. In the A. O. A. C. method of calculation, a tolerance of 3 per cent is allowed on all results for

---

[45] Elsdon and Stubbs, *Analyst* 61, 382 (1936).
[46] Sutton, Markland, Barraclough, and Chapman, *Analyst* 75, 42 (1950).

added water determined on the basis of the average freezing point depression of 0.550° C. However, in this recommended method of calculation it is not considered necessary to deduct the tolerance figure from results showing added water in excess of 3 per cent.

Paley and Tzall [47] in a study of the freezing points of 1450 samples of fluid market milk found only 4 samples having freezing points of −0.550° C., 3 had −0.551° C., and 2 greater than this representing only 0.65 per cent of the total. Some 40 per cent had depressions within the tolerance set by the A. O. A. C. method. The greatest number, representing 9 per cent, had freezing points of −0.531° C. These investigators suggest that since the average depression value of 0.55 is one based on work performed in the 1920's and on market milk of that period, a new average value should be adopted based on market milk of the present time.

SELECTED REFERENCES

Bates, *Polarimetry, Saccharimetry, and the Sugars; Nat. Bur. Standards (U.S.), Circ.* **C440**, 1942.
Bigelow, National Canners Association, *Bulletin No. 27 L* (1934).
Central Scientific Co., Bulletin No. 104 (1937).
Clark, *Determination of Hydrogen Ions.* Williams & Wilkins, Baltimore, 1920.
Findlay, *Practical Methods of Physical Chemistry.* Longmans, London, 1933.
Fryer and Weston, *Oils, Fats, and Waxes.* Cambridge (1920).
Glasstone, *Textbook of Physical Chemistry.* Van Nostrand, New York, 1948.
Hawk, Oser, and Summerson, *Practical Physiological Chemistry*, 13th ed. Blakiston, Philadelphia, 1954.
Landolt-Long, *The Optical Rotation of Organic Compounds* (1902).
Leeds and Northrup, Bulletin No. 755 (1930).
Leeds and Northrup, Catalog No. EN.-96 (1937).
Leeds and Northrup, Note Book No. 3 (1931).
Mahin, *Quantitative Analysis*, McGraw-Hill, New York, 1924.
Perley, *Measurement and Control of Hydrogen Ion Concentration in Industrial Plants,* Leeds and Northrup 1936.
Radley and Grant, *Fluorescence Analysis.* Van Nostrand, New York, 1933.
Reilly, Rae, and Wheeler, *Physico-Chemical Methods,* Van Nostrand, New York, 1925.
Weissberger, *Technique of Organic Chemistry—Physical Methods,* 2nd ed. Interscience, New York, 1949.

---

[47] Paley and Tzall, *N. Y. State Assoc. Milk Sanitarians 27th Conf.*, October, 1950.

CHAPTER III

COLORING MATTERS IN FOODS

THE addition of coloring matter to foods has several objectives, some desirable, others undesirable. First and foremost is the use of color to increase the attractiveness and thus at times the palatability of a food; second, is the addition of color in the processing of foods to achieve uniformity of product; and third, is the employment of color in an attempt to conceal damage or inferiority. That this is so is clear from the fact that there is additional cost in not only the dye, itself, but also in the process of adding the dye. Since manufacturers compete with other manufacturers no unnecessary costs may be permitted. Where color is added for a fraudulent purpose, governmental agencies do not tolerate their use. Color in such materials as confections, where the color is added, primarily, in order to increase the attractiveness, is not deemed objectionable. On the other hand, artificially coloring tomato paste, or cake, or orange drink, or French ice cream or similar products where color should naturally be present because of normal colored components is definitely objectionable.

CLASSIFICATION

Food coloring matters may be placed into 3 classes and in the order of their importance:

1. Coal-tar dyes or artificial coloring matters consisting of
   a. Permitted and certified water soluble acid dyes and oil soluble dyes
   b. Nonpermitted and noncertified water soluble acid dyes, water soluble basic dyes and oil soluble dyes
2. Natural or vegetable colors
3. Mineral colors

The latter are generally termed pigments. Thus, for instance, lampblack, charcoal, and other forms of carbon are pigments used to color food. Other mineral colors are talc, ochres, ultramarine blue, Prussian blue, and umbers. Their use is decreasing. Examples of vegetable colors are annatto and turmeric. An example of an animal color is carmine from cochineal.

The poisonous character of some coal-tar colors was recognized soon after their introduction in 1858. Weyl[1] in 1888 wrote a small book concerning the coal-tar dyes with special reference to their injurious qualities and the restriction of their use. This was brought to the attention of our country by Henry Leffman. This was re-emphasized by Lieber[2] in 1904, who, while advocating the use of synthetic colors in foods, pointed to the necessity of limiting this to specified colors.

It should be pointed out, however, that natural colors have to be processed also before they can be added to foods. In that processing they may become contaminated. There is no assurance that just because a color is obtained from animal or vegetable sources that it is pure.

Certification.—Hesse[3] in 1912 recommended that:

1. Coal-tar dyes should not be used indiscriminately in foods.

2. Only specified coal-tar dyes should be used.

3. Only listed and certified dyes should be used. These rules are equally good today.

The synthetic coal-tar colors that are used in the United States with governmental permission are known as certified colors. Often they are termed permitted colors. In order to be used, however, with this governmental permission they must undergo a procedure known as certification. This procedure consists in the submission by a manufacturer of suitable samples so that the government may ascertain by chemical, biochemical, toxicological, and medical analysis if the dye is free of deleterious substances. As ordinarily manufactured for textile, or other industrial and commercial purposes, synthetic coal-tar dyes often contain impurities. Some of these impurities may be harmless; others, on the other hand, may be deleterious and may even be toxic. These impurities may not detract from the value of the dye for industrial use, but they are highly objectionable in a substance prepared for human consumption. Nearly all artificial colors at one stage or another in their manufacture are treated with sulfuric acid or nitric acid, both of which are frequently contaminated with arsenic. Indeed, one of the most notorious cases of food poisoning in history was that of the arsenic poisoning cases in Manchester, England, in 1900, which were traced to the presence of arsenic in

---

[1] Weyl, Leffman, *Coal-Tar Colors.* Blakiston, Philadelphia, 1892.

[2] Lieber, *Use of Coal-Tar Colors in Foods.* Lieber, New York, 1904.

[3] Hesse, *U. S. Dept. Agr., Bur. Chem. Bull.* 147 (1912).

beer made with commercial glucose that was manufactured with arsenic contaminated sulfuric acid. Sulz[4] pointed out in 1888 that coal-tar dyes were frequently contaminated with arsenic. Not only may the acids used in the manufacture of these dyes contain harmful metals but the vessels in which the dyes are made may also contain these elements. Therefore, unless precautions are taken in the manufacture of the dyes, they may be seriously contaminated with arsenic and other harmful compounds. Special apparatus made of wood, glass, enamel or noble-metal lined equipment must be used to observe these precautions.

In the manufacture of organic compounds, such as these dyes, it is difficult to control the reaction so that only one substance is formed. Harmful intermediates—that is, the ingredients used in making dyes—may remain in the dye or other harmful organic compounds may be produced by reactions other than the main or desired reaction. Even if some substances are formed which are comparatively harmless they are undesirable in food dyes because of their unknown character.

To obtain harmless dyes, the standards of purity set for certified colors necessitate, as has been noted, special precautions in their manufacture and purification, in order that appreciable quantities of objectionable substances may not be present in the finished products. Because of these factors it is necessary for some agency with an adequate scientific staff to perform the clinical, medical, physiological and biological, chemical, and physical-chemical tests necessary to select from the hundreds of coal-tar dyes available those that may be used with safety in foods. The agency charged with these functions under the Federal Food, Drug, and Cosmetic Act is the Food and Drug Administration, Department of Health, Education, and Welfare. Certification by the Food and Drug Administration implies not only that the dye itself is harmless, but that it is not contaminated with poisonous substances. This agency has set rigid specifications for each permitted coal-tar color. The maximum limit for all FD&C colors for arsenic (as As_2O_3) is 0.00014 per cent and for lead is 0.001 per cent. Only a trace of other heavy metals precipitable as sulfides is permitted and no barium lakes are permitted.

The specifications set for volatile matter, water insoluble matter, ether extract, chlorides and sulfates mixed oxides, subsidiary dyes, uncombined intermediates and pure dye are listed for each color in the *Coal-Tar Regulations* of the Food and Drug Administration[5] and by Jacobs in *Synthetic Food Adjuncts.*

[4] Sulz, *Treatise on Beverages.* Dick & Fitzgerald, New York, 1888.
[5] *U. S. Food Drug Admin. S.R.A., F.D.&C.* 3, September, 1940, as amended through 1957.

The use of color of any kind to conceal damage or inferiority in a food product is defined by the United States Federal Food, Drug, and Cosmetic Act as an adulteration and, when damage or inferiority is concealed, the employment of artificial color is not permissible, even though certified colors are used and their presence is declared on the label. In general, where colors are legitimately used in foods and beverages a statement on the label of the presence of artificial color is required.

The Certified Colors.—The original list of 7 permitted colors— namely, amaranth, ponceau 3R, erythrosine, orange I, naphthol yellow S, light green SF yellowish, and indigo carmine, was selected after a critical study of the reports of pharmacological tests on the more important dyestuffs. The number was slowly increased until 15 dyes were on the permitted list, but only after appropriate pharmacological and toxicological tests had proved them to be harmless. These were, in addition to those previously mentioned, ponceau SX, sunset yellow FCF, tartrazine, yellow AB, yellow OB, Guinea green B, fast green FCF, and brilliant blue FCF. Additional certifications raised this number and there were 18 synthetic coal-tar colors permitted for use in food in 1944. In 1953 FD&C Violet No. 1 (Acid Violet 6B) was added to the group of colors permitted in foods.

In 1955 FD&C Orange No. 1 (Orange I), FD&C Orange No. 2 (Orange SS), and FD&C Red No. 32 (Oil Red XO) were deleted from the certified list of colors permissible for use in or on foods, but the cutoff date of Oil Red XO for use on oranges was extended to March, 1959. In 1957 an order was tentatively promulgated removing FD&C Yellow Nos. 1, 2, 3, and 4 from the certified list, but this order was suspended pending further hearings.

After the Food, Drug, and Cosmetic Act was passed, the number of dyes permitted were given specific letters and numbers for identification. The procedure of certification was extended so that it could be used for certification of two other large categories of permitted dyes in addition to that for foods, making a total of three groups. The first group comprises the ones with which we have been dealing. These are classified FD&C colors, signifying that they are certified for use in foods, drugs and cosmetics. They are all soluble in either water or oils, for no insoluble colorants may be used for coloring foods. However, water insoluble calcium and aluminum lakes of the FD&C colors may be used for external coloring purposes such as Easter eggs in shells. The second group are classified D&C colors to signify that they are certified for use in drugs and cosmetics but not in foods. There were 69 colors placed in this category. The third group are classified Ext. D&C colors to signify that they are certified for restricted use in preparations such as externally applied drugs and cosmetics. There were 29 synthetic dyes placed in this category. The decertified food colors were added to this group.

TABLE 4. CERTIFIED FOOD COLORS

Trade Name	FD&C Color	No.	"Colour Index" No.
Brilliant Blue FCF	Blue	1	...
Indigotine (Indigo Carmine)	Blue	2	1180
Guinea Green B	Green	1	666
Light Green SF Yellowish	Green	2	670
Fast Green FCF	Green	3	...
Ponceau 3R	Red	1	80
Amaranth	Red	2	184
Erythrosine	Red	3	773
Ponceau SX	Red	4	...
Acid Violet 6B	Violet	1	697
*Naphthol Yellow S	Yellow	1	10
*Naphthol Yellow S, potassium salt	Yellow	2	10
*Yellow AB	Yellow	3	22
*Yellow OB	Yellow	4	61
Tartrazine	Yellow	5	640
Sunset Yellow FCF	Yellow	6	...

* Considered for decertification in 1957.

It is to be noted, however, that there is a general restriction on the use of any and all coal-tar dyes, certified as well as uncertified. Their use is prohibited in any product applied to the area of the eye. The *Coal-Tar Color Regulations* of the Food and Drug Administration [5] state:

"The authorization contained in these regulations for the certification of coal-tar colors shall not be considered to authorize the certification of any coal-tar color for use in any article which is applied to the area of the eye. A coal-tar color used in any such article which is so applied shall be considered to be from a batch that has not been certified in accordance with these regulations, even though such color is from a batch that has been certified for other use."

"The term 'area of the eye' means the area enclosed within the circumference of the supra-orbital ridge and the infra-orbital ridge, including the eyebrow, the skin below the eyebrow, the eyelids and the eyelashes, the conjunctival sac of the eye, the eyeball, and the soft areolar tissue that lies within the perimeter of the infra-orbital ridge."

[5] *U. S. Food Drug Admin. S.R.A., F.D.&C.* **3**, September, 1940.

The letters and numbers appearing after or in a given trade name of a dye are the manufacturer's code. In fact, many manufacturers have their own trade name for a given dye. Thus, ponceau 3R is also known as ponceau N3R, ponceau FRRR, scarlet S3R, cumidine scarlet, cumidine red and cumidine ponceau. For the synonyms and trade names of many of the dyes it is well to consult the Colour Index.[6]

It is well to note the interpretation placed by the Food and Drug Administration on the use of color in certain items. Thus, if color is added to a processed food like mayonnaise or sausage and the food conforms in all other respects to the standards set by that agency, the Food and Drug Administration does not deem it to be an adulteration, provided the words "color added" are placed conspicuously on the label of the product. One of the reasons adduced for this permission is that there has been no diminution of nutrients. In other words, the sausage contains the proper amount of meat and no excess of filler or water. The mayonnaise contains the required amount of oil and eggs. The reason given by the manufacturer is that unless he is permitted to use color, he cannot obtain a uniformly colored product. Sometimes the meat he uses is dark, sometimes it is light. Customers, they say, come to expect a certain shade. If they do not get that, they begin to think that something is wrong with the product. Therefore, since the raw materials they use are not uniform, color must be used in order to produce a uniform product.

Certain additional factors concerning the certification of colors and the interpretation on the use of color by the Food and Drug Administration are to be noted. Flavoring mixtures containing coal-tar coloring which are to be used primarily to color as well as to flavor the food to which they are added require certification.[7] However, if the dye is added simply to color the flavoring essence, or a food product such as gelatin dessert, or soft drink powders, certification is not necessary. The dry color manufacturer [8] may dilute a certified color at the order of an interstate customer and submit, in the name of the customer, a sample of the mixture for certification before bulk shipment is made.[9]

The term "Food Color Added" is considered inappropriate by the aforementioned Agency to indicate the presence of artificial color because this term might be interpreted erroneously to mean the color is obtained from a food. The term preferred is "Artificial Color Added" or "Color Added." [10]

---

[6] Rowe, ed., *Colour Index.* Society of Dyers and Colourists, Bradford, England, 1924.

[7] *U. S. Food Drug Admin. F.D.&C. Act Trade Correspondence* TC-171, March 14, 1940.

[8] *U. S. Food Drug Admin. F.D.&C. Act Trade Correspondence* TC-219, March 21, 1940.

[9] *U. S. Food Drug Admin. F.D.&C. Act Trade Correspondence* TC-182, March 15, 1940.

[10] *U. S. Food Drug Admin. F.D.&C. Act Trade Correspondence* TC-167, March 14, 1940.

Pure Dye Percentage

Pure dye percentage is defined as the percentage by weight of coal-tar dye in a color. The Food and Drug Administration requires that brilliant blue FCF, Guinea green B, and light green SF yellowish contain at least 82 per cent pure dye in the straight certified color. Most commercial products of this group are prepared to contain 86-89 per cent of pure dye in the straight color. The remaining water-soluble, coal-tar colors are required to contain at least 85 per cent pure dye for certification. Most commercial straight certified dye colors in this group contain 90-92 per cent. The remaining components are principally salt and water. The oil soluble dyes are required to contain a much higher percentage of straight dye for certification—namely 98 per cent for orange SS, 95 per cent for oil red XO, and 99 per cent for yellow AB and OB.

It is important for the purchaser to note the percentage of pure dye given on the label of any product, for some products, which are ostensibly less expensive than others, actually may cost more for the amount of dye incorporated because they contain less dye per unit cost.

Water Soluble Food Colors

All of the water-soluble dyes belong to the class of dyes known as acid dyes. They are, as can be seen from their formulas, sulfonates or salts of color acids. They generally can be stripped from the food in which they are used by soaking in ammonia water.

Blue Colors.—*Brilliant Blue FCF, FD&C Blue No. 1.*—

Brilliant blue FCF ($C_{37}H_{34}N_2O_9S_3Na_2$) is the disodium salt of dibenzyldiethyldiaminotriphenylcarbinol trisulfonic acid anhydride (disodium salt of 4-([4-(N-ethyl-p-sulfobenzylamino)-phenyl]-(2-sulfoniumphenyl)-methylene)-[1-(N-ethyl-N-p-sulfobenzyl)-$\Delta^{2,5}$-cyclohexadienimine]), and thus belongs to the triphenylmethane group of dyes. The dye is prepared by condensing 1 mole of benzaldehyde o-sulfonic acid with 2 moles of benzylethylaniline sulfonic acid, followed by oxidation and conversion into the disodium salt. Brilliant blue FCF is a bronze-purple powder, dissolving in water, giving a greenish-blue solution; it is moderately soluble in 95 per cent alcohol and is soluble in glycerol and glycols. This coal-tar color is fairly fast to light and has good resistance to the action of acetic acid. To 10 per cent hydrochloric acid its resistance is moderate, the dye turning greener in hue. Its resistance to 30 per cent hydrochloric acid is poor, the dye turning greener or yellower in hue. Brilliant blue FCF is moderately resistant to 10 per cent sodium hydroxide solution, whereas 30 per cent sodium hydroxide solution causes a change in shade to wine red on standing. Alkali, in general, causes this change on warming. Its resistance to reducing agents is better than that of the azo colors. Thus invert sugar will decolorize amaranth but will leave brilliant blue FCF practically unaffected. It has little resistance to oxidizing agents. This dye is unaffected by contact with copper and aluminum in either plain water or acid solution.

Indigotine, FD&C Blue No. 2.—

Indigotine (indigo carmine) ($C_{16}H_8N_2O_8S_2Na_2$) is the disodium salt of 5,5'-indigotindisulfonic acid. The dye belongs to the indigoid group of coloring matters. It is prepared by the following series of chemical changes: naphthalene → phthalic anhydride → anthranilic acid → phenylglycine-o-carboxylic acid → indoxyl carboxylic acid → indigo. Synthetic indigo, which is sulfonated with concentrated or slightly fuming sulfuric acid, has almost entirely replaced the natural product because it is much more homogeneous in composition. Indigotine is a blue, brown, reddish powder, easily soluble in water, giving a blue solution; it is only sparingly soluble in 95 per cent alcohol, but is readily soluble in glycerol and the glycols. Indigotine has very poor resistance to light and is thus a fugitive color. Acids have little effect on it, for its resistance to acetic acid is very good, to 10 per cent hydrochloric acid it is good, and to hydrochloric acid it is moderate. Indigotine has only a moderate fastness rating to 10 per cent sodium hydroxide solution and a very poor rating to a 30 per cent solution of this reagent. Its resistance to oxidizing agents is poor but its fastness to reducing agents is moderate. Indigotine is practically unaffected by contact with copper and aluminum in either plain water or acid solution.

Green Colors.—*Guinea Green B, FD&C Green No. 1.*—

Guinea green B ($C_{37}H_{35}N_2O_6S_2Na$) is the monosodium salt of diben-zyldiethyldiaminotriphenylcarbinol disulfonic acid anhydride (4-[4-(*N*-ethyl-*p*-sulfobenzylamino)-diphenylmethylene]-[1-(*N*-ethyl-*N*-*p*-sulfoniumbenzyl)-$\Delta^{2,5}$-cyclohexadienimine]). It is the product formed by the condensation of benzylethylaniline sulfonic acid with benzal-dehyde, followed by oxidation of the dibenzyldiethylaminotriphenyl-methane sulfonic acid formed and conversion to the monosulfonate. It belongs to the triphenylmethane group of dyes. It is a dull, dark green powder, easily soluble in water, giving a green solution; it is more soluble in 95 per cent alcohol than any of the other FD&C colors, for about 4.5 g. will dissolve in 100 ml. of 95 per cent alcohol. It is also readily soluble in glycerol and glycols. This coal-tar color is not fast to light but has good resistance to acetic acid and 10 per cent hydrochloric acid. However, its resistance to 30 per cent hydro-chloric acid is poor for the dye turns yellower in hue and the color acid may precipitate. The resistance of Guinea green B to alkalis is very poor; hence it is rapidly destroyed by 10 per cent sodium hydroxide and 30 per cent sodium hydroxide. In commercial prac-tice, an alkaline pH will nearly always result in the decolorization of the dye because of the small quantities used for coloring purposes. Its resistance to oxidizing agents is poor, and its fastness rating to reducing agents is analogous to that of brilliant blue FCF. Guinea green B is not affected by aluminum in either plain water or acid solution, but copper generally causes an oily film to appear.

Light Green SF Yellowish, FD&C Green No. 2.—

Light green SF yellowish ($C_{37}H_{34}N_2O_9S_3Na_2$) is the disodium salt of dibenzyldiethyldiaminotriphenylcarbinol trisulfonic acid anhydride;

that is, it is the disodium salt of 4-([4-(N-ethyl-p-sulfobenzylamino)-phenyl]-(4-sulfoniumphenyl)-methylene)-[1-(N-ethyl-N-p-sulfoben-zyl)-$\Delta^{2,5}$-cyclohexadienimine]. It belongs to the triphenylmethane group of dyes and is prepared by condensing benzylethylaniline with benzaldehyde, and then sulfonating the product, with fuming sulfuric acid to the trisulfonate, followed by oxidation and conversion into the disodium salt. The product is a reddish-brown powder, easily soluble in water, giving a green solution; it is moderately soluble in 95 per cent alcohol but is readily soluble in water-alcohol mixtures and in glycerol and glycols. The fastness ratings of light green SF yellowish are entirely analogous to those of Guinea green B. However, 30 per cent hydrochloric acid, to which it has poor resistance, makes the dye tinctorially weaker, whereas 30 per cent sodium hydroxide solution not only makes the dye tinctorially weaker but also turns it yellower in hue.

Fast Green FCF, FD&C Green No. 3.—

Fast green FCF ($C_{37}H_{34}N_2O_{10}S_3Na_2$) is similar to Guinea green B, differing only in that p-hydroxy-o-sulfobenzaldehyde is substituted for benzaldehyde in its preparation, and thus this dye also belongs to the triphenylmethane group of dyes. It is the disodium salt of 4-([4-(N-ethyl-sulfobenzylamino)-phenyl]-(4-hydroxy-2-sulfonium-phenyl)-methylene)-[1-(N-ethyl-N-p-sulfobenzyl)-$\Delta^{2,5}$-cyclohexadieni-mine]. The dye is prepared by condensing 1 mole of the aldehyde with 2 moles of benzylethylaniline monosulfonic acid, followed by oxidation and conversion into the disodium salt. Fast green FCF is a reddish or brownish-violet powder, easily soluble in water, giving a bluish-green solution; it is slightly soluble in 95 per cent alcohol but is much more soluble in water-alcohol mixtures. It is soluble in glycerol and glycols. Fast green FCF has fastness ratings analogous to both Guinea green B and light green SF yellowish. It is, however, much faster to light but has poorer resistance to 30 per cent hydrochloric acid. Fast green FCF has poorer resistance than the aforementioned dyes to alkali only in the sense that a change to a violet tint results. It has better resistance in that it does not decolorize as rapidly as do weak solutions of Guinea green B and light green SF yellowish. Fast green FCF turns brownish in the presence of copper and loses some color in the presence of aluminum in acid solutions but is little affected by contact with these metals in plain water.

Orange Color.—*Orange I* (decertified for food use in 1955).—

$$NaO_3S \underset{}{\bigcirc} - N = N - \underset{\bigcirc}{\bigcirc} OH$$

Orange I ($C_{16}H_{11}N_2O_4SNa$) is the monosodium salt of 4-p-sulfophenylazo-1-naphthol. It is prepared by diazotizing 1 mole of sulfanilic acid and coupling it with 1 mole of 1-naphthol, and this dye belongs to the monazo group of dyes. The product is a reddish-brown powder soluble in water, giving an orange-red solution; it is only moderately soluble in 95 per cent alcohol (approximately 0.3 g. in 100 ml. of alcohol) forming an orange solution; it is soluble in glycerol and the glycols. FD&C Orange No. 1 is moderately fast to light but has a tendency to become darker or turn muddy in appearance upon exposure to light. It has a good fastness rating to acetic acid, but is only moderately resistant to 10 per cent hydrochloric acid while 30 per cent hydrochloric acid destroys the dye or makes it colorless. However, relatively concentrated solutions are first turned violet and then brown or a violet precipitate may form. It has moderate resistance to 10 per cent and poor resistance to 30 per cent sodium hydroxide solution, the color of the dye being changed to blood-red. Its fastness rating to reducing agents is very poor although it has fair resistance to oxidizing agents. In contact with copper in plain water, orange I turns cloudy and redder in shade becoming tinctorially weaker; in contact with copper in acid solutions, the dye turns muddy and its color is completely destroyed. Aluminum has little effect in water solutions turning the dye slightly redder, but in acid solutions the result is practically the same as with copper.

Red Colors.—*Ponceau 3R, FD&C Red No. 1.*—

$$H_3C \underset{H_3C}{\overset{CH_3}{\bigcirc}} - N = N - \underset{SO_3Na}{\overset{HO \quad SO_3Na}{\bigcirc}}$$

Ponceau 3R ($C_{19}H_{16}N_2O_7S_2Na_2$) is the disodium salt of 1-pseudo-cumylazo-2-naphthol-3,6-disulfonic acid. It is the product formed by diazotizing 1 mole of pseudocumidine and coupling it with 1 mole of 2-naphthol-3,6-disulfonic acid (R salt) and belongs to the monazo group of dyes. Ponceau 3R is a dark red powder which dissolves readily in water to yield a poppy-red solution; it is slightly soluble in 95 per cent alcohol; it dissolves easily in glycerol and the glycols. Ponceau 3R has a very good rating of fastness to light, a good rating of fastness to glacial and 10 per cent acetic acid, to 10 and 30 per cent hydrochloric acid, and to 10 per cent sodium hydroxide solution but only a fair rating of fastness to 30 per cent sodium hydroxide solution with a tendency for its hue to become duller or darker. Its solution becomes hazy or cloudy in 5 per cent alum solution. Ponceau 3R has only a fair rating of fastness to oxidizing agents and is very easily reduced. Aqueous solutions of ponceau 3R turn slightly bluer in contact with copper. Although acid solutions of this dye also turn bluer in shade in contact with this metal, they also become cloudy and duller. Aluminum has little effect on water solutions of this dye but makes acid solutions slightly bluer and tinctorially weaker.

Amaranth, FD&C Red No. 2.—

Amaranth ($C_{20}H_{11}N_2O_{10}S_3Na_3$) is the trisodium salt of 1-(4-sulfo-1-naphthylazo)-2-naphthol-3,6-disulfonic acid. This dye belongs to the monazo group of dyes. It is prepared by diazotizing 1 mole of naphthionic acid and coupling it with 1 mole of 2-naphthol-3,6-disulfonic acid (R salt). Amaranth is a reddish-brown powder which dissolves readily in water to yield a magenta-red or bluish-red solution; it is also sparingly soluble in 95 per cent alcohol; it is easily soluble in the glycols and glycerol. Amaranth is only moderately fast to light, otherwise its fastness properties are similar to those of ponceau 3R, although ferrous sulfate makes it turn darker and alum has little effect. Both aqueous and acid solutions of amaranth become dark brown and cloudy in contact with copper. Aqueous solutions in contact with aluminum turn slightly yellower but tinctorial loss is evident in acid solutions.

Erythrosine, FD&C Red No. 3.—

Erythrosine ($C_{20}H_6O_5I_4Na_2\cdot H_2O$) is the disodium salt of 9-*o*-carboxy-phenyl-6-hydroxy-2,4,5,7-tetraido-3-isoxanthone. It belongs to the xanthane group of dyes and is formed by iodinating fluorescein which is a dye made by condensing phthalic anhydride with resorcinol in aqueous or alcoholic solution. It is a brown powder, soluble in 95 per cent alcohol, giving a red solution with a slight fluorescence. It dissolves in water to give a cherry-red solution without fluorescence and is readily soluble in glycerol and the glycols. Erythrosine has only fair resistance to light. The dye is readily precipitated in acid solutions; thus it cannot be used in beverages like soda pop which are made with citric, tartaric or lactic acid. It has very good fastness rating to 10 per cent sodium hydroxide solution. Alum and ferrous sulfate precipitate the dye. It has fair resistance to oxidizing agents and is not reduced as readily as the azo dyes. Aqueous solutions in contact with copper are little affected.

Ponceau SX, FD&C Red No. 4.—

Ponceau SX ($C_{18}H_{14}N_2O_7S_2Na_2$) is the sodium salt of 2-(5-sulfo-2,4-xylylazo)-1-naphthol-4-sulfonic acid. It belongs to the monazo group of dyes and is the product formed by diazotizing 1 mole of 1-amino-2,4-dimethylbenzene-5-sulfonic acid and coupling it with 1 mole of naphthionic acid. It is a red powder easily soluble in water, giving an orange-red solution, and it is slightly soluble in 95 per cent alcohol but is readily soluble in glycerol and the glycols. The fastness properties of ponceau SX are similar to those of ponceau 3R. It is slightly faster to 10 per cent acetic acid, and 30 per cent sodium hydroxide

solution makes it turn yellower in hue. While it is precipitated by
5 per cent alum solution, aqueous and acid solutions of the dye are
little affected in contact with aluminum. Copper, on the other hand,
turns both water and acid solutions of the dye darker and the shade
is duller and yellower.

Violet Color.—*Acid Violet 6B, FD&C Violet No. 1.*—As men-

tioned in a previous section, in 1950 the promulgation of FD&C
Violet No. 1 as an additional certified dye was considered by the
Food and Drug Administration [*Federal Register* 15, 2814 (1950)].
Acid violet 6B ($C_{39}H_{40}N_3O_6S_2Na$) is the monosodium salt of 4-([4-
(N-ethyl-p-sulfobenzylamino)-phenyl]-[4-(N-ethyl-p-sulfoniumben-
zylamino)-phenyl]-methylene)-(N,N-dimethyl-$\Delta^{2,3}$-cyclohexadieni-
mine). It belongs to the triphenylmethane group of dyes. Acid violet
6B, also known as benzyl violet 4B, gives a bright violet. It is soluble
in water, glycerol, glycols, and alcohols. It is insoluble in oils and
ethers. It has fair fastness to light and moderate fastness to acetic
acid. Acid violet 6B is shaken out along with Guinea green and
orange I in the schematic separations by immiscible solvents detailed
in this chapter. This dye is virtually the same as D&C Violet No. 1.

Yellow Colors.—*Naphthol Yellow S,* (decertification ordered in 195

Naphthol yellow S ($C_{10}H_4N_2O_8SNa_2$) is the disodium salt of 2,4-dini-
tro-1-naphthol-7-sulfonic acid, and thus belongs to the nitro group
of coloring matters. It is the product resulting from the nitration
of the tri- or disulfonic acids of 1-naphthol or from the nitroso compound of the 2,7-disulfonic acid. The sodium salt of naphthol yellow

S is a light yellow or orange-yellow powder, and, being a nitro compound, burns readily, accompanied by scintillation. It is easily soluble in water, giving a bright yellow solution, but is only slightly soluble in 95 per cent alcohol. It is also soluble in glycerol and the glycols. Naphthol yellow S has moderate resistance to light and good resistance to acetic acid, and 10 per cent sodium hydroxide solution. Naphthol yellow S has its tinctorial value greatly reduced by acid. In dilute solution it is practically decolorized. It has moderate resistance to 30 per cent hydrochloric acid, in more concentrated solutions, the latter reagent turning the dye much greener in hue. This coal-tar color has fair resistance to both oxidizing and reducing agents. Both aqueous and acid solutions of this dye are turned redder in contact with aluminum and copper.

Naphthol Yellow S, Potassium Salt (decertification ordered in 1957).—

Naphthol yellow S (potassium salt) ($C_{10}H_4N_2O_8SK_2$) is prepared in similar manner as the salt of naphthol yellow S, but potassium compounds are substituted for sodium compounds in its preparation. Its solubility and its fastness ratings are entirely analogous to those of sodium salt.

Tartrazine, FD&C Yellow No. 5.—

Tartrazine ($C_{16}H_9N_4O_9S_2Na_3$) is the trisodium salt of 3-carboxy-5-hydroxy-1-sulfophenyl-4-*p*-sulfophenylazopyrazole. The principal method of preparation consists in treating 2 moles of phenylhydrazine-*p*-sulfonic acid with 1 mole of dioxytartaric acid. An alternative method of preparation is to couple diazotized sulfanilic acid and oxalacetic ether, condense the product with phenylhydrazine-*p*-sulfonic acid, and then hydrolyze the ester with sodium hydroxide solution. Tartrazine is an orange-yellow powder easily soluble in water, giving a golden-yellow solution; it is only slightly soluble in 95 per cent alcohol but is readily soluble in glycerol and glycols. Tartrazine has good resistance to light, acetic acid, hydrochloric acid, and 10 per cent sodium hydroxide solutions. However, toward 30 per cent sodium hydroxide solution its resistance is only fair and it turns redder in hue. It is reduced readily and has only a fair fastness rating to oxidizing agents. Ferrous sulfate solution tends to make the hue of this dye duller. Aqueous solutions of tartrazine are not affected by aluminum. Copper may turn the color slightly redder.

Sunset Yellow FCF, FD&C Yellow No. 6.—

$$NaO_3S\langle\bigcirc\rangle - N = N - \overset{OH}{\underset{SO_3Na}{\bigcirc\bigcirc}}$$

Sunset yellow FCF ($C_{16}H_{10}N_2O_7S_2Na_2$) is the disodium salt of 1-*p*-sulfophenylazo-2-naphthol-6-sulfonic acid. It is the product obtained by diazotizing 1 mole of sulfanilic acid and coupling it with 1 mole of 2-naphthol-6-sulfonic acid (Schaeffer's acid). It belongs to the monazo group of dyes. Sunset yellow FCF is an orange powder, easily soluble in water, giving an orange-yellow solution. It is only slightly soluble in 95 per cent alcohol but is readily soluble in glycerol and glycols. The fastness ratings of sunset yellow FCF are similar to those of tartrazine. However, it has only moderate fastness to ferrous sulfate solutions. Acid and aqueous solutions of this dye become dark brown, opaque, and the shade duller in contact with copper. Aluminum has less effect, for the solutions only become slightly redder.

OIL-SOLUBLE COLORS

The oil soluble dyes formerly permitted for foods belong to the azo group. They are insoluble in water principally because they have no salt-forming radicals. They are generally soluble in most of the common fat-dissolving organic solvents, such as chloroform, ether, petroleum ether, benzene, etc., but are much more soluble in aromatic hydrocarbons than they are in aliphatic hydrocarbons. They are also moderately soluble in alcohol.

Orange Color.—*Orange SS*, (see page 105).—

Orange SS ($C_{17}H_{14}N_2O$), 1-*o*-tolylazo-2-naphthol, is the product obtained by diazotizing 1 mole of *o*-toluidine and coupling it with 1 mole of 2-naphthol. It thus belongs to the monazo type of dyes. It is one of the newer permitted food dyes. Orange SS is an orange powder insoluble in water but soluble in 95 per cent alcohol, giving an orange color. It is soluble in oils and is only slightly soluble in glycerol and glycols. Orange SS has only fair fastness to light. It has very poor resistance to reducing agents and poor resistance to oxidizing agents.

Red Color.—*Oil Red XO*, (see page 105).—

Oil red XO ($C_{18}H_{16}N_2O$), 1-xylylazo-2-naphthol, belongs to the monazo group of dyes and is prepared by diazotizing 1 mole of xylidine mixture (from which the meta component has been partially removed) with 1 mole of 2-naphthol. Oil red XO is a brownish-red powder which is soluble in oils. It is insoluble in water but is moderately soluble in alcohol, giving an orange-red color. It is also slightly soluble in glycerol and glycols. Oil red XO has only fair resistance to light and to oxidizing agents. It is very easily reduced.

Yellow Colors.—*Yellow AB*, (see page 105).—

Yellow AB ($C_{16}H_{13}N_3$), 1-phenylazo-2-naphthylamine, is prepared by diazotizing 1 mole of aniline and coupling it with 1 mole of 2-naphthylamine. It belongs to the monazo group of dyes. Yellow AB is an orange powder insoluble in water, and slightly soluble in 95 per cent alcohol, giving an orange-yellow color. It is readily soluble in fats and oils and slightly soluble in glycerol and glycols. It is one of the principal coloring matters used for butter and margarine. Yellow AB has only poor fastness to light and oxidizing agents. Its resistance to reducing agents is very poor.

Yellow OB (see page 105).—

Yellow OB ($C_{17}H_{15}N_3$), 1-*o*-tolylazo-2-naphthylamine is prepared by diazotizing 1 mole of *o*-toluidine and coupling it with 1 mole of 2-naphthylamine. It belongs to the monazo group of dyes. Yellow OB is an orange powder insoluble in water and only slightly soluble in 95 per cent alcohol, giving an orange-yellow solution. It is readily soluble in oils and fats but is only slightly soluble in glycerol and glycols. It is also one of the principal colors used for coloring butter and margarine. The fastness ratings of yellow OB are identical with those of yellow AB.

Short Method for Water Soluble Acid Dyes

There are hundreds of dyes listed in Schultz, Julius, and Green and in the Colour Index, but the food analyst actually need look only among a very small number. The reason is that certified dyes are comparatively inexpensive because of increase in volume of manufacture. Furthermore, a small amount of artificial dye goes a long way. Consequently, only seldom does the analyst meet non-permitted colors. In fact the analyst has little means of concluding whether a certified or noncertified but permitted color has been used in a food product except by analyzing the dry color which in most cases can be obtained from the manufacturer of the food product. The dry color may then be carefully analyzed for purity. The use of coal-tar colors follows the price scale very closely. For example, as the price of eggs goes up, the tendency for use of yellow shades of coal-tar dyes in processed foods also increases.

Water soluble acid dyes have the property of dyeing wool in acid solutions. The dye may then be stripped from the wool by boiling in ammoniacal solution and subsequently may be redyed on wool in acid solution. A short general method for detecting water-soluble acid coal-tar colors based on these properties is the following.

Procedure.—Comminute 20-100 g. of the foodstuff according to the amount of color and mix with about 300-500 ml. of water, or if the material is soluble, dissolve the same proportion, if possible, in 300-500 ml. of hot water. Make slightly alkaline with ammonia, allow to stand for 1 or 2 hours so that the dye may be stripped from the foodstuff. Filter or decant through cheesecloth or through a wire gauze made into a cone to fit a funnel. Make the filtrate slightly acid with hydrochloric acid, adding sufficient water to adjust the volume to about 500 ml. Add a 6 to 8-in. length of white wool yarn and bring the solution to a boil. Boil cautiously from 10 to 30 minutes. Remove the wool and wash thoroughly with cold water. Place in a beaker, add 25 ml. of water, 5 drops of strong ammonium hydroxide, and boil to strip the color. Remove the wool, dilute the solution to 350 ml., and make acid by adding 10 drops of hydrochloric acid. ⅃ l a fresh 6-in. strip of white wool yarn and boil cautiously for 10-⋯ minutes. Remove the wool, wash thoroughly with cold water, and ⋯en press dry. If the wool is colored, the food product probably contains a coal-tar dye . Perform the spot tests with concentrated hydrochloric acid, concentrated sulfuric acid, 10 per cent sodium hydroxide solution, and 12 per cent ammonium hydroxide solution. Compare the reactions with Table 6 and identify the dye. If the spot tests are inconclusive, the wool has probably been dyed with a mixture of colors and one must proceed with the method of mixtures detailed below. It is possible at times that vegetable colors such as archil or turmeric may dye, strip, and redye, and cognizance of such possibility must be taken before drawing conclusions. If orange I is suspected, in this short method, do not dilute the solution containing the stripped color of the first piece of wool. Make the 25 ml. barely acid with hydrochloric acid and dye on wool by warming alone, otherwise the orange I may be destroyed.

If a mixture of dyes is suspected, it is convenient to dye eight to ten strips of wool directly, then strip the pieces with ammonia water in a small volume and proceed as detailed in Table 5 "Scheme for Separation of Permitted Coal-tar Water-soluble Dyes."

Systematic Examination of Water-Soluble Dyes

If the short method is inconclusive and there is reason to suspect that the coloring matter is not natural to the foodstuff, a more comprehensive treatment is necessary. These methods make use of the law of partition which states that a solute will distribute itself between a mixture of immiscible solvents at constant temperature in a constant ratio dependent on the respective solubility of the solute in each solvent. By the addition of salt or acid the dyes are made more or less soluble in the aqueous layer and correspondingly less or more soluble in the other solvent. By the addition of gasoline or carbon tetrachloride, the dyes are made less soluble in the immiscible solvent and more soluble in water.

Preparation of Sample.—Prepare a water solution of the coloring matter and strain as directed above. If the material is not soluble, comminute or grind and to 20-100 g. add 100 ml. of hot water, make slightly ammoniacal, allow to stand in a warm place for 2 hours and decant through a wire skimmer or sieve. Solid materials may be digested with 70 per cent alcohol containing 1 to 5 per cent ammonium hydroxide solution, although there is little advantage over steeping in ammoniacal water solution except to obtain traces of color. The alcohol, added or in the sample, must be removed by evaporation on a steam bath.

The details for the systematic examination of water-soluble dyes by separation by immiscible solvents follow the methods of the Association of Official Agricultural Chemists.[11] The numbers in parentheses and brackets following the name of a dye represent in the first instance the number of that dye as listed in "A Systematic Survey of the Organic Colouring Matters" (1904), by Arthur G. Green, while the second number designates the number as listed in the Society of Dyers and Colourists' "Colour Index" 1st edition, January, 1924. In 1957 new editions of Green and the "Colour Index" became available but reference to the old numbers is retained.

In the following methods, where multiple extractions with a supernatant solvent are to be made, Jacobs-Singer separatory extraction flasks are most convenient for this purpose.

Basic Dyes.—Most basic dyes may be separated from mixtures by making the mixture alkaline with 10 per cent sodium hydroxide solution and shaking with ether. Use the sample prepared as directed above. Separate the ether layer which may or may not be colored; wash it twice with a few milliliters of water to remove excess of alkali; and shake with acetic acid (1:18). This will take up any dye present and form a colored solution. Although this treatment may, to some extent, alter the common basic colors, it can be used for the detection of methyl violet B (451) [680], magenta 448 [677], bismarck brown (197) [331], malachite green (427) [657], and rhodamine B (504) [749]. With care auramine (425) [655] also may be separated

11 *Methods of Analysis, A.O.A.C.* Washington, 1945.

in this way, although it is quickly decomposed on standing in alkaline solution.

Acid Dyes.—The following short procedure is convenient for the examination of mixtures of acid dyes. Make the prepared sample strongly acid by adding half its volume of hydrochloric acid and shake 3 times with portions of amyl alcohol. Separate and combine the amyl alcohol solutions and wash by shaking with successive portions of half its volume of water, reserving the portions in separate bottles. Because of the varying acid content of the amyl alcohol these washings will show a regular decrease in acidity, and the coloring matters will appear in maximum quantity in the different fractions according to their respective solubilities.

Approximate Normality	Dye Washed Out
1.0	Ponceau 6R (108) [186]
0.25	Amaranth (107) [184] Brilliant scarlet (106) [185] Tartrazine (94) [640] Sunset yellow FCF Orange G (14) [27] Soluble blue (480) [707]
0.625 - 0.0039	Palatine scarlet (53) [77] Ponceau 2R (55) [79] Ponceau 3R (56) [80] Ponceau SX Naphthol yellow S (4) [10] Cochineal (706) [1239] Crystal ponceau (64) [89] Azorubine A (103) [179]
Practically all acid removed	Orange I (85) [150] Orange II (86) [151] Croceine orange (13) [26]

Orange IV (88) [143] and metanil yellow (95) [138] are less readily washed out.

Finally the unsulfonated coloring matters, such as erythrosine G (516) [772], erythrosine B (517) [773], and the rose bengals (520) [777] and 523 [779] are removed very slowly by water or not at all unless the solvent is diluted with gasoline and the dyes are removed with water containing a few drops of ammonium hydroxide solution. Acid yellow (8) [16] and brilliant yellow S (89) [144] are not very uniform in composition. They are partially taken up by amyl alcohol from acid solution and appear chiefly in the first washings. Indigo-tine (692) [1180] behaves somewhat similarly.

Transfer the colored washings to beakers and dilute with a large amount of water. Add a strip of wool and heat to boiling. Boil for a few minutes and then remove the wool, wash thoroughly with cold water and perform the spot tests to verify the identification of the dye.

Separation of the Coal-Tar Water-Soluble Dyes

The following directions and those given in the scheme are general and the proportions of amyl alcohol, dichlorohydrin, and other reagents used should be varied with the amount of color present in the aqueous solution.

To the solution containing the color add sufficient 25 per cent salt solution to make the concentration about 10 per cent and 1 part of acetic acid to every 7 parts of solution. Extract with three 50-ml. portions of amyl alcohol. Draw off the lower layer and reserve for further treatment. Wash the amyl alcohol extract in rotation with 25-ml. portions of 5 per cent salt solution until the washings are colorless or nearly so. Add the washings to the original aqueous solution. Dilute the amyl alcohol extract with an equal volume of gasoline or petroleum ether and wash with 25-ml. portions of water until all color is extracted. The coloring matters obtained are orange I and Guinea green B. For their separation see below. Treat the amyl alcohol gasoline solution with 10-ml. portions of 0.1 N sodium hydroxide solution or with 10-ml. portions of ammonium hydroxide solution (1:9) to remove erythrosine. Acidify the original solution and washings, from which the 3 named dyes were removed, with hydrochloric acid (1 volume acid to 40 volumes of solution) and extract in 50-ml. volumes with three 50-ml. portions of amyl alcohol. Reserve the lower aqueous layer for further treatment. Wash the amyl alcohol extract with 25-ml. portions of 0.25 N hydrochloric acid until the washings are colorless or nearly so. Combine the washings with the aqueous solution above. Extract the amyl alcohol with several 25-ml. portions of water until all color is extracted. The coloring matters obtained are ponceau 3R, ponceau SX, and naphthol yellow S. For their separation see below. Treat the original solution and washings, from which the 6 named dyes were removed, in 50-ml. volumes with three 50-ml. portions of α-dichloro-hydrin. Reserve the upper aqueous layer for further treatment. Wash the dichlorohydrin extract in rotation with several 20-ml. portions of 25 per cent salt solution. Combine the washings with the

TABLE 5. SCHEME FOR SEPARATION OF COAL-TAR WATER-SOLUBLE DYES

Original color solution made up to 10% salt solution with NaCl and containing 1 volume of HOAc to 7 volumes of solution. Extract in 50-ml. volumes with three 50-ml. portions of amyl alcohol.

Amyl alcohol layer	Aqueous layer and washings
Amyl alcohol layer. Wash with successive 25-ml. portions of 5% NaCl solution until colorless and return washings to original solution. This layer now may contain orange I, guinea green B, erythrosine. Add an equal volume of gasoline and wash with 25-ml. portions of water.	Aqueous layer and washings. Acidify with HCl, one volume of acid to 40 volumes of solution, and extract in 50-ml. portions with three 50-ml. portions of amyl alcohol.

Amyl alcohol layer subdivision:

Aqueous extract contains orange I and Guinea green B, and acid violet 6B.	Shake amyl alcohol-gasoline layer with 0.1 N NaOH or NH$_4$OH (1:9). Aqueous extract contains erythrosine.

Aqueous layer and washings subdivision:

Amyl alcohol layer. Wash with 25-ml. portions of 0.25 N HCl and return washings to original solution. Wash amyl alcohol layer with 25-ml. portions of water. Aqueous layer may now contain ponceau 3R, ponceau SX, naphthol yellow S.	Aqueous layer and washings. Extract in 50-ml. volumes with three 50-ml. portions of α-dichlorohydrin.

α-Dichlorohydrin subdivision:

α-Dichlorohydrin layer. Wash with 20-ml. portions of 25% NaCl solution and return washings to original solution. Dilute with 2 volumes of CCl$_4$ and wash with water until all color is extracted. Aqueous layer may now contain light green SF yellowish, fast green FCF and brilliant blue FCF.	Aqueous layer and washings. Acidify further with HCl, 1 volume acid to 40 volumes of solution. Extract in 50-ml. volumes with three 50-ml. portions of amyl alcohol. Discard aqueous layer. Wash out dyes from amyl alcohol layer with 25-ml. portions of water. Aqueous layer may now contain the remaining coloring matters, indigotine, amaranth, tartrazine and sunset yellow FCF.

For separation of remaining groups refer to the separations detailed in the text.

aqueous solution above. Dilute the dichlorohydrin extract with 2 volumes of carbon tetrachloride and extract with several 25-ml. portions of water until all color is extracted. The coloring matters obtained are light green SF yellowish, fast green FCF, and brilliant blue FCF. For their separation see below. Further acidify the original solution and washings, from which the 9 named dyes were removed, with hydrochloric acid (1 volume of acid to 40 volumes of solution), and extract in 50-ml. volumes with three 50-ml. portions of amyl alcohol. If the color intensity of the solution was not too strong, all coloring matter should have been extracted by the solvent. Discard the lower colorless or nearly colorless layer and wash out the dyes from the amyl alcohol extract in rotation with several 25-ml. portions of water until all color is extracted. The coloring matters obtained are indigotine, amaranth, tartrazine, and sunset yellow FCF. For their separation see below.

Orange I and Guinea Green B.—Extract the combined colors with two 20-ml. portions of dichlorohydrin. Discard the colorless upper aqueous layer, dilute the solvent with 2 volumes of carbon tetrachloride, and extract orange I in rotation with several 10-ml. portions of water, and Guinea green B with several 10-ml. portions of 25 per cent alcohol.

Ponceau 3R, Ponceau SX, and Naphthol Yellow S.—Acidify the combined colors with hydrochloric acid (1 part acid to 10 parts of solution), and extract the naphthol yellow S with two 20-ml. portions of washed ethyl acetate or amyl acetate. Ponceau 3R and ponceau SX are not extracted appreciably and remain in. the aqueous layer. Wash the solvent with 5-ml. portions of N hydrochloric acid to remove traces of the ponceaus. Naphthol yellow S is removed from the combined ethyl acetate or amyl acetate with 5-ml. portions of ammonium hydroxide solution (1:9). Extract the remaining ponceau solution with 20-ml. portions of amyl alcohol and wash out excess of acid twice with a few ml. portions of water. Dilute the amyl alcohol with an equal volume of gasoline, or petroleum ether, and remove the color with small volumes of water. Treat 10 ml. of this solution with 1 ml. of hydrochloric acid, 2 ml. of strong bromine water and lastly 3 ml. of saturated hydrazine sulfate solution and immediately pour into a test tube containing 10 ml. of 2 N sodium carbonate solution and 2 drops of 1 per cent alcoholic α-naphthol. A light orange solution indicates ponceau 3R. A deep brownish red solution indicates ponceau SX. Add to the solution 5 ml. of ether, mix well, and draw off the lower aqueous layer which, if colored, contains ponceau SX. To the ethereal extract add an equal volume of hydrochloric acid.

The formation of a purplish solution confirms the presence of ponceau 3R.

Light Green SF Yellowish, Fast Green FCF and Brilliant Blue FCF.—Treat the combined colors with an equal volume of 2 N sodium carbonate solution and extract in 25-ml. volumes with two 50-ml. portions of *n*-butyl alcohol. Draw off the lower aqueous layer containing the fast green FCF and wash out the last traces from the solvent with 25-ml. portions of 2 N sodium carbonate solution. Reserve the washings and add to the aqueous solution for confirmatory tests. Light green SF yellowish is colorless in the solvent, whereas brilliant blue FCF imparts a bluish green to it. To prove the presence of light green SF yellowish in the presence of brilliant blue FCF proceed as follows: Dilute the solvent with an equal volume of gasoline and remove color with small portions of water. Treat 20 ml. of solution with 4 ml. of 10 per cent sodium hydroxide solution and boil for 5 minutes. Brilliant blue FCF is changed to a red phase, while light green SF yellowish is changed to a yellow. Acidify with 10 ml. of glacial acetic acid, which changes brilliant blue FCF to a violet and light green SF yellowish to a green. Treat with about 3 g. of zinc dust and heat until the solution is decolorized. Filter, make slightly alkaline with ammonium hydroxide, and later make acid with acetic acid. Bring to a boil. In the presence of light green SF yellowish a deep green solution is formed while brilliant blue FCF remains colorless.

Indigotine, Amaranth, Tartrazine, and Sunset Yellow FCF.—To separate the indigotine, heat a small portion of the solution, which should be neutral or faintly acid, to boiling and add a few crystals of sodium hyposulfite, $Na_2S_2O_4$, until all the dyes are reduced. On adding a few drops of glacial acetic acid and shaking with air the indigotine is quickly restored, while amaranth, tartrazine, and sunset yellow FCF are destroyed. If a positive test for indigotine is obtained, add to the remainder of the mixed dye solution several decigrams of urea, heat, and while the mixture is boiling add 1 or 2 drops of 10 per cent sodium nitrite solution. Indigotine is converted to the pale yellow isatine sulfonate, while amaranth, tartrazine, and sunset yellow FCF are but little affected. Acidify the resultant mixture with sulfuric acid (1:4), using 1 part of the dilute acid to 10 parts of solution. Extract in 25-ml. portions with three 50-ml. portions of *n*-butyl alcohol. Draw off the lower layer and pass successively through all the funnels. Reserve the aqueous layer if colored, if not colored, discard. Prepare the following solution: 13.5 ml. of sulfuric acid, 100 g. of anhydrous sodium sulfate, and sufficient water to make 1 liter. Extract the *n*-butyl alcohol successively with 25-ml. portions of the solution until washings are colorless. Reserve them for amaranth and tartrazine. Dilute the n-butyl alcohol with an equal volume of gasoline and remove sunset yellow FCF with water. Confirm with dyeing tests and wet reactions.

Acidify the reserved solution with hydrochloric acid (1 volume of acid to 20 of solution) and extract with two 30-ml. portions of amyl alcohol. This will extract both amaranth and tartrazine while the isatine compound, being less readily extracted, remains in the lower layer and is discarded. Remove the coloring matter with several 10-ml. portions of water. To a portion of the solution add 5 drops of ammonium hydroxide solution and a few crystals of sodium hyposulfite, $Na_2S_2O_4$. This treatment will destroy amaranth completely, leaving tartrazine practically unaltered. Add an excess of hydrochloric acid and speedily extract the dye with a small amount of amyl alcohol, from which solution tartrazine can be removed with 0.25 N hydrochloric acid. Treat another 10-ml. portion of the neutral dye solution in a test tube with 2 ml. of 20 per cent ammonium chloride and 1 ml. of 25 per cent potassium cyanide solution and heat in a boiling water bath for 5 minutes. Cool rapidly and acidify with 2 ml. of hydrochloric acid and extract with 10 ml. of amyl alcohol. [Caution]. Draw off the lower layer and discard. Remove the tartrazine with 5-ml. portions of 0.25 N hydrochloric acid, amaranth is converted to a lower sulfonated dye, and is not removed at that acid concentration. Dilute the solvent with an equal volume of gasoline or petroleum ether and extract the dye with small volumes of water. Amaranth is modified to a brownish red dye.

IDENTIFICATION OF DYES BY SPOT TESTS

Transfer the separated coloring matter to wool, or silk in the case of oil soluble dyes, by boiling as directed above. Cut off four small pieces from the strip of dyed wool yarn or silk thread and place in the depressions of a white porcelain spot plate. Moisten the pieces with concentrated hydrochloric acid, concentrated sulfuric acid, 10 per cent sodium hydroxide solution, and 12 per cent ammonium hydroxide solution, respectively. At the same time, repeat with four pieces of known fiber of similar concentration. Compare the reaction with the reactions listed in Table 6.

COMMON MIXTURES

The common mixtures of certified colors used as food colors have been discussed in detail by Jacobs.[12] In this discussion it is shown

---

[12] Jacobs, *Synthetic Food Adjuncts*. Van Nostrand, New York, 1947.

TABLE 6. COLOR CHANGES PRODUCED ON DYED FIBERS BY VARIOUS REAGENTS [a,b]

Coloring Matter	Concd. HCl	Concd. H₂SO₄	10% NaOH	12% NH₄OH
Rhodamine B	Orange	Yellow	Bluer	Bluer
Rose Bengal	Almost decolorized	Orange	No change	No change
Archil	Red	Reddish-brown	Violet	Violet
Magenta	Yellowish-brown	Yellowish-brown	Decolorized	Paler
Acid Magenta	Almost decolorized	Yellow	Decolorized	Decolorized
Palatine Red	Darker	Blue	Dull brown	Little change
Bordeaux B	Violet	Blue	Brick red	Little change
Amaranth	Slightly darker	Violet to brownish	Dull brownish to orange red	Little change
Azorubine A	Little change	Violet	Red	Red
Erythrosine	Orange-yellow	Orange-yellow	No change	No change
Ponceau 6RB	Blue	Blue	Dull violet-red	Little change
Ponceau 6R	Violet-red	Violet	Brown	Orange-red
Crystal Ponceau	Red	Violet	Dull brown	Little change
Ponceau 3R	Little change	Little change	Dull orange	Little change
Ponceau SX	Deeper red	Deeper red	Orange-yellow	Orange-yellow
Sudan III†	Violet, then brown	Green	Violet-red	Little change
Safranine	Greenish-blue	Green	Red	Red
Brilliant Scarlet	Red	Violet red	Yellowish brown	Orange-red
Ponceau 2R	Little change	Little change	Brownish yellow	No change
Palatine Scarlet	Darker	Violet red	Brownish yellow	No change
Erythrosine G	Yellow-orange	Yellow-orange	No change	No change
Sudan II†	Red	Violet red	Little change	No change
Sudan I†	Orange-red	Red	Redder	No change
Cochineal	Little change	Little change	Violet-red	Violet red
Bismark Brown	Redder, darker	Browner	Yellower	Yellower
Bismark Brown R	Redder, darker	Browner	Yellower	Yellower
Orange I	Violet	Violet	Red, dark	Red, dark
Orange II	Red	Red	Dull red	No change
Croceine Orange	Orange-red	Orange	Slightly darker	No change
Orange G	Little change	Orange	Dull brownish-red	No change
Acid Violet 6B	Brownish yellow	Dark brownish yellow	Yellow	More bluish

† Oil-soluble.

[a] Mathewson, *U. S. Dept. Agr., Bull.* **448** (1917).

[b] *Methods of Analysis, A.O.A.C.* Washington, 1945.

TABLE 6—*Continued*

Coloring Matter	Concd. HCl	Concd. H₂SO₄	10% NaOH	12% NH₄OH
Orange SS†	Cherry red	Cherry red	Slightly yellow	No change
Oil Red XO†	Cherry red	Cherry red	Slightly yellow	No change
Yellow OB†	Red	Violet	Little change	No change
Yellow AB†	Red	Violet	Little change	No change
Sudan G†	Orange-yellow	Brownish yellow	Orange-yellow	Orange-yellow
Butter Yellow†	Violet-red	Orange-yellow	No change	No change
Aniline Yellow†	Violet-red	Orange-yellow	Little change	No change
Aminoazo-*o*-toluene†	Dull orange	Orange-yellow	Little change	No change
Fluoresceine	Little change	Little change	Green fluorescent	Green fluorescent
Metanil Yellow	Violet red	Violet	No change	No change
Azoflavine	Violet red	Violet red	Dull brown	Little change
Acid Yellow	Red	Orange	Little change	No change
Brilliant Yellow S	Violet red	Violet red	Little change	Little change
Tartrazine	Slightly darker	Slightly darker	Little change	Little change
Sunset Yellow FCF	Redder	Redder	Browner	Little change
Naphthol Yellow S	Almost decolorized	Very pale, dull brown	No change	No change
Auramine	Decolorized	Almost decolorized	Decolorized	No change
Turmeric	Red	Reddish-brown	Orange	Paler
Quinoline Yellow	Slightly darker	Brownish-yellow	Slightly paler	Orange
Naphthol Green B	Yellowish	Brownish-yellow	No change	Little change
Guinea Green B	Pale orange-yellow	Yellow-brown	Decolorized	No change
Light Green SF Yellowish	Pale orange-yellow	Yellow-brown	Decolorized	Decolorized
Fast Green FCF	Orange	Green to brown	Blue	Blue
Brilliant Blue FCF	Yellow	Yellow-brown	No change	No change
Night Green 2B	Pale orange-yellow	Almost decolorized	Decolorized	Paler
Malachite Green	Almost decolorized	Pale, dull yellow or brown	Decolorized	Decolorized
Erioglaucine A	Yellow	Green to brown	Slightly darker	Little change
Patent Blue A	Pale orange-yellow	Green to brown	Little change	Little change
Soluble Blue	Paler	Brown	Pale red	Almost decolorized
Indigotine	Slightly darker	Darker	Greenish-yellow	Greenish-blue
Formyl Violet	Pale orange-yellow	Pale, dull orange	Decolorized	Decolorized
Methyl Violet	Yellowish	Yellowish	Decolorized	Almost decolorized
Nigrosine, Soluble	Dull bluish	Dull greenish	Brownish red, paler	Pale reddish

† Oil-soluble.

that most of the mixtures consist of two primary colors, a large number are composed of three primary colors, and there are a few formulas consisting of four, five, and even six dyes so that the foregoing method is merely to be used as a general scheme, the analyst, skipping those parts that unquestionably do not enter into the method for the moment. The following are some of the combinations that are more common.

Egg shade:	tartrazine	+ orange I	
	tartrazine	+ sunset yellow FCF	
Greens:	tartrazine	+ indigotine	(pistachio)
	tartrazine	+ brilliant blue FCF	(lemon)
	tartrazine	+ fast green FCF	(lemon)
	tartrazine	+ guinea green B	(lemon)
	tartrazine	+ light green SF yellowish	(lemon)
Orange:	tartrazine	+ orange I	
	tartrazine	+ ponceau 3R	
	tartrazine	+ sunset yellow FCF	
	tartrazine	+ ponceau 3R + amaranth	
Brown:	ponceau 3R	+ orange I	(chocolate and
	ponceau SX	+ orange I	frankfurters)
	orange I	+ indigotine + amaranth	
		+ tartrazine	
	indigotine	+ light green SF yellowish	
		+ ponceau 3R	
Yellow:	tartrazine	+ sunset yellow FCF	
Red:	amaranth	+ orange I + ponceau 3R	
	amaranth	+ ponceau 3R	
	ponceau SX	+ amaranth	
	erythrosine	+ amaranth	
Violet:	amaranth	+ indigotine	(grape color
	amaranth	+ brilliant blue FCF	in jams,
	amaranth	+ fast green FCF	jelly, etc.)
	amaranth	+ ponceau 3R + orange I	

The use of orange I is declining because a muddy color or precipitate develops in the presence of iron. The use of a particular dye or any other combination of dyes is very often a matter of price. Thus, erythrosine is seldom used because it is comparatively expensive; amaranth is used instead. The use of coal-tar colors in food products is also discussed in detail by Jacobs.[12]

A simple means for extracting water-soluble coloring matters from a food product is to take up the dye on five to ten 6-in. pieces of white wool yarn, stripping in a small amount of water with the aid of ammonia and then going on with the systematic method.

Tartrazine

The following method for the detection of tartrazine [13] in alimentary pastes, cakes, and the like is applicable if tartrazine is the sole dye present.

Procedure.—Place 800 ml. of cold water and 5 ml. of concentrated ammonium hydroxide solution into a liter Erlenmeyer flask. Add 200 g. of the unground sample. Stopper the flask and shake at intervals for 3 to 4 hours, which is sufficient time to permit the material to disintegrate. Dislodge the material from the bottom with a glass rod. Transfer to centrifuge bottles, centrifuge, and decant the clear supernatant fluid into a liter flask. To precipitate nearly all the protein matter, add a 50 per cent magnesium sulfate solution, prepared by dissolving 50 g. $MgSO_4 \cdot 7H_2O$ in 100 ml. of water, 10 ml. of 12 per cent silicotungstic acid, 10 ml. of concentrated hydrochloric acid, shake thoroughly, and allow to stand for 1 hour. Transfer again to centrifuge bottles, centrifuge, and decant the supernatant solution into a liter casserole. If only a small amount of color is present, the solution may appear to be colorless.

Put 4 pieces of washed wool into the solution and heat on a steam bath until the dye is adsorbed on the fiber. Usually this requires evaporation to about half the original volume. Remove the wool, wash thoroughly with water, and place it into a 125-ml. casserole. Add about 25 ml. of water, a few drops of concentrated ammonium hydroxide solution, and warm on the steam bath to elute the color. Discard the wool and evaporate the alkaline solution to dryness. To the residue add 25 ml. of water to dissolve the dye and filter. About 40 per cent of the dye originally present is recovered. Place 15 ml. into a 50-ml. casserole, acidify slightly with hydrochloric acid, and add a small piece of washed wool to adsorb the dye. Perform the spot tests detailed in this chapter and compare the color changes with a standard piece of wool of approximately the same intensity.

To the remaining 10 ml. add 2 drops of hydrochloric acid and an excess of bromine water. If a precipitate forms at this stage because of the presence of soluble proteins, centrifuge or filter off the precipitate. Destroy the excess bromine with about 1 ml. of a saturated solution of hydrazine sulfate and add immediately 1 drop or 1 per cent alcoholic a-naphthol in 5 ml. of 10 per cent aqueous sodium carbonate solution. The formation of an orange or pink coloration indicates the presence of tartrazine.

---

[13] Biletch, personal communication, 1947.

Short Method for Oil-Soluble and Natural Coloring Matters

Grind such materials as alimentary pastes, comminute other materials, and mix thoroughly foodstuffs similar to mayonnaise and salad dressing. Place about 50-75 g. of the material in a bottle or beaker, if pasty, and wet with 10-15 ml. of alcohol. This aids in the breaking of emulsions and in the extraction of color. Add 100 to 150 ml. of ethyl ether and allow to stand for a few hours or overnight. Filter through filter paper. In the case of pasty materials, pour the supernatant ether layer through the filter. Wash the residue with ether, pouring the washings through the filter. Catch the filtrate in an evaporating dish and evaporate off the ether and alcohol. The residue in the dish is fat or oil and color. Take up in petroleum ether or gasoline. Place one portion in a test tube and shake with a few milliliters of 13 N sulfuric acid. Allow to stand and separate, or centrifuge. A pink or orange color in the acid layer shows the presence of a coal-tar dye. This color may be due to yellow OB or AB or some other oil-soluble coal-tar color. The probability is that if coal-tar color is present it will be due to yellow OB or AB. If such color is present, or, if this short test is inconclusive, proceed with the methods detailed under the sections, "Separation of Yellow AB and OB" and "Separation of Oil-soluble Dyes."

Return the other portion to the evaporating dish and evaporate off the petroleum ether or gasoline. Take up the residue in ethyl ether and place a portion in a test tube. Add a few milliliters of 10 per cent sodium hydroxide solution and shake. Allow to stand or centrifuge. If the sodium hydroxide layer is colored, a natural coloring matter is present. This may be due to annatto or turmeric. Draw off the sodium hydroxide layer, if colored, with the aid of a small separatory funnel and divide into two portions. Dilute one portion with an equal amount of water and allow a piece of absorbent cotton to steep in the solution for a few hours or overnight. If the cotton after washing gently in water remains colored a straw hue which is turned pink or purple by a drop of stannous chloride solution, annato is confirmed. To the other portion, add sufficient 4 per cent boric acid in strong hydrochloric acid to make the mixture acid. A red color confirms the presence of turmeric.

If traces of color are present, the sodium hydroxide layer will scarcely be colored. Therefore this rapid test is not completely applicable for traces of color. The change from straw to pink, undergone by annatto, is probably due to the acid of the stannous chloride solution and not to the stannous chloride, itself, for hydrochloric acid, acetic acid, and sulfuric acid, all give this color reaction, although stannous chloride is not present. Some authorities believe that the purple color produced is due to the stannous chloride.

Evaporate off the ether in the remaining portion and take up in chloroform. Place 0.1 ml. in a micro test tube and add 9 drops of trichloroacetic acid solution (9:1). An intense blue color indicates the presence of carotene.[14] To verify, add to another 0.1 ml. portion 3-4 drops of antimony trichloride. Again a blue color develops in the presence of carotene. A red color at this point in either of these two tests indicates ergosterol or some similar sterol.

SEPARATION OF YELLOW AB AND YELLOW OB

The following details are given by the A. O. A. C. for the separation of yellow AB and OB. Extract a gasoline solution of these dyes 3 times with half its volume of 13 N sulfuric acid. Shake each acid extract successively with 2 portions of equal volume of low-boiling gasoline or petroleum ether, using the same 2 portions for each acid portion. Extract each of the two latter gasoline portions with 20 ml. of 13 N sulfuric acid, using the same acid portion successively for both gasoline portions. Finally, extract the second of these gasoline portions with another 20-ml. portion of 13 N sulfuric acid. The original gasoline solution has now been shaken with acid 3 times, the next gasoline portion 4 times, and the third 5 times. Combine the acid extracts, dilute with water, re-extract with low boiling gasoline, and evaporate the solvent. Yellow AB will be found in a practically pure state. Combine the gasoline solutions, original and subsequent solutions left after the acid washings, wash with small portions of water to remove excess of acid, and evaporate the solvent. The yellow OB will remain as a residue. While this method is not absolutely quantitative, it is sufficiently accurate to make a separation of either of the dyes with comparatively little contamination from the other.

The following color test may be applied to the separated dyes to confirm their identity: Shake 5 ml. of a neutral gasoline solution of the dye in a test tube with 5 ml. of a mixture of 1 part of 40 per cent formaldehyde solution and 4 parts of acetic anhydride. Both coloring matters are extracted by the acetic anhydride, yellow AB giving in a few seconds a red colored solution, and yellow OB, under the same conditions, giving an orange colored solution.

Because the structure of these oil-soluble yellows differs only by a methyl group, their reactions are generally similar. As an alterna-

---

[14] Levine and Bein, *Proc. Soc. Exptl. Biol. Med.* 32, 335 (1934).

tive method [15] of differentiation, proceed as follows: To 1 ml. of an alcoholic solution of the dye (0.005-0.01 per cent) add 0.1 ml. of 40 per cent formaldehyde solution, 0.1 ml. of concentrated sulfuric acid, and 8.0 ml. of water. Yellow AB gives a red color which is unaltered by the addition of an excess of strong ammonium hydroxide solution and which is somewhat intensified by the addition of an excess of glacial acetic acid. Yellow OB treated similarly produces a yellow coloration.

Copper-Pyridine Test.—To 1 ml. of an alcoholic solution of the dye (0.005-0.01 per cent) add 0.1 ml. of copper-pyridine reagent consisting of 0.5 g. of copper sulfate, $CuSO_4 \cdot 5H_2O$, 1.0 ml. of pyridine, and complete to 10 ml. volume with water. Yellow AB gives a pink color turning to a purple on addition of strong ammonia. Yellow OB treated similarly becomes colorless or slightly bluish.

SEPARATION OF OIL-SOLUBLE DYES

If nonpermitted oil-soluble coal-tar dye is suspected, one may proceed according to the following method from the A. O. A. C.[16] or as outlined in a text dealing exclusively with identification of dyes. Prepare an alcoholic solution of the dye by applying one of the following methods to the oil or fat obtained by extraction with ether or gasoline if the nature of the substance requires it:

(a) Shake the oil or melted fat with an equal volume of alcohol, 90 per cent by volume, and wash the alcoholic extract with several portions of gasoline to free the coloring matter from fats. The alcohol, after separation, will contain aniline yellow, butter yellow, aminoazotoluene, auramine, sudans, yellow OB, yellow AB, etc., if present.

(b) Saponify 20 to 200 g. of the oil or fat with 0.5 N alcoholic potassium hydroxide solution, remove most of the alcohol on the steam bath, and extract the soap with ether or gasoline. Remove the dyes from the solvent with 10 ml. portions of a mixture containing 1 part of hydrochloric acid and 5 parts of glacial acetic acid. Most of the common dyes are removed by this treatment, although the digestion with strong alkali may cause some decomposition and make the extraction rather troublesome.

(c) Dilute 20 to 200 g. of the oil or melted fat with 1-2 volumes of gasoline and shake out successively with 2-4 per cent potassium hydroxide or sodium hydroxide solution, hydrochloric acid (1:3), and phosphoric acid-sulfuric acid mixture, prepared by mixing 85 per

---

[15] Jablonski, in Jacobs, Ed. *Chemistry and Technology of Food and Food Products.* Interscience, New York, 1944.

[16] *Methods of Analysis, A.O.A.C.* Washington, 1945.

cent phosphoric acid with about 10-20 per cent sulfuric acid, by volume.

Reagent	Dye Extracted
2—4% KOH	Sudan G (10) [23] Annatto (709) [1241]
HCl (1:3)	Aniline yellow (7) [15] ⎱ become orange red Aminoazotoluene [17] ⎰ Butter yellow (16) [19]—becomes cherry red
H₃PO₄—H₂SO₄ mixt.	Sudan I (11) [24] Sudan II (49) [73] Sudan III (143) [248] Sudan IV [258] Benzeneazo-β-naphthylamine [22] and homologues

Benzeneazo-β-naphthylamine (—) [22] and homologues readily undergo chemical changes in the strongly acid mixtures. The procedure is not very suitable in the presence of auramine, but this dye is seldom found in oils. Neutralize the alkaline and the dilute hydrochloric acid solutions, dilute the phosphoric acid mixture and partially neutralize, cooling the liquid during this operation. Extract the dyes by shaking with ether or gasoline.

For the direct dyeing test use the alcoholic solution obtained as directed under (a). Evaporate to dryness the ether or gasoline solutions obtained as directed under (b) and (c) and dissolve the residue in 10-20 ml. of 95 per cent alcohol. To the alcoholic solution add some strands of white silk and a little water and evaporate on a steam bath until the alcohol has been removed or the dye is taken up by the silk. The dyeing test is sometimes unsatisfactory, and in all cases a small portion of the alcoholic solution should be tested by treating with an equal volume of hydrochloric acid and stannous chloride solution. The common oil soluble coal tar dyes are rendered more red or blue by the acid and are decolorized by the reducing agent. Most of the natural coloring matters become slightly paler with the acid and are little changed by the stannous chloride solution.

The dyed strands of silk may then be identified by performing the spot tests previously outlined, checking by using known strands of about equal intensity and referring to the color reactions listed in Table 6.

Particular care should be taken in the detection of aniline yellow, aminoazotoluene, and butter yellow, *Colour Index* Nos. 15, 17, and 19, for it has been demonstrated that these nonpermitted colors may cause carcinoma in rodents.

Color on Citrus Fruits

In addition to the use of ethylene gas, 2,4-dichlorophenoxyacetic acid, and similar agents to develop a "riper" color and to hasten "ripening" (these agents probably act by decolorizing the chlorophyll and thus apparently ripening the fruit by giving it its ripe color) of oranges and other citrus fruits, the use of coal-tar colors to augment the desirability and salability of oranges was a common practice. Some oranges even though ripe still retain a green color and other oranges that are off-color or spotty or otherwise considered second grade, may be made to appear the equal of best grade oranges by judicious use of color.

The addition of color to the skin of oranges may easily be detected by a method based on the solution of the dye in chloroform, dyeing the color on silk from 70 per cent alcohol solution and subsequent identification of the color by means of spot tests.

Procedure.—Place the orange in a funnel. Wash with a stream of chloroform directed from a wash bottle, rotating the orange at intervals so that the entire orange is washed with the chloroform. Catch the washings in a 400-ml. beaker. Wash two or three more oranges in a similar manner and catch the washings in the same beaker. Evaporate the chloroform solution to dryness on a water bath. Dissolve the color in 70 per cent alcohol and dye the color on silk by steeping the silk in the alcohol solution of the color on a water bath. Replenish the alcohol from time to time by the addition of 70 per cent alcohol. Remove the silk and wash thoroughly with water. Dry and apply spot tests.

A more exhaustive method is to dry the peel, extract the oil and color with ether or gasoline, and then proceed as directed in the section, "separation of oil-soluble colors."

Method for Detection of Artificial Color in Olive Oil

There are various methods for the separation of artificial coloring matters from oils, but the following method is a special method [17] designed to extract the artificial coloring matter of olive oil and to separate a mixture of yellow OB or AB and quinazarin green SS.

Procedure.—Transfer approximately 5 g. of oil to a 300-ml. Erlenmeyer flask. Add 30 ml. of 95 per cent alcohol and 5 ml. of 50 per

---

[17] Jacobs, unpublished work, 1940.

cent aqueous sodium hydroxide solution. Adjust an air condenser and reflux for 1 hour. Transfer the saponified mixture to an extraction cylinder (unsaponifiable residue type) or separatory funnel, or a Jacobs-Singer separatory flask, and make up to 60 ml. with water washings of the Erlenmeyer flask. Extract 4 times with 45 to 50-ml. portions of petroleum ether, drawing off the ether layer as closely as possible into a liter beaker.

Repeat the entire process with three to five more 5-g. portions of oil. Combine all the petroleum ether extracts and evaporate to about 200-250 ml. Transfer the ether to a separatory funnel and wash the beaker twice with small portions of petroleum ether adding the washings to the separatory funnel. Wash the petroleum ether extract with three or four 25-ml. portions of 10 per cent alcohol by volume. Discard the 10 per cent alcohol washings.

Extract the petroleum ether layer with successive 5-ml. portions of 20 N sulfuric acid until no more color is noted in the acid layer. Transfer the acid layers to a separatory funnel containing 150-200 ml. of water. Reserve this extract for the separation of yellow OB or AB described below.

Wash the petroleum ether layer with successive 25-ml. portions of water until the water washings are neutral to litmus. Evaporate the petroleum ether to a volume of about 25-50 ml. Return to a separatory funnel and extract with successive 3 to 5-ml. portions of syrupy phosphoric acid-sulfuric acid mixture (5 parts 85 per cent H_3PO_4-1 part concentrated H_2SO_4). Combine the phosphoric acid-sulfuric acid mixture extracts and wash 2 or 3 times with 15 to 20-ml. portions of petroleum ether to free the acid extract of lipoid substances. Discard these ether washings. Dilute the acid mixture extract with 150 ml. of water and extract with three 50-ml. portions of petroleum ether. The color will now go back to the ether layer. Draw off the acid layer and discard. Wash the ether layer with water until the water washings are free of acid. Take up the residue in alcohol. Dye on silk if sufficient color is present or keep in a vial. An separate a mixture of yellow OB or AB and quinazarin green SS.

Extract the diluted 20 N sulfuric acid extracts with three 50-ml. portions of petroleum ether. Combine the ether layers and wash with successive 25-ml. portions of water until the water is neutral to litmus. Evaporate the ether layer to dryness. Take up in 95 per cent alcohol and dye on silk. This color may be yellow OB or AB. The acetic anhydride-formaldehyde test for the differentiation between these colors may be performed on this extract.

Separation of Natural Coloring Matters

Natural coloring matters show little tendency to dye animal fibers. However, they may be extracted by various solvents, re-extracted with water, and subjected to color tests with divers reagents on portions of the aqueous solution. From neutral solutions ether extracts carotene, xanthophyll, the pigments found in leaves, fats and oils, egg yolk, carrot, etc., the coloring matter of tomatoes and paprika, and green chlorophyll. The coloring matter remains in the ether solution on shaking with normal sodium hydroxide solution or normal hydrochloric acid, no apparent change taking place, although chemically the substances may be altered more or less by this treatment.

Annatto, turmeric, alkanet, and the coloring matter of red dyewoods and other woods and flavone colorings matters are extracted by ether from acid solution almost completely. The coloring matters of this group are easily removed from the ether by shaking with alkaline solutions but they are rather readily decomposed.

Amyl alcohol extracts largely the coloring matters of logwood, archil, saffron, and cochineal, from slightly acid solutions. Amyl alcohol extracts in relatively small proportions caramel and the anthocyans constituting the red coloring matter of the most common fruits.

Carotene

One type of sophistication encountered is the addition of carotene to alimentary pastes and other foods for the purpose of increasing the color. Since alimentary pastes normally contain carotenoid pigments, the problem becomes one of determining added carotene. This is done by estimating total carotenoids and comparing this value with known values for the carotene content of alimentary pastes.

Briefly the method follows the following line. The color and fat are extracted from a sample. A known quantity is saponified, and the total carotenoid coloring is extracted by a solvent. Xanthophyll, lutein, and the like are extracted by a solvent immiscible in the first solvent, leaving the carotene. This may then be estimated colorimetrically [18] and as detailed below.

Wall and Kelley Method.—In the Wall and Kelley [19] procedure and also that of Moore and Ely,[20,21] a Waring blendor or similar de-

[18] Munsey, *J. Assoc. Official Agr. Chem.* 21, 626, 664 (1938).
[19] Wall and Kelley, *Ind. Eng. Chem., Anal. Ed.* 15, 18 (1943).
[20] Moore and Ely, *Ind. Eng. Chem., Anal. Ed.* 13, 600 (1941).
[21] Derby and DeWitt, *J. Assoc. Official Agr. Chem.* 31, 704 (1948).

vice is used for the extraction of the carotenoid pigments with cold solvents.

Preparation of Sample.—Grind the sample, such as dehydrated alfalfa meal to pass a 40-mesh sieve, place in a closed container, and store in a refrigerator. Remove portions and weigh immediately prior to assay.

Procedure.—Extract a 2-5 g. sample with 200 ml. of a 5:3 mixture of 95 per cent ethyl alcohol and petroleum ether (boiling point 35-65° C.) in a Waring blendor for 10 minutes. Transfer the extract and the finely divided material to a fritted glass filter connected to a suction flask and wash with alternate portions of alcohol and petroleum ether until a colorless filtrate is obtained. Remove the alcohol from the extracts by washing with water in a separatory funnel, and concentrate the petroleum ether extracts on a steam bath. Dry by filtration through anhydrous sodium sulfate into a 100-ml. volumetric flask, and adjust to volume with petroleum ether.

Chromatography.—Prepare a column composed of a 1:3 mixture of magnesium oxide (Westvaco #2641) and Hyflo-Super Cel (Fisher) in an absorption tube measuring 22 x 100 mm. fitted with a fritted disk of medium porosity. Pack the adsorbent to a depth of approximately 75 mm., and add a 1-cm. layer of anhydrous sodium sulfate. Add the extract solution. Develop the chromatogram and elute the carotene with a 5 per cent solution of acetone in petroleum ether. Adjust the eluate to a desired volume with petroleum ether.

Optical Measurement.—Determine the optical density of the purified pigment at 5 mμ intervals between 400 and 500 mμ.

Modified A. O. A. C. Method.—This method [22] depends upon a hot extraction of the carotenoid pigments from plant materials with low-boiling solvents, such as hydrocarbons or a mixture of hydrocarbons and a polar solvent.

Procedure.—Extract a 2-4 g. sample in a Bailey-Walker apparatus for 1 hour with 30 ml. of a 30 per cent solution of acetone in petroleum ether. Transfer the extract to a 100-ml. flask and make up to volume with petroleum ether. Proceed with the method as detailed above.

Alcoholic Potash Digestion.—In this variation the plant material is digested with hot alcoholic potassium hydroxide solution, which treatment is followed by extraction with petroleum ether or other solvents.

Procedure.—Place a 2-5-g. portion of the sample in an Erlenmeyer flask with 100 ml. of 12 per cent alcoholic potassium hydroxide

22 *J. Assoc. Official Agr. Chem.* **31**, 111 (1948).

solution, and digest on a steam bath for 30 minutes. Hold the volume constant throughout the digestion by the addition of 95 per cent alcohol. Cool to room temperature, transfer to a 500-ml. separatory funnel, rinse the Erlenmeyer flask with 100 ml. of water, and add the rinsings to the separatory funnel. Extract the resulting mixture with 50-ml. portions of petroleum ether until a colorless extract is obtained. This requires about 5 to 6 extractions. Combine the extracts in another separatory funnel, add 200 ml. of water, but take care not to shake or swirl the contents. After 2 minutes discard the aqueous layer. Add another 100 ml. of water and swirl gently. Allow 2 minutes for separation, draw off the aqueous layer, add 25 ml. of water, and shake vigorously for 30 seconds. Continue extractions with 25-ml. portions of water until the washings are neutral to phenolphthalein indicator solution. After the final washing, allow the funnel to stand for 5 minutes, draw off any residual water, and concentrate the petroleum ether layer under vacuum on a steam bath. Filter through anhydrous sodium sulfate into a 100-ml. volumetric flask and make up to volume with petroleum ether.

BEET DYE IN TOMATO PRODUCTS

Occasionally beet juice or the coloring matter of beets, betanin, is used to intensify the color of tomato products like tomato sauce, puree, and catsup. This sophistication may be detected by utilizing the differences in solubility of the normal pigments of tomato and that of betanin.[23]

Procedure.—Extract the normal coloring matter of tomato products from an aqueous suspension or solution with amyl alcohol. Betanin remains in the aqueous solution. Put in a piece of tannin mordanted cotton and dye the betanin on the cotton. A terra cotta shade is produced.

CARAMEL

Caramel is an artificial color, according to the Food and Drug Administration,[24] made by the artifice of partially breaking down sugar. It is used widely for the coloring of imitation vanilla flavors, "cream" flavors, and the like.

To prepare a relatively small batch of caramel or sugar coloring[25] fuse 10 lbs. of sugar in a wide low-form type of vessel like an iron frying pan. Heat 5 pints of water to boiling and add it to the

---

[23] Jablonski, personal communication, 1949.

[24] *U. S. Food Drug Admin. F.D.&C. Act Trade Correspondence* TC-203, March 21, 1940.

[25] *Pharmaceutical Recipe Book.* Am. Pharm. Assoc., 1929.

fused sugar which caramelizes during the fusion process. Boil until a syrup is obtained.

An alternative procedure is to stir the fused mass until it is brown, remove it from the heat and add sufficient water to dissolve the caramelized sugar. A paste may be made by the addition of commercial glucose.

Marsh Test.—Caramel and some coal tar colors are insoluble in the Marsh reagent which consists of 100 ml. of pure amyl alcohol, 3 ml. of syrupy phosphoric acid, and 3 ml. of water. Thus this reagent provides a means of detecting caramel. The test applied to alcoholic beverages can be modified for use in nonalcoholic beverages and other food products.

Procedure.—To 6 ml. of a nonalcoholic sample add 6 ml. of alcohol and mix. Add 15 ml. of the Marsh reagent and shake the mixture gently for 2 minutes, avoiding the formation of an emulsion. Allow the layers to separate completely. A colored aqueous layer may indicate the presence of caramel or some coal-tar colors which can be identified by the methods detailed in the preceding sections.

Mathers[26] Test.—Place 10 ml. of the filtered sample into a Babcock cream bottle (see Chapter VII) or any convenient small centrifuge bottle and add 1 ml. of a pectin solution prepared by dissolving 1 g. of pectin in 75 ml. of water and adding 25 ml. of alcohol for preserving it. Shake the pectin solution thoroughly before using it. Add 3 drops of concentrated hydrochloric acid and fill the bottle with alcohol. Shake well, centrifuge for 5 to 10 minutes and decant the supernatant liquid from the gelatinous residue. Dissolve the latter in 10 ml. of water and swirl thoroughly. Add 3 drops of concentrated hydrochloric acid and sufficient alcohol to fill the bottle. Shake thoroughly and centrifuge. Decant the supernatant fluid. Repeat this treatment until the upper alcoholic layer is clear and colorless. After the final decantation of the water-white supernatant alcohol, dissolve the gelatinous residue in 10 ml. of hot water. A clear brown solution indicates caramel coloring.

To confirm the presence of caramel, add 1 ml. of a reagent prepared by dissolving 1 g. of 2,4-dinitrophenylhydrazine in 7.5 ml. of concentrated sulfuric acid and bringing up the volume to 75 ml. with 95 per cent alcohol. This reagent should be preserved in an amber glass-stoppered bottle in which it is stable for months. Place the bottle in a beaker of boiling water for 30 minutes. Set aside for 30 minutes to cool. In the presence of substantial quantities of caramel a precipitate forms almost immediately. Lesser quantities give a precipitate in about 15 minutes. Even the smallest amount of cara-

---

[26] Valaer, *J. Assoc. Official Agr. Chem.* **30,** 331 (1947).

mel will show a precipitate under the conditions detailed and if caramel is absent no precipitate will form. In order to verify the formation of a precipitate, filter the hot test solution through a small filter and wash any residue with hot water. A reddish brown precipitate will be clearly seen on the filter. This precipitate although amorphous is characteristic and may be examined with a low-power microscope for confirmation.

GROUP SEPARATION OF MIXTURES OF FD&C, D&C, AND EXT. D&C COLORS

Koch Immiscible Solvent Method.—As mentioned previously, the use of immiscible solvents for the separation of colors was comprehensively investigated by Mathewson.[27,28] Ruggli and Jensen [29] used chromatographic methods and reported considerable success. Koch [30] devised a systematic group separation of mixtures of FD&C, D&C, and Ext. D&C colors. Although the latter two groups are not to be used as food colors, they may at times be used in this manner, and the method of Koch will serve to separate them. As noted in the instance of mixtures of food colors, the condition will seldom arise in which the mixture will contain more than a few colors so that, although the groups are large, the number of colors actually isolated in any group will be a very small number.

Procedure.—Dissolve 5-15 mg. of the mixture in 24 ml. of water, and make alkaline with 6 ml. of 10 per cent sodium hydroxide solution. If the unknown is insoluble in water, boil a similar sized sample with 30 ml. of a 2 per cent sodium hydroxide solution for 1 minute, cool, and filter.

Extract the solution or filtrate with 30 ml. of ether, repeating the extraction with a second and a third portion of ether, if necessary. Wash the extracts with 20-ml. volumes of 0.5 per cent sodium hydroxide solution, passing the washings through the separatory funnels in the original sequence. Discard the washings if they are lightly tinted, but reserve them if they are deeply colored (cf. "Washings" below).

Acidify the color solution, from which the alkaline ether extracts have now been removed, with 10 ml. of acetic acid, and extract with one or more 30-ml. portions of ether. Wash the extracts with 20-ml. volumes of a 0.5 per cent acetic acid solution, discarding or reserving the washings according to the intensity of the color. The ether extracts contain:

27 Mathewson, *U. S. Dept. Agr. Bull.* 448 (1917).
28 Mathewson, *Am. Dyestuff Reporter* 22, 721 (1933).
29 Ruggli and Jensen, *Helv. Chim. Acta* 18, 624 (1935); 19, 64 (1936).
30 Koch, *J. Assoc. Official Agr. Chem.* 26, 246 (1943).

Group 1.—Alkaline Ether Extract.—Rhodamine B (D&C Red Nos. 19, 20, 37).

Rhodamine B is practicaly decolorized by alkali; it bleeds out of the ether layer with a 0.5 per cent sodium hydroxide solution, and it is washed out of the ether by a 2.5 per cent acetic acid solution, the color being restored. Rhodamine B stearate is extracted also in Group 8.

Group 2.—Acid Ether Extract.—
    Erythrosine (FD&C Red No. 3, D&C Red No. 3)
    Fluorescein (D&C Yellow Nos. 7, 8, 9)
    Tetrabromofluorescein (D&C Red Nos. 21, 22, 23)
    Tetrachlorofluorescein (D&C Red Nos. 24, 25, 26)
    Tetrachlorotetrabromofluorescein (D&C Red Nos. 27, 28)
    Bluish Orange T.R. (D&C Red No. 29)
    Dibromofluorescein (D&C Orange Nos. 5, 6, 7)
    Dichlorofluorescein (D&C Orange Nos. 8, 9)
    Diiodofluorescein (D&C Orange Nos. 10, 11, 12, 13)
    Dichlorotetraiodofluorescein (Ext. D&C Red Nos. 4, 5, 6)
    Orange T.R. (D&C Orange No. 14)
    Alizarin (D&C Orange No. 15)
    Dibromodiiodofluorescein (D&C Orange No. 16)

Dilute the color solution, from which groups 1 and 2 have been extracted, to 60 ml. with a 20 per cent salt solution. Extract with 30-ml. portions of amyl alcohol. Wash the amyl alcohol extracts with a 5 per cent salt solution, reserving or discarding the washings as usual. The amyl alcohol extract contains:

Group 3.—
    Orange I
    Guinea Green B (FD&C Green No. 1, D&C Green No. 1)
    Quinoline Yellow W.S. (D&C Yellow No. 10)
    Orange II (D&C Orange No. 4)
    Lithol Rubin B (D&C Red Nos. 6, 7)
    Deep Maroon (D&C Red No. 34)
    Resorcin Brown (D&C Brown No. 1)
    Alizarin Cyanine Green F (D&C Green No. 5)
    Alizarin Astrol B (D&C Blue No. 5)
    Patent Blue (D&C Blue Nos. 7, 8)
    Acid Violet 6B (FD&C Violet No. 1)
    Metanil Yellow (Ext. D&C Yellow Nos. 1, 2)
    Methylene Blue (Ext. D&C Blue Nos. 1, 2)

Violamine R (Ext. D&C Red No. 3)
Alizarin Carmine (Ext. D&C Red No. 7)
Coomassie Black (Ext. D&C Black No. 1)
Anthraquinone Violet (Ext. D&C Violet No. 1)
Alizurol Purple (Ext. D&C Violet No. 2)
Fast Light Yellow (Ext. D&C Yellow No. 3)

All the dyes in Group 3 are washed out of the ether layer with a 1 + 9 ammonia solution. Alizarin is changed to a purple shade by this treatment. Alizarin and brilliant lake red R are extracted also in Group 7. Red lake D may bleed strongly into Group 2.

Quinoline yellow bleeds into other groups. Lithol Rubin is yellow in alkali and red in acid. Guinea green B is decolorized by alkali. Metanil yellow is changed to a violet by acid. Methylene blue turns violet after boiling with a 2 per cent alkali solution and may bleed into Groups 1 and 2. Red lake C may bleed into this group, but the dye will not easily wash out of the amyl alcohol with water. Acid violet 6B is decolorized by alkali. Anthraquinone violet may bleed into Groups 1 and 2. Patent blue is not wholly extracted in Group 3, but is completely removed in Group 4. It turns violet on boiling with dilute alkali. Deep maroon is slightly soluble in a 2 per cent alkali solution. To remove the dyes of Groups 3, 5, and 6 from the amyl alcohol, dilute with petroleum benzine, and wash with water. Naphthol blue black (D&C Black No. 1), when present in small quantities, will completely extract in Group 3 and may not wash out with a 5 per cent salt solution. When present in large amounts, the color will completely extract with the amyl alcohol in Group 3 and will largely wash out with a 5 per cent salt solution, leaving a subsidiary (?) dye in the organic solvent.

Extract the remaining color solution with 15-ml. portions of a 1 + 2 carbon tetrachloride-dichlorohydrin mixture. Wash the extracts with a 25 per cent salt solution. Reserve or discard the washings in the usual manner. The extract contains:

Group 4.—
Brilliant Blue FCF (FD&C Blue No. 1, D&C Blue No. 1)
Light Green SF Yellowish (FD&C Green No. 2, D&C Green Nos. 2, 4)
Fast Green FCF (FD&C Green No. 3, D&C Green No. 3)
Alphazurine FG (D&C Blue No. 4)

Brilliant blue and alphazurine FG turn a violet-red shade when boiled with a 2 per cent sodium hydroxide solution. Light green is decolorized by alkali. Fast green is changed to a purplish blue by alkali. To remove the dyes of Group 4 from the organic solvent, dilute with carbon tetrachloride and wash with water.

Acidify the color solution, from which groups 1, 2, 3, and 4 have now been removed, with 1.4 ml. of hydrochloric acid for every 40 ml. of solution, and extract with one or more 30-ml. volumes of amyl alcohol. Wash the extracts with 0.125 N hydrochloric acid. Reserve or discard the washings as usual. The amyl alcohol contains:

Group 5.—
  Naphthol Yellow S
  Ponceau 3R (FD&C Red No. 1, D&C Red No. 1)
  Ponceau SX (FD&C Red No. 4, D&C Red No. 4)
  Ponceau 2R (D&C Red No. 5)
  Naphthol Blue Black (D&C Black No. 1)
  Croceine Scarlet MOO (Ext. D&C Red No. 13)

Naphthol yellow is nearly decolorized by acid; with sodium hydrosulfite in ammoniacal solution, a rose-red color is produced. Ponceau 2R and ponceau 3R are easily destroyed by alkaline peroxide, whereas ponceau SX is not. Naphthol blue black bleeds into Groups 3 and 4; in Group 4 the 25 per cent salt solution precipitates the dye out of the organic solvent. Croceine scarlet MOO is dark red in the presence of alkali.

Add excess hydrochloric acid to the remaining dye solution and extract with 30-ml. portions of amyl alcohol. The residual solution should be colorless. The amyl alcohol contains:

Group 6.—
  Tartrazine (FD&C Yellow No. 5, D&C Yellow No. 5)
  Sunset Yellow FCF (FD&C Yellow No. 6, D&C Yellow No. 6)
  Amaranth (FD&C Red No. 2, D&C Red No. 2)
  Indigotine (FD&C Blue No. 2, D&C Blue No. 2)
  Orange G (D&C Orange No. 3)
  Acid Fuchsin D (D&C Red No. 33)
  Amidonaphthol Red 6B (Ext. D&C Red No. 1)
  Pigment Scarlet NA (Ext. D&C Red No. 2)
  Alizarin Saphirol (Ext. D&C Blue No. 4)
  Naphthol Green B (Ext. D&C Green No. 1)

Alizarin saphirol extracts with difficulty.
If the unknown was insoluble or slightly soluble in water:

Boil a new sample with 30 ml. of an acid mixture containing 1 part of concentrated acetic acid plus 1 part of 1 + 1 hydrochloric acid for 30 seconds. Cool, and filter. Add 30 ml. of ether to the filtrate, agitate in the extraction funnel, and then dilute with 30 ml. of water. Agitate again. Extract with 2 more volumes of ether to complete the extraction. Wash the combined ether extracts with N. hydrochloric acid. Discard washings. The ether contains:

Group 7.—
    Lithol Red (D&C Red Nos. 10, 11; 12, 13)
    Lake Red C (D&C Red Nos. 8, 9)
    Lake Red D (D&C Red Nos. 14, 15, 16)
    Brilliant Lake Red R (D&C Red No. 31)
    Alizarin (D&C Orange No. 15)

Group 7.—Lithol red will not wash out of the ether layer with water or with 1 + 9 ammonia. The ether solution of red lake C will bleed slightly when washed with normal hydrochloric acid, and the color will wash out completely with water. Red lake D. and brilliant lake red R will wash out of the ether layer with 1 + 9 ammonia. Group 2 will bleed into this group and will wash out of the ether layer with 1 + 9 ammonia. Dyes of Group 8 may bleed occasionally and may be mistaken for lithol red. Deep maroon is extracted with ether, washes out very slowly with N hydrochloric acid and nearly wholly with water, leaving a residual dye that is extractable with 1 + 9 ammonia.

Digest a 10-25 mg. sample of the unknown with hot benzene, and filter. The benzene contains:

Group 8.—
    Yellow AB
    Yellow OB
    Orange SS
    Oil Red XO, Sudan Red II
    Toney Red (D&C Red No. 17)
    Oil Red OS (D&C Red No. 18)
    Helindone Pink CN (D&C Red No. 30)
    Alizarin Green SS (D&C Green No. 6)
    Alizurol Purple SS (D&C Violet No. 2)
    Rhodamine B Stearate (D&C Red No. 37)
    Toluidine Red (D&C Red No. 35)
    Chlorinated-*p*-nitraniline Red (D&C Red No. 36)
    Hexyl Blue (Ext. D&C Blue No. 5)

Hansa Yellow (Ext. D&C Yellow No. 5)
Hansa Orange (Ext. D&C Orange No. 1)
Quinoline Yellow SS (D&C Yellow No. 11)

Group 8.—Colors of Group 2 may bleed into the benzene but are removed with dilute alkali. Helindone pink has a yellowish fluorescence in benzene solution. Microscopic examination of the crystal structure and the melting point determination will be of great assistance in the identification of the dyes in Group 8.

Groups 2-6 and Group 8 contain dyes of the food-color class, and the procedure of the A.O.A.C. described in other sections of this chapter is used if the sample is in this category.

Washings.—If, in the opinion of the analyst, the washings contain dyes other than those found in the group analysis, add excess hydrochloric acid and extract with amyl alcohol. Discard the residual washings if they are colorless, but if they are colored add solid salt to bring the salt concentration to 12.5 per cent and extract with portions of carbon tetrachloride-dichlorohydrin mixture. Dilute the amyl alcohol with petroleum benzine and remove the colors with water. Evaporate the water solution to dryness, and repeat the group analysis, 1-6. Dilute the carbon tetrachloride-dichlorohydrin extract with an equal volume of carbon tetrachloride, and remove the coloring matters with water. Evaporate the water solution to dryness, and test for the dyes of Group 4.

MINERAL COLORS

The use of mineral colors in foods is very small, primarily because many of them are poisonous. These mineral colors are called pigments. Lakes are organic substances combined with metals or metallic salts and are considered with mineral colors because of their metallic content. Examples of pigments are lampblack or other forms of carbon, prussian blue, talc, ochres, umbers, and ultramarine blue. These may all be identified by the usual tests for metals and other elements as for example carbon in lampblack and sulfur in ultramarine blue. The analysis of these materials will be discussed in some detail in the section on metals in foods. It must be noted that certain natural coloring matters, such as chlorophyll derivatives, contain metallic elements as magnesium.

For special tests for the water soluble acid and basic dyes, the oil-soluble coal-tar colors and the natural and mineral coloring matters, the reader is referred to any standard text on the analysis and identification of dyes. Some of these will be found in the bibliography at the end of the chapter. In these texts the reader will also find complete methods for the analysis of the dry dye powder.

SELECTED REFERENCES

Food and Drug Administration, Service and Regulatory Announcements, Food, Drug, and Cosmetics No. 3, *Coal-Tar Color Regulations.* Washington, D. C., 1940, as amended to 1957.

Green, *Analysis of Dyestuffs.* Lippincott, Philadelphia, 1916.

Green, *Organic Coloring Matters.* Macmillan, New York, 1908.

Jablonski, in Jacobs, ed., *Chemistry and Technology of Food and Food Products.* Vol. I. Interscience, New York, 1944.

Lieber, *Use of Coal-Tar Colors in Foods.* Lieber, New York, 1904.

Mathewson, *U. S. Dept. Agr., Bull.* **448** (1917).

Methods of Analysis. A.O.A.C., Washington, 1945.

Peacock, "*Application Properties of the Certified 'Coal-Tar' Colors.*" *Calco Tech. Bull. No. 715,* American Cyanamid, Bound Brook, N. J., 1945.

Rowe, ed., *Colour Index.* Society of Dyers and Colourists. Bradford, Yorkshire, 1924.

Weyl, and Leffmann, *Coal-Tar Colors.* Blakiston, Philadelphia, 1892.

Woodman, *Food Analysis.* McGraw-Hill, New York, 1942.

CHAPTER IV

CHEMICAL PRESERVATIVES IN FOODS [1,2]

THE use of chemical preservatives in foods by man is a very ancient practice, for long ago man learned to preserve his food by the use of various chemicals. In all likelihood, the discovery of the use of salt was accidental and was related to the finding of salt-encrusted carcasses in the deserts of Asia. The preservation of eggs by the Chinese by dipping them in water glass is a very old method.

About 50 years ago, the use of chemical preservatives in food was on the increase. Leach [3] mentions about a dozen mixtures which were commercially available. With the passage of the Wiley Food and Drug Act of 1906, however, and as a result of Wiley's active campaigning, the use of these agents became less common. Advances in other methods of preservation, production, and sanitation also tended to decrease their use. In the fifth decade of the 20th century, however, another change took place, and the use of chemical preservatives in food and food processing took an upward turn. This undoubtedly was a result of the increased use of chemicals for the processing and fortification of foods.

As Monier-Williams [4] points out, chemical preservatives have the advantage of continuing to exert their preserving effect even though the food is exposed to air at ordinary temperatures and the treatment of food in this way may be a more economical method than the application of heat or cold. Food preserved with chemicals, however, will remain sound only for a limited period of time, for the growth of microorganisms is merely retarded and not entirely prevented. The greater the amount of preservative added, the longer will the decomposition be delayed.

The mechanism by which chemical preservatives act has not been entirely elucidated, but it is possible that food which is not heavily contaminated with bacteria may be preserved by chemicals because these chemicals prolong the lag phase of bacterial growth. Inter-

---

[1] See also Jacobs, ed., *Chemistry and Technology of Food and Food Products*, 2nd ed. Interscience, New York, 1951.

[2] See also Jacobs, *Synthetic Food Adjuncts*. Van Nostrand, New York, 1947.

[3] Leach, *Food Inspection and Analysis*, 1st ed. Wiley, New York, 1906.

[4] Monier-Williams, *Chemistry in Relation to Food*. Chem. Ind. Pamphlets, Ernest Benn, London, 1924.

ference with chain reactions may be another explanation of the action of chemical preservatives.

The use of chemicals to preserve food, although economical, permits both housewife and commercial processor to handle and utilize fruits, vegetables, meats, and other foodstuffs in a much more unsanitary and careless manner than would be possible without the use of such preservatives. Chemical preservation should never be used as a substitute for cleanliness in the processing of food. For this reason alone it is unwise to subscribe unqualifiedly to the contention that the unrestricted and indiscriminate use of chemical preservatives is entirely harmless. In addition, there is insufficient knowledge concerning their complete mode of action. Consequently, the use of chemicals for the preservation of foods should be rigidly controlled.

It must be stressed that the use of preservatives will not improve the quality of inferior material nor will preservatives, once spoilage has set in, enable a processor to make a wholesome product out of a polluted one for the products of spoilage are still present in the foodstuff.

Definition.—Preservatives are ordinarily defined as substances which have antiseptic properties under the conditions of use—that is, they are substances which inhibit the growth of microorganisms without necessarily destroying them. Effective inhibition of microbiological growth prevents spoilage of foods. However, this definition is too limited from a practical aspect. Spoilage may occur which has no relationship to the growth of microorganisms; for instance, there is spoilage which is attributed to oxidation or to the action of autolytic enzymes. Since substances which will prevent such spoilage must also be considered preservatives, preservatives are more generally defined as chemical agents which serve to retard, hinder, or mask undesirable changes in food.

Many of the substances used as preservatives may in themselves be harmless or relatively harmless. Among these are sugar, salt, nitrates, vinegar, organic fruit acids, wood smoke, hops, and alcohol—all used in the preservation of foods. One would not ordinarily consider them as chemical preservatives, although they do have a bactericidal, germicidal, or antifermentative action. Chenowith[5] classes these materials as common or kitchen preservatives in contradistinction to chemical preservatives. Sherman[6] classes sugar, vinegar, wood smoke, and salt as materials with condimental properties as opposed to sodium benzoate, salicylic acid, boric acid, borax,

[5] Chenowith, *Food Preservation*. Wiley, New York, 1930.
[6] Sherman, *Food Products*. Macmillan, New York, 1933.

sulfur dioxide, formaldehyde, etc., which he terms noncondimental agents. The word condiment originally meant a preservative.

In conformance with such concepts, the term preservative for foods is defined by the British Food and Drug Act of 1928 as:

" 'Preservative' means any substance which is capable of inhibiting, retarding, or arresting the process of fermentation, acidification, or other decomposition of food or of masking any of the evidences of any such process; but does not include common salt (sodium chloride), saltpeter (sodium or potassium nitrate), sugars, acetic acid, or vinegar, alcohol or potable spirits, spices, essential oils or any substance added to the food by the process of curing known as smoking.''

In a broader sense, however, if these materials and others like them, as, for example, lactose or lactic acid, are used to retard damage or to conceal inferiority, strict, interpretation would imply that these, too, are chemical preservatives rather than kitchen preservatives. Thus, where lactic acid instead of vinegar is added to white horse-radish colored with beet juice in order to preserve the red color of the beet juice, because lactic acid has less action on this color than has acetic acid, it becomes difficult to decide whether an inferiority is concealed or not. In the same light, if lactose or milk powder, or some similar material has been added to frozen egg yolks so that the yolks when thawed will not yield a gummy mass, the question arises whether or not the lactose or milk powder has been added because it is an inexpensive filler or is an antifermentative agent rather than an aid in the thawing step.

In the United States as far as federal regulation is concerned no separate standards for food preservatives have been proposed. In unstandardized foods, they are covered by the general provisions of the Food, Drug and Cosmetic Act of 1938—that is, preservatives are prohibited if the preservative is a poison. In standardized foods they may be used only if recognized as an optional ingredient. However, many of the States and some municipalities have rigid prohibitions about the use of preservatives. The Federal Food, Drug, and Cosmetic Act requires the label declaration of chemical preservatives in food.

Unscrupulous dealers and manufacturers will have no compunction about the use and nondeclaration of preservatives; therefore, it behooves the analyst, even though the general use of antiseptics has declined in some respects, to be on the lookout for such substances and to consider the possibility that new type preservatives may be employed. As in the use of color, the use of preservatives very often follows the price scale—that is, as the price of food materials rises, the use of preservatives increases.

In some instances, as, for example, benzoates in cranberries, sulfur dioxide in horse-radish, or boric acid in fruits, the preservative may be a natural component. In these cases, quantitative determinations have to be made, and only if the preservative content exceeds the known natural limits may the analyst conclude that preservative has been added.

The reasons for the prohibition, declaration, and limitation of preservatives are clear. First, they may be used to conceal damage or inferiority, as in the case of the use of sulfites in meats; second, they may be unknown to the consumer, as in the case of undeclared benzoates in a foodstuff; and third, if maximum limits of preservative content are not enforced, the cumulative effect might be harmful.

In times of stress or of food shortage, every effort must be made to conserve food supplies. Consequently, the use of preservatives must be carefully re-examined so that those substances which are physiologically and toxicologically harmless may be used. However, the consideration as to whether or not they are being used as substitutes for cleanliness or for utilization of inferior material, as previously noted, must be factors to be evaluated.

For example, certain preservatives were permitted in Germany. Regulations concerning them were published in 1932. These preservatives had to comply with the standards of the German Pharmacopoeia. The sale of mixtures of preservatives other than those mixtures appearing in the official list were prohibited, as was the sale of preservatives mixed with other substances, with the exception of mixtures containing salt, sugar, tartaric acid, citric acid, and mixtures of the ethyl and propyl esters of p-hydroxybenzoic acid with sodium carbonate. Preserved foodstuffs sold in packages had to be labeled "chemically preserved" or "chemically preserved with boric acid" when boric acid was present. The approved preservatives were: Ethyl and propyl esters of p-hydroxybenzoic acid including their sodium compounds and their mixtures, hexamethylenetetramine, hydrogen peroxide, · benzoic acid, formic acid (25% solution), and sulfurous acid. Details concerning the use of these as permitted in Germany are given by Jacobs.[7]

Classification.—Chemical preservatives may be classified in a number of ways. For instance, they may be grouped into (1) inorganic preservatives, (2) organic preservatives, and (3) sweeteners. In the first group, among the principal agents used are nitrates, nitrites, sulfites and sulfurous acid, borates, iodates, free chlorine, hypochlorites, and peroxides. In the second group, the principal

---

[7] Jacobs, ed., *Chemistry and Technology of Food and Food Products*, Vol. III, 2nd ed. Interscience, New York, 1951.

agents are benzoates, formaldehyde, salicylates, formic acid, esters of *p*-hydroxybenzoic acid, propionic acid and its sodium and calcium salts, thiourea and many others. In the third group, the principal sweetening agents are saccharin and dulcin.

Chemical preservatives may also be classified according to their use or action—namely, as antiseptics, germicides, fungistats, mycostats, antioxidants, neutralizers, stabilizers, emulsifiers, and coating agents, and in other categories. For the purposes of this chapter, it is preferable to discuss chemical preservatives under the type of classification mentioned above. In this text we will consider the analytical chemistry of food preservatives. For other information the reader is referred to Jacobs.[7,8]

BORIC ACID AND BORATES

Boric acid and turmeric react to yield a characteristic red color. The production of this color forms the base of the following test for borates as it did for turmeric in Chapter III.

Turmeric Test.—Prepare strips of turmeric paper by soaking filter paper in an alcoholic extract of ground turmeric. Allow the paper to dry spontaneously and then cut into strips.

Procedure.—If the sample is a solid, prepare a paste of the material with water. Otherwise use the sample as is and add 7 ml. of hydrochloric acid for every 100 g. of sample used. Stir vigorously, and dip into the mixture a piece of the prepared turmeric paper. Allow the paper to dry spontaneously. If a red color develops which is changed by ammonium hydroxide solution to a dark blue green and back again to red by acid, boric acid is present.

Quantitative Determination.—The volumetric method [9] for the determination of boric acid, after ashing, in foodstuffs is liable to considerable errors when applied to samples of food containing large proportions of fat, or phosphates or small proportions of boric acid such as those found naturally in certain fruits. The removal of phosphate requires great care, and the results obtained by titration are sometimes vitiated by the accidental presence of phosphates or carbonates. Avoidance of these errors may be made by making use of the ready esterification of boric acid with methyl alcohol, distillation of the methyl borate formed, and final titration after the addition of mannitol.

Procedure.—Boil the acidified ash with methyl alcohol in a 300-ml. Kjeldahl flask, and carry the vapors over into a vertical, double-sur-

---

[8] Jacobs, *Synthetic Food Adjuncts.* Van Nostrand, New York, 1947.
[9] Alcock, *Analyst* **62**, 522 (1937).

face condenser to which the Kjeldahl is inclined at a 30° angle. The exit tube of the condenser is elongated and dips below the surface of alkaline methyl alcohol in a conical flask. The flask stands on a tripod and gauze and is strongly heated by a burner. It is closed by a rubber stopper which carries, besides the condenser exit, another wide glass tube passing through a second hole in the stopper of the Kjeldahl flask down into the methyl alcohol. The ascending limb of this tube must be lagged, which may be conveniently done by enclosing it in a rubber tube. Immediately above the stopper of the conical flask is a side tube sealed on to the condenser outlet. This runs up close to the condenser and carries a trap bulb and a thistle funnel. This tube serves to maintain atmospheric pressure inside the apparatus, to prevent possible losses through bumping in the receiver flask, and to make possible the washing out of the lower part of the condenser outlet.

Moisten from 40 to 50 g., or less when large amounts of boric acid are anticipated, with 10 ml. of 2 N sodium hydroxide solution. If much fat is present, remove it by washing with petroleum ether, decanting the petroleum ether layer, and repeating the process until nearly fat free. Evaporate the water and, if petroleum ether was used, the petroleum ether on a steam bath, and ash the sample. There is no need to burn away all the carbon. Transfer the ash to the Kjeldahl flask with as little water as possible. Finally wash the dish with a few milliliters of dilute sulfuric acid, and dissolve the ash in the flask in a further quantity of acid by warming. This removes most of the carbon dioxide and dissolves the lumps of ash. Add methyl red indicator solution and run in 30 per cent sodium hydroxide solution until the color changes to yellow. Concentrate the liquid to 1 or 2 ml. over a burner, with continuous stirring. After cooling, add 60 ml. of methyl alcohol and 1 ml. of methyl red solution. Add sulfuric acid drop by drop until, after shaking, the solution is strongly acid to the indicator. Attach the flask to the apparatus, and also the receiving flask containing 0.5 ml. of N sodium hydroxide solution and a few drops of phenolphthalein solution. Heat the Kjeldahl flask and, when sufficient alcohol has collected in the receiving flask, heat strongly, so that the vapor bubbles vigorously through the acid liquid in the Kjeldahl flask. Adjust the flame beneath that flask so that 15 to 20 ml. of methyl alcohol remains. If during the distillation the color of the phenolphthalein in the conical flask is discharged, add a further 0.5 ml. of the sodium hydroxide solution through the thistle funnel.

After 30 minutes replace the flame beneath the Kjeldahl flask by a beaker of cold water, and distill as much of the methyl alcohol as possible up into the Kjeldahl flask for subsequent recovery. Remove the receiving flask, and wash the condenser tube, both inside and out, into it with water. Boil the remaining methyl alcohol off, make the solution just acid to methyl red with 0.1 N sulfuric acid, and boil for an additional 5 minutes to remove carbon dioxide. After cooling, readjust the acidity with 0.05 N sodium hydroxide solution until it is just not acid to methyl red. Add more phenolphthalein and, after addition of 1 g. of mannitol, carry the titration to the phenolphthalein change. Make a blank determination with water in place of a distillate. This result is usually of the order of 0.1 ml. of 0.05 N sodium hydroxide solution. One ml. of sodium hydroxide solution is equivalent to 0.0031 g. boric acid. A volume of glycerol, neutral to phenolphthalein, equal to the volume of the solution to be titrated may be substituted for the mannitol.

SULFUR DIOXIDE AND SULFITES

The presence of sulfites as such in a material is not easy to detect qualitatively within the foodstuff because of the possible presence of other reducing agents. However, the following rapid method for the determination of sulfur dioxide modified from that of Alesi,[10] overcomes some of these difficulties. This method is based on the bleaching action of sulfur dioxide on iodine. The sulfur dioxide is carried over the iodine in a foam and current of gas caused by carbon dioxide.

Procedure.—Place 20 to 50 g. of the food material rubbed up to a paste with 20 ml. of water, if solid, in a wide-mouth, 8-ounce bottle. Add 20 ml. of a solution containing 2 per cent sodium hydroxide solution and 3 per cent sodium carbonate solution. Then acidify with 30 ml. of hydrochloric acid (1 : 1). When the foam has decreased, suspend in the bottle a piece of starch-potassium iodide paper slightly blued with iodine vapor or by dipping into a dilute solution of chloramine-T. The starch potassium iodide paper is suspended by inserting it into a slit made in a cork, provided with a vent, that fits the wide mouth bottle. If sulfur dioxide is present the paper is decolorized. Some measure of the amount of sulfur dioxide present is obtained by the rapidity with which the starch-iodide paper is decolorized.

Rapid Quantitative Method.—The determination of sulfur dioxide or sulfites is generally made on a quantitative basis. The following method is comparatively short, practical, and does away with the sometimes unnecessary refinements of distillation in an atmosphere

[10] Alesi, *Ind. ital. conserve aliment.* 11, 47 (1936).

of carbon dioxide or some other inert gas. Exactly the same setup is used as that needed for a Kjeldahl ammonia distillation.

Procedure.—Transfer 50-100 g. of the material, comparatively accurately weighed, to an 800-ml. Kjeldahl flask. Add 250 ml. of water and 15 ml. of 20 per cent phosphoric acid to make certain that the mixture is distinctly acid. Connect the flask by means of a trap to a condenser equipped with an adapter dipping into and under the level of a dilute solution of iodine in a beaker. Distill over about 150 ml. If the iodine is decolorized by the liberated and distilled sulfurous acid, sulfites are present. It is possible that other materials may cause the decolorization. Add more iodine solution immediately and continue the distillation until no further decolorization takes place. Care must be taken to see that the distilled solution does not foam over or that the iodine solution is sucked back. Remove the beaker, wash the adapter, catching the washings in the beaker, and boil off the excess iodine. Make acid with hydrochloric acid and add 10 per cent barium chloride solution until no further precipitation takes place. If no precipitate is formed, sulfites are absent. If a precipitate is formed, digest on a hot plate for an hour or leave in a warm place over night and then filter through ashless filter paper. Wash thoroughly with water. Transfer the filter paper and precipitate to a weighed quartz or platinum crucible previously heated to the same temperature as that to which the ash will be heated and burn to a white ash in a muffle. Remove from the muffle, place in a desiccator to cool, and weigh when cool. The additional weight is due to barium sulfate. Calculate the amount of sulfur dioxide present and report in parts per million of the original sample.

Care must be taken in the interpretation of results, especially for low amounts of sulfur dioxide. If any doubt arises a more precise determination in an oxygen free atmosphere should be made. Such a method is that of Monier-Williams [11] which is applicable in the presence of other volatile sulfur compounds but is not in the presence of nitrates and nitrites.

Monier-Williams Method.— Connect an 800-ml. Kjeldahl flask to a sloping reflux condenser, as illustrated in Fig. 45, the lower end of which is cut off at an angle. Pass carbon dioxide from a generator or tank through a sodium carbonate solution. Connect the upper end of the reflux condenser to the bottom of a small flask which is connected in turn with a Peligot tube. One Peligot tube has been found to be sufficient to catch traces of sulfurous acid swept through the small receiving flask. The receiving flask contains 15 ml. of pure neutral 3

---

[11] Monier-Williams, Ministry of Health, *Reports of Public Health and Medical Subjects, No. 43.* London (1927).

per cent hydrogen peroxide solution, and the Peligot tube contains 5 ml. of the same reagent. Hydrogen peroxide generally contains free sulfuric acid. In order to free the peroxide from sulfate, start with 30 per cent hydrogen peroxide solution, dilute somewhat, and neutralize with barium hydroxide solution, using bromophenol blue solution as indicator. After the reagent has settled in the cold, filter from the barium sulfate, and determine its strength by a permanganate titration, and finally adjust to a 3 per cent strength. The bromophenol

Fig. 45. Monier-Williams Apparatus

blue indicator in the hydrogen peroxide remains unaffected for some time.

Connect the apparatus, introduce into the flask 300 ml. of water and 20 ml. of hydrochloric acid, and boil for a short time in a current of carbon dioxide. Then add the food to be tested, adapting the procedure to the sort of food. Add liquids directly by means of a dropping funnel and solids by rapid transfer directly to the Kjeldahl flask. After introducing the food, boil the mixture for 1 hour, or 1.5 hours in the case of dried fruits, in a slow current of carbon dioxide, stopping the flow of water in the condenser just before the end of the distillation. This will cause the condenser to become hot and drive over the residual traces of sulfur dioxide retained in the condenser. When the delivery tube just above the receiving flask becomes hot to the touch, remove the stopper connecting the delivery tube to the upper end of the condenser.

Wash the contents of the Peligot tube into the receiving flask and titrate the liquid at room temperature with 0.1 N sodium hydroxide solution, using bromophenol blue as indicator. The sodium hydroxide must be standardized with this indicator. Bromophenol blue is unaffected by carbon dioxide and also gives a distinct color change in cold hydrogen peroxide solution. One ml. of 0.1 N sodium hydroxide solution is equivalent to 3.2 mg. of sulfur dioxide, so that titration of small quantities of sulfur dioxide requiring less than 0.5 ml. of sodium hydroxide is not accurate. A gravimetric determination may be made after titration, the precipitation of barium sulfate being carried out at room temperature. After allowing the supernatant liquid to settle, filter, and wash the residual barium sulfate 3 times by decantation with boiling water. Determine a blank on the reagents, both by titration and gravimetrically, and correct the results accordingly.

Determination of Sulfites in the Presence of Nitrates and Nitrites. —Sulfur dioxide may not be recovered if nitrates or nitrites are present because of the oxidizing action of the nitrates and nitrites in the presence of the hot dilute acid necessary for the Monier-Williams method.[12] To overcome this, the nitrate and nitrite may be reduced while leaving the sulfites unattacked. This may be done as in the following method.

Procedure.—Dissolve 5 g. of hydrazine sulfate in 100 ml. of hot water and transfer to the Kjeldahl flask of the Monier-Williams apparatus. Boil the liquid to expel air, and cool in a current of carbon dioxide. Add twenty ml. of a 20 per cent sodium hydroxide solution through the tap-funnel, if used, and shake the flask thoroughly. Transfer a suitable quantity of the liquid or solid to be tested and boil the contents of the flask in a current of carbon dioxide for 5 minutes. Remove the flame from the flask and add 10 ml. of phosphoric acid, sp. gr. 1.75, and a small quantity of pumice powder through the tap-funnel. Boil the contents of the flask in the current of carbon dioxide and distill the sulfur dioxide and collect in hydrogen peroxide in the usual way. In applying this process to brine and meat pickles, it is desirable to dilute the reduced solution with boiled water before the addition of phosphoric acid.

12 Sherratt, *Analyst* 62, 267 (1937).

Mathers Method.—In this method,[12a] the sulfur dioxide is trapped in neutral lead acetate solution and after optical examination, if desired, it is acidified and the sulfite is estimated iodometrically.

Procedure.—Transfer the sample to a distillation set-up, dilute if necessary to 50 ml., add some boiling stones and 50 ml. of 5 per cent sulfuric acid. Place 50 ml. of 1 per cent neutral lead acetate solution into a graduated cylinder and use this as the receiver. Adjust the condenser tip or adapter to dip two inches below the surface of the liquid in the receiver. Collect about 50 ml. of distillate, make up to 100 ml., and after determining the turbidity in a Coleman spectrophotometer at 600 mμ if desired, acidify the suspension with hydrochloric acid. Titrate with 0.02 N iodine solution. Be sure to wash the adapter or condenser with the acid for the titration.

If free sulfur dioxide is to be estimated, do not acidify with sulfuric acid before distillation. Subtract this result from the total sulfur dioxide determined as detailed to obtain the combined sulfur dioxide.

FLUORIDES

Etching Test.—Hydrofluoric acid reacts with glass and therefore can be used for etching glass. This property is the basis of the following test for the detection of soluble fluorides in the absence of silica.

Procedure.—Boil 150 ml. of the sample or of an aqueous extract of the sample. Add 5 ml. of 10 per cent potassium sulfate solution and 10 ml. of 10 per cent barium acetate solution. Centrifuge to collect the precipitate of barium fluoride. Filter on a small filter. Transfer to a dish and ash.

Prepare a microscope slide or similar glass plate by cleaning and dipping into a hot mixture of equal parts of carnauba wax or wax of similar melting point and paraffin and allow to cool. Make a characteristic mark through the wax. Place the ash in a small flat-bottomed lead dish and add a few drops of sulfuric acid. Cover with another lead dish having an orifice of about 1 cm. Cover the orifice with the glass plate so that the characteristic mark is directly over the orifice and warm, keeping the glass plate cool so that the wax will not melt. If fluorides are present in sufficient quantity, the glass will be distinctly etched where the characteristic mark was made. Smaller quantities must be determined by a far more sensitive method to be discussed in the chapter on inorganic determinations, Chapter XVIII.

[12a] Mathers, *J. Assoc. Official Agr. Chem.* **32**, 745 (1949).

Drop Test.—To determine insoluble fluorides and soluble fluorides in the presence of silica, a somewhat different procedure must be followed. The hydrogen fluoride formed reacts with the silica to yield gaseous silicon fluoride which reacts with water to form gelatinous silica and hydrofluosilicic acid.

Procedure.—Add enough lime water to about 200 g. of the sample to make it alkaline and evaporate to dryness. Ash and leach the ash with water and acetic acid to decompose carbonates. Filter, re-ignite the residue, and re-extract with acetic acid, (1 : 1). The filtrate contains any boric acid that might be present and which may be tested for as detailed above. The residue contains calcium fluoride and silicates.

Ash the filter and residue in a platinum crucible. Add some precipitated silica and 1 ml. of sulfuric acid. Cover with a watch glass plate from which a drop of water is suspended. Heat on a bath for an hour at 70-80° C. If a gelatinous precipitate of hydrated silica resulting from the reaction between the silicon fluoride and water is formed, fluorides are present.

$$2CaF_2 + 2H_2SO_4 + SiO_2 \rightarrow 2CaSO_4 + SiF_4 + 2H_2O$$

$$3SiF_4 + 4H_2O \rightarrow Si(OH)_4 + 2H_2SiF_6$$

If both borates and fluorides are present, the original substance probably contained a fluoborate. If the gelatinous precipitate is given without the addition of silica, fluosilicate probably was present in the original sample.

IODATES

The literature of recent years mentions the use of calcium and potassium iodate as a preservative. A simple method for the detection of these substances is based on the iodide-iodate reaction.

Procedure.—Make an acid serum of the sample by the addition of sufficient dilute hydrochloric acid to an aqueous mixture or solution of the sample to make it slightly acid, and filter. Add to the filtrate potassium iodide solution and starch solution and allow to stand for a few minutes. The following reaction runs with the formation of free iodine and the subsequent production of blue color with the starch solution.

$$KIO_3 + HCl \rightleftarrows HIO_3 + KCl \text{ and } KI + HCl \rightleftarrows HI + KCl$$

$$HIO_3 + 5HI \rightarrow 3H_2O + 3I_2$$

Landolt Reaction.—An alternative method is a variation of the famous Landolt reaction. Make an acid serum of the sample and add starch. Add a solution of a sulfite drop by drop until a slight excess of sulfite is present. When the excess of sulfite is used up, the formation of a blue color indicates the presence of an iodate. This color is formed by the following series of reactions.

$$2HIO_3 + 5H_2SO_3 \rightarrow I_2 + 5H_2SO_4 + H_2O$$

$$I_2 + H_2SO_3 + H_2O \rightarrow 2HI + H_2SO_4$$

$$HIO_3 + 5HI \rightarrow 3H_2O + 3I_2$$

Only when the excess of sulfite is used up will the blue color with the starch appear.

FREE CHLORINE

The presence of free chlorine in a substance may be detected in the following manner. To an acid solution of the substance add a solution of potassium iodide and starch solution. The chlorine displaces the iodide, forming free iodine. A blue color indicates the presence of free chlorine. Starch-iodide paper serves equally well at times, especially if the color is likely to be masked by the material being analyzed.

OXIDIZING AGENTS

Diphenylamine Reaction.—Hypochlorites, nitrites, nitrates, etc., may be detected by the diphenylamine reaction. Dissolve 1 g. of diphenylamine in 100 ml. of sulfuric acid. Add a few drops of the reagent to a small portion of an aqueous extract of the material in a porcelain dish. An intense blue color at the junction shows the presence of an oxidizing agent.

A more sensitive test is obtained if diphenylbenzidine is used instead of diphenylamine. Wood, Illing, and Fletcher [13] found that this reagent will detect 0.1 ppm. of nitrite in milk.

Free chlorine in small amounts is not detected by the diphenylamine reagent and should be tested for as directed above.

PEROXIDES

The use of hydrogen peroxide, in particular, and peroxides, in general, caused a great deal of interest a number of years ago when their presence in chocolate flavored drinks was demonstrated. The usual test for peroxides in milk is to pour 10 drops of a solution of

---

[13] Wood, Illing, and Fletcher, *Analyst* **58**, 149 (1933).

vanadic pentoxide or of ammonium vanadate in 100 ml. of concentrated sulfuric acid down the side of the test tube containing the sample. A reddish color indicates the presence of peroxides.

In the case of chocolate flavored drinks, or chocolate flavored milk, this color is easily masked. The following test devised by Jacobs [14] overcomes this difficulty.

Procedure.—To 10 ml. of the sample add 1 ml. of a 40 per cent solution of trichloroacetic acid and, after allowing to stand for 5 minutes, filter. To 3 ml. of the serum produced, add 2-5 drops of a solution of titanium oxide in concentrated sulfuric acid. A fairly persistent yellow color indicates the presence of peroxide. This may be confirmed by adding to another 3-ml. portion of the trichloroacetic serum a few drops of vanadium pentoxide-sulfuric acid solution, 1 g. of vanadium pentoxide dissolved in 100 ml. sulfuric acid (1:3). A fairly persistent reddish color verifies the presence of peroxides.

Contrary to the general conception, peroxides do not disappear readily on addition to milk products. Thus, the author obtained positive tests for peroxides as long as one month after its addition to samples by the above test.

ORGANIC PRESERVATIVES

There are hundreds of organic substances which have been suggested as food preservatives. Many of them are sold under trade names. The more common trade names and the chemical names of many organic preservatives are tabulated by Jacobs.[15] Many of these organic materials have actually been used in food materials and some of them are permitted in some countries, as, for example, Germany, although no extensive work has been performed to establish their harmlessness. Fortunately they need to be used in such small quantities that this danger is lessened although not eliminated. The price and efficiency of these organic materials limit their use to a very small group.

The general means of detecting organic preservatives and sweetening agents falls into two main classes:

First, if volatile, they may be distilled and subsequently identified, for example, formaldehyde and formic acid.

Secondly, if soluble in a water immiscible solvent, they may be extracted by that solvent and again subsequently identified, for example, benzoic acid, salicylic acid, saccharin, and many others.

---

[14] Jacobs, unpublished work (1936).
[15] Jacobs, *Synthetic Food Adjuncts.* Van Nostrand, New York, 1947.

FORMALDEHYDE

There are many well-known tests for formaldehyde. Some of these may be performed directly on the food sample, whereas others can only be made on an alcoholic extract or on the distillate from phosphoric acid solution of the food material. Among these tests are the sulfuric acid-ferric chloride reaction of Hehner, the hydrochloric acid reaction of Leach, the phenylhydrazine hydrochloride reaction, the phenylhydrazine hydrochloride plus sodium nitroprusside reaction, etc. The following tests are adequate.

Schiff's Test.—*Fuchsin-Sulfite Reagent.*—Dissolve 0.5 g. pure fuchsin in 400 ml. of warm water. Cool, add 2 g. of anhydrous sodium bisulfite, and stir until dissolved. Add 4 ml. of concentrated sulfuric acid. Transfer to a brown bottle and allow to stand at least overnight before use. This reagent will not deteriorate over long periods of time if kept in a refrigerator.

Rosaniline-Sulfite Reagent.—Dissolve 0.2 g. of rosaniline, or an equivalent weight of its salt, in 120 ml. of hot water, cool, and add this to a solution of 2 g. of sodium bisulfite in 20 ml. of water. Finally, add 2 ml. of concentrated hydrochloric acid and dilute the entire mixture to 200 ml. This solution should become colorless or nearly so after standing. If it is protected from air, no deterioration occurs.

Procedure.—To about 10 ml. of the sample, made acid with at least 7 ml. of hydrochloric acid to 100 ml. of sample or prepared solution or mixture, add 1 ml. of either the fuchsin-sulfite reagent or the rosaniline-sulfite reagent, and allow to stand. If a pink to deep violet color develops on standing, formaldehyde is present. This test can be performed directly on most food products; however, if the color may be masked, or it is desired to make the test more sensitive, it should be performed with 10 ml. of a distillate.

Chromotropic Acid Reaction.—When a solution containing formaldehyde is heated with an acid solution of chromotropic acid, 1,8-dihydroxynaphthalene-3,6-disulfonic acid or its sodium salt, a magenta to violet colored solution is formed.

Procedure.—To a few drops of a distillate, or a colorless solution, if that is available, add 5 ml. of freshly prepared chromotropic acid solution, prepared by dissolving 50 mg. of chromotropic acid or its sodium salt in 100 ml. of 75 per cent sulfuric acid, and heat in a water bath for 10 minutes at 60° C.

Hehner Test.—When a solution of formaldehyde in milk is used to overlay some concentrated sulfuric acid containing an oxidizing agent such as ferric ions, a violet color develops at the junction. This reaction depends on the presence of tryptophan as a constituent of the protein component. Fulton [16, 17] developed two modifications of this test, the first a zone test and the second a uniform color test. Bromine set free by the sulfuric acid is used as the oxidant.

Procedure A.—Dilute 5 ml. of concentrated sulfuric acid with 1 ml. of water, cool, and transfer 3 ml. of this diluted acid to a test tube. Add a small crystal of potassium bromide. Shake, then overlay at once with 1 ml. or slightly more of the milk to be tested. If formaldehyde is present, a violet zone develops quickly. The test is sensitive to 1 part of formaldehyde in 1 million parts of milk. The reaction is strongest with 1 part in 50,000.

If no violet appears by the time the acid has become orange-yellow, the test is negative. Because of the prior dilution of the acid, the color forms in a fairly broad zone rather than in a narrow ring. The escape of bromine and the formation of hydrobromic acid tend to mix the solutions so that the color spreads throughout the tube, being strong violet in the bottom, deep red at the center, and purple at the top. For good results the concentration of formaldehyde should not exceed 1 part in 1000.

Procedure. B.—Dilute 8 ml. of concentrated sulfuric acid with 5 ml. of water, cool, and place 4 ml. of this mixture in a test tube. Add 1 ml. of milk and mix, cooling under running water. A clear and practically colorless solution should result unless high concentrations of formaldehyde are present. The curd first formed is dissolved.

Prepare a bromine oxidizing solution by mixing equal volumes of concentrated sulfuric acid and saturated bromine water, and cool. Add 0.5 ml. of the oxidizing solution to the sulfuric acid-milk mixture and shake. A violet color develops at once in the presence of formaldehyde ranging to light purplish pink for low concentrations of formaldehyde. The blank has only a faint pink tinge, probably due to trace formation of formaldehyde from the carbohydrate of the milk by the action of the strong acid.

Weinberger [18] Dimetol Method.—Prepare a 5-10 per cent solution of dimethylcyclohexanedione in alcohol. This reagent is also known as methone, dimetol, and dimethylhydroresorcin. To an aqueous solution of the aldehyde add sodium chloride and neutralize or make faintly acid with acetic acid. Add a few drops of the reagent to the cold solution and stir vigorously. If formaldehyde or any other aldehyde is present a precipitate will form. Filter off the precipitate and recrystallize from hot water or alcohol and determine the melting point. Dimetolformaldehyde melts at 187° C.

[16] Fulton, *Ind. Eng. Chem., Anal. Ed.*, 3, 199 (1931).
[17] Oakley, *J. Assoc. Official Agr. Chem.*, 28, 296 (1945).
[18] Weinberger, *Ind. Eng. Chem., Anal. Ed.* 3, 365 (1931).

FORMIC ACID

Steam distill about 50 g. of the sample made acid with 15 ml. of 20 per cent phosphoric acid until about 200 ml. of the distillate has been collected. Test the distillate for formaldehyde by one of the above tests. If formaldehyde is absent, add to another portion of the distillate sulfuric acid (1 : 4) and some magnesium filings. The hydrogen produced will reduce the formic acid, if present, to formaldehyde which may then be detected by one of the above methods.

ETHER EXTRACTIVE PRESERVATIVES

Braverman [19] Method.—This method is a rapid, simple procedure for the detection of ether extractive preservatives. It avoids the formation of troublesome emulsions that are often encountered in other methods. Interfering substances, such as proteins and fats, are removed by precipitation with copper sulfate solution and sodium hydroxide solution. The preservatives are then extracted with ethyl ether and subsequently identified.

Procedure.—To 50 g. of the sample in a beaker, if pasty or solid, add 50 ml. of water and mix thoroughly. If the sample is liquid, use 100 ml. directly. Now add 5 ml. of 10 per cent sodium hydroxide solution and stir well. Add 10 ml. of 35 per cent copper sulfate solution, 350 g. $CuSO_4 \cdot 5H_2O$ dissolved in water and made to a liter, and stir well. Place on a hot plate and bring to a boil. Stir vigorously and filter at once through a filter directly into a separatory funnel. Cool under running water. Add 5 ml. of hydrochloric acid and mix. Add 75-100 ml. of ethyl ether and shake. Allow to separate. Draw off the water layer and discard. Wash the ether layer 3 successive times with 5-ml. portions of water, drawing off each washing before the addition of the subsequent one. Transfer the ether layer to an evaporating dish, and evaporate off the ether. If no crystalline residue remains, preservatives are absent. If a crystalline extractive remains, it may be benzoic acid, salicylic acid, methyl *p*-hydroxybenzoate and analogues, furoic acid, saccharin, dulcin, etc., or some fruit extractive.

Benzoic acid may be verified by its odor, crystalline appearance, by subliming the residue, by the formation of ethyl benzoate with its characteristic fruity odor on boiling an alcoholic solution of the sublimate with concentrated sulfuric acid, by the formation of a flesh colored precipitate on the addition of ferric chloride solution to an aqueous solution of the sublimate.

---

[19] Braverman, personal communication (1930).

Salicylic acid may be identified by the formation of a deep violet color with ferric chloride solution. Other substances also give colors with ferric chloride solution, but they do not have the shade given by salicylic acid.

In case of doubt, apply the Jorissen [20] test. To an aqueous solution of the residue add 4-5 drops of 10 per cent potassium nitrite or sodium nitrite solution, 4-5 drops of acetic acid, and 1 drop of 1 per cent copper sulfate solution. Boil. If a blood red color is produced, salicylic acid is present. Phenol will give the same reaction.

Saccharin may be verified by its sweet taste and by transforming it to salicylic acid, which may then be identified as directed above. To do this, transfer the residue by means of ether to a nickel dish and evaporate off the ether. Add 10 ml. of 10 per cent sodium hydroxide and evaporate to dryness. Bake for an hour or so and fuse cautiously over a Bunsen flame. Leach with water and transfer to a separatory funnel. Make acid with hydrochloric acid. Extract with ether and wash as directed above. Transfer the ether layer to an evaporating dish and evaporate off the ether. Test the residue for salicylic acid. If the test for salicylic acid was negative before fusion and positive after fusion, saccharin was present in the original sample.

To identify other preservative substances—which it might be said, in passing, are seldom used in the United States, probably because of price and because they are looked upon with disfavor—purify the ether extract either by sublimation or by recrystallization from a suitable solvent. Determine the melting point [21] and the crystalline structure by means of a microscope. If these tests are insufficient to identify the compound, the analyst may proceed as directed under the subsequent section, "separation of organic preservatives and sweetening agents" which may give a clue to the identity of the substance. Otherwise the analyst must proceed according to a systematic method of organic analysis, as is described in Mulliken, "Identification of Pure Organic Compounds," or some similar text.

Lactic Acid

It has been mentioned in the foregoing that lactic acid is sometimes used as a preservative. It is also worth while to note that lactic acid has been recommended in place of vinegar in a number of food

[20] Jorissen, *Bull. acad. roy. sci. let. Belf.* [3] 3, 259 (1882).
[21] Jansen, *Chem. Weekblad* 33, 1 (1936).

products. A modification of a method used by Palm [22] may be used for its identification.

Procedure.—Place an aliquot of an aqueous solution or mixture of the sample in the lower section of a Jacobs-Singer separatory flask. Make acid with sulfuric acid and then stopper with the upper section. Add water to the connecting joint and then extract with 3 successive portions of ethyl ether. Plural separatory funnels or a liquid-liquid continuous extraction device may also be used. Decant the successive ether layers into an evaporating dish and evaporate to dryness. Take up the residue in a small portion of water and transfer to a test tube. Add 1 ml. of 10 per cent neutral lead acetate solution and filter through a small filter. Add 1-2 drops more of 10 per cent neutral lead acetate solution and, if no precipitate forms, add a few drops of alcoholic ammonia. If lactic acid is present, it will precipitate as a heavy granular precipitate of basic lead lactate, $3PbO \cdot 2C_3H_6O_3$. If a precipitate forms on the further addition of neutral lead acetate solution, more need be added and the mixture filtered again before the addition of the alcoholic ammonia.

Methods for the quantitative determination of lactic acid are detailed in Chapter XX.

ABRASTOL

Sangle-Ferriere Test.—Abrastol, asaprol, is the calcium salt of β-naphthol-α-sulfonic acid. Sangle-Ferriere [23] recommends the following method. Boil 200 ml. of the sample with 8 ml. of hydrochloric acid for an hour under a reflux condenser. This treatment converts the abrastol to β-naphthol. Transfer the refluxed sample to a separatory funnel and extract with 10 ml. of chloroform. Draw off the chloroform into a test tube, add a few drops of 0.5 N potassium hydroxide solution, and place in a boiling water bath. The formation of a deep blue color which changes to green and then to yellow indicates the presence of β-naphthol, which in turn indicates the presence of abrastol.

Sinibaldi Test.—Sinibaldi [24] uses the following procedure. Make 50 ml. of the sample alkaline with a few drops of ammonium hydroxide solution and extract with 10 ml. of amyl alcohol, adding ethyl alcohol if an emulsion forms. A Jacobs-Singer separatory flask may be used conveniently. Decant the amyl alcohol, filter if turbid, and evaporate to dryness. Add to the residue 2 ml. of nitric acid (1 : 1),

22 Palm, *Z. anal. Chem.* **33**, 16 (1894).
23 Sangle-Ferriere, *Compt. rend.* **117**, 796 (1893).
24 Sinibaldi, *Mon. sci.* **7**, 842 (1893).

heat on a water bath until half of the liquid is evaporated, and transfer to a test tube with the addition of 1 ml. of water. Add about 0.2 g. of crystallized ferrous sulfate and an excess of ammonium hydroxide solution, dropwise with constant shaking. If the resultant precipitate is of a reddish color, dissolve it in a few drops of sulfuric acid, and add crystallized ferrous sulfate and ammonium hydroxide as before. As soon as a dark colored or greenish precipitate is obtained, introduce 5 ml. of 95 per cent alcohol, dissolve the precipitate in sulfuric acid, shake well and filter. In the absence of abrastol a colorless or light yellow liquid is produced, while a red color is produced in the presence of 0.01 g. of abrastol.

SEPARATION OF ORGANIC PRESERVATIVES

According to Fischer,[25] the following method may be used to separate and identify some of the more important organic preservatives and sweetening substances. These substances are Nipagin, ethyl p-hydroxybenzoate, Nipasol, benzoic acid, o-chlorobenzoic acid, salicylic acid, cinnamic acid, p-hydroxybenzoic acid, dulcin, p-chlorobenzoic acid, and saccharin. The extraction for all of these materials is carried out as directed in the preceding sections of the chapter with ethyl ether; the ethyl ether extract is then shaken out with an aqueous alkali solution. The dulcin remains in the ether layer; all the other substances go into the aqueous alkaline layer. Separate the aqueous layer and acidify. Shake out thoroughly with petroleum ether. The petroleum ether extract contains all the aromatic preservatives except p-hydroxybenzoic acid and saccharin, which are insoluble in petroleum ether.

These two substances may be removed from the water layer by means of ethyl ether. The solubility of o-chlorobenzoic acid in petroleum ether is small, but the repeated use of this solvent removes the acid quantitatively. Dulcin is recovered from the original ether layer. It is redissolved in hot water, filtered, and again extracted with ether and recovered by evaporation of the ether extract.

The other ether and petroleum ether extracts are evaporated and the residues are sublimed. The identification of the crystals obtained follows by determinations of the melting point. The p-hydroxybenzoic acid will sublime more easily than will saccharin.

In a simple mixture of benzoic acid and saccharin, the benzoic acid may be separated from the saccharin by subliming the benzoic acid by heating on a water bath. The saccharin remains as a residue.

---

[25] Fischer, Z. Untersuch. Lebensm. 67, 161 (1934).

An alternative method of procedure is to dry the ether extract, obtained as described in the preceding sections, over anhydrous sodium sulfate and then evaporate the solvent below 40° C. The extract is then sublimed in an apparatus consisting of a long aluminum plate,[26] which is heated at one end, while at intervals along the plate are depressions in which portions of the substance are placed and covered by a watch glass. A thermometer near each depression registers the temperature at this point.

p-Hydroxybenzoic acid methyl ester gives a red color with Millon's reagent in warm acid solution. It sublimes at about 70° C. If it is necessary to cool the receiver during this operation, the crystals obtained are in the metastable condition and melt at 110° C., the stable type melt at 126° C. Saponification with 5 ml. of 2 per cent potassium hydroxide solution produces methyl alcohol, which may be separated by distillation and subsequently identified.

p-Hydroxybenzoic acid propyl ester and ethyl ester have melting points of 97° and 116° C., respectively. The sublimate may be saponified by boiling for 1 hour with 2 ml. of 10 per cent potassium hydroxide solution and 4 ml. of water under a reflux condenser. Then 4 ml. is distilled and the alcohols in the distillate may be identified. The melting point of the combination of acids produced from Nipacombin-A, the sodium compound of a 6:4 mixture of the propyl and ethyl esters, is 95° C.

p-Hydroxybenzoic acid sublimes at 135° C. and has a melting point of 213° to 214° C. Copper sulfate produces small, bright blue crystals when added to the warm acid.

p-Chlorobenzoic acid sublimes at 95° C. and melts at 236° C. The ortho compound sublimes at 75° C. and melts at 142° C. A mixture of the extract heated, on a water bath for 20 minutes, with 0.25 ml. of sulfuric acid and a crystal of potassium nitrate followed by the addition of 2 ml. of ammonia water and then 1 ml. of 2 per cent solution of hydroxylamine hydrate yields a green color at the junction of the liquids, if this compound is present.

Cinnamic acid melts at 133° C. and sublimes at 90° C. To identify it, make the ether extract alkaline, evaporate, and re-extract with acidified ether in the presence of a little alcohol to prevent emulsification Wash the extract 3 times with water and shake with 0.33 N potassium hydroxide solution. Remove the ether from the separated water layer by warming. Add a 1 per cent solution of potassium permanganate when the solution is cool, and the benzaldehyde produced may then be recognized by its odor. The reaction is sensitive to 1 mg.

26 Jansen, *Chem. Weekblad* 33, 1 (1936).

of cinnamic acid, but is lower if the sodium salt is used. The purified acid may be estimated by solution in a known amount of 0.1 N sodium hydroxide, the excess of which is back-titrated with hydrochloric acid. The benzaldehyde produced may be confirmed by the following reactions: (1) To the solution containing the benzaldehyde add 1 drop of a solution of phenol and 2 ml. of sulfuric acid. A hard resinous mass is produced on warming. Cool the mixture, dilute with 10 ml. of water, make alkaline with 20 per cent potassium hydroxide solution, benzaldehyde gives a violet color which may be extracted by shaking with acidified ether. (2) To the oxidized liquid containing the benzaldehyde, add twice its volume of a solution of dimethylaniline in sulfuric acid, warm the mixture to 150° C., and dilute with an equal volume of water. Malachite green separates on the addition of potassium dichromate and sodium acetate.

Anisic acid is converted to p-hydroxybenzoic acid and methyl iodide by the action of hydriodic acid. It may be separated from p-hydroxybenzoic acid by extraction with chloroform, in which only anisic acid is soluble.

Volatile Fatty Acids in Bakery Products [27]

It has been demonstrated that fatty acids containing from 2 to 14 carbon atoms, particularly propionic, butyric, caproic, caprylic, etc., acids are effective mold inhibitors. This action is discussed by Jacobs.[28] Propionic acid and its salts are fairly widely used in bakery products for the prevention of ropiness and of mold growth. It is necessary at times to establish the concentration of mold inhibitor used.

Preparation of Sample.—Air-Dried Bread.—Cut a loaf or half a loaf of bread into slices 2-3 mm. thick, spread the slices out on paper, and allow to dry in a warm room until it is sufficiently dry to grind adequately in a mill. Grind the entire sample to pass a 20-mesh sieve, mix thoroughly, and store in an airtight container.

Fresh Bread or Cake.—For analysis of the fresh product or for analysis of cake, which is often difficult to air-dry without spoilage, pass the sample through a meat grinder, equipped with a ⅛-in. hole plate, and reduce to a finely divided condition by rubbing through an 8-mesh sieve. Proceed with the analysis promptly.

Procedure.—Transfer 25 g. of air-dried or 35 g. of fresh sample to a 250-ml. volumetric flask. Add 100 ml. of water and mix by swirling until all particles are wet and any lumps are completely broken

[27] J. Assoc. Official Agr. Chem. 31, 99 (1948).
[28] Jacobs, Synthetic Food Adjuncts. Van Nostrand, New York, 1947.

up. Add 25 ml. of N sulfuric acid, shake for 2 minutes and let stand for a half hour, shaking occasionally to stir up the particles. Do not allow bread to clog in the neck of flask. This can be prevented by avoiding too vigorous shaking. Wash down with a small amount of water if necessary. Add 15 ml. of 20 per cent phosphotungstic acid (W/V), shake for 2 minutes and make to volume. Transfer the mixture to a centrifuge bottle and centrifuge for 10 minutes at 1000-1500 r.p.m. Disregard turbidity. Decant the supernatant which should be about 180 ml. and pipette 150 ml. to a 500-ml. flask equipped with a water condenser for refluxing. Add 1.0 g. silver sulfate and heat for 5 minutes after boiling begins. Cool the flask under running water to room temperature—reflux connected—wash down the condenser, transfer the contents of flask to a 200-ml. volumetric flask and make to volume. Mix in a 400-ml. beaker with 3-5 g. of Filter-Cel, stir, and filter using S.S. #589 15 cm. Pour back to give best possible clearness. Difficulty in obtaining a clear filtrate indicates insufficient silver sulfate. Test for excess silver by allowing a few drops of filtrate to flow into a test tube containing about 5 ml. of 5 per cent sodium chloride solution in nitric acid (1:3). If excess of silver is not indicated, add more silver sulfate (0.2 g.), shake and filter on a new paper.

Transfer 150 ml. of the chloride-free filtrate to the standard distillation flask and proceed as directed in Chapter XX in the determination detailed for volatile acids in dried eggs.

Collect only the two initial portions of distillates (50 ml. and 200 ml.) unless the ratio of their titers is less than the standard C ratio for propionic acid by more than 0.1. In this case collect two additional 200-ml. portions of distillate for calculation of acids higher than propionic.

Determine formic acid in composite of distillates and correct the titrations for the titer contributed by this acid. Using the prescribed aliquots, the results in terms of milligrams of acid per 100 g. of sample is calculated from the determined ml. of 0.01 N acid in the distillation by means of the following factors: Formic acid—4.09; acetic acid—5.33; propionic acid—6.58.

MONOCHLOROACETIC ACID

Monochloroacetic acid, $CH_2ClCOOH$, has been banned by the Food and Drug Administration[29] as a preservative for foods because it

---

[29] *U. S. Food Drug Admin. F.D.&C. Act Trade Correspondence* TC-277, December 29, 1945.

was found that the acute toxicity of this substance is comparable to that of such recognized poisons as mercuric chloride, phenol, and strychnine. It therefore concludes that a substance which exhibits this order of toxicity has no place in foods and that monochloroacetic acid will be considered an adulteration no matter what amount is added.

Argentometric Method.—Monochloroacetic acid in concentrations of the order of 5 to 150 mg. in 150 ml. of carbonated beverages and fruit juices may be estimated argentometrically [30,31] by prior hydrolysis of the chlorinated preservative.

$$CH_2ClCOOH + NaOH \rightarrow CH_2OHCOOH + NaCl$$

Reagent.—Ammonium Thiocyanate Solution.—Dissolve 4.03 g. of ammonium thiocyanate in water and dilute to 1 liter. Standardize against pure sodium chloride solution, containing 3.093 g. per liter. This solution contains 1.8762 g. of chlorine which is equivalent to 5 g. of monochloroacetic acid for the latter contains 1.8764 g. of chlorine.

Procedure.—Add 3 ml. of sulfuric acid to 100 ml. of the sample. Transfer to a separatory funnel and extract successively with three 100-ml. portions of ether. Combine the ether extracts and wash successively with two 30-ml. portions of 1 N sodium hydroxide solution. Combine the sodium hydroxide layers and digest on the steam bath for 2 hours or boil under reflux for 0.5 hour. Add 50 ml. of water, 15 ml. of nitric acid, and a known volume of silver nitrate solution, prepared by dissolving 9 g. of silver nitrate in water and diluting to 1 liter, shake for 0.5 to 1 minute; add 5 ml. of a saturated solution of hydrazine sulfate, to remove any nitrous acid; 5 ml. of a saturated solution of ferric ammonium alum; 1 ml. of nitrobenzene for each 0.05 g. of chloride, if desired; and titrate the excess silver with the standard ammonium thiocyanate solution. Titrate an equal volume of silver nitrate solution in the same way. The difference in the titrations is equivalent to the amount of monochloroacetic acid.

Pyridine Method.—A quantitative method for monochloroacetic acid is described by Ramsey and Patterson.[32, 33] It is an empirical gravimetric method based on the reaction of monochloroacetic acid with an excess of pyridine to yield a mixture of the neutral and basic salts of pyridine betaine which are insoluble in pyridine.

[30] Wilson, *J. Assoc. Official Agr. Chem.* 25, 145 (1942).
[31] *J. Assoc. Official Agr. Chem.* 31, 104 (1948); 32, 97 (1949).
[32] Ramsey and Patterson, *J. Assoc. Official Agr. Chem.* 29, 100 (1946).
[33] *J. Assoc. Official Agr. Chem.* 32, 99 (1949).

THIOUREA

In 1933, it was shown that thiourea prevented browning in cut fruit. In 1937 and 1943 patents were obtained for use of this chemical for this purpose. Subsequent work disclosed that thiourea could be employed for the prevention of mold on wheat and for the protection of oranges against stem-rot. However, it was demonstrated in 1941 that thiourea had a depressing effect on the thyroid for it inhibits the production of thyroxine. For this reason thiourea is not considered suitable for use in foods.

Detection of Thiourea in Orange Juice.[34]—*Pentacyanoammonioferrate Test.*—*Reagent.*—Dissolve 10 g. of sodium nitrosoferricyanide (nitroprusside) in 40 ml. of concentrated ammonium hydroxide solution (sp. gr. 0.88) and keep at about 0° C. until all the nitrosoferricyanide has decomposed. This is shown when a few drops of the mixture no longer give a red color when added to a solution of creatinine in N sodium hydroxide solution. Decomposition is complete by the end of 24 hours. Remove the precipitate by filtration and precipitate the residual pentacyanoammonioferrate in the solution by addition of absolute ethyl alcohol until no further precipitate appears. Collect the precipitate, wash with absolute ethyl alcohol until free of ammonia, and dry *in vacuo* over sulfuric acid. Keep the solid reagent in a desiccator over calcium chloride in the dark.

Prepare a 1 per cent solution of the solid in distilled water, expose it to light and air for a day, and then store in a brown glass bottle in the dark. The reagent is now ready for use, gains in potency for several weeks, and can be kept for about six months.

Preparation of Sample.—Extract a volume of juice with about two thirds its volume of ethyl ether, centrifuge, and separate lower layer. Stir in some Filter Cel and filter with suction. Keep the vacuum on for a shirt time and agitate to remove most of the ether.

Procedure.—To about 5 ml. of the extracted sample add 5 drops of the above reagent Note the color. If a blue color does not develop, add about 0.1 N iodine solution a drop at a time, shaking after each drop. Usually, about 5 drops are necessary to develop maximum color (blue green). Excess iodine tends to reduce the color.

Test with Grote's Reagent.[35]—*Reagent.*—Dissolve 0.5 g. of sodium nitroprusside, $Na_2(NO)Fe(CN)_5 \cdot 2H_2O$, in 10 ml. of water in a 25-ml. volumetric flask. Weigh out 1 g. of sodium bicarbonate and 0.5 g. of hydroxylamine hydrochloride, $NH_2OH \cdot HCl$, and mix the two solids in a small beaker or dish by gently grinding and crushing any lumps

---

34 *J. Assoc. Official Agr. Chem.* 31, 104 (1948).
35 *J. Assoc. Official Agr. Chem.* 32, 100 (1949).

in the material with a small pestle or flattened glass rod. Transfer the entire amount of solid to the nitroprusside solution with the help of a short-stemmed funnel and a brush. Allow the flask to stand without stirring until the rapid evolution of carbon dioxide subsides and then swirl to dissolve any of the sodium bicarbonate remaining. When the reaction virtually ceases, add 0.10 ml. of bromine, after which an additional evolution of gas occurs. When swirling no longer produces effervescence, make up to 25 ml. with water and filter.

Test the effectiveness of the reagent in the following manner: To a mixture of 5 ml. of a solution containing 0.25 mg. of thiourea, 5 ml. of water and 1 drop of acetic acid, add 1 ml. of diluted Grote's reagent. A strong blue color should develop in 5 minutes. If this color does not develop, it is necessary to prepare the reagent again.

Permit the reagent to stand at room temperature for 5 to 10 hours for aging. The reagent should have a mahogany brown color. If it has a greenish cast, it will probably lose its effectiveness in a short time. The reagent, if adequate, may be stored in a refrigerator for several weeks.

Dilute Grote's Reagent.—Dilute 2 ml. of the above reagent with 8 ml. of water and use 1 ml. for the test. The dilute reagent is stable for about 1 day.

Procedure.—Use extracted sample prepared as directed above. To about 5-10 ml. of prepared sample add 0.02 N iodine solution dropwise until a drop remains and does not disappear for some time. Add a milliliter or so of diluted Grote's reagent. A blue green or blue color develops rather gradually in the presence of thiourea.

Methods for the quantitative estimation of thiourea in oranges and orange juice[35] and in frozen peaches[36] have been described.

2-AMINOPYRIDINE IN ORANGES

Another chemical preservative, 2-aminopyridine,

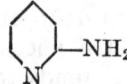

has been proposed as a fungicide for the control of stem-rot in oranges. Because of its toxic nature it is considered objectionable if it penetrates the flesh of oranges given this treatment. Winkler[37] has devised a method to detect its presence in the flesh of oranges.

[36] *J. Assoc. Official Agr. Chem.* 31, 102 (1948).
[37] Winkler, *J. Assoc. Official Agr. Chem.* 31, 760 (1948).

The method is based on the extraction of 2-aminopyridine as the free base by ether with subsequent estimation by direct titration, or nitrogen determination, or by ultraviolet absorption measurements.

Preparation of Sample.—Obtain the juice of oranges with a reamer, strain, mix in a blendor, transfer 150 ml. to a centrifuge bottle, add 50 ml. ether, shake well, and centrifuge. Remove the aqueous layer with the aid of a syphon, add calcined magnesium oxide to adjust the pH to 7.5-8.0, add Filter Cel, mix, centrifuge, decant into a suction flask, and remove the ether with suction. Extract 50 ml. of the clear liquid 4 times with 100-ml. portions of washed ether or extract 100 ml. with a Palkin or Matchett-Levine type of continuous extractor (see Chapter I). Only 95 per cent of the 2-aminopyridine is extracted by the separatory funnel variation.

Titrimetric Procedure.—To the combined ether extracts, add 40 ml. of water and 5-10 ml. accurately measured, of 0.02 N hydrochloric acid. Shake, allow to separate, swirl, draw off the aqueous layer, swirl, draw off the remainder of the aqueous layer. Repeat with additional 40-ml. portions of water. Combine the aqueous layers and, if a spectrophotometric determination is to be made, make up the volume to 110 ml. using 10 ml. for that determination.

To 100-ml. aliquot add 3 drops of bromocresol green indicator solution and titrate with 0.02 N sodium hydroxide solution to a blue color. Bring the solution to pH 4.4 with 0.02 N hydrochloric acid using about 110 ml. of pH 4.4 buffer with 3 drops of bromocresol green as a control. Titrate 10 ml. of 0.02 N hydrochloric acid to the same end point. Obtain the amount of acid consumed by 2-aminopyridine by subtracting the acid equivalent of the alkali from the total acid used. From the volume of acid consumed, subtract a reagent blank of 0.08 ml. for the procedure involving the separatory funnel extraction and of 0.15 ml. for the continuous extraction procedure, to obtain the acid necessary to react with 2-amino-pyridine. Multiply the corrected titer in milliliters of 0.02 N acid by 1.98 to obtain the milligrams of 2-aminopyridine in 50 ml. of the sample obtained by the separatory funnel extraction and by 1.882 for the 100 ml. obtained by the continuous extractor, and also by 1.1 if 100-ml. aliquot of 110 ml. was used. Ten ml. of 0.02 N hydrochloric acid is equivalent to 18.82 mg. of 2-aminopyridine.

Nitrogen Procedure.—Shake out the combined ether extracts once with 40 ml. of 0.02 N hydrochloric acid and twice with 40 ml. of water or use the titrated sample. Add 4 ml. of concentrated sulfuric acid, evaporate to 25-50 ml., transfer to a 125-ml. Erlenmeyer flask, add 1.5 g. of anhydrous sodium sulfate, and 0.1 g. mercuric oxide, boil down to sulfur trioxide fumes, insert a small, short-stemmed funnel, continue the digestion for 30 minutes until the mixture is clear and colorless. Cool, transfer to a 200-ml. Erlenmeyer flask with 140 ml. of water, add 15 ml. of 50 per cent sodium hydroxide solution without mixing, 5 ml. of sodium thiosulfate solution, connect to the customary Kjeldahl trap and condenser and distill. Catch the distillate in a measured quantity of 0.02 N hydrochloric acid containing methyl red as indicator. Titrate the excess hydrochloric acid with 0.02 N sodium hydroxide solution. Conduct a blank determination and subtract. Each milliliter of 0.02 N hydrochloric acid is equal to 0.2801 mg. of nitrogen or 0.941 mg. of 2-aminopyridine.

Spectrophotometric Procedure.—To confirm that 2-aminopyridine is the material being estimated, place a portion of the reserved aliquot in a quartz cell and make readings in a spectrophotometer at intervals of 5 or 10 mμ between 230 and 340 mμ, using 0.02 N hydrochloric acid as a blank. From the concentration obtained as above, convert extinction readings to corresponding coefficients in terms of a 0.01 per cent concentration and 1-cm. cell. Compare with a standard 2-aminopyridine solution containing 5 mg. per 100 ml. There is a minimum at 253 mμ and a maximum at about 287 mμ.

QUATERNARY AMMONIUM COMPOUNDS

Within the past decades, there has been interest displayed in the use of quaternary ammonium compounds as food preservatives. Many of these chemicals are surface-active agents. A large number of these surface-active agents have been developed for use as detergents and are generally classified into three groups: (1) the cationic compounds, such as the alkylbenzyldimethylammonium chlorides in which the hydrophobic group is in the cation; (2) the anionic compounds, such as sodium lauryl sulfate in which the hydrophobic group is in the anion; and (3) the unionized compounds, such as the polyethers and polyglycerol esters. The detergents of the cationic group are more effective germicidal agents than the ionic group, and for this reason this is the group in which the food analyst is most interested. The structure, toxicity, trade names, and use of these compounds are discussed by Jacobs.[38, 39]

The work of Woodard and Calvery [40] indicates that many of these substances are relatively toxic. Unpublished data of the author on fish appear to substantiate the work of Woodard and Calvery, for

[38] Jacobs, *Synthetic Food Adjuncts.* Van Nostrand, New York, 1947.

[39] Jacobs, *Chemistry and Technology of Food and Food Products*, Vol. III, 2nd ed. Interscience, New York, 1951.

[40] Woodward and Calvery, *Proc. Sci. Section Toilet Goods Assoc.* No. 3, 1, (1935).

none of eight guppies survived concentrations of the order of 4 p.p.m. for more than 2 days except one female guppy which lived 9 days.

Auerbach Method.—The alkylbenzyldimethylammonium chlorides sold under a number of trade names such as Zephiran, Roccal, BTC, and many other names, form one of the more important groups of quaternary compounds. These substances form colored salts with bromophenol blue (tetrabromophenolsulfonphthalein) [41] and with bromothymol blue (dibromothymolsulfonphthalein) which are readily extracted from alkaline aqueous solutions by a number of organic solvents, especially the chlorinated solvents. Thus bromophenol blue, either in its acid form or as the sodium salt, is insoluble in ethylene dichloride; nor does it form salts extractable from alkaline solution with the common primary, secondary, and tertiary amines and alkaloids. In the Auerbach method,[42, 43] the quaternary ammonium-dye salt in carbonate solution is extracted with ethylene dichloride, and the intensity of the color of the extract is measured with the aid of a photoelectric colorimeter.

Preparation of Solution and Extraction.—*Milk.*—Transfer 25 ml. of milk, with the aid of a pipette, to a 250-ml. volumetric flask containing 10 mg. of bromophenol blue and stir until the indicator is dissolved. Add gradually with stirring 50 ml. of acetone, and then 1 ml. of hydrochloric acid dropwise. This generally produces a bright yellow color; if such a color is not obtained, add more hydrochloric acid until it is produced and then 0.2 to 0.3 ml. in excess. Dilute to volume with acetone, adding it gradually and with stirring. Stopper, mix, allow to stand 30 minutes, and filter through a fluted filter. Transfer 200 ml. of the filtrate to a separatory funnel and add 200 ml. of water. Extract the aqueous acetone mixture with three successive 50-ml. portions of petroleum ether. When the phases separate, filter each petroleum ether layer through a filter paper which should be reserved for the procedure followed if a deep color is produced. Transfer the aqueous acetone layer to a beaker, evaporate on the steam bath under a current of air until odor of acetone is not noticeable and the volume of the mixture is reduced to 100 ml. or less. Cool, transfer to a separatory funnel with water, reserving the evaporating beaker, add 3 to 5 ml. of hydrochloric acid, and proceed as detailed below.

Fruit Juices in Bottled Beverages.—Mix thoroughly and measure out 50 ml. of the sample in a cylinder. Filter through a 7-cm. Büchner funnel and dilute with water to make the volume 100 ml. (solu-

41 Auerbach, *Ind. Eng. Chem., Anal. Ed.,* 15, 492 (1943).
42 Wilson, *J. Assoc. Official Agr. Chem.* 29, 33 (1946).
43 *J. Assoc. Official Agr. Chem.* 31 (194

tion A). Place the filter paper into a 400-ml. beaker and extract with small portions of alcohol until no more color is extracted and the paper remains white. Transfer the alcoholic extract to a 500-ml. distillation flask, add 10 mg. of bromophenol blue, 2 ml. of hydrochloric acid (1:1), and 100 ml. of water. Steam distill collecting a volume of distillate which is at least 100 ml. greater than the volume of alcohol in the extract. Cool the residue, transfer to a separatory funnel, and wash with 40-ml., 30-ml., and 30-ml. portions of petroleum ether, and proceed with the method as detailed below.

Transfer a suitable aliquot, say 5 or 10 ml., of solution A into a separatory funnel, add 3 ml. of bromophenol blue solution (prepared by dissolving 40 mg. of bromophenol blue solid reagent in warm water, cooling, and diluting to 100 ml.), 1 ml. of hydrochloric acid (1:1), and proceed with the method as detailed below:

Fruit Juices.—Transfer 20 ml. of fruit juice, with a pipette, to a 50-ml. centrifuge tube. Centrifuge for 15 minutes and decant the liquid into a 500-ml. flask equipped for steam distillation. Add 10 ml. of bromophenol blue, 2 ml. of hydrochloric acid (1:1), 80 to 100 ml. of water, and steam distill, collecting 100 ml. of distillate. Cool the residue, transfer to a separatory funnel, wash with 40 ml., 30 ml., and 30 ml. of petroleum ether, and proceed with the method detailed below.

Add 30 ml. of alcohol to the pulp in the centrifuge tube, stir with a rod, allow to stand for 10 minutes, centrifuge for 5 minutes, and pour off the supernatant liquid into a 500-ml. steam distillation flask. Extract the pulp twice more, using 15 to 20 ml. of alcohol and then proceed as above, but collect 200 ml. of distillate and then continue with the method.

Soda Pop.—Transfer 50 ml. of decarbonated soda pop or carbonated beverage to a separatory funnel, add 3 ml. of bromophenol blue solution (see above), 1 ml. of hydrochloric acid (1:1), and proceed with the method.

Table Syrup.—Weigh out 20 g. of the sample and transfer with water to a 100-ml. volumetric flask. Make to volume with water. Transfer an aliquot to a separatory funnel, add 3 ml. of bromophenol blue solution (40 mg. of the dye per 100 ml. of water), 1 ml. of hydrochloric acid (1:1), and proceed with the method.

Mayonnaise and Salad Dressings.—Place 10 g. of the sample in a 250-ml. beaker, add 100 ml. of acetone, and mix with a rod. If a layer of undissolved oil separates on the bottom of the beaker, add sufficient acetone to dissolve it. If gummy material separates, knead it with the rod to press out the oil. Filter with the aid of a Büchner funnel using a 589 SS white ribbon filter or equal. Wash 2 or 3 times with 15-20 ml. of acetone. Transfer the acetone to a separatory funnel, and wash out the suction flask with an equal volume of water adding the washings to the separatory funnel. Wash the suction flask with 75 ml. of petroleum ether, transferring the petroleum ether to the separatory funnel. Shake for 1 minute, allow to separate, draw off the aqueous acetone layer, and discard the petroleum ether phase. Repeat the extraction of the undesired fat with two additional 50-ml. portions of petroleum ether. Transfer the washed aqueous acetone layer to a beaker and evaporate on a steam bath until the odor of acetone is gone, and the volume has been reduced to less than 75 ml. Cool the solution, transfer to a 100-ml. volumetric flask, complete to volume with water, and mix. Transfer 25 ml. to a separatory funnel, add 3 ml. of bromophenol blue solution, add 1 ml. of hydrochloric acid (1:1), and proceed as directed below.

Other Products.—Depending upon whether the food product contains fatty material or not, the quaternary ammonium-dye salt can be prepared by a variation of the method detailed above.

Procedure.—Add 50 ml. of ethylene chloride, with the aid of a pipette equipped with a safety pipetter, to the separatory funnel containing the samples prepared as detailed above, and shake for 3 to 4 minutes. Allow to stand until the layers separate and are clear, and draw off the lower layer into a second separatory funnel containing 10 ml. of 1 per cent sodium carbonate solution, 10 g. Na_2CO_3 per liter, and shake for 3 to 4 minutes. When the two layers separate, observe the lower layer. If it is blue, a quaternary ammonium compound is present. The depth of color indicates whether or not it can be estimated photometrically. Draw off the lower layer into a glass-stoppered flask containing 1-2 g. of anhydrous granular sodium sulfate, allow to stand for 30 minutes, and read in an appropriate cell using an adequate light filter adequate for 610 mμ.

If the color is too deep, acidify the contents of the second separatory funnel with 1 to 2 ml. of hydrochloric acid (1:1), shake until the contents become yellow, and transfer back to the first separatory funnel. Add a second 50-ml. portion of ethylene dichloride to the first separatory funnel, shake 3 to 4 minutes, and allow to stand until the lower layer is clear.

Estimate a suitable volume of the extract to be used as an aliquot. Transfer this aliquot with a pipette to a 50-ml. volumetric flask, and fill to the mark with ethylene dichloride. Then continue with the method as detailed above.

The clarified colored solution can be decanted into a Klett-Summerson photocolorimeter tube and read in the instrument, using a No. 54 filter, or an equivalent device and filter may be used.

Standard Curve.—A standard curve can be prepared. Standardize a 1 per cent solution of the quaternary ammonium preparation, if obtainable, by the ferricyanide method (see below). Then ascertain the maximum and minimum quantities of the material that will produce, in 50 ml. of ethylene dichloride using the method detailed, colors having densities within the range of the instrument used. Prepare a set of three or more standards containing, in 50 ml., appropriate quantities of the quaternary covering the range. With a wedge photometer, standards containing 0.0, 0.1, 0.2, and 0.25 mg. per 50 ml. are adequate.

Ferricyanide Method.—This method [42, 43, 44] is based on the precipitation of the quaternary ammonium compound as a salt of ferricyanic acid. The excess ferricyanide is then estimated iodometrically.

$$3 \begin{bmatrix} C_nH_{2n+1} & CH_3 \\ & N-Cl \\ \bigcirc\ CH_2 & CH_3 \end{bmatrix} + K_3Fe(CN)_6 \longrightarrow \begin{bmatrix} C_nH_{2n+1} & CH_3 \\ & N \\ \bigcirc\ CH_2 & CH_3 \end{bmatrix}_3 Fe(CN)_6 + 3KCl$$

Since each molecule of ferricyanide liberates 1 atom of iodine, each oxidation equivalent corresponds to 3 molecules of the quaternary ammonium base.

Reagents.—Sodium Thiosulfate Solution, 0.02 *N*.—Dissolve 5 g. of crystallized sodium thiosulfate, $Na_2S_2O_3 \cdot 5H_2O$ in water and standardize as detailed in Chapter IX. One ml. of 0.02 N thiosulfate solution is equivalent to 0.02142 g. of alkylbenzyldimethyl ammonium chlorides, average molecular weight 357.

Buffer Solution.—Dissolve 130 g. of sodium acetate in water, add 42 ml. of acetic acid, and make up to 500 ml.

Approximation of Quaternary Content.—Place 1 ml. of buffer solution, 2 ml. of ferricyanide solution, and 20 ml. of water into each of 4 small Erlenmeyer flasks. Transfer to each of these flasks with a pipette, 0.5, 1.0, 2.0, and 4.0 respectively, of the quaternary sample, mix, and observe. The approximate content is indicated in Table 7.

Procedure.—Transfer with a pipette an aliquot of the quaternary solution containing 5 g. to a 100-ml. Kohlrausch flask. Dilute if necessary to make the volume 50 ml. Add 5 ml. of buffer and mix. Add with a pipette 30 ml. of potassium ferricyanide solution, prepared by dissolving 6.6 g. $K_3Fe(CN)_6$ in water and diluting to 1 liter, and rotate the flask during the addition of the reagent. Fill to the mark with water and mix. After 0.5 hour, filter, and discard the first 10-15 ml. of the filtrate. Transfer 50 ml. with a pipette to a 500-ml.

---

[44] *New and Non-Official Remedies.* Am. Med. Assoc., 1946.

Erlenmeyer flask, add 100 ml. of water and 1 to 2 g. of solid potassium iodide. Swirl until the salt dissolves. Add 10 ml. of hydrochloric acid (1:1), stir, and allow to stand for 2 minutes. Add 10 ml. of zinc sulfate solution, prepared by dissolving 20 g. of $ZnSO_4 \cdot 7H_2O$ in 180 ml. of water. Mix and titrate with the standard thiosulfate solution adding 1 ml. of starch indicator solution when the iodine has been reduced to a light yellow. Run a blank determination, using water instead of the sample. Calculate the concentration of the quaternary from the difference in the titrations.

TABLE 7. APPROXIMATION OF CONTENT OF ALKYLDIMETHYLBENZYLAMMONIUM CHLORIDE (MOL. WT. 357)

	Sample Added			
Per Cent	A 0.5 Ml.	B 1.0 Ml.	C 2.0 Ml.	D 4.0 Ml.
8.4 or more......	No ppt.	No ppt.	No ppt.	No ppt.
5	Ppt.	No ppt.	No ppt.	No ppt.
2.5	Ppt.	Ppt.	No ppt.	No ppt.
1.25	Ppt.	Ppt.	Ppt.	No ppt.
1 or less	Ppt.	Ppt.	Ppt.	Ppt.

SELECTED REFERENCES

Jacobs, ed., *Chemistry and Technology of Food and Food Products,* 2nd ed. Interscience, New York, 1951.

Jacobs, *Synthetic Food Adjuncts.* Van Nostrand, New York, 1947.

Leach, *Food Inspection and Analysis.* Wiley, New York, 1920.

Mellor, *Modern Inorganic Chemistry.* Longmans, London, 1927.

Methods of Analysis, A.O.A.C., Washington, 1945.

Mulliken, *Identification of Pure Organic Compounds,* Wiley, New York, 1916.

Thurston, *Pharmaceutical and Food Analysis.* Van Nostrand, New York, 1922.

Woodman, *Food Analysis.* McGraw-Hill, New York, 1931.

CHAPTER V

METALS IN FOODS

THE determination of metals in foods resolves itself into the problem of determining those metals in the presence of organic material. These procedures present, therefore, no greater difficulty than such estimations ordinarily present. In general, the organic material must be destroyed or removed in some manner before the estimation of the metal is made, and then the usual qualitative or quantitative method applied. Sometimes it is possible to run a procedure for a metal in the presence of organic material without great loss in sensitivity, for example, the Reinsch test for arsenic.

A text of this type cannot go into an exhaustive survey of methods for metals. Furthermore, the food analyst is mainly interested in those metals that are harmful and may come into contact with foodstuffs for the reasons detailed in Chapter VI. The food analyst is also interested in the determination of metals which are micronutrients or are of nutritional significance, such as iron, cobalt, nickel, manganese, copper, zinc, and aluminum, and in the identification of the plating of equipment and utensils used in the food industry for the processing and storage of foods or used in the home for the cooking and storage of foods. His interest is further conditioned by the factor that the presence of anything more than a trace of some metals is generally illegal, and only approximate quantitative analysis need be made where comparatively large quantities of metal are present.

The food analyst is interested in the following metals: arsenic, lead, mercury, copper, zinc, chromium, cadmium, antimony, aluminum, manganese, tin, nickel and, rarely thallium. The methods for some of these metals will be detailed, and procedures for the others will be outlined. Selenium, sodium, potassium, iodine, and other elements will be discussed in Chapter XVIII, "Inorganic Determinations." Authorities are more or less agreed that iron, aluminum, nickel, chromium, silver, and gold are nontoxic; copper, tin, and zinc are moderately toxic; and lead, antimony, cadmium, mercury, arsenic, and thallium are highly toxic.

The role of arsenic in foodstuffs is of historic importance. The numerous cases of arsenic poisoning in Manchester, England, in 1900, traced to the presence of arsenic in beer made with commercial glucose that was manufactured with contaminated sulfuric acid, brought the subject of metallic contamination in food sharply to the public. In recent years, the use of arsenical, mercurial, and lead-bearing and other metallic insecticides and fungicides to preserve the growing food supply from destruction by insect and mold pests has increased enormously. The realization of danger of epidemic proportions lying in the use of these metallic insecticides and fungicides without proper removal before the food is sold for consumption has developed a great deal of interest in methods for the detection and determination of very small amounts of these metals.

PREPARATION OF ASH

There are two general methods for the destruction of interfering organic matter. The first is called the "wet ash" or acid-digestion method, and the second is ordinary ashing by means of heat with, or without, the aid of an "ash aid" mixture or of an alkaline fixitive for volatile metals, such as arsenic or mercury. Acid digestion is to be preferred for arsenic, mercury, and tin.

Wet Ash or Acid Digestion.—Depending on the type of foodstuff and whether the metallic contamination is throughout the product as might be possible in the case of fish, or whether it is exclusively on the outside as in the case of insecticide on fruits, weigh a representative portion of the product, generally 100 to 200 g., sometimes varying from 5 g. to 5 lbs., according to the metal content, and peel, if possible. Place the weighed portion or the peelings in one or more 800-ml. Pyrex Kjeldahl flasks. Add 50 ml. of nitric acid and then carefully add 20 ml. of sulfuric acid. Heat cautiously so that no excessive foaming takes place. Add nitric acid in small portions until all the organic matter is destroyed. This point is reached when no further darkening of the solution occurs on continued heating after the production of a clear solution and copious fumes of sulfur trioxide. Cool, add 75 ml. of water and 25 ml. of a saturated solution of ammonium oxalate to aid in the expulsion of nitrogen fumes. Evaporate again to the appearance of sulfur trioxide fumes. Cool, dilute with water, transfer to a 500-ml. or liter volumetric flask, and make to volume. Use aliquot portions for the analyses detailed below.

Methods for wet ashing with nitric and perchloric acids are described by various investigators.[1,2] Great care must be exercised in the use of perchloric acid to avoid explosions.

---

[1] Gerritz, *Ind. Eng. Chem., Anal. Ed.* **7**, 167 (1935).
[2] Gieseking, Snider, and Getz, *Ind. Eng. Chem., Anal. Ed.* **7**, 185 (1935).

A method for wet ashing with potassium permanganate and sulfuric acid is described in this chapter in connection with procedures for the determination of mercury.

Ash by Ignition.—Weigh a representative portion of the material to be analyzed and transfer to a porcelain dish or casserole of convenient size. This will be from 5 to 200 g. depending on the amount of metal in the food. Add 2 to 5 ml. of a solution of aluminum nitrate and calcium nitrate, 40 g. of $Al(NO_3)_3 \cdot 9H_2O$ + 20 g. of $Ca(NO_3)_2 \cdot 4H_2O$ in 100 ml. of water or an excess of lime or magnesia. Dry in a thermostatically controlled oven at 100° C. Char the material, controlling swelling, if any, by playing the flame from a glass jet over the sample. Ash in a muffle overnight, if possible, at not over 450° C. If a clean ash cannot be obtained, cool the dish and add more ash-aid solution, that is, aluminum nitrate-calcium nitrate solution, or 2-3 ml. nitric acid, if permissible, dry and re-ash. Do not add nitric acid if an alkaline ash is necessary, as for example, if arsenic is to be estimated Dissolve in an appropriate solvent when the ashing process is complete, and proceed with one of the methods detailed below.

HEAVY METALS

In general, the qualitative demonstration of the presence of heavy metals in foodstuffs can be made by using the group sulfide precipitations.

Procedure.—Transfer the wet ash to a beaker, dilute, and neutralize. Add 1 ml. of hydrochloric acid (1:3) to each 10 ml. of solution and warm to 50° C. Add an equal volume of saturated hydrogen sulfide solution and allow to stand for 10 minutes at 35° C. If no color or precipitate is produced, the following metals may be considered absent in more than traces: silver, arsenic, antimony, tin, copper, mercury, bismuth, thallium, and cadmium. Lead may remain behind as $PbSO_4$, lead sulfate, in the wet ash process; hence, if any residue remains, it should be investigated separately with ammonium acetate solution.

If the mixture of sample solution and hydrogen sulfide is now filtered and made alkaline with ammonium hydroxide, any precipitate which forms may be due to one of the following: aluminum, chromium, zinc, manganese, iron, cobalt, and nickel. This portion of the test is not of much value because many food materials contain iron, which would give the test.

ARSENIC

Reinsch Test.—The Reinsch test is a simple, although not very sensitive, one. It is based on the deposition of arsenic from solution as a copper arsenide. This test may very often be applied directly without previous destruction of organic material.

Procedure.—Place 200 ml. of the liquid food or beverage, or of a mixture of water and the solid food, in a casserole or similar container, and acidify with 1 ml. of arsenic-free hydrochloric acid. Then evaporate to one-half its volume. Add 15 ml. more of hydrochloric acid and also a piece of pure burnished copper foil. Keep the liquid simmering for an hour and replace the water lost by evaporation from time to time. If at the end of this time the copper foil remains bright, arsenic is absent. If the copper has a black or brown deposit, remove it and wash well with water, alcohol, and ether, and dry. Place the foil in a subliming tube and heat over a low flame. If a sublimate is present, examine under a microscope. Arsenic forms tetrahedral crystals in contradistinction to mercury. Antimony, silver, and bismuth will also give a deposit but will not sublime.

Gutzeit Method.—The Gutzeit method is based on the liberation of arsine from an arsenic solution. The arsine subsequently reduces mercuric bromide on a prepared strip of paper with the production of stains. The stain, if the method is followed in detail, is proportional to the amount of arsenic.

Preparation of Generator.—Prepare a generator, Fig. 46, as follows: Use a 2-oz. wide-mouth bottle. Equip the bottle by means of a perforated stopper with a glass tube 1 cm. in diameter and 6-7 cm. long, with an additional constricted end to facilitate connection. Place a small wad of glass wool in the constricted bottom end of the tube and add 3.5 to 4 g. of 30-mesh clean sand. Moisten the sand with 10 per cent lead acetate solution, and remove the excess by light suction. The lead acetate is used to remove any hydrogen sulfide that might be generated along with the arsine and thus vitiate results, if permitted to reduce the mercuric bromide. Connect the tube by means of a rubber stopper with a narrow glass tube 2.6 to 2.7 mm. in internal diameter, and 10 to 12 cm. long, and place in this tube a strip of mercuric bromide paper.

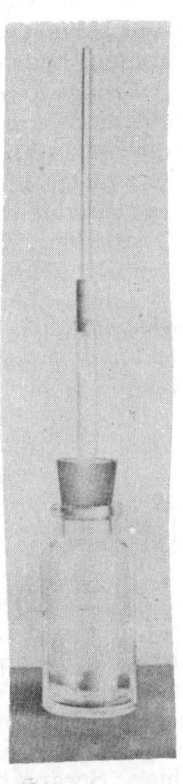

FIG. 46. Gutzeit Generator

Preparation of Test Paper.—These strips may be made by cutting paper, similar to Whatman No. 40, into strips exactly 2.5 mm. wide and about 12 cm. long. Soak the strips for 1 hour or longer in a fresh

3 to 6 per cent solution of mercuric bromide in 95 per cent alcohol. Dry and use within 2 days. For approximately quantitative work, these strips may be stored in a stoppered blackened tube.

Goldstone Modification.—The Gutzeit method for arsenic is an empirical one. It requires strict adherence to all details. To overcome one source of error, namely, the uneven evolution of hydrogen, Goldstone [3] suggested the use of short zinc rods treated so that a constant surface area would be exposed and thus a relatively even evolution of hydrogen would result.

Preparation of Zinc Rods.—Clamp a 15- by 125-mm. Pyrex test tube on a stand in a vertical position, place a 6-in. stick of arsenic-free zinc into the tube, and heat carefully with a Bunsen burner until the zinc melts and fills the entire tube. As an alternative procedure, melt the zinc in a beaker and pour the molten metal into a preheated Pyrex tube. Tap the tube to dislodge any air pockets that may have formed and allow the mass to solidify gradually, playing the flame on the upper portion in order to make this section solidify last. This precaution prevents the formation of a hollow core attributable to the contraction of the metal as it solidifies and insures a solid, uniform cylinder of zinc metal. Allow to cool and remove the cylinder by breaking the test tube. Cut the cylinder into lengths slightly less than the diameter of the generating bottle with the aid of a hack saw, grind the ends smooth with an emery wheel, and coat with the wax composition, described below. Lengths of 1.5 in. are suitable and generally last for 15 determinations before they become too short.

Rub some magnesium carbonate into gum arabic paste, coat the plane ends of the short rods with the paste, and allow to dry. Dip one end of the short rods into a beaker of molten wax prepared from three parts of paraffin and one part of Acrowax C, withdraw, allow to harden, and repeat the operation on the other end of the rod, covering the entire surface with a layer of wax about $\frac{1}{16}$ inch thick. It may be necessary to repeat the coating in order to get it sufficiently thick. Scrape the plane ends free of the wax and soak in water to remove the paste coating. Activate the uncoated ends of the rods with stannous chloride, as directed below, and store under water acidified with a drop of concentrated hydrochloric acid. Since after each arsenic determination the plane surfaces remain activated, the initial activation is the only one necessary. As the zinc is dissolved during a series of determinations, the protruding collar of wax should be scraped off.

[3] Goldstone, *Ind. Eng. Chem., Anal. Ed.* **18,** 797 (1946).

Instead of preparing strips each time a determination is to be made, Goldstone [3] suggests that a sheet of 32 strips be cut into 9-cm. lengths, which may be suspended permanently in the alcoholic mercuric bromide solution stored in a 10-ml. glass-stoppered cylinder. Withdraw the strips as required, press immediately between filter paper, and permit the strips to dry in air for 0.5 hour before use.

Determine the acid in an aliquot of the solution prepared from the wet ash as described above. Place aliquots, not to exceed 30 ml., depending on the amount of arsenic trioxide, 0.01 to 0.03 mg., in the Gutzeit generator. If the aliquot contains only hydrochloric acid, add sufficient hydrochloric acid to make a total volume of 5 ml. If it contains sulfuric acid, add sufficient arsenic-free 25 per cent sodium hydroxide to exactly neutralize it and add 5 ml. of hydrochloric acid; or add sufficient hydrochloric acid to the sulfuric acid in the aliquot to make a total volume of 5 ml. Cool, if necessary, and add 5 ml. of potassium iodide solution (15 g. of potassium iodide dissolved in water and made up to 100 ml.) and 4 drops of stannous chloride solution (40 g. of arsenic-free stannous chloride, $SnCl_2 \cdot 2H_2O$, in hydrochloric acid made up to 100 ml. with hydrochloric acid). Add a piece of activated zinc (prepared by placing the zinc in contact with hydrochloric acid (1:3), to which has been added 2 ml. of the stannous chloride reagent and allowing the action to proceed for 15 minutes), 10 to 15 g. in weight or 2 to 5 g. of granulated zinc, center the strip of mercuric bromide paper in its tube, and set the tubes in position.

Immerse the apparatus in a water bath kept at 20-25° C. to within 1 in. of the top of the narrow tube and allow the evolution of the arsine to proceed for 1 hour to 1.5 hours. Remove the strip and average the length of the stains on both sides in millimeters. Locate the length of the unknown on a standard graph and read off on the abscissa the quantity of arsenic present. The graph may be made by running known quantities of arsenic by the above method, using length of stain as ordinates and milligrams of arsenic trioxide as abscissas. Many authorities advise against the use of a standard graph on the ground that one cannot be prepared. They advise the running of a series of controls with every unknown determination.

All the reagents used in this determination should be arsenic free. However, as a precaution, it is best to run blanks on the reagents. In some cases the test may be made without previous destruction of organic material but the results obtained are probably only approximate. An aliquot containing 0.02 to 0.025 mg. of arsenic trioxide is considered optimum for reading the stain.

Clarke [4] says that general experience has made it plain that not one of the various modifications of the Gutzeit method can be used by the average analyst with the assurance or even probability that his results will be accurate unless he attains considerable experience in its use.

Molybdenum Blue Method.—The molybdenum blue method [5-13] for the estimation of arsenic is one of the most sensitive methods that can be used for this purpose. Phosphorus reacts with ammonium molybdate to form a complex molybdiphosphate. This may subsequently be reduced with the formation of a complex molybdenum compound strongly colored blue. Arsenic undergoes an entirely analogous reaction with the formation of an intensely colored blue complex. This reaction of arsenic and its use in methods for the estimation of arsenic have been discussed by a number of investigators.

The arsenic is put into solution by methods previously detailed. It is evolved as arsine, which is trapped and oxidized by bromine water or by sodium hypobromite solution. Ammonium molybdate is added, and the color of molybdenum blue is developed by the use of hydrazine sulfate, $N_2H_4 \cdot H_2SO_4$.

Procedure.—Make an acid digestion of the material to be analyzed as directed on page 183. Prepare a Gutzeit generator in the usual way (page 185); however, instead of the tube containing the mercuric bromide test paper, attach another tube leading the generated gases to a trapping device containing 3 ml. of sodium hypobromite solution (3 ml. half-saturated bromine water plus 1 ml. 0.5 N sodium hydroxide solution), as shown in Fig. 47, or use two ordinary small gas-washing devices. It is better to have two trapping devices in series, and the second need contain only water. Treat the arsenic test solution in the same way as in the Gutzeit method.

Allow the generation of arsine to proceed as directed in the Gutzeit method. After generation is complete, transfer the contents of the traps to a graduated colorimeter tube, Nessler tube, or volumetric flask. Wash the trap with six 2-ml. portion of water, delivering the water to the trapping device with a 2-ml. pipette. Use a rubber-bulb aspirator to blow the wash solutions out of the trap into

[4] Clarke, *J. Assoc. Official Agr. Chem.* 11, 438 (1938).
[5] Deniges, *Compt. rend.* 171, 802 (1920).
[6] Atkins and Wilson, *Biochem. J.* 20, 1225 (1926).
[7] Maechling and Flinn, *J. Lab. Clin. Med.* 15, 779 (1930).
[8] Deemer and Schricker, *J. Assoc. Official Agr. Chem.* 16, 226 (1933).
[9] Zinzadze, *Ind. Eng. Chem., Anal. Ed.* 7, 227, 230 (1935).
[10] Snell and Snell, *Colorimetric Methods of Analysis.* Van Nostrand, New York, 1936.
[11] Chaney and Magnuson, *Ind. Eng. Chem., Anal. Ed.* 12, 691 (1940).
[12] Jacobs and Nagler, *Ind. Eng. Chem., Anal. Ed.* 14, 442 (1942).
[13] Ruchhoft, Placak, and Schott, *U. S. Pub. Health Service, Reprint* 2527 (1943).

the collection vessel. Press the aspirator bulb gently in this step. Add exactly 5 ml. of 2 N sulfuric acid and stir; add 1 ml. of ammonium molybdate reagent (page 753) and shake. Add 1 ml. of the half-saturated hydrazine sulfate solution and swirl, make to a volume of 25 ml., and allow to stand for 0.5 hour for full development of the blue color. Compare with standards or a standard treated in a similar way at the same time.

Preparation of Standards.—Prepare the standards or standard from the diluted stock standard arsenious oxide solution (page 190). Add 3 ml. of sodium hypobromite solution to the aliquot or aliquots selected, dilute to 15 ml. with water, add exactly 5 ml. of 2 N sulfuric acid, and stir. Add 1 ml. of the molybdate reagent, stir, add 1 ml. of half-saturated hydrazine sulfate solution, and stir. Make up to the same volume as the test solution. Run a blank on all the reagents as a check.

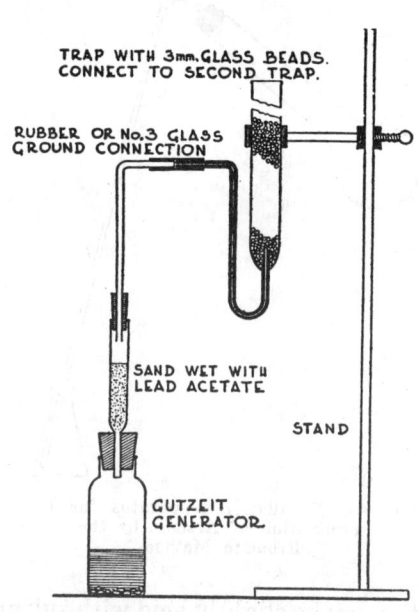

TRAP WITH 3mm. GLASS BEADS. CONNECT TO SECOND TRAP.

RUBBER OR No.3 GLASS GROUND CONNECTION

SAND WET WITH LEAD ACETATE

STAND

GUTZEIT GENERATOR

If a final volume of 25 ml. is to be used in making the comparisons, use exactly 5 ml. of 2 N sulfuric acid, in order to have the proper acidity for the development of the molybdenum blue color. If less than this quantity of acid is used, the blank may itself be reduced. If more than this quantity of acid is used, the development of the blue complex will be delayed.

FIG. 47. Apparatus for Jacobs-Nagler Method

Bromate Method.—The bromate method [14] for the determination of arsenic is applicable when the amount of arsenic trioxide to be estimated is of the order of 0.35 mg. The method is based on placing the arsenic into solution by means of the wet method, or acid digestion. Then the arsenic is distilled as arsenious chloride, $AsCl_3$, along with hydrogen chloride. The distillate is titrated with standard bromate solution using methyl orange as indicator.

The distillation apparatus (Fig. 48) consists of an 800-ml. Kjeldahl flask, a tube, and a 300-ml. wide-mouth flask. To make the tube,

---

[14] *Methods of Analysis, A.O.A.C.* Washington, 1945.

bend a 10-15 mm. glass tube to an acute angle of about 70°. Draw the longer arm, which is about 15-20 in. long, down to an orifice of about 3 mm. Fit the shorter arm, which is about 4 in. long, with a rubber stopper, which has previously been boiled in 10 per cent sodium hydroxide solution for about 15 minutes, and then in hydrochloric acid for 15 minutes in order to remove most of the sulfur compounds which might be distilled and react with the bromate solution.

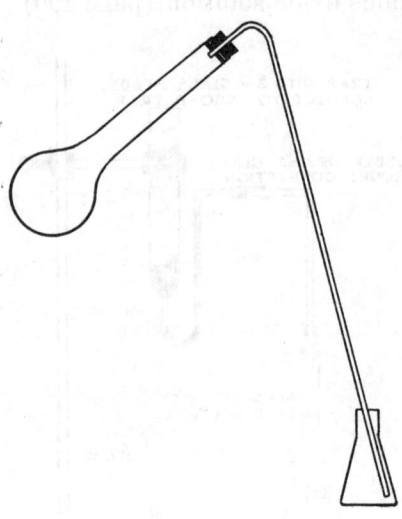

Preparation of Reagents.— Standard Potassium Bromate Solution.—Dissolve 0.1823 g. of potassium bromate, $KBrO_3$, in water and dilute to 1 liter. One ml. of this solution is equivalent to 0.324 mg. of arsenic trioxide, As_2O_3. Standardize by titration against standard arsenious oxide solution, making the titration at about 90° C. and in the presence of about 100 ml. of water and 25 ml. of hydrochloric acid, in order to simulate the conditions under which the unknown samples will be titrated. One ml. of the bromate solution should be equivalent to 1 ml. of the arsenious oxide solution.

Standard Arsenious Oxide Solution.—Dissolve 0.3241 g. of arsenic trioxide, As_2O_3, in 25 ml. of 10 per cent sodium hydroxide solu-

FIG. 48. Distillation Apparatus for Determination of Arsenic by the Bromate Method

tion, make slightly acid with sulfuric acid (1:6), and dilute with water to 1 liter.

Hydrazine Sulfate-Sodium Bromide Solution.—Dissolve 20 g. of hydrazine sulfate and 20 g. of sodium bromide in 1 liter of hydrochloric acid (1:4).

Procedure.—Proceed with the wet-ash digestion using exactly 20 or 25 ml. of concentrated sulfuric acid at the beginning of the digestion. After the digestion is complete, add 50 ml. of water and 25 ml. of saturated ammonium oxalate solution containing 50 g. of urea per liter, and boil until white sulfur trioxide fumes extend up into the neck of the flask to decompose oxalates and urea completely.

Add 25 ml. of water and cool to room temperature. Place 100 ml. of water into the 300-ml., wide-mouth flask. Add to the mixture in the Kjeldahl flask 20 g. of sodium chloride, not iodized, and 25 ml. of the hydrazine sulfate-sodium bromide solution. Connect the distilling apparatus. Heat the Kjeldahl flask over a small well-protected flame and catch the distillate in the water in the wide-mouth flask. The heating generates hydrogen chloride gas, which carries over the arsenious chloride with it. The absorption of the evolved hydrogen chloride gas by the water causes a rise in temperature, by means of which rise the progress of the distillation can be followed. Adjust the flame so that the temperature of the distillate solution will rise to 90° C. in 9-11 minutes, and then discontinue the distillation. The residual mixture in the Kjeldahl flask should not be less than 55 ml. If the distillation proceeds further, or a larger quantity of sulfuric acid than that specified is used in the digestion, sulfur dioxide may be distilled. This is titrated as arsenious oxide.

Titrate the distillate at once with the bromate solution, using 3 drops of methyl orange indicator. Single drops of indicator, but not exceeding 3, may be added during titration as the red color fades. Toward the end of the titration add the bromate solution very slowly and with constant agitation to prevent local excess. The end point is reached when a single drop of the bromate just destroys the final tinge of red color. To determine when this point has been reached use a similar wide-mouth flask of clear water for comparison. The end point must not be exceeded, as the action of the indicator is not reversible and back titrations are not reliable. At the proper end point, the red color produced by 2 additional drops of methyl orange indicator should persist for at least 1 minute. Correct the results for the volume of bromate used in a blank determination using 5 g. of pure sucrose and the same quantities of reagents, as well as the same distillation procedure. The blank titration should not exceed 0.7 ml. of bromate solution and variations in the blank should not exceed 0.1 ml. when chemicals from the same lot are used. If doubt arises, run a Gutzeit determination on an aliquot.

Iodometric Method.—Cassil and Wichmann [15] describe a rapid titrimetric method for the determination of arsenic in microgram quantities. An acid digestion is performed on the material. Then the arsenic is evolved in a special generator as arsine, which is trapped in a mercuric chloride solution contained in a special tube made of methyl methacrylate resin. The liberated arsine is absorbed quantitatively by the mercuric chloride solution, forming mercury arsenides. The arsenides are oxidized by the excess mercuric chloride with the formation of mercurous chloride and arsenious acid.

---

[15] Cassil and Wichmann, *J. Assoc. Official Agr. Chem.* **22**, 436 (1939).

The arsenious acid may then be oxidized to arsenic acid with 0.001 N iodine solution.

Chlorometric Method.—Arsenic may also be determined by chlorometry, using the method of Goldstone and Jacobs [16] as detailed on page 195 for the estimation of antimony.

ANTIMONY

Antimony may get into foods that are cooked or processed in enamelware, the enamel of which was made with antimony compounds,.or it may contaminate foods that are covered with antimony bearing tin foil. Acids, such as citric, may at times extract sufficient antimony to have an emetic or even more harmful effect.

It is the opinion of the author that some of the published work concerning the analytical recovery of antimony is unreliable. It has been shown that antimony is lost as antimony pentachloride, $SbCl_5$, when hydrochloric acid solutions of antimony are boiled. Antimony chlorides behave differently from arsenic chlorides. Thus the lower valent arsenious chloride, $AsCl_3$, boils at a lower temperature (130.2° C.) than the higher valent arsenic pentachloride; whereas the lower valent antimonious chloride, $SbCl_3$, boils at a higher temperature (223° C.) than the corresponding higher valent antimony pentachloride, $SbCl_5$, boiling at 140° C. The pentachloride dissociates slowly at its boiling temperature, yielding free chlorine and the trichloride. The trichloride has an appreciable volatilization in hydrochloric acid solutions at temperatures as low as 110° C.

Some investigators use the acid-digestion method for the recovery of antimony from organic materials, from mixtures with organic materials, and to free it from the cellulosic materials used to trap the fume or dust while sampling. Goldstone [17] and also Jacobs [18] have shown that antimony is "lost" in any acid digestion containing an oxidizing medium such as nitric acid, hydrogen peroxide, etc., or when acid solutions of antimony containing an oxidizing substance such as nitric acid, chlorine, bromine, or hydrogen peroxide are boiled or evaporated.

Maren pointed out that an investigation of the valency state of antimony following a typical sulfuric acid-nitric acid digestion showed that about 35 per cent was in the trivalent state, 15 per cent was in the quinquevalent state, and 50 per cent was apparently "lost," that is, was in neither the trivalent nor quinquevalent state.

---

[16] Goldstone and Jacobs, *Ind. Eng. Chem., Anal. Ed.* 16, 206 (1944).
[17] Goldstone, personal communication, 1941.
[18] Jacobs, work performed, 1941.

It was shown many years ago that nitric acid oxidation of antimony is never complete to the quinquevalent state [19] and that there is evidence of a quadrivalent antimony.[20]

These sources of error must be taken into consideration in using any procedure for the determination of antimony.

It may be detected qualitatively by the Reinsch and other tests. The Reinsch test is detailed under the section, "Arsenic." It may be estimated quantitatively as directed by Bamford.[21]

Sulfide Method.—In this method the antimony is thrown down as antimony sulfide and is then determined colorimetrically against standards prepared in a similar manner. The antimony sulfide colloid is stabilized by the use of a solution of gum arabic or ghatti.

Cut up finely a weighed portion of the samples and mix in a silica dish with sufficient magnesium oxide to give a definitely alkaline reaction. Cover the material with a saturated solution of magnesium nitrate, $Mg(NO_3)_2$. In general, 35 to 40 ml. of this solution are sufficient for 100 g. of animal matter. Heat the mixture on a sand-bath, with frequent stirring until the material has dried, charred, and begun to whiten. Crush the charred mass with a pestle and heat strongly if necessary over a blowpipe flame. The ash should be white. If not, cool, mix with a concentrated solution of ammonium nitrate, and reheat until free of nitrates. Moisten the ash, when cold, with water and sufficient hydrochloric acid to dissolve the magnesium oxide and to give a definitely acid reaction.

Dilute the solution with water and treat with hydrogen sulfide Filter off the precipitate, wash in the usual way, and dissolve in a minimum quantity of hot concentrated hydrochloric acid. Dilute this solution with water, refilter, add 1 ml. of a 5 per cent solution of gum ghatti or gum arabic per 100 ml. of the liquid to hold the precipitated antimony sulfide in suspension, and make the solution up to a definite volume. Pass in hydrogen sulfide again and compare the color produced with that of a standard of approximately equal concentration with the aid of a colorimeter, or transfer to a Nessler tube and compare with standards in similar tubes.

Preparation of Standard.—Prepare the standard by diluting 1 ml. of a 5 per cent tartar emetic solution [potassium antimonyl tartrate, $2K(SbO)C_4H_4O_6·H_2O$], slightly acid and mixed with gum arabic or gum ghatti solution, to 1 liter. Treat an appropriate aliquot with sufficient gum solution to give the same concentration as that in the

[19] Mellor, *Treatise on Inorganic Chemistry.* Longmans, London, 1927.
[20] Maren, *Bull. Johns Hopkins Hosp.* **77**, 338 (1945).
[21] Bamford, *Analyst* **59**, 101 (1934).

unknown and make up to definite volume. It is then saturated with hydrogen sulfide and used as the standard.

Separation from Interferences.—Where the antimony-bearing food is also lead bearing, the metals must be separated before estimation of the antimony can proceed. This may be done by the usual polysulfide method. Prepare a solution of the sample as described under lead. Pass in hydrogen sulfide until the solution is saturated. Add an equal volume of water and again saturate the solution. Filter the sulfides preferably through a sintered-glass filter, such as a Jena-glass filter No. 11 GA or an equivalent Pyrex type. Dissolve the antimony and any tin, if present, with five applications of 5 ml. each of warm polysulfide reagent, prepared as directed under the Wichmann-Clifford method for lead. Wash the filter 4 times with 3 per cent sodium sulfate solution, 3 g. of anhydrous sodium sulfate, Na_2SO_4, in 100 ml. of water. Combine all the polysulfide filtrates and the wash solutions. The precipitate on the filter may be estimated for lead as described. Neutralize the combined filtrates slowly with hydrochloric acid, dilute, adding the acid dropwise near the neutral point, finally add 2 ml. of dilute acid in excess. Filter off the precipitate. Redissolve in a minimum quantity of hot concentrated hydrochloric acid. Dilute with water and filter if necessary. From this point, if tin is not present, proceed as directed above, estimating antimony sulfide in the presence of a stabilizing gum, or use the following bromate method.

If tin is present, the sulfides are dissolved in concentrated sulfuric acid, the tin forming a stannic compound and the antimony forming an antimonious compound. By the use of a standard solution of potassium permanganate, the antimonious ion may be oxidized to the antimonic ion, the amount of antimony being estimated from the relationship that

$$5Sb_2(SO_4)_3 + 4KMnO_4 + 24H_2O \rightarrow 10H_3SbO_4 + 2K_2SO_4 + 9H_2SO_4 + 4MnSO_4$$

Tin may subsequently be estimated by reducing it to the stannous form by heating with antimony metal and subsequently titrating with standard iodine solution, which oxidizes the tin back to the stannic form.

$$SnCl_2 + 4HCl + I_2 \rightarrow H_2SnCl_6 + 2HI$$

Bromate Method.—In this method [22] antimony is precipitated as the sulfide. It is separated from interferences by the use of polysulfide and then is reprecipitated as the sulfide. After dissolving the sulfide in concentrated hydrochloric acid, the antimony is reduced

[22] Anderson, *Ind. Eng. Chem., Anal. Ed.* 11, 224 (1939).

to the antimonious state by means of sulfite and is estimated by oxidation to the antimonic state by standard bromate solution.

Procedure.—Place the sample in a porcelain crucible, add 1 ml. of saturated sodium carbonate solution and 0.5 g. of magnesium oxide. Dry the mixture in an oven and carefully ash at low red heat, first over a low flame and finally in a muffle oven. Dissolve the ash in 25 ml. of hot, dilute hydrochloric acid. Filter the solution and wash the crucible and filter paper, catching the filtrate and washings in a wide-mouth flask. Pass hydrogen sulfide gas through the solution for 0.5 hour and allow the precipitate to settle overnight. Then filter through paper and wash with hydrogen sulfide water. Return the precipitate with the filter paper to the flask and warm with a solution of sodium polysulfide. Filter and wash well with hot water, catching the filtrate and washings in a 250-ml. beaker. Acidify the filtrate with hydrochloric acid and again allow to stand overnight. Filter the precipitate through a Gooch crucible equipped with an asbestos mat and wash with hydrogen sulfide water. Place the Gooch crucible back into the precipitation beaker; break up the asbestos mat with a glass rod and dissolve the sulfide in 20 ml. of boiling hydrochloric acid (1:1). Filter and wash with hot water. Evaporate the filtrate to a volume less than 50 ml. Transfer to a 50-ml. volumetric flask and make to volume.

Transfer an aliquot portion of this solution to a 100-ml. Erlenmeyer flask. Add 5 ml. of concentrated hydrochloric acid and 20 mg. of sodium sulfite. Boil the solution to remove the sulfite. Titrate the hot solution with 0.005 N potassium bromate solution (0.1392 g. $KBrO_3$ dissolved in water and made up to 1 liter) with a micro- or semimicroburette, using 2-3 drops of 1 per cent methyl orange solution as indicator. One ml. of 0.005 N potassium bromate solution is equivalent to 0.3044 mg. of antimony. Run a blank determination. The blank titration should not exceed 0.3 ml. of 0.005 N potassium bromate solution.

Chlorometric Method.—"Chlorometry" is the term used to designate the quantitative estimation of various substances by use of a standard hypochlorite solution, in a manner entirely analogous to iodometry and bromometry. It is a popular misconception that, because sodium hypochlorite is a highly reactive substance, it is too unstable to be used as a standard titrimetric reagent. Undoubtedly one reason for the lack of enthusiasm among analysts for the use of sodium hypochlorite as a titrimetric reagent is the apparent difficulty of preparing such solutions. It may be prepared in the following simple manner.[23]

---

[23] Goldstone and Jacobs, *Ind. Eng. Chem., Anal. Ed.* 16, 206 (1944).

Preparation of Standard Sodium Hypochlorite Solution.—Transfer 8.0 ml. of a commercial preparation of sodium hypochlorite solution containing 5 per cent of available chlorine to a glass-stoppered brown-glass bottle, and dilute with water to about 2 liters. If necessary, add sufficient sodium hydroxide (1 g.) to raise the pH to about 12.5, the optimum pH for stability. To ascertain if the proper pH has been reached, the customary colorimetric methods for the determination of pH in the range 12 to 14 may be used. Obtain the titer of the solution by titration against a primary standard of sodium arsenite made as follows:

Weigh 0.2473 g. of arsenious oxide (arsenic trioxide, As_2O_3, National Bureau of Standards) and dissolve in 25 ml. of 10 per cent sodium hydroxide solution. Transfer to a 1-liter volumetric flask, make slightly acid with sulfuric acid (1 to 6), and dilute with water to 1 liter. This solution is 0.005 N.

The solution of sodium hypochlorite made as directed above is generally somewhat stronger than 0.005 N. Its exact titer can be determined by titration against the standard arsenite solution. Its normality may be adjusted to exactly 0.005 N by the usual procedure.

Titrimetric Procedure.—Transfer a known aliquot of standard arsenite solution to a 125-ml. Erlenmeyer flask or a 150-ml. beaker: a 4-ml. aliquot if a microburette is to be used for the standard hypochlorite solution and a 5-ml. aliquot if a semimicroburette is to be used. A standard solution of tartar emetic [potassium antimonyl tartrate, $2K(SbO)C_4H_4O_6 \cdot H_2O$] containing 1 mg. of antimony per 10 ml. of solution may also be used. Add 5 ml. of concentrated hydrochloric acid and adjust the volume of the solution to 35 to 40 ml. by adding distilled water. Fill a micro- or semimicroburette with the standard hypochlorite solution. Add 1 drop of 0.05 per cent methyl orange indicator solution to the test solution and titrate directly with the sodium hypochlorite solution. Add another drop of methyl orange indicator solution near the end point and continue the titration until the color of the methyl orange is destroyed. Make a blank titration using exactly the same volume of hydrochloric acid, water, and 2 drops of methyl orange indicator solution, replacing the volume of arsenite or antimony test solution by additional distilled water. The blank should run about 0.12 to 0.14 ml.

Antimony solutions prepared from samples treated as described in previous paragraphs, particularly on page 193, may be estimated titrimetrically as detailed above.

Several precautions must, however, be observed in using sodium hypochlorite solution as a titrimetric reagent. It must be preserved in brown, glass-stoppered bottles. It may be kept at room temperature without deterioration over considerable periods of time. Keeping the solution at lower temperatures is perhaps preferable.

The optimum conditions for the titrations are a volume of at least 35 to 40 ml. with an acid concentration equivalent to 5 ml. of concentrated hydrochloric acid.

Rhodamine B Colorimetric Method.—In this method [24-27] all the antimony in the intermediate or unreactive state (see page 192) is either oxidized to the quinquevalent state by use of perchloric acid at the end of the digestion, or is reduced to trivalent antimony by sulfur dioxide following the destruction of organic matter by the acid digestion and subsequently oxidized to the quinquevalent state by ceric sulfate in the presence of hydrochloric acid. A lake is prepared using rhodamine B; this lake is extracted by a suitable solvent, and the color is estimated colorimetrically or photometrically. If perchloric acid is used in the digestion,[28] it is not necessary to use the sulfur dioxide reduction or the ceric sulfate oxidation.

Procedure.—Add 2 drops of 60 per cent perchloric acid to the water-white acid digest and heat until fumes of sulfur trioxide are evolved. If charring or yellow occurs, it is necessary to add additional perchloric acid but not over a total of 0.5 ml. when 10 ml. of 18 N sulfuric acid is used initially. Cool, add 3 ml. of water, and heat until fumes are evolved. Cool again and place in a cold-water bath. Add 5 ml. of 6 N hydrochloric acid.

Benzene Extraction.—Add 8 ml. of 3 N phosphoric acid, 70 ml. of concentrated acid diluted to 1 liter, and 5 ml. of 0.02 per cent rhodamine B solution, 0.20 g. of the dye dissolved in water and diluted to 1 liter. Shake the flask and cool again if necessary. The benzene extraction must now be performed without delay. Transfer to a separatory funnel. Rinse the digestion flask with 10 ml. of benzene and transfer the benzene to the separatory funnel. Shake 150-200 times, draw off the lower aqueous layer, and transfer the benzene phase to a tube. Allow to stand and settle. The color is stable at this point. Transfer 6 to 8 ml. to a cuvette and read at 565 mμ or use a green filter.

Isopropyl Ether Extraction.—After the addition of the hydrochloric acid add 13 ml. of water and transfer to a separatory funnel. Add 15 ml. of isopropyl ether to the digestion flask, rinse, and transfer to the separatory funnel. Shake about 100 times, discard the aqueous layer. Add 5 ml. of 0.02 per cent rhodamine B solution.

[24] Fredrick, *Ind. Eng. Chem., Anal. Ed.* **13**, 992 (1941).
[25] Maren, *Bull. Johns Hopkins Hosp.* **77**, 338 (1945).
[26] Maren, *Anal. Chem.* **19**, 487 (1947).
[27] Webster and Fairhall, *J. Ind. Hyg. Toxicol.* **27**, 183 (1945).
[28] Freedman, *Anal. Chem.* **19**, 502 (1947).

Shake again 150 times and, after settling, discard the aqueous layer. Transfer the ether layer to a tube. Read immediately at 545 mμ or use a green filter.

Standard Solutions.—Weigh accurately 0.1000 g. of chemically pure antimony and add 25 ml. of concentrated sulfuric acid. Heat until the metal dissolves. Cool and dilute to 1 liter. It is stable and contains 100 micrograms of antimony per ml. It can be diluted to give working standards.

In preparing a standard curve, add known amounts of antimony up to 40 micrograms to 5 ml. of sulfuric acid. Make an acid digestion with nitric acid. Treat with perchloric acid as detailed and then proceed with the remainder of the analysis.

LEAD

Lead bearing sprays and powders were employed extensively as insecticides. Modern procedure determines lead in preference to arsenic because fruit or other food materials so sprayed or coated may be freed more easily from arsenic than lead. Lead is considered far more toxic than formerly and the determination of small amounts is important in biochemical estimations as an indication of lead poisoning. Lead sometimes gets into foods which are processed in lead lined tanks or pipes or that are stored in leaded containers or in containers that are soldered.

Sulfide Method.—The following method based on the precipitation of lead as the sulfide is a comparatively simple one and is adaptable to all types of foodstuffs.

Procedure.—Ash an appropriate amount of the material in a porcelain or silica dish, and after obtaining a clean ash as directed in a preceding section, dissolve the ash in 10 ml. of water and 1 ml. of nitric acid. Filter and make the colorless filtrate alkaline with ammonia. The phosphate precipitate will contain practically all of the lead and the copper will be in the filtrate. Filter, wash well, and dissolve the precipitate in 5 ml. of dilute acetic acid. Make up to 50 ml. in a Nessler tube. Add 5 ml. of hydrogen sulfide solution and match the color with standard lead solutions treated the same way. The lead sulfide precipitate may be stabilized by the use of gums, arabic or ghatti, and may then be estimated colorimetrically against a standard as detailed in the method for antimony.

Dithizone Method.—With the introduction by Fischer [29] of the dithizone method for the determination of lead, numerous variations of this method, among which may be mentioned the colorimetric,

---

[29] Fischer, *Z. angew. Chem.* **42**, 1025 (1929).

mixed colors, and photometric modifications, have appeared.[30-44] These methods have the ability of detecting very small quantities of lead and are based on the formation of a red precipitate of a lead-dithizone complex, which is soluble in chloroform or carbon tetrachloride when an ammoniacal cyanide solution of dithizone is added to a solution containing lead.

Dithizone is the short name for diphenylthiocarbazone. It is the type of reagent which is best used for estimation of low concentrations. It forms green solutions in chloroform. The lead complex has a red color and is soluble in chloroform but is practically insoluble in

$$\text{\LARGE $\langle\!\bigcirc\!\rangle$}-\underset{\underset{\text{H}}{|}}{\text{N}}-\underset{\underset{\text{H}}{|}}{\text{N}}-\underset{\overset{\text{S}}{\|}}{\text{C}}-\text{N}=\text{N}-\text{\LARGE $\langle\!\bigcirc\!\rangle$}$$

Diphenylthiocarbazone

dilute ammonia, whereas dithizone itself is soluble in this ·solvent. Upon these factors, the various methods for the isolation and subsequent determination of lead depend. The nature of the reaction that takes place between dithizone and a metallic salt and the structure of the resulting compound are not definitely known but it is probable that the hydrogen of the NH group adjacent to the phenyl group is replaced by a metal ion and an inner keto complex is formed with the nitrogen of the azo group adjacent to the phenyl group. An enol complex may also be formed in which an additional metal ion is attached to the sulfur atom with a double bond between the central carbon atom and a nitrogen.

Interferences.—Dithizone is not a specific reagent for lead, for it will form colored compounds with 14 other metals. Even in the presence of excess potassium cyanide, stannous tin, bismuth, and thal-

[30] *Methods for Determining Lead in Air and in Biological Materials.* Am. Pub. Health Assoc., New York, 1955.

[31] Wichmann, Murray, Harris, Clifford, Loughrey, and Vorhes, *J. Assoc. Official Agr. Chem.* **17**, 108 (1934).

[32] Vorhes and Clifford, *J. Assoc. Official Agr. Chem.* **17**, 139 (1934).

[33] Winter, Robinson, Lamb, and Miller, *Ind. Eng. Chem., Anal Ed.* **7**, 265 (1935).

[34] Wilkins, Willoughby, and Kraemer, *Ind. Eng. Chem., Anal. Ed.* **7**, 33 (1935).

[35] Willoughby, Wilkins, and Kraemer, *Ind. Eng. Chem., Anal. Ed.* **7**, 285 (1935).

[36] Clifford and Wichmann, *J. Assoc. Official Agr. Chem.* **19**, 130 (1936).

[37] Cholak, Hubbard, McNary, and Story, *Ind. Eng. Chem., Anal. Ed.* **9**, 488 (1937).

[38] Hubbard, *Ind. Eng. Chem., Anal. Ed.* **9**, 493 (1937).

[39] Fischer and Leopoldi, *Z. anal. Chem.* **119**, 161 (1940).

[40] Bambach, *Ind. Eng. Chem., Anal. Ed.* **11**, 400 (1939); **12**, 63 (1940).

[41] Bambach and Burkey, *Ind. Eng. Chem., Anal. Ed.* **14**, 904 (1942).

[42] Schultz and Goldberg, *Ind. Eng. Chem., Anal. Ed.* **15**, 155 (1943).

[43] Gómez, *Rev. Sanidad Asistencia Social (Venezuela)* **11**, 477 (1946).

[44] Jacobs, in Cantarow and Trumper, *Lead Poisoning.* Williams & Wilkins, Baltimore, 1944.

lium interfere. Bismuth is eliminated as an interference by an extra
dithizone extraction [42] from the lead solution before its final estimation by extracting a nitric acid solution of the two metals, which has
been adjusted to a pH of 3.5, with a chloroform solution of dithizone.
Table 8 shows the pH at which dithizone will extract different metal
ions.

TABLE 8. SEPARATIONS WITH DITHIZONE IN CHLOROFORM [45]

pH of Aqueous Solution	Metal Ions Extracted
Less than 2	Noble metals plus Hg
2-3	Cu, Bi, Sn^{++}
4-7	Zn, Cd, Pb, Tl, and all the above
7-10	All the above. Washing with 0.04 ammonia solution removes Sn^{++}. Addition of KCN leaves only Pb, Tl, or Bi, if not previously removed.

[45] Hibbard, *Ind. Eng. Chem., Anal. Ed.* 9, 127 (1937).

Both thallous thallium and stannous tin are converted to the
thallic and stannic states during the evaporation step in the preparation of the sample by oxidation with nitric acid. This minimizes the
possibility of their extraction by dithizone.

If interferences, such as tin, bismuth, or thallium, are likely to be
present, it may be better to use the Wichmann-Clifford electrolytic
method rather than any of the dithizone methods.

Purification.—Commercial diphenylthiocarbazone generally must
be purified before use. Dissolve about 1 g. of the commercial reagent
in 50 to 75 ml. of chloroform and filter if insoluble material remains.
Shake out in a Jacobs-Singer separatory flask, an apparatus designed
to permit multiple extractions with a solvent of lighter specific gravity without disturbing the solvent layer of higher specific gravity,
with four 100-ml. portions of metal-free, redistilled ammonium hydroxide solution (1 : 99). Dithizone passes into the aqueous layer to
give an orange solution. Filter the aqueous extracts into a large separatory funnel through a pledget of cotton inserted in the stem of a
funnel. Acidify slightly with dilute hydrochloric acid and extract the
precipitated dithizone with two or three 20-ml. portions of chloroform. Combine the extracts in a Jacobs-Singer separatory flask and
wash 2 or 3 times with water. Pour off into a beaker and evaporate
the chloroform with gentle heat on the steam bath, avoiding spattering as the solution goes to dryness. Remove the last traces of moisture by heating for 1 hour at not over 50° C. in vacuo. Store the dry
reagent in the dark in a tightly stoppered bottle. Make up the reagent

solutions for extraction to contain approximately 100, 50, and 10 mg. per liter in redistilled chloroform. A stock solution of dithizone in chloroform containing 1 mg. per ml. will keep a long time and is convenient for use in making dilutions.

Rapid Method for Lead Spray on Fruits.—One of these methods is the Vorhes-Clifford [46] rapid method for the determination of lead spray on fruits.

Reagent.—Sodium Oleate Solution.—To 45 ml. of 30 per cent sodium hydroxide, in a 1.5 liter beaker add 400 ml. of water. Add slowly while heating and stirring, 90 g. (by difference from a separatory funnel) of oleic acid. Heat the mixture on a steam bath until the soap is entirely dissolved, cool, dilute to 1 liter and filter.

Procedure.—Weigh 1400 g., equivalent to 10 or more apples or pears, or 350 g. of cherries, pull or cut out the stems and expose the junction of stem and fruit. Trim off the sepals. Allow stems and sepals to fall into a large funnel inserted in the neck of a 500 ml. volumetric flask. To 25 ml. of 30 per cent sodium hydroxide solution in a 600 ml. beaker, add 175 ml. of water and 25 ml. of sodium oleate reagent and bring to a gentle boil. Impale each fruit on a pointed glass rod and immerse in the alkaline solution, rotate slowly until the skin begins to check, remove to the funnel, and rinse with the aid of a wash bottle containing hot nitric acid (1 : 49) or hot hydrochloric acid (3 : 97), if arsenic is to be determined also, being careful to flush out the stem and calyx ends thoroughly and allowing the rinse acid to flow over the stems and sepals in the funnel. When all the fruit has thus been treated, cool the alkaline solution and add it through the funnel to the acid solution in the flask. Rinse the beaker and funnel with any remaining acid and with water, using the entire 250 ml. of rinse acid. Cool and make to volume. In a flask place 10 ml. of concentrated nitric acid or hydrochloric acid to conform to the kind of acid used in rinsing. Thoroughly mix the sample solution, withdraw 100 ml. by means of a pipette and add to the acid in the flask, while swirling vigorously. Filter on a rapid filter. If the first portion of the filtrate is cloudy, return to the filter until a clear filtrate is obtained.

Transfer 20 ml. portions of the filtrate to each of three small glass stoppered bottles. First add 10 ml. of the ammonia-citric acid-potassium cyanide solution [dissolve 10 g. of potassium or sodium cyanide, phosphate-free, and 10 g. of citric acid in 500 ml. of ammonium hydroxide (sp. gr. 0.90) and dilute to 1 liter.] to each bottle. To one bottle add 20 ml. of standard dithizone solution, 20 mg. of purified

---

[46] Vorhes and Clifford, *J. Assoc. Official Agr. Chem.* **17**, 130 (1934).

dithizone dissolved in 1 liter of chloroform, and to the other two bottles 20 ml. of clear chloroform. Shake the flasks vigorously for 1 minute. Transfer the contents of the bottles to Nessler tubes and allow the layers to separate. With a tube of clear chloroform backing the sample tube which contains the dithizone and one sample tube containing chloroform backing each of two standard tubes, compare the color in the lower layer of the sample with that of the standards. If the range is exceeded, repeat with a smaller aliquot of the filtrate making up to 20 ml. with the "blank" solution.

Standards.—The standards for comparison are prepared as follows: Place into each of two 1 liter volumetric flasks, 47.5 ml. of 30 per cent sodium hydroxide. According to the rinse acid, add 100 ml. of concentrated nitric or 104.6 ml. of concentrated hydrochloric acid to each flask. To one of the flasks add 7.27 mg. of lead, from a stock solution containing 2 mg. lead [3.197 mg. lead nitrate, $Pb(NO_3)_2$] per ml. in 1 per cent nitric acid. Mark this flask "standard" and the other "blank." Dilute both solutions to volume and mix. By a combination of the two solutions in suitable proportions the equivalent of any lead load from 0 to 0.02 grain/lb. may be obtained. The following tabulation gives the quantities of "standard" and "blank" to be added to the Nessler tubes for each interval. They are conveniently measured into the tubes by means of a burette.

Grain/lb.	Standard Ml.	Blank Ml.
0.000	0.0	20.0
0.002	2.0	18.0
0.004	4.0	16.0
0.006	6.0	14.0
0.008	8.0	12.0
0.010	10.0	10.0
0.012	12.0	8.0
0.014	14.0	6.0
0.016	16.0	4.0
0.018	18.0	2.0
0.020	20.0	0.0

Then add to each tube 10 ml. of the ammonia-citric acid-potassium cyanide solution, followed by 20 ml. of standard dithizone solution. Shake vigorously for 1 minute, and allow the layers to separate. The color of the lower layer is used as the standard for comparison.

Mixed-color Photometric Method.—In this method [47] lead is extracted with an excess of dithizone in chloroform solution but the

---

[47] Clifford and Wichmann, *J. Assoc. Official Agr. Chem.* **19**, 130 (1936).

excess is allowed to partition between the aqueous and chloroform phases and thus modify the color of the extract according to the relative amounts of lead and dithizone. Because, according to this proportion, a series of colors from red to green could be arranged with intermediate crimsons, purples, and blues, Clifford and Wichmann [47] termed this procedure a mixed-color method. If the extraction is made under definite conditions of volume and strength of dithizone solution and volume and strength and pH of aqueous fraction, the mixed color obtained is definite, reproducible, and, provided excess dithizone is present, depends only upon the amount of lead present.

The transmission spectra of the two components in the dithizone extract, namely, lead dithizonate and free dithizone, show a marked difference in their ability to absorb light of a wave length of 510 $m\mu$, for the red-lead complex absorbs strongly and the free dithizone transmits freely. Consequently when the absorption of light of this wave length by the individuals of a standard color series, measured through suitable cell length, is determined photometrically, a practically linear relation is observed between the amounts of lead and the absorption coefficient.

Preparation of Reagents.—(1) Ammonia-Cyanide Solution.—Add 75 ml. of concentrated ammonium hydroxide solution, sp. gr. 0.9, to 100 ml. of 10 per cent potassium cyanide solution and make up to 500 ml. with distilled water.

(2) Standard Dithizone Solution.—Dissolve 0.125 g. of purified dithizone (see page 200) in chloroform in a 250-ml. volumetric flask and complete to volume with chloroform. Each milliliter is equivalent to 0.5 mg. of dithizone. The standard solutions listed in Table 9 may be prepared from this solution.

(3) Standard Lead Solutions.—Solution A.—Weigh out accurately 1.598 g. of recrystallized lead nitrate, $Pb(NO_3)_2$, and transfer to a 1-liter volumetric flask. Add 10 ml. of concentrated nitric acid and a few milliliters of redistilled water to dissolve the salt. Make to volume with water and mix. This solution contains 1 mg. of lead per milliliter. It is stable and should be used as the standard lead stock solution. It should be discarded if any cloud or sediment appears. Solution B.—Dilute 100 ml. of solution A to 1 liter with water and then dilute 50 ml. of this to 500 ml. with nitric acid (1 : 1,000). One milliliter of this solution contains 0.01 mg. of lead. This solution may be used for standardizing the dithizone in the 0-100 and 0-200 microgram lead ranges. Solution C.—Dilute 50 ml. of solution B to 500 ml. with nitric acid (1 : 1,000). This solution may be used for standard-

izing the dithizone for the 0-5, 0-10, 0-20, and 0-50 microgram lead ranges. Prepare solutions B and C as needed.

Standardization of Dithizone Solutions.—The appropriate volumes and concentrations of solutions specified for the various ranges of lead content and the cell length are given in Table 9.

TABLE 9. DITHIZONE CONCENTRATIONS AND CELL LENGTHS FOR
VARIOUS LEAD RANGES

Lead Ranges, μg.	Dithizone Concentration, Mg./Liter	Volume, Ml.	Cell Length, Ins.
0-5	4	5	2
0-10	4	10	2
0-20	8	10	1
0-50	8	25	1
0-100	10	30	½
0-200	20	30	½

Transfer with the aid of pipettes the required volumes of standard lead solution, 1 ml. of which equals some simple fraction or multiple of 1 microgram of lead, to a series of separatory funnels. Add sufficient nitric acid (1 : 1,000) to bring the volume to 50 ml. For so-called zero lead, use 50 ml. of nitric acid (1 : 1,000). Saturate each mixture with 2 ml. of chloroform by shaking. Allow to stand for a few minutes, swirling the funnel to carry down any globules of chloroform clinging to the side, and draw off the chloroform layer completely, being careful not to draw off any of the aqueous layer. Remove any chloroform in the stem with a pledget of cotton, or use filter paper. Add 10 ml. of the ammonia-cyanide mixture and mix. Immediately add the appropriate volume of dithizone solution as given in Table 9 and shake for 1 minute. Allow to stand for 2 minutes, and then filter the chloroform extract through lead-free filter paper inserted gently into the neck of a dry fifty-ml. Pyrex Florence or similar flask in order to avoid loss of chloroform by evaporation. Rinse the proper absorption cell with a small volume of the dithizone extract and then fill the cell almost to the top of the vent with the extract. Set the cell in the trough of the photometer. Take the average of 5 or 10 readings, which seldom vary more than 2 mm. Plot scale readings against micrograms of lead on a large scale graph.

Procedure.[48]—Evaporate the sample to dryness on a steam bath or hot plate. Transfer the dish to an electric muffle oven and gradu-

[48] Gant, *Lead Poisoning.* Industrial Health, Chicago, 1939.

ally raise the temperature to 500° C. Ash overnight at this temperature. Remove the dish, cool, wash down the sides with 2 ml. of concentrated nitric acid, and evaporate to dryness as before. Replace in the muffle furnace and heat for about 30 minutes or until a white ash is obtained. Remove the dish, cool, add 10 ml. of concentrated hydrochloric acid, and evaporate to dryness on a steam bath or hot plate. Add another 10-ml. portion of concentrated hydrochloric acid and again evaporate to dryness. Remove the dish and, while it is still hot, add 2 ml. of hydrochloric acid and about 20 ml. of hot water to dissolve the ash completely. Transfer to a 250-ml. separatory funnel. Add 10 ml. of 50 per cent citric acid solution to the dish, add a small quantity of hot water, swirl gently, and add to the separatory funnel. Rinse the dish three times with hot water and add the washings to the funnel. Mix the contents of the separatory funnel and add 2 to 3 drops of metacresol purple indicator solution. Adjust the pH to 8.5 with concentrated ammonium hydroxide with the aid of a burette and cool. Generally 7 to 8 ml. is needed. Add 5 ml. of 10 per cent potassium cyanide solution to the dish, rinse into the separatory funnel with water, and mix. The total volume should be about 100 to 125 ml. Add 5-ml. portions of the dithizone solution (20 mg. dithizone per liter of chloroform), shaking between additions until the chloroform extract assumes a purple color. Allow to stand for a few minutes and swirl to shake down the chloroform globules. Draw off the chloroform phase into a 125-ml. separatory funnel containing 50 ml. of nitric acid (1 : 1,000) but permit a drop or two of chloroform to remain in the first funnel. Repeat the extraction of the aqueous phase with 20 ml. of the dithizone solution and combine the chloroform extracts. Shake for 1 minute to strip the lead from the dithizone complex. Discard all but 2 to 3 ml. of the dithizone solution, dilute with 2 to 3 ml. of chloroform and shake for 2 minutes. If the dithizone retains its original green color, bismuth is absent. A trace of bismuth will give the dithizone solution a dirty purple or iridescent blue color, a larger amount of this metal yields a yellowish brown. If bismuth is present, extract repeatedly with excess dithizone, shaking for 2 minutes between extractions until the dithizone retains its original color. Discard the dithizone layer and wash the aqueous portion with successive 2- to 3-ml. portions of chloroform until free from dithizone. Shake down globules of chloroform and allow to stand for a few minutes. Draw off the chloroform layer completely, being careful not to draw off any of the aqueous layer.

Add 10 ml. of the ammonia-cyanide mixture and mix. Add the appropriate volume of standardized dithizone solution (see Table 9) and shake for 1 minute. Allow to stand for 2 minutes and filter through specially prepared filter papers inserted directly into the neck of a 50-ml. Pyrex Florence or similar flask. The lead-free filter papers are prepared by soaking overnight in nitric acid (1 : 100) and then washing with large volumes of water with the aid of a Büchner funnel until free of acid. Rinse out the proper cell with a small amount of the filtered extract and fill it almost to the top. Determine the absorption coefficient, using the standardized dithizone with the same cell used in making the standard curve, and read the amount of lead from the curve or calculate from the factor of the dithizone solution as detailed by the A. O. A. C.[49]

Simple color matching may be made without the use of a photometer by making a series of 10 standards as detailed on page 202, but drawing off the dithizone layers into a series of tubes, vials, or Nessler tubes. The unknown is treated in a similar way and drawn off into a similar tube or vial. View longitudinally for ranges up to 20 micrograms in the flat-bottomed vials and transversely for higher ranges in Nessler tubes. If the range is exceeded, use a smaller aliquot, or re-extract with nitric acid reagent and make standards covering a higher range.

Bambach-Burkey Modification.[50]—*Preparation of Reagents.*—(1) Ammonium Citrate Solution.—Dissolve 400 g. of citric acid in water and add sufficient ammonium hydroxide solution to make the solution alkaline to phenol red. Dilute the solution to 1 liter with water and extract with successive portions of a chloroform solution of dithizone until the dithizone solution retains its original green color. Remove the excess dithizone by repeated washes with chloroform. Sodium citrate may be used in place of the citric acid and ammonium hydroxide. It should be purified in the same way.

(2) Hydroxylamine Hydrochloride Solution.—Dissolve 20 g. of hydroxylamine hydrochloride, $NH_2OH \cdot HCl$, in sufficient water to make 65 ml. of solution. Add a few drops of metacresol purple indicator solution. Add concentrated ammonium hydroxide solution until a yellow color is obtained. Add sufficient 4 per cent aqueous sodium diethyldithiocarbamate solution to combine with all the lead and most of the other metals present and leave an excess of reagent. Extract the excess and the metallorganic complexes with chloroform. Test for complete removal by shaking a portion of the chloroform extract with a dilute aqueous solution of a copper salt. Add distilled hydro-

[49] *Methods of Analysis*, A.O.A.C. Washington, 1945.
[50] Bambach and Burkey, *Ind. Eng. Chem., Anal. Ed.* 14, 904 (1942).

chloric acid until the indicator turns pink, and complete the volume to 100 ml. with water.

(3) Potassium Cyanide Solution.—Dissolve 50 g. of potassium cyanide in sufficient water to make 100 ml. Extract repeatedly with a chloroform solution of dithizone until all the lead has been removed. Excess dithizone in the aqueous layer can be removed by successive washes with chloroform. Dilute with water to give a 10 per cent solution.

(4) Dithizone Extraction Solution.—Shake 1 liter of chloroform with 100 ml. of water containing about 0.5 g. of hydroxylamine hydrochloride, which has been made alkaline to phenol red with ammonium hydroxide. Drain off the chloroform and dissolve 30 mg. of dithizone in it. Add 5 ml. of alcohol to this solution if it is to be kept for several days. Shake the quantity of dithizone that is to be used with 100 ml. of hydrochloric acid (1 : 100) just before use.

(5) Standard Dithizone Solutions.—Prepare chloroform as directed in the preceding paragraph. Filter the chloroform through dry filter paper into a Pyrex bottle equipped with a glass stopper and shield from light with wrapping paper or a wooden box. Dissolve dithizone in the following ratios in the chloroform: 5 mg. per liter for the 0-10-microgram range of lead; 10 mg. per liter for the 0-50-microgram range; and 20 mg. per liter for the 0-100-microgram range. Add 5 ml. of absolute alcohol per liter and hold in a refrigerator.

(6) Buffer Solution.—Transfer 9.1 ml. of reagent nitric acid to a 1-liter volumetric flask and dilute to about 500 ml. with water. Add bromophenol blue indicator and adjust the pH to 3.4 with ammonium hydroxide solution. Add 50 ml. of double-strength Clark and Lubs potassium acid phthalate-hydrochloric acid buffer (pH 3.4) prepared by diluting 50 ml. of 0.2 M potassium acid phthalate and 9.95 ml. of 0.2 M hydrochloric acid to 100 ml., and dilute the entire mixture to 1 liter.

Procedure.—Transfer 15 ml. of ammonium citrate solution to a separatory funnel. Add the proper aliquot of the sample being analyzed and 1 ml. of hydroxylamine hydrochloride solution, mix, and make the mixture alkaline to phenol red with ammonium hydroxide solution. Add 5 ml. of potassium cyanide solution.

Start the extraction of lead with 5 ml. of dithizone extraction solution and note the color to assist in choosing the proper standard dithizone solution to be used later. Less than 10 micrograms of lead is indicated by a greenish-blue color. Add another 5-ml. portion of dithizone extraction solution. Shake, allow to separate, and again note the color to ascertain whether the quantity of lead is greater than 50 or 100 micrograms. Drain off the dithizone and continue the extraction with successive 5-ml. portion of dithizone solutions, noting the color in each instance before draining, until all of the lead is extracted. Wash the combined dithizone extracts with 50 ml. of water, and wash the water with 5 ml. of chloroform. This wash should be green in color; if it is not, the presence of more lead or of zinc is indicated. Add a drop of potassium cyanide solution and shake the funnel again. If the chloroform layer does not become green, the water should be washed at least once with dithizone extraction solution. Add all chloroform washings to the dithizone extract and discard the aqueous layer. Strip the lead from the dithizone extract by shaking with 50 ml. of buffer solution (pH 3.4); if the dithizone solution does not return to its original color, bismuth is present. Drain the dithizone solution from the separatory funnel. If more than 100 micrograms of lead is indicated, discard an aliquot portion of the buffer solution, sufficient to bring the quantity within the 100 microgram range, and make up to 50 ml. again with buffer solution. If bismuth is indicated, shake the buffer solution with one 5-ml. portion of dithizone solution. Drain, add 5 ml. of chloroform, and shake. Allow to stand until the supernatant drop of chloroform evaporates and draw off as much chloroform as possible without permitting any of the aqueous phase to enter the bore of the stopcock.

Estimation of Lead.—Do not allow direct sunlight to strike the solutions. Add the proper standard dithizone solution, described above, to the separatory funnel containing the lead in the buffer solution, using 10 ml. of the 0-10-microgram solution and 25 ml. of the other solutions. Add 7 ml. of ammonia-cyanide mixture and shake immediately for 1 minute. Do not release the pressure that is developed through the stopcock, but permit the gases to escape through the stopper. Flush the stem of the separatory funnel with 2 ml. of the 0-10 microgram standard and with 10 ml. of the other standards and dry the stem. Rinse the photometer cells twice with the test solution, but since the 0-10 microgram cell will hold the entire 8 ml. remaining of the test solution, these cells should be cleaned and dried with acetone after each determination. Use a 5-cm. (2-in. cell for the 0-10-microgram solution; a 1.25-cm. (0.5-in.) cell for the 100-microgram solution; and a 2.5 cm. (1-in.) cell for the 0-50-microgram solution. Read in a photometer and refer to a calibration curve prepared as directed below.

Prepare a fresh lead standard by taking an aliquot of the lead standard solutions prepared as directed above and adjust the pH to 3.4 by addition of dilute ammonium hydroxide. Add the proper amount of 3.4 pH buffer and dilute the mixture to a known volume. Add measured quantities of this prepared standard lead solution to separatory funnels, adjust the volume to 50 ml. with additional buffer solution, and add the dithizone solution to be standardized. Read in a photometer as described above and prepare a calibration curve.

One-color Extraction Method.[51]—In this method the lead is extracted with a small excess of dithizone in chloroform solution and the excess dithizone is removed from the combined extracts by washing with dilute ammonia-potassium cyanide solution. The amount of lead in the extract is then estimated colorimetrically by a comparison of the red color of the lead-dithizone complex.

Preparation of Reagents.—(1) Prepare a 5 per cent solution of ammonium citrate from citric acid by the addition of ammonium hydroxide until just alkaline to litmus paper.

(2) Lead Extractive Solution.—Mix 15 ml. of 10 per cent potassium cyanide solution, 10 g. of potassium cyanide dissolved in water and made up to 100 ml., and 20 ml. of ammonium citrate solution with 53 ml. of concentrated ammonium hydroxide solution, sp. gr. 0.9, and then add 450 ml. of water. This solution is used to neutralize the excess nitric acid and provide the proper pH of from 9.5 to 10.

(3) Dithizone Extractive Solution.—Dilute 5 ml. of 10 per cent potassium cyanide solution and 15 ml. of concentrated ammonium hydroxide to 500 ml. with water.

(4) Standard Lead Solution.—Dissolve 1.598 g. of recrystallized lead nitrate in 0.1 per cent nitric acid and make up to 1 liter with this solvent. One ml. of this solution contains 1 mg. of lead. Dilute 10 ml. of this solution to 1 liter. One ml. of this dilution equals 0.01 mg. of lead.

Procedure.—Prepare a nitric acid solution of the ash of the sample so that 100 ml. of the test solution contains 10 ml. of nitric acid. Remove a 5-ml. aliquot containing approximately 0.5 ml. of concentrated nitric acid. It is important to add exactly 10 ml. of nitric acid and no more. Too little acid will cause the destruction of the dithizone reagent. Too great an amount will lower the pH below 9.7, which is the alkalinity to which the lead extractive solution has been adjusted. This was shown to be the optimum pH by Clifford and Wichmann.[52] The reagents have been adjusted to care for all the commonly interfering metallic ions, but the procedure must nevertheless be closely followed.

Place 15 ml. of the lead extractive solution containing the potas-

---

[51] Harrold, Meek, and Holden, *J. Ind. Hyg.* **18**, 725 (1936).
[52] Clifford and Wichmann, *J. Assoc. Official Agr. Chem.* **19**, 130 (1936).

sium cyanide, ammonium citrate, ammonium hydroxide, and water into a Squibb or pear-shaped separatory funnel. Add the 5-ml. aliquot of the unknown with the aid of a standard pipette. Add dithizone solution, 25 mg. dithizone per liter of chloroform, from a semimicro burette, in 0.3-ml. portions, shaking the separatory funnel after each addition until a slight purple tinge is noticed in the chloroform layer. This purple tinge shows that uncombined dithizone is now present. The dithizone in the chloroform layer turns a bright cherry red when shaken with solutions containing lead. Add chloroform from a burette in sufficient amount so that the total of dithizone-chloroform solution and chloroform is equal to exactly 10 ml. Shake for not more than 10 seconds. Allow the layers to separate. Drain the chloroform layer into another Squibb separatory funnel containing 20 ml. of the dithizone extractive solution, consisting of potassium cyanide and ammonium hydroxide. Shake the chloroform layer with the 20 ml. of dithizone extractive solution, if too great an excess of dithizone has been added. Repeat if necessary to remove the excess dithizone. It has been shown that two extractions are usually sufficient to remove any excess. The color of the resulting chloroform solution is a bright, clear cherry red. The drop at the top of the aqueous layer may be brought down to the rest of the chloroform by repeatedly tapping and shaking with a slight rocking motion. Transfer the chloroform layer into a test tube or comparator tube, first wiping the inside of the stem of the separatory funnel with a cotton swab or pipe cleaner to remove moisture. The test tube or comparator tube should then be stoppered.

Three standard solutions containing 0.005, 0.01, and 0.015 mg. of lead, respectively, are made at the same time that the unknown is prepared for analysis. These standards, after the lead extraction and the removal of excess dithizone, are placed in tubes similar to those used for the unknown. By placing the unknown sample between the two standards that are nearest the unknown in shade, and holding it up in front of a standard source of white light, one is able to determine the lead content within 0.001 mg. This does not hold when more than 0.1 mg. of lead is present in the sample taken for analysis. Best results are obtained when the aliquot taken contains less than 0.04 mg. of lead.

To prepare the 0.005-mg. standard, add 5 ml. of the dilute lead nitrate standard solution to 40 ml. of water and 5 ml. of concentrated nitric acid. Then, after this solution has been thoroughly mixed, take a 5-ml. aliquot and treat it exactly as has been described for the unknown. Use proportionately larger amounts of the dilute lead standard solution for the other standard comparison solutions.

To make up standards for comparison containing 0.045 mg. or more of lead, use the concentrated lead nitrate solution containing 1.0 mg. of lead per ml. Thus, for making an 0.07 mg. standard, add 0.7 ml. of the concentrated lead nitrate standard solution to 5 ml. of nitric acid and dilute to 50 ml. with water. Take 5 ml. of this as an aliquot for analysis. The resultant pH is between 9.5 and 10 under conditions as stated.

When large amounts of iron are present in samples analyzed for lead by the colorimetric dithizone method, fading will occur unless a small amount of hydroxylamine hydrochloride is present as an inhibitor.

Add 2 to 4 drops of a saturated aqueous solution of hydroxylamine hydrochloride to each sample and standard prior to the addition of dithizone. When 200 to 300 times as much iron as lead is present, extract the lead with an excess quantity of very strong dithizone solution (approximately 100 mg. per liter) and then strip the excess with the dithizone extractive solution. The final color is developed after adding 2 drops of hydroxylamine hydrochloride to the solution, which is brought to a pH of 9.5 to 10.0 by the addition of a known quantity of standard dithizone solution, as in the procedure above.

Titrimetric Method.—In this method [53, 54] the lead is separated from a given solution by means of dithizone and the resulting lead-dithizone complex is then isolated. The latter is freed of lead by washing with acid. The chloroform solution of dithizone remaining is mixed with some dilute cyanide solution, which removes most of the dithizone from the chloroform, imparting a brown color to the aqueous layer. A lead solution is added from a burette to this mixture until all the dithizone has been reconverted to lead dithizonate, as indicated by (1) the disappearance of the brown color in the aqueous layer, and (2) the absence of a red color when the aqueous layer is mixed with chloroform and additional lead solution. As the final titration is carried out directly with a known standard lead solution, it eliminates the necessity for special precautions in the handling of the dithizone.

Preparation of Reagents.—(1) Dithizone Solution.—Dissolve 40 mg. of dithizone in 400 ml. of chloroform and filter into a 500-ml. Pyrex separatory funnel. Add 50 ml. of water containing 2 ml. of 25 per cent hydroxylamine hydrochloride solution and shake. Keep in a cool dark place and withdraw the chloroform solution as needed.

[53] Horwitt and Cowgill, *J. Biol. Chem.* 119, 553 (1937).
[54] Moskowitz and Burke, *N. Y. State Ind. Bull.* 17, 492 (1938); *J. Ind. Hyg. Toxicol.* 20, 457 (1938).

The acid aqueous layer not only prevents the oxidation of the dithizone but also extracts any lead that might be present.

(2) Potassium Cyanide Solution, 0.5 per cent. Prepare when needed by diluting 25 ml. of a freshly prepared 10 per cent solution of potassium cyanide to 500 ml. with water. It is important that this solution be lead-free. To insure this, place 100 ml. of 10 per cent potassium cyanide solution into a separatory funnel and extract with 2 ml. of chloroform containing 2 drops of dithizone solution. If a pink color appears in the chloroform layer, withdraw it and repeat the extraction until the chloroform layer is colorless. The slight excess of dithizone which remains in the 10 per cent potassium cyanide solution is insignificant because the amounts which remain after dilution to form the 0.5 per cent solution of potassium cyanide are not detectable.

(3) Standard Lead Solution. Dissolve 1.598 g. of recrystallized lead nitrate with the aid of 1 ml. of nitric acid in a volumetric flask and dilute to 100 ml. This solution, which contains 10 mg. of lead per ml., is stable. By diluting 10 ml. of this solution to 100 ml. and then in turn diluting 10 ml. of the latter to 1 liter, a solution containing 0.01 mg. of lead per milliliter is prepared.

(4) Sodium Citrate, 20 per cent. To 800 ml. of this solution add 8 ml. of 10 per cent potassium cyanide and extract in a 1-liter separatory funnel with 15-ml. portions of dithizone solution until the citrate mixture is free of lead. Wash twice with 25-ml. portions of chloroform, acidify with 4 ml. of 20 per cent hydrochloric acid, and complete the extraction of the excess dithizone with 20-ml. portions of chloroform.

Procedure.—Place the specimen in a silica evaporating dish, evaporate to dryness, if necessary, dry, and ignite in a muffle oven at about 475° C. Remove the dish from the muffle and, if a clean ash is not obtained, add 2 ml. of nitric acid, evaporate to dryness, and ignite again in the muffle. Place the dish on a hot plate, carefully add 15 ml. of 20 per cent hydrochloric acid (1 : 1), and heat until the ash is dissolved. Wash the contents into a 125-ml. separatory funnel with about 20 ml. of hot water. Add 10 ml. of 20 per cent sodium citrate solution and 3 ml. of ammonium hydroxide to the silica dish, mix, and transfer to the separatory funnel with enough water to make a total volume of about 75 ml. Cool, add 1 ml. of 25 per cent hydroxylamine hydrochloride solution, 1 drop of phenol red, and bring to a pH 8.0 with ammonium hydroxide delivered from a Pyrex burette. Cool.

Add 0.5 ml. of 10 per cent potassium cyanide solution drop by drop, shaking between additions, and immediately extract with 0.5 ml.

of dithizone solution and 4 ml. of chloroform. If, after shaking, the chloroform layer does not contain a noticeable excess of uncombined dithizone, add 0.5-ml. portions of dithizone solution, shaking between additions until the green excess becomes evident. Transfer the chloroform phase to another separatory funnel and repeat the extraction of the aqueous phase twice with 0.2-ml. portions of dithizone in 2-ml. of chloroform. To the combined chloroform solutions add an amount of 0.5 per cent potassium cyanide equal to 1.5 times the volume of the chloroform solution and shake for 10 seconds. Transfer the chloroform layer to another separatory funnel and wash the aqueous cyanide solution with 1 ml. of chloroform. Combine the chloroform solutions and again extract with 1.5 volumes of 0.5 per cent potassium cyanide solution. The extraction with cyanide removes the uncombined dithizone unless a very large excess has been used, in which case the extraction with 0.5 per cent potassium cyanide solution is continued until the absence of color in the aqueous phase indicates that the dithizone excess has been removed.

Remove any lead which may have dissolved in the aqueous layer by extraction with 2 ml. of chloroform. Separate the lead from the red dithizone complex by shaking for 15 seconds with 2 volumes of 1.0 per cent hydrochloric acid. Withdraw the green chloroform layer and then extract the acid aqueous solution with 1 ml. of chloroform to recover the last traces of dithizone. Combine the chloroform fractions.

Add to the dithizone solution 0.5 of its volume of 0.5 per cent potassium cyanide and shake. Most of the dithizone goes into the aqueous layer, giving that mixture a brown color. Add the standard lead solution (0.01 mg. per ml.) from a burette a drop at a time, shaking between additions, until only a very faint color remains in the aqueous layer. This is evidence that practically all of the dithizone has combined with lead and gone into the chloroform phase. Discard the red chloroform layer and wash the aqueous layer with chloroform, 2 ml. at a time, until the chloroform layer remains colorless after shaking. Add 1 or 2 drops of the lead solution and shake for 5 seconds. Draw off the pink chloroform solution and continue the extraction with 2-ml. portions of chloroform plus 1 or 2 drops of lead solution until further addition of lead gives no pink color to the chloroform solution after shaking. The end point is a slight pink in the chloroform solution; extraction with 1 drop more results in a colorless solution. In order to facilitate the titration, a solution of the lead-dithizone complex containing a small amount of the order of 1 or 2 drops of the lead solution in 2 ml. of chloroform is kept for comparison. When the color obtained after an addition of lead solution is less than that given by 1 drop of the lead solution, the end point has been attained. It is suggested that the analyst unaccustomed to this analysis add 2 drops (equivalent to about 0.0006 mg.) at a time until his eyes become accustomed to the change.

For ease of manipulation, the aliquot taken for analysis should contain less than 0.050 mg. of lead. A rough estimate of the quantity of lead in the aliquot can be made by observing the volume of dithizone solution used to produce an excess. One ml. of dithizone solution will combine with approximately 0.040 mg. of lead.

The number of milliliters of standard lead solution used in the titration multiplied by 0.01 gives the quantity of lead in milligrams in the aliquot taken for analysis.

"Mush" Method.—An alternative method for placing lead into solution for analysis is termed the mush method. Run the food material through a lead-free sheering food grinder 3 times. Place 200 g. of the material in an 800-ml. beaker, dilute to 300 ml., and add 40 ml. of strong nitric acid. Bring the mixture to a boil and stir until the initial foaming ceases and a comparatively smooth mixture results. Cool, transfer to a volumetric flask, make to volume, and filter. Place 250 ml. of the filtrate in a separatory funnel and extract with dithizone solution. Re-extract the lead with 1 per cent nitric acid from the dithizone solution and determine by a suitable method.

Wichmann-Clifford Method.[55]—The method is based on the electrolytic separation of lead as the peroxide and its titration by iodometric means. The lead is deposited on the anodic, positive pole, by the use of a low electric current. Tin, antimony, bismuth, and manganese interfere with the deposition and must, therefore, be removed. Samples are ashed and precipitated with hydrogen sulfide, using copper as a collector for the lead. The sulfides are filtered, washed with hot polysulfide solution, and finally with sodium sulfate solution. The lead and copper sulfides remaining are then dissolved in hot nitric acid, neutralized with ammonium hydroxide, and made up to 2 per cent acid, with nitric acid. Potassium dichromate solution is added, the solution heated, electrolyzed, and the lead deposited as the peroxide, PbO_2. It is then washed thoroughly and removed from the anode, with a sodium acetate acidic solution. Potassium iodide is added and the liberated iodine titrated with 0.001 N sodium thiosulfate solution, using starch as an indicator.

Reagents.—(1) Polysulfide Solution.—Dissolve 480 g. of sodium sulfide, $Na_2S \cdot 9H_2O$ and 40 g. of sodium hydroxide, NaOH, in water.

---

[55] Wichmann and Clifford, J. Assoc. Official Agr. Chem. 17, 123 (1934).

Add 16 g. of powdered sulfur, shake until the sulfur dissolves, filter and dilute to 1 liter.

(2) Sodium Thiosulfate Solution.—Dissolve 24.85 g. of sodium thiosulfate, $Na_2S_2O_3 \cdot 5H_2O$ in one liter of carbon dioxide free water. Protect the solution with a soda lime tube and thiosulfate trap and allow it to stand for about two weeks. Prepare approximately 0.001 N solution by diluting this reagent with carbon dioxide free water in the ratio of 1:100. Standardize this dilute solution by running known quantities of lead nitrate of the order of 2 to 4 mg. or 0.2 to 0.5 mg. respectively. Prepare the dilute solutions at least every other day, but standardize the solution daily.

(3) Standard Lead Solution.—Lead nitrate may be recrystallized from water to obtain a reasonably pure salt. Dissolve 20 to 50 g. of C. P. lead nitrate in a minimum amount of hot water and cool with stirring. Filter the crystals with suction on a small Büchner funnel, redissolve and repeat the crystallization. Dry the crystals at 100-110° C. to constant weight. Cool in a desiccator and preserve in a tightly stoppered bottle. The product has no water of crystallization and is not appreciably hygroscopic. Prepare a solution containing 10 mg. of lead per ml. in about 1 per cent nitric acid and from this solution make weaker dilutions as needed. Because lead tends to precipitate, probably as a silicate, from very dilute solutions, the weaker dilutions should not be used over long periods of time.

Procedure.—Prepare an ash of an adequate amount of material as directed in the section, "Ash by Ignition." Add 30 ml. of hydrochloric acid (1:1), and heat to boiling. Filter and wash with hot water into a 250-ml. beaker. If a large quantity of unburnt carbon remains, return the residue and filter into the casserole or ashing dish and re-ash. Rinse the ashing dish and then extract the residue with 20 ml. of hydrochloric-citric acid mixture (1:1), the hydrochloric acid containing 20 per cent citric acid. Filter the extract into the same beaker. Place a few pellets of sodium hydroxide in the ashing dish and add 1 to 2 ml. of hot water. Allow the syrupy solution to wet the inside of the dish completely and heat until nearly dry. Take up the residue in a little water and run the alkaline wash directly into the filtrate, so as not to redissolve the silica that may be on the filter. Finally rinse the casserole or ashing dish with a few milliliters of hot hydrochloric acid (1:1) followed by 1 to 2 washings with hot water.

Cool the filtrate. Add 1 ml. of thymol blue, 40 mg. thymol blue indicator in 100 ml. of water, and add ammonium hydroxide solution from a burette until the color changes from reddish orange to a distinct yellow, at which the pH is 2.8. To obtain a more accurate pH, add 4 drops of bromophenol blue, 40 mg. bromophenol blue indicator in 100 ml. of water, and continue adding ammonium hydroxide until the color changes through olive green to a purple color, which is obtained at a pH of 3.8 or greater, The pH range, for the precipitation of lead sulfide, without coprecipitation of iron sulfide, is from 2.5 to 3.4. The color of the solution using these indicators is an incipient purple to olive green at this pH. Adjustment to the correct pH point must always be made from the acid side to prevent precipitation of alkaline earths, aluminum, iron, hydrates, and phosphates.

Add 5 ml. of copper sulfate solution, 10 g. $CuSO_4 \cdot 5H_2O$ dissolved in water and made up to a liter, and pass hydrogen sulfide into the cold solution for 3 minutes and filter the sulfides immediately, preferably on a Jena Glass Filter No. 11 GA, or an equivalent type. This type of filter may be cleaned with sulfuric acid, hydrochloric acid, or 10 per cent sodium hydroxide, followed by reverse flushing with hot water. A light mat of asbestos should cover the filter to prevent the filter from clogging. Dissolve any tin, antimony, or arsenic sulfides from the filter with 5 applications of 5 ml. each of warm polysulfide reagent, prepared as directed above. Wash the filter 4 times with 3 per cent sodium sulfate solution, 3 g. anhydrous Na_2SO_4 in 100 ml. water.

Dissolve the sulfides retained with 5 ml. of hot nitric acid catching the filtrate in a beaker mounted under the funnel, and rinse the filter thoroughly with hot water. Boil the solution until the sulfur is coagulated or oxidized. Neutralize the solution with ammonium hydroxide, add 2 ml. of concentrated nitric acid, and bring the volume to 100 ml. with water. These volumes are preferable when electrodes $1'' \times 5/16'' \times 5''$ overall are used.

The conditions of electrolysis should be carefully regulated. Low acidity, constant speed of the revolving anode, constant elevated temperature, and low current density counteract the interference of phosphates, arsenic, and traces of chlorides, and insure the complete deposition of small amounts of lead as the peroxide. Low acidities are essential for the amounts of lead usually found in foods, circa 0 to 10 mg. of lead. Higher acidities may be used for larger quantities of the metal. Low current densities are also essential. High acidities and high current densities promote the production of nitrites which seriously interfere with the determination of small amounts of lead.

Before electrolyzing, heat the anode to red heat in a Bunsen burner. Heat the solution to about 75° C. over a gas flame and then add 1 ml. of potassium dichromate solution, 100 g. $K_2Cr_2O_7$ dissolved in water and made up to 1 liter, to suppress nitrite formation. Start the current and adjust to 75 to 80 milliamperes and electrolyze, while anode is revolving at 450 r.p.m. Maintain the temperature between 70 to 80° C. Electrolytic apparatus designed so that all the conditions specified may be controlled are available.

The time of electrolysis may vary according to conditions. Efficient stirring shortens the time so that, with proper equipment, 15 minutes is usually long enough. If the speed is not great enough, stir the solution well, and if more than 5 to 10 mg. of lead is expected, increase the time to 20-25 minutes.

If possible arrange to remove the acid by siphoning, at the same time adding water to keep the level of the solution above the deposit on the anode. The acid is entirely removed only when the current drops to zero.

Place the anode in a small flat bottom tube, add 5 ml. of stripping solution, consisting of 5 ml. of acetate mixture (to 20 ml. of a saturated sodium acetate solution add 10 ml. of glacial acetic acid and make up to 100 ml. with water), and 1 ml. of potassium iodide solution, 2 g. of KI dissolved in water and made up to 100 ml. Add a few drops of starch solution and titrate the liberated iodine, with 0.001 N sodium thiosulfate solution using a microburette. Using the electrode as a stirrer, sight through the entire depth of liquid to detect the delicate end point, which is reached when the last faint starch iodine color just disappears. A white base support aids in ascertaining the end point. If the quantity of lead is high, of the order of 1 to 5 mg., which may be noted from the dark appearance of the anode, use a double quantity of the reagents and 0.005 N sodium thiosulfate solution. The starch solution should not be added until the end point is approached. This may be detected when the yellow color of the solution begins to disappear. Calculate the quantity of lead determined from the factor of the thiosulfate solution, previously standardized against pure lead nitrate. The reagents used should be as pure as possible. Every precaution should be taken to have clean apparatus. As a precautionary measure, it is wise to run blanks with every series of determinations.

Combination Method.—As an illustration of a method [56] combining the information of the preceding methods, namely, the mush digestion, dithizone extraction, electrolytic deposition of lead peroixde and iodometric estimation, the following method is appended.

Procedure.—Place 10 g. of the material or other suitable dry weight in a 400-ml. beaker, moisten with water, in order to prevent

---

[56] Cassil and Smith, *Am. J. Pub. Health* **26**, 902 (1936).

combustion when heated with nitric acid, and add 50 ml. of nitric acid. Allow the contents to stand, with a watch glass over the beaker, until the initial action has subsided. Place the beaker over a flame and boil for about 0.5 hour or until the material is well mushed. Dilute the mushed solution when cool, transfer to a 300-ml. volumetric flask, make to the mark, allow to stand for 15 minutes, shaking at short intervals, filter, and take a suitable aliquot, say 200 ml. for the analysis. Add to the aliquot in a separatory funnel 20 ml. of citric acid, 20 g. citric acid dissolved in water and made up to 100 ml., and 8 g. of sodium hexametaphosphate. Make this solution just alkaline with ammonium hydroxide, and add 10 ml. of potassium cyanide solution,. 10 g. potassium cyanide dissolved in water and made up to 100 ml. Extract with small portions of dithizone solution, 50 mg. dithizone in 100 ml. chloroform, until the color of one portion remains unchanged. Drain the successive chloroform extracts into a smaller separatory funnel containing 20 ml. of ammonium hydroxide solution (1:99), and shake as a means of washing the chloroform. When the extraction is complete, drain the combined portions of dithizone-chloroform solution, containing the lead, into a 150-ml. beaker and evaporate to dryness over a steam bath. After the chloroform has been completely evaporated, add 2 ml. of concentrated nitric acid, place a watch-glass over the beaker, and boil until the gases evolved are colorless. Dilute to about 100 ml., heat to 75-80° C., add 2 ml. of potassium dichromate solution, and electrolyze, maintaining the stated temperature as directed in the preceding method. Dissolve the lead peroxide in a mixture of 2 ml. of potassium iodide, 2 g. dissolved in water and made up to 100 ml., and 4 ml. of acid sodium acetate solution, consisting of 20 ml. of saturated sodium acetate solution, 10 ml. of glacial acetic acid and 70 ml. of water. Titrate with approximately 0.001 N sodium thiosulfate solution that has been previously standardized against a known amount of lead as has been described in the foregoing. The very dilute thiosulfate solution may be protected from decomposition for several weeks by adding 1 per cent of amyl alcohol to the boiled water with which it is prepared.

If interferences are likely to be present, as for instance, if canned goods are being examined and tin be possibly present, it is better to proceed with the Wichmann-Clifford method.

Polarographic Method.—A polarographic method for the estimation of lead has been detailed in Chapter II.

Mercury

This metal generally comes in contact with food products because it is a component of some insecticides and fungicides. It is also the metallic constituent of many organic antiseptics and coloring matters, as for instance, merthiolate and mercurochrome.

Reinsch Test.—Mercury may be detected directly in foodstuffs by applying the Reinsch test. This is carried out in a manner similar to the test for arsenic. A piece of pure burnished copper is immersed in a mixture of the material to be analyzed, with 1/5 its volume of hydrochloric acid, and the mixture is allowed to simmer for a number of hours. A bright lustrous mirror is formed on the copper in the presence of mercury. The piece of foil is washed with water, alcohol, and ether, dried, placed in a subliming tube and heated. The mercury will deposit in the cool part of the tube and may be identified under the microscope.

Electrolytic Method.—If small quantities are present, the mercury may be electrolytically determined, either in a nitric acid electrolyte, or using an alkaline sulfide electrolyte. The mercury may be deposited on a gauze cathode while employing rotating anodes with 2.5 amperes of current. The cathode is washed with water and dried in a desiccator. Do not dry mercury deposits in an oven. The cathode may be placed in a subliming tube and then heated. The mercury will sublime and deposit in the cool part of the tube. A more detailed method is the following.

Procedure.[57]—Transfer the sample to a 500-ml. Kjeldahl flask and boil off the alcohol very rapidly for 5 to 10 minutes. Cool, add 10 ml. of concentrated sulfuric acid and then 4 g. of potassium permanganate. Wash down the neck of the flask with water. Allow the sample to digest at a temperature just below boiling for 2 hours. Decolorize with oxalic acid, for which 3 g. is generally required, warming if necessary. If the reaction mixture is not cold, add the oxalic acid in small amounts during this step. Transfer the solution to a 250-ml. wide-mouth glass-stoppered centrifuge bottle, add 1 ml. of 0.5 per cent copper sulfate solution, and then pass in hydrogen sulfide for about 0.5 hour. Stopper the flask and permit the precipitate to settle overnight. Wash by centrifuging. Pass chlorine gas into the centrifuge bottle containing the washed mercury and copper sulfides after the addition of 5 ml. of water. Solution is generally effected in 15 minutes. Aspirate air through the flask in order to remove the excess chlorine.

Transfer this solution to a 50-ml. beaker. Add 2 ml. of a saturated solution of oxalic acid and 5 ml. of a saturated ammonium oxalate solution. Plate out the mercury, using a pure gold cathode, 1 by 3 cm. and ¼ mm. thick. Keep the voltage at about 1.3-1.5 volts. At this

[57] *Methods of Analysis, A.O.A.C.* Washington, 1945.

voltage, 18 to 24 hours is required for complete deposition. The gold electrode can be easily made in the laboratory. Cut a piece of gold foil to the specified dimensions and weld it to a platinum wire by heating the wire and foil in position on an anvil with a Bunsen flame, finally tapping gently with a small hammer. Care must be taken not to melt the gold by excessive heating. Wash the electrode upon which both copper and mercury have deposited with water, alcohol, and ether, successively, dry in a desiccator, and weigh on a micro balance.

After weighing, place the electrode in a Pyrex combustion tube, pass through a stream of hydrogen, heat the tube carefully, and drive off the mercury. Cool in a desiccator and weigh again on a micro balance. The difference in weight represents mercury. Never dry the original mercury deposit in an oven.

Dithizone Method.—Dithizone, diphenylthiocarbazone, may also be used for the estimation of mercury.[58,59] This method is based on the following principles. When a dilute acid solution containing mercury and other metals is shaken with a chloroform or carbon tetrachloride solution of dithizone, the normal green color of dithizone solution changes to a bright orange yellow color attributable to the formation of a soluble organic mercury complex, which approaches two molecules of dithizone to one atom of mercury. (One mg. of mercury reacts with 2.6 mg. of dithizone.) The yellow color persists as long as the mercury is in excess. When sufficient dithizone is added to react with all the mercury present, any excess of the reagent turns the solution green or red or reddish violet, depending upon whether traces of copper are present in the mixture. The fact that mercury under proper conditions will react with dithizone first is the basis of a determination of mercury by titration. High concentrations of copper must be removed before the titration of the mercury. This may be done by the addition of potassium iodide, for in the presence of iodide the copper is extracted with dithizone, whereas mercury remains in the aqueous solution. Mercury cannot be extracted or titrated with dithizone in acid solution when iodides are present but it can be extracted in ammoniacal solution. It can also be extracted from an acid solution containing iodides by the use of sodium diethyldithiocarbamate and chloroform as the extractant.

Procedure.—Digest the sample under a reflux condenser with about 25 ml. of concentrated nitric acid, 2 ml. of concentrated sulfuric acid, and sufficient potassium permanganate until the organic matter is completely destroyed. Add more permanganate and nitric acid if necessary. Remove the excess permanganate and the man-

---

[58] H. Fischer, *Z. angew. Chem.* **42**, 1025 (1929).
[59] W. O. Winkler, *J. Assoc. Official Agr. Chem.* **18**, 638 (1935).

ganese dioxide by the dropwise addition of 30 per cent hydrogen peroxide. Expel the dissolved oxygen by boiling. Cool, add about 0.5 g. of hydroxylamine hydrochloride, and extract the solution by shaking with successive portions of a chloroform solution by dithizone, containing 25 mg. per liter, until it is present in excess, that is, the green color of the dithizone predominates over the yellow of the mercury complex.

Treat the chloroform extract with 50 ml. of water at 50-60° C., 2 ml. of 5 per cent potassium permanganate solution, and 2 ml. of sulfuric acid (1:1). The mercury passes intờ the aqueous layer. Withdraw the chloroform layer and discard. To the aqueous solution, add sufficient 10 per cent sodium or potassium nitrite solution to react with the excess permanganate. Destroy the free nitrous acid left by the addition of about 0.5 g. of hydroxylamine hydrochloride and heat just to boiling.

If large amounts of copper are present, the mercury may be inactivated by the addition of iodide ion. On re-extracting with dithizone, the copper is removed and the mercury is left in the aqueous solution. After the excess iodide is destroyed or the solution made ammoniacal, the mercury may be estimated by the titration procedure described below. Small amounts of copper will not interfere with the analysis for mercury, since only a small percentage of the copper present will be extracted with dithizone from a solution at a pH 2, which is sufficiently acid to permit complete extraction of the mercury present.

Titrate the cool solution in a separatory funnel with a carbon tetrachloride solution of dithizone containing about 1.25 mg. dithizone per liter. Add small portions of this solution until it is present in excess. Use standard mercuric nitrate solution, containing exactly 10 mg. of mercury per liter, for back-titration of the excess. Determine the exact strength of the dithizone solution each time it is used by treating a solution containing 0.1 mg. of mercury as mercuric nitrate in the same manner as described for a test sample. All the reagents used must be as pure as it is possible to procure them, and blank determinations on water alone must be performed to determine the mercury content of the water and the reagents used in the analytical procedure. Subtract the mercury found in the blank from that found in the test sample to ascertain the quantity actually present in the material analyzed.

CADMIUM

Cadmium may occur in food products that have been produced or processed in cadmium plated vessels or molds. It is considered poisonous in small amounts. Cadmium may be detected in the wet ash solution after the total expulsion of nitric acid by precipitation as the sulfide, by separation from copper, if necessary, and reprecipitation as the sulfide.

Sulfide Method.—A method for the detection and estimation of cadmium is based on the separation of cadmium from other metals as the sulfide, with its subsequent estimation by the amount of yellow color produced in a solution containing cadmium when viewed under a mercury-arc lamp.[60] In concentrations of less than 0.1 mg. of cadmium in 50 ml. of solution, differences in the yellow color of cadmium sulfide are indistinguishable in ordinary light, while under the quartz mercury-vapor lamp the yellow color is perceptible in concentrations as low as 0.01 mg. per 50 ml. of solution.

Procedure.—Add sufficient nitric acid to the sample to cover it and heat gently. After the solid material has dissolved, add 10 ml. of concentrated sulfuric acid and add, when necessary, small amounts of nitric acid until oxidation is complete. The method described for arsenic may also be used. Dilute to 75 ml. and add the equivalent of 0.5 mg. of copper and 2 g. of sodium citrate. The copper is added to act as an entrainer or collector. Neutralize the acid solution for the first precipitation with ammonium hydroxide and adjust the concentration of hydrogen ion to pH 3 by means of the indicators thymol blue (thymolsulfonphthalein) and bromophenol blue (tetrabromophenolsulfonphthalein). Saturate with hydrogen sulfide solution for five to ten minutes, add 1 drop of 5 per cent aluminum chloride solution, and allow the solution to stand for 6 to 12 hours. Filter, dissolve the precipitate in nitric acid and hydrochloric acid, and carefully evaporate to dryness. Repeat the precipitation as sulfide twice more, omitting the addition of sodium citrate the last time and adjusting the pH to 2 by means of dilute potassium hydroxide. Carefully evaporate the final solution of chloride to dryness, dissolve it in water, and make up to a convenient exact volume in a volumetric flask. Transfer an aliquot portion of this prepared solution to a Nessler tube for the final reading. To each tube add 5 drops of 10 per cent potassium cyanide, water, 5 ml. of hydrogen sulfide water, and make to volume. Mix thoroughly and compare under a flood of ultraviolet light with standards similarly prepared. The solution should exhibit a bright clear yellow color under the mercury arc. Dark or turbid solutions may indicate incomplete removal of iron.

For larger quantities of cadmium, the following method may be used. Neutralize the wet-ash solution, make slightly acid but suffi-

---

[60] Fairhall and Prodan, *J. Am. Chem. Soc.* **53**, 1321 (1931).

ciently so as to hold all the zinc in solution, and pass in hydrogen sulfide. Yellow cadmium sulfide is precipitated. If copper is present, it may be separated as follows: Redissolve the sulfides in either sulfuric or hydrochloric acid. Add an excess of sulfurous acid to make certain that no oxidizing medium exists and then add N ammonium thiocyanate solution. Copper precipitates as the dimeric cuprous thiocyanate, $Cu_2(SCN)_2$. Filter, wash with cold water. Collect the filtrate and the washings. Precipitate the cadmium in the filtrate with hydrogen sulfide and estimate in a manner similar to lead by comparing with standard cadmium sulfide precipitates in Nessler tubes. This method has a large error, at times, since cadmium sulfide is often contaminated with a basic salt in the hydrogen sulfide precipitation.

In the presence of copper, 1-(2-quinolyl)-4-allylthiosemicarbazide [61] may be used as a precipitant for cadmium. One ml. of a saturated 50 per cent alcohol solution of this reagent with 10 ml. of solution gives a precipitate with 1 p.p.m. of cadmium in the presence of potassium iodide. Zinc, nickel, cobalt, sulfate, and ammonia interfere. The metals may be eliminated by the usual sulfide separation, the sulfates with barium and the ammonia by evaporation.

A method for the determination of cadmium using dithizone is given by Sandell.[62,63] Other methods for the determination of cadmium with the aid of organic reagents such as anthranilic acid quinaldinic acid, naphthoquinoline, and others are detailed by Prodinger.[64]

Detection of Cadmium on Plated Ware.—Goldstone [65] devised a method for the detection of cadmium on plated ware which is performed easily in the field. This method is a refinement and adaptation of the laboratory method proposed by Coleman. It depends on the precipitation of yellow cadmium sulfide in the presence of excess cyanide.

Reagents.—(1) Ammonia-Sodium Nitrate Reagent.—Dissolve 200 ml. of ammonia water (28 per cent) and 100 g. of sodium nitrate in water and dilute with water to 1 liter.

(2) Sodium Sulfide Reagent.—Dissolve 100 g. of sodium sulfide in water and dilute with water to 1 liter.

[61] Scott and Adams, *J. Am. Chem. Soc.* 57, 2541 (1935).

[62] Sandell, *Ind. Eng. Chem., Anal. Ed.* 11, 364 (1939).

[63] Sandell, *Colorimetric Determination of Traces of Metals.* Interscience, New York, 1944.

[64] Prodinger, *Organic Reagents Used in Quantitative Inorganic Analysis. Elsevier,* New York, 1940.

[65] Goldstone, personal communication, 1940.

(3) Potassium Cyanide Reagent.—Dissolve 100 g. of potassium cyanide in water and dilute with water to 1 liter.

Procedure.—To a small pinch of the metal scrapings in a test tube, add 3 ml. of the ammonia-sodium nitrate reagent; bring the mixture to a boil over a flame and allow to stand for a minute or two. Pour the clear supernatant liquid into another test tube, add 1 ml. of the cyanide reagent and, after shaking, add 1 drop of sodium sulfide reagent. This produces a canary yellow precipitate if cadmium is present. The metals, iron, tin, antimony, arsenic, silver, copper, nickel, chromium, zinc, and aluminum do not interfere. In the case of zinc and aluminum, a whitish gray precipitate is formed which is readily distinguishable from the canary yellow color of cadmium sulfide. If cadmium is present in addition to any of these metals, it is instantly detected. The only metals which do interfere are lead and mercury, but these are rarely, if ever, used as plating metals under these conditions.

THALLIUM

Thallium may sometimes contaminate food materials through use as an insecticide and rodenticide. Thallium may be detected spectroscopically by the green color it gives to the flame. It can be precipitated from sodium carbonate solution, in the presence of potassium cyanide, by ammonium sulfide. The precipitated thallous sulfide is soluble in hot 10 per cent sulfuric acid and may be reprecipitated as the chloride (thallous chloride, $TlCl$), as thallous iodide in solutions neutralized with sodium carbonate, and as thallous chromate, Tl_2CrO_4, from neutral solutions.

Small amounts of thallium may be estimated by a method similar to that used for copper by the liberation of iodine from thallic chloride, $TlCl_3$, by the addition of potassium iodide with the formation of thallous iodide, TlI, and free iodine which is subsequently titrated by standard sodium thiosulfate solution. A variation depending on an ether extraction is detailed below.

Thallium may also be estimated volumetrically by oxidation from the thallous to the thallic state in hydrochloric acid solution by the use of standard potassium permanganate or bromate.[66,67]

Extraction-Titration Method.[68, 69]—Destroy organic matter with concentrated hydrochloric acid and potassium chlorate, or with 4 N hydrochloric acid and potassium chlorate, using 0.1 N potassium per-

[66] Marshall, *J. Soc. Chem. Ind.* **19**, 994 (1900).
[67] Zintl and Rienaecker, *Z, anorg. Chem.* **153**, 276 (1926).
[68] Noyes, Bray, and Spear, *J. Am. Chem. Soc.* **30**, 516, 559 (1908).
[69] Reith and Gerritsma, *Rec. trav. chim.* **65**, 770 (1946).

manganate solution as a catalyst, or with nitric and sulfuric acids in the usual manner. In the latter instance it is necessary to add some free chlorine as, for instance, by use of a heated solution of potassium chlorate in 4 N hydrochloric acid.

Transfer the sample containing free chlorine to a Jacobs-Singer separatory flask. Check the reaction with starch-iodide paper. Add an equal volume of ether. Shake vigorously, allow the layers to separate, and draw off the ether layer into a separatory funnel. Add 1 to 2 ml. of sulfur dioxide water to the ether layer and shake vigorously, until the aqueous layer no longer reacts with starch-iodide paper. Adjust the volume of the aqueous layer to about 5 ml. and draw off into an evaporating dish. Shake out the ether layer with 2 ml. of water and add this washing to the evaporating dish. Repeat the extraction of the sample an additional two times with ether. Extract the second and third ether extractions successively with sulfur dioxide water and water. Add each aqueous extract and wash to the evaporating dish, making a total of six additions.

Evaporate the combined aqueous sulfur dioxide extracts on a steam bath in a hood. Transfer, with the aid of a glass rod and a few drops of nitric acid, to a 50- by 18-mm. Pyrex glass tube, add 0.2 ml. of concentrated sulfuric acid, and digest in the customary manner. Wash the evaporating dish with drops of nitric acid, adding the washings to the digestion tube. The digestion may be considered complete when the sulfuric acid remains colorless or a light yellow.

Add 0.8 ml. of water, mix, cool, and filter with suction through a micro filter of sintered glass into precipitation tube 40 by 10 mm. Adjust the volume to 1.8 ml., add 0.1 ml. of a freshly prepared saturated solution of sodium sulfite, $Na_2SO_3 \cdot 7H_2O$, and mix with a glass rod. Add 0.2 ml. of 10 per cent potassium iodide solution and mix. An orange-yellow precipitate indicates thallium. Rinse off the rod and allow the covered tube to stand for 12 to 18 hours in the dark. Centrifuge at 1,500 rpm. for 5 minutes, pour off the supernatant liquid with the aid of a glass rod, and wash the precipitate with 2 ml. of 50 per cent alcohol, stirring the precipitate with the rod, which is rinsed off with a few drops of alcohol. Centrifuge and decant. Wash again with 2 ml. of 90 per cent alcohol. At this point the precipitate may be estimated gravimetrically by the usual micro gravimetric methods, the factor Tl/TlI being 0.6160.

Titrimetric Procedure.—Dry the tube. In the range of 10 to 25 micrograms of thallium add 0.1 ml. of glacial acetic acid and a small drop of bromine. Shake for a moment every 5 minutes until no solid particles are visible and allow to stand an additional 15 minutes.

Transfer the contents of the tube to a 25-ml. flask with not more than 2 ml. of water. Heat until the mixture is light yellow, allow to cool, add 2 M sodium formate solution, prepared by dissolving 24.2 g. of sodium formate, $HCOONa \cdot 3H_2O$, in water and diluting to 100 ml., until the solution is colorless, and then add an excess of 0.2 ml. Mix carefully and moisten the walls of the flask. Allow to stand 5 minutes. Add 2 ml. of 30 per cent sodium chloride solution, 1 drop of 10 per cent potassium iodide solution, 0.2 ml. of 4 N sulfuric acid, and 5 drops of 0.2 per cent starch-indicator solution. Titrate with 0.01 N sodium thiosulfate solution.

Comparison solutions of thallium must be standardized because many salts are of dubious purity. Dissolve 131 mg. of thallous carbonate, Tl_2CO_3, or an equivalent amount of another salt in water and dilute to 100 ml. This is approximately 1 mg. of thallium per ml. Transfer 1 ml. of this solution to a 100-ml. flask, add 10 ml. of water, 0.3 ml. of glacial acetic acid, and sufficient bromine water to give a yellow color and 2 drops in excess. Allow to stand for 15 minutes, and remove the excess bromine with 2 M sodium formate solution. Allow to stand an additional 5 minutes, add 20 ml. of 30 per cent sodium chloride solution, 0.5 ml. of 10 per cent potassium iodide solution, 1 ml. of 4 N sulfuric acid solution, 2 ml. of 0.2 per cent starch-indicator solution, and titrate with 0.01 N sodium thiosulfate solution. One milliliter of the latter is equivalent to 1.022 mg. of thallium.

Lead, mercury, copper, arsenic, antimony, bismuth, and iron do not interfere in this method.

ZINC

Zinc salts are sometimes used instead of copper salts to increase the green color of vegetables. The metal sometimes contaminates powdered eggs that have been dried in galvanized iron containers, because the eggs are scraped from the container and some zinc gets scraped into the food material. If acid foods are processed in galvanized containers or pass through zinc-lined vats or pipes, the food is likely to pick up some of the metal. In these procedures the zinc is separated from other metals by first precipitating the metals having sulfides, insoluble in relatively high acid concentration, the zinc remaining in the filtrate, and then estimating the zinc by precipitating it as zinc sulfide in a buffered faintly acid solution, thus separating the metal from those other metals whose sulfides are soluble in faintly acid solution.

Ferrocyanide Method.—Small quantities of zinc may be estimated by the use of the ferrocyanide method.[70,71] The zinc is freed from iron by the use of cupferron or phosphate. It is precipitated as the sulfide with the addition of copper, if necessary, to act as a collector. The sulfides are redissolved and the zinc is separated from the copper by the usual sulfide separation. Zinc is then converted to the chloride and estimated with ferrocyanide, either by titration using uranium acetate as an external indicator, or nephelometrically.

Preparation of Reagents.—Potassium Ferrocyanide Solution.— Dissolve 3.464 g. of recrystallized potassium ferrocyanide, $K_4Fe(CN)_6 \cdot 3H_2O$, in water and dilute to 1 liter. Allow to stand for a day or two and filter from any residue. This solution may be standardized by running a titration against known amounts of standardized zinc chloride solution as described in the method. One ml. of this solution is equivalent to 1 mg. of zinc.

Standard Zinc Chloride Solution.—Dissolve 1.2446 g. of ignited zinc oxide in a slight excess of hydrochloric acid (1:1). Dilute to 1 liter with water. One ml. is equivalent to 1 mg. of zinc.

Uranium Acetate Solution.—Dissolve 40 g. of uranium acetate [uranyl acetate, $UO_2(C_2H_3O_2)_2 \cdot 2H_2O$] in 800-900 ml. of water. Allow to stand for several days. Filter from any residue into a volumetric flask and dilute to 1 liter.

Procedure.—Adjust the volume of the prepared hydrochloric acid solution of zinc to about 75 ml. containing 10 to 15 ml. of hydrochloric acid (1:1). To precipitate iron, add to the cold solution an excess of an aqueous solution of cupferron (ammonium nitrosophenylhydroxylamine). The iron is completely precipitated when white crystals of the cupferron are noticed on further addition of the reagent. If a colloid forms, shake and stir until it is flocculated. Filter. Partly neutralize with sodium hydroxide solution and add ammonium acetate until the free hydrochloric acid is replaced by acetic acid, as is shown by using methyl orange as indicator. Add 0.5 mg. of copper as copper nitrate to act as a collector and saturate with hydrogen sulfide. Filter. Wash the precipitate with water and finally with hot alcohol. Dissolve the precipitate from the filter with alternate washings of concentrated nitric acid and hot water, catching the filtrate in the original sulfide precipitation vessel, until completely dissolved. Evaporate to dryness. Add 1 ml. of sulfuric acid (1:2) and 2 ml. of nitric acid. Heat in a hood until all traces of organic material have been oxidized and the excess sulfuric acid has been driven off. Dissolve the residue in 5 ml. of hydrochloric acid (1:1) and 20 ml. of

70 Fairhall, *J. Ind. Hyg.* 8, 165 (1926).
71 Drinker, Fehnel, and Marsh, *J. Biol. Chem.* 72, 375 (1927).

water. Heat to boiling and titrate with the standard potassium ferrocyanide solution, using a spot plate and uranium acetate as an external indicator.

Nephelometric Method.—As an alternative method [72-75] the following may be used. Iron is not precipitated but is held in solution by sodium citrate. The mixed sulfides of copper and zinc are precipitated as before but the ionic strength of the salts in the final solution is adjusted so that the sensitivity of the reaction with ferrocyanide is increased.

Procedure.—Adjust the volume of the prepared hydrochloric acid solution of the zinc to about 75 ml., add 5 g. of sodium citrate, 2 mg. of copper as copper sulfate, and a drop of thymol blue indicator. Add dilute potassium hydroxide solution until the solution becomes yellow and then add a drop of bromophenol blue. If the solution is bluish at this point, add dilute acid until the yellow color is just restored. Saturate the cold solution with hydrogen sulfide, filter and wash well to free from iron salts. Dissolve the combined sulfides in nitric acid and hydrochloric acid, dissolve the residue in hydrochloric acid, and adjust the pH of the solution as described above, omitting the use of sodium citrate. This makes the color changes sharper because of the absence of the citrate buffer. Saturate the cold solution with hydrogen sulfide. Filter and wash well. Dissolve the sulfides in 5 ml. of hydrochloric acid (1:1) and 20 ml. of water. Slowly saturate the cold solution with hydrogen sulfide and filter. Copper alone is precipitated and the zinc is in the filtrate. Evaporate the filtrate to dryness and dissolve the residue in 4 to 5 drops of hydrochloric acid (1:1) and a little water. If necessary warm slightly before the addition of water. Transfer to a 25-ml. volumetric flask and make to volume. To an appropriate aliquot, usually 5 or 10 ml. of this solution, add 10 ml. of 0.1341 N potassium hydroxide, standardized against potassium hydrogen phthalate. Carefully neutralize the excess potassium hydroxide with 0.1 N hydrochloric acid, using phenolphthalein as indicator, and then add exactly 1 ml. of acid in excess. Transfer to a 50-ml. Nessler tube. Dilute with water to 45 ml., add 1 ml. of 2 per cent potassium ferrocyanide solution, mix thoroughly, and make to volume. The solution is 0.002 N with respect to acid, and 0.0268 M with respect to potassium chloride. The nephelometric standards in Nessler tubes should be prepared in exactly the same

[72] Fairhall and Richardson, *J. Am. Chem. Soc.* **52**, 938 (1930).

[73] Bartow and Weigle, *Ind. Eng. Chem.* **24**, 463 (1932).

[74] *Standard Methods of Water Analysis* (8th ed.), Am. Pub. Health Assoc., New York, 1936.

[75] Ouzdina and Blajek, *Chimie & Industrie* **37**, 1096 (1936).

way to insure the ionic strength to be the same in the standards and in the unknown. The most suitable range for comparison is that of standards containing 0.20 to 0.25 mg. per 50 ml. matched against solutions of the unknown of nearly the same opacity. The standards for comparison should vary from 0.25 to 0.50 mg. of zinc in steps of 0.05 mg. They may be prepared from ignited zinc oxide as directed in a preceding paragraph.

Rapid Turbidometric Method.—Ash 50 g. of the material in a porcelain crucible, dissolve the ash in 25 ml. of dilute hydrochloric acid, filter, and wash. Pass hydrogen sulfide through the filtrate to remove metals of the first groups, allow to stand overnight, filter, and wash. Boil the solution to remove hydrogen sulfide, neutralize with ammonium hydroxide solution and buffer the solution with 1 ml. of glacial acetic acid, 5 ml. of 50 per cent ammonium acetate solution, and 5 ml. of 20 per cent ammonium chloride solution. Pass hydrogen sulfide through the solution for 0.5 hour and let stand overnight. Filter the precipitate through a hardened paper, taking care that a clear filtrate is obtained.

Return the precipitate and paper to the beaker and boil with dilute hydrochloric acid to dissolve the zinc sulfide, add bromine water to oxidize the iron, and neutralize with excess of ammonium hydroxide solution. Cool, filter, and wash well. If much iron is present, the iron hydroxide must be redissolved and reprecipitated to extract any adsorbed zinc.

Combine the filtrates. Evaporate the filtrate and make up to volume of 50 ml. filtering to obtain a clear solution if necessary. To 5 and 10 ml. aliquots in 50-ml. Nessler tubes add 5 ml. of concentrated hydrochloric acid and 1 ml. of 3.5 per cent potassium ferrocyanide solution and make up to the 50-ml. mark. Compare the turbidity of the liquids against known standards using amounts of zinc from 0.1 mg. to 0.5 mg. for comparison.

Dithizone Method.—The dithizone method [76, 77] provides a fairly rapid means of estimating zinc. The zinc may be brought into solution as described for lead on page 212. The solution is made alkaline with ammonia; a chloroform solution of dithizone is added, the mixture is shaken, and then permitted to stand and separate. If zinc is present, it combines with the dithizone in chloroform and colors it red. The intensity of the color is proportional to the amount of zinc, which should be kept within the range of 0.001 to 0.010 mg. In the method detailed below, the zinc is extracted with the lead, the total amount of dithizone is estimated titrimetrically, and the

[76] Hibbard, *Ind. Eng. Chem., Anal. Ed.* 9, 127 (1937).
[77] Moskowitz and Burke, *N. Y. State Ind. Bull.* 17, 492 (1938).

amount of zinc is calculated by subtracting the volume used in the lead determination.

Reagents.—The reagents are prepared as directed in the method for lead, page 211.

Procedure.—Transfer an aliquot of the solution prepared as directed on page 212, containing less than 25 micrograms of zinc or the equivalent of zinc and lead, to a 125-ml. separatory funnel. Do not add potassium cyanide solution. Note, however, that if copper is present, it will be extracted along with the zinc and lead, and consequently the method should be modified as detailed below.

Add 5 ml. of chloroform and small portions of dithizone. Shake and continue the addition of dithizone until it is present in excess. Draw off the chloroform layer and wash the aqueous phase with a small portion of chloroform containing a few drops of dithizone solution. Combine the chloroform layers.

Wash the chloroform layer at least twice with 3 volumes of dilute ammonium hydroxide solution in each washing. To break the emulsion that may form with the first washing, draw off the clear portion of the chloroform, as formed on standing. Add 1 ml. of chloroform and invert the separatory funnel several times gently. Again draw off the clear chloroform layer as it forms. After the major portion of chloroform has been withdrawn, add an additional 1 to 2 ml. and repeat as above. Avoid too many washings, since some free dithizone will distribute itself between the aqueous and chloroform phases. The chloroform extract should be bright red and should have no free dithizone. The last wash of ammonia water should be colorless.

Shake the chloroform layer with 2 volumes of 1 per cent hydrochloric acid vigorously in order to break the zinc and lead complexes. Draw off the dithizone solution, add half its volume of 0.5 per cent potassium cyanide solution, and titrate with the same standard lead solution as used for the determination of lead (page 213) in exactly the same way. Subtract the volume of standard lead solution equivalent to the amount of lead found in an equal aliquot as that taken for the zinc determination from the volume found above, representing both zinc and lead. Multiply this difference by 3.15 to give the quantity of zinc in micrograms present in the aliquot taken for analysis.

Blank analyses should be run using the water and reagents used in the analysis, and the result should be subtracted from the results obtained in the regular analysis to get the corrected results.

Copper Interference.—To remove copper—which will be an interference in the lead and zinc determination, since copper will also be extracted by dithizone in the absence of cyanide—the test solution must be treated with successive acid and alkaline washes.

Shake the chloroform extract with twice its volume of 1 per cent hydrochloric acid. Repeat the acid extraction. If copper is to be estimated (page 234) retain the chloroform layer, otherwise it may be discarded. Adjust the pH to 8 with ammonium hydroxide solution (1:1), using phenol red as the indicator. Re-extract the zinc, lead, and any residual copper with an excess of dithizone solution and repeat the acid extraction. Adjust the pH to 8, extract again with excess dithizone solution, and wash out the excess dithizone with dilute ammonium hydroxide solution. Decompose the zinc and lead complexes with hydrochloric acid, and titrate the liberated dithizone with standard lead solution in the presence of potassium cyanide.

Sulfide Method.—The foregoing methods have been detailed for the estimation of small quantities of zinc. Larger quantities of zinc may be estimated by the sulfide method.[78] In this procedure the zinc is separated from the other metals by first precipitating the metals having sulfides, insoluble in relatively high acid concentration, the zinc remaining in the filtrate, and then determining the zinc by precipitating it as zinc sulfide in a buffered faintly acid solution, thus separating the metal from those other metals whose sulfides are soluble in faintly acid solution.

Procedure.—Boil the filtrate from the copper determination to expel the hydrogen sulfide and reduce the volume to 250 ml. Add a drop of methyl orange indicator, 5 g. of ammonium chloride and make alkaline with ammonium hydroxide solution. Add hydrochloric acid (1:9) dropwise to faintly acid reaction, add 10-15 ml. of sodium or ammonium acetate solution, 50 g. of salt made up to 100 ml. with water, and pass in hydrogen sulfide until precipitation is complete. Allow the precipitate to settle, filter, and wash twice with hydrogen sulfide water. Dissolve the precipitate on the filter with a little hydrochloric acid (1:3), wash the filter with water, boil the filtrate and washings to expel hydrogen sulfide, and cool. Add a distinct excess of bromine water. Add 5 g. of ammonium chloride and ammonium hydroxide solution until the bromine disappears. Add hydrochloric acid (1:3) dropwise until the bromine color just reappears. Then add 10-15 ml. of sodium or ammonium acetate solution and 0.5 ml. of ferric chloride solution, 10 g. $FeCl_3 \cdot 6H_2O$ in 100 ml. water, or enough to precipitate the phosphates. Boil until all the iron is precipitated. Filter while hot and wash the precipitate with water containing a little sodium acetate. Pass hydrogen sulfide into the

[78] *Methods of Analysis,* A.O.A.C. Washington, 1945.

combined filtrate and washings until all the zinc sulfide, which should be pure white, is precipitated. Filter through a weighed, prepared Gooch crucible and wash with hydrogen sulfide-ammonium nitrate water. Dry the crucible, ignite at a bright red heat, cool and weigh as zinc oxide, ZnO. Calculate the weight of metallic zinc, using the factor, 0.8034.

Zinc sulfide sometimes forms colloidal precipitates which will not flocculate and pass through the filter. Caldwell and Moyer [79] recommend the addition of a solution containing 0.5 to 2 mg. of gelatin of very low ash content. The gelatin solution will produce instantaneous and complete flocculation of as much as 0.3 g. zinc sulfide in 300 ml. of solution.

COPPER

Copper often occurs in foods that have been processed in copper kettles. The copper is oxidized to copper carbonate which is soluble in organic acids. Thus tomato products often are contaminated with copper. Copper salts are sometimes added to intensify the green color of chlorophyll. The views on copper contamination have undergone changes, and it is now accepted that minute amounts of this metal are needed for proper human metabolism.

Iodide Thiosulfate Method.—In the methods detailed below the copper is separated from the other metals by means of a sulfide precipitation and it is subsequently estimated by iodine liberated in the cupric-cuprous iodide reaction.

Procedure.—Weigh 100 g. of the sample into a porcelain dish, about 9-10 cm., and add 5 ml. of a mixture of 5 ml. of sulfuric acid and 95 ml. of 95 per cent alcohol. Burn to a white ash. Add 100 ml. of water after cooling and hydrochloric acid until slightly acid. Try to dissolve all of the ash. Filter and if necessary burn again. Pass in hydrogen sulfide for 15 to 20 minutes. Filter through quantitative paper. Wash. Place the filter paper plus the precipitate directly into a 100-ml. flask, a squat flask with a wide mouth, customarily called a "fat" flask, is preferable and add 3-4 ml. of sulfuric acid and 6 ml. of nitric acid. Place glass hooks on the flask and cover with a watch glass. If the resultant mixture is dark, add nitric acid until it is clear. Evaporate to 1 or 2 ml., cool, add 30 ml. of water and an excess of bromine water. Place the flask on the steam bath until the solution is colorless, after which cool and add ammonium hydroxide solution. In case iron is present, filter and then evaporate off the ammonium hydroxide solution. Make acid with acetic acid

[79] Caldwell and Moyer, *J. Am. Chem. Soc.* 57, 2372 (1935).

and titrate with 0.01 N sodium thiosulfate solution in the presence of about 5 g. potassium iodide and starch solution. Factor on 0.01 N thiosulfate solution divided by 2 multiplied by the titration equals the mg. of copper per 100 g.

The A. O. A. C.[80] gives the following alternative method. Dissolve the ash in hydrochloric acid, neutralize with ammonium hydroxide, add 5 ml. of sulfuric acid, dilute to 200 ml., and boil for 1 minute. Add cautiously 10 ml. hot, saturated solution of sodium thiosulfate and continue boiling for 5 minutes. Filter the precipitate, wash 6 times with hot water, and reserve the filtrate for the zinc determination. Fold the filter paper, place in a crucible, and ignite in a muffle at 500° C. Treat the residue with 1 ml. nitric acid (2:5) and dry on a steam bath. Add 5 ml. of water and evaporate to dryness on steam bath. Add 20 ml. of water and an excess of ammonium hydroxide and heat until the copper salts dissolve. Transfer to a 100-ml. flask. Make acid to litmus with acetic acid (1:1) and add 1 ml. in excess. Boil for 1 minute and cool to room temperature. Add 2 g. of potassium iodide, dissolved in enough water to make the final solution 50 ml., and titrate the free iodine with 0.01 or 0.005 N sodium thiosulfate solution until the end point is nearly reached. Add 2 ml. of 1 per cent starch solution and continue titrating until the color is discharged.

Potassium Ethyl Xanthate Method.—Very small amounts of copper may be determined conveniently by the potassium ethyl xanthate colorimetric method. The copper, separated from other metals as the sulfide, is dissolved in nitric acid and a 0.1 per cent solution of potassium ethyl xanthate is added. The color developed is compared against standards treated the same way.

Procedure.—Separate the copper from other metals as the sulfide. Dissolve in a drop of nitric acid, if possible, otherwise keep the nitric acid down to a minimum. Transfer to a 50-ml. volumetric flask and make to volume. Transfer a 5-ml. aliquot to a Nessler tube containing 10 ml. of a freshly prepared 0.1 per cent solution of potassium ethyl xanthate. Dilute to 25 ml. and mix. Place 10 ml. of the ethyl xanthate reagent into another Nessler tube. Dilute to 15 ml. Add from a 10-ml. semimicroburette, a drop at a time, while continually stirring, a standard copper solution containing 0.1 mg. of copper per ml. To prepare this standard, dissolve 0.3928 g. of copper sulfate, $CuSO_4 \cdot 5H_2O$, in water, transfer to a 1-liter volumetric flask, make to volume, and mix. Continue the addition and stirring until the color in the tube containing the standard copper solution apparently

[80] *Methods of Analysis, A.O.A.C.* Washington, 1945.

matches the color of the unknown. Adjust the volume to 25 ml. and the color to match the test solution as closely as possible. Compute the quantity of copper from the volume of standard copper solution used.

Sodium Diethyldithiocarbamate Method.[81,82]—After the copper has been separated as the sulfide, dissolve in a minimum amount of nitric acid. Evaporate almost to dryness to drive off excess acid if necessary. Dissolve in water, transfer to a volumetric flask, and make to volume. Take a 50-ml. aliquot, add 5 ml. of ammonium hydroxide solution (1:5), and filter if a precipitate forms. Transfer to a Nessler tube and add 5 ml. of a 0.1 per cent solution of sodium diethyldithiocarbamate [1 g. of sodium diethyldithiocarbamate, $N(C_2H_5)_2 \cdot CS_2Na$, dissolved in water and diluted to 1 liter]. Compare the color produced within 1 hour with that of standards treated the same way. The standards may be prepared by diluting 25 ml. of the 0.1 mg. of copper per ml. standard of the potassium ethyl xanthate method to 250 ml. This yields a solution containing 0.01 mg. copper per ml. Convenient standards contain from 0.005 to 0.05 mg. of copper.

Dithizone Method.—Copper may often be present along with lead and zinc. It may be estimated in such mixtures by the dithizone method.

Procedure.—Wash the initial extract of lead, zinc, and copper dithizonates (see page 231) with dilute ammonium hydroxide solution to remove the excess free dithizone. Treat with 1 per cent hydrochloric acid and retain the aqueous layer for the estimation of zinc and lead. Shake the chloroform solution with half its volume of 0.5 per cent potassium cyanide solution and titrate with the standard lead solution (page 213). The presence of potassium cyanide makes it unnecessary to decompose the copper-dithizone complex by acid since this is done by the cyanide. The difference between the volume of lead solution used in this titration and that used for the titration of zinc and lead together, multiplied by 3.07, equals the quantity of copper, in micrograms, present in the aliquot taken for the analysis.

TIN

Sulfide Method.—Canned food products sometimes contain tin, due to the action of fruit or organic acids, which exert a solvent action on the tin. It is best to obtain the tin for quantitative determination by the wet ash process, for heat ashing often yields low

[81] Callan and Henderson, *Analyst* **54**, 650 (1929).
[82] Haddock and Evers, *Analyst* **57**, 495 (1932).

results. The tin is precipitated as stannous sulfide and separated from the sulfides insoluble in polysulfide by solution in polysulfide and filtration. The metal is then reprecipitated as the sulfide, and estimated as the oxide after roasting.

Procedure.—Add 200 ml. of water to the digested sample and transfer to a 600-ml. beaker. Rinse the Kjeldahl flask with 3 portions of boiling water, making a total volume of approximately 400 ml. Cool, and add ammonium hydroxide until just alkaline, then 5 ml. of hydrochloric acid or 5 ml. of sulfuric acid (1:3) for each 100 ml. of solution. Place the beaker, covered, on a hot plate. Heat to about 95° C. and pass in a slow stream of hydrogen sulfide for an hour. Digest at 95° C. for an hour and allow to stand 30 minutes longer. Filter, and wash the precipitate of stannous sulfide alternately with 3 portions each of wash solution and hot water. The wash solution consists of 100 ml. of saturated ammonium acetate solution, 50 ml. of glacial acetic acid, and 850 ml. of water. Transfer the filter and precipitate to a 50-ml. beaker, add 10-20 ml. of ammonium polysulfide, heat to boiling, and filter. Repeat the digestion with ammonium polysulfide and the filtration twice, and then wash the filter with hot water. Acidify the combined filtrate and washings with acetic acid (1:9), digest on a hot plate for an hour, allow to stand overnight, and filter through a double 11 cm. filter. Wash alternately with two portions each of the wash solution and hot water and dry thoroughly in a weighed porcelain crucible. Ignite over a Bunsen flame, very gently at first to burn off filter paper and to convert the sulfide to oxide, then partly cover the crucible and heat strongly over a large Meker burner. Weigh as stannic oxide, SnO_2 and calculate to metallic tin, using the factor, 0.7877.

Tin may also be determined colorimetrically by dissolving the purified stannous sulfide in 2.5 ml. of hydrochloric acid. Place this solution in a test tube fitted with a cork and delivery tube. Add a small piece of zinc and, when it is dissolved, pass in carbon dioxide to replace the air, add 2 ml. of 0.2 per cent dinitrodiphenylamine-sulfoxide in 0.1 N sodium hydroxide solution. Boil the mixture for a few minutes and dilute to 100 ml. Add a few drops of ferric chloride solution. The violet color so obtained may be matched against standard solutions of tin treated the same way.

Thioglycolic Acid Method.—In this method [83] the tin is separated as stannous sulfide, it is redissolved in sodium hydroxide solution, and after being made acid with hydrochloric acid, the color obtained with a reagent containing thioglycolic acid is compared against standards.

[83] De Giacomi, *Analyst* **65**, 216 (1940).

Procedure.—Convert the tin to stannous sulfide as described above and filter. Digest the paper and precipitate with 10 ml. of 10 per cent sodium hydroxide solution on the steam bath for at least 10 minutes. Filter and wash well. Make the solution just acid by adding concentrated hydrochloric acid, add 2 drops of thioglycolic acid and dilute to 100 ml. with water. Take an aliquot of 5 ml. in a boiling tube with 5 ml. of water, 0.5 ml. of concentrated hydrochloric acid and 0.5 ml. of a reagent containing 0.1 g. dithiol [84] and 0.25 ml. of thioglycolic acid dissolved in 50 ml. of 1 per cent sodium hydroxide solution. Immerse in a bath of boiling water for 30 seconds, allow to stand for one minute, and compare with standards treated similarly.

The reagent is best kept in an atmosphere of hydrogen but should be rejected as soon as a white precipitate of disulfide appears. It seldom keeps longer than two weeks.

SELECTED REFERENCES

Curtman, *Qualitative Chemical Analysis*. Macmillan, New York.

Deposition Manual. Wilkens-Anderson, Chicago, 1936.

Furman, *Scott's Standard Methods of Chemical Analysis*. Van Nostrand, New York, 1939.

Jacobs, *Analytical Chemistry of Industrial Poisons*. Interscience, New York, 1949.

McAlpine Soule, *Prescott and Johnson's Qualitative Analysis*. Van Nostrand, New York, 1933.

Methods of Analysis, A.O.A.C., 8th ed. Washington, 1955.

Monier-Williams, *Trace Elements in Food*. John Wiley & Sons, New York, 1949.

Sandell, *Colorimetric Determination of Traces of Metals*. Interscience, New York, 1944.

---

[84] 4-Methyl-1, 2-dimercaptobenzene.

CHAPTER VI

CHEMICAL FOOD POISONING

Food poisoning is a common occurrence, yet most of these cases are of bacterial origin rather than of chemical origin. One of the distinctions between chemical types of food poisoning and food poisoning of bacterial origin is that mass poisonings with chemicals are rare, although not unknown. Thus in Cuena, Ecuador, in May, 1957, about a thousand persons were poisoned by eating bread made from arsenic-contaminated flour. There were, however, no fatalities. Another difference in bacterial and chemical types of food poisoning is that in the latter the onset of illness is generally much more sudden.

That illness has occurred because of the presence of metallic and nonmetallic chemical contaminants in ingested foods is unquestionable. In some instances the symptoms exhibited simulate so closely those developed in bacterial infections that poisoning by these substances is not suspected until after a chemical investigation has been made. The fact that foods containing metallic and nonmetallic contamination may be deleterious and injurious to health has been recognized in the Food, Drug, and Cosmetic Act of 1938.

Among the metals as noted in the preceding chapter which have been incriminated in causing illness as a result of the ingestion of food may be mentioned: arsenic, lead, antimony, cadmium, selenium, mercury, and thallium. Some investigators attribute illness to the ingestion of tin, zinc, and copper. Inorganic compounds which have caused poisoning are hydrocyanic acid, fluorides, sulfur dioxide, ammonia, and in general, rodenticides, insecticides, and fungicides. Among the organic compounds frequently mentioned as the cause of food poisoning we may consider phenol, formaldehyde, glucosides, methyl chloride, organic refrigerants, organic insecticides, and diethylene glycol and other glycols. The role of arsenic as a poison in foodstuffs is, as has been mentioned, of historic importance.

ROLE OF POISONS

The exact role that metallic and nonmetallic compounds play in poisoning has not been completely evaluated. We do know, however, that the inhalation and injection of poisonous compounds as a general rule produce far more serious effects than the ingestion of these compounds. This is so because a poison which is inhaled or injected enters the blood stream and is very rapidly pumped over the entire body. On the other hand, a poison which is ingested must first undergo the regular processes of digestion before it is absorbed by the blood.

The damage arising from the ingestion of a poison may be local or remote, depending upon whether the material is a protoplasmic poison like hydrogen cyanide, whether it is caustic in its action, or whether it is absorbed into the blood stream to be carried to other points to exert its evil effect or be deposited to exert its effect at some later time.

Acute and Chronic Poisoning

Poisoning attributable to the ingestion of contaminated foods also falls into the general classification of acute and chronic poisoning. The first is induced by relatively massive doses of a harmful substance. The latter is the result of repeated small doses. A typical example of an attack of acute poisoning attributable to the ingestion of a harmful compound is that of lemonade containing antimony. Typical examples of chronic poisoning are those in which food or water contaminated with lead is repeatedly ingested and the repeated consumption of seleniferous plants.

Although the effects of acute poisoning are more readily apparent, as evidenced by sudden gastro-intestinal distress and other severe symptoms, the deep-seated effects of slow and chronic poisoning are often much more damaging. The chances of recovery from acute poisoning, if it is not lethal, are greater than the chances of recovery from chronic poisoning.

The degree of toxicity of the compounds ingested along with foods is dependent upon a number of factors among which may be mentioned the chemical combination of the compound ingested, its solubility in the fluids of the body, the length of time in contact with body fluids, the quantity ingested, and the quantity present in the circulation at a given time. The susceptibility of a particular individual is also important. For instance, the ingestion of metals in the form of their compounds is generally more toxic than the ingestion of the metal itself. Thus zinc chloride is far more caustic than zinc metal. Litharge, lead monoxide (PbO), is far more soluble in blood serum than in water; 1,152 mg. of litharge is soluble in a liter of serum, whereas a liter of water will only dissolve 17.1 mg. Metals in chemical combination with proteins are generally less toxic than in simple inorganic combinations. Indeed, toxicologists have noted the modifying action that proteins exert in cases of lead poisoning. Flinn and Inouye [1] noted that animals tolerated much larger quantities of harmful metals when ingested along with their food.

OCCURRENCE OF METALLIC AND NONMETALLIC CONTAMINANTS IN FOOD

Metals, nonmetallic, and organic contaminants may be present in foods in small amounts and even in poisonous amounts, naturally. Indeed, some foods are toxic because of the poisonous amount of substance that they contain. In addition to the normal amount of metal present in a foodstuff, we may find deleterious quantities of metal or other injurious substance because of a variety of forms of contact. Poisonous metal or nonmetallic compounds may be present in a foodstuff because of

(1) Their presence as natural components;
(2) The use of insecticides, rodenticides, fungicides, germicides, antiseptics, and other economic poisons;
(3) Solution of metal or compound from the utensils or container in which a food is prepared or processed, or stored;
(4) Contact during a manufacturing process, or during fumigation, insect extermination, or cleaning;
(5) Deliberate use for a fraudulent purpose;
(6) Deliberate use as a component in a processing step;
(7) Accident;
(8) Criminal intent.

Poisons as Natural Components of Foods.—Some authorities are reluctant to admit the possibility that some foods naturally contain poisonous quantities of metallic components for to many the terms "natural" and "pure" are synonymous. They may readily agree to the poisonous compounds contained by mushrooms, rayless goldenrod, etc., but are not so sure of poisonous amounts of metallic contaminants. An excellent example of the presence of toxic amounts of metal in vegetation is the cause of the illness known as "alkali disease." This is a broad term commonly used in the West for a number of disorders which are often associated with an illness caused by the consumption of seleniferous vegetation. The existence of large areas of land containing sufficient selenium to produce toxic vegetation has been demonstrated in the area surrounding the Black Hills, in western Colorado, in portions of the valleys of Uncom-

---

[1] Flinn and Inouye, *J. Am. Med. Assoc.* **90**, 1010 (1928).

pahgre, Gunnison, and Colorado rivers, in a portion of western Kansas, and in certain portions of Montana.

Other examples that may be cited are the presence of relatively large amounts of oxalic acid in spinach and rhubarb leaves. The amount of oxalic acid present may range from that sufficient to make the calcium content of these plants nutritionally valueless to toxic amounts. Fatal cases arising from the consumption of almonds and rangoon beans containing excessive amounts of glucosides which yielded cyanides or hydrolysis are further instances of foods naturally containing poisonous amounts of chemicals.

Use of Insecticides, Fungicides, Germicides, and Rodenticides.— Fruits and vegetables are important to man. They are also important to rats, to insects, to molds, and to other microorganisms. There is an eternal war being waged to see which group, man, rats, insects, or microorganisms, will win these materials for food. In order to prevent the attack of pests, fruit, vegetables, and other food products must be protected. One of the means of protection is the use of rodenticides, insecticides, fungicides, germicides, and antiseptics. These are, generally, chemical substances that kill or inhibit the growth of rats, insects, fungi, bacteria, and other microorganisms. Some of these substances are cheaper than others; some have better properties than others. An ideal rodenticide, insecticide, or fungicide would probably be one that would be perfectly harmless to man, that would prevent rodent attack, insect infestation, or the attack of other organisms at the proper time, and when that danger was over would be capable of being easily removed, or washed completely away in a simple manner. Unfortunately killing rodents, insects, and preventing the action of other organisms is not a simple matter. Most rodenticides are poisonous to human beings. There are some agents lethal to insects and apparently harmless to man. This harmlessness is more often apparent than real, for instance, the insecticide pyrethrum is stated to be harmless to man, but it actually causes allergic conditions in many who use it. There are holders of exterminator permits in the City of New York who refuse to use pyrethrum because of its effect upon their respiratory tract.

There are other agents that are toxic both to insects and to man. Some agents will kill some insects readily and other insects slowly or not at all. For these reasons the agents which will kill most insects are preferred as the means of combatting these pests. Unfortunately these are the type that are poisonous to human beings. They consist in the main of lead arsenate and copper arsenate. Other insecticides and fungicides using selenium and fluorine are also used. In more recent years the insecticide DDT, dichlorodiphenyltrichloromethane, technically 2,2-bis-(p-chlorophenyl)-1,1,1-trichloroethane, has been used on a large scale but this substance, it has been established, causes toxic effects in human beings, although under proper conditions of use no danger exists.

The spray residue problem is serious, but while it is not the author's wish to minimize the seriousness of the problem, it is not his wish to exaggerate it either. Spray residue is the term applied as a class name to the insecticide or fungicide that remains on the fruit or vegetable after the rains, washing, and other cleaning process that the fruit or vegetable undergoes.

Often rodenticides and insecticides are the cause of accidental poisonings. Some discussion will follow in that section below.

Solution of Metal or Compound from Utensils.—It is well known that foods may become contaminated with metal by solution of the metal from the utensils in which it is made or from the container in which the food is stored. Thus illness has been caused by eating ices prepared in cadmium plate forms; by drinking lemonade prepared in enamelware having antimony opacifiers; by eating meat prepared in galvanized iron utensils; etc.

In most instances where a food is badly contaminated by a metal because of solution of that metal, the food is acid in reaction. Some metals such as cadmium, iron, and zinc are readily soluble in acids, even organic acids. Others such as copper are not. However, some of these insoluble metals form compounds which are readily soluble in acid. For instance, copper kettles form a layer of copper carbonate. This compound is easily soluble in weak organic acids.

Enamelware sometimes contains antimony opacifiers. Some of the antimony compounds used for this purpose are more readily soluble than others in acids. Sufficient antimony may actually be dissolved by such an acid as citric or tartaric to cause illness.

Solution may not necessarily take place only in acid media. Alkali will dissolve some metals such as aluminum and zinc. Lead and its compounds are comparatively easily soluble in salts such as ammonium acetate. Very likely there are similar salts in foodstuffs which have this power.

Metals may also dissolve because of local electrolytic effects. This is probably one of the causes for the etching and solution of tin and iron by foods packed in tin cans. The application of lacquer to the tin can inhibits and diminishes this electrochemical effect but does not entirely eliminate it because the food may have limited contacts with iron and tin surfaces where minute abrasions in the enamel coating unavoidably expose these metals to the food. The amounts of these metallic salts acquired by the food will depend upon the character of the food. In general, acid foods tend to take up more metal especially after the can is opened and air can also act on the container.

The amounts of tin and iron salts normally present in commercially canned foods are generally without significance as far as possible hazards to consumer health is concerned.[2]

An interesting case arose as follows: Some refrigerator units had been replated with cadmium, a metal which is prohibited by the Sanitary Code of the City of New York for use in the manufacture, sale, or keeping of any drink, beverage, or food, any tap, faucet, fountain refrigerator, utensil, vessel, etc. In one of these units a microscopic orifice permitted the refrigerant sulfur dioxide to escape from the expansion coils. The sulfur dioxide dissolved in the surrounding ice and the sulfurous acid formed dissolved the cadmium. This dripped into the water being frozen for ice cubes. When the ice cubes were used for their usual purpose, the people consuming the beverage were made ill.

Another case of interest concerned copper poisoning. In the summer of 1945, four girls employed by a small radio manufacturing company became ill and complained of nausea and vomiting. An investigation disclosed that after taking a soft drink from a dispensing machine at the plant the four girls developed these symptoms in 5 to 15 minutes.

This machine had two compartments, one for the syrup and the other for the water. The latter compartment was lined with copper. A tank of compressed carbon dioxide gas was part of the unit. Upon the insertion of a coin, there was a flow of water from the water compartment and this was charged with the carbonic acid of the tank. Simultaneously, some of the syrup flowed into the glass. A valve intended to protect the water compartment from the entrance of carbonic acid gas was found to be defective, and as a result carbon dioxide entered the water compartment. The acid mixture was sufficient to dissolve some of the copper so that the water actually turned a greenish color. A chemical analysis of the water showed the presence of 85 p.p.m. copper; thus a glass of this drink contained a quantity sufficient to cause toxic symptoms.

By Contact During a Manufacturing Process.—Metals and compounds of a poisonous nature may become incorporated into foods during a manufacturing process, or during fumigation, insect extermination, or cleaning. Solution of the metal in such instances plays the same role as described in the previous section.

---

[2] *Canned Food Reference Manual,* American Can Co., New York, 1939.

A situation illustrative of the type where a food poisoning occurred due to fumigation follows. In 1937, a suspected food poisoning outbreak involving several members of a family who had partaken of uncooked raisins, among other articles of food, was investigated by the Federal Food and Drug Administration. The circumstances were such that suspicion was directed toward the raisins as the possible cause of the illness. A chemical examination of the remaining portion of the package of raisins disclosed that the fruit contained approximately 3,000 p.p.m. of hydrocyanic acid.

In the search for an explanation for this condition, it was learned that a fumigation had been carried out under emergency conditions on packaged raisins which had been stored for an unusually long period pending settlement of a maritime dispute. The raisins involved in the food poisoning outbreak were a portion of a very large lot which had been fumigated in this manner. The method of fumigation employed and the amount of hydrocyanic acid used were presumed, based on previous experience, to yield satisfactory results. The suggestion was made that the failure of an atomizing nozzle to function properly resulted in wetting some of the packages with liquid hydrocyanic acid. Because of the packaging and possibly because of other factors, conditions arose which prevented the evaporation of the fumigant which would normally be expected during subsequent aeration.

Cyanides may also contaminate foodstuffs when they are used as silverware cleaning polishes. These polishes are generally forbidden but are sometimes used in spite of such prohibitions.

Through Deliberate Use for a Fraudulent Purpose.—Poisonous compounds may become incorporated into foods through deliberate use for a fraudulent purpose. The user may not, of course, know that he is using a poison. As examples of the use of substances for fraudulent purposes we may mention the use of lake dyes for coloring foods; aminoazotoluene

as a color for butter (this substance has been shown to have carcinogenetic properties), other carcinogenetic agents that may sometimes be used are aminoazobenzene, $C_6H_5N:NC_6H_4NH_2$, (CI 15, aniline yellow), dimethylaminoazobenzene or butter yellow (CI 19, N,N-dimethyl-p-aminoazobenzene) N-methyl-p-aminoazobenzene, and 4'-methyl-4-aminoazobenzene; (actually CI 19 has been isolated from several food products such as salad dressings and mayonnaise) ; mineral pigments such as chrome yellow, Prussian blue and ochres, although such use is very rare in these times; potassium aluminum sulfate and similar astringents to give tang to pickles; and the addition of copper or zinc salts to vegetables such as green peas or fruit as citron to intensify the green color of chlorophyll.

Through Deliberate Use as a Component in a Processing Step.—Diethylene glycol was used as a solvent in many artificial fruit flavors. It was not until the many deaths which were attributed to the use of Elixir Sulfanilamide in which diethylene glycol was used as the solvent for sulfanilamide brought the dangers of this solvent so dramatically to the attention of the country that it was realized that it was widely used in foods and had even been recommended for such use.

In this category we can place the use of chemical preservatives such as formaldehyde, monochloroacetic acid, borates, chlorine, sodium benzoate, salicylic acid, the esters of p-hydroxybenzoic acid and the like. Particular attention should be called to monochloroacetic acid. The use of this material is generally considered illegal, and poisonings have been traced to its use in an orange beverage base. The use of some of these substances is permitted in certain countries when declared. Excessive amounts may, however, cause illness, and they should be kept to a minimum whenever used. Attention should be called to the increasing use of the quaternary ammonium compounds such as the alkylbenzyldimethyl ammonium chlorides, Roccal, Zephiran, and similar surface active agents. Some of these are considered toxic, and their use in foods will probably be prohibited. Nitrogen trichloride, a bleaching and maturing agent for flour, apparently causes epileptiform seizures in dogs.

One other example in this category is the improper use of mineral oil as a food ingredient. This material has no food value and causes illness through its deleterious effect on the assimilation of vitamins A, D, K, and carotene.

By Accident.—One may consider the contamination of raisins during fumigation as cited in a previous section an illustration of accidental commercial contamination. A more striking illustration is that of household contamination. In one instance, a woman mistook a white insecticide powder containing sodium fluoride for flour and used this powder in mixing a batter for potato pancakes. She consumed about eight and died within 24 hours. Two others ate a few of the pancakes, were made ill but fortunately recovered. Serious cases of fluoride poisoning resulting in death occurred in an asylum in Oregon and in a family in Baltimore.

The City of New York in an attempt to reduce such accidents was the first to introduce the practice of coloring insecticides nile blue or microcline green.

A case occurred in New York City in which 11 men were made seriously ill as a result of eating oatmeal contaminated with sodium nitrite. Salt shakers had been accidentally refilled with a meat curing composition containing 92 per cent of sodium nitrite. Several salt shakers contained sodium chloride only. Some had faint traces of sodium nitrite. Others contained as much as 0.137 per cent sodium nitrite. Some of the salt shakers probably had been partly filled with the curing compound. The 11 men who became ill unfortunately used the shakers containing large amounts of the nitrite. The ingestion of the nitrite in the shakers plus the nitrite in the oatmeal caused the toxic symptoms.

From time to time instances occur in which persons, often children, become ill or die from the ingestion of a rodenticide like zinc phosphide or sodium fluoroacetate.

Criminal or Malicious Intent.—While we know well enough that poisons have been added to foods with criminal intent, such cases are really the province of the forensic chemist. Since any such criminal addition, especially of the common poisons, will in some measure simulate the contamination of a food from the sources previously mentioned, it is not necessary to dwell longer on this mode of contamination.

DETECTION AND DETERMINATION

In Chapters III, IV, and V concerning methods for the detection and estimation of coloring matters, preservatives, and metals in foods, methods have been given in detail for the detection and determination of various poisons. These methods have great value. They are, however, often time-consuming and, when an emergency arises, do not lend themselves to the rapid evaluation of a food as to its toxic quality. For this reason a systematic method for the chemical analysis of foods from a toxicological point of view is valuable.

Classification of Poisons.—The various substances poisonous to man have been classified into different groups by several toxicologists. Most place such poisons into four groups, namely, (1) volatile, (2) metallic, (3) alkaloidal, and (4) nonalkaloidal poisons. Gettler uses a much more elaborate classification: (1) poisonous gases: carbon monoxide, hydrogen cyanide, hydrogen sulfide, nitrogen oxides, etc.; (2) volatile poisons: acetone, alcohols, aniline, benzene, camphor, carbon disulfide, chloral hydrate, chloroform, croton oil, cyanides, formaldehyde, nitrobenzene, phenols, phosphorus, pyridine, ricin oil, etc.; (3) acid-ether soluble poisons: acetanilide, antipyrine, barbiturates, benzoic acid, caffeine, cantharadin, colchicine, polyhydroxy phenols, oxalic acid, phenacetin, picric acid, picrotoxin, salicylates, santonin, sulfonal, etc.; (4) alkaline-ether soluble poisons: aconitine group, atropine group, poisonous amines, cinchonine, cocaine group, codeine, delphinium group, emerine group, hydrastine group, lobeline, narcotine, nicotine group, pilocarpine, piperazine group, quinine, strychnine group, taxine, veratrine, yohimbine, etc.; (5) ammonia-ether or ammonia-chloroform soluble poisons: apomorphine, morphine, papaverine, theobromine, theophylline, etc.; (6) metallic poisons: antimony, arsenic, barium, chromium, copper, lead, mercury, radium, thallium, uranium, zinc, etc.; (7) mineral acids and alkalis: hydrochloric acid, nitric acid, sulfuric acid, ammonium hydroxide, sodium hydroxide, potassium hydroxide, sodium carbonate, etc.; (8) halogens and their salts: fluorides, bromides, iodides; (9) salts of oxy-acids: borates, chlorates, nitrates, nitrites, etc.; (10) poisons isolated by special methods: anthraquinine derivatives, ergot, glucosides, muscarine, saponins, nitroglycerine, strophanthin, etc.

The routine toxicological examination made by Goldstone [3] on foods and allied products provides for the detection of the following substances:

Volatile	Metallic	Alkaloidal	Anions
Cyanides	Arsenic	Atropine	Borates
Yellow (white)	Antimony	Brucine	Oxalates
phosphorus	Bismuth	Cocaine	Fluorides
Zinc phosphide	Mercury	Codeine	
Phenol	Lead	Heroin	
Methanol	Cadmium	Morphine	
Formaldehyde	Thallium	Pilocarpine	
	Zinc	Strychnine	
	Barium		

A schematic procedure, in general falls short of being all-inclusive. It can provide only for the detection of a limited number of all the foreign substances which might possibly find their way into a food product, but experience shows that when contamination does occur, it is confined almost invariably to a relatively small number of toxic substances, generally the active principles of commercial insecticides, rodenticides, or disinfectants which are usually found

[3] Goldstone, *Anal. Chem.* **21**, 781 (1949).

in a household, retail establishment, or processing plant and which get into the food as detailed in a prior section.

Sometimes the preliminary organoleptic examination offers a clue to the presence of an unusual contaminant, whose characteristic odor, appearance, or taste is easily recognizable. Analytical procedure is then focused on direct isolation and detection of the suspected substance. The history of the case, which should always be available to the analyst, may also provide a valuable base for analytical attack.

Aside from the obvious saving of time and expense resulting from the use of a systematic analytical procedure, it is of especial value where the size of the sample is limited to only a few small scraps left over from the meal supposedly responsible for the toxic symptoms.

Goldstone Method for Chemical Food Poisons [3]

Organoleptic Examination.—*Appearance.*—Examine the material with a magnifying glass for the presence of extraneous substances such as glass fragments, wood splinters, mold, or powdery substance not characteristic of the type of food under consideration. Evidence of fermentation or decomposition may indicate bacterial contamination as the toxic causative agent. In the case of canned goods take note of the physical condition of the container. Defects such as spring or swell, rust marks, pin-hole leaks or corrosion of internal metal surface may be relevant.

Odor.—The presence of toxic substances such as yellow phosphorus, cyanides, phenols, and industrial solvents, even in very low concentration is often revealed by their characteristic odors. Rancidity and sour odors commonly result from oxidative degradation and, at times, from bacterial contamination.

Taste.—It is inadvisable because of personal health hazard, and for esthetic reasons, to taste specimens, but if the sample is from a bulk source, the hazard is greatly reduced and tasting may occasionally provide a clue.

pH Value.—Macerate 0.5 g. of sample in 2 ml. of distilled water in a test tube. Allow the suspended matter to settle, decant a few drops of the supernatant liquid on to a white spot plate, and add one drop of universal indicator. Compare against standard buffer solutions to which one drop of indicator has been added.

Phenol.—*Reagents.*—Gibbs' Reagent.[4]—Dissolve 0.1 g. of 2,6-di-

[4] Gibbs, *J. Biol. Chem.* **72,** 649 (1927).

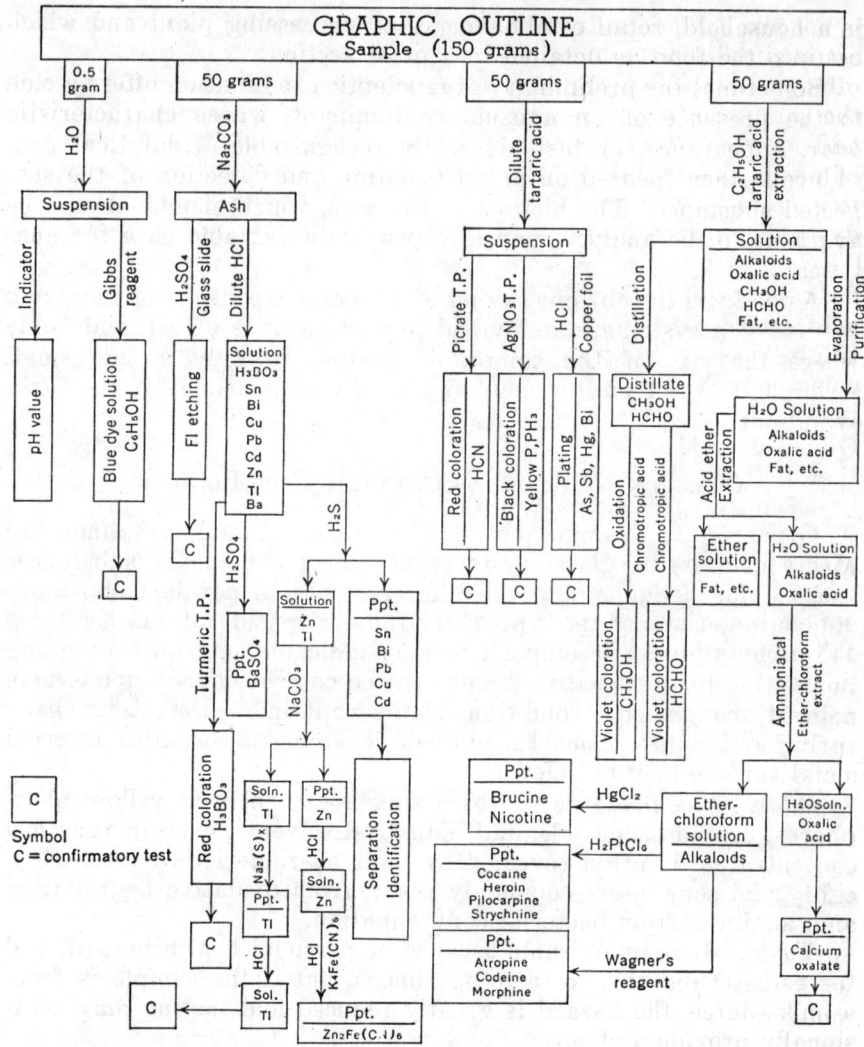

bromoquinonechloroimide in 25 ml. of 95 per cent ethyl alcohol. Stored in a brown glass-stoppered bottle in refrigerator, it remains stable for a week. Discard when solution turns brown.

Sodium Borate Buffer Solution.—Prepare a saturated solution of sodium biborate in water.

Procedure.—To the residue remaining in the test tube in the pH value determination add 5 ml. of sodium borate buffer solution, shake, test with litmus paper to check the alkalinity. If the solution is not alkaline add a drop or two of sodium hydroxide solution, never ammonium hydroxide. Allow to settle and decant or filter into another test tube. A perfectly clear solution is not essential. Divide the solution between two test tubes, and to one serving as a control add one drop of dilute phenol solution prepared by dissolving 100 mg. of phenol in 100 ml. of water. Now add 3 drops of Gibbs' reagent to each tube and shake. The formation of the deep blue indophenol dye indicates the presence of phenol.

Confirmatory Test.—Transfer 10-25 g. of sample to a distilling flask, cover with 75 ml. of water, make slightly acid with sulfuric acid, and distill 50 ml. into a separatory funnel. Shake out the distillate with two 25-ml. portions of ethyl ether and evaporate the combined ether extracts slowly on a steam bath. Take up the residue in 5 ml. of sodium borate buffer solution, then add three drops of Gibbs' reagent. The presence of phenol is confirmed by the formation of the deep blue indophenol dye.

Cyanides, Yellow Phosphorus, and Zinc Phosphide.—*Reagents.*—Ammonium Molybdate Solution.—Dissolve 10 g. of molybdic acid, MoO_3, in a mixture of 15 ml. of ammonium hydroxide solution (sp. gr. 0.90) and 27 ml. of water. Cool and pour slowly with constant stirring into a cool mixture of 49 ml. of nitric acid (sp. gr. 1.42) and 115 ml. of water. Keep in a warm place for several days or until a portion heated to 40° C. deposits no yellow precipitate of ammonium molybdiphosphate. Decant solution from any sediment and preserve in a glass-stoppered bottle.

Picric Acid Test Paper.—Wet a sheet of filter paper with a saturated water solution of picric acid and allow excess liquid to drain. Air-dry and cut into strips 1 x 7 cm.

Procedure.—Macerate 50 g. of sample in 50 ml. of water in a 250-ml. Erlenmeyer flask and add 10 ml. of tartaric acid solution. Suspend over the surface of the liquid a test paper strip moistened with a drop of silver nitrate solution, prepared by dissolving 10 g. in water and diluting to 100 ml., and a picric acid test paper strip moistened with a drop of sodium carbonate solution. Warm the mixture for 15 minutes at 50° C. on a steam bath. Presence of hydrocyanic acid is indicated by a red rose coloring of the picric acid paper. Blackening of the silver nitrate paper *may* indicate the presence of yellow phosphorus or zinc phosphide. Volatile substances such as formaldehyde, formic acid, and hydrogen sulfide also blacken silver nitrate; hence a positive paper strip test must be confirmed by the distillation test.

Confirmatory Test.—Suspend 25 g. of sample in 200 ml. of water, make slightly acid with sulfuric acid, and distill in a dark room using an ordinary distilling apparatus with an upright condenser, or preferably that described by McNally.[5] Presence of yellow phosphorus is confirmed by the appearance of a luminous ring in the upper part of the condenser. Minute amounts of hydrocyanic acid and yellow phosphorus will reveal their presence by characteristic odors if the condenser exit is *cautiously* smelled. If zinc phosphide is present phosphine is generated and this also has a very characteristic odor. Allow the vapors to condense into a flask containing a few milliliters of dilute nitric acid. Phosphorus and phosphine are oxidized to phosphoric acid which is precipitated as the yellow molybdiphosphate on the addition of ammonium molybdate solution.

Reinsch Test for Arsenic, Antimony, Bismuth, Mercury.—Add 20 ml. of concentrated hydrochloric acid to the Erlenmeyer flask containing the material used in the paper strip test for phosphorus and cyanide. Drop in a strip of clean, burnished copper foil about 1 cm. square and allow to simmer on a hot plate for one-half hour. The presence of arsenic, antimony, or bismuth is indicated by deposition of a bluish-black plating of the reduced metal on the copper foil. Mercury salts deposit a shiny silvery plating of the free metal.

Confirmatory Tests.—Carefully wash the plated copper foil with water, dry with alcohol and ether, place in a small dry test tube and cautiously heat over a small flame. Arsenic, antimony, and mercury are deposited on the cooler area of the inner surface of the tube directly above the copper foil. Examine through a low powered microscope. Arsenic, as the trioxide, is deposited in characteristic octahedral crystalline form, mercury as a mass of minute opaque globules, and antimony as an uncharacteristic amorphous smudge. Bismuth does not volatilize, but its presence can later be confirmed in the ash of the sample.

Ashing.—Transfer 50 g. of sample to a porcelain crucible, wet down with 1 ml. of saturated sodium carbonate solution, dry in an oven, carefully burn over a Bunsen flame and complete to a grey or white ash at 500° C. in a muffle furnace.

Fluorides.—*Apparatus.*—Fluoride Etching Crucible.—A satisfactory crucible for performing the etching test for fluorides is not available commercially but may easily be fashioned in the laboratory.

Place a 30-ml., tall-form, porcelain crucible on a wire gauze over a Bunsen burner and melt sufficient printer's linotype metal to almost fill the crucible. After cooling remove the mold by gently tapping the overturned crucible. Clamp the mold in a turning lathe and bore out to shape and approximate dimensions shown in Fig. 49. Place the top edge against a piece of sandpaper set on a flat surface and

---

[5] McNally, *Toxicology*. Industrial Medicine, Chicago, 1937.

rub until smooth. For the cover, hammer a slug of lead on a flat metal surface to a thickness of 1 mm. and cut to size with a pair of shears. Smooth with sandpaper and punch a hole in center.

Clean after each test by soaking in hot alkali solution and scrubbing with a wad of steel wool.

Procedure.—Transfer a portion of the ash to a metal etching test crucible, moisten with a drop of water, and cautiously add concentrated sulfuric acid, drop by drop, until effervescence ceases. Wet the top edge of the crucible with sulfuric acid to form a seal, place cover on and set a glass microscope slide over hole in cover. Heat the crucible on a hot plate for 1 hour, wash and dry the slide and examine its surface for any etching produced by the generation of hydrofluoric acid. When the etching is very light due to small concentration of fluorine, breathing on it will render it more distinct, or its roughness may be felt by gently scratching with the fingernail.

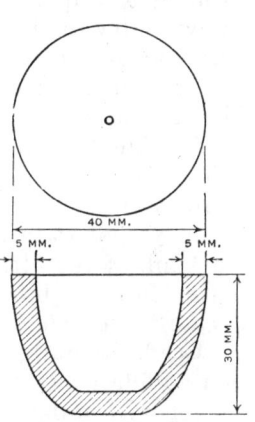

FIG. 49. Fluoride Etching Crucible and Cover

Hydrofluoric acid in the presence of a borate forms a volatile borofluoride which does not have an etching action on glass. Hence, when boron is present its interference must be eliminated by a separation of calcium borate from calcium fluoride based on the insolubility of the latter in dilute acetic acid.[6]

In a similar manner, hydrofluoric acid reacts with silica to form the volatile fluosilicic acid, and this too lacks the ability to etch a glass surface. Here, however, it is unnecessary to make a separation, for the formation of fluosilicic acid serves as the basis of an even more sensitive test for a soluble fluoride than the etching test, for which it may be substituted or used as a confirmatory test, in the manner adapted by Gettler and Ellerbrook[7] for the detection of fluorine in tissues.

Confirmatory Test.—Transfer a pinch of the dry ash to a 5-ml. porcelain crucible, mix with an equal amount of powdered glass or silica, cautiously add a few drops of concentrated sulfuric acid, and immediately cover with a microscope slide from the under surface of which is suspended a small drop of sodium chloride solution. Place the crucible on a hot plate maintained at a temperature of 150° C. and put a drop of cold water on the upper surface of the slide di-

---

[6] Furman, *Scott's Standard Methods of Chemical Analysis.* Van Nostrand, New York, 1939.

[7] Gettler and Ellerbrook, *Am. J. Med. Sci.* 197, 625 (1939).

rectly over the suspended drop, to retard evaporation. After 5 minutes heating remove the slide, allow the suspended drop to dry in the air, and examine under a microscope (450 magnifications) for six-pointed stars or hexagonal crystals of sodium fluosilicate. Ten micrograms of fluorine under this treatment should be detected without difficulty.

Borates.—Dissolve the remainder of the ash in 20 ml. of water and make slightly acid to litmus paper with hydrochloric acid. Moisten a strip of turmeric test paper with a drop of the solution and allow to dry in air. Presence of a borate is indicated by a cherry red coloring of the test paper, which changes to a dark blue-green on wetting with a drop of ammonium hydroxide and is restored to the blue-green by acid.

Soluble Barium Salts.—Add 5 ml. of concentrated hydrochloric acid to the solution of the ash obtained in the test for borates, and evaporate to dryness on a steam bath. Redissolve the residue in 25 ml. of hot water, filter, and wash into a 125-ml. Erlenmeyer flask. To 1 ml. of the filtrate in a test tube add a few drops of dilute sulfuric acid. A white precipitate of barium sulfate indicates the presence of a soluble barium salt.

Lead, Bismuth, Copper, and Cadmium.—Pass a current of hydrogen sulfide gas through the filtrate obtained in the test for barium salts for 15 minutes and allow to stand until the precipitate coagulates. Pass through a paper filter, wash, and retain the filtrate for the detection of zinc and thallium. The precipitate, which may contain tin, lead, copper, bismuth, and cadmium is subjected to the procedure for separating and identifying metals of this group detailed in Chapter V.

Zinc.—Boil the filtrate obtained in the separation of the copper group to remove hydrogen sulfide and make alkaline with saturated aqueous sodium carbonate solution, precipitating zinc, iron, and the alkaline earth metals. Filter, wash with water, and retain the filtrate for the detection of thallium. Return precipitate and paper to the original flask, add 25 ml. of dilute hydrochloric acid, 5 ml. of bromine water to oxidize any iron present, boil to remove excess bromine, and add an excess of ammonium hydroxide to precipitate iron and alkaline earth phosphates. The zinc remains in solution. Cool, filter into a Nessler tube, add 10 ml. of hydrochloric acid (1:1), then 1 ml. of 3.5 per cent potassium ferrocyanide solution and dilute to 100 ml. The presence of zinc is indicated by the formation of a white precipitate of zinc ferrocyanide. Since many foodstuffs, particularly the proteins, contain small amounts of zinc, it is advisable for the analyst to run a control along with the sample to accustom himself to estimate roughly small concentrations of zinc. Pipette 2 ml. of standard zinc chloride solution, add the reagents, and compare the turbidities in both tubes. Prepare the standard zinc chloride solution by dissolving exactly 100 mg. of chemically pure zinc in hydrochloric acid and dilute to 1 liter. Each milliliter of this solution is equivalent to 0.1 mg. of zinc.

Thallium.—*Reagent.*—Sodium Polysulfide Solution.—Dissolve 48 g. of sodium sulfide, $Na_2S \cdot 9H_2O$, and 4 g. of sodium hydroxide in water, add 1.6 g. of powdered sulfur, shake until sulfur dissolves, filter, and dilute to 100 ml.

Procedure.—Add 1 ml. of sodium polysulfide solution to the alkaline filtrate obtained in the detection of zinc. A brown precipitate of thallium sulfide indicates the presence of thallium. Let stand until the precipitate coagulates, filter, and wash, and dissolve the precipitate in a few cubic centimeters of dilute sulfuric acid, boil to remove hydrogen sulfide, cool, and neutralize exactly with ammonium hydroxide, using litmus paper as indicator.

Confirmatory Test.—Divide the solution into two equal portions in test tubes. Make one portion slightly alkaline with ammonium hydroxide and add 1 ml. of 10 per cent potassium iodide solution. A yellow crystalline precipitate of thallium iodide is formed. To the other test tube add 1 ml. of 5 per cent potassium chromate solution, producing a yellow precipitate of thallium chromate.

Alkaloids.—*Reagents.*—Mercuric Chloride Solution.—Dissolve 5 g. of mercuric chloride in water and dilute to 100 ml. Platinic Chloride Solution.—Dissolve 5 g. of platinic chloride, $H_2PtCl_6 \cdot 6H_2O$, in water and dilute to 100 ml. Wagner's Reagent.—Dissolve 1.27 g. of iodine and 2 g. of potassium iodide in water and dilute to 100 ml.

Procedure.—Transfer 50 g. of sample to a Florence flask, macerate in 50 ml. of 80 per cent alcohol, add 5 ml. of tartaric acid solution and reflux on a steam bath for 1 hour. Connect the flask with a suitable condenser and distill over 5 ml. of liquid, retaining this for the detection of methanol and formaldehyde if the nature of the material is such that the presence of either or both of these substances is suspected. Filter the mash, wash well with 80 per cent ethyl alcohol, and evaporate the filtrate on a steam bath to a volume of 5 ml. Slowly add 50 ml. of 95 per cent ethyl alcohol, stirring and breaking up any clumps formed with a glass rod, again filter and wash with ethyl alcohol. Evaporate on a steam bath to remove the alcohol, dilute to a volume of 25 ml. with water and transfer to a separatory funnel. Extract with three 25-ml. portions of ethyl ether, wash the combined ether extracts twice with 2-ml. portions of water, and add the washings to the original water solution. This ether extract from acid solution may contain in addition to fat such compounds as salicylic acid, acetylsalicylic acid, barbituric acid, acetanilid, chloral hydrate, DDT, phenols, and organic solvents, which may be tested for if their presence is suspected.

Make the water solution distinctly alkaline to litmus paper with ammonium hydroxide solution and extract with three 25-ml. portions of ethyl ether, followed by two extractions with 25-ml. portions of chloroform. Retain the water solution for the detection of oxalates. After washing the combined solvent extracts with several 2-ml. portions of water, pass through a dry paper filter, evaporate slowly on a steam bath to remove the solvents, and take up the residue in four drops of water.

CHARACTERISTICS OF MICROCHEMICAL TESTS FOR ALKALOIDS [8]

	Reagent	Description of Crystals
Atropine	Wagner's	Rods and triangular plates, singly and in groups
Brucine	Mercuric Chloride	Transparent, rectangular plates and rosettes of thin plates
Cocaine	Platinic Chloride	Delicate feathery crystals
Codeine	Wagner's	Red-brown precipitate, crystallizing in yellow blades, extending in branches
Heroin	Platinic Chloride	Spherical clusters of golden yellow needles, around a nucleus
Morphine	Wagner's	Heavy red-brown precipitate in shining overlapping plates extending in branches
Nicotine	Mercuric Chloride	Radiating transparent blades in slight excess of H_2SO_4. Feather-like blades with HCl.
Pilocarpine	Platinic Chloride	Layers of thin, yellow triangular plates
Strychnine	Platinic Chloride	Clusters of wedge-shaped needles moving about in the field

Of the common alkaloids this residue may contain atropine, brucine, codeine, heroin, morphine, cocaine, nicotine, pilocarpine, and strychnine. With a glass rod transfer three separate drops to a glass microscope slide. To the first drop by means of a glass rod add a drop of Wagner's reagent, to the second a drop of mercuric chloride solution, and to the third a drop of platinic chloride solution. The formation of a precipitate in any of the drops indicates the presence of an alkaloid. For identification, examine the slide under a microscope without stirring or covering (100-150 magnifications) and compare crystal characteristics with known controls prepared in the same manner, and also with comparison chart. Confirmatory tests for specific alkaloids may then be applied.

8 *Methods of Analysis, A.O.A.C.* Washington, 1945.

Oxalates.—Make the water solution obtained in the extraction of alkaloids slightly acid with hydrochloric acid and warm on a steam bath to remove residual ether and chloroform. Add 1 ml. of calcium chloride solution, prepared by dissolving 10 g. $CaCl_2$ in water and diluting to 100 ml., make ammoniacal, filter, and wash the precipitate with water. Redissolve the precipitate in 5 ml. of hot dilute hydrochloric.acid, filter, wash with water, and evaporate the filtrate almost to dryness on a steam bath. Cool, take up in 25 ml. of 95 per cent alcohol, and then add 25 ml. of ether. Filter through paper and repeat if necessary until a clear solution is obtained. Evaporate off the ether and alcohol on a steam bath and take up residue in a few milliliters of water. Add 1 ml. of calcium chloride solution and make ammoniacal. The presence of oxalic acid is indicated by the formation of a white, silky precipitate of calcium oxalate.

Confirmatory Test.—Filter the precipitate, wash with water, and dissolve in a small amount of hot dilute hydrochloric acid. The decolorization of a drop of potassium permanganate solution added confirms the presence of oxalic acid.

Methyl Alcohol and Formaldehyde.—*Reagent.*—Chromotropic Acid Solution.—Dissolve 5 mg. of chromotropic acid, 1,8-dihydroxy-naphthalene-3,6-disulfonic acid, in 10 ml. of a mixture of 9 ml. of concentrated sulfuric acid and 4 ml. of water.

Procedure.—Transfer 1 drop of the distillate obtained in the extraction of alkaloids to each of two test tubes. To the first tube add a drop of water, a drop of phosphoric acid solution, a drop of potassium permanganate solution, let stand for 1 minute, then add sodium bisulfite solution drop by drop until the permanganate color is discharged. If a brown color remains add another drop of phosphoric acid solution. To both tubes now add 5 ml. of freshly prepared chromotropic acid solution and heat in a water bath at 60° C. for 10 minutes. The appearance of a violet color in both tubes indicates the presence of formaldehyde and possibly methyl alcohol. If the color appears only in the tube oxidized with permanaganate, then methyl alcohol alone is present.

DETECTION OF POISONS IN MILK AND WATER

The following scheme was devised by the author [9] for the rapid detection of poisons in water and milk. Some of the tests detailed have been described in prior sections of the text but are included here for the sake of continuity.

[9] Jacobs, *Technical Manual for Chemical Detection and Decontamination of Warfare Agents.* Dept. Health, New York, 1951.

ORGANOLEPTIC ANALYSIS

It must be noted that any scheme for the detection of poisons in milk or food in the field has serious limitations. Most of these tests are applicable, of course, to water. Negative tests are to be regarded as conclusive unless special information is available. Doubtful and positive tests must be confirmed in the laboratory. A mobile or plant laboratory would be excellent for this scheme, but a relatively small kit can be arranged. Small test tubes should be used for the tests.

An organoleptic or sensory analysis is one which is made by the use of our senses, principally sight, odor, and taste.

A description of the tests for milk is given but, as explained, most of these tests are applicable to water and food with but slight modification.

All personnel handling milk should be on the alert for any deviation from the normal of the appearance, odor, or taste of the milk at any stage of its handling and manipulation. If an off-color, or off-odor of the milk, milk sediment, etc., or off-taste of the milk itself is noted, a report to a superior should be made immediately. Personnel should smell and taste cautiously.

In order to make a smell test, sniff once or twice gently and try to recall the odor. In order to make a taste test, roll a small amount of milk over the tongue, but do not swallow. Gross contamination of milk will often be readily detected by these simple tests.

The value of organoleptic analysis as an aid in the detection of poisons should neither be underestimated or overestimated. Thus, for instance, a poisonous substance added to milk might be sufficiently acid to curdle the milk and thus will readily be detected. Or another poison might have sufficient coloring power to give the milk an off-color. Such an addition can also be readily detected by close observation. These are examples of noting the appearance of the milk.

So too, if a poison such as a phenol or hydrogen cyanide, or nitrobenzene having a characteristic odor is added to milk, attention to the off-odor will assist in the detection of the added poisonous material.

Many alkaloidal and nonalkaloidal poisons have an intensely bitter taste even when diluted. Any bitter taste or other off-taste in milk should be noticed. Among the poisons which have a very bitter taste are picrotoxin, colchicine, picric acid, veronal, strychnine, brucine, and similar substances.

CHEMICAL EXAMINATION

pH, Excess Acid or Alkali.—The normal pH value of milk is about 6.6, with a range of 6.5 to about 6.7. Any marked variation from these values is to be regarded with suspicion. An extreme low range of 6.3 has at times been noted. Marked acidities may cause the milk to curdle.

Procedure.—To obtain the pH value, place 0.5 ml. of milk in a small test tube or, better, a spot plate, add 2 drops of bromothymol blue indicator solution, and compare the color against prepared buffers, known good milk, or by the use of a chart. If the milk is normal it will have a slightly greenish-yellow color; if abnormal, it will have a yellow color on the acid side and a green or blue on the alkaline side.

Inorganic Poisons.—White (Yellow) Phosphorus and Cyanides.—Place 50 ml. of the sample in an Erlenmeyer flask and add 10 ml. of 10 per cent tartaric acid solution. Suspend above the surface of the liquid test strips of (1) filter paper moistened with a drop of silver nitrate solution, (2) picric acid paper moistened with a drop of saturated sodium carbonate solution. Warm the mixture to 40°-50° C. for 15 minutes on the steam bath. The papers may be suspended by attaching them with the aid of a rubber band to a glass stopper which fits loosely in the mouth of the flask.

The presence of cyanides is indicated by a rose color produced on the picric acid paper. Sulfur dioxide and other reducing agents may interfere.

The benzidine-copper acetate test paper is better for testing for cyanides, but it is more difficult to prepare.

Blackening of the silver nitrate test paper may indicate the presence of yellow phosphorus. Volatile reducing substances such as formaldehyde and formic acid also reduce silver nitrate as well as hydrogen sulfide. A positive test must be confirmed in the laboratory.

Borates.—Take 10 ml. of milk and transfer to a test tube. Add 0.7 ml. of concentrated hydrochloric acid. Stir vigorously, and dip into the mixture a piece of prepared turmeric test paper. If a red color develops which is changed by ammonium hydroxide solution to a dark blue green and back again to red by acid, boric acid is present.

Oxidizing Agents (Chlorine, Hypochlorites, Nitrites, Peroxides, Nitrates, Iodates, etc.)—Place 3 ml. of milk in a test tube. Add 2 drops of hydrochloric acid and mix. Add 1 ml. of 7 per cent potassium iodide solution and 0.5 ml. of 1 per cent starch solution, mix and wait a few moments. A blue color indicates the presence of oxidizing agents. Starch-iodide test paper serves equally well.

Place a few drops of milk in a porcelain dish. Add a drop or so of diphenylamine reagent, 1 g. of diphenylamine in 100 ml. of concentrated sulfuric acid. An intense blue color at the junction shows the presence of an oxidizing agent. Diphenylbenzidine may be used instead of diphenylamine.

Fluorides.—The presence of fluorides may be detected by the use of sodium zirconium alizarinate paper.

Reagent.—Dissolve 0.87 g. of zirconium nitrate $Zr(NO_3)_4 \cdot 5H_2O$, in 100 ml. of water, and 0.17 g. of sodium alizarinate in 100 ml. of water. Place each solution in a large watch glass or evaporating dish. Steep filter paper first in the dye solution and then place the dyed paper in the zirconium nitrate solution. Allow to dry and cut into strips.

Procedure.—Acidify the suspected water or milk with hydrochloric acid (1:1) and filter the milk. Test with the test paper. If the paper changes from pink to yellow, fluorides are present. Bleaching agents bleach the pink and yellow color and therefore interfere.

Heavy Metals.—The heavy metals as a group (arsenic, antimony, lead, mercury, cadmium, and bismuth) may be detected in the following manner.

Prepare an acid serum by the addition of trichloroacetic acid to the milk, diluted with an equal volume of water. Allow the precipitate to settle and filter. Add 1 ml. of hydrochloric acid (1 : 3) to each 10 ml. of test solution, and warm to 50° C. Add an equal volume of saturated hydrogen sulfide water and allow to stand at 35° C. for 10 minutes. If no color is produced, the above-mentioned metals may be considered absent in more than traces.

If the mixture is now made alkaline (if a precipitate was formed above, the solution must be filtered), a precipitate indicates the presence of zinc, cobalt, nickel, iron, aluminum, chromium, and thallium. To test for the latter which is the most important, redissolve in nitric acid and free of hydrogen sulfide by boiling, precipitate iron, the zinc-nickel group, and the alkaline earths by boiling with sodium carbonate solution, filter, and add ammonium sulfide. A brown precipitate indicates thallium. Verify by dissolving the brown precipitate in sulfuric acid, boil off the hydrogen sulfide, add potassium iodide, a yellow crystalline precipitate indicates thallium.

Arsenic and Antimony.—Arsenic and antimony may also be detected as follows:

Dilute 25 ml. of milk with an equal volume of water. Add 1 ml. of concentrated hydrochloric acid. Allow the precipitate to coagulate and filter into a 4-oz., wide-mouth bottle. Add 4 ml. more of concentrated hydrochloric acid, 5 ml. of 15 per cent potassium iodide solution, and 4 drops of stannous chloride solution, 40 g. of arsenic-free $SnCl_2 \cdot 2H_2O$, in hydrochloric acid made up to 100 ml. with concentrated hydrochloric acid. Equip the 4-oz., wide-mouth bottle with a guard tube containing sand wet with saturated lead acetate solution, and then attach a delivery tube leading to a test tube. Allow the delivery tube to dip into 2-3 ml. of 2 per cent silver nitrate solution. Add 2-5 g. of 20-mesh granulated zinc, stopper the bottle with the guard tube and allow the generation of gases to proceed for 20-30 minutes. A darkening of the silver nitrate solution or black precipitate beginning at the jet of the delivery tube and extending through the solution is evidence of the presence of arsenic or antimony.

Selenium and Tellurium.—Dilute 25 ml. of milk with an equal volume of water. Add 1 ml. of concentrated hydrochloric acid. Allow the precipitate to coagulate and filter a portion into a test tube; the remainder can be used for the test for arsenic and antimony. Add an equal volume of saturated sulfur dioxide solution and some hydroxylamine hydrochloride solution. A reddish precipitate indicates elementary selenium; a blackish precipitate indicates elementary tellurium.

Organic Poisons.—Common Organic Poisons.—Phenols.—Phenols as a group can readily be detected in milk by use of the Gibbs reagent. Place 3 ml. of milk and 2 ml. of borate buffer solution (pH 9.6) into a test tube and shake. Add 3 drops of an alcoholic solution of 2,6-dibromoquinonechloroimide. A deep blue color indicates the presence of a phenol. If there is any possibility that the color is masked, shake out with 1 ml. of normal or isobutyl alcohol. A blue color in the supernatant layer indicates the presence of phenol, cresol, lysol, tricresyl-o-phosphate, etc. The latter is not very soluble in water but sufficient is dissolved to give a test with the Gibbs reagent. Gross quantities of these poisons would give the milk a pronounced odor.

Formaldehyde.—Place 5 ml. of the milk in a test tube. Add at least 0.5 ml. of concentrated hydrochloric acid. Add 0.5 ml. of Schiff's fuchsin-sulfite or rosaniline-sulfite reagent, and allow to stand. A pink to deep violet color develops on allowing the mixture to stand. Gross quantities of formaldehyde would be noticeable by the odor.

Oxalates.—Add 50 ml. of ethyl alcohol to 25 ml. of milk and allow to stand for a moment. Add a drop of concentrated hydrochloric acid and allow to stand. Filter, transfer the filtrate to a separatory funnel, and shake out with two successive portions of ethyl ether. Evaporate the ether layers to dryness, take up residue with 1 ml. of water, and filter into a small test tube. Add 1 ml. of saturated calcium sulfate solution, boil the solution and make slightly ammoniacal. The presence of oxalate is indicated by a white precipitate of calcium oxalate.

The test is difficult to perform in the field because it requires extractions but could easily be performed in a small plant laboratory or in a mobile unit.

Nonalkaloidal and Alkaloidal Poisons.—It is difficult to test for these substances in the field. Furthermore, it would be very unlikely that these substances would occur as contaminants because of price and unavailability. However, if it were desired to make a test for the presence of these substances, the following procedure can be used.

Procedure.—To 50 ml. of milk made acid with tartaric acid add 100 ml. of alcohol. Allow to stand and filter. Triturate the residue with 25 ml. more of alcohol and filter. Evaporate the alcohol until a syrup is obtained. Dissolve in water and extract with ether (Extract I). Make alkaline with sodium hydroxide solution and extract with ethyl ether (Extract II). Make neutral with hydrochloric acid and again alkaline with ammonia, and extract with ethyl ether (Extract III). Extract the last mixture with chloroform. (Extract IV). Evaporate each extract and note if there is any suspicious residue, such as a crystalline, or very bitter-tasting or colored residue. If the test is positive further work must be performed in a well-equipped laboratory in order to identify the substance and verify the presence of a poison.

Glucosides and Saponins.—These substances might be very difficult to detect. However, glucosides of the *Digitalis* and *Strophanthin* groups would be rather difficult to obtain because of price and availability. Saponins might be detected by the persistent foam they would impart to the milk. Another test that might be useful would be to note the color imparted to a sulfuric acid serum of the milk. Sulfuric acid-molybdic acid (Froehde's reagent) gives colors with various saponins and may be tried.

SELECTED REFERENCES

Brooks and Alyea, *Poisons.* Van Nostrand, New York, 1946.
Furman, *Scott's Standard Methods of Chemical Analysis.* Van Nostrand, New York, 1943.
Goldstone, *Anal. Chem.* 21, 781 (1949).
Gonzales, Vance, and Helpern, *Legal Medicine and Toxicology.* Appleton-Century, New York, 1937.
Jacobs, *Analytical Chemistry of Industrial Poisons.* Interscience, New York, 1949
Jacobs, *Technical Manual for Chemical Detection and Decontamination of Warfare Agents.* Department of Health, New York, 1951.

CHAPTER VII

MILK AND CREAM

MILK and cream, alone or in combination with other foods, comprise about one-sixth of the weight of food eaten by an average American family. Still the United States does not consume as much milk as some European countries do, on a per capita basis.

Milk is the whole, fresh lacteal secretion obtained by the complete milking of one or more healthy cows, excluding that obtained within 15 days before and 5 days after calving, or such longer period as may be necessary to render the milk practically colostrum free. The name, milk, unqualified, means cow's milk. It consists largely of water, milk fat, lactose or milk sugar, protein and mineral matter. These are probably present in some form of combination such as fat-protein, protein-mineral matter. The three main characteristic components are milk or butter fat, casein, and lactose.

Milk is one of the most important foods in the human diet because it has many components present in very small quantities that are essential to growth and well-being. That this is true follows from the fact that mammalian animals can live and thrive for weeks and months without the addition of other foods. There are other foods that contain these materials also but are not as easily assimilable as milk. Some of the minor components, but very important, are lactalbumin, lactoglobulin, lactic acid, sodium, potassium, calcium, magnesium, chlorides, phosphates, citrates, iodine, cholesterol, lecithin, enzymes, and the vitamins A, B_1, B_2, and C. The chemistry of milk, cream, and dairy products is discussed by Jacobs.[1]

COMPOSITION

Davies [2] gives as the proximate analysis of milk, based on the analysis of thousands of samples by various investigators of United States, England, Germany, and Scotland:

Fat	3.71%
Solids not fat	8.99%
Total solids	12.7 %

---

[1] Jacobs, *Chemistry and Technology of Food and Food Products*, 2nd ed. Interscience, New York, 1951.

[2] Davies, *Chemistry of Milk*. Van Nostrand, New York, 1936.

The maximum, minimum and average percentage composition of the more important components of milk as compiled by Davies from the work of many investigators and that of the New York Agricultural Experimental Station at Geneva are given in the following table:

TABLE 10. COMPOSITION OF MILK

	Maximum	Minimum	Average	N. Y. State	Andrade[3]
Water ....,....	90.0	82.0	87.3	87.1	87.0
Fat	7.8	2.3	3.67	3.9	4.2
Casein	....	1.5	2.86	2.5	....
Albumin	....	0.5	0.56	0.7	....
Total Protein	4.5	2.0	3.42	....	3.27
Lactose	6.0	3.5	4.78	5.1	4.78
Ash	0.9	0.6	0.73	0.7	0.7
Total Solids	18.0	10.0	12.69	12.9	12.98
Solids-not-Fat........	10.6	7.5	8.77	9.0	8.77

[3] Andrade, *Estudios sobre la leche.* Caracas, Venezuela, 1940.

A more complete description of the composition of milk can be obtained from Table 11.

TABLE 11. COMPOSITION OF MILK

Fraction	Component
Lipid	Consists of fat composed of the mixed triglycerides of the following fatty acids in order of importance: oleic, palmitic, myristic, stearic, and butyric; of lesser importance—caproic, caprylic, lauric, capric, decenoic, tetradecenoic, hexadecenoic; and arachidonic. In addition, the oil-soluble vitamins, cholesterol, xanthophyll, phospholipids, cephalin, and lecithin are present
Protein	Casein, lactalbumin, and lactoglobulin. These proteins contain some 20 or more amino acids including all of the essential ones
Carbohydrate	Lactose or milk sugar
Mineral	Sodium, potassium, calcium, and phosphorus, chlorine, and sulfur in substantial quantities. Magnesium, copper, iron, zinc, manganese, and iodine in smaller amounts. Carbonates are also present
Vitamin	Vitamin A (and its precursors such as carotene), thiamine, ascorbic acid, vitamins D, vitamins E, riboflavin, niacin, pantothenic acid, and pyridoxine are present
Enzyme	Phosphatase, amylase, lipase, catalase, peroxidase, galactase, reductase
Other organic	Citric and lactic acid, creatine, creatinine, urea, choline
Gas	Carbon dioxide, oxygen, and nitrogen
Water	

Oleic, followed closely by palmitic, is the predominant fatty acid. Myristic, stearic, and butyric acids follow in order. The remainder may be considered the minor fatty constituents.[4] Casein, albumin, and globulin constitute about 93 per cent of the nitrogeneous components of milk, and the remainder, which includes amino acids and amides, phosphatides, purine substances, ammonia, choline thiocyanate and riboflavin are to be regarded as the minor nitrogeneous components.[5]

Variation.—The composition of milk varies with the breed of cow, as illustrated in Table 12, the time of year, the time of day, the portions of any one milking, the individuality of the cow, the age of the cow, the period of lactation, feeding, and other factors. Jersey, Guernsey, and Ayrshire give milk richer in fat than Holstein and Dutch Belt. In the fall and early winter richer milk is obtained thon in the spring and early summer. The evening milking yields richer milk than the morning milking. The first portions or "fore" milk of milk drawn in the milking process are poorer than the last portions or "strippings." The reader is referred to texts on dairying for complete information.

TABLE 12. VARIATION IN COMPOSITION OF MILK ATTRIBUTABLE TO BREED

	Fat	Lactose	Protein	Ash	Water
Jersey	5.43	4.85	3.96	0.75	85.01
Guernsey	5.16	4.80	3.92	0.75	85.37
Ayrshire	4.09	4.57	3.27	0.69	87.38
Shorthorns	3.91	4.80	3.27	0.73	87.29
British Friesians	3.63	4.62	3.11	0.71	87.93
Dutch Belt	3.60	5.00	2.62	0.68	87.97
Holstein	3.39	4.89	2.99	0.69	88.04

Breed of Cow.—The food analyst is, however, little concerned, except theoretically and from the investigational viewpoint, with these variations except with that of breed of cow, for his main interest lies in market milk, in which most of the variations disappear into an average due to mixing of herd milk on a large scale. The breed of cow has a large effect, as can be seen from Table 12. Thus, for example, in the New York City milk shed, that is, the area surrounding New York City which supplies it with milk, probably 70

4 Hilditch, *Analyst* **62**, 250 (1937).
5 Bushill, Lampitt, and Filmer, *Analyst* **62**, 260 (1937).

per cent of the cows are Holstein. These give milk of low fat content, circa 3.35 per cent, whereas the usual average is much nearer 3.7.

Adequate consideration must be given to the fact that much of the data on milk in the literature are results obtained from few samples. Thus it is well known that, although the minimum standard for milk fat percentage in many States of the United States and also in England is 3 per cent, milk obtained from some herds actually falls below that figure. Hence the minimum selected is not one which actually occurs but is one which has been adopted as a health measure.

Some other significant factors concerning milk are the following. The yield of milk is inversely proportional to the energy value—that is, the calorific value—of the total solids per unit weight of milk. In other words, the greater the total volume of milk given by any one cow, the lower the total solid content.

Milk possesses the same order of osmotic pressure as blood and, since the osmotic pressure of blood is and must remain practically constant, it follows that the osmotic pressure of milk should also be practically constant. The osmotic pressure is dependent on the number of dissolved particles in a solvent. The freezing point also depends on the number of particles dissolved in a solvent and there is a mathematical relationship between these factors and other colligative properties such as the vapor pressure and the boiling point. Consequently the freezing point of milk also varies slightly and may be considered an index of a normal milk. In Chapter II, "Physical Chemical Methods," the freezing point depression determination or the cryoscopic method is detailed. This method may be used to estimate added water in milk and cream.

Milks of Mammals.—The average composition of the milks of different mammals is tabulated in Table 13.

Composite Sample

It is best to have fresh, single samples for the analysis of milk. It is, however, not always possible nor practical to obtain or hold such samples; consequently composite samples are secured, particularly of market milk or milk used for manufacturing purposes. A composite sample of milk may be defined as a sample prepared from proportionate amounts of each delivery from a single source of supply.

In making or taking a composite sample, a proportionate amount, but not less than 10 ml., of each volume of milk sampled is transferred with the aid of a sampling tube or dipper to the composite sample jar. This jar should be adequately labeled and should be capable of being tightly stoppered and properly sealed. Such composite samples should not be collected for periods exceeding fifteen days.

TABLE 13. AVERAGE COMPOSITION OF MILKS OF MAMMALS [a]

Species	No. of Samples	Water, %	Protein.[b] %	Fat, %	Lactose, %	Ash, %
Human	1154	87.43	1.63	3.75	6.98	0.21
Human		87.68	1.05	4.37	6.79	0.18
Cow	1998[c]	86.21	3.77	4.45	4.86	0.72
Cow	208[d]	87.90	3.13	3.65	4.50	0.72
Goat	326[c]	87.14	3.71	4.09	4.20	0.78
Ewe	2[e]	82.90	5.44	6.24	4.29	0.85
Egyptian buffalo	61	82.09	4.16	7.96	4.86	0.78
Chinese buffalo	30[e]	76.80	6.04	12.60	3.70	0.86
Philippine carabao	19[e]	78.46	5.88	10.35	4.32	0.84
Camel		87.61	2.98	5.38	3.26	0.70
Mare	104	89.04	2.69	1.59	6.14	0.51
Mare	...	90.23	2.30	0.78	6.42	0.44
Ass	...	89.70	2.10	1.50	6.40	0.30
Reindeer	...	63.30	10.30	22.46	2.50	1.44

[a] Wright, Deysher, and Cary, *U. S. Dept. Agr., Yearbook, Separate* 1704 (1940).

[b] 6.38 × nitrogen.

[c] 198 whole lactation periods from 14 Ayrshire, 16 Guernsey, 19 Holstein, and 15 Jersey pure bred cows and from 66 Guernsey-Holstein cross-bred cows.

[d] 208 New York herd samples.

[e] Number of animals used for samples.

Composite samples as well as single samples, if not analyzed within a day or two, even when properly refrigerated, should be preserved with a preservative such as formaldehyde, mercuric chloride, or potassium dichromate. Mercuric chloride or corrosive sublimate is the chemical most commonly employed. Where such preservatives are objectionable or cannot be used because of interference with a subsequent determination, the sample should be kept cold by proper icing but should not be frozen.

Samples should not be permitted to freeze and should be stirred gently with a rotary motion each day to prevent the accumulation of a cream layer which is difficult to disperse. This also helps to distribute the preservative.

SPECIFIC GRAVITY

Milk is a fat-water emulsion, consequently its specific gravity is a function of the specific gravity of the fat and of that of the water solution. The specific gravity of the fat is about 0.93 and that of the solids-not-fat is 1.5. Hence, as the fat content of the milk increases, the specific gravity decreases and, conversely, as the solids-not-fat content increases, the specific gravity of the milk also increases. The actual specific gravity found is some function of the two. Milk normally has a specific gravity which varies in the range 1.027-1.035 with an average value of 1.032 at 60° F., although the average value used for calibrating lactometers is 1.029.

The specific gravity of milk is usually obtained by means of a lactometer although it may also be obtained with a pyknometer. The New York Board of Health lactometer is a hydrometer graduated with arbitrary scale divisions from 80° to 120°, in which 100° equals 1.029, the average specific gravity of milk. This lactometer is calibrated at 60° F. and should be read at that temperature, otherwise corrections need be made, by adding 2° of the arbitrary scale for every 5° F. rise and subtracting 2° of the arbitrary scale for every 5° F. below 60° F.

Another type of lactometer is the Quevenne with a scale divided into 25 equal parts from 15 to 40 in which 29 equals 1.029, the average specific gravity of milk. This instrument is also calibrated at 60° F. and should be read at that temperature. It may be corrected for readings not made at this temperature by adding 0.1 to the reading for each degree F. above 60° F, or subtracting 0.1 for each degree F. below 60° F.

Procedure.—Stir the milk by passing it from vessel to vessel, taking care to occlude as little air as possible, and place in a suitable cylinder. Immerse the lactometer in the fluid and allow it to rise to its proper level. This is done in order to overcome surface tension and viscosity effects. Take the reading and the temperature, and correct the reading for temperature difference. If a New York Board of Health lactometer is used the reading must be multiplied by 0.29 to convert to the Quevenne scale. The specific gravity may be obtained from the Quevenne reading by placing 1.0 before that scale reading, in other words, the readings form the hundredths and thousandths place after 1.0.

FAT

There are a number of purposes in making fat determinations in milk. These have been listed by Matt [6] as serving (1) as the method for payment of milk on the basis of the fat content; (2) to detect losses of fat in the manufacture of butter and cheese and thus implying the ability to prevent such losses; (3) as the means for the detection of watering and skimming of milk both by regulatory agencies and by purchasers of bulk shipments; and (4) as the means

---

[6] Matt, *Kimble Manual for Sampling and Testing Milk*. Kimble, Vineland, N. J., 1942.

of detecting cows yielding milk of low fat content, thus implying the ability to select better stock by weeding out poorer cows.

There are many methods for the determination of fat in milk. Of these, the Babcock and Gerber methods are the quickest and simplest. The Roese-Gottlieb is the best of the longer and more accurate methods.

Babcock Method.—This method depends on the solution of all components of milk except fat and lipoid bodies in sulfuric acid, and the subsequent estimation of the fat by centrifuging into a graduated narrow neck of a special flask as the supernatant layer over the heavier layer of sulfuric acid.

Procedure.—Measure 18 g. of milk at 60° F. from a properly mixed sample into a standard State branded milk test bottle, Fig. 50, by using a 17.6-ml. standard State branded pipette, add 17.5 ml. of standard commercial sulfuric acid, specific gravity 1.813, which is best and avoids charring the fat layer, and shake until all the curd has disappeared. Continue the shaking for about one-half minute longer. Before mixing, the milk and acid should have a temperature of about 60° F., if not, the amount of acid must be adjusted to give the proper rate of color development.

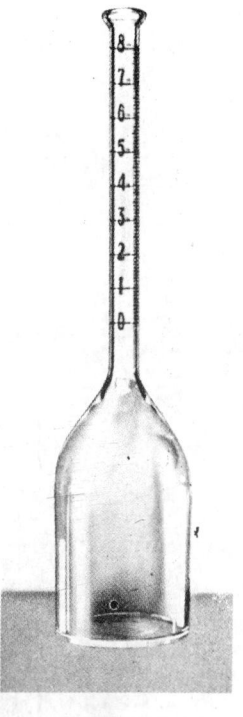

Fig. 50. Babcock Milk Flask

Place the test bottles in the Babcock centrifuge and whirl at the proper speed for 5 minutes. Fill the bottles with hot water, having a temperature of at least 200° F. to the bottom of the neck. Whirl for 2 minutes, fill with hot water at 200° F. to the top of the graduations, and whirl again for 1 minute. If a large number of samples are to be read, place these samples in a water bath at 135° F. to 140° F. for 5 minutes. Read the per cent fat by measuring from the lowest point of the fat column to the highest point of the meniscus at the top of the fat column. Discard all results that do not have a clear fat column. Some centrifuge machines are equipped with heating apparatus. When using such machines, the temperature of the water added should be not less than 160° F. The Babcock method makes use of standard pipettes and bottles whose specifications are rigidly drawn.[7]

---

[7] Y. State Dept. Agr. Markets, Circ. 505 (1935).

Gerber Method.—This method depends on the solution of all milk components other than fat in sulfuric acid using amyl alcohol to help break the milk emulsion and prevent charring of the fat layer. The amyl alcohol should be pure and should be tested by running

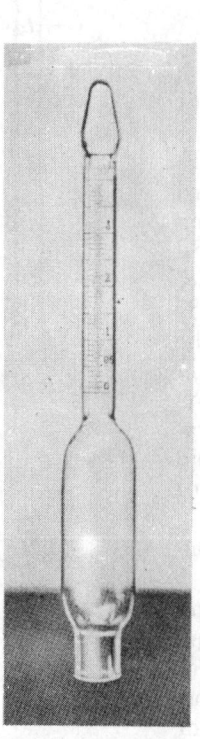

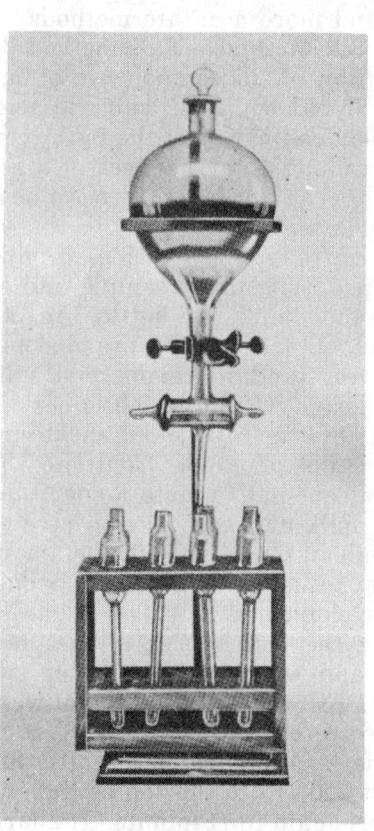

Fig. 51. Gerber Milk
Butyrometer

Fig. 52. 10 cc. Automatic Burette
and Butyrometers.

a control exactly as detailed in the method, using water instead of milk. The fat reading should be zero. The amyl alcohol reacts with the sulfuric acid forming an ester which is completely soluble in the sulfuric acid and hence has no effect on the fat result. The fat is subsequently estimated by centrifuging, with the lipoids forming the supernatant layer in the capillary graduated portion of a butyrometer, Fig. 51. This method has the marked advantage that centrifuging only once is necessary. If the control test with the amyl alcohol gives an apparent fat reading, it must be rejected. Only amyl alcohol giving no fat reading may be used in this method.

Procedure.—Measure 10 ml. of sulfuric acid (sp. gr. 1.82) into the milk butyrometer, Fig. 52. Carefully add exactly 11 ml. of the milk sample equivalent to 11.33 g. with an 11-ml. pipette and add 1 ml. of amyl alcohol. The temperature of acid and milk should be near 60° F. When ready to mix, insert the stopper and shake. When the milk curd is completely dissolved, invert the bottle several times to mix the acid remaining in the neck of the bottle with the rest of the mixture. Place in a Gerber centrifuge or in adapters in a Babcock centrifuge and whirl for 5 minutes at the proper speed for the machine that is being used. The machine, if heated, should be at a temperature of about 160° F. A heated machine usually gives more satisfactory results. Remove and read immediately. By adjusting the stopper, the bottom of the fat column can be made to coincide with the zero or maximum division on the scale. The extreme lower part of the upper surface or meniscus of the fat column is read. When it is not possible to adjust the lower surface of the fat column to zero it may be adjusted to any other whole per cent mark and the proper calculation made when reading the test. The reading gives the per cent fat directly. The Gerber method makes use of standard pipettes and butyrometers whose specifications are rigidly drawn. This equipment is also known as State branded glassware.[8]

Roese-Gottlieb Method.—The Roese-Gottlieb method replaces successfully the more tedious continuous extraction methods such as the Adam's coil method. It depends on the use of ammonia to soften the curd of the milk, of ethyl alcohol to break the milk emulsion and the fat-protein combination, and of mixed ethers to extract the fat. The alcohol assists the ethyl ether in coming in contact with the fat. The petroleum ether is used to decrease the solubility of water and alcohol in the ethyl ether and of course thus decrease the solubility of salts in the ether layer. The petroleum ether also decreases the solubility of ethyl ether in the water layer. The fat thus extracted is subsequently estimated by weighing. In performing these extractions, the Jacobs-Singer separatory flask or Mojonnier extraction tube are to be preferred as far superior to the unwieldy Röhrig tube or similar apparatus.

Procedure.—Transfer 10 g. of the milk sample to the lower section of a Jacobs-Singer separatory flask, Fig. 15, either by weighing directly into the flask, or by weighing by difference with the aid of a Mojonnier 10 g. pipette and carriage, Fig. 53, or by means of a

[8] *N. Y. State Dept. Agr. Markets, Circ.* **515** (1936).

calibrated delivery weight pipette. A Mojonnier extraction tube, Fig. 16, may also be used. Stopper the Jacobs-Singer separatory flask with the upper section. Add 1.25 ml. of ammonium hydroxide, or 2 ml. if the sample is sour, and mix thoroughly. Add 11 ml. of 95 per cent alcohol and mix well. Add 25 ml. of ethyl ether, shake vigorously for 30 seconds, add 25 ml. of petroleum ether and shake again for 30 seconds. Let stand for 20 minutes, or until the upper layer is perfectly clear, or centrifuge. Draw off as much as possible of the mixed ether layer into a tared "fat" flask, so designated because it is squat in form, through a small, quick-acting filter. Again extract the liquid remaining in the tube or flask, this time with 5 ml.

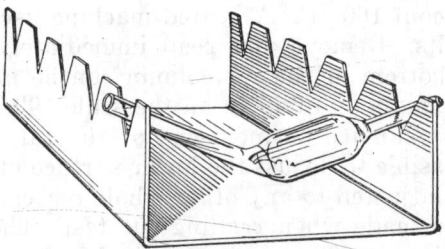

FIG. 53. Mojonnier Weight Pipette and Carriage
(Courtesy of Mojonnier Brothers)

of 95 per cent alcohol, and 15 ml. of each ether, shake vigorously 30 seconds after each addition and allow to settle. Draw off the clear solution through the small filter into the same tared flask. Repeat the extraction with another 15-ml. portion of each ether, shake vigorously 30 seconds after each addition and allow to settle. Draw off the clear solution through the small filter into the same tared flask. Add water to the separatory flask or to the extraction tube with the aid of a wash bottle until the level of the water layer reaches the middle of the constriction of the extraction tube or the connecting joint of the separatory flask. Only a few milliliters will be necessary. Draw off the remaining ether solution through the small filter into the same tared flask as carefully as possible. Evaporate off the ether in the tared flask after each extraction slowly, on a steam bath, while the subsequent extraction is allowed to settle. Wipe off the outside of the flask and place in an oven thermostatically controlled at 100° to 105° C. Weigh the flask with a similar flask as a counterpoise after cooling in a desiccator. If a 10 g. calibrated weight pipette was used, the weight of the fat in the flask divided by 10 and multiplied by 100 equals the per cent fat in the milk sample.

Total Solids

Weigh into a tared flat-bottomed dish, containing 10 to 15 g. of pure dry sand, 5 g. of milk with the aid of a pipette or by difference. Place in an oven thermostatically controlled at 99° C. for 4 hours. Place into a desiccator and weigh quickly when cool; report the increase in weight as total solids.

Total Protein

Determine the nitrogen in 5 g. of the milk added to an 800-ml. Kjeldahl digestion flask by means of a weight pipette, as described in the Roese-Gottlieb method, by the Kjeldahl-Gunning-Arnold method detailed in Chapter I. Multiply the percentage of nitrogen by 6.38 to obtain the equivalent percentage of milk proteins.

Casein

Nitrogen Method.—The casein is separated from albumin and other proteins by precipitation with acetic acid and is subsequently estimated by determining the amount of nitrogen in that precipitate.

Procedure.—To 10 g. of milk add 90 ml. of water at 40-42° C. and then 1.5 ml. of 10 per cent acetic acid. Stir and allow to stand 3-5 minutes. Decant on a filter, wash by decantation with cold water and transfer the precipitate to the filter. Wash twice on the filter. Determine the nitrogen in the washed precipitate and filter by the Kjeldahl-Gunning-Arnold method and multiply by 6.38 to obtain the percentage of casein. This determination must be made on the fresh sample. The milk should be preserved with formaldehyde, 1 part to 2,500 parts of milk, if the analysis is not made shortly after receipt.

Formol Titration of Casein.—The following method devised by McDowall and McDowell [9] is based on the fact that the addition of formaldehyde to a protein endows the protein with acidic properties. This method differs from the Walker formaldehyde-volumetric casein test in that the casein is separated from the milk by precipitation, and is redissolved in alkali before the addition of formaldehyde.

Procedure.—Dilute 20 ml. of milk with 100 ml. of water at 42° C., but not higher in order to avoid coagulating soluble protein, in a 150-ml. beaker. Add at once 1.5 ml. of 1.67 N acetic acid (10 per cent), and then stir gently by rotating the stirring rod 4 times in the beaker. After allowing the beaker to stand about 20 minutes, add 4.5 ml. of 0.25 N sodium acetate solution and, after stirring gently, allow to stand for at least an hour.[10]

---

[9] McDowall and McDowell, *Analyst* **61**, 824 (1936).
[10] Moir, *Analyst* **56**, 147 (1931).

Decant the mixture through filter paper under gentle suction on a Büchner 6-cm. funnel. Wash the precipitate with water and allow to settle. Again decant the liquid through the filter paper. It is advisable to disconnect the suction pump as soon as all the liquid has passed through the filter, otherwise any casein that has passed over on to the filter will form an impervious layer. Repeat the washing and decantation a second time and finally transfer all of the precipitate to the funnel. Suction should cease well before the precipitate is dry. Return the precipitate and filter to the original beaker. Invert the funnel and wash with a little water to remove adhering particles. Add 4 to 5 ml. of 0.1 N sodium hydroxide solution. The total volume including the precipitate should now be about 20 ml. Place the beaker in a boiling water bath for 5 minutes and shake occasionally until all the casein is dissolved and the fat emulsified. Cool the milky solution to 21-24° C. Add 1 ml. of 1 per cent phenolphthalein and 0.1 N sodium hydroxide until an end point is reached matching that of 20 ml. of milk tinted with a few drops of 0.01 per cent aqueous rosaniline acetate solution. Use a portion of the sample of milk being analyzed, if possible, for the control. Add 4 ml. of 40 per cent formaldehyde of analytical quality and continue the titration with 0.1 N sodium hydroxide until the same end point is reached. The number of milliliters of 0.1 N sodium hydroxide used in the second titration multiplied by the average factor 0.92 gives the percentage casein in the sample. The factor actually varies from 0.89 to 0.94, but the factor recommended gives results accurate to ±0.05 per cent.

ALBUMIN

Exactly neutralize the filtrate obtained in the method for casein described in the section, "Casein-Nitrogen Method," with 10 per cent sodium hydroxide solution, add 0.3 ml. of acetic acid (1:9), and heat on a steam bath until the albumin is completely precipitated. Collect the precipitate on a filter, wash with cold water, determine nitrogen as directed in the Kjeldahl-Gunning-Arnold method, and multiply by 6.38 to obtain the equivalent of albumin.

MINOR NITROGENEOUS COMPONENTS [11]

A trichloroacetic acid precipitation will give the most satisfactory results for precipitation of the proteins. Ammonia may be estimated by steam distillation under reduced pressure. Urea may be determined by a modification of the urease method and creatinine may

---

[11] Bushill, Lampitt, and Filmer, *Analyst* 62, 260 (1937).

be estimated by a modification of the colorimetric method with picric acid, after the influence of lactose has been eliminated. Nonprotein nitrogen may be determined on the trichloroacetic acid filtrate.

The total nitrogen content [12] of milk is about 0.5 per cent, and of that total casein comprises about 76.1 per cent. Nonprotein nitrogen is 5.9 per cent of the total nitrogen.

ASH

The ash content of milk is one of its most constant characteristics. Although the literature indicates that the ash of milk varies in the range 0.6-0.9 per cent, it seldom falls below 0.68 or rises above 0.74 per cent. The average value is very close to 0.7 per cent.

Procedure.—Add to a tared porcelain dish by means of a weight pipette 20 g. of the sample. Add 6 ml. of nitric acid and evaporate to dryness. Ignite at a temperature below redness until the ash is free from carbon. Cool in a desiccator, weigh, and report the increase in weight as ash.

ACIDITY

Normal market milk has a titratable acidity which averages about 0.15-0.16 per cent expressed in terms of lactic acid. The range for normal milk from individual cows extends from 0.10 to 0.22 per cent, but normal commercial milk practically never exceeds the average value. This acidity is largely due to the casein content (0.05-0.08 per cent) and the phosphate content. Carbon dioxide (0.01-0.02 per cent) citrates (0.01 per cent), and albumin (less than 0.01 per cent) also contribute to it.

Procedure.—Place 17.6 ml. of milk by means of a Babcock pipette into a 125-ml. flask and dilute with an equal volume of water, recently boiled and cooled, washing out the pipette with this water. Titrate with 0.1 N sodium hydroxide solution, using 0.5 ml. of phenolphthalein indicator. The number of milliliters of 0.1 N sodium hydroxide required divided by 20 gives the percentage of lactic acid.

Instead of using 0.1 N sodium hydroxide solution, it is fairly common to use a 0.02 N sodium hydroxide solution to obtain a greater titrating volume. When 0.02 N sodium hydroxide is used, divide the burette reading by 100 to obtain the percentage of acidity expressed as lactic acid.

Rapid Test.—*Reagent.*—Prepare a 0.02 N sodium hydroxide solution containing phenolphthalein indicator. This may readily be done by diluting 200 ml. of 0.1 N sodium hydroxide solution with 790 ml.

---

[12] Wright, Deysher, and Cary, *U. S. Dept. Agr., Yearbook Separate* **1704** (1940).

of water and 10 ml. of phenolphthalein indicator solution. **Each milliliter is equivalent to 0.01 per cent lactic acid.**

Procedure.—With the aid of a 17.6-ml. pipette, transfer 17.6 ml. of milk to a porcelain dish or white cup. Add by means of a pipette or burette exactly 18 ml. of 0.02 N sodium hydroxide solution, and mix. If the mixture remains pink, the milk contains less than 0.18 per cent acidity; and if the mixture turns white, more than 0.18 per cent acid is present.

LACTOSE

Dilute 25 g. of the sample with 400 ml. of water in a 500-ml. volumetric flask, add 10 ml. of copper sulfate solution, 34.639 g. of copper sulfate, $CuSO_4 \cdot 5H_2O$, dissolved in water, diluted to 500 ml. and filtered through an asbestos mat, and 8.8 ml. of 0.5 N sodium hydroxide solution or an equivalent amount of potassium hydroxide solution. The alkali added should be sufficient to precipitate completely the copper as hydroxide from 1 volume of the copper sulfate solution. Fill the flask to the 500 ml. mark, mix, filter through a dry filter, and determine lactose in an aliquot by one of the copper reduction methods described in the chapter on sugars, Chapter X.

ADDED WATER

To 1 volume of copper sulfate solution, 72.5 g. of copper sulfate, $CuSO_4 \cdot 5H_2O$, per liter, adjusted if necessary to read 36 at 20° C. on the scale of the immersion refractometer, or to a specific gravity of 1.0443 at 20° C./4° C., add four volumes of milk. Shake well and filter. Determine the refractometer reading of the clear serum— that is, the filtrate—at 20° C. A reading below 36 indicates added water.

A much more sensitive method for the determination of added water in milk has been detailed in Chapter II, under the section of freezing point methods. Such methods are also known as cryoscopic methods.

RAPID METHOD

A fairly complete means of determining the chemical quality of milk is to combine the use of the lactometer for estimating the specific gravity and the Babcock method for estimating the per cent fat content. Then by the use of these two factors and a mathematical formula or table the total solids may be ascertained. The milk sample or samples are allowed to come to approximately 60° F., the lactometer readings are taken with a New York Board of Health lactometer, as described above, and recorded. The fat is then ascertained by the Babcock method. These two results may then be substituted in the formula of Hehner and Richmond [13] modified for the New York Board of Health lactometer readings to give the calculated total solids.

$$T.S. = \frac{L \times 0.29}{4} + 1.2F + 0.14$$

in which

$T.S.$ = per cent total solids
L = New York Board of Health lactometer reading
F = per cent fat

Let us assume that the lactometer reading was 107, corrected to 60° F., and that the fat content was 3.8 per cent; then by substitution

$$T.S. = \frac{107 \times 0.29}{4} + [1.2 \times 3.8] + 0.14$$

then

$$T.S. = 12.46$$

TABLE 14.—PERCENTAGE OF TOTAL SOLIDS IN MILK. LACTOMETER READINGS ACCORDING TO NEW YORK BOARD OF HEALTH LACTOMETER. ALL READINGS TO BE CORRECTED TO 60° F.

	95	96	97	98	99	100	101	102	103	104	105	106	107	
2.8	10.25	10.32	10.39	10.46	10.54	10.61	10.68	10.76	10.83	10.90	10.97	11.05	11.12	2.8
2.9	10.37	10.44	10.51	10.58	10.66	10.73	10.80	10.88	10.95	11.02	11.09	11.17	11.24	2.9
3.0	10.49	10.56	10.63	10.70	10.78	10.85	10.92	11.00	11.07	11.14	11.21	11.29	11.36	3.0
3.1	10.61	10.68	10.75	10.82	10.90	10.97	11.04	11.12	11.19	11.26	11.33	11.41	11.48	3.1
3.2	10.73	10.80	10.87	10.94	11.02	11.09	11.16	11.24	11.31	11.38	11.45	11.53	11.60	3.2
3.3	10.85	10.92	10.99	11.06	11.14	11.21	11.28	11.36	11.43	11.50	11.57	11.65	11.72	3.3
3.4	10.97	11.04	11.11	11.18	11.26	11.33	11.40	11.48	11.55	11.62	11.69	11.77	11.84	3.4
3.5	11.09	11.16	11.23	11.30	11.38	11.45	11.52	11.60	11.67	11.74	11.81	11.89	11.96	3.5
3.6	11.21	11.28	11.35	11.42	11.50	11.57	11.64	11.72	11.79	11.86	11.93	12.01	12.08	3.6
3.7	11.33	11.40	11.47	11.54	11.62	11.69	11.76	11.84	11.91	11.98	12.05	12.13	12.20	3.7
3.8	11.45	11.52	11.59	11.66	11.74	11.81	11.88	11.96	12.03	12.10	12.17	12.25	12.32	3.8
3.9	11.57	11.64	11.71	11.78	11.86	11.93	12.00	12.08	12.15	12.22	12.29	12.37	12.44	3.9
4.0	11.69	11.76	11.83	11.90	11.98	12.05	12.12	12.20	12.27	12.34	12.41	12.49	12.56	4.0

	108	109	110	111	112	113	114	115	116	117	118	119	120	
2.8	11.19	11.26	11.34	11.41	11.48	11.55	11.63	11.70	11.77	11.81	11.92	11.99	12.06	2.8
2.9	11.31	11.38	11.46	11.53	11.60	11.67	11.75	11.82	11.89	11.96	12.04	12.11	12.18	2.9
3.0	11.43	11.50	11.58	11.65	11.72	11.79	11.87	11.94	12.01	12.08	12.16	12.23	12.30	3.0
3.1	11.55	11.62	11.70	11.77	11.84	11.91	11.99	12.06	12.13	12.20	12.28	12.35	12.42	3.1
3.2	11.67	11.74	11.82	11.89	11.96	12.03	12.11	12.18	12.25	12.32	12.40	12.47	12.54	3.2
3.3	11.79	11.86	11.94	12.01	12.08	12.15	12.23	12.30	12.37	12.44	12.52	12.59	12.66	3.3
3.4	11.91	11.98	12.06	12.13	12.20	12.27	12.35	12.42	12.49	12.56	12.64	12.71	12.78	3.4
3.5	12.03	12.10	12.18	12.25	12.32	12.39	12.47	12.54	12.61	12.68	12.76	12.83	12.90	3.5
3.6	12.15	12.22	12.30	12.37	12.44	12.51	12.59	12.66	12.73	12.80	12.88	12.95	13.02	3.6
3.7	12.27	12.34	12.42	12.49	12.56	12.63	12.71	12.78	12.85	12.92	13.00	13.07	13.14	3.7
3.8	12.39	12.46	12.54	12.61	12.68	12.75	12.83	12.90	12.97	13.04	13.12	13.19	13.26	3.8
3.9	12.51	12.58	12.66	12.73	12.80	12.87	12.95	13.02	13.09	13.16	13.24	13.31	13.38	3.9
4.0	12.63	12.70	12.78	12.85	12.92	12.99	13.07	13.14	13.21	13.28	13.36	13.43	13.50	4.0

PER CENT OF BUTTERFAT

[13] Hehner and Richmond, *Analyst* 17, 170 (1892).

Other formulas have been suggested by Babcock and other investigators. One of these that agrees closely with the total solids of milk of the New York City milk shed determined gravimetrically and, undoubtedly, is applicable to herd milk in general, is the following:

$$T.S. = 1.2[F - 3.0] + 0.07[L - 100] + 10.89$$

in which

$T.S.$ = per cent total solids
L = New York Board of Health lactometer reading at 60° F.
F = per cent fat

Thus, for the above example, we have by substitution

$$T.S. = 1.2[3.8 - 3.0] + 0.07[107 - 100] + 10.89$$

then

$$T.S. = 12.34$$

GELATIN AND OTHER THICKENING AGENTS

A rare type of adulteration of milk is the addition of gelatin, or the addition of other thickening agents. These are much more likely to occur in cream or other milk products and will be discussed fully under those sections.

PRESERVATIVES

The detection and estimation of preservatives have been fully detailed in Chapter IV, "Chemical Preservatives in Foods." These methods may be applied successfully to milk.

FOREIGN FAT

A milk containing foreign fat is generally considered adulterated. Even if the foreign fat is added as the "so-called" solvent for oil soluble vitamins, it is deemed an adulteration. Foreign fat may be detected in a manner similar to that detailed in the section under cream.

CHLORIDE TEST

A test based on the fact that infected udder tissues allow the salts of the blood plasma to filter through into the freshly secreted milk was developed by Hammer and Bailey.[14] An increase over the normal amount of chloride in the freshly drawn milk therefore indicates an abnormal condition in the udder. In a similar respect, an increase in the chloride content of market milk indicates milk that comes

---

[14] Hammer and Bailey, *Iowa State Coll. Agr. Mech. Arts, Bull.* 41 (1917).

from a poor source or milk that has been tampered with and possibly reconstituted. Thus the use of salt butter in reconstitution would very likely raise the chloride content far above normal.

Procedure.—Place 10 ml. of milk in a flask and dilute with 40 ml. of water. Add 8 to 10 drops of 10 per cent potassium chromate solution, 10 g. K_2CrO_4 dissolved in water and diluted to 100 ml. Titrate the mixture with 0.1 N silver nitrate solution. One milliliter of 0.1 N silver nitrate solution is equivalent to 3.55 mg. chlorine.

A shorter method was suggested by Hayden.[15] Measure accurately 5 ml. of a silver nitrate solution, 1.3415 g. of silver nitrate dissolved in a liter of water, into a test tube. Add 2 drops of a 10 per cent potassium chromate solution and then exactly 1 ml. of milk. If the milk contains more than 0.14 per cent chlorides, a yellow color will develop in the tube, because all the silver has been precipitated as the chloride. This yellow color indicates abnormal milk because, from general experience, it is assumed that the normal content of chlorides varies from 0.09 to 0.14 per cent. If the milk contains less than the stated amount of chloride the tube would have a color that varies from reddish to brownish depending upon the amount of silver chromate precipitated.

HEAT STABILITY TESTS

Alcohol Coagulation Test.—A simple test [16] to indicate whether a given milk will be able to withstand the heat of the condensing and sterilizing processes without becoming curdy is the alcohol coagulation test.

Procedure.—Transfer 5 ml. of milk to a test tube and rapidly add 5 ml. of 70-75 per cent alcohol. Close the test tube with the thumb and invert twice. Examine for coagulation. If no coagulation occurs, the glass walls of the test tube are clear. A fine precipitate leaves a cloud on the walls of the tube while in marked coagulation small lumps are produced.

Phosphate Coagulation Test.—This test is not designed to indicate the sanitary quality of milk but rather to assist in eliminating milks which have poor stability under sterilizing conditions.

Reagent.—Dissolve 68.1 g. of monopotassium dihydrogen phosphate, KH_2PO_4, in water and dilute to 1 liter.

Procedure.—Transfer 10 ml. of the milk sample to a tube and add 1 ml. of the phosphate solution. Mix thoroughly and immerse in a boiling water bath for 5 minutes. Cool and observe for coagulation which, if visible, indicates poor heat stability.

---

[15] Hayden, *Cornell Veterinarian* **22**, 277 (1932).
[16] Ramsdell, Johnson, Jr., and Evans, *J. Dairy Sci.* **14**, 93 (1931).

Pasteurized Milk

Many cities and towns do not permit the sale of raw or improperly pasteurized milk because such milk may constitute a health hazard. The United States Public Health Service defines pasteurized milk as milk every particle of which has been subjected to a temperature of 161° F. for 15 seconds or to a temperature not lower than 143° F. for not less than 30 minutes and then promptly cooled to 50° F. or lower. In England and other countries, however, milk must be held for a half hour at 145° to 150° F. for proper pasteurization. In New York City and in many other communities in the United States, milk in order to be properly pasteurized must be treated by processes conforming to the requirements of the United States Public Health Service. By improper pasteurization is meant, then, a milk which has not been held at 143° F. for the proper length of time, namely, 30 minutes, or one in which the temperature of pasteurization was less than 142° to 143° F., or one which was heated to less than 161° F. or held at that temperature for less than 15 seconds.

The correct process of pasteurization entails expense and an unscrupulous distributor may improperly pasteurize milk, or may add raw milk to standardize, his product, that is, adjust the chemical composition by addition of milk, in order to lower costs. Many times milk is improperly pasteurized due to carelessness or poor equipment. At any rate, this is an important problem in the regulation and control of the sale of milk. This problem has been attacked from the fact that many, if not all of the enzymes in milk are heat labile at the temperature of pasteurization and because of the holding-time in the pasteurization process. This is the basis of many of the older as well as newer tests for heated and pasteurized milk.

The Schardinger test,[17] based on the destruction of the peroxidase enzyme in the pasteurization process and the Leahy [18] test based on the assumption that amylase enzyme is completely inactivated by heating milk at 143° F. for 30 minutes have been completely superseded by the phosphatase tests because these tests and also that of Gould [19] and the older one of Rothenfusser [20] do not, for unknown reasons, always distinguish between pasteurized and improperly pasteurized milk and sometimes do not even distinguish between pasteurized and raw milk. Hence, caution must be observed in interpreting results with these procedures.

[17] Schardinger, Z. Nahr. Genussm. 5, 1113 (1902).
[18] Leahy, Intern. Assoc. Dairy Milk Inspectors, Ann. Rept. 23, 93 (1934).
[19] Gould, J. Dairy Sci. 15, 230 (1932).
[20] Rothenfusser, Z. Untersuch. Lebensm. 60, 94 (1930).

Kay and Graham Phosphatase Method.—The inadequacy of most other tests in distinguishing between pasteurized and improperly pasteurized milk led to the development of the phosphatase test of Kay and Graham.[21] This test is based on the distinction produced by the complete or incomplete destruction of the enzyme phosphatase by heating. Inorganic phosphate and phenol are liberated from a buffered solution of disodium phenylphosphate by the action of phosphatase which is present in raw or improperly pasteurized milk, or in mixtures containing raw or improperly pasteurized milk. The phenol liberated is estimated by the use of the Folin and Ciocalteu reagent.

Reagents.—Sodium Veronal—Disodium Phenylphosphate Buffer. —Dissolve 11.54 g. of sodium barbital or sodium veronal and 1.09 g. disodium phenylphosphate in water saturated with chloroform and make up to a liter with water saturated with chloroform. The disodium phenylphosphate may sometimes be contaminated with phenol. It may be purified from any such contamination by washing with ethyl ether until the washings give no test for phenol with 2,6-dibromoquinonechloroimide. To perform this test, add 10 ml. of water to the last of the ether washings and evaporate off the ether. Adjust the pH to 9.6 by the addition of ½ ml. borax buffer and add a few drops of 2,6-dibromoquinonechloroimide solution, 100 mg. dissolved in 25 ml. of 95 per cent alcohol. No development of a blue color indicates absence of phenol. To prepare the borax buffer, dissolve 15 g. of anhydrous sodium tetraborate powder in 900 ml. of warm water. Add 3.27 g. sodium hydroxide in the form of a 20-40 per cent solution and make up to a liter after cooling to room temperature. Five ml. of this buffer added to 100 ml. of water should produce a pH of 9.6.

Folin and Ciocalteu Reagent.—Dissolve 100 g. of sodium tungstate, $Na_2WO_4 \cdot 2H_2O$ and 25 g. of sodium molybdate, $Na_2MoO_4 \cdot 2H_2O$ in 700 ml. of water in a 1,500-ml. flask connected by a joint to a reflux condenser. Add 50 ml. syrupy 85 per cent phosphoric acid and 100 ml. of hydrochloric acid. Reflux the mixture gently for 10 hours. After this time, cool, add 150 g. lithium sulfate, Li_2SO_4, 50 ml. water and 4-6 drops of liquid bromine. Boil the mixture under a hood without the condenser for 15 minutes to boil off excess bromine. Cool, dilute to 1 liter, and filter. The finished reagent should have a golden yellow color with no greenish tint. Dilute with 2 volumes of water as needed for use.

Short Test Procedure.—Place 10 ml. of the sodium veronal-disodium phenylphosphate buffer in four test tubes of 25-ml. capacity.

---

[21] Kay and Graham, *J. Dairy Research* **6**, 191 (1935).

To two tubes, which act as controls, add 4.5 ml. of diluted Folin phenol reagent. To all four tubes add 0.5 ml. of milk and mix well. Incubate the two tubes without the Folin phenol reagent at 47° C. ± 2° in a water bath for 10 minutes. After this time remove from the bath, cool, add 4.5 ml. diluted Folin phenol reagent. Allow to stand for 3 minutes. Filter all four tubes. To 10 ml. of the filtrate, add 2 ml. of 14 per cent sodium carbonate solution. Mix, place in a boiling water bath for 15 minutes, and filter. Pasteurization or non-pasteurization is ascertained by the depth of the blue color produced.

Long Test Procedure.—To 0.5 ml. of milk add 10 ml. of the sodium veronal-disodium phenylphosphate buffer plus 2 drops of chloroform and incubate for 24 hours at 37° to 38° C. To the above add 4.5 ml. of Folin phenol reagent, properly diluted. Let stand for at least 3 minutes and filter. Pipette 10 ml. into a test tube and add 2 ml. of 14 per cent sodium carbonate solution. Heat in a boiling water bath for at least 5 minutes and filter. Run in duplicate, and always run controls which are not incubated and to which diluted Folin reagent is added immediately. Compare the color produced with standards. A light blue indicates a pasteurized milk. A dark blue indicates an under-pasteurized milk. A purplish color indicates a raw milk. Any color exceeding 2.3 Lovibond units of blue indicates milk not properly pasteurized. Gilcreas and Davis [22] recommend the production of blue colors by Folin's reagent with known phenol solutions as standards instead of Lovibond blue units. They found that a phenol value of 0.037 mg. per 0.5 ml. of sample examined or less indicates adequate pasteurization.

The Kay and Graham test was designed for milk pasteurized according to British standards, namely, at 145° to 150° F. for one-half hour. However, by suitable modification the phosphatase test can be adjusted to suit American standards. This is done by diminishing the length of time of incubation. If a given milk has been claimed to be heated for one-half hour at 142°-143° F., it should not give more than 2.3 Lovibond blue units after 2½ hours incubation at 37° C. If it gives a greater color than this at this temperature of incubation then the milk is not properly pasteurized. On the other hand, the test may be made more sensitive for American pasteurization procedure by keeping the 24-hour incubation period and increasing the maximum allowable Lovibond blue units obtained to 4.

Modified Kay and Graham Method, Dye Reagents for Phenol.— The essential feature of the Kay and Graham method is the co-

---

[22] Gilcreas and Davis, *Annual Proceedings Intern. Assoc. Milk Sanitarians*, p. 15 (1936).

liberation of phosphate and phenol. They estimate the phenol by means of the Folin and Ciocalteu reagent. The use of dye intermediates like the Gibbs' reagent,[23] 2,6-dibromoquinonechloroimide and of p-dimethylphenylenediamine [18] as a means for detecting the liberated phenol directly within the milk have been developed. p-Nitrosodimethylaniline [24] may also be used but must be reduced to the diamine by zinc dust prior to use. Other similar reagents may be used. The use of these reagents has advantages for rapid and not too sensitive work.

Procedure.—To 10 ml. of Kay and Graham sodium veronal-disodium phenylphosphate buffer solution, add ½ ml. of milk, mix, stopper, and maintain at about 37° C. for about 5 to 10 minutes. At the end of this period add 4 drops of a solution of 2,6-dibromoquinonechloroimide, 100 mg. of the dye intermediate reagent in 25 ml. of 95 per cent alcohol. Allow 5 to 10 minutes for the color to develop. A blue color, grayed over by the milk, indicates an improperly pasteurized milk. Properly pasteurized milk will remain unchanged or may develop a faint blue shade.

Preferably, the colors developed may be extracted by solvents.[25] After allowing the color to develop, add 3 ml. of isobutyl alcohol, n-butyl alcohol, amyl alcohol, or ethyl acetate and mix gently. Any blue color that was formed in the milk by the reaction between the phenol liberated by the phosphatase enzyme and the imide, with the formation of an indophenol, will be extracted by one of the aforementioned solvents. In isobutyl alcohol, n-butyl alcohol and amyl alcohol, the color will remain blue, in ethyl acetate it will change to red. A deep blue or red color indicates a raw or improperly pasteurized milk. A properly pasteurized milk will not color the solvent layer, or will impart a mere tinge of red or blue, as the case may be.

The details of the test with p-dimethylphenylenediamine reagent are somewhat different. To 10 ml. of the Kay and Graham sodium veronal-disodium phenylphosphate buffer add ½ ml. of milk and allow to stand for 5 to 10 minutes at about 37° C. for about 5 to 10 minutes. At the end of this period add 4 drops of 5 per cent sodium bicarbonate solution, 5 g. sodium bicarbonate dissolved in water and made up to 100 ml., four drops of 0.1 per cent diamine reagent, 0.1 g. p-dimethylphenylenediamine dissolved in 100 ml. of water and then sodium hypochlorite solution containing 0.05 per cent of available chlorine until the pink color first produced changes to colorless or blue. The sodium hypochlorite solution may be prepared by dissolv-

[23] Gibbs, *J. Biol. Chem.* 72, 649 (1927).
[24] Houghton and Pelley, *Analyst* 62, 117, (1937).
[25] Jacobs, unpublished work, 1937.

ing 60 g. of sodium carbonate and 40 g. of bleaching powder in 400 ml. of water, filtering, and diluting the filtrate so that it contains 0.05 per cent of available chlorine. Allow the milk reaction mixture to stand for a few minutes for the color to develop and then shake out the dye formed with either chloroform or carbon tetrachloride. A blue color in the lower solvent layer indicates raw or improperly pasteurized milk or a mixture of the two.

Sanders and Sager Test.—Since the introduction of the phosphatase test by Kay and Graham in 1935, many variations of this test have been proposed and a number are detailed in this book. Difficulty has been encountered in applying the original test and some of its modifications to determine the adequacy of pasteurization of products other than milk and even of milk itself, at times. The Sanders and Sager modification [26, 27] by making use of a buffer, the pH of which can be varied to provide optima for various milk products, and of zinc-copper protein precipitants has overcome this difficulty. It is the pasteurization test accepted by the Food and Drug Administration.

Reagents.—(1) *Buffers.*—(a) Barium Borate-Hydroxide Buffer. —Dissolve 25.0 g. of barium hydroxide, $Ba(OH)_2 \cdot 8H_2O$—fresh, not deteriorated, in water and dilute to 500 ml. In another flask or cylinder dissolve 11.0 g. of boric acid, H_3BO_3, and dilute to 500 ml. Warm each to 50° C., mix the two together, stir, cool to approximately 20° C., filter, and stopper the filtrate tightly (pH 10.6).

The buffer thus prepared is designated as the 25-11 buffer, the figures indicating the grams per liter of each of the respective reagents. Modifications in the quantities of these two reagents, necessary in preparing the appropriate buffers for testing various products, are indicated in Table 15.

(b) Color Development Buffer.—Dissolve 6.0 g. of sodium metaborate $(NaBO_2)$ and 20 g. of sodium chloride in water and dilute to 1 liter with water (pH 9.8).

(c) Color Dilution Buffer.—Dilute 100 ml. of color development buffer 1-b to 1 liter with water.

(d) Standard Borax Buffer.—0.01-molar, for checking pH meter —pH 9.18 at 25° C.—Dissolve 0.9603 g. of pure borax (Bureau of Standards Sample 187) in water (distilled recently or freshly boiled and cooled) and dilute to 250 ml. Keep tightly stoppered.

(2) *Buffer Substrates.*—(a) For Evaluating Pasteurization.— Dissolve 0.10 g. of phenol-free crystalline disodium phenyl phos-

[26] Sanders and Sager, *J. Dairy Sci.* **30**, 909 (1947)
[27] *Federal Register* **14**, 1960 (1949).

phate in 100 ml. of the appropriate (Tables 15 and 21) barium borate-hydroxide buffer 1-a.

(b) For Quantitative Results with Raw Milk and Raw-Milk Products.—Dissolve 0.20 g. of the phenol-free crystalline disodium phenyl phosphate in 100 ml. of the appropriate (Tables 15 and 21) barium borate-hydroxide buffer 1-a.

(3) *Protein Precipitants.*—(a) Zinc-copper Precipitant for Milk.—Dissolve 3.0 g. of zinc sulfate, $ZnSO_4 \cdot 7H_2O$, and 0.6 g. of copper sulfate, $CuSO_4 \cdot 5H_2O$, in water and dilute to 100 ml. with water. The precipitant thus prepared is designated as the 3.0-0.6 precipitant.

TABLE 15. PHOSPHATASE TEST MODIFICATIONS FOR VARIOUS DAIRY PRODUCTS OTHER THAN CHEESE

Product	Quantity of Sample	Buffer for optimum pH (9.85-10.20)	Precipitant	Criterion, Experimental, Phenol Equivalents[a]
Milk:				
Fresh	1 ml.	25-11[b] (5+5)[c]	3.0-0.6[d]	2 γ/0.5 ml.
Old or slightly sour.....	1 ml.	25-11	6.0	2 γ/0.5 ml.
Cream:				
Fresh ..,...........	1 ml. or 1 g.	25-11 (5+5)	3.0-0.6	2 γ/0.5 ml. or 0.5 g.
Old or slightly sour.....	1 ml. or 1 g.	25-11 (8+2)	4.5	2 γ/0.5 ml. or 0.5 g.
Ice cream mix...........	1 ml.	25-11 (8+2)	4.5-0.1	2 γ/0.5 ml.
Sherbet mix	1 ml.	25-11 (5+5)	3.0-0.6	2 γ/0.5 ml.
Chocolate drink	1 ml.	25-11 (8+2)	4.5-0.1	2 γ/0.5 ml.
Butter	1 g.	18-8	6.0	2 γ/0.5 g.
Sweet buttermilk	1 ml.	25-11 (5+5)	6.0[e]	2 γ/0.5 ml.
Cultured buttermilk and fermented drinks:				
Medium acid	1 ml.	25-11		2 γ/0.5 ml.
Very acid, pH < 4.5..	1 ml.	26-11	6.0	2 γ/0.5 ml.
Goats' milk[f]	3 ml.	27-11	7.5-0.1	0 γ/1.5 ml.
Cheese whey	1 ml.	25-11 (5+5)	3.0-0.6	2 γ/0.5 ml.

[a] Values higher than those shown indicate under-pasteurization.
[b] Grams $Ba(OH)_2 \cdot 8 H_2O$ and H_3BO_3, respectively, per l.
[c] Five parts of 25-11 buffer plus 5 parts of water.
[d] Grams $ZnSO_4 \cdot 7 H_2O$ and $CuSO_4 \cdot 5 H_2O$, respectively, per 100 ml.
[e] Grams $ZnSO_4 \cdot 7 H_2O$ per 100 ml.
[f] 4-hour incubation period; use 7.0 ml. of filtrate and add 3.0 ml. of color development buffer 1-b.

(4) 2,6-Dibromoquinonechloroimide Solution (Gibbs' Reagent).—Dissolve 40 mg. of 2,6-dibromoquinonechloroimide powder in 10 ml. of absolute ethyl or methyl alcohol and transfer to a dark-colored dropper bottle. This reagent remains stable for at least a month if kept in the ice tray of a refrigerator. Do not use it after it begins to turn brown.

(5) *Other Reagents.*—(a) Copper Sulfate, 0.05 per cent, for Standards.—Dissolve 0.05 g. of copper sulfate in water and dilute to 100 ml.

(b) Butyl Alcohol.—Specify *n*-butyl alcohol, boiling point 116°- 118° C. To adjust the pH, mix 50 ml. of the color development buffer 1-b with a liter of the alcohol.

(6) *Phenol Standards.*—(a) Stock Solution.—Weigh accurately 1.0 g. of pure phenol, transfer to a liter volumetric flask, dilute to a liter with water, and mix. One ml. contains 1 mg. of phenol. Use this stock solution to prepare standard solutions. It is stable for several months in the refrigerator.

(b) *Preparation of Standards.*—Dilute 10.0 ml. of the stock solution 6-a to a liter with water, and mix. One ml. contains 10γ or *units* of phenol. Use this standard solution to prepare more dilute standard solutions; for example, dilute 5, 10, 30, and 50 ml. to 100 ml. with water to prepare standard solutions containing 0.5, 1.0, 3.0, and 5.0 γ or units of phenol per ml., respectively. Keep standard solutions in the refrigerator.

In a similar manner, prepare from the stock solution as many more concentrated standard solutions as may be needed, containing, for example, 20, 30, and 40 units per ml.

Measure appropriate quantities of the phenol standard solutions into a series of tubes (preferably graduated at 5.0 and 10.0 ml.) to provide a suitable range of standards as needed, containing 0 (control or blank), 0.5, 1.0, 3.0, 5.0, 10.0, etc., to 30 or 40 units. To increase the brightness of the blue color and improve the stability of the standards, add 1.0 ml. of 0.05 per cent copper sulfate solution 5-a to each.

Add 5.0 ml. of color dilution buffer 1-c and add water to bring the volume to 10.0 ml. Add four drops (0.08 ml.) of 2,6-dibromoquinonechloroimide solution, mix, and allow to develop for 30 minutes at room temperature. If the butyl alcohol extraction method is to be used in the test, extract the standards as described under "*Procedure.*"

Read the color intensities with a photometer, subtract the value of the blank from the value of each phenol standard, and prepare a standard curve (straight line). When the standards are to be used for visual comparisons, they should be stored in a refrigerator.

Photometric Determination.—To read the color in aqueous solution, use a filter with maximum light transmission in the region of 610 mμ wave length.

To read the color in butyl alcohol, extract the color as described above and centrifuge the sample for 5 minutes to break the emulsion and to remove the moisture suspended in the alcohol layer. A Babcock centrifuge can be adapted for this purpose by making special tube holders as follows: Slice a section 0.25 in. thick from a rubber stopper of suitable diameter to fit in the bottom of the centrifuge cup. Glue together two cork stoppers of appropriate diameter, bore through the center a hole of proper size to hold the tube snugly, and insert the double cork section into the cup. After centrifuging, remove nearly all of the butyl alcohol by means of a pipette with a rubber bulb on the top end. Filter the alcohol into the photometer cell and read with a filter with maximum light transmission in the region of 650 mμ wave length.

If more than approximately 4 ml. of butyl alcohol is required for the photometer used, conduct the test in a larger tube and extract the color, in both the test and the standards, with the necessary quantity of butyl alcohol rather than with 5 ml. specified above.

Procedure.—Transfer with the aid of a pipette a 1.0-ml. sample (preferably two) into a tube and an additional 1.0 ml. into another tube as a control or blank. In testing cream, the sample may be weighed (1.0 g.) if desired; in testing goats' milk, use a 3.0-ml. aliquot of the sample (Table 15). Heat the *blank* to at least 85° C. for 1 minute in a beaker of boiling water with the beaker covered so that the entire tube is heated to approximately this temperature, and cool to room temperature. From this point, treat the blank and the test in a similar manner. Add 10.0 ml. of the appropriate barium buffer substrate 2-a or 2-b (Table 15), stopper the tube, and mix. Incubate in a water bath at 37°-38° C. for 1 hour, mixing or shaking the contents occasionally. Place in a beaker of boiling water for nearly a minute, heating to approximately 85° C. (use a thermometer in another tube containing the same volume of liquid). Cool to room temperature. Pipette in 1.0 ml. of the zinc precipitant 3-a.

Filter (5-cm. funnel, 9-cm. Whatman No. 42 or No. 2 paper recommended), and collect 5.0 ml. of filtrate in a tube, preferably graduated at 5.0 and 10.0 ml. Add 5.0 ml. of color development buffer 1-b (pH of mixture, 9.3-9.4). Add 2 drops of 2,6-dibromoquinonechloroimide solution, mix, and allow the color to develop for 30 minutes at room temperature.

Determine the amount of blue color by either of two methods.—

(a) *With a photometer.* Read the color intensity of the blank and that of the test, subtract the reading of the blank from that of the test, and convert the result into phenol equivalents by reference to the standard curve described under "Phenol Standards." The butyl alcohol extraction method ordinarily is unnecessary when using a photometer.

(b) *With visual standards.* For quantitative results in borderline instances, *e.g.*, tests yielding 0.5 to 5 units of color, extract with butyl alcohol 5-b. Add 5.0 ml. of the alcohol and invert the tube slowly several times; centrifuge if necessary to increase the clearness of the alcohol layer, and compare the blue color with the colors of standards in the alcohol.

With samples yielding more than 5 units, compare the colors in aqueous tests with those of aqueous standards.

Calculation and Evaluation of Result.—When using 1.0 ml. of fluid sample and adding 11.0 ml. of liquid (total liquid 12.0 ml., 5.0 ml. of filtrate used), multiply the value of the reading by 1.2 to convert it to phenol equivalents per 0.5 ml. of sample. If desired, the result may be converted to phenol equivalents per 1 ml. by multiplying by 2.4. Evaluate the result by comparing it with the criteria of pasteurization in Table 15.

HOMOGENIZED MILK

Homogenized milk is the milk product that has been mechanically treated in such a manner as to alter its physical properties, with particular reference to the condition and appearance of the fat globules. The U. S. Public Health Service defines homogenized milk as milk which has been treated in such manner as to insure breakup of the fat globules to such an extent that after 48 hours' storage no visible cream separation occurs on the milk, and the fat percentage of the top 100 ml. of milk in a quart bottle, or of proportionate volumes in containers of other sizes, does not differ by more than 5 per cent of itself from the fat percentage of the remaining milk as determined after thorough mixing.

In 1950, approximately 70 per cent of the 3,225,000 quarts of milk sold daily in New York City was homogenized milk.

Fat may be estimated by the Babcock method as detailed under milk, but both the homogenized milk and the acid should be at about 70° F., the acid should be added in small portions, and only about 16 ml. of acid should be used.

Formerly it was considered sufficient evidence of homogenization to note the regularity of particle size under the microscope due to the belief that the colloid mill or other homogenizing machine produced an emulsion in which the particles are of equal size. This method has proved unsuccessful. Lampitt and Bogod [28] measured the degree of homogenization by estimating the residual fat in definite strata of milk after centrifuging at 2000 r. p. m. for 2 minutes.

---

[28] Lampitt and Bogod, *Chemie et Industrie* Spc. No. 1004-09, April (1934).

Cream Line Method.—This method is based on the observation that a completely homogenized milk will show no cream line on centrifuging at moderate speeds.

Procedure.—Place 20 ml. of milk in a test tube, 15 × 150 mm., and whirl in a centrifuge at a speed near 1200 r. p. m. for a period of 2 minutes. Measure the cream line produced with a millimeter scale and average the measurement. An unadulterated pasteurized milk gives an average cream line of 10 mm. with a minimum of 8 mm. A fully homogenized milk gives a 0 to 1 mm. cream line.

Curd Tension.—This test measures the resistance of a knife to passage through a curd produced by the action of pepsin on milk. There are many processes for altering the properties of curd in milk in order to make the milk more digestible. One of the methods for measuring whether or not a milk has been treated to alter its curd is the Hill test modified by Otting and Quilligan.[29]

Reagent.—Make up 6 g. of scale pepsin, 1 to 3,000 [This ratio implies the pepsin is of such strength that it will digest not less than 3,000 times its own weight of coagulated and disintegrated egg albumin.], and 3.5 ml. of 0.1 N hydrochloric acid to 100 ml. with water. This pepsin solution has a pH of 1.66 determined electrometrically.

Procedure.—Place 10 ml. of the pepsin solution in a suitable jar, brought to a temperature of 95° F. by placing in a water bath. Bring 100 ml. of the milk to the same temperature and add to the jar containing the pepsin and the Hill curd knife. Rotate the jar and contents for an instant, place it back in the water bath, and hold at this temperature for 10 minutes, after which determine the curd tension.

Most milk has a curd tension varying from 60 to 250 g. That is, there is a resistance equivalent to that weight in pulling the Hill curd knife through the curd. A soft curd milk, often a homogenized milk, will have a curd tension of zero to 30 g.

SOFT CURD MILK

Until relatively recent times it seldom occurred to anyone that cow's milk might be unsuitable for some human beings, particularly infants. The milk of cows produces, according to some investigators, a tough, leathery, tenacious mass when acted upon by gastric juice. Human milk yields a fine, flocculent curd under the same conditions. Because of these differences in curd structure, it is believed by some authorities that the former is more indigestible than the latter.

There are a number of methods available for the preparation of

---

[29] Otting and Quilligan, *Milk Dealer*, p. 36, August (1934).

soft curd milk. One of these methods, namely, that of homogenization, has been discussed. Another method is to heat milk considerably above the temperature used for pasteurization. Such treatment produces a soft curd milk. Heat treatment probably accounts for the soft curd produced with evaporated milk or boiled normal milk.

A third method is to gather a herd of cows each one of which has the individual characteristic of producing milk of low curd tension. Since this characteristic property generally remains constant such herds can be assembled so that the herd milk produced has a curd tension of less than 30 g.

The base-exchange treatment of milk is another method for the production of soft curd milk. By this process, in which milk acidified with citric acid is passed through a zeolite filter, about 20 per cent of the calcium is removed with a slight alteration in the phosphorus content and salt content of the milk.

Milk of low curd tension may be obtained by the addition of the sodium salts of hexametaphosphate, pyrophosphate, metaphosphate, and citrate. Another method is to treat milk with a proteolytic enzyme such as trypsin. Still another method is to subject milk to intense high-frequency vibration in an oscillating device. No pressure is used. This type is known as sonic soft curd milk.

The methods of analysis to differentiate a soft curd milk from other types of milk consist of making a curd tension determination (discussed under the section "Homogenized Milk" in this chapter) and analyses of the calcium concentration in the case of zeolite treated milk. Other types must be analyzed for specific components such as added phosphate, citrate, etc.

RECONSTITUTED MILK

Reconstituted, reconstructed, or recombined milk is a product resulting from the recombining of milk components with water and which complies with the standards of milk fat and solids-not-fat of normal milk. Remade, reconstituted, or reconstructed milk is milk that has been made from milk powder either skimmed or whole, from cream, from evaporated or frozen milk, or from any combination of these products and butter or butter oil. In general, it is any type of milk manufactured by any process other than the normal production by cows.

In order to produce a properly reconstructed milk it is generally homogenized. Such milk is generally made according to a formula and, if the milk is to comply with the standards of normal milk, each ingredient must be fully incorporated. Often the sole ingredients are butter or butter oil, skim milk powder, and water. In order for such a mixture to remain in suspension, it must be properly homogenized, otherwise a sediment will settle out.

During World War II it was demonstrated that milk could be reconstituted from several source materials to give a relatively palatable beverage. This reconstituted, or reconstructed milk as it is sometimes called, often had virtually the same appearance as normal pasteurized milk but did not have the same odor, taste, or flavor. One characteristic of reconstituted milk is that it is homogenized either by peptization methods or by means of a homogenizer.

For many years the use of homogenized milk was prohibited by many communities because it was feared that reconstituted milk might be substituted in whole or in part for normal milk and the detection of homogenization served as a convenient means of differentiating reconstituted milk from normal milk. However, with the acceptance of homogenized milk as an approved article of commerce, this simple method of differentiation was no longer available. In addition, reconstituted milk is now an accepted article of commerce in certain communities, and very likely as methods are developed for its differentiation from fresh milk it will become a permitted milk product in communities that forbid its sale.

The *Milk Ordinance and Code* of the U. S. Public Health defines reconstituted milk as follows: Reconstituted or recombined milk is a product resulting from the recombining of milk constituents with water, and which complies with the standards for milk fat and solids not fat as defined. Reconstituted cream or recombined cream is a product resulting from the combination of dried cream, butter, or butterfat with cream, milk, skim milk, or water.

The Louisiana State Sanitary Code defines reconstituted milk as a product resulting from the recombining of milk constituents with fluid milk or water and which complies with the standards for milk, milk fat, and solids-not-fat of milk as defined.

The use of reconstituted milk was greatly expanded during World War II for there was a marked shortage of safe fluid milk both abroad and in areas of our own country. To overcome this, the Quartermaster Corps of the U. S. Army issued specifications for non-fat and whole dried milk solids as well as concentrated whole and skim milk to be used in making reconstituted milk.

The status of reconstituted milk was surveyed by Hauser and King.[30] The legal aspects of its sale in 1947 in various States is tabulated in Table 16.

---

[30] Hauser and King, *Am. J. Pub. Health* **37**, 1284 (1947).

The tests detailed in the preceding section merely distinguish between homogenized and nonhomogenized milk and do not necessarily imply reconstitution. Tests for reconstituted milk depend on the estimation of components and properties that are heat labile or are likely to change due to the manner of processing or due to age. Some of the factors and properties that change readily are taste, odor, flavor, general appearance, vitamin content, ammonia content, sulfhydryl and volatile sulfur content,[31,32] fluorescent properties [33] and others. If, for example, butter has been used to reconstitute milk and the butter was colored, the milk will contain color. If milk powder, either whole or skimmed, is used and the milk is not properly homogenized, the sediment will generally be greater and the milk will not filter easily through a Fisher cotton filter pad placed in a Gooch crucible, or some similar arrangement.

TABLE 16. STATE REGULATIONS GOVERNING SALE OF RECONSTITUTED MILK [a]

Illegal	U. S. P. H. S. Milk Ordinance and Code	State Enforcement Through Local Ordinances	State Milk Ordinance	No State Regulations	Unclassified
California	Colorado	N. Dakota	Alabama	Delaware	Missouri
Florida	Georgia	S. Dakota	Connecticut	Indiana	Oregon
Massachusetts	Kentucky	Virginia	Idaho	Iowa	
Michigan	Nevada	Texas	Illinois	New Jersey	
New York	S. Carolina	Washington	Louisiana	New Hampshire	
Ohio	Tennessee		Maine	Pennsylvania	
Rhode Island	W. Virginia		Maryland	Utah	
Wisconsin			Minnesota		
			Mississippi		
			Nebraska		
			New Mexico		
			Vermont		
			Wyoming		

[a] After Hauser and King, Am. J. Pub. Health 37, 1284 (1947).

If all of the gross characteristics of a milk are apparently normal, which will indeed be difficult in the matter of taste and odor in a fully reconstituted milk, reconstitution may be determined by the estimation of a heat labile component such as vitamin C or by ascertaining the quantity of ammonia which increases with age. Butter and most skim milk powders have little vitamin C because most of the vitamin

---

[31] Beck and Urack, Z. Untersuch. Lebensm. 45, 399 (1933).
[32] Lea, Analyst 71, 227 (1946).
[33] Radley, Analyst 58, 527 (1933).

is lost in the various processes for the production of the material. Vitamin C, or rather the iodine reducing materials of milk, may be estimated by the following procedure.

Iodine Reduction.—To 100 ml. of milk in a small beaker, add 8 ml. of 40 per cent trichloroacetic acid. Stir thoroughly, allow to stand 5 minutes, and transfer to a centrifuge bottle. Centrifuge at 1400 r. p. m. for 5 minutes. Filter the supernatant liquid through a medium fast retentive filter paper. Refilter. Transfer 50 ml. of the filtrate to a small flask, add 1 to 2 ml. of starch solution, and titrate with 0.01 N iodine solution which has been carefully standardized, to a blue color.

The above method is purely empirical and does not really give only the vitamin C content, because all substances easily reduced are included. Nevertheless, the result is indicative of the quality of the milk tested. This test should not be performed on old milk, for such results would have little meaning. To illustrate the differences in reducing power of various milks the average vitamin C content of approximately two hundred samples are given:

Raw milk ....0.028 mg. per ml.
Pasteurized milk ....0.023 mg. per ml.
Reconstituted milk ....0.009 mg. per ml.

These averages are based on the assumption that 1 ml. of 0.01 N iodine solution is equivalent to 0.88 mg. of ascorbic acid. The lower and upper limits of vitamin content overlap in the case of raw and pasteurized milk, but a milk containing 25 per cent reconstituted milk is suspicious and one containing 50 per cent is easily detected.

Ammonia Content.—The quantity of ammonia increases markedly with age and quality of milk.[34, 35] The relationship is illustrated by the following data:

Best quality ....0.1 to 0.12 mg. per 100 ml.
Medium quality ....0.15 to 0.18 mg. per 100 ml.
Poor quality ....0.22 to 0.23 mg. per 100 ml.

Methods for the determination of ammonia are detailed elsewhere in the text.

Evenson Method.—This method[36, 37] for the detection of reconstituted milk is based on the production of a yellow color in the curd of remade milk.

[34] Burstein and Frum, Z. Untersuch. Lebensm. 69, 421 (1935).
[35] Kluge, Z. Untersuch. Lebensm. 71, 232 (1936).
[36] Evenson, Intern. Assoc. Dairy Milk Inspectors, Ann. Rept. 12, 354 (1923).
[37] King and Schonest, Assoc. Food Drug Officials U. S., Quart. Bull. 8, 136 (1944).

Procedure.—To 25 ml. of milk add 25 ml. of water and warm to 30° C. Add 4 ml. of 10 per cent acetic acid. Add 200 ml. of water and allow the curd to settle. Filter through bolting cloth or silk, wash 3 or 4 times with water and wash the curd back into the beaker. Wash 3 or 4 times by decantation, filter again, and wash. Place the curd in test tubes or vials or in small beakers and add 10 ml. of 5 per cent sodium hydroxide solution. After 2 hours, a yellow color appears in the case of remade milk. It is best to run a control on known milk along with the sample to be tested and compare the colors produced.

Milk reconstituted from milk powder made by the spray wheel system or from unsweetened condensed or concentrated skim milk gave negative or at best only doubtful positive Evenson reactions.

Peroxidase Test.—There are methods (Storch test [38] and Arnold test [39]) in the literature designed to detect ''heated milk,'' that is, milk that has been heated for any length of time over pasteurization temperatures (143° F.—30 minutes holding method; 160° F.—15 seconds high temperature—short time method). These methods are based on the demonstration that the peroxidase enzyme or enzymes are destroyed.

There are 20-25 units of peroxidase enzyme in a liter of milk according to Willstatter and Stoll,[40, 41] that is, milk has a purpurogallin number of 0.020-0.025. This enzyme is inactivated by heating above 70° C., by decreasing the pH to less than 4 and it is irreversibly inactivated at pH 10. In addition, it is apparently inactivated by the methods used at the present time (1950) for the dehydration of milk.

The method adopted by the author [42] for the routine differentiation of reconstructed milk from normal approved and homogenized milk is a modification of the Arakawa method.[43]

Reagents.—Guaiac-Resin Reagent.—Dissolve 0.3-1 g. of guaiac resin in absolute alcohol. Add 0.02 g. of arsenic trioxide, 0.6 g. of glacial acetic acid, and 1.3 g. of sodium acetate. Mix and add sufficient absolute alcohol to make the volume 100 ml.

Guaiacol-Peroxide Reagent.—Dissolve 2 ml. of guaiacol in absolute alcohol. Add 0.1 ml. of 3 per cent hydrogen peroxide solution and make to 100 ml. with absolute alcohol.

Phosphate Buffer pH 7.8.—Dissolve 0.92 g. of sodium dihydrogen

[38] Storch, *Z. Untersuch. Nahr. u. Genussm.* 2, 239 (1899).
[39] Jensen, *Essentials of Milk Hygiene.* J. B. Lippincott, Philadelphia.
[40] Willstatter and Stoll, *Ann.* 416, 21 (1918).
[41] Davies, *Chemistry of Milk.* Van Nostrand, New York, 1939.
[42] Jacobs, work performed in 1946.
[43] Tayura Arakawa, *Tohoku J. Exptl. Med.* 16, 83 (1930).

phosphate, $NaH_2PO_4 \cdot H_2O$ ($M/15$) in water and dilute to 100 ml. Dissolve 32.24 g. of disodium hydrogen phosphate, $Na_2HPO_4 \cdot 12H_2O$ (0.1 M) in water and dilute to 900 ml. Mix both solutions.

Procedure.—To 1 drop of milk in a test tube, add 3 ml. of phosphate buffer, pH 7.8, 1 ml. of guaiac-resin reagent, and mix. Run 1 ml. of guaiacol-peroxide reagent down the side of the tube so that this reagent overlays the mixture in the tube. Note the color at the interface within 5-10 seconds. In the case of normal milk, a deep blue color appears almost instantaneously; in the case of reconstituted milk, the interface remains colorless or a very faint blue is produced after 10 seconds. Mix the contents of the tube by shaking. A blue color spreads throughout the body of the mixture with normal milks but reconstituted milks remain colorless or virtually colorless.

Certified Milk

Milk which is produced under exceptionally rigid conditions of cleanliness and sanitation so that its original bacteria count is very low has been marketed under the name of certified milk. Certified milk-raw is milk which conforms with the requirements of the American Association of Medical Milk Commissions. Certified milk-pasteurized is certified milk-raw which has been pasteurized, cooled, and bottled in a milk plant conforming with the requirements for Grade A pasteurized milk. Formerly, the Sanitary Code of the City of New York required that certified milk-raw contain not more than 10,000 bacteria per ml., whereas certified milk-pasteurized contain not more than 500 bacteria per ml.

Certified milk may be analyzed by the methods previously detailed in this chapter.

Vitamin D Milk

Vitamin D milk is milk in which the vitamin D content has been increased by a method and in an amount approved by public health authorities. It was first introduced in 1932, primarily because ordinary market milk does not contain sufficient vitamin D to be a completely protective food as far as this vitamin is concerned. In 1943, about 310,000 qts. of vitamin D milk were sold per day in New York City. This represented about 10 per cent of the total volume of household fluid milk sold daily in that city. By 1950, approximately 70 per cent of the 3,225,000 qts. of milk sold daily in New York City was both homogenized and enriched with vitamin D.

Vitamin D milk may be analyzed by the methods detailed in this chapter. The vitamin D content is estimated by bioassays.

(a) *Irradiated Milk.*—There are a number of different types of vitamin D milk marketed. One type, namely, irradiated milk, is produced by the direct irradiation of milk with ultraviolet light. Since milk is known to contain cholesterol, irradiated milk contains the cholesterol type of vitamin D.

(b) *Metabolized Vitamin D Milk.*—Another type of vitamin D milk is known as metabolized vitamin D milk. It is produced by feeding the cow irradiated yeast which is known to contain ergosterol. In view of the basic difference in types of vitamin D in the milk one should expect a difference in potency of these milks when tested on rats and on chickens. There is little practical difference between the two milks, unit for unit, in respect to their effectiveness for rachitic infants.

(c) *Fortified Vitamin D Milk.*—Another type of vitamin D milk is fortified vitamin D milk. It is made by adding to milk a vitamin D concentrate from cod-liver oil or other fish-liver oils. The potency of fortified milk is adjusted by producers at approximately 400 U. S. P. Units per quart and it contains the cholesterol type of vitamin D.

The term fortified vitamin D milk has been extended to cover the use of other concentrates such as other fish-liver oils or irradiated ergosterol. The vitamin D concentrate is emulsified in an oil such as cottonseed oil or in cream and is mixed with the milk to be fortified.

Vitamin Fortified Milk

Some milk is prepared on a commercial scale which has been fortified with vitamins other than vitamin D. These are known as multivitamin milks. They are generally sold under trade names and are enriched with an adult's requirement of some of the vitamins. Vitamins A, B_1, B_2, C, nicotinic acid amide, calcium pantothenate, and vitamin D are used.

CREAM

Cream or sweet cream is the portion of milk, rich in butterfat, that is separated from the remainder of the milk by means of centrifugal force, or that rises to the surface of milk on standing. By far most of the cream sold is the product obtained by means of centrifugal separators. The most important component of cream is, therefore, butterfat.

The different states, cities, and communities of the United States have different standards for grades of cream. The Food and Drug Administration definitions [44] are the following.

---

[44] *U. S. Food Drug Admin., S.R.A., F.D.&C. 2,* Part 18, May, 1953.

Cream Class of Food; Identity.—Cream is the class of food which is the sweet, fatty liquid or semiliquid separated from milk, with or without the addition thereto and intimate admixture therewith of sweet milk or sweet skim milk. It may be pasteurized, and, if it contains less than 30 per cent of milk fat, it may be homogenized. It contains not less than 18 per cent of milk fat, as determined by the Roese-Gottlieb Method. The word "milk" as used herein means cow's milk.

Light Cream, Coffee Cream, Table Cream; Identity.—Light cream, coffee cream, table cream, conforms to the definition and standard of identity prescribed for the cream class of food except that it contains less than 30 per cent of milk fat, as determined by the method referred to above.

TABLE 17. COMPOSITION OF CREAM

Type	Fat, %	Solids-not-fat, %	Total Solids, %
	16.02	7.43	23.45
	17.01	7.34	24.35
Light	18.00	7.25	25.25
	19.04	7.16	26.20
	20.03	7.07	27.10
	21.02	6.98	28.05
	22.00	6.90	28.90
Medium	23.00	6.81	29.81
	24.00	6.72	30.72
	25.00	6.63	31.63
	26.00	6.54	32.54
	27.00	6.46	33.46
	28.00	6.37	34.37
	29.00	6.28	35.28
Whipping	30.00	6.19	36.19
	31.00	6.10	37.10
	32.00	6.09	38.09
	33.00	5.92	38.92
	34.00	5.83	39.83
	35.00	5.75	40.75
Heavy	36.00	5.66	41.66
	37.00	5.57	42.57
	38.00	5.48	43.48
	39.00	5.39	44.39
	40.00	5.30	45.30
	45.00	4.86	49.86
	50.00	4.42	54.42
	55.00	3.98	53.98

Whipping Cream Class of Food; Identity.—Whipping cream is the class of food which conforms to the definition and standard of identity prescribed for the cream class of food except that it contains not less than 30 per cent of milk fat, as determined by the method referred to above.

Light Whipping Cream; Identity.—Light whipping·cream conforms to the definition and standard of identity prescribed for the whipping cream class of food except that it contains less than 36 per cent of milk fat, as determined by the method referred to above.

Heavy Cream, Heavy Whipping Cream; Identity.—Heavy cream, heavy whipping cream, conforms to the definition and standard of identity prescribed for the whipping cream class except that it contains not less than 36 per cent of milk fat, as determined by the method referred to above.

The minimum requirements for fat as given in the Sanitary Code of the City of New York are 18 per cent for light cream, 23 per cent for medium cream, and 36 per cent for heavy cream.

FAT

The principles underlying the method for estimation of fat in cream are similar to those discussed in the sections on milk.

Cold samples should be warmed to 100°-110° F. in order to insure homogeneity of the sample on mixing but overheating should be rigorously avoided to prevent oiling off or separation of the fat.

Babcock Method.—On cream-testing scales which are in proper working condition and of proper sensitiveness weigh 9 g., if a 9-g. bottle is to be used, Fig. 54, or 18 g. if an 18-g. bottle is to be used, of the properly mixed sample into a standard cream test bottle. If 9 g. of cream have been weighed, add 9 ml. of water at about 60° F. rinsing down the neck of the bottle with this water. If 18 g. of cream have been weighed, no water need be added. Add 17.5 ml. of sulfuric acid, specific gravity 1.82 at a temperature of 60° F.

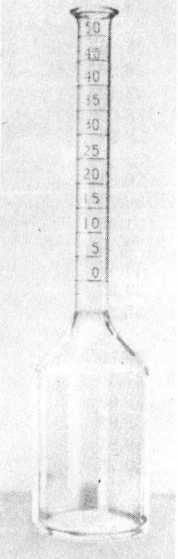

FIG. 54. Babcock Cream Flask

After shaking, if the proper amount of acid of the correct strength has been used, the mixture of cream, water, and acid or of cream and acid should be chocolate brown in color or the color of coffee to which cream has been added. After mixing the sample, whirl in a Babcock centrifuge at the proper speed for 5, 2, and 1 minutes, respectively, filling the bottles with water at a temperature of about 200° F. to the bottom of the neck after the first whirling and to near the top graduation after the second whirling. The centrifuge machine, if heated, should be at a temperature of about 160° F.

If a large number of samples are to be read they should be placed in a tempering bath at 135°-140° F., leaving them in the bath for 5 minutes before reading. Add a few drops of meniscus remover by allowing it to run down the inside wall of the bottle and spread over the fat. Read at once, preferably by using dividers. Read the cream test by measuring the fat from the lowest point of the fat column to the line which divides the fat column from the meniscus remover or leveler, observing all the precautions mentioned in the directions for estimating the percentage fat by the Babcock method in milk.

Meniscus remover may be made by adding an oil soluble dye to Russian mineral oil until a deep red or blue color is obtained. For instance, dissolve 1 oz. of National Oil Red C in 1 qt. of warm light mineral oil and dilute to 5 gal. with the same solvent. Where little of meniscus remover is required proportionately smaller quantities may be prepared. The fat column may also be read without any aids as in the analysis of milk, but in this case from the lower surface to the bottom of the upper meniscus.

The fat column of all tests should be clear, translucent, and have a golden yellow to amber color. All tests which are milky or foggy or show the presence of curd, charred matter, or other foreign material in or below the fat column, or of which the reading is indistinct or uncertain should be rejected. Only test bottles that have the proper specifications should be used.[45]

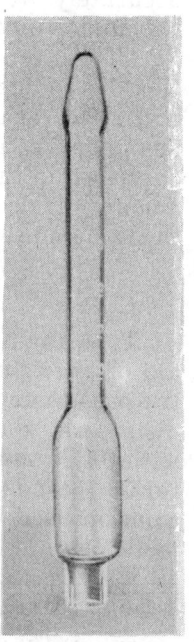

Fig. 55. Cream Butyrometer

Gerber Method.—Measure 10 ml. of sulfuric acid, specific gravity 1.82 into the cream butyrometer, Fig. 55, and balance on a sensitive scale, properly protected against drafts. Weigh in, very carefully, 5 g. of the properly prepared cream sample, or add by means of a calibrated syringe 5 g. of the cream, add 5 ml. of water and 1 ml. of amyl alcohol, insert the stopper and shake, inverting the bottle sev-

[45] *N. Y. State Dept. Agr. Markets, Circ.* **505** (1935).

eral times after the curd is all dissolved, as in the case of milk. Centrifuge for 5 minutes at the proper speed, remove, and read the same as for milk. The fat column of the finished test should be clear, translucent, and should have a golden yellow to amber color. All uncertain tests should be rejected. Only standard butyrometers [46] should be used.

Roese-Gottlieb Method.—Transfer approximately 5 g. of cream accurately weighed, by weighing by difference with the aid of a Mojannier pipette and a carriage to a Mojonnier extraction tube or a Jacobs-Singer separatory flask, or weigh directly into the lower section of the Jacobs-Singer separatory flask, about 5 g. of the sample, noting the exact weight. Add 5 to 6 ml. of water, according to the amount of sample being used, warm to 60° C., and then proceed as described in the method for milk. Care must be taken to add the full amount of alcohol. Divide the weight of fat found by the weight of the sample used and multiply by 100 to obtain the percentage fat.

Acidity

Transfer 8 ml. of the cream sample to a porcelain dish with the aid of an 8.8-ml. pipette or an accurately calibrated syringe. Rinse the pipette or syringe with 8.8 ml. of water and add the rinse to the dish. Add 1 ml. of phenolphthalein indicator solution and titrate with 0.1 N sodium hydroxide solution until a pink color which lasts for 30 seconds is obtained. Divide the number of milliliters of alkali solution used by 10 to obtain the percentage of acidity in terms of lactic acid.

Determination of Other Components of Cream

Total solids, lactose, proteins, and ash may be estimated as described previously under milk. Preservatives may be detected and estimated in a manner similar to those given in the chapter on preservatives, Chapter IV. Added coloring matters may be ascertained as detailed in the chapter on coloring matter in food, Chapter III.

Thickening Agents

Thickening agents are substances added to food and food products to increase the viscosity of those products. Closely allied to thickening agents are emulsifying and stabilizing substances. In ice cream and cheese manufacture, these agents form an important part of the production procedure. In other milk products, the use of these

---

[46] *N. Y. State Dept. Agr. Markets, Circ.* 515 (1936).

materials is of doubtful value. Gelatin, sucrate of lime, starch, gums, and pectins are among the substances generally used.

Within the past decade a number of synthetic thickening agents have been suggested for use in foods. These are discussed by Jacobs.[47] Some of these thickening agents, as, for instance, cellulose methyl ether, have no food value.

Gelatin.—Gelatin has for many years been used as a thickening agent in dairy products. In ice cream it has been accepted because it improves the texture as well as the standing qualities of this product. However, in milk, cream, and like products, its use has generally been frowned upon because of a tendency to use it to conceal inferiority. The methods of Stokes, Jacobs and Jaffe, and Richardson and Tarassuk are detailed. The method of Stokes is adequate for milk and sweet cream, that of Jacobs and Jaffe is adequate for sour milk products, generally. The Richardson and Tarassuk method is much longer than either of the others but is very valuable where doubtful results are obtained by the other methods.

Stokes Method.[48]—The method mainly used for the detection of gelatin in dairy products has been that of Stokes. It depends on the precipitation of the milk proteins by acid mercuric nitrate and the subsequent detection of gelatin by the gelatin-picrate reaction.

Procedure.—To 10 ml. of milk or cream or milk product, add an equal volume of acid mercuric nitrate solution, mercury dissolved in twice its weight of nitric acid and this solution diluted to 25 times its volume with water. Shake the mixture, add 20 ml. of water, stir again, allow to stand 5 minutes, and filter. If much gelatin is present, the filtrate will be opalescent and cannot be obtained clear. To a portion of the filtrate contained in a test tube, add an equal volume of saturated aqueous picric acid solution. A yellow precipitate will be produced in the presence of any considerable amount of gelatin, whereas smaller quantities will be indicated by a cloudiness. In the absence of gelatin, the filtrate will remain perfectly clear.

Sour cream, cultured milks, and sour dairy products and those made with rennet invariably give precipitates with picric acid when the above detailed test is performed. The A. O. A. C. recommends that the characteristics of the gelatin-picrate precipitate be taken into consideration in deciding whether or not gelatin is present. The gelatin-picrate precipitate is a fine granular one which does not settle rapidly, whereas that due to rennet and proteins is a flocculent, rapidly precipitating type.

---

[47] Jacobs, *Synthetic Food Adjuncts.* Van Nostrand, New York, 1947.
[48] Stokes, *Analyst* 22, 220 (1897).

Jacobs and Jaffe Method.[49]—In view of the limitations of the Stokes method, these investigators devised a different method for the detection of gelatin. Basic lead nitrate is used as a protein precipitant, and calcined charcoal is used to adsorb the pseudo-gelatins formed in the souring process. The basic lead nitrate reagent consists of two solutions added separately to the solution or mixture to be clarified.

Procedure.—To 10 ml. of milk or milk product, add 3 ml. of lead nitrate solution, 250 g. lead nitrate dissolved in water and made up to 500 ml., and stir. Add 3 ml. of sodium hydroxide solution, 25 g. sodium hydroxide dissolved in water and made up to 500 ml. and stir. Add 5 ml. of water and stir, add 0.1 g. of calcined charcoal and stir thoroughly, allow to stand for 5 minutes, and filter. To 3 ml. of the filtrate add 2 drops of nitric acid and then a few drops of freshly or recently prepared 5 per cent tannic acid solution. In the presence of gelatin there is a white or brownish voluminous precipitate. In the absence of gelatin the solution remains perfectly clear. As a confirmatory test, add to a portion of the filtrate (no addition of nitric acid is now necessary) an equal volume of freshly filtered saturated aqueous picric acid solution. In the case of considerable quantities of gelatin, there is a heavy precipitate of gelatin-picrate. In the case of smaller quantities there is a turbidity which develops within 2 minutes. In the absence of gelatin, the filtrate will remain perfectly clear even on standing. For the tannic acid test the addition of nitric acid is essential, for otherwise tannic acid will always give a precipitate. For the picric acid test there is no need to use acid. The addition of acid in this case reduces the sensitivity of the test, because gelatin-picrate is somewhat soluble in nitric acid.

The basic lead nitrate test will give a good test for gelatin for one part of gelatin in 2,000 of the milk product. The test is thus less sensitive but more definite than the official or Stokes method. The Jacobs and Jaffe method gives a precipitate with milk products made with rennet as does the Stokes procedure. Richardson and Tarassuk [50] have modified both methods so that this difficulty is overcome by the use of trichloroacetic acid as a secondary protein precipitant.

Richardson and Tarassuk Method.—To 10 ml. of the filtrate obtained from the Stokes or the Jacobs and Jaffe procedures or to an aliquot of these filtrates add ½ volume of saturated picric acid solution. Observe the mixture for clearness and type of precipitate. Chill the remainder of the filtrates and add ½ volume of 20 per cent tri-

---

[49] Jacobs and Jaffe, *Ind. Eng. Chem., Anal. Ed.* 4, 418 (1932).
[50] Richardson and Tarassuk, *J. Assoc. Official Agr. Chem.* 17, 315 (1934).

chloroacetic acid solution, shake well, and allow to stand at 8-10° C. for about 16 hours, with occasional shaking, particularly during the early part of the period. Observe the filtrate-trichloroacetic acid mixture for clearness and type of precipitate. Filter cold, using a medium fast filter paper. Add ½ volume of saturated picric acid to this last filtrate. Observe. The purpose of observing carefully lies in the explanation given by Richardson and Tarassuk that the gelatin-picrate precipitate is a fine granular one, which does not settle rapidly. That attributable to rennet and proteins is a flocculent, rapidly settling type. These tests may be used and should be used to supplement each other and thus definitely prove the presence or absence of gelatin.

Starch.—This substance is probably seldom used as a thickening agent and may be easily detected by the starch-iodine reaction. If the amount of starch used is small, the blue color may be masked and the test should be performed on an acetic acid or preferably a trichloroacetic acid serum of the milk product made by adding 1 ml. of 40 per cent trichloroacetic acid to 10 ml. of milk product or 1 ml. of glacial acetic acid to 100 ml. of diluted sample, allowing to stand and filtering after the curd has settled. The iodine solution is added to a portion of the filtrate and the color produced is observed.

Sucrate of Lime or Sucrose.—Lythgoe [51] Method.—In this method any sucrose present is inverted by acid to form dextrose and levulose, whereas the lactose present is hydrolyzed to dextrose and galactose. Levulose being a ketose will preferentially reduce the molybdate reagent. It is interesting to note that the ketose reagent of Fischl, which is detailed in Chapter X, will also be reduced if levulose is present.

Procedure.—To 25 ml. of cream add 10 ml. of 5 per cent uranium acetate solution, shake well, allow to stand for 5 minutes, and filter. To 10 ml. of the clear filtrate or to as much of the filtrate as is obtained, add a mixture of 2 ml. of saturated ammonium molybdate solution and 8 ml. of dilute hydrochloric acid, 1 volume of acid of 1.12 specific gravity to 7 volumes of water, and place in a water bath at 80° C. for 5 minutes. If sucrose is present the solution will be a prussian blue color, which should be compared with the standard prussian blue color solution. This is prepared by adding a few drops of potassium ferrocyanide and 5 drops of 10 per cent hydrochloric acid to 20 ml. of water containing 1 ml. of a 0.1 per cent ferric chloride solution. Occasionally a sample of pure milk will be found which

[51] Lythgoe, *U. S. Dept. Agr., Bur. Chem. Bull.* 132 (1910).

will give an apparent test but to a much less degree than the standard. Moreover, the color in this case can be removed by filtration, leaving a green filtrate, whereas the color due to sugar or sucrate of lime is not thus removed.

Seliwanoff Resorcinol Reaction.—To 5 ml. of Seliwanoff reagent which is prepared by dissolving 0.05 g. of resorcinol in 100 ml. of dilute hydrochloric acid (1:1) in a test tube, add 1 ml. of a hydrochloric acid serum of the cream or milk product, made by adding hydrochloric acid to milk, allowing to stand and filtering. Heat to boiling preferably in a boiling water bath. A positive reaction is indicated by the production of a red color and the separation of a brown-red precipitate which is soluble in alcohol with the formation of a striking red color.

Gums.—The use of common gums such as tragacanth, karaya, etc., as thickening agents and binders for dairy products has increased a great deal especially in ice cream and cheese products. In cream such substances are considered adulterants. A complete schematic method for their identification is detailed in the chapter on gums, Chapter XI. They may be detected by one of the following methods. These methods depend on the precipitation of proteins by some protein precipitant which does not affect gums, filtering the gum solution and then precipitating the gum with alcohol. These methods should be used to supplement one another.

Patrick [52] Method.—To the sample containing the gum, add half its volume of water and boil for a few minutes. Then add 2 ml. of 10 per cent acetic acid for every 50 ml. of sample, heat to boiling, and add 3 teaspoonfuls of kieselguhr for every 50 ml. of the mixture. Filter on a plaited filter and discard the precipitate. Precipitate the gum from the filtrate by the addition of 12 ml. of 95 per cent alcohol for every 3 ml. of filtrate. Add 3 ml. of a mixture of 95 ml. of 95 per cent alcohol and 5 ml. of concentrated hydrochloric acid for every 3 ml. of filtrate. The acidified alcohol dissolves the milk proteins that have not been previously precipitated. A flocculent or stringy precipitate insoluble in the acid-alcohol mixture and remaining insoluble upon the further addition of 3 ml. of water for every 3 ml. of filtrate shows the presence of gums.

Jacobs Method.—To 10 ml. of the sample or to 10 ml. of a paste of the sample and water add 1 to 2 ml. of 40 per cent trichloroacetic acid solution. Allow to stand for 5 minutes, centrifuge and filter or filter directly as desired. If the filtrate does not come through clear, pass through the filter again. To 2 ml. of the filtrate add 10 ml. of

[52] Patrick, *U. S. Dept. Agr., Bur. Chem. Bull.* **116** (1908).

95 per cent alcohol, stopper and shake. Add a drop of methyl red indicator solution and then 2 to 3 drops of strong ammonium hydroxide solution or sufficient to make the mixture alkaline. Stopper and shake, wait for the protein precipitate to develop. Add 2 to 3 drops of hydrochloric acid and mix. A flocculent or stringy precipitate persisting at this point indicates the presence of gums in the original sample.

All of the gums are soluble in the trichloroacetic acid solution. Some of the split proteins are also soluble. Some of the gums are precipitated on the addition of the alcohol. After neutralization with ammonium hydroxide, all of the gums and the split proteins are precipitated and, on the addition of the hydrochloric acid, all of the proteins are dissolved leaving only the gums as the precipitant. Gelatin, some starches and dextrins are also soluble in the trichloroacetic acid mixture. They are also insoluble upon the addition of ammonium hydroxide and the further addition of hydrochloric acid does not dissolve them. The separation and distinction of gums from these substances is discussed in the chapter on gums, Chapter XI.

RECONSTITUTED CREAM

A discussion of the reconstitution and homogenization of milk was given in this chapter. Most of the discussion applies equally well to cream. That is, the taste, odor, appearance, sediment, cream line after dilution with water and centrifuging, peroxidase content and the so-called added color may all be altered by reconstituting. Hence, the alteration of these properties serves as means for the detection of reconstructed cream.

Evenson Method.—The details of this method[53] for re-made cream are somewhat different than in milk because the fat of the cream must be removed.

Procedure.—To 15 ml. of cream add 15 ml. of water and warm to 35° C. Precipitate the curd with 2 ml. of 10 per cent acetic acid. Filter and wash. Remove the fat by washing first with 25 to 40 ml. of 95 per cent alcohol and then with 100 to 150 ml. of ethyl ether. Wash thoroughly with water. Test for complete removal of lactose by applying the Molisch test. Add a few drops of the Molisch reagent, a 15 per cent alcoholic solution of α-naphthol to a small portion of the wash water and underlay with sulfuric acid. A pink to violet coloration shows the presence of carbohydrate. Continue washing with water until the carbohydrate test is negative. Place the washed curd in test tubes, vials, or small beakers and add 10 ml. of 5 per cent

[53] Evenson, *Intern. Assoc. Dairy Milk Inspectors, Ann. Rept.* 12, 354 (1923).

sodium hydroxide solution. A yellow coloration of the curd indicates a remade cream. It is best to run a control on known samples of cream.

Peroxidase Method.—The peroxidase method for the detection of reconstituted cream has been detailed in a previous section in this chapter.

Other Methods.—Harral[54] uses the formaldehyde titration of milk proteins to detect reconstituted cream prepared from dried milk. The nitrogen content and the formol titration of a sample are estimated. The ratio of nitrogen to formol titration is calculated. The ratio is highly increased by reconstitution because of the destruction of available amino acids.

Letzig[55] detects thickening agents and possible reconstruction by use of the viscosimeter. The addition of pectin and other thickening agents produces an increase in the relative viscosity of the serum over the normal upper limits of these products.

If 5 ml. or 5 g. of cream are shaken with 5 ml. of a mixture of equal parts of benzene and alcohol and the mixture is centrifuged for a short period, reconstituted cream will separate readily its butter fat forming an amber upper layer, whereas fresh cream will remain as an emulsion with an appreciable separation of the fat layer. After strong centrifuging, a reconstituted mixture shows three distinct layers; whereas, on the other hand, a fresh cream mixture shows only two layers, namely, the aqueous lower layer and an upper opaque stratum.[56]

If, instead of the aforementioned solvents, acetone is used, the following results are obtained:[57]

1. Natural cream gives no definite separation.
2. Artificial cream made from
   a. New milk yields an extensive clear·lower layer.
   b. Skimmed dried milk yields four layers consisting of a small amber upper layer, a lower opaque stratum, a clear liquid with some coagula, and a little sediment as the bottom or fourth layer.

It is best to run controls with fresh cream side by side with the sample being tested.

[54] Harral, *Analyst* **58**, 604 (1933).
[55] Letzig, *Z. Untersuch. Lebensm.* **72**, 312 (1936).
[56] Richardson, *Analyst* **53**, 335 (1928).
[57] Richardson, *Analyst* **58**, 686 (1933).

FOREIGN FAT

An imitation cream may sometimes be made from foreign fat, such as coconut oil, hydrogenated fats, etc., and sometimes the fat content of ordinary cream may be raised by the addition of such materials. These practices are considered adulterations and they are forbidden under the Filled Milk Act. They may be detected by an examination of the fat of the sample.

Procedure.—If the cream is sweet, allow it to sour and filter through a large coarse filter with the addition of an equal volume of water. When drained, scrape off the curd and fat from the filter paper and transfer to a beaker. Place the beaker in an oven thermostatically controlled at 100 to 105° C. until the curd chars and the fat is dry. Filter the fat through cotton in a small funnel into a suitable container and then proceed as directed in the chapter on fats and oils, Chapter IX. Sour cream may be filtered directly with the aid of additional water.

Preferentially dry the curd and fat in a vacuum oven, transfer to a continuous extractor, extract the fat with ether, evaporate off the ether, and dry the fat at 70 to 75° C. in an oven thermostatically controlled. Then proceed with the examination of the fat.

SOUR CREAM

Sour cream is defined as cream the acidity of which is more than 0.20 per cent expressed as lactic acid. It is sometimes called cultured cream. Formerly sour cream was made by permitting sweet cream to sour naturally. Most of the commercial product is prepared by inoculation of standardized storage cream with a bacterial culture.

The methods of analysis for sour cream are those detailed for cream. To overcome difficulties in weighing out samples for analysis because of the high-viscosity of this food product, the apparatus devised by the author, shown in Fig. 56, is helpful. It merely consists of two tubes, one adjusted to reach the bottom of the sample bottle and bent so that it will easily fit a Babcock flask and the other tube adjusted above the level of the

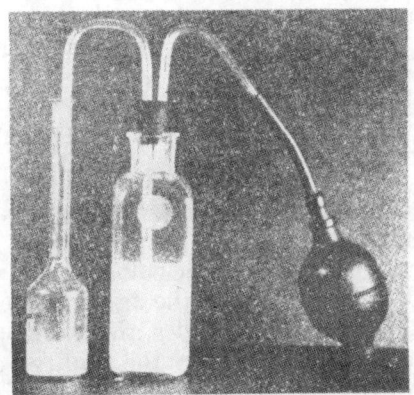

FIG. 56. Aspirator Device

cream to which is attached an aspirator bulb. Pressure on the aspirator bulb forces the cream up the exit tube into the analysis flask.

PLASTIC CREAM

Plastic cream is chilled cream of 65-83 per cent butterfat content, generally, however, covering the range of 80-83 per cent butterfat. It is true cream for, although its butterfat content is similar in quantity to that of butter, the fat of plastic cream is dispersed throughout its water, in contradistinction to butter, in which the water is dispersed in the fat. In other words, plastic cream, like milk or cream, is an oil-in-water emulsion, whereas butter is a water-in-oil emulsion.

Plastic cream is made by an intense centrifugal treatment which skims out nine-tenths of the serum and serum solids. Fluid cream of any fat content may be turned into plastic cream by heating the cream and passing it through a special centrifugal separator. It may then be pasteurized. The process may be used in the salvage of poor and dirty cream, for the extreme method of separation removes the insoluble filth as a residue of bowl-cake sediment.

Plastic cream may be distinguished from butter in that it is crumbly in texture whereas butter is greasy. It is shipped in cartons and tubs, like butter, as it is not fluid. One of its chief uses is the manufacture of a type of cream cheese. This product is also recommended for the manufacture of reconstituted cream and whipping cream.

Plastic cream may be analyzed by the methods detailed for butter in the next chapter.

WHIPPED CREAM

Light cream that is neither pasteurized nor aged will not whip with ease. Formerly whipping cream was defined as cream which contains not less than 30 per cent of milk fat. After aging, light cream will whip, although using a cream of higher fat content will yield better results. By the aging process the acidity is slightly increased. The acid subsequently affects the proteins, namely, casein and lactalbumin, giving them the ability to form the gelatinous consistency which is necessary to entrap the air while the cream is being whipped. This may be arranged more quickly by the addition of 0.1 to 0.3 per cent of commercial lactic acid or by the addition of calcium sucrate or sucrate of lime (Viscogen). The addition of these substances is prohibited by most communities.

Another method for the manufacture of whipped cream is that of impregnation with nitrous oxide. Other gases have been used, but it has been found that none of these can give the high overrun or swell in volume that is. obtained with nitrous oxide. This type of whipped cream is distributed at times in cylinders under pressure, or it may be made at the dispensing counter with the aid of cylinders of nitrous oxide or by use of siphoning apparatus containing bulbs of the gas. By the use of this gas an overrun or increase in volume of as much as 450 per cent can be obtained. With the siphon arrangement single portions of whipped cream can be served.

Products resembling whipped cream are sometimes made by whipping, that is, agitating with the incorporation of air, mixtures of milk powder with some stabilizing agents such as acacia, tragacanth, or other type of gum or gelatin.

The addition of lactic acid may be detected by estimation of the lactic acid content as described in Chapter XX and subsequent comparison with known lactic acid values of cream. Calcium sucrate may be detected as detailed in the foregoing sections.

To detect the presence of nitrous oxide in whipped cream, place the sample in a flask equipped with a stopper having an exit tube. Destroy the foam by the addition of caprylic alcohol and evacuate the flask trapping the nitrous oxide in a gas collector tube. The gas may then be identified by methods detailed by Jacobs.[58]

To detect the presence of carbon dioxide, proceed as above but pass the gas through barium hydroxide solution to precipitate barium carbonate. Some carbon dioxide will normally be present but a marked increase is indicative of the use of this gas in preparing the whipped cream.

The presence of gums may be detected as detailed in previous sections of this chapter and in Chapter XI.

SELECTED REFERENCES

Allen's Commercial Organic Analysis, 5th ed., Vol. IX. Blakiston, Philadelphia, 1933.

Burke, *Practical Dairy Tests.* Olsen, 1935.

Davies, *Chemistry of Milk.* Van Nostrand, New York, 1939.

Eckles, Combs, and Macy, *Milk and Milk Products.* McGraw-Hill, New York, 1936.

Eckles, *Dairy Cattle and Milk Production.* Macmillan, New York, 1918.

Jacobs, ed., *Chemistry and Technology of Food and Food Products,* 2nd ed. Interscience, New York, 1951.

---

[58] Jacobs, *Analytical Chemistry of Industrial Poisons, Hazards, and Solvents.* Interscience, New York, 1949.

Leach-Winton, *Food Inspection and Analysis*. Wiley, New York, 1920.
Methods of Analysis A.O.A.C. Washington, 1945.
Ling, *Textbook of Dairy Chemistry*. Wiley, New York, 1930.
Mojonnier and Troy, *Technical Control of Dairy Products*, Mojonnier Bros., Chicago, 1925.
Rogers, *Fundamentals of Dairy Science*. Reinhold, 2nd Ed., New York, 1935.
Standard Methods of Milk Analysis. Am. Pub. Health Assoc., 9th ed., New York, 1948.
Standard Methods of Water Analysis, Am. Pub. Health Assoc., 9th ed., New York, 1946.
Wolf, *The Human Fuel*. Chapman and Grimes, 1936.
Woodman, *Food Analysis*. McGraw-Hill, New York, 1941.

CHAPTER VIII

MILK PRODUCTS

BUTTER

When cream is churned the fat droplets coalesce and form progressively larger clusters of fat globules. These grains finally break away from the surrounding liquid and form a semisolid or plastic material called butter. It is made exclusively from milk or cream, or both with or without the addition of common salt and with or without the addition of coloring matter. It must contain not less than 80 per cent butterfat.

Either sweet or sour cream may be used for the manufacture of butter. The cream is "neutralized," that is, its acidity is adjusted by the addition of calcium hydroxide, magnesium hydroxide, calcium caronate, sodium carbonate, sodium bicarbonate, etc. After pasteurization, starters are added so that the cream can be ripened. The cream is then churned in which process the colloidal system is inverted from the oil-in-water system of cream to the water-in-oil system of butter. Washing, salting, and working the butter mass prepares it for the market.

Butter is generally sold as salt or sweet butter and in the form of tub, cube, print, or whipped butter. Whipped butter is a more recent departure in butter making. It is the product of the process of incorporating air into the butter mass by means of mechanical agitation or whipping. This produces a product which will spread as easily in the winter as in the summer. It also gives the butter a light yellowish color, a fluffy appearance and a creamy taste. Since butter is generally sold by weight, the incorporation of air makes little difference in its sale in contradistinction to ice cream which is sold by volume and in which the incorporation of air makes a large difference. In the whipping of butter, the size of the butterfat particles is reduced. Thus in unwhipped butter the size of the particles range from 6-7 to 15-20 μ, while in whipped butter the particle-size range is 1-2 to 6-7 μ.

The chief component of butter is, of course, butterfat, which comprises at least 80 per cent of the food material. Thus the average composition of 672 samples [1] was 80.9 per cent fat, 16.1 per cent moisture, and 2.3 per cent salt. Table 18 gives some representative values of butter compositions.

TABLE 18. COMPOSITION OF CREAMERY BUTTER

	No. of Samples	Fat, %	Protein (Curd), %	Lactose, %	Salt, %	Water, %
Maximum	695[a]	86.91	3.42	..	5.26	16.83
Minimum	..	77.64	0.20	..	0.92	10.52
Average	..	82.41	1.18	..	2.51	13.90
Average	573[b]	81.12	0.98	..	2.44	15.46
Average[c]	..	81.0	0.6	0.4	2.5	15.5
Whipped sweet	..	80.1	..	..	0.7[d]	19.2

[a] Thomburg, *Chicago Daily Produce*, 10, 14 (1925).
[b] Thompson, Shaw, and Norton, *U. S. Dept. Agr., Bur. Animal Industry Bull.* 149 (1912)
[c] Chatfield and Adams, *U. S. Dept. Agr., Circ.* 549 (1940).
[d] Curd and salt.

MOISTURE

Weigh 3 g. of the properly prepared sample into a tared, low, flat-bottomed dish and place in a thermostatically controlled oven at 100 to 105° C. for 4 hours. At the end of this period place the dish in a desiccator and, when cool, weigh. The loss in weight divided by the weight of the original sample taken for analysis is deemed to be moisture.

FAT BY DIFFERENCE

Prepare a Gooch crucible with a thin pad of asbestos. Wash, dry, ignite, cool, and weigh. Transfer the contents of the dish from the moisture determination to the Gooch with the aid of a wash bottle containing petroleum ether, maintaining suction on the crucible during the washing. Carefully transfer all of the curd and salt to the crucible with the aid of the ether. Continue washing with the ether until all of the fat has been washed out. Suck dry and place in the oven for 1 hour. Remove the crucible, place in a desiccator, and weigh when cool. The increase in weight is the curd plus salt. The percentage of moisture plus the percentage of curd and salt subtracted from 100 gives the percentage fat.

[1] *North Dakota State Laboratories Dept., Bull.* 67 (1942).

Fat

The percentage fat may be determined directly by placing the dish and contents from the moisture determination in a Soxhlet continuous extractor or some similar apparatus, extracting the fat with petroleum or ethyl ether, receiving the ether in a tared flask, and then evaporating off the petroleum or ethyl ether extract. The flask is dried, cooled, and weighed. The increase in weight is calculated as the percentage fat.

Paley Method.—More rapid methods for determining fat in butter are available. Thus, for instance, a Gerber butyrometer is available for butterfat determinations in butter. In addition, variations of the Babcock method using Babcock test bottles have also been proposed but these have disadvantages.

Fig. 57. Paley-Butter Test Bottle

The Paley-type Babcock test bottle (Fig. 57) has been designed to overcome these difficulties. It is constructed so that the graduated tube is at the side of the body. This permits an opening in the bottle which is large enough to place viscous or lumpy material directly into the flask. The material being analyzed can thus be weighed directly into the bottle on a balance or cream scale. This large opening is closed with a rubber stopper while the calibrated tube is open so that acid or water may be added as in the regular Babcock test. The rubber stopper permits one to raise or lower the fat column in the graduated tube by pressing the stopper in or alternatively drawing the stopper out slightly.

Procedure.—Weigh out 9 g. of butter properly prepared to the consistency of a mayonnaise into an 86 per cent in 0.2 per cent Paley bottle. Melt butter in a warm place until the fat separates and visible curd appears. Add 10 ml. of hot water and mix thoroughly. Add about 10 ml. of Babcock sulfuric acid. Mix well and let stand for 5 minutes. Centrifuge for 5 minutes in a hot centrifuge, then shake very well. Add water almost to the neck of the bottle as for milk and shake vigorously until all visible curd is dissolved. Centrifuge for 3 minutes and shake vigorously. Add water to about the 80 per cent mark. Centrifuge for 10 minutes. After the machine has stopped, permit the bottles to remain in the machine with the heater on for 5 minutes. Place bottle in water bath of about 140° F. to permit fat to be brought to about that temperature, adjust bottom of fat column to zero, replace bottle in same water bath for 5 minutes, then adjust bottom of fat column to zero again, add meniscus remover and read.

SALT

Weigh 10 g. of butter into a beaker. Transfer with the aid of hot water to a separatory funnel. Allow the layers to separate and draw off the water layer into a flask. Do not allow any of the fat to pass through the funnel. Repeat the extraction 10 to 15 times with 20 ml. of hot water and collect the extractions in the same flask. Rinse the original beaker with each portion of wash water before adding to the separatory funnel. Titrate the washings, which will contain practically all of the salt, with standard silver nitrate solution, using potassium chromate as the indicator.

JACOBS RAPID CRUCIBLE METHOD

The moisture, fat, curd plus salt, and salt are all determined on one weighed portion of butter by the use of a specially prepared crucible and technical carbon tetrachloride as a solvent. The method uses a Gooch crucible containing a thin pad of asbestos over which a layer three-eighths of an inch thick of finely ground alundum is placed. The asbestos pad must be thick enough to prevent any of the alundum from passing through, yet thin enough to allow a free flow of the solvent. The crucible is ignited in a muffle oven and then cooled. It is washed well with water, alcohol, and ether and dried in a constant temperature oven. By using technical carbon tetrachloride as the solvent, the collected washings from the fat-by-difference determinations may be saved. The carbon tetrachloride washings may then be purified by distillation and re-used.

Procedure.—Soften the sample of butter, contained in a mason jar, but do not melt, in a steam oven or by some other suitable means and mix thoroughly with a long and broad spatula so that a homogeneous mass is obtained. Weigh 1 to 1.5 g. of the butter accurately into a tared crucible prepared as directed above. Heat the crucible for an hour and a half at 100 to 105° C., place in a desiccator to cool, and weigh at the end of 20 minutes. Calculate the loss in weight as per cent moisture.

Place the crucible in an 8-oz. bottle having a mouth wide enough to hold the crucible and wash with 160 ml. of carbon tetrachloride. Fill the crucible very carefully almost to the top with the solvent and allow to drain completely before adding any more of the solvent. Splashing must be avoided, otherwise the results will be vitiated. At the completion of the washing apply slight suction to the crucible to drain it as completely as possible. Place it in a constant temperature oven and dry at 100 to 105° C. for 1 hour, place in a desiccator to cool, and weigh. The gain in weight over that of the crucible, itself, is calculated as per cent curd plus salt. The sum of the per cent moisture and the curd plus salt subtracted from 100 gives the per cent butterfat.

Ignite the crucible at just below redness, place in a desiccator, and weigh when cool. The loss in weight may be calculated as casein.

Wash the crucible well with water and catch the washings in a flask. Titrate the salt solution with standard silver nitrate solution using potassium chromate solution as indicator, and obtain the percentage of salt.

TOTAL ACIDS IN BUTTER [2]

Weigh 50 g. of butter into each of two 250-ml. centrifuge bottles, add 10 ml. of water, and warm the mixture gently to melt the butter. Neutralize the contents of the bottles with N sodium hydroxide solution using phenophthalein solution for the indicator. Extract the fat with 50 ml. each of ether and petroleum ether. Make the residues in each bottle acid with sulfuric acid and add sodium tungstate. Transfer to a lactic acid extractor '(see Chapter XX) and extract for 3 hours. Combine the extracts from the two 50-g. portions, add 50 ml. of water, and titrate the material with 0.1 N sodium hydroxide solution. This gives the total acidity for 100 g. of butter.

FOREIGN FAT

The butterfat may be prepared for examination for adulteration with fats foreign to milk by placing the butter in a beaker and heating the beaker and its contents in an oven at 100 to 105° C. until it is dry and the curd is charred. Or the butter may be melted in a tall cylinder at about 60° C. for a few hours until the fat and water layers are completely separated and then the supernatant fat is filtered through cotton in a small funnel into a suitable container. The physical and chemical constants of the fat may then be determined as directed in the chapter on oils and fats, Chapter IX.

RENOVATED BUTTER AND OLEOMARGARINE

The *foam test,* also known as the *spoon test* for distinguishing between butter on the one hand, and renovated butter and oleomar-

[2] Hillig and Ahlmann, *J. Assoc. Official Agr. Chem.* 31, 739 (1948).

garine on the other, was originally intended as a household test. It is just as indicative in the laboratory. Heat 2 to 3 g. of the sample in either a spoon or a dish over a small flame. True butter will foam copiously, whereas process butter will bump and sputter like hot grease, with little or no foaming. Oleomargarine behaves like process butter, but chemical tests will determine whether the sample is oleo-margarine or butter.

Another simple test used to distinguish between true butter and reworked butter and margarine is the following. Melt 50 to 100 g. of the sample at 50° C. The curd from butter will settle, leaving a clear supernatant fat, in the case of renovated butter, the super-natant fat remains more or less turbid.

BIACETYL AND ACETYL METHYL CARBINOL

The compound responsible for most of the flavor in butter is biacetyl. Entering into the butter during manufacture are both biacetyl (diacetyl, dimethylketone, 2,3-butandione, $CH_3COCOCH_3$), and acetyl methyl carbinol (acetoin, 3-hydroxy-2-butanone, CH_3-$CHOHCOCH_3$), which are natural flavoring components of butter. The biacetyl arises from 2 sources,

1. Bacterial action on the acetyl methyl carbinol
2. Auto-oxidation of the same compound.

Well-ripened cream, that is of 0.6 per cent lactic acidity from a good starter, contains 5 to 10 p.p.m. of biacetyl and 100 to 200 p.p.m. of carbinol. A starter is a pure bacterial culture used to initiate an industrial fermentation. However, the resulting butter will contain of these substances, calculated as biacetyl, less than 1 to 2 p.p.m. in slightly flavored butter [3] and as much as 4 p.p.m. in very highly fla-vored butter. A greater content of total biacetyl may be viewed with suspicion.

Butter that has an off-taste and odor is sometimes reflavored by the addition of biacetyl. This is an adulteration, according to the definition of butter, because nothing except salt or color may be added. This addition is sometimes done during the process of whipping, that is, during the process of incorporating air into the butter.

Vizern and Guillot Method.—Vizern and Guillot give the follow-ing method [4] for the detection of biacetyl. To 50 g. of sample in a round-bottom, 250-ml. distillation flask, add 20 ml. 95 per cent alcohol,

---

[3] Davies, *Dairy Ind.* 1, 165 (1936).
[4] Vizern and Guillot, *Ann. fals.* 25, 45 (1932).

insulate the neck of the flask to avoid excessive fractionation. Distill by immersing the flask in a calcium chloride bath at 115 to 120° C. and using a vertical condenser, collect 20 ml. of the distillate. Transfer the distillate to a porcelain dish and rinse the receiver with 5 ml. of water. Add successively 1 ml. of 10 per cent hydroxylamine hydrochloride solution and 1.7 ml. of N aqueous sodium hydroxide solution. Stir for 1 minute. Add 1 ml. of 10 per cent nickel sulfate solution and then 0.6 ml. of N acetic acid, drop by drop with constant stirring. Evaporate the alcohol on a water bath. In the presence of biacetyl, a characteristic red zone of nickel dimethylglyoxime adheres to the dish when the contents of the dish have been reduced to about 2 to 3 ml. According to Vizern and Guillot, pure butter, free from improver gives a negative test but, according to more recent investigations, pure butters do have a slight amount of biacetyl as noted above.

This test may be made more sensitive and far more characteristic by continuing the evaporation to dryness. Then extract the residue with 3 successive portions of chloroform and filter each extract through a small filter, catching the filtrate in another porcelain dish. Evaporate to dryness either spontaneously or on a water bath at low temperature. If biacetyl was present in the original sample, a series of red rings resulting from nickel dimethylglyoxime will be found in the porcelain dish. With continued experience the approximate amount of biacetyl may be estimated from the depth of color in the porcelain dish attributable to the rings.

Modified Barnicoat Method.—The following Barnicoat [5] method, modified by Jacobs, may be used to determine the amount of biacetyl quantitatively in butter and like products.

Procedure.—Weigh 400 g. of butter into a liter Florence flask or better into a liter flask equippel with a stopper. The stopper has two tubes arranged for steam distillation one of which tubes has a Polenske or other trap attachment. Place the flask into an oil bath and add 75 ml. of 0.1 N sulfuric acid and enough salt to saturate the solution. It is best to have the flask almost completely immersed in the oil at a temperature of 115° to 120° C. Connect the flask with a source of steam and to an upright condenser by means of a Polenske trap or with the above-mentioned connections. Have the tip of the condenser dip into or have an adapter connected to the condenser dip into a mixture of 4 ml. of 20 per cent hydroxylamine hydrochloride solution, 4 ml. of 20 per cent sodium acetate, and 2 ml. of 5 per cent nickel sulfate. Steam distill until 150 to 200 ml. has been carried over, stirring the distillate and the fixing solution occa-

[5] Barnicoat, *Analyst* **60**, 653 (1935).

sionally. Evaporate the distillate on a hot plate or steam bath at 80° C. until a volume of 10 ml. is obtained. Allow to stand and cool. Transfer the remaining solution and precipitate to a small separatory funnel, washing the vessel in which the evaporation was performed with two 5-ml. portions of water. Extract the nickel dimethylglyoxime with 4 successive portions of chloroform, 15 ml. for the first portion and then with 10-ml. portions. Wash the vessel in which the evaporation was carried out with each successive portion of chloroform before transferring to the separatory funnel. Allow each chloroform extract to separate completely and then draw off the lower layer, passing it through a very small filter, catching the filtrate in a flat-bottomed crystallization dish or small evaporating dish that has been previously dried and weighed. After the four portions of chloroform have passed through the filter paper wash the filter with one more 5-ml. portion of the solvent. Evaporate the chloroform spontaneously or on a water bath at low heat. Dry in a thermostatically controlled oven at 100 to 105° C. for an hour, place in a desiccator to cool and weigh. The weight is due to nickel dimethylglyoxime, and the various relationships may be computed as follows:

1. Nickel dimethylglyoxime × 0.596 equals biacetyl
2. Nickel dimethylglyoxime × 0.610 equals acetyl methyl carbinol
3. Distillation with ferric chloride yields both biacetyl and acetyl methyl carbinol and is computed as total biacetyl
4. Distillation in carbon dioxide atmosphere yields only biacetyl
5. The results (3) minus (4) give acetyl methyl carbinol

The reactions taking place are the following:

$$CH_3COCOCH_3 + H_2NOH \rightarrow CH_3\overset{\overset{NOH}{||}}{C}-COCH_3 + H_2O$$

$$CH_3\overset{\overset{NOH}{||}}{C}-COCH_3 + H_2NOH \rightarrow CH_3\overset{\overset{NOH}{||}}{C}-\overset{\overset{NOH}{||}}{C}CH_3 + H_2O$$

$$2CH_3-\overset{\overset{NOH}{||}}{C}-\overset{\overset{NOH}{||}}{C}-CH_3 + NiCl_2 \rightarrow [\ CH_3\overset{\overset{NOH}{||}}{C}-\overset{\overset{NO}{||}}{C}-CH_3\ ]_2\ Ni + 2HCl$$

The nickel dimethylglyoxime forms a yellow solution in chloroform and may be estimated colorimetrically by comparing with known amounts of the nickel organic compound dissolved in the same solvent. Nickel dimethylglyoxime is also soluble in tetrachloroethane which may be used as the solvent in colorimetric determinations.

PASTEURIZATION

Phosphatase Test.—The Sanders and Sager method may be used to check whether or not butter has been made from pasteurized milk or has been pasteurized.

Procedure.—Weigh, on a piece of wax paper about 1 × 1 in., on a balance, a 1.0-g. sample (preferably two) and insert the paper with the sample into the tube. Similarly, weigh another sample and place in a tube as a control or blank. Heat the *blank* to the temperature indicated under "Cheese" and cool to room temperature. From this point, treat the blank and the test in a similar manner. Add 10.0 ml. of barium buffer substrate 2-a or 2-b as described on page 282 (prepared with 18-8 barium buffer, Table 15), stopper the tube, and mix.

Follow the directions given for the corresponding steps under "Cheese," mixing the contents frequently and thoroughly during incubation, substituting the appropriate zinc precipitant (Table 15) and, for merely detecting under-pasteurization, using 2 rather than 4 drops of 2,6-dibromoquinonechloroimide solution.

Calculation and Evaluation of Result.—When using 1.0 g. of butter and adding 11.0 ml. of liquid, multiply the value of the reading by 1.1 to convert the result to phenol equivalents per 0.5 g. of butter. Evaluate the result by comparing it with the criterion of pasteurization in Table 15.

BUTTER OIL AND GHEE

Butter oil is the product made by clarifying butter by means of melting and centrifuging. It contains practically no water or any of the other components of butter, and, consequently, is almost entirely butterfat.

Ghee is a very similar product. The butter made from buffalo milk is clarified by boiling and filtration so that a product resembling an oil is obtained. This product will not turn rancid as easily as butter.

The methods used for the analysis of butter oil and ghee are analogous to those detailed for butter.

CHEESE

Cheese is the food product made from the separated curd obtained by coagulating the casein of milk, skimmed milk, or milk enriched with cream. The coagulation is accomplished by means of rennet or other suitable enzyme, lactic fermentation, or by a combination of the two. The curd may be modified by heat, pressure, ripening ferments, special molds, or suitable seasoning. Certain varieties, such as for example Roquefort cheese, which is made from the milk of sheep, are made from the milk of animals other than the cow. Cheese unqualified generally means American Cheddar cheese. There are hundreds of names for cheese, native and foreign, although there are only about 18 varieties. These may all be classified under the headings, whole milk cheese, part skim milk cheese, skim milk cheese, pasteurized cheese and process cheese, or as hard or soft curd cheese.

TABLE 19. COMPOSITION OF VARIETIES OF CHEESE

Variety	Water	Fat	Protein	Ash
Brick	42.5	30.7	21.1	3.0
Caciocavallo	35.0	22.0	34.3	7.0
Camembert	47.9	26.3	22.2	4.1
Cheddar	36.8	33.8	23.7	5.6
Cottage	69.8	1.0	23.3	1.9
Cream	42.7	39.9	14.5	1.9
Edam	38.1	22.7	30.9	6.2
Emmentaler and Swiss....	33.0	30.5	30.4	4.2
Gorgonzola	37.3	34.7	25.2	3.8
Gouda	38.1	24.5	29.6	6.1
Gruyère	30.0	28.2	33.0	4.0
Limburger	54.8	19.6	21.3	5.2
Munster	40.6	31.0	22.2	4.6
Neufchatel	52.1	23.5	19.3	5.0
Parmesan	17.0	22.7	49.4	7.6
Pecorino	29.8	30.5	33.5	6.2
Romano	29.6	27.7	31.2	8.7
Roquefort	38.7	32.2	21.4	6.1
Sap sago	47.8	2.0	41.6	11.9
Stilton	33.6	31.2	29.0	3.0

The analyses given in the table have been abstracted from *U. S. Dept. Agr. Bull.* 608, "Varieties of Cheese" (1932). Analyses for process cheese and for filled cheese should follow closely the varietal name.

The classifications other than process cheese are self-explanatory. In this generation, the processing of cheese has become a large industry. Process cheese is the modified cheese made by comminuting and mixing one or more lots of cheese into a homogeneous, plastic mass, with the aid of heat, with or without the addition of water and with the incorporation of not more than 3 per cent of a suitable emulsifying agent. The name process cheese unqualified applies to process Cheddar cheese and those with a varietal name correspond to the process variety indicated by that name. These types of cheese must conform to the limits set for the variety of cheese.

Table 19 gives the percentage composition of some of the more well-known varieties of cheese.

Cheese is adulterated in a number of ways. It may be substandard in the limits set for fat and protein. It may contain excessive amounts of water, or it may contain foreign fat. It may, if it be a process cheese, contain excessive amounts of water and emulsifying agent or binder. Many times cheese is misbranded not only as to brand name but also as to method of processing. These adulterations and misbranded statements are generally capable of detection by chemical methods. Some of these will be detailed.

DEFINITIONS

The Food and Drug Administration [6,7] proposed a number of definitions and standards of identity for various types of cheese, processed cheese, and cheese foods and spreads. These are tabulated in Table 20. Abstracts of some of these follow.

Cheddar Cheese, Cheese; Identity.—(a) Cheddar cheese, cheese, is the food prepared from milk by the procedure set forth in paragraph (b) below. It contains not more than 39 per cent of moisture, and its solids contain not less than 50 per cent of milk fat.

(b) Milk, which may be pasteurized and which may be warmed, is subjected to the action of harmless lactic-acid-producing bacteria, present in such milk or added thereto. Harmless artificial coloring may be added. Sufficient rennet is added to set the milk to a semisolid mass. The mass is cut, stirred, and heated with continued stirring, so as to promote the separation of whey and curd. The whey is drained off and the curd is matted into a cohesive mass. The mass is cut into slabs which are piled and handled so as to promote the drainage of whey and the development of acidity. The slabs are then cut into pieces, which may be rinsed by sprinkling or pouring water over them, with free and continuous drainage; but the duration of such rinsing is so limited that only the whey on the surface of such pieces is removed. The curd is salted, stirred, further drained, and pressed into forms.

(c) Determine the moisture content. Subtract the per cent of moisture found from 100; divide the remainder into the per cent of milk fat found; the quotient multiplied by 100 is considered to be the per cent of milk fat contained in the solids.

---

[6] *Federal Register* 12, 1192 (1947); 14, 1960 (1949).

[7] *U. S. Food Drug Admin., S.R.A., F.D.&C.* 2, Part 19, June, 1952, and amendments through 1957.

TABLE 20. CHEESE FAT AND MOISTURE STANDARDS [a]

Type	Fat[b] % Contained in Solids (not less than)	Moisture % (not more than)	Aging (Curing) (for not less than)	Remarks
Cheddar	50	39	60 days not less than 35° F.	
Washed curd cheese (soaked curd cheese)	50	42	60 days not less than 35° F.	
Colby	50	40	60 days not less than 35° F.	
Granular (stirred curd)	50	39	60 days not less than 35° F.	
Swiss and emmentaler	43	41	60 days not less than 35° F.	
Gruyere	45	39	90 days	
Brick	50	44	60 days not less than 35° F.	
Muenster	50	46	......	
Edam	40	45	60 days not less than 35° F.	
Gouda	46	45	60 days not less than 35° F.	
Blue	50	46	60 days	
Gorgonzola	50	42	90 days	
Roquefort	50	45	60 days	
Limburger	50	50	60 days not less than 35° F.	
Monterey	50	44	......	
High-moisture jack	50	more than 44 but less than 50	......	
Provolone, pasta filata	45	45	90 days not < 35° F.	made in loaves 14-17 lb.
Cacciocavallo	42	40	14 months	
Parmesan, reggiano	32	32	14 months	
Romano	38	34	5 months	
Asiago, fresh and soft	50	45	60 days	

320

TABLE 20—*Continued*

Type	Fat[b] % Contained in Solids (not less than)	Moisture % (not more than)	Aging (Curing) (for not less than)	Remarks
Asiago, medium	45	35	6 months	
Asiago, old	42	32	12 months	
Cream	33 (in finished cheese)	55		
Neufchatel	Not less than 20 but less than 33 (in finished cheese)	65		
Cottage		80		
Creamed cottage	4 (in finished cheese)	80		
Cook		80		
Sap Sago		38		
Hard	50	39	60 days not less than 35° F.	
Semisoft	50	More than 39 but less than 50	60 days not less than 35° F.	
Semisoft part skim	Not less than 45 but less than 50	50	60 days not less than 35° F.	
Soft ripened	50		60 days not less than 35° F.	
Spiced	50		60 days not less than 35° F.	Spices or oils not less than 0.015 oz./lb.
Part-skim spices	Less than 50 but not less than 20			
Hard grating	32	34	6 months	Coated with blue colored paraffin or other blue coating
Skim milk, cheese for manuf.		50		

Subtract the % moisture found from 100, divide the remainder into the % milk fat found; the quotient multiplied by 100 shall be considered to be the % milk fat contained in the solids.

[a] *Federal Security Agency, Food Drug. Admin. S. R. A., F. D. & C. 2,* Part 19, June, 1952.
[b] *Federal Register* 14, 1960 (1949).

321

TABLE 20—Continued

Type	Fat[b] % Contained in Solids (not less than)	Moisture % (not more than)	Aging (Curing) (for not less than)	Remarks
Pasteurized and pasteurized blended	Not less than min. of variety, but not less than 47	Not more than 1% over variety but not more than 43		
Washed curd		40		
Colby		40		
Swiss	43	44		
Gruyère	45	44		
Limburger		51		
With fruits, vegetables, or meats	May be 1% less than variety of pasteurized	May be 1% more than variety of pasteurized		Meats comprise not less than 10% of weight of finished food
Pasteurized process cheese food	23	44		Cheese content not less than 51%
with fruits, vegetables, nuts, or meats	22	44		Cheese content not less than 51% and meat not less than 10% of weight of food
Pasteurized process cheese spread	20	More than 44 but less than 60		Cheese content not less than 51%
with fruits, nuts or meats		More than 44 but less than 60		Meat not less than 10% of weight
Pasteurized cheese spread		More than 44 but less than 60		No emulsifier
Pasteurized process pimento	49	41		Cheddar cheese not less than 75% of wt., pimento 0.2% of wt.

322

The milk may be adjusted by the separation of part of the fat or the addition of cream or skim milk.

Washed Curd Cheese, Soaked Curd Cheese; Identity.—(a) Washed curd cheese, soaked curd cheese, is the food prepared from milk by the procedure set forth in paragraph (b) below. It contains not more than 42 per cent of moisture, and its solids contain not less than 50 per cent of milk fat, as determined by the methods detailed for moisture and fat.

(b) Milk, which may be pasteurized and which may be warmed, is subjected to the action of harmless lactic-acid-producing bacteria, present in such milk or added thereto. Harmless artificial coloring may be added. Sufficient rennet is added to set the milk to a semi-solid mass. The mass is cut, stirred, and heated with continued stirring, so as to promote the separation of whey and curd. The whey is drained off and the curd is matted into a cohesive mass. The mass is cut into slabs which are piled and handled so as to promote the drainage of whey and the development of acidity. The slabs are then cut into pieces, cooled in water, and soaked therein until the whey is partly extracted and water is absorbed. The curd is drained, salted, stirred, and pressed into forms.

The milk may be adjusted by the separation of part of the fat therefrom or the addition thereto of cream or skim milk.

Colby Cheese; Identity.—(a) Colby cheese is the food prepared from milk by the procedure set forth in paragraph (b) below. It contains not more than 40 per cent of moisture, and its solids contain not less than 50 per cent of milk fat, as determined by the methods detailed for moisture and fat.

(b) Milk, which may be pasteurized and which may be warmed, is subjected to the action of harmless lactic-acid-producing bacteria, present in such milk or added thereto. Harmless artificial coloring may be added. Sufficient rennet is added to set the milk to a semi-solid mass. The mass is cut, stirred, and heated with continued stirring, so as to promote the separation of whey and curd. A part of the whey is drained off and the curd is cooled by adding water, the stirring being continued so as to prevent the pieces of curd from matting. The curd is drained, salted, stirred, further drained, and pressed into forms.

This milk may be adjusted by the separation of part of the fat therefrom or the addition thereto of cream or skim milk.

Cream Cheese; Identity.—(a) Cream cheese is the soft uncured cheese prepared by the procedure set forth in paragraph (b) below. The finished cream cheese contains not less than 33 per cent of milk fat and not more than 55 per cent of moisture, as determined, respectively, by the methods detailed for moisture and fat.

(b) (1) Cream or a mixture of cream with one or more of the dairy ingredients specified in (3) below is pasteurized and may be homogenized. To such cream or mixture harmless lactic-acid-producing bacteria, with or without rennet, are added, and it is held until it becomes coagulated. The coagulated mass may be warmed; it may be stirred; it is then drained. The curd may be pressed, chilled, worked, seasoned with salt; it may be heated, with or without added cream or one or more of the dairy ingredients specified in (3) below or both, until it becomes fluid, and it may then be homogenized or otherwise mixed.

(2) In the preparation of cream cheese one or any mixture of two or more of the optional ingredients gum karaya, gum tragacanth, carob bean gum, gelatin, or algin may be used; but the quantity of any such ingredient or mixture is such that the total weight of the solids contained therein is not more than 0.5 per cent of the weight of the finished cream cheese.

(3) The dairy ingredients referred to (1) above are milk, skim milk, concentrated milk, concentrated skim milk, and nonfat dry milk solids. If concentrated milk, concentrated skim milk, or nonfat dry milk solids are used, water may be added in a quantity not in excess of that removed when the milk or skim milk was concentrated or dried.

Neufchatel Cheese; Identity.—(a) Neufchatel cheese is the soft uncured cheese prepared by the procedure set forth in paragraph (b) below. The finished neufchatel cheese contains not less than 20 per cent but less than 33 per cent of milk fat and not more than 65 per cent of moisture, as determined by methods detailed in this chapter.

(b) (1) Milk or a mixture of cream with one or more of the dairy ingredients specified in (3) below or a mixture of concentrated milk with milk or with water not in excess of that removed when the milk was concentrated is pasteurized and may be homogenized. To such milk or mixture harmless lactic-acid-producing bacteria, with or without rennet, are added and it is held until it becomes coagulated. The coagulated mass may be warmed; it may be stirred; it is then drained. The curd may be pressed, chilled, worked, seasoned with salt; it may be heated, with or without added cream or one or more of the dairy ingredients specified in (3) or both, until it becomes fluid, and it may then be homogenized or otherwise mixed.

(2) In the preparation of neufchatel cheese one or any mixture of two or more of the optional ingredients gum karaya, gum traga-

canth, carob bean gum, gelatin, or algin may be used; but the quantity of any such ingredient or mixture is such that the total weight of the solids contained therein is not more than 0.5 per cent of the weight of the finished neufchatel cheese.

(3) The dairy ingredients referred to in subparagraph (1) of this paragraph are milk, skim milk, concentrated milk, concentrated skim milk, and nonfat dry milk solids. If concentrated milk, concentrated skim milk, or nonfat dry milk solids is used, water may be added in a quantity not in excess of that removed when the milk or skim milk was concentrated or dried.

Cottage Cheese; Identity.—(a) Cottage cheese is the soft uncured cheese prepared by the procedure set forth in (b) below. The finished cottage cheese contains not more than 80 per cent of moisture.

(b) (1) One or more of the dairy ingredients specified in (2) below is pasteurized; calcium chloride may be added in a quantity of not more than 0.02 per cent (calculated as anhydrous calcium chloride) of the weight of the mix; harmless lactic-acid-producing bacteria, with or without rennet, are added and it is held until it becomes coagulated. The coagulated mass may be cut; it may be warmed; it may be stirred; it is then drained. The curd may be washed with water and further drained; it may be pressed, chilled, worked, seasoned with salt.

(2) The dairy ingredients referred to (1) above are sweet skim milk, concentrated skim milk, and nonfat dry milk solids. If concentrated skim milk or nonfat dry milk solids is used, water may be added in a quantity not in excess of that removed when the skim milk was concentrated or dried.

Creamed Cottage Cheese; Identity.—Creamed cottage cheese is the soft uncured cheese prepared by mixing cottage cheese with pasteurized cream or a pasteurized mixture of cream with milk or skim milk or both. Such cream or mixture is used in such quantity that the milk fat added thereby is not less than 4 per cent by weight of the finished creamed cottage cheese. The finished creamed cottage cheese contains not more than 80 per cent of moisture.

MOISTURE

Weigh 4 to 5 g. of the properly prepared sample into a tared flat-bottomed dish containing a small stirring rod and sand. Rub the cheese and sand carefully together and place the dish in a constant temperature oven overnight at 100 to 105° C. Remove the dish from the oven, place in a desiccator to cool, and weigh. The loss in weight is calculated as per cent moisture.

FAT

Weigh accurately into the lower section of a Jacobs-Singer separatory flask about 1 g. of the sample and add 9 ml. of water and 1 ml. of ammonium hydroxide. Warm on a steam bath or hot plate, regulated at a low temperature. Stir by shaking until the curd is completely softened. Add ½ ml. of hydrochloric acid and stir. Add 10 ml. more of hydrochloric acid and a pinch of sand. Boil the mixture gently for 5 minutes. Cool the mixture and stopper with the upper section of the separatory flask. Add water to the middle of the connecting joint, and shake. Add 25 ml. of ethyl ether and, after stoppering the flask, shake thoroughly. Add 25 ml. of petroleum ether, again shaking vigorously. Allow the layers to separate and draw off the ether layer into a tared fat flask. From this point proceed as directed in the Roese-Gottlieb method for the determination of fat from the point of drawing off the ether layer into a tared fat flask.

As an alternative procedure, weigh into a tall form 100-ml. beaker, 1 g. of the sample and add 9 ml. of water and 1 ml. of ammonium hydroxide. Warm on a hot plate, stirring with a glass rod, until the curd is completely softened. Add ½ ml. of hydrochloric acid and stir. Add 10 ml. more of hydrochloric acid and a pinch of sand. Cover the beaker with a watch glass on glass hooks and boil the mixture gently for 5 minutes. Cool the mixture and transfer to a Mojonnier tube or to a Jacobs-Singer separatory flask. Rinse the tall form beaker with sufficient water to bring the level of the water up to the middle of the constriction in the Mojonnier extraction tube or the middle of the connecting joint of the Jacobs-Singer separatory flask. Then rinse the beaker and watch glass and glass hooks with 25 ml. of ethyl ether and add the washings to the extraction tube or flask. Stopper and shake thoroughly. Repeat with 25 ml. of petroleum ether. From this point proceed as directed above.

Gerber Method.—The fat of a soft curd cheese may easily be determined by the following modification of the Gerber method. Weigh or counterpoise a scoop, Fig. 58, held in a one hole rubber stopper and weigh into the scoop exactly 2.5 g. of the cheese. Place the scoop into a cheese butyrometer, Fig. 58, and add 9 ml. of 4 per cent borax solution and 1 ml. of amyl alcohol. Heat to 70° C. until the curd is completely disintegrated. Cool slightly and add carefully 10 ml. or less if necessary of sulfuric acid, specific gravity 1.82. Insert the small stopper and then mix thoroughly by inverting the butyrometer several times. Whirl for 10 minutes at the proper speed, remove the butyrometer, and read the fat percentage directly as directed under milk and cream. The specifications for the cheese buty-rometer are also rigidly drawn.[8]

Paley Method.—Using a cream scale or balance, weigh into the bottle (with 20 per cent, 9 g., in ²⁄₁₀ per cent bottle, or 50 per cent, 9 g., in ½ per cent bottle), through the large opening, 9 g. of a representative sample. Use a narrow spatula for transferring sample into bottle, or a stirring rod if the material is cream cheese or cottage cheese. Wash the inside of the mouth of the bottle with about 10 ml. of hot water. Insert stopper securely into the opening. Without delay, very carefully add 17.5 ml. of sulfuric acid (specific gravity 1.82-1.83 at 68° F.), 1 to 2 ml. at a time, through the reading tube, mixing thoroughly after each addition until the material assumes a chocolate brown color and no

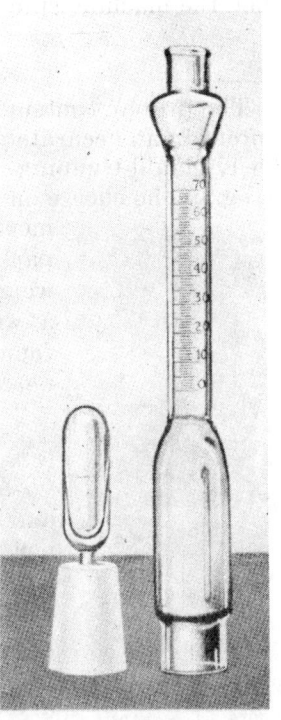

Fig. 58. Gerber Cheese Butyrometer and Scoop

lumps or particles of cheese are inside. Let bottle stand for 5 minutes, shaking well every minute or two. Place bottle in a centrifuge such as is used in milk analysis by the Babcock Method and centrifuge for 5 minutes. Add hot water until liquid level is at base of neck and centrifuge again for 2 minutes. Shake thoroughly. Add hot water again until entire fat column is in the reading tube, and centrifuge for 10 minutes. Place in water bath at 140° F., for 5 minutes, add meniscus remover and read direct, or manipulate the stopper until the bottom of the fat column is exactly at the zero mark, then add meniscus remover and read to junction of the meniscus remover and the butter fat.

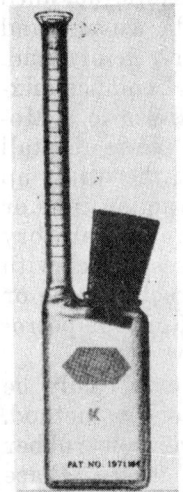

Fig. 59. Paley 20% Cheese Test Bottle (Courtesy of Kimble Glass)

After final addition of water, the Paley test bottles should be placed in the centrifuge so that the stopper faces the center of the

---

[8] *N. Y. State Dept. Agr. Markets, Circ.* **515** (1936).

machine, that is, the reading tube must be nearest to the periphery of the machine. The fat will then flow completely in the reading tube when the machine spins.

PROTEIN

The protein content may be estimated by determining the nitrogen content in an accurately weighed portion of the cheese, as directed in the Kjeldahl-Gunning-Arnold method given in Chapter I. Weigh 2 to 3 g. of the cheese on ashless filter paper as described in the above-mentioned method and then the contents and filter paper are transferred to the Kjeldahl flask or weigh the cheese by difference into the flask from a weighing bottle. The percentage of nitrogen found multiplied by the factor, 6.38, equals the percentage of protein.

FIG. 60. Paley 50% Cheese Test Bottle (Courtesy of Kimble Glass)

OTHER COMPONENTS

Ash, preservatives, coloring matters, acidity, and chlorides may be estimated in the usual manner. For the determination of tartaric acid, citric acid, and phosphates that are used as emulsifying agents in the manufacture of process cheese, refer to the appropriate chapters. For the estimation of lactose, refer to Chapters VII and X. For the detection of gums in cheese, the reader may use one of the methods outlined under cream or in the chapter on gums, Chapter XI. The fat may be extracted for examination in a manner entirely analogous to the one detailed in the following section concerning ice cream.

PASTEURIZATION

Sanders and Sager Phosphatase Method.—The basis for this method [9] and most of the reagents have been discussed and detailed in Chapter VII in the section of Pasteurized Milk.

Sampling.—Hard Cheese.—Take a sample from the interior with a clean Roquefort trier, place in a small tube, stopper the tube, and keep it in a refrigerator.

Soft and Semi-soft Ripened Cheese.—Harden the cheese by chilling it in the freezing chamber of a refrigerator. Take special pre-

---

[9] Sanders and Sager, *J. Dairy Sci.* **30**, 909 (1947).

cautions to avoid contaminating the sample with phosphatase that may be present on the surface; use either of the following methods for sampling:

a. Cut a portion from the end of the loaf or from the side of the cheese, extending in at least 2 in. if possible or to a point somewhat beyond the center in the case of a small cheese. Cut a slit about 0.5 in. deep at least halfway around the portion and midway between the top and bottom. Break the portion into two parts, pulling it apart so that it breaks on a line with the slit, being careful not to contaminate the freshly-exposed, broken surface. Remove the sample from the freshly-exposed surface at or near the center of the cheese.

b. Remove the surface of the area to be sampled, that is, the end and the adjacent sides, with a clean knife or spatula, to a depth of 0.25 in. Clean the instrument and hands with hot water and phenol-free soap and wipe them dry. Remove the freshly-exposed surface to a similar or greater depth, and repeat the cleaning. Then take the sample from the center of the freshly-exposed area, preferably at or near the center of the cheese in the case of a small cheese.

Process Cheese, Spreads, Butter, and Other Non-fluid Products. —Take the sample from beneath the surface with a clean knife or spatula.

Additional Reagents.—Zinc-copper Precipitant for Unripened Cheese.—Dissolve 6.0 g. of zinc sulfate and 0.1 g. of copper sulfate in water and dilute to 100 ml. with water. This precipitant is designated as the 6.0-0.1 precipitant.

Zinc Precipitant for Ripened Cheese.—Dissolve 6.0 g. of zinc sulfate in water and dilute to 100 ml. with water. This precipitant is designated as the 6.0 precipitant.

The quantities of the respective reagents to use in preparing the precipitants for testing other products, not mentioned under "Protein precipitants" above, are indicated in Tables 15 and 21.

Procedure.—Weigh, on a clean balance pan or watch glass, a 0.50-g. sample (or preferably two samples) and place in a culture tube 16 or 18 × 150 mm. Similarly, weigh another sample and place in a tube or a control or blank. If the cheese is sticky, weigh the sample on a piece of wax paper about 1 × 1 in. and insert the paper with the sample into the tube. Macerate the blank and the test with a glass rod about 8 × 180 mm.

Add to the *blank* 1.0 ml. of the appropriate barium buffer 1-a (without substrate added) (Table 21), macerate with the rod, leave the rod in the tube, heat for about a minute to at least 85° C. in a beaker of boiling water with the beaker covered so that the entire

TABLE 21. PHOSPHATASE TEST MODIFICATIONS FOR DIFFERENT KINDS OF CHEESE AND CHEESE OF DIFFERENT AGES [9]

Kind of Cheese	Age or Extent of Curing; Other Details	Buffer for Opt. pH (9.85-10.20)	Precipitant	Criterion, Experimental, Phenol Equivalent[a]
				$\gamma/0.25$ g.
Cheddar, granular,	< 1 wk.	25-11[b]	6.0-0.1[c]	3
stirred curd, hard	1 wk.-1.5 mo.	25-11	6.0[d]	3
cheese	1.5-4 mo.	26-11	6.0	3
	> 4 mo.	27-11	6.0	3
Washed curd, soaked	< 1 wk.	25-11	6.0-0.1	3
curd, Colby	1 wk.-2 mo.	25-11	6.0	3
	> 2 mo.	26-11	6.0	3
Swiss, Gruyère	< 1 wk.	25-11	6.0-0.1	3
	1 wk.-1 mo.	25-11	6.0	3
	1-3 mo.	26-11	6.0	3
	> 3 mo.	27-11	6.0	3
Brick, Muenster	< 1 wk.	25-11	6.0-0.1	3
	1 wk.-1 mo.	25-11	6.0	3
	1-2 mo.	25-11	6.0	3
	> 2 mo.	26-11	6.0	3
Edam, Gouda	< 1 wk.	25-11	6.0-0.1	3
	1 wk.-2 mo.	25-11	6.0	3
	2-4 mo.	26-11	6.0	3
	> 4 mo.	27-11	6.0	3
Blue mold, blue	< 1 wk.	25-11	6.0-0.1	3
	1 wk.-1 mo.	26-11	6.0	3
	1-4.5 mo.	27-11	6.0	3
	> 4.5 mo.	28-11	6.0	3
Camembert, Limburger	< 1 wk.	25-11	6.0-0.1	4
	1 wk.-1 mo.	25-11	6.0	4
	1-2 mo.	26-11	6.0	4
	> 2 mo.	27-11	6.0	4
Monterey	< 1 wk.	25-11	6.0-0.1	3
	1 wk.-2 mo.	25-11	6.0	3
	> 2 mo.	26-11	6.0	3
High moisture Jack	< 1 wk.	25-11	6.0-0.1	3
	1 wk.-2.5 mo.	25-11	6.0	3
	> 2.5 mo.	26-11	6.0	3
Provolone, pasta filata	< 1 wk.	25-11	6.0-0.1	3
	1 wk.-1 mo.	25-11	6.0	3
	1-3 mo.	26-11	6.0	3
	> 3 mo.	27-11	6.0	3
Parmesan, reggiano,	< 1 wk.	25-11	6.0-0.1	3
monte, modena,	1 wk.-2 mo.	26-11	6.0	3
Romano, asiago old	2-6 mo.	27-11	6.0	3
	6 mo.-1 yr.	28-11	6.0	3
	> 1 yr.	29-11	6.0	3
Asiago fresh	Same as Cheddar	.....	......	..

TABLE 21—*Continued*

Kind of Cheese	Age or Extent of Curing; Other Details	Buffer for Opt. pH (9.85-10.20)	Precipitant	Criterion, Experimental, Phenol Equivalent[a]
Asiago medium	< 1 wk.	25-11	6.0-0.1	3
	1 wk.-1 mo.	25-11	6.0	3
	1-3 mo.	26-11	6.0	3
	> 3 mo.	27-11	6.0	3
Gorgonzola	Same as blue	.....	...,....	
				$\gamma/0.25$ g.
Cottage,[e] cook cheese, koch kaese	Dry	25-11	6.0-0.1	0
	Moist	25-11(8+2)[f]	4.5-0.1	0
Cream cheese	......	25-11(7+3)	4.5-0.1	3
Semi-soft cheese	< 1 wk.	25-11	6.0-0.1	3
	1 wk.-1 mo.	25-11	6.0	3
	>1 mo.	26-11	6.0	3
Soft ripened cheese	< 1 wk.	25-11	6.0-0.1	4
	1 wk.-1 mo.	25-11	6.0	4
	> 1 mo.	26-11	6.0	4
Nokkelost, kuminost, sage cheese	< 1 wk.	25-11		3
	1 wk.-1.5 mo.	25-11	6.0-0.1	3
	1.5-4 mo.	26-11	6.0	3
	> 4 mo.	27-11	6.0	3
Pasteurized process, pasteurized process, pimento, pasteurized process with fruits, meats, etc.	Soft, mild	25-11	6.0	3
	Medium firm	26-11	6.0	3
	Firm, sharp (including Swiss, Gruyère)	27-11	6.0	3
Pasteurized process cheese foods; pasteurized process cheese foods with fruits, meats, etc.	Same as pasteurized process	....	....	....
Pasteurized process cheese spreads; pasteurized process cheese spreads with fruits, meats, etc.	Soft, high moisture, incl. cream spreads	25-11	6.0	3
	Less soft., incl. Blue	26-11	6.0	3
Cold pack, club; cold pack cheese foods; cold pack cheese foods with fruits, meats, etc.	Mild to medium flavored, soft	26-11	6.0	3
	Sharp, firm	27-11	6.0	3

[a] Values higher than those shown indicate under-pasteurization.
[b] Grams $Ba(OH)_2 \cdot 8H_2O$ and H_3BO_3, respectively, per 1.
[c] Grams $ZnSO_4 \cdot 7H_2O$ and $CuSO_4 \cdot 5H_2O$, respectively, per 100 ml.
[d] Grams $ZnSO_4 \cdot 7H_2O$ per 100 ml.
[e] See also more sensitive modification in text—alternative.
[f] Eight parts of 25-11 buffer plus 2 parts of water.

tube is heated to approximately 85° C., cool to room temperature, and macerate again with the rod.

Add to the *test* 1.0 ml. of the appropriate barium buffer substrate 2-a or 2-b (Table 21), and macerate.

From this point, treat the blank and the test in a similar manner.

Add 9.0 ml. of the appropriate barium buffer substrate 2-a or 2-b (total, 10.0 ml. added), and mix. The rod may be left in the tube during incubation; if it is removed at this point, cut a piece of filter paper approximately 1 × 1 in., wrap and hold it tightly around the rod, rotate the rod while withdrawing it from within the tube so as to wipe the rod clean, insert the paper with the adhering fat into the tube, and stopper the tube.

Incubate in a water bath at 37-38° C. for 1 hour, mixing or shaking the contents occasionally. Place in a beaker of boiling water for nearly a minute, heating to approximately 85° C. (use a thermometer in another tube containing the same volume of liquid), and cool to room temperature. Pipette in 1.0 ml. of the zinc precipitant for ripened cheese, or the zinc-copper precipitant for unripened cheese, and mix thoroughly (pH of mixture, 9.0-9.1). Filter (5-cm. funnel, 9-cm. Whatman No. 42 or No. 2 paper recommended), and collect 5.0 ml. of filtrate in a tube, preferably graduated at 5.0 and 10.0 ml. Add 5.0 ml. of color development buffer 1-b (pH of mixture, 9.3-9.4). Add four drops of 2,6-dibromoquinonechloroimide solution, mix, and allow the color to develop for 30 minutes at room temperature. For merely detecting under-pasteurization in testing unripened cheese, 2 drops are sufficient, provided the visual standards likewise are prepared with 2 drops. Determine the intensity of blue color by either of two methods:

(a) *With a photometer.*—Read the color intensity of the blank and that of the test, subtract the reading of the blank from that of the test, and convert the result into phenol equivalents by reference to the standard curve described under "Phenol Standards." The *n*-butyl alcohol extraction method ordinarily is unnecessary when using a photometer.

(b) *With visual standards.*—For quantitative results in borderline instances, for example, tests yielding 0.5 to 5 units of color, extract with butyl alcohol 5-b. Add 5.0 ml. of the alcohol and invert the tube slowly several times, centrifuge if necessary to increase the clearness of the alcohol layer, and compare the blue color with the colors of standards in the alcohol.

With samples yielding more than 5 units, compare the colors in aqueous tests with those of aqueous standards.

Dilution Method for Quantitative Results.—In tests that are observed to be strongly positive during color development—for example, 20 units or more—in which 4 drops of 2,6-dibromoquinonechloroimide solution may be much less than sufficient to combine with all of the phenol, pipette an appropriate proportion of the contents into another tube, make up to 10.0 ml. with color dilution buffer 1-c, and add 2 drops more of 2,6-dibromoquinonechloroimide solution in the case of an unripened product or 4 drops in the case of a ripened product. With each test, dilute and treat the blank in the corresponding manner. Dilute each strongly positive test thus until the final color is within the range of the visual standards or photometer. Allow 30 minutes for color development after the last addition of 2,6-dibromoquinonechloroimide solution, and make the reading at the end of the 30-minute period. To correct, multiply by 2 for a $5 + 5$ dilution, by 10 for a $1 + 9$ dilution, and by 50 for a $1 + 9$ followed by a $2 + 8$ dilution.

Alternatively, to reduce the amount of yellow off color, add 2 instead of 4 drops of 2,6-dibromoquinonechloroimide solution after each dilution, and allow the color to develop. Then test the completeness of color development by adding a third drop. Repeat the dilution procedure until the addition of an extra drop does not cause any further increase in the amount of blue color.

Calculation and Evaluation of Result.—When using 0.5 g. of solid sample and adding a total of 11.0 ml. of liquid, multiply the value of the reading by 1.1 to convert it to units of color or phenol equivalents per 0.25 g. of cheese. The result, if desired, may be converted to phenol equivalents per 1 g. by multiplying by 4.4. Evaluate the result by comparing it with the criteria of pasteurization in Table 21.

EVAPORATED MILK

Evaporated milk is the product resulting from the evaporation of a considerable portion of the water from milk, or from milk with the adjustment, if necessary, of the ratio of fat to nonfat solids by the addition or abstraction of cream. Maximum, minimum, and average compositions of this product are given in Table 22.

Evaporated milk is defined by the Food and Drug Administration [10] as follows.

Evaporated milk is the liquid food made by evaporating sweet milk to such point that it contains not less than 7.9 per cent of milk fat and not less than 25.9 per cent of total milk solids. It may contain one or both of the following optional ingredients:

---

[10] *U. S. Food Drug Admin., S.R.A., F.D.&C.* 2, Part 18, May, 1953.

(1) Disodium phosphate or sodium citrate or both, or calcium chloride, added in a total quantity of not more than 0.1 per cent by weight of the finished evaporated milk.

(2) Vitamin D in such quantity as increases the total vitamin D content to not less than 25 U. S. P. units per fluid ounce of finished evaporated milk.

It may be homogenized. It is sealed in a container and so processed by heat as to prevent spoilage.

The milk may be adjusted, before or after evaporation, by the addition or abstraction of cream or sweet skim milk, or by the addition of concentrated sweet skim milk.

Vitamin D content may be increased by the application of radiant energy or by the addition of a concentrate of vitamin D (with any accompanying vitamin A when such vitamin D in such concentrate is obtained from natural sources) dissolved in a food oil; but if such oil is not milk fat the quantity thereof added is not more than 0.01 per cent of the weight of the finished evaporated milk.

TABLE 22. COMPOSITION OF EVAPORATED MILK

	Fat, %	Protein, %	Lactose, %	Ash, %	Total Solids, %	Water, %
Maximum[a, b]	8.90	7.08	10.84	1.75	27.40	74.57
Minimum[a, b]	7.90	6.38	9.38	1.34	25.43	72.60
Average[a, b]	8.22	6.71	10.13	1.55	26.37	73.63
Maximum[c]	8.1	...	...	...	27.7	74.4
Minimum[c]	7.9	...	...	...	25.6	72.3
Average[c, d]	8.02	...	...	...	26.5	73.5
Maximum[e]	8.0	7.2	...	1.7	27.6	74.6
Minimum[e]	7.8	6.3	...	1.5	25.4	72.4
Average[e, f]	7.84	6.8	...	1.6	26.1	73.9

[a] Street, *Connecticut Agr. Expt. Sta., Bull.* 213 (1919).
[b] Rogers, *Fundamentals of Dairy Science.* Chemical Catalog Co., New York, 1928.
[c] *North Dakota State Laboratories Dept., Bull.* 67 (1942).
[d] 12 commercial samples.
[e] *North Dakota State Regulatory Dept., Bull.* 47 (1935).
[f] 19 commercial samples.

After dilution with water the sample may be analyzed in a manner similar to milk with a correction for the dilution in calculating results.

Scrape, if necessary, the contents of the can into a dish large enough to hold all the sample, and mix sufficiently to make the sample homogeneous. If the material has separated out, the sample may be warmed and passed through a hand pressure colloidal mixer, commonly known as a mechanical cow. Weigh 40 g. into a flask, add 60 g. of water, and proceed with the determinations as directed under milk correcting the results obtained for dilution.

RAPID METHOD FOR THE ANALYSIS OF EVAPORATED MILK

Weigh 40 g. of the properly mixed sample into a beaker and add 80 g. of water. Stir thoroughly until a homogeneous mixture is obtained. Transfer to a cylinder and take the lactometer reading in the usual way at 60° F. Determine the fat in exactly the same way as that detailed in the Babcock method for milk, except that slightly more acid need be added. Calculate the total solids according to one of the formulas given in the section "Rapid Method" under milk. Multiply the fat and calculated total solids results by 3 to obtain the percentage in the original material.

FAT

The fat concentration of evaporated milk can be determined by the Gerber method using ice-cream butyrometers, Fig. 62, weighing a 5-g. sample or by use of the Pennsylvania method described under the section, "Ice Cream," in this chapter.

Procedure.—Weigh directly 6 g. of the sample into a 9-g., 20-per cent ice-cream-test bottle or into an 18-g. milk-test bottle and proceed with the method as detailed. If a 9-g., 20-per cent ice-cream-test bottle was used, multiply the reading by 1.5 to obtain the fat percentage. If an 18-g. milk-test bottle was used, multiply the percentage by 3 to obtain the percentage of fat.

CONCENTRATED MILK

Concentrated milk or plain condensed milk should not be confused with sweetened condensed milk sold in cans to consumers. It is whole milk from which part of the water is removed by evaporation in a vacuum pan. It is generally sold in bulk to bakeries and other establishments which use milk in large quantities as an ingredient in their manufactured products.

Under an official standard of identity, the Food and Drug Administration requires concentrated milk to conform to the definition of evaporated milk except that it need not be sterilized nor need it be packed in hermetically sealed containers.

The methods of analysis are the same as those detailed for evaporated milk.

CONDENSED MILK

Condensed milks may actually be divided into two large groups of unsweetened condensed and sweetened condensed milk. The former is generally designated as evaporated milk which has been discussed in a previous paragraph. Condensed milk is generally taken to mean the sweetened product and, if an unsweetened condensed milk is sold in bulk, it is designated as plain condensed milk or concentrated milk. Thus evaporated milk and condensed milk are two different products from both composition and method of production.

Condensed milk, sweetened condensed milk is the product resulting from the evaporation of a considerable portion of the water from milk to which sugar or dextrose or both have been added. The composition of sweetened condensed milk, as given by a typical analysis, is: fat, 8.5 per cent; protein, 8.1 per cent; lactose and sugar, 54.7 per cent; ash, 1.7 per cent; water, 27.0 per cent.

Sweetened Condensed Milk; Identity.—Sweetened condensed milk is the liquid or semi-liquid food made by evaporating a mixture of sweet milk and refined sugar (sucrose) or any combination of refined sugar (sucrose) and refined corn sugar (dextrose) to such point that the finished sweetened condensed milk contains not less than 28.0 per cent of total milk solids and not less than 8.5 per cent of milk fat. The quantity of refined sugar (sucrose) or combination of such sugar and refined corn sugar (dextrose) used is sufficient to prevent spoilage.

The milk may be adjusted, before or after evaporation, by the addition or abstraction of cream or sweet skim milk, or the addition of concentrated sweet skim milk.

Condensed Milks which Contain Corn Syrup; Identity.—Condensed milks which contain corn syrup are the foods each of which conforms to the definition and standard of identity prescribed for sweetened condensed milk above, except that corn syrup or a mixture of corn syrup and sugar is used instead of sugar or a mixture of sugar and dextrose. For the purpose of this definition the term "corn syrup" means a clarified and concentrated aqueous solution of the products obtained by the incomplete hydrolysis of cornstarch, and includes dried corn syrup; the solids of such corn syrup contain not less than 40 per cent by weight of reducing sugars, calculated as anhydrous dextrose.

Fat

The concentraton of fat in sweetened condensed milk can also be estimated by use of the Pennsylvania method detailed in the section, "Ice Cream," in this chapter.

Procedure.—Balance two beakers on opposite sides of a trip scale or cream scale. Place about 1 oz. of condensed milk in one beaker. Exactly counterbalance with an equal weight of water. Add the water to the sweetened condensed milk and mix until homogeneous. Weigh 9 g. of this mixture into a 9-g., 20 per cent ice-cream-test bottle and proceed with the method as detailed. Multiply the result by 2 since the condensed milk comprises only half of the portion analyzed.

ANALYSIS

Warm the contents of the can, scrape into a dish, and mix thoroughly until the mass is homogeneous. Weigh 100 g. of the sample into a 500-ml. volumetric flask, then dilute to the mark with water and mix. The usual determinations are made in a manner similar to that of milk. For the determination of sucrose the reader is referred to the chapter on sugars, Chapter X. The fat content of sweetened condensed milk may be determined by the Gerber method using ice-cream butyrometers, Fig. 62, or as above. Condensed milk is diluted one-half with water before being weighed into the butyrometer.

SKIM MILK

Skim milk is defined as milk from which a sufficient portion of the fat has been removed to reduce the fat content below that of 3.25 per cent. More specifically skim milk is the product obtained in the separation of cream from milk. Davies prefers to differentiate between skim milk and separated milk. The former is obtained by permitting milk to stand and then skimming off the cream and the latter is the product issuing from the skim-milk screw when cream is produced by use of a centrifugal separator.

A representative composition of skim milk is the following: 0.2 per cent fat, 3.5 per cent protein, 5.0 per cent lactose, 0.8 per cent ash, 9.5 per cent total solids, and 90.5 per cent water. The similarity to the composition of buttermilk is evident (see below). Hand-skimmed milk will generally contain more fat than separator-skimmed milk. The fat content of hand-skimmed milk is of the order of 0.75 per cent, otherwise the composition is about the same.

FAT

American Association Method.—This is a modified Babcock method in which *n*-butyl alcohol is used to assist a lesser volume of sulfuric acid in disintegrating the colloidal mixture.

Procedure.—Transfer to a regular Babcock 18-g., 0.5-per cent or 0.25-per cent skim-milk-test bottle, 2 ml. of *n*-butyl alcohol from a burette, add 9 ml. of skim milk, and mix well. Add 7-9 ml. of sulfuric acid, specific gravity 1.82-1.83, and mix well. Centrifuge 6 minutes and add water to the base of the neck of the bottle. Centrifuge again for 2 minutes and add water to bring the fat into the neck of the bottle. Centrifuge a third time for 2 minutes and place the bottle in a water bath at 135-140° F. for 5 minutes. Multiply the reading by 2 to obtain the percentage of fat.

Pennsylvania Method.—The amount of fat in skim milk can also be determined by use of the Pennsylvania method detailed in the section, "Ice Cream," of this chapter.

Procedure.—Transfer 17.5 ml. into a skim-milk-test bottle, or milk-test bottle and proceed with the method.

BUTTERMILK

Buttermilk was was formerly defined as the product that remains when fat is removed from milk or cream in the process of making butter by churning. If sweet cream is used for churning, the buttermilk produced does not differ from ordinary skimmed milk. The amount of buttermilk produced in the United States is very large because for every 100 lbs. of butter, approximately 166 lbs. of buttermilk are obtained. This means there is available about 4,000,000,000 lbs. of buttermilk yearly. Actually little of this is used for human consumption. Buttermilk contains most of the nutrients of milk except the butterfat.

TABLE 23. COMPOSITION OF BUTTERMILK

Type	Fat, %	Protein, %	Lactose, %	Ash, %	Total Solids, %	Water, %
Genuine[a]	0.5	3.0	5.3	0.7	9.5	90.5
Genuine[b]	0.5	3.5	4.6	0.7	9.3	90.7
Cultured[b]	0.2	3.5	5.0	0.8	9.5	90.5

[a] Rogers, *U. S. Dept. Agr., Bull.* 319 (1916).
[b] Chatfield and Adams, *U. S. Dept. Agr., Circ.* 549 (1940).

Most of the buttermilk available commercially for food purposes is cultured buttermilk made by souring pasteurized skimmed or partly skimmed milk, or reconstituted milk made from skim-milk powder and water, by the addition of an ordinary butter starter such as *Streptococcus lactis* or some similar type. It is preferable to limit the maximum fat content in cultured buttermilk to a low value so that people who must be on a fat-free diet may be able to rely on buttermilk as part of that diet.

Buttermilk may be analyzed in a manner similar to other milk products. Fat can be determined by either the Roese-Gottlieb method, the Gerber method, or the methods detailed under skim milk and whey. Because of the viscosity only gravimetric determinations of total solids can be made. The other determinations follow the line previously described.

CONDENSED BUTTERMILK

Condensed buttermilk is the semisolid product prepared by evaporating fluid buttermilk, usually in a vacuum pan, until it has a heavy consistency. It contains enough acid to prevent spoilage.

Condensed buttermilk may be analyzed by the methods detailed for evaporated and condensed milk.

WHEY

Whey is the serum which remains after the coagulation of the casein when cheese is manufactured. Thus, for instance, about 90 lbs. of whey and about 10 lbs. of cheese are obtained from 100 lbs. of milk in the manufacture of Cheddar cheese. Whey consists principally of the water, lactose, lactalbumin, and most of the ash of the original milk. It has slightly more fat than buttermilk. A representative composition contains 0.3 per cent fat, 1.0 per cent protein, 5.1 per cent lactose, 0.6 per cent ash, 7.0 per cent total solids, and 93 per cent water. The composition may vary considerably, however, depending on the type of cheese manufactured.

Fat

Minnesota Method.—*Reagents.*—Use the reagents detailed for this method in the section on "Ice Cream,' in this chapter.

Procedure.—Transfer 9 ml. of whey to 18-g., 0.5-per cent or 0.25-per cent skim milk-test bottles, add 10 ml. of Minnesota reagent, and shake thoroughly. Digest in a water bath at 160-178° F. for 6 to 7 minutes and shake several times during this interval. Centrifuge 5 minutes and add water at 135-140° F. to bring the contents of the flask to the base of the neck. Centrifuge again for 2 minutes and add water at 135-140° F. to bring the fat column into the graduated column of the neck. Centrifuge for a third time for 1 minute and place the tests in a water bath at 135-140° F. for 5 minutes. Multiply the reading by 2 to obtain the fat percentage.

Pennsylvania Method.—Transfer 17.5 ml. of whey into a test bottle and proceed with the method as detailed under the section, "Ice Cream," in this chapter.

FERMENTED MILK

Milk is fermented in order to preserve it. A number of organisms have been used to ferment milk. One of the most common is *Lactobacillus bulgaricus*. This has been mentioned in connection with the discussion of buttermilk and sour cream. Buttermilk is a fermented milk.

TABLE 24. COMPOSITION OF FERMENTED MILKS [a]

Product	Fat, %	Pro-tein,[b] %	Lactose. %	Ash, %	Lactic acid, %	Alcohol, %	Water, %
Kefir
2 days old	3.6	3.1	3.7	0.6	0.7	0.2	...
4 days old	3.6	3.1	2.2	0.6	0.8	0.8	...
6 days old	3.6	3.1	1.7	0.6	0.9	1.1	...
Koumiss
1 day old	1.2	2.0	1.6	0.35	0.8	2.7	91.4
8 days old	1.1	1.8	0.5	0.35	1.1	2.9	92.1
22 days old	1.3	1.8	0.23	0.35	1.3	3.0	92.1

[a] Rogers, *U. S. Dept. Agr., Bull.* 319 (1916).
[b] Some denaturation of protein takes place.

Kefir, a product of southeastern Russia and the Caucasus, is a fermented milk made from the milk of various animals—principally sheep, goats, and cows. To prepare this drink, a characteristic starter which consists of "seeds" or Kefir grains is used.

Koumiss, another drink which originated in Russia, is a fermented milk made from mares' milk. It is fermented by milk-souring types of bacteria and yeasts. Both Kefir and Koumiss are limpid, slightly acid, and distinctly alcoholic drinks.

Yogurt, a Turkish product, and Matzoon, an Armenian product, are fermented milks that are sufficiently curdled to form a thick, plastic mass which is decidedly acid and which contains comparatively little alcohol. Similar products are Egyptian Leban, Indian Dadhi, and Sardinian Gioddu.

Acidophilus milk is a fermentation product resulting from the inoculation of milk with *L. acidophilus* bacteria. It is very similar in appearance to buttermilk.

The composition of some of these products is given in Table 24.

Fermented milks may be analyzed by the methods detailed for milk, cream, and sour cream, depending on the approximate fat content and viscosity.

FILLED MILK

Filled milk is defined by the Filled Milk Act of 1923 as any milk, cream, or skimmed milk, whether or not condensed, evaporated, concentrated, powdered, dried, or desiccated, to which has been added, or which has been blended or compounded with, any fat or oil other than milk fat, so that the resulting product is in imitation or semblance of milk, cream, or skimmed milk, whether or not condensed, evaporated, concentrated, powdered, dried, or desiccated. By its terms the Filled Milk Act prohibits the interstate distribution of any combination of milk, cream, or skimmed milk, with any foreign fat or oil so as to resemble or imitate milk or skimmed milk in any form. This legislation was enacted after a Congressional hearing. Congress determined and declared, "that filled milk, as herein defined, is an adulterated article of food injurious to public health and its sale constitutes a fraud upon the public."

A more vicious adulteration of an analogous type is the substitution, not of an edible oil such as lard, or oleomargarine, but the substitution of mineral oil. Instances have been found in which this relatively nutritionally inert material has been incorporated in cheese.

The character of the fat or oil present in filled milk products may be elucidated by isolating the fat by methods detailed in Chapter VII under the section, "Cream," and then performing the various tests detailed in Chapter IX.

IMITATION MILK

A type of imitation milk is a mixture made from soybean. One of the methods of making this product is to boil the beans, beat them to a pulp, and then suspend the vegetable protein in water. This process yields a milky mixture which appears to have many of the properties of milk. It is used a great deal by the Chinese. An analogous type of imitation vegetable milk is made from tara or Payo meal.

MILK BEVERAGE

A milk beverage or a skim-milk beverage is a food mixture or confection which is prepared from milk or skim milk as the case may be and to which has been added a syrup or flavor consisting of wholesòme ingredients. Among the most common of these are chocolate milk and chocolate drink. Chocolate milk is the beverage made by the addition of chocolate or chocolate flavors to whole milk and contains not less milk fat than does whole milk. Chocolate drink or chocolate-flavored drink generally consists of fluid skim milk or skim-milk powder and water to which have been added cocoa powder, sugar, a stabilizer such as a gum or other thickening agent like tapioca flour, and salt.

DRIED MILK PRODUCTS

Dried milk products are the products resulting from the removal of water from milk and milk products. The principal dried milk products are dried whole milk, dried skim milk, dried buttermilk, dried whey, and dried cream. These products are also known as whole-milk powder, skim-milk powder, buttermilk powder, powdered cream or cream powder, and whey powder. All of them are generally made by use of spray or drum drying methods but whole-milk powder is more commonly prepared by the spray process and buttermilk powder by the drum method. Actually little dried cream is manufactured. The ratios of the components of all of these products are the same as those of the milk or other milk products from which they are prepared. The composition of several of these products is given in Table 25.

TABLE 25. COMPOSITION OF DRIED MILK PRODUCTS

Product	Fat, %	Protein, %	Lactose, %	Ash, %	Total Solids, %	Water, %
Whole milk[a]	26.7	25.8	38.0	6.0	96.5	3.5
Skim milk[a]	1.0	35.6	52.0	7.9	96.5	3.5
	1.0	37.0	49.0	9.0	96.0	4.0
Cream[b]	65.1	13.4	17.9	2.9	99.3	0.7
Buttermilk[b]	5.9	38.7	39.9	7.7	98.1	1.9
Malted milk[a]	8.5	14.6	70.7[c]	3.6	97.4	2.6
	7.5	13.0	73.5	3.0	97.0	3.0

[a] Chatfield and Adams, U. S. Dept. Agr., Circ. 549 (1940).
[b] Eckles, Combs, and Macy, Milk and Milk Products. McGraw-Hill, New York, 1936.
[c] Total carbohydrate.

Dry skim milk is identified as the product obtained by drying sweet skim milk. It must not contain more than 5 per cent of water. Dry whole milk must contain not less than 26 per cent milk fat and not more than 5 per cent of water.

The fat content of this milk product may be estimated by the Gerber method in a manner similar to that of evaporated milk. Weigh accurately the amount of powder taken and the amount of water used to dissolve it and then proceed as directed under evaporated milk, making the proper correction for dilution.

MOISTURE

Weigh 50 g. of the dried milk product into a 300-ml., round-bottom or 300-ml. Erlenmeyer flask, preferably equipped with a neck. Add 75-100 ml. of toluene and proceed with the Immiscible Solvent Distillation Method as detailed in Chapter I.

SOLUBILITY INDEX

In this method [11] the amount of sediment remaining in a tube after centrifugation of a suspension of 10 g. of dry-milk solids in water is measured. This is an empirical method, consequently the procedure must be strictly followed.

Apparatus.—Equip a 50-ml. conical sediment tube graduated in 0.1-ml. divisions from 0 to 1 ml.; in 0.2-ml. divisions from 1 ml. to 2 ml.; in 0.5-ml. divisions from 2 ml. to 10 ml.; and in 1.0-ml. divisions from 10 to 20 ml.; and with the 50-ml. mark at least 0.5 in. below the top of the tube, with a siphon arrangement.

Procedure.—Weigh 10 g. of dry milk solids into an 18- or 20-oz. tumbler having approximately a 2-in. base and containing 100 ml. of water at 75° F. Immediately agitate vigorously for exactly 90 seconds with a Dumore electric mixer. Allow the mixture to stand, for a period not exceeding 25 minutes, until the foam has broken away from the liquid and remove the foam with a spoon or equivalent device. Stir the mixture thoroughly and rapidly and pour 50 ml. immediately into the 50-ml. centrifuge tube. Place at once in the centrifuge and whirl for 5 minutes at the speed used for Babcock determinations. Insert the siphon arrangement so that it is at least 2 ml. above the level of any sediment and siphon off the supernatant liquid taking care not to draw over any of the sediment. Add 25 ml. of water at 75° F. and shake to dislodge the sediment using a wire if necessary. Then bring up to the 50-ml. mark with additional water. Mix the contents thoroughly and then centrifuge again for 5 minutes. Read the volume of sediment in milliliters by holding the tube toward a strong light.

[11] Turnbow, Tracy, and Raffetto, *The Ice Cream Industry.* Wiley, New York, 1947.

MALTED-MILK POWDER

Malted milk is the product made by the combination of whole milk with the liquid separated from a mash of ground barley malt and wheat flour, with or without the addition of sodium chloride, sodium bicarbonate, and potassium bicarbonate, in such a manner as to secure the full enzymic action of the malt extract, and by removing water. The manufacture of malted milk powder involves the preparation of a malt-flour infusion by the process of mashing and the evaporation to dryness of the infusion-milk mixture. The ground malt is used to make malt flour. Malted milk should contain 7.5 per cent of butterfat as a minimum and not more than 3.5 per cent water.

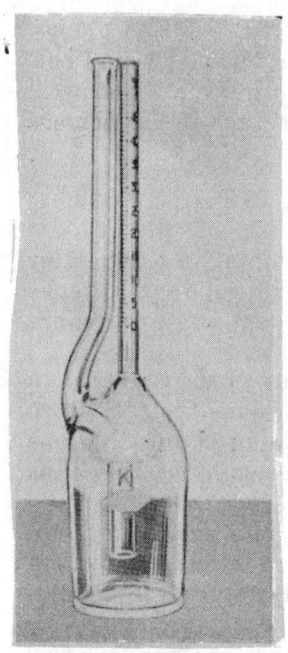

FIG. 61. Babcock Skim Milk Flask

The composition of a representative powder is given in Table 25. This powder is used in combination with milk or water to make a beverage usually by agitation with an electrical mixer commonly spoken of as malted milk. Often ice cream, cream, and whipped cream are added.

This product may be adulterated by the substitution of skim milk for whole milk in the aforementioned process, by mixing skim-milk powder with the malted-milk powder, by adding starch, sugar, and other substances. Many adulterations of malted milk are easily detected by microscopical examination with the aid of comparative photomicrographs identifying malted milk and its allied products. The chemical quality of this product may, of course, be determined by suitable means.

FAT

Modified Babcock Method.—The fat content of malted milk powder may be estimated by a modified Babcock method. Transfer 1 g. of the powder, accurately weighed, to a tall form 100-ml. beaker containing 5 ml. of water. Stir this mixture thoroughly and then add 5 ml. of 85 per cent lactic acid. Bring the mixture to a boil and continue heating until the powder is completely dissolved. Transfer to a skim-milk Babcock flask, washing the beaker with a total of 10 ml. of water and adding the washings to the flask, Fig. 61. Add 2 ml.

of amyl alcohol or 1 ml. of n-butyl alcohol and 1 ml. of amyl alcohol. Add 17.6 ml. of a mixture of 200 ml. of concentrated sulfuric acid and 30 ml. of water, adding the acid carefully to the water. Stir as in the Babcock milk method and centrifuge for 5, 2, and 1 minutes, adding hot water at the end of the first and second whirlings as previously described. Read the percentage of fat by assuming full-scale reading equals 90 mg., or on basis of 1-g. sample 9 per cent. Therefore each division is equivalent to 0.18 per cent or 1.8 mg. fat.

Roese-Gottlieb Method.—The fat content may also be estimated by the Roese-Gottlieb method. Weigh into the lower section of a Jacobs-Singer separatory flask 1 g. of the powdered sample. Add 11 ml. of water and shake until a thick paste is formed. Add 2 ml. of ammonium hydroxide, boil for 5 minutes and cool. Add 11 ml. of ethyl alcohol and bring the volume of the mixture up to the middle of the connecting joint. Add 25 ml. of ethyl ether and proceed with the method as detailed under milk.

Alternatively weigh 1 g. of the powdered sample into a tall form 100-ml. beaker. Add 1 ml. of water and stir to a thick paste. Add 5 ml. more of water, stir thoroughly, add 2 ml. of ammonium hydroxide, boil for 5 minutes, cool and transfer to a Mojonnier extraction tube or a Jacobs-Singer separatory flask. Wash the beaker with a total of 5 ml. of water and add the washings to the extraction tube or flask. Proceed with the method as detailed immediately above and as under milk using the first portions of the solvent to wash the beaker.

Moisture may be determined by the vacuum oven method in the usual manner, or it may be estimated by drying at a lower temperature for a longer period of time. Protein, ash, carbohydrate, and other components may be ascertained by the methods detailed in this text.

MALTED-MILK DRINK

Sometimes it is necessary for the analyst to find out whether or not malted-milk powder has been added to a drink sold as a malted-milk drink. Malt contains dextrins, for, as the malt is prepared from barley or other grain, higher dextrins, malto-dextrins, etc., must be formed as intermediates in the formation of maltose. Hence, maltose and malto-dextrins are characteristic components of malted-milk powder.

Based on the foregoing characteristic property, the following test for the detection of malted-milk powder in malted-milk drink was devised by Jacobs.[12] To 10 ml. of the malted milk drink, add 5 ml.

---

[12] Jacobs, work performed in 1936.

of 5 per cent uranium acetate solution, 5 g. $UO_2(OAc)_2 \cdot 2H_2O$ dissolved in water, made up to 100 ml. and centrifuged. Allow the mixture to stand for 5 minutes after stirring thoroughly. Filter and repass the first 2 ml. through the filter again. To 2 ml. of the filtrate, add two drops of hydrochloric acid. Then add 20 ml. of 95 per cent alcohol and mix. The presence of dextrins is indicated by a marked persistent white turbidity. In the case of large amounts, a flocculent precipitate may form. A positive test does not prove the presence of malted-milk powder in the drink, for if the drink were made, let us say, with commercial glucose, a positive test for dextrins would also be obtained. A negative test indicates the probable absence of malted-milk powder in the drink. Gums and pectins do not interfere because these substances are either precipitable by uranium acetate in the presence of milk or do not yield a white turbidity on the addition of alcohol. Chocolate flavor syrup may interfere because such syrups often contain commercial glucose and dextrins.

ICE CREAM

Ice cream is the frozen product made from a combination of milk products and two or more of the following ingredients: eggs, water, and sugar with harmless flavoring and harmless coloring, with or without stabilizer, and in the manufacture of which freezing is accompanied by agitation of the ingredients. The milk products generally used are one or more of the following: cream, butter, milk, evaporated milk, skimmed milk, condensed milk, sweetened condensed milk, condensed skimmed milk, sweetened condensed skimmed milk, dried milk, dried skimmed milk. Ice cream formerly was considered more nearly a confection than a staple article of our diet but for many people, especially children, it is a regular item of their daily fare. Consequently its manufacture and sale should be as rigidly controlled as is milk.

Ice-cream manufacture presents one possible type of adulteration that has not been encountered previously in this text. It is made by agitating while it is being frozen. This process incorporates a great deal of air and increases the volume of the ice-cream mix, the term applied to the mixture of ingredients before freezing, usually from 90 per cent to 110 per cent. Since ice cream is sold by volume rather than by weight, it is obvious that it is necessary to limit the incorporation of air. This is generally done by having a minimum value of total food solids per gallon of ice cream. The other types of adulteration are not different from that encountered in the previously described milk products. The addition of yellow coloring matter to ice creams labeled "French ice cream," "Custard," or "Frozen Custard" which are required to contain eggs is considered an adulteration also.

TABLE 26. INGREDIENTS OF ICE-CREAM MIXES FOR 10% ICE CREAM ON 1000-LB. BASIS

18% cream	366.1	415.6	210.7*	491.5	385.5	500.8
Evaporated milk	379.7	....	....	....	....	....
Whole Milk	109.2	220.4	416.7	....	311.5	259.2
Plain condensed whole milk	....	219.0	....	....	....	....
Condensed skim milk	....	....	227.6	....	....	....
Sweet condensed whole milk	....	....	....	....	238.0	....
Sweet condensed skim milk	....	....	....	285.0	....	163.9
Sugar	140.0	140.0	140.0	10.3	40.0	71.1
Gelatin	5.0	5.0	5.0	5.0	5.0	5.0
Water	....	....	....	208.2	....	....
Total	1000.0	1000.0	1000.0	1000.0	1000.0	1000.0
Skim milk powder	45.8	110.0	....	....		
Butter	89.6	119.2	68.9	117.8		
Whole milk	719.6	....	1224.0	....		
Skim milk	....	....	....	1174.0		
Sugar	140.0	140.0	140.0	140.0		
Gelatin	5.0	5.0	5.7	5.7		
Water	....	625.8	....	....		
Total	1000.0	1000.0	1438.6	1437.5		
Water to be removed by vacuum pan	....	....	438.6	437.5		

These formulas are extracted from the extensive tables given by Mojonnier and Troy.[13]
* Forty per cent cream.

FAT

Preparation of Sample.—Ordinarily in the preparation of an ice-cream sample for analysis it is customary to pass the melted ice cream through cheesecloth or equivalent straining device to free the ice cream of admixed nuts and fruits. Where analyses are performed for fat on such samples to ascertain if they conform to the fat requirements of certain codes and a lower fat specification is often allowed because of the incorporation of added food material such as fruits or nuts, the fat result obtained will be high.

To obtain a correct fat result, Maack and Tracy [14] suggest the

[13] Mojonnier and Troy, *Technical Control of Dairy Products.* Mojonnier Bros., Chicago, 1925.
[14] Maack and Tracy, *J. Milk Tech.* 3, 123 (1940).

use of a malted-milk mixer to disintegrate the particulate matter of the sample and make it relatively homogeneous. Chocolate chip and mint stick ice cream can also be treated in this manner, for, although the chocolate and mint dissolve or mix with the ice cream, they do so more quickly in the mixer. Samples can then be weighed out as explained in previous sections.

In chocolate covered vanilla ice cream, it is a common practice to remove the chocolate coating while the ice cream is still frozen and then proceed with the analysis as if the ice cream were straight vanilla.

Prucha [15] also points out that ice-cream products containing nuts, fruits, etc., can be made smooth, homogeneous, and uniform by comminuting them in a malted milk mixer or similar device.

When this procedure has been followed, the fat in the comminuted ingredients will be included in the fat determinations so that a determination of total fat rather than of milk fat is obtained. This procedure is analogous to the inclusion of cocoa fat commonly encountered in the determination of fat in chocolate ice cream.

This procedure is not applicable in those communities where laws or regulations permit lower milk-fat content in ice cream to which ingredients of an insoluble nature have been added.

Procedure.—Fill the mixer not more than one-third full with the melted sample. This requires 4-5 oz. by weight of sample. Operate the mixer until the insoluble particles are broken down into a finely divided state. The soft fruits are quickly comminuted. The harder products must be treated longer to secure the desired breakup.

To prevent churning, warm the sample above the melting point of the fat (to about 40° C. or 104° F.) before comminuting. Samples from properly homogenized mixes do not churn out. Cool the comminuted samples and collect in 400-ml. beakers.

They are now in condition for pipetting for the examination for the fat.

Gerber Method.—The fat content of ice cream is easily determined by the Gerber method. This method applies equally well to ice cream containing cocoa powder or chocolate flavor. Allow the ice cream to come to room temperature by standing or warming slightly. Measure 10 ml. of sulfuric acid (to 13 parts of water add 87 parts of sulfuric acid specific gravity 1.82, for all ice cream except chocolate, for which, to 6 parts of water 94 parts of the acid are added) into the ice cream butyrometer, Fig. 62, which is balanced on a suitable balance. Weigh 5 g. of the properly prepared sample

---

[15] Prucha, *Am. J. Pub. Health* **33**, 595 (1943).

carefully into the tube. Add from 4.5 to 5.5 ml. of water according to the volume of the butyrometer and 1 ml. of amyl alcohol. Stopper and shake until all of the curd is dissolved and then mix the acid remaining in the neck by inverting several times. Centrifuge for 6 minutes. Remove the butyrometer and read in the same manner as directed under milk.

Modified Babcock Methods.—The Babcock method is not suitable for the determination of fat in ice cream because the sulfuric acid reagent reacts with the sugar in the ice cream producing so much charred matter that a good fat reading cannot be obtained. Because Gerber-type butyrometers were not readily available and were more expensive than Babcock-test bottles, a number of variations of the Babcock test have been developed which overcome the difficulty mentioned. In these variations special alkaline and acid reagents are used. As a conclusion from personal experience with thousands of samples, the author is of the opinion that the Gerber method is the preferred rapid method for the estimation of fat in ice cream.

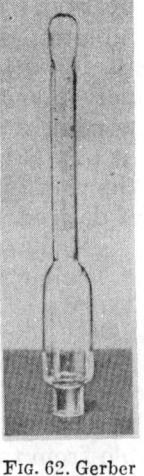

Fig. 62. Gerber Ice Cream Butyrometer

Pennsylvania [16] **Method.**—Sulfuric Acid Reagent.— Prepare a diluted sulfuric acid reagent by adding 5.5 parts by volume of commercial sulfuric acid, specific gravity 1.82-1.83 to 1 part by volume of water. The diluted sulfuric acid will have a specific gravity of 1.72-1.74.

Procedure.—Weigh a 9-g. sample into an ice-cream test bottle. Add 2 ml. of 28-29 per cent ammonium hydroxide solution and mix with a rotatory motion. Add 3 ml. of *n*-butyl alcohol and mix thoroughly. Add 17.5 ml. of the prepared sulfuric acid reagent and mix until a dark color is obtained. Place promptly in a centrifuge and whirl for 5 minutes. Add sufficient hot water at 135-140° F. to raise the liquid level to the base of the neck of the test bottle but do not mix the contents of the flask to avoid interference with clean fat separation. Centrifuge again for 2 minutes and add sufficient hot water to bring the fat into the graduated portion of the neck of the bottle. Whirl again for 1 minute. Place the flasks in a water bath at 135-140° F. for 5 minutes and read the percentage of fat. Add meniscus remover if preferred. Multiply the reading by 2 to obtain the percentage of fat if a milk-test bottle was used.

Minnesota [17] **Method.**—*Reagent.*—Dissolve 110 g. of sodium carbonate and 200 g. sodium salicylate in water and make up to 1 liter

16 *Pennsylvania State College Agr. Exp. Sta., Bull.* **412** (1941).
17 Petersen and Herreid, *Minnesota Agr. Exp. Sta., Bull.* **63** (1929).

with water. To this solution add 30 ml. of 50 per cent sodium hydroxide solution and 100 ml. of *n*-butyl alcohol.

Procedure.—Weigh a 9-g. sample into an ice-cream- or milk-test bottle. Add 15 ml. of the Minnesota reagent. Shake thoroughly and place the flask in a water bath at 180° F. to boiling until the fat appears in a clear layer at the top. This generally requires about 12-15 minutes. Shake the flask several times to facilitate the movement of the fat. Place in a centrifuge and whirl for 0.5 minute. Add sufficient hot water to the flask to bring the fat column up into the graduated portion of the neck. Centrifuge again for 0.5 minute. Place in a bath at 133-137° F. for 5 minutes and read the percentage of fat. Multiply the result by 2 if a milk-test bottle was used. Add meniscus remover if desired.

Modified Minnesota [18] **Method.**—*Reagent.*—To a mixture of 645 g. sodium salicylate, 355 g. potassium carbonate and 165 g. sodium hydroxide add 3 liters of water. After the chemicals have dissolved, add 1 liter isopropyl alcohol. Store in cork or rubber-stoppered glass bottles.

Sampling.—Sample frozen ice cream from the container or can by removing about ½-1 in. of the upper surface and taking the sample from the freshly uncovered lower layer. Sample wrapped, sliced ice cream by removing the wrapper from any slice which is representative of the total volume. Samples of stick ice cream are obtained by trimming off outer chocolate or other coating. Ice cream mix is sampled from a storage vat after thorough agitation or from a can after thorough stirring.

Melt the frozen product in a water bath (temperature about 38° C. or 100° F.) by long storage at 5-10° C. (40-50° F.) or at room temperature. Keep the sample covered during the melting to prevent evaporation of water from the sample. Thoroughly mix the melted sample or liquid mix by pouring back and forth from one container to another just before it is weighed into the test bottles. Use care to incorporate any accumulated foam layer.[17]

Procedure.—Preferably run each determination in duplicate. Weigh 9 g. of prepared sample into a 20-per cent ice-cream-test bottle. Add 15 ml. of the reagent. Shake thoroughly. Digest for 12-15 minutes in a gently boiling water bath, having the bottles in a rack held at least 2½ in. above the bottom of the bath. Shake the mixture in the test bottle vigorously at the time when at least half the contents of the bottle have turned dark brown (usually about 2½ minutes after placing them in the water bath). Shake vigorously again about

[18] Martin, *Am. J. Pub. Health* **33,** 597 (1943).

1 minute later. Some care may be necessary when starting to shake the bottles the second time, as the isopropyl alcohol in the reagent may boil off through the neck of the bottle, taking with it some of the mixture.

Centrifuge the test bottles for 0.5 minute at the speed used for the regular Babcock test. Add hot water (55-65° C. or 130-150° F.) to float the milk fat well up into the neck of the test bottle. Centrifuge for 0.5 minute. Place the test bottles in a water bath at 55-60° C. or 130-140° F. and leave for 5 minutes only, because prolonged contact may entail loss of fat by saponification.

Tests may be read with the use of a colored reading fluid, or a reading may be taken from the bottom of the upper meniscus to the bottom of the lower meniscus. If a colored reading fluid is used, immediately before reading each test allow just a few drops of the fluid to flow gently down the inside of the neck of the bottle; the liquid must not be dropped onto the surface of the fat column. Hold the bottles in a level position and read as one would read a Babcock cream test, measuring from the bottom of the lower meniscus to the sharp line of demarcation between the colored mineral oil and the fat. To secure accurate readings, apply divider points to the smooth side of the bottle neck using care to prevent slipping of lower point of divider. When adjusting the lower point of the dividers, keep the eyes on a level with that point; and when adjusting the upper point, raise the eye accordingly. Report the average of the duplicate determinations.

Illinois Method.—In this method an alkaline reagent is also used for breaking the milk-fat emulsion, dissolving the proteins, and releasing the fat.

Procedure.—Weigh 9 g. of sample into a milk- or mix-test bottle. Add with the aid of a safety pipetter or from a burette 2.5 ml. of a mixture of 75 ml. of 28-29 per cent ammonium hydroxide solution, 35 ml. of n-butyl alcohol, and 15 ml. of 95 per cent alcohol, and mix thoroughly. Add 9-10 ml. of a reagent consisting of a mixture of 200 g. trisodium phosphate, 150 g. sodium acetate, and 1 liter of water, and again mix thoroughly. Place the flask in a water bath and heat the bath to boiling for several minutes. Shake the flask from time to time while heating. When the fat has risen to the top as a separate layer, which will require from 15-30 minutes for most ice cream and about 30-45 minutes for chocolate ice cream, place the flask in the centrifuge and whirl for 5 minutes. Add water to the base of the neck and whirl for 2 minutes. Add sufficient water to bring the fat into the neck of the flask and centrifuge again for 1 minute and then place in a water bath at 135-140° F. for 5 minutes.

Read the percentage of fat from the bottom of the column to the top of the meniscus and multiply the reading by 2 if a milk-test bottle was used.

Nebraska Method.—*Reagents.*—Sulfuric Acid-Alcohol Solution. —Pour 100 ml. of sulfuric acid, specific gravity 1.82-1.83 very carefully and gradually down the side of a beaker into 100 ml. of 95 per cent ethyl alcohol. Mix cautiously and allow to cool to room temperature. The reagent prepared with pure ethyl alcohol is stable. If denatured alcohol is used, as for instance Formula No. 30, a brown color is produced on standing and it is best to prepare the reagent fresh.

Procedure.—Weigh 9 g. of the sample into an ice-cream or milk-test bottle. Add 5 ml. of a mixture of 90 ml. of *n*-butyl alcohol and 10 ml. of ammonium hydroxide solution 28-29 per cent, and mix thoroughly. Add 30 ml. of the acid reagent but not more than required to keep the level below the neck of the flask. Shake thoroughly until all the curd is dissolved. This should occur before the contents turn dark. Heat the flask in a water bath at 175-180° F. for 15 minutes and shake at least 3 times in this interval. Centrifuge for 5 minutes and shake again. Add water at 180° F. if necessary to bring the level up to the base of the neck. Centrifuge for 3 minutes and again shake. Add water at 180° F. to bring the fat within the graduated portion of the neck and centrifuge for 1 minute. Place the flask in a water bath for 5 minutes at 135-140° F. and read the percentage of fat. If a milk-test bottle was used multiply the reading by 2.

Acetic Acid Method.—Weigh 9 g. of sample into an ice cream or milk-test flask. Add 8 ml. of glacial acetic acid and mix thoroughly. Before the acid settles add 9 ml. of sulfuric acid, specific gravity 1.82-1.83, in 6 equal portions, shaking after each addition and then proceed with the method as detailed for milk. If a milk-test bottle was used, multiply the reading by 2 to obtain the percentage of fat.

Roese-Gottlieb Method.—Transfer about 5 g. of ice cream accurately weighed by difference with the aid of a Mojonnier pipette and a carriage to a Jacobs-Singer separatory flask or a Mojonnier extraction tube, or weigh directly into the lower section of a Jacobs-Singer separatory flask. Add 5-6 ml. of water and 2 ml. of ammonium hydroxide, warm to 60° C., stir thoroughly, and then proceed as in the method described for milk. Care must be taken to add the full amount of alcohol. Divide the weight of the fat found by the exact weight of the sample transferred and multiply by 100 to obtain the percentage fat.

Acidity

The acidity of ice-cream mix and ice cream may be determined in a manner wholly analogous to that detailed for milk.

Weigh 9 g. of sample into a porcelain dish and dilute with an equal volume of water. Titrate as described under milk with 0.1 N sodium hydroxide solution. Divide the result by 10 to obtain the acidity in per cent, in terms of lactic acid.

Total solids, proteins, ash, preservatives, and coloring matters may all be determined by methods similar to those described in prior sections of the text. Sugars must be determined by the methods for mixtures detailed in Chapter X. "Stabilizing agents" is just another word for emulsifying agents and thickening agents and these substances may be detected in a manner entirely analogous to the procedures discussed in the sections on cream.

Total Food Solids per Given Volume

The sample must be received in the original container in a frozen condition, or in a container of known volume which has been completely filled with the frozen product.

Procedure.—Weigh the ice cream and its container as received in a solid condition to the nearest gram or tenth of a gram. Transfer the ice cream to some other sealed or covered container and clean and dry the original container. Weigh. The weight of the ice cream, plus its container, less the weight of the container, equals the net weight of the ice cream.

Determine the total solids of the ice cream by transferring about 5 g. of the ice cream which has been allowed to come to room temperature, accurately weighed with the aid of a Mojonnier pipette and a carriage, to a tared flat-bottomed dish containing sand. Place in the oven to dry. The gain in weight of the dish is the total solids. This gain in weight divided by the weight transferred to the dish and multiplied by 100 is the per cent total solids.

The net weight of the ice cream, multiplied by the per cent total solids, equals the weight of the total food solids in the ice cream. This weight, divided by the volume of the original container, gives the total food solids per given volume. If the volume of the container is accepted as the volume stated on the label, the actual volume of the container need not be determined. If it is desired to obtain the exact volume, it may be done as detailed in the first chapter by measurement. If the stated volume is given in pints, quarts, or gallons, the weights which were taken in grams must be multiplied by the factor, 0.0022, in order to convert to pounds. If the original volume was a pint, the ratio should be multiplied by 8/8; if the original volume was a quart, the ratio should be multiplied by 4/4 to bring the expression in the form of pounds/gallon. Any other weight/volume relationship is multiplied by similar factors to arrive at the pounds/gallon expression which is the customary ratio.

These calculations may be summarized as follows:

Gross weight in g. — weight of container in g. = net weight in g.
Net weight in g. × % total solids = weight of total solids in g.
Weight of total solids in g. × 0.0022 = weight of solids in lbs.

$$\text{weight of solids in lbs. divided by original vol.} = \frac{\text{lbs.}}{\text{given volume}}$$

$$\frac{\text{Lbs.}}{\text{Given volume}} \times \text{appropriate factor} = \frac{\text{lbs.}}{\text{gallon}}$$

Thus, if in a one pint ice cream sample we find:

Gross weight ....258 g.
Container .... 27 g.
Net weight .... 231 g.
Total solids 37.2%

then

$$231 \times 0.372 = 85.9 \text{ g.}$$

$$85.9 \times 0.0022 = 0.189 \text{ lbs.}$$

$$\frac{0.189}{\text{pint}} \times \frac{8}{8} = \frac{1.51 \text{ lbs.}}{8 \text{ pints}} = \frac{1.51 \text{ lbs.}}{\text{gallon}}$$

OVERRUN

Overrun is the increase in volume resulting from the amount of air incorporated into the ice cream during the agitation and freezing process. It is generally expressed as percentage overrun.

For the proper regulation of overrun, both bulk and packaged frozen desserts must be examined. Public health and other regulatory laboratories are ordinarily interested in determining the amount of food solids per gallon in the final product which may be reduced below the legal standard when overrun is not carefully controlled.

It is an accepted procedure to consider the weight and the volume of the melted, de-aerated frozen dessert as the weight and volume, respectively, of mix from which the frozen dessert has been made. However, the volume of the melt may contain variable and unknown amounts of air, tenaciously held by the stabilizer even after long heating, thus making the volume of the melt greater than that of the original mix. In such cases, the overrun calculation will give too low a figure. Lucas [19] describes a method overcoming these difficulties.

Sampling.—Select packaged samples in the original unbroken package directly from the hardening room of the manufacturing plant or in the storage cabinet of the dealer; immediately store them in an insulated container with dry ice, and hold them frozen in the container until they are examined in the laboratory. Optionally, an equivalent refrigeration practice may be followed.

In sampling bulk frozen dessert certain determinations must be made in the field. First weigh the original container, usually of 1, 2½, or 5 gal. capacity, together with its contents. (If the weight is less than 15 lbs., the scale used should have a tolerance no greater than ⅜ oz.; if the weight is between 15 and 50 lbs., the tolerance should not exceed ¾ oz.) If the weight of the container is not already known, remove the contents and determine the weight of the dry, empty container. (The contents should be handled in a sanitary manner so that they may be returned to a subsequent batch of mix for refreezing.) The difference between these two weighings is the net weight of the given volume of frozen dessert.

Record the net weight of the frozen dessert that was in the container and the capacity of that container on the information sheet submitted with the sample. Then place a sample of no less than 8 oz. in a stoppered bottle to be taken or shipped to the laboratory.

Volume of Frozen Dessert.—To determine the volume of packaged frozen desserts follow one of the three procedures given below. The determination of volume of bulk frozen desserts has been made in the field as specified above. See also Chapter I.

1. Determine the capacity of the container up to the level of the fill of the contents by measuring the dimensions of the container; if cylindrical, the measurement should include the inside diameter and the depth from the level of the fill; if tapering measure the diameter at the level of the fill and at the bottom, using the mean of these measurements as the average (mid) diameter. Calculate the volume from the formula

$$\text{Volume} = \pi r^2 h$$

where π is 3.1416, r is the radius, and h is the height.

2. A quicker way to determine the volume is to fill the container with water at 20° C. (68° F.) up to the level of the fill, and then measure this volume of water. In case it is not practical to empty the original container, measure the volume of several similar containers to the same fill, and average their volumes.

[19] Lucas, *Am. J. Pub. Health* **33**, 595 (1943).

3. The volume of a wrapped slice or brick of frozen dessert can be calculated from its measured dimensions.

Weight of Frozen Dessert.—

1. Weigh the product, the volume of which has been determined on a balance sensitive to 0.1 g.

2. In the case of wrapped slices or bricks, weigh the frozen dessert in its immediate container or carton. Remove all traces of product from the container, dry the container thoroughly, and weigh it. The difference in the two weights represents the net weight of the frozen dessert.

*Volume and Weight of Mix from Melted Frozen Dessert.—*Weigh about 130 g. of well-mixed frozen dessert (weighed accurately to the nearest centigram) into a tared 250-ml. volumetric flask. (A 250-ml. sugar flask is convenient because of its wider neck.) Heat contents to 50° C. (120° F.) for 3 minutes to expel air. Add 10 g. of *n*-amyl alcohol, specific gravity 0.817/20° C., or capryl alcohol, specific gravity 0.827/20° C., to break the foam. Adjust temperature of contents to 20° C. (68° F.). Fill flask to mark with distilled water at 20° C. (68° F.). Dry outside of flask and weigh.

Calculate the volume of this known weight of mix (i e., melted and de-aerated frozen dessert) as follows:

$$\text{(weight of contents of 250 ml. flask)} - \text{(weight of frozen dessert + amyl alcohol)} = \text{g. water added}$$

$$\frac{\text{g. water}}{0.998} = \text{ml. water}$$

$$250 \text{ ml.} - (\text{ml. water} + 12.24 \text{ ml. amyl alcohol}) = \text{ml. mix}$$

$$\frac{\text{g. mix}}{\text{ml. mix}} = \text{density of mix}$$

The weight of any volume of mix equals its density times its volume.

Calculation of Overrun.—

$$\frac{\text{(Wt. unit volume of mix)} - \text{(wt. same volume of frozen dessert)}}{\text{(Wt. same volume of frozen dessert)}} \times 100 = \% \text{ overrun}$$

A more rapid method for the determination of overrun is that of Benkendorf.[20]

*Procedure.—*Press an ice cream sampler of exactly 50-ml. volume, a short, wide tube open at both ends, into the frozen ice cream, until it is below the surface. Allow it to remain until chilled and remove by working from side to side while drawing out. Trim off

---

[20] Benkendorf, *Wisconsin Exp. Sta., Bull.* 241 (1914).

the ice cream protruding from both ends with a sharp knife and wipe the exterior dry. Transfer the ice cream to a beaker, or place the ice cream directly into a funnel resting in a 250-ml. volumetric flask. Transfer the ice cream to the flask with the aid of exactly 200 ml. of warm water and the aforementioned funnel. Add 1 to 2 ml. of ether to destroy the foam and then fill the flask to the 250-ml. mark by adding water from a burette. Record the exact amount of additional water and ether added. The percentage overrun may be calculated by the following series of expressions:

50 ml. (volume of ice cream) − volume of ether and water = volume of ice cream before agitation and freezing

$$\frac{\text{volume of ether and water}}{\text{volume of ice cream before agitation and freezing}} = \% \text{ overrun}$$

For example,

50 ml. − (22 ml. water + 2 ml. ether) = 26 ml.

$$\frac{24 \text{ ml.}}{26 \text{ ml.}} = 92.3\% \text{ overrun}$$

Foreign Fat

To a quart of ice cream add an equal volume of water in a large beaker or Erlenmeyer flask and bring the mixture to a boil. Add 25 ml. of 20 per cent copper sulfate solution and filter hot through a large Buchner funnel with the aid of suction. Drain and dry the precipitate in a vacuum oven. Mix with an equal amount of anhydrous copper sulfate, place in a large thimble and extract with petroleum ether in a Soxhlet extractor as described in Chapter I, or extract with 3 successive portions of petroleum ether. Evaporate off the petroleum ether, dry the fat in a constant temperature oven at 75° C. and proceed to examine the fat as directed in the chapter on fats and oils, Chapter IX.

Milk Sherbet

In some states another frozen product made from the same ingredients as ice cream but containing only a small percentage of milk solids is allowed to be sold under the name of milk sherbet. In New York State, milk sherbet must not contain more than 5 per cent of milk solids. This food product is analyzed exactly as is ice cream. However, it is more important, or rather, more often necessary to determine the casein content in order to control the maximum limits of milk solids.

ICES

Ices are another type of frozen product but ices are not permitted to contain any milk solids. They are made from sugar and water, flavor, color, and stabilizer. They must be rigidly controlled as to milk content for religious, dietary, and medicinal reasons, as well as for the prevention of fraud.

ACIDITY

A method similar to that used for ice cream mix can be used to estimate the acidity of water ices.

Procedure.—Allow the ice to melt, mix thoroughly, and transfer 8.8 ml. with the aid of a pipette or calibrated syringe to a porcelain dish. Add 1 ml. of phenolphthalein indicator solution and titrate with 0.1 N sodium hydroxide solution to a pink color which persists for 30 seconds. To express the result in terms of lactic acid divide the volume of standard alkali used by 10. To express the acidity in terms of hydrated citric acid multiply the percentage of acidity as lactic acid by 7/9.

DETECTION OF MILK POWDER IN FOODS

The use of skim milk powder and even whole milk powder as a binder or filler in different food products has increased in recent years. Standards have been established to govern this practice by governmental agencies. In some instances their use has been totally prohibited in certain food products as, for example, in New York State, in ices and ice sherbets, in order to prevent substitution in place of a better product such as milk sherbet or even ice cream. In meat products, their use is sometimes prohibited for dietetic or religious reasons. Milk powder as a binder in sausages, a filler in ices, an anti-crystallizing agent in frozen eggs, a component of milk chocolate and confectionery, are illustrations of the varied use of this product.

Milk has three main characteristic components, milk fat, casein, and milk sugar or lactose. The detection of any one of these in a food would be proof of the addition of milk product to that food. In the case of a food containing an oil or fat, the addition of whole milk powder would be evidenced by the alteration in the constants of the fat or oil of that food. Thus, for example, the Reichert-Meissl value would be raised. In food products which contain no fat, the problem of the detection of added milk product resolves itself to one of the identification of casein and lactose. The method for lactose depends upon the complete precipitation of all proteins, split-proteins, gelatin and pseudo-gelatins by tannic acid and neutral lead acetate, identification by means of Fehling's, Barfoed's, Tauber's and mucic acid tests, and estimation by the usual methods. The method for casein depends upon its solubility in 3 per cent sodium oxalate solution, whereas other proteins including vitellin are insoluble. The casein is subsequently estimated by the usual methods.

Jacobs Method.—*Preparation of Sample.* (1) *Meats.*—Comminute the meat or other meat product such as sausage and weigh 100 g. into a 600-ml. beaker, add 200 ml. of water, heat to boiling, and boil gently for 5 minutes. Stir the contents of the beaker frequently during this and other extractions to prevent bumping. Remove the beaker and, stirring constantly, cool under cold running water. Place in a refrigerator until the fats have solidified and filter by decantation through a Büchner funnel fitted with a rapid filter paper. Repeat the entire extraction with two 150-ml. portions of water, following each detail. Finally transfer the residue to the filter and press dry. Transfer the filtrate to an evaporating dish or to a beaker and evaporate on a steam bath or a hot plate regulated at low heat to a volume of 25 ml. From this point proceed as detailed in the method.

(2) *Eggs.*—Weigh, by difference, approximately 40 g. of the frozen egg yolks or other egg products into a liter volumetric flask containing 600 ml. of water, mix gently, fill to the mark with water, and shake gently. Filter through an 18½ cm. fluted filter paper. If the filtrate is cloudy, allow the filtration to proceed until drops of the filtrate become clear. Return the cloudy filtrate to the filter and wash the receiving flask twice with clear filtrate, returning the washings to the filter. Evaporate 500 ml. of the filtrate on a steam bath and proceed as directed in the method.

(3) *Ices and Sherbets.*—Use 25 g. and proceed with the method.

(4) *Other Food Products.*—Prepare as directed above a hot water extract, or cold water extract, or use the sample unchanged according to the character of the sample.

Procedure for Lactose.—Transfer the 25 ml. prepared as directed above to a 100-ml. volumetric flask with not more than 25 ml. of water. Mix thoroughly. Add 10 ml. of 20 per cent freshly prepared tannic acid solution and shake the flask. Add 5 ml. more. Fill to the mark after allowing to stand for half an hour. Mix gently and centrifuge or allow to stand over night. Filter through a dry, small filter, discarding the first 15 ml. Transfer a 50-ml. aliquot to a 100-ml. volumetric flask and add 10 ml. of saturated neutral lead acetate solution. Make to volume. Allow to stand or centrifuge. Filter through a small, dry filter, discard the first portion, and delead by precipitating lead as lead sulfate with anhydrous potassium sulfate. Allow to settle. Filter through a small, dry filter and discard the first portion. Test the filtrate for reducing sugars by the Fehling or Benedict test as directed in the chapter on sugars, Chapter X. If no positive test is obtained, lactose as well as other reducing sugars are not present. If a positive test is obtained, test the filtrate with Barfoed's or Tauber's reagent, as detailed in Chapter X. If no positive test is obtained with these reagents, the reduction in the case of the Fehling test is due to a disaccharide. If the Tauber or Barfoed test is also positive, a reducing monosaccharide is present and the filtrate should be tested for lactose by applying the mucic acid test.

Take 25 ml. of the deleaded filtrate, dilute to 100 ml., add 20 ml. of nitric acid, and evaporate to 20 ml. on the steam bath or on a low heat hot plate. Dilute again to 100 ml., add 20 ml. more of nitric acid, and evaporate to 20 ml. as before. Allow to stand for 24 hours. Add 10 ml. of water and allow to stand another 24 hours, to permit the mucic acid to crystallize. Pass through a small filter and wash the mucic acid crystals with 30 ml. of water to remove the nitric acid. Return the filter and contents to the original beaker. Add 30 ml. of ammonium carbonate solution (consisting of 1 part ammonium carbonate, 19 parts of water, and 1 part of ammonium hydroxide) and heat the mixture in a water bath at 80° C. for 15 minutes with constant stirring. Wash the filter paper and the contents several times with hot water by decantation, passing the washings through a filter paper, to which finally transfer the residue. Evaporate the filtrate to dryness on a water bath, avoid unnecessary heating which causes decomposition, add 5 ml. of nitric acid (specific gravity 1.15), stir the mixture thoroughly and allow to stand for 30 minutes. Collect the precipitated mucic acid on a weighed Gooch crucible, wash with 10-15 ml. water, then with 60 ml. ·95 per cent alcohol and then a number of times with ether. Dry in a constant temperature oven for 3 hours and weigh. Multiply the weight of mucic acid by 1.33 to convert to galactose and then convert to lactose monohydrate by multiplying by 2.

The lactose may be estimated by gravimetric or volumetric reduction methods on a portion of the deleaded filtrate, as detailed in Chapter X. If a monosaccharide is present, as is entirely possible in the case of ices, the methods for mixtures should be used. The above method may be used but it is empirical and requires the factor 1.33 because only 75 per cent of the galactose present is recovered as mucic acid, and hence only part of the lactose is recovered also. The method is best used for qualitative identification because only lactose and galactose of the sugars give the mucic acid test.

Maltose is also a reducing disaccharide, hence care must be taken in interpreting the reducing test. It does not yield mucic acid.

Procedure for Casein.—Weigh 22.5 g. of the original sample into a 500-ml. Erlenmeyer flask and add 250 ml. of 3 per cent sodium oxalate solution. Boil gently for 5 minutes or allow to stand for 4 hours shaking frequently. Add 5 g. of light magnesium carbonate or 1 g. of Filter Cel, allow to cool and filter. Take 100 ml. of the filtrate, add 2 ml. of glacial acetic acid and dilute to 200 ml. Allow to stand for 1 hour or over night. Filter, return the filter paper and the precipitate to the beaker in which the acetic acid precipitation was made, add 250 ml. of 3 per cent sodium oxalate solution, boil gently for 5 minutes or allow to stand for 4 hours, shaking frequently, add 5 g. of light magnesium carbonate, allow to cool, and filter. Take 100 ml. of this filtrate, add 2 ml. of glacial acetic acid, dilute to 200 ml., and allow to stand. If a dense turbidity is formed, deep enough to obscure a stirring rod held in back of the beaker, which later develops into a curdy white precipitate, casein is present.

To obtain quantitative results, determine nitrogen in an aliquot of the filtrate from the second boiling with sodium oxalate solution and then determine the nitrogen in an aliquot of the filtrate after the addition of acetic acid. The difference in the nitrogen determinations, after corrections for dilution and the aliquots taken, multiplied by 6.38 equals casein. Casein × 1.25 equals milk protein.

Casein may also be estimated by the McDowall and McDowell method as described in the section "formal titration of casein," Chapter VII.

Lactosazone Method.—To detect dried skimmed milk in meat products Kerr [21] recommends the following method. It is based on the principle that the lactose is adsorbed on the charcoal and is subsequently liberated by boiling the charcoal in acetic acid solution. The lactosazone is then formed in sodium acetate solution by boiling with phenylhydrazine hydrochloride and the crystals that are formed are identified under the microscope. This is only a qualitative test.

Procedure.—To 25 g. of the finely divided meat in a 250-ml. beaker, add 50 ml. of water. Thoroughly break up the meat with a glass rod and boil the mixture for a few seconds. Filter through a wet filter paper and add to 25 ml. of the filtrate 1 g. of good absorbent charcoal. Mix by shaking, boil for a few seconds, cool thoroughly and shake at intervals for 10 minutes. Filter through a small paper

21 Kerr, *J. Assoc. Official Agr. Chem.* **19**, 410 (1936).

or use a filter pump. When the charcoal has completely drained, transfer it to a porcelain dish containing 10 ml. of water and 1 ml. of glacial acetic acid. This is best done by opening the paper, holding it by the clean half, and moving it about in the liquid. The greater part of the charcoal is thus removed from the paper. Stir the charcoal with a glass rod and transfer the mixture to a boiling tube. Heat to boiling for about 10 seconds and filter the hot solution through a small paper into a test tube containing 0.5 to 1 g. of solid phenylhydrazine hydrochloride and 2 g. of solid sodium acetate. Mix thoroughly and filter from any insoluble oily residue. Place the tube in a boiling water bath and leave it there for 45 minutes. Remove the tube and allow it to stand at room temperature for at least 1 hour and preferably longer. Pipette off a little of the deposit, if any, and examine it on a slide under the microscope.

Lactosazone crystallizes in characteristic clumps with projecting spines ("hedgehog" crystals). Recrystallize by filtering through a small paper, washing with a small amount of water, and then passing about 4 ml. of boiling water through the paper into a clean tube. Boil the filtrate and pass through the paper 2 or 3 times, boiling between every filtration. On allowing the solution to stand, typical crystals of the osazone separate out. Filter, dry, and take the melting point (200° C.). Care must be taken that the phenylhydrazine hydrochloride solution does not become too concentrated during the boiling process, otherwise lactosazone crystals may not separate out.

Milk Solids in Bread.—This method is based on the determination of lactose and the estimation of butterfat using the Reichert-Meissl number, and applies to skim-milk solids as well as whole-milk solids. It is recommended by Hoffman, Schweitzer, and Dalby [22] as a reasonably rapid and practical method for this determination.

Remove the crust of the bread, air-dry the crumb, then grind sufficiently to pass a 20-mesh sieve. Digest 50 g. of the prepared material in 400 ml. of water at about 40° C. for 3 hours, and transfer the mixture to a large centrifuge tube. Centrifuge and decant the liquid portion into a 1-liter volumetric flask. Wash the residue 4 times using 75 ml. of water each time, and separate solids by centrifuging. Decant after each washing and add the liquid portion to the first extract. Add 35 g. of baker's compressed yeast, suspended in a small amount of water, 0.5 g. of ammonium sulfate, and 0.2 g. of sodium bisulfite, and let stand over night at room temperature stoppered, but with a vent for the escape of carbon dioxide. The ammonium sulfate is used as a yeast stimulant, and the sodium bisulfite retards bacterial action.

---

[22] Hoffman, Schweitzer, and Dalby, *Ind. Eng. Chem., Anal. Ed.* **8,** 298 (1936).

After standing over night add 20 ml. of copper sulfate solution (regular Fehling's A), and add sufficient sodium hydroxide solution to give a definite blue color and clarify the solution. Make up to volume in the liter flask, shake, and filter through a good filter paper. Take 50 ml. of the filtrate and determine lactose, using the Munson and Walker gravimetric method as directed in Chapter X.

Fifty ml. of the filtrate are equivalent to 2.500 g. of bread after correction for the yeast is made. The fat-free milk solids are calculated from the percentage of lactose found, since skim-milk powders average 50 per cent lactose with only slight variations. Therefore twice the percentage of lactose found (after calculation to the dry basis) is equal to the percentage of fat-free solids on the dry basis of the bread.

Extract the fat necessary for the Reichert-Meissl number determination by placing 200-300 g. of finely ground air-dried bread, depending upon the fat content, in a 2-liter flask containing 1000 ml. of water and 30 ml. of hydrochloric acid. Digest the mass by boiling 1 hour or until it shows good digestion, and add 10 g. of Filter Cel. Filter through a Büchner funnel containing a filter paper upon which is a thin pad of Filter Cel. Apply suction until the mass is fairly dried. Transfer the residue to a beaker, stir with ether, and filter again through Filter Cel into a dry flask. Evaporate the ether, and, if the oil is clear, it is ready for the Reichert-Meissl number determination. Use about 5 g. of fat and determine the Reichert-Meissl value as detailed in Chapter IX on Oils and Fats.

The average oil content of flour on the dry basis is 0.7 per cent. This oil has a Reichert-Meissl value of 1, for which allowance must be made in estimating the amount of butterfat in the bread. If other shortening is present, an allowance must be made for the Reichert-Meissl number of the shortening used. The normal variation in the Reichert-Meissl number of butterfat must also be considered in making the calculation.

The amount of butterfat that would be present if the bread contained whole-milk solids is calculated from the percentage of fat-free milk solids found. The Reichert-Meissl number, after correction for fats other than butterfat present, determines whether the amount of butterfat calculated is actually present. If the Reichert-Meisel number does not indicate any butterfat, skim-milk solids were used in the manufacture of the bread. If the Reichert-Meissl number indicated only part of the butterfat necessary to balance the skim-milk solids in the ratio of skim-milk solids to butterfat in whole-milk solids, then partially skimmed milk was used. The factor 0.4115 multiplied by the percentage of skim-milk or fat-free solids gives the amount of butterfat necessary to balance the skim-milk solids. The Reichert-Meissl number alone without a determination of lactose makes an estimation of the milk solids present uncertain.

It may be repeated that the effect of the shortening used in the bread is not to be underestimated, and therefore more reliance must be placed on the lactose determination.

SELECTED REFERENCES

Andrade, *Estudios sobre la leche*. Caracas, Venezuela, 1940.

Barthel, *Untersuchung von Milch und Molkereiprodukten*. Parey, Berlin, 1920.

Burke, *Practical Dairy Tests*. Olsen, 1935.

Davies, *Chemistry of Milk*. Chapman and Hall, London, 1939.

Eckles, Combs, and Macy, *Milk and Milk Products*. McGraw-Hill, New York, 1936.

Federal Register, 14, 1960 (1949).

Hunziker, *Condensed Milk and Milk Powder*. La Grange, Ill., 1946.

Jacobs, ed., *Chemistry and Technology of Food and Food Products*, 2nd ed. Interscience, New York, 1951.

Ling, *Textbook of Dairy Chemistry*. Wiley, New York, 1930.

Methods of Analysis, A.O.A.C. Washington, 1945.

Mojonnier, and Troy, *Technical Control of Dairy Products*. Mojonnier Bros., Chicago, 1925.

N. Y. State Dept. Agr. Markets, *Circ.* 481 (1934) ; *Circ.* 505 (1935).

Roadhouse and Henderson, *The Market-Milk Industry*. McGraw-Hill, New York, 1941.

Rogers, *Fundamentals of Dairy Science*. Chemical Catalog Co., New York, 1928.

Totman, McKay, and Larsen, *Butter*. Wiley, New York, 1939.

Turnbow, Tracy, and Raffeto, *The Ice Cream Industry*. Wiley, New York, 1947.

U. S. Food Drug Admin., S.R.A., F.D.&C. 2, Part 18, May, 1953; Part 19, June, 1952, as amended May, 1955.

CHAPTER IX

OILS AND FATS

EDIBLE oils and fats are the mixed triglycerides of fatty acids, so treated as to be wholesome foods. They are found in the fruits, nuts, seeds, and roots of plants and vegetables. In animals, fats are present throughout the entire body but especially so in the adipose tissue and bone marrow. Edible fats and oils, chemically, are mixtures of fatty acid esters of the trihydroxy alcohol, glycerol. Table 27 details the most common fatty acids of edible oils and fats.

TABLE 27. MORE IMPORTANT ACIDS OF EDIBLE OILS AND FATS

Acid	Formula	Occurrence	Melting Pt., ° C.	Boiling Pt., ° C.
Saturated Acids				
Acetic	CH_3COOH	Oil of spindle tree	16.6	118.1
n-Butyric	$CH_3(CH_2)_2COOH$	Butter	—7.9	163.5
Isovaleric	$(CH_3)_2CHCH_2COOH$	Oils of dolphin and porpoise	—37.6	176.7
n-Caproic	$CH_3(CH_2)_4COOH$	Butter, coconut oil, palm-nut oil	—1.5	205
n-Caprylic	$CH_3(CH_2)_6COOH$	Butter, coconut oil, palm-nut oil	16	237.5
Capric	$CH_3(CH_2)_8COOH$	Cow and goat milk, coconut oil	31.5	268-270
Lauric	$CH_3(CH_2)_{10}COOH$	Milk, spermaceti, laurel oil, palm-kernel oil	44	225^{100}
Myristic	$CH_3(CH_2)_{12}COOH$	Nutmeg oil, milk and vegetable fats; lard and cod-liver oil	58	250.5^{100}
Palmitic	$CH_3(CH_2)_{14}COOH$	Most animal and vegetable fats	64	339-356 d.
Stearic	$CH_3(CH_2)_{16}COOH$	Most animal and vegetable fats	69.4	383
Arachidic	$CH_3(CH_2)_{18}COOH$	Peanut oil	76.3	328
Behenic	$CH_3(CH_2)_{20}COOH$	Oil of behen, butterfat	80.7 (84)	306^{60}
Lignoceric	$CH_3(CH_2)_{22}COOH$	Peanut oil, sphingomyelin, ground-nut oil	81	...

TABLE 27.—(*Continued*)

Acid	Formula	Occurrence	Melting Pt., ° C.	Boiling Pt., ° C.
		Acids with One Double Bond		
Hypogeic		Peanut and maize oils	33	236^{15}
Palmitoleic	$CH_3(CH_2)_5CH=CH(CH_2)_7-COOH$	Seal oil		
Physetoleic		Sperm and seal oils		
Oleic	$CH_3(CH_2)_7CH=CH(CH_2)_7-COOH$	Most fats and oils	14	286^{10}
Rapic		Rape or colza oil		
Gadoleic	$CH_3(CH_2)_9CH=CH(CH_2)_7-COOH$	Herring, sperm and cod-liver oils		
Erucic	$CH_3(CH_2)_7CH=CH-(CH_2)_{11}COOH$	Rapeseed, mustard seed, cod-liver oils	33.5 (31-32)	281^{13}
		Acids with Two or More Double Bonds		
Linoleic	$CH_3(CH_2)_4CH=CHCH_2-CH=CH(CH_2)_7COOH$	Cottonseed, linseed, poppyseed oils	—11	230^{16}
Linolenic	$CH_3CH_2CH=CHCH_2-CH=CHCH_2CH=CH-(CH_2)_7COOH$	Linseed, hemp, perilla oils	....	$230\text{-}232^{17}$
Clupanodonic	$C_{22}H_{34}O_2$	Cod-liver, herring, sardine, whale oils	Less than —78	236^{5}
Arachidonic	$C_{20}H_{32}O_2$	Liver tissue of pigs		

Fats and oils obtained from various sources differ from one another in their physical and chemical properties because they contain varying amounts of different mixed esters. Some of these esters are solid, some liquid, some volatile, some saturated, and some unsaturated compounds. Therefore, each ester influences the physical and chemical properties of an oil or fat in some measure according to the amount of that ester in the fat or oil. These differences are the bases of tests for their identification.

Fats and oils cannot be sharply differentiated, and difference in terminology is mainly attributable to temperature and environment. Hence, the terms fat and oil are almost interchangeable. At low enough temperatures all oils and fats are solid and at high enough temperatures all fats are liquid. For instance, it is common to speak of coconut oil but since this product melts in the range 22-26° C., in temperate zones it is generally a solid, particularly in cold weather, and thus might equally well be called coconut fat. In the torrid zone, it is almost invariably a liquid and consequently derives its common name of coconut oil. A fat is, at ordinary temperatures, a mixture of liquid and solid esters of glycerol. These give it the property of a plastic mass.

TABLE 28. RELATIVE OCCURRENCE OF FATTY ACIDS IN COMMON
EDIBLE FATS AND OILS [a]

Fatty Acid	Occurrence (in terms of % by weight)		
	0-5	5-25	Over 25
Butyric	Butter	None	None
Caproic	Butter	None	None
Caprylic	Butter	Coconut, palm-kernel	None
Capric	Butter	Coconut, palm-kernel	None
Lauric	Butter	None	Coconut, palm-kernel
Myristic	Many oils and fats	Butter, coconut, palm-kernel	None
Palmitic	Rapeseed	Many oils and fats	Palm, lard, tallow
Stearic	Many oils and fats	Lard, tallow	Cocoa butter
Arachidic	Many oils and fats	None	None
Behenic	Peanut, rapeseed	None	None
Lignoceric	Peanut, rapeseed	None	None
Palmitoleic	Probably many oils	Fish, whale	None
Oleic	None	Coconut, palm-kernel, fish	Most oils and fats
Gadoleic	Probably many oils	Fish	None
Erucic	None	None	Rapeseed
Linoleic	Coconut, palm-kernel, cocoa butter	Olive, rapeseed, palm, lard, fish, whale	Peanut, cotton sunflower, corn soybean
Linolenic	Soybean, rapeseed	None	None
Arachidonic, clupanodonic, etc.	Lard	Whale	Fish

[a] Bailey in Jacobs, ed., *Chemistry and Technology of Food and Food Products.* Interscience, New York, 1944.

It must be understood that the glycerides of fats and oils are not simple esters where all three hydroxy groups of glycerol have condensed with the same acid radical, but are rather mixed glycerides, in which different acid radicals are attached to the same glycerol molecule.

TABLE 29. COMPONENT FATTY ACIDS OF BUTTERFAT [a]

Acid	Per Cent Weight
Butyric	2.6— 3.5
Caproic	1.3— 1.9
Caprylic	0.7— 1.6
Capric	1.8— 3.6
Lauric	3.2— 5.7
Myristic	6.9—11.1
Palmitic	22.8—29.1
Stearic	6.5—12.5
Arachidic (?)	0.6— 0.9
Oleic	31.3—41.3
Linoleic	3.6— 5.1

[a] Hilditch, *Analyst* 62, 250 (1937).

Edible oils are generally classified according to drying properties as follows:

1. Nondrying vegetable oil of the olive oil type, namely, olive, almond, peach kernel, and peanut.

2. Nondrying vegetable oil of rape oil type, namely, rape and mustard seed.

3. Nondrying oil of animal type, namely, lard.

4. Semidrying vegetable oil, namely, cottonseed, corn, sesame, and soya.

5. Drying vegetable oil, namely, sunflower, poppy, safflower.

6. Fish and marine animal oil, namely, cod liver, menhaden, whale.

7. Vegetable fat, namely, coconut, cocoa butter, cottonseed stearin, palm.

8. Animal fat, namely, butterfat, beef tallow, lard, mutton tallow.

A drying oil is one which has the property of drying, or of being oxidized—that is, becoming thick, viscous, and forming a membrane on being exposed to air. The terms semidrying and nondrying refer to slower drying and lack of ability to dry, in the oils so classified.

Edible oils and fats can also be classified according to source as follows:

I. Plants

1. Seeds of annuals.—Corn, cottonseed, peanut, rapeseed, sesame, soybean, and sunflower oils.

2. Fruit coats of perennials.—Olive and palm oils.
3. Seeds of perennials.—Coconut, cocoa butter, and palm kernel oils, including babassu, cohune and the like.

II. Animals
1. Milk of domestic.—Butterfat.
2. Bodies of domestic.—Beef tallow and its derivatives, oleostearine, oleo oil, and oleostock, lard and lard oil, and mutton tallow.
3. Bodies of marine.—Fish oils from sardine, menhaden, and the like, and whale oil.

The analysis of oils and fats is concerned not, as in other foods and food products, with percentage composition but rather with the physical and chemical properties which serve as a basis for identification and assay, and also with an evaluation of the suitability of a given oil or fat for a given purpose. Mixtures of oils and fats will have properties of the individual oils or fats comprising the mixture. The chemical and physical properties of an oil or fat vary between certain limits and, because of the comparatively small variation, are called *constants*. Some of the more important physical constants are specific gravity, refractive index, and melting point. Some of the more important chemical constants are the iodine value, saponification value, Reichert-Meissl value, Polenske value, free fatty acids and unsaponifiable residue. Table 30 contains the numerical values of the constants of the more important edible oils.

The literature contains many tables of these constants in which the range for specific constants is wide. This is due to the collection by authors of the results of analyses of oils from many sources. It is a fact, however, that because of modern methods or processing oils on a large scale, the range of variation of constants is actually very small. Table 31 gives the constants of some commercial oils. This small variation of commercial oils is also shown by Bailey.[1]

It is not necessary to determine every constant for every oil. Thus the purity of olive oil can almost always be determined by ascertaining the iodine value, refractive index, saponification value, added color and by performing a series of qualitative tests for added oils foreign to olive oil. Similarly, the purity of butterfat may be learned by determining the Reichert-Meissel, Polenske, Kirschner, and saponification value and also the melting point and refractive index. These constants will, in general, disclose whether fat foreign to butter has been added. It would, for example, be superfluous to make

---

[1] Jacobs, ed., *Chemistry and Technology of Food and Food Products.* Interscience, New York, 1944.

TABLE 30. CONSTANTS OF OILS AND FATS

Name	Sp. Gr. $\frac{15^\circ C}{15^\circ C}$	Saponification Value	Iodine Value	Acid Value	Refractive Index at 25° C.	Unsaponifiable Matter	Reichert-Meissl Value	Hehner Value
Almond	0.914-0.921	183-207	93-103	0.5-3.5	1.4593-1.4646[a]	0.75	0.5	96.0
Beef tallow	0.895	196-200	35-42	0.25	1.449-1.452[b]	.....	0.2-0.5	95.0-96.0
Butterfat	0.930-0.940	210-230	26-28	0.45-35.4	1.4555-1.4578[a]	0.3-0.5	17.0-34.5	86.5-89.5
Coconut	0.926	253-262	6-10	2.5-10	1.4477-1.4495[a]	0.2	6.6-8.4	82.3-90.5
Cocoa butter	0.950-0.974	192-202	33-42	1.1-1.9	1.4537-1.4580[a]	.....	0.3-1	94-95
Cod liver	0.922-0.931	171-190	135-175	5.6	1.4758-1.4783	0.5-2.7	0.2	95.4
Corn	0.921-0.928	187-193	111-128	1.4-2.0	1.4733-1.477	1.5-2.8	4.3	93-95
Cottonseed	0.920-0.925	191-196	103-115	0.6-0.9	1.4743-1.4752[c]	1.1	0.95	95-96
Cottonseed stearin	0.918-0.923	194-195	89-103	.....	.....[a]	.....	.....	96
Lard oil	0.913-0.916	193-198	62-82	1.5	1.4607[a]	0.6	.....	97
Lard	0.934-0.938	195-203	47-66	0.5-0.8	1.450-1.454[a]	.....	0.2-0.6	93-95
Menhaden	0.923-0.933	188-193	148-185	5-8	1.4787	0.6-1.4	1.2	.....
Mutton tallow	0.937-0.953	192-196	32-61	1.7-14	1.4545-1.4582[a]	.....	0.3	95-96
Olive	0.915-0.920	185-196	79-90	0.3-1.0	1.4657-1.4667	0.4-1.0	0.6-1.5	95
Palm	0.921-0.924	196-204	49-59	10	1.4603-1.4639[a]	.....	0.9-1.9	94-97
Peanut	0.917-0.926	186-194	85-100	0.8	1.4620-1.4653[a]	0.5-0.9	0.5	95-96
Poppy seed	0.924-0.926	190-195	128-141	2.5	1.4739-1.4743	0.4	0.6	95-96
Rape	0.913-0.917	168-179	94-105	0.4-1.0	1.4649-1.4659[a]	1.5	0.8	94-96
Sesame	0.921-0.925	188-193	103-117	9.8	1.4704-1.4717	0.9-1.3	1.1	95
Soya	0.924-0.927	189-194	122-134	0.3-1.8	1.4723-1.4756	1.3-1.5	0.5-2.8	93-95
Sunflower	0.924-0.926	188-194	120-136	11.2	1.4659-1.4721[a]	0.3	0.5	95
Tea seed	0.911-0.927	188-196	88-90	.....	1.4707	.....	0.6	.....
White mustard	0.912-0.916	171-174	94-98	5.4	1.4649	.....	.....	96-97

[a] At 40° C.
[b] At 60°C.
[c] At 15°C.

a Reichert-Meissl determination for ascertaining the adulteration of olive oil.

TABLE 31. CONSTANTS OF COMMERCIAL OILS AND FATS

Oil or Fat	Iodine Value		Saponifi- cation Number	Refractive Index	
	Range	Mode		Range	Mode
Corn	115-121	116	190-195	1.4720-25[a]	1.4721
Cottonseed	103-110	109	190-195	1.4704.12[a]	1.4710
Olive	80-86	83.5	190-195	1.4671-77[a]	1.4675
Peanut	91-95	94	190-195	1.4682-92[a]	1.4690
Rapeseed	100-105	102	175-180	1.4705-10[a]	1.4708
Sesame	100-108	107	190-195	1.4704-12[a]	1.4710
Soybean	120-125	124	190-195	1.4727-35[a]	1.4730
Beef tallow	35-40	38	190-198	1.4595-605[b]	1.4598
Butterfat	26-28	26	215-225	1.4538-60[b]	1.4550
Coconut	8-12	11	255-260	1.4490-98[b]	1.4495
Lard	55-62	60	190-198	1.4570-80[b]	1.4575

[a] At 25° C.
[b] At 40° C.

One fat and one oil, namely, butterfat and olive oil, have assumed a far greater importance in the eyes of the public than any other oil or fat. Butterfat being an animal fat and a milk product is of unquestionable value as a food. Olive oil, on the other hand, has not only been credited with food value, in which it is of no more value than any other edible oil, but has, it is claimed, medicinal and therapeutic powers. It has a distinct flavor which some consumers prefer. Hence, this oil has in the belief of these people a greater monetary value. Consequently adulteration with a cheaper oil is a fraud. The importance of olive oil as an article of commerce is easily seen from the fact that 104,788,000 lbs. of the edible oil were imported in 1939. This value dropped to a low of 1,105,000 in 1943 during World War II but for the fiscal year 1947-8, a total of 24,731,000 lbs. were imported. It is true that virgin olive oil is obtained pure directly from the precursor, the olive, just as butterfat may be obtained pure directly from its precursor, milk or cream. Other oils and fats must be refined before being acceptable for food use, for example, cottonseed oil.

SPECIFIC GRAVITY

The specific gravity of oils and fats may be determined by the methods described in Chapter I. The tables of specific gravity of oils are generally given at 15.5° C. Many fats are not liquid at this temperature and hence the specific gravity must be determined at a higher temperature and this figure must then be corrected for expansion. This may be done as follows.

$$\text{sp. gr.} \frac{t° \text{ C.}}{15.5° \text{ C.}} = \frac{W_1}{W}$$

$$\text{sp. gr.} \frac{15.5° \text{ C.}}{15.5° \text{ C.}} = \frac{W_1}{W} + 0.00064 t_1° \qquad \text{For all temperatures above 15.5° C.}$$

W_1 = weight of oil
W = weight of an equal volume of water
$t°$ = temperature at which oil is weighed
$t_1°$ = $t° - 15.5°$.

The value 0.00064 is an average coefficient of expansion for all oils.

REFRACTIVE INDEX

The refractive index of oils is generally taken at 25° C. and of fats at 40° C. This may be done as described in Chapter II. The correction to be applied for readings taken at a temperature greater than 40° C. is +0.00038 times the number of degrees over 40° C. A corresponding correction may be applied for readings taken over 25° C. by using the same correction factor.

MELTING POINT

Introduce a very small portion of fat into a capillary tube by drawing up the melted fat into the tube by suction. Place the capillary on ice for at least 30 minutes and preferably overnight. Remove the capillary and attach it to a thermometer graduated to 0.2° by means of rubber bands. Suspend the thermometer in a beaker containing water and determine the point of incipient fusion at which the fat is translucent, and the point of complete fusion at which the fat is transparent in the usual way.

TITER TEST

It is difficult to obtain the solidifying point of an oil or fat. This is much more easily performed on the fatty acids derived from the oil or fat. The procedure followed in obtaining the fatty acids and then determining the solidifying point of these mixed fatty acids is termed the titer test. This constant is more characteristic than that of the melting point. However, the labor and time involved in the method is not commensurate with the value of the result obtained, except in certain instances. The conditions of the test have been standardized by the Committee on Analysis of Commercial Fats and Oils of the American Chemical Society.[2]

PHYSICAL ASPECTS

The viscosity, surface tension, optical rotation, fluorescence, and absorption spectra of oils and fats may be obtained as detailed in Chapter II. These determinations are being made with increasing frequency and may become important means of determining the purity of an oil or fat.

Virgin olive oil gives a yellow fluorescence, probably attributable to carotene. The oxidation induced by long exposure to air causes a change from yellow to bluish-green or violet tones. A violet fluorescence is particularly noticeable in most oils, as, for example, cottonseed, soya, palm kernel, rape seed, peanut, etc., which have been refined under pressure or obtained by solvent extraction, the yellow color that is normally present probably being destroyed by the processing. The addition of annatto or carotene causes the reappearance of the yellow fluorescence.

Butter and margarine both fluoresce, butter with a yellowish color and margarine with a strong blue. Stadler [3] recommends that the fat be dissolved in petroleum ether, in which solvent a 15 per cent adulteration of butter with margarine can be detected by the strong bluish fluorescence. Fat separated from artificial cream appears pure white with a bluish fluorescence, whereas genuine cream yields fat which gives a striking yellow color.

Cocoa butter does not have much fluorescence unless refined or purified by a solvent extraction process.

IODINE VALUE

The unsaturated glycerides of an oil or fat have the ability to absorb a definite amount of iodine, especially when aided by a carrier such as iodine chloride or iodine bromide, and thus form saturated compounds. The quantity of iodine absorbed is a measure of the unsaturation of an oil or fat. The iodine value is generally expressed as the number of grams of iodine absorbed by 100 g. of the oil. The two methods usually employed for the estimation of the iodine value are the Hanus method using iodine bromide as the carrier and the Wijs method using iodine chloride as the carrier. The preparation

---

[2] *Ind. Eng. Chem., Anal. Ed.* 12, 383 (1940).
[3] Stadler, *Z. Untersuch. Lebensm.* 55, 404 (1928).

of iodine bromide solution is easier than the Wijs reagent. There is some difference in the iodine values obtained by these two methods, but the difference is not greater than the variation in the iodine values of the oils or fats themselves.

Hanus Iodine Value.—Iodine Bromide Reagent.—Dissolve 13.2 g. of iodine in 1 liter of glacial acetic acid. Add small portions of warmed glacial acetic acid to the iodine. When the iodine is completely dissolved and the mixture is cool, add enough bromine to double the halogen content, usually about 3 ml. is sufficient. Dissolve the iodine in one portion of the liter of glacial acetic acid and the bromine in another smaller portion. Estimate the strength of each solution by titrating a small portion with potassium iodide solution and standard thiosulfate. With the aid of these results the exact amount of the bromine solution to be added to the iodine solution may be calculated.

Although the iodine bromide reagent is relatively stable, it should be checked from time to time with an oil of known iodine value and discarded if it appears to give abnormal results and a fresh reagent should be prepared. Such deterioration is more readily detected with oils of high iodine value.

Procedure.—Weigh about 0.1 to 0.5 g. of the oil or fat, according to the unsaturation, into a small capsule. Record the exact weight. Transfer the capsule and its contents to a 250-ml. glass-stoppered bottle. Add 10 ml. of chloroform to dissolve the sample and then 25 ml. of the Hanus iodine solution with the aid of a safety pipette,[4] and allow to stand for 30 minutes with occasional shaking. A large excess of iodine should always be present, at least 60 per cent. At the end of this period, add 10 ml. of 15 per cent potassium iodide solution, shake thoroughly, and then wash down the sides of the bottle and the stopper with 100 ml. of freshly boiled and cooled water. Titrate with the standard 0.1 N sodium thiosulfate solution, adding it with constant shaking until the yellow color of the iodine has almost disappeared. Add 2 ml. of a 1 per cent starch solution to act as an indicator and continue the titration with the thiosulfate. When the blue color has almost disappeared, stopper the bottle and shake vigorously so that any iodine remaining in the chloroform layer will pass into the potassium iodide solution. Complete the titration. Run two blank determinations on an equal portion of Hanus solution. The number of milliliters of 0.1 N sodium thiosulfate solution required by the blank less the quantity used in the determination gives the thiosulfate equivalent of the iodine absorbed

---

[4] Singer and Jacobs, *Anal. Chem.* **20**, 496 (1948).

by the oil or fat. Calculate the grams of iodine absorbed by 100 g. of the fat or oil. This is the Hanus iodine value.

A convenient method of weighing oils for the determination of the iodine value is the following: Take an ordinary glass slide, place 5 or 6 glass capsules on it, and obtain the tare weight. Add successively with the aid of individual glass tubes, inside diameter about 2 mm., the requisite number of drops of oil. Weigh after each specimen has been sampled in this manner. Six or seven weighings thus give the weight of each of the five or six samples.

Wijs Iodine Value.—Iodine Chloride Solution.—Dissolve 13 g. of iodine in 1 liter of glacial acetic acid. Pass in sufficient washed and dried chlorine until the halogen content is almost but not quite doubled. This may be ascertained by thiosulfate titrations. An excess of iodine is permissible, but an excess of chlorine must be avoided.

Procedure.—Weigh accurately from 0.1 to 0.5 g. of the oil or fat, according to the unsaturation of the oil or fat, into a small capsule and transfer the capsule and its contents to a 250-ml. glass-stoppered bottle. Add 15 ml. of chloroform or carbon tetrachloride to dissolve the oil or fat and then 25 ml. of the Wijs iodine solution, with the aid of a safety pipette, allowing the pipette used to deliver the iodine solution to drain for a definite time. Place the bottle in a dark place and allow to stand for 30 minutes. At the end of this period, add 20 ml. of 15 per cent potassium iodide solution, stopper the bottle and shake thoroughly, and wash down the sides of the bottle and the stopper with 100 ml. of recently boiled and cooled water. Titrate with standard 0.1 N sodium thiosulfate solution, adding the reagent with constant shaking until the yellow color of the iodine has almost disappeared. Add 2 ml. of a 1 per cent starch solution and continue the titration. When the blue color has almost disappeared, stopper the bottle and shake vigorously so that any iodine remaining in the organic solvent layer will pass into the water layer. Complete the titration. Run two blank determinations on an equal portion of the Wijs reagent, allowing the pipette to drain for the same length of time as for the unknown. The number of milliliters of 0.1 N sodium thiosulfate solution required by the blank less the quantity used in the determination gives the thiosulfate equivalent of the iodine absorbed by the oil or fat. Calculate the grams of iodine absorbed by 100 g. of the fat or oil. This is the Wijs iodine value.

THIOCYANOGEN VALUE

Thiocyanogen (CNS)$_2$ may be regarded as a pseudo-halogen in a manner analogous to ammonium being regarded as a pseudo-alkali metal. It may be obtained by the action of bromine on lead thiocyanate in ethereal solution. It forms colorless crystals melting at $-3°$ to $-2°$ C., and reacts as a typical halogen, hence it will be absorbed by unsaturated glycerides as is iodine. By means of the thiocyanogen method and a series of mathematical formulas derived from the theoretical thiocyanogen and iodine values, the percentage of oleic, linoleic, and saturated acids in an oil or fat may be estimated, provided other unsaturated acids are not present in large amounts.

It was believed that thiocyanogen was added stoichiometrically at each double bond. It has, however, been shown that the actual absorption of thiocyanogen is not exactly theoretical and, consequently, empirical thiocyanogen values have been set for the various glycerides and fatty acids. These empirical values are used in calculating the composition of mixtures of fatty acids or of fats and oils.[5]

Preparation of Lead Thiocyanate.—Dissolve 250 g. of neutral lead acetate $[Pb(CH_3COO)_2 \cdot 3H_2O]$, in 500 ml. of water. Dissolve 250 g. of potassium thiocyanate in 500 ml. of water. Add the lead acetate solution to the potassium thiocyanate solution slowly and with stirring. Filter off the precipitated lead thiocyanate on a Büchner funnel and wash successively with water, alcohol, and ether. Dry the lead thiocyanate as much as possible by drawing air through it. Remove from the funnel and dry on a watch glass in a phosphorus pentoxide desiccator for 8 to 10 days before using. This lead thiocyanate should be a greenish or yellowish white crystalline material; if it is at all discolored it must be discarded. Precipitated lead thiocyanate may be kept for a period not exceeding two months.

Preparation of Acetic Acid.—Acetic acid is conveniently dehydrated by refluxing with acetic anhydride. In a 3-liter Florence flask, with a large test tube set in the neck and through which cold water is passed to serve as a condenser, place 2 liters of glacial acetic acid (99.5 to 100.0 per cent) and 100 ml. of acetic anhydride (90 to 100 per cent). Reflux this mixture over an oil bath for 3 hours at approximately 135° C. After the anhydrous acid has cooled to room temperature place it in a cleaned and dried glass-stoppered bottle.

Preparation of an 0.2 N Solution of Thiocyanogen.—For the preparation of 1 liter of solution, suspend 50 g. of the dry lead thiocyanate in 500 ml. of anhydrous acetic acid; dissolve 5.1 ml. of bromine in another 500 ml. of acid. Two glass-stoppered acid bottles of 2 or 3 liter capacity, which have previously been cleaned and dried, should be used for this purpose. Add the bromine solution to the lead thio-

[5] Riemenschneider, Swift, and Sando, *Oil & Soap* 18, 203 (1941).

cyanate suspension slowly, in small portions, and shake vigorously, between each addition, until the solution is completely decolorized. After all the bromine has been added, allow the precipitated lead bromide and the excess lead thiocyanate to settle out, then filter the solution as rapidly as possible. Use a 13-cm. Büchner funnel and qualitative filter paper together with two 2-liter pressure flasks for the filtration. These should have been previously dried for 1 hour at 105° C.

Filter the entire solution by suction into the one flask. Transfer the funnel, containing the paper and some cake, to the second flask and refilter the solution. It should be perfectly clear upon the second filtration. Store the solution in glass-stoppered brown bottles and keep in a cool place, 15.5-21° C. (60-70° F.).

If it is convenient, the following method for the preparation of the thiocyanogen solution can be used to advantage. Suspend 50 g. of the dry lead thiocyanate in 600 ml. of anhydrous acetic acid in a round-bottomed 2-liter flask, equipped with a mechanical stirrer and dropping tube. Slowly add with agitation 5.1 ml. of bromine suspended in 200 ml. of the dry acid in the dropping tube. The acetic acid-bromine solution should be added at a rate such that the liquid in the reaction flask remains only faintly tinged with brown. When the entire bromine-acetic acid solution has been added, rinse the dropping tube out with an additional 200 ml. of the dry acid which is added immediately to the reaction mixture. When the bromine has all reacted, as indicated by the color of the reaction mixture, stop the agitation, allow the precipitated lead bromide to settle, and filter the thiocyanogen solution as described above.

Procedure.—Weigh accurately 0.1 to 0.3 g. of the oil as above and transfer to a dry 125-ml. glass-stoppered flask. Add from a safety pipette 25 ml. of thiocyanogen solution and allow to stand for 24 hours in the dark. The storage place should be from 18-21° C. (65-70° F.) in temperature and should not exceed 21° C. for any length of time. The size of the sample is governed largely by the expected thiocyanogen absorption. The excess thiocyanogen should be at least 100 per cent and preferably 150 per cent of the amount absorbed by the oil, although a greater excess seems to do no harm. At the end of 24 hours, add 1 g. of dry powdered potassium iodide to the flask and swirl the flask rapidly for 2 minutes. It is advisable to agitate the blank determination for 3 minutes. Mechanical agitation such as is employed at times for iodine values is found very satisfactory for thiocyanogen values. Then add 30 ml. of water and titrate the liberated iodine with 0.1 N sodium thiosulfate solution, using starch as an indicator. At least three blanks should be run with the samples. The solution should also be titrated at the beginning of the 24-hour period. If the drop is more than 0.2 ml. on the blank titrations, the solution is decomposing too rapidly and erratic and low figures will be the result.

An iodine value must be determined by the regular Wijs method as a factor in the calculation.

$$TV = \frac{(\text{blank titr.} - \text{titr. of sample}) \times (N \text{ of } Na_2S_2O_3) \times 12.69}{\text{weight of sample}}$$

in which

TV = thiocyanogen value of the oil
N = normality

or the calculation may be expressed as follows:

$$TV = \frac{(\text{blank titr.} - \text{titr. of sample}) \times Na_2S_2O_3 \text{ factor } (IV) \times 100}{\text{weight of sample}}$$

in which the $Na_2S_2O_3$ factor (IV) is expressed as grams of I_2 per ml.

TV = thiocyanogen value of the oil
IV = iodine value of oil Wijs method.

This index makes it possible to calculate the composition of an oil composed of the glycerides of oleic, linoleic, and a group of saturated acids. Hence the method may, in the main, be applied to most of the common edible oils. From the theoretical values of the thiocyanogen and iodine values of the glycerides the following equations may be set up.

	Theoretical Iodine Value	Thiocyanogen Value
Linoleic glycerides (L)	173.20	90.59
Oleic glycerides (O)	86.01	86.01
Saturated glycerides (S)	0	0

then placing

$$L + O + S = 100$$
$$173.2L + 86.0\ O + S = IV \times 100$$
$$90.59L + 86.0\ O + S = TV \times 100$$

we have a set of simultaneous equations that may be solved.

Calculation of Fat Composition.—The following calculations may be used when the iodine number and thiocyanogen number are determined on the fat directly and it is desired to express the percentages of the various acids as glycerides. In these formulas the derivation of which is shown above, no unsaturation greater than linoleic is assumed to be present:

$$L = 1.210 \ (IV - TV)$$
$$O = 2.438 \ (TV \div 1.274 \ IV)$$
$$S = 100\% - (L + O).$$

in which

IV = iodine number of the oil
TV = thiocyanogen number of the oil
L = per cent of linoleic glycerides
O = per cent of oleic glycerides
S = per cent of saturated glycerides S also includes the percentage of unsaponifiable matter.

The factors and formulas for calculations with the free fatty acids are:

	IV	TV
Linoleic acid (x)	181.0	96.3
Oleic acid (y)	89.9	89.9
Saturated acids (z)	0	0

$$x + y + z = 95.5\%$$
$$181.0x + 89.9y + 0z = IV$$
$$96.3x + 89.9y + 0z = TV$$

$$x \ (\% \text{ linoleic acid}) = 1.181 \ (IV - TV)$$
$$y \ (\% \text{ oleic acid}) = 2.377 \ (TV - 1.265 IV)$$
$$z \ (\% \text{ satd. acids + unsapon. matter}) = 95.5 - (x + y).$$

In these equations the assumption is made that the fatty acids plus the unsaponifiable matter amount to 95.5 per cent of the fat or oil. In those instances in which the unsaponifiable matter is greater than 1 per cent, it is better to prepare the fatty acids and separate the unsaponifiable matter. The fat acids are then used for the determination of the thiocyanogen number in place of the original sample.

The thiocyanogen method can also be applied to oils which contain linolenic acid, but it is necessary, in this instance, to know the percentage of both the unsaturated and saturated acids if the quantities of the unsaturated acids in the sample are to be calculated.

Acids	IV	TV
Linolenic (Ln)	273.7	167.3
Linoleic (L)	181.0	96.3
Oleic (O)	89.9	89.9

$$273.7 \ Ln + 181.0 \ L + 89.9 \ O = IV \qquad (1)$$
$$167.3 \ Ln + \ 96.3 \ L + 89.9 \ O = TV \qquad (2)$$
$$Ln + L + O = 95.5 - (\text{satd. acids + unsapon. matter}) \qquad (3)$$

From these equations, the following may be derived

$$106.4 \ Ln + 84.7 \ L = 100(IV - TV) \tag{4}$$
$$106.4 \ Ln + 8.58 \ L = 137.47TV - 123.59(95.5 - \text{satd. acids} +$$
$$\text{unsapon. matter}) \tag{5}$$

The percentage of linoleic acid is calculated by substituting the proper values for IV and TV and also that for saturated acids plus unsaponifiable matter, simplifying, and subtracting (5) from (4) and solving for L. The percentage of linolenic acid can then be calculated by substituting this value of L in either equations (4) or (5) and then solving for Ln. The percentage of oleic acid is obtained by substituting the values calculated for L and Ln in either (1) or (2) and solving for O.

All glassware and chemicals used in the preparation or handling of thiocyanogen solutions must be absolutely free from water. The glassware should be scrupulously cleaned with cleaning solution, water, alcohol, and ether and then dried for 1-2 hours in an oven at 105° C. The thiocyanogen solution should not be exposed to air, heat, or light for any length of time. The 0.2 N thiocyanogen solution cannot be used after its decomposition exceeds 0.2 ml. of 0.1 N sodium thiosulfate solution for 25 ml. over a period of 24 hours. This rate of decomposition should not be exceeded in less than 7 days.

SAPONIFICATION VALUE

Saponification is the hydrolysis of esters. This is generally done by boiling the esters with alkali. Oils and fats are the fatty acid esters of the trihydroxy alcohol, glycerol. The saponification value or Koettstorfer value is the number of milligrams of potassium hydroxide required to saponify 1 g. of oil or fat. Since 1 g. of an oil or fat containing glycerides of the low molecular weight fatty acids will have more molecules than an oil or fat containing glycerides of the high molecular weight, the number of milligrams of potassium hydroxide required to saponify the oil or fat will be greater in the former than in the latter case. Thus the saponification value of butter is higher than that of beef tallow because the former contains the low weight fatty acid, butyric acid.

Reagent.—Purify a portion of alcohol by refluxing 1.2 liters of 95 per cent alcohol over 10 g. of potassium hydroxide and 6 g. of aluminum foil for about 30 minutes. Distill by heating on a steam bath, discard the first 50 ml. of the distillate. Dissolve 40 g. of potassium hydroxide in one liter of the distillate and keep the solution in a glass-stoppered bottle in the dark.

Procedure.—Weigh accurately 5 g. of the oil or fat into a 300-ml. Erlenmeyer flask. This may be done with the aid of a Mojonnier pipette and carriage, by difference, or by some other suitable means. Add 50 ml. of the alcoholic potassium hydroxide with a pipette, allowing the pipette to drain for a definite time. Connect the flask with an air condenser and reflux until the fat is completely saponified, which point is reached when a solution free of fat globules is obtained. Cool, titrate the excess potassium hydroxide with 0.5 N hydrochloric acid using phenolphthalein as indicator. Run a blank along with the sample, using the same 50-ml. pipette and allowing the pipette to drain for the same length of time. Subtract the number of milliliters of 0.5 N hydrochloric acid used in the determination from that used in the blank. The remainder is the number of milliliters of 0.5 N hydrochloric acid equivalent to the potassium hydroxide used to saponify the oil or fat. Calculate the number of milligrams used for one g. of oil or fat. This number is the saponification value.

To calculate the saponification value, the following formula may be used.

$$\text{Saponification value} = \frac{(a-b) \times 28.05}{\text{weight of oil}}$$

where

a = the number of ml. of 0.5 N hydrochloric acid required for the blank,

and

b = the number of ml. of 0.5 N hydrochloric acid required for the determination.

Hehner Value

Most of the fatty acids of an oil or fat are insoluble in water. In the case of those fatty bodies having fatty acids of low molecular weight which are relatively soluble in water, such as butter, coconut, and palm kernel oils, the percentage of insoluble acids will be lower. The Hehner value is a number expressing the percentage of insoluble fatty acids plus unsaponifiable matter in an oil or fat.

Procedure.—Evaporate off the alcohol in the residue from the saponification value determination, dissolve the soaps completely in sufficient warm water, and transfer to a beaker. Add 10 ml. of hydrochloric acid, specific gravity 1.12, and heat the mixture until all the fatty acids rise to the top as an oil. Cool and place the beaker in a refrigerator. Weigh a thick sheet of filter paper. Wet it thoroughly and place it in a Büchner funnel. Filter the liquid and transfer the cake of fatty acids to the filter. Wash out all the fatty acids from the beaker by using hot water and freezing the oil formed before filtering. Wash the fatty acids on the filter thoroughly with cold water. Cool the funnel. Transfer the filter and fatty acids to a weighed beaker and dry at 100 to 105° C. Cool and weigh. Calculate the percentage of insoluble fatty acids. This is the Hehner value.

REICHERT-MEISSL AND POLENSKE VALUES

Some of the fatty acids obtained from fats and oils are volatile with steam and some are not. The volatile fatty acids consist mainly of butyric, caproic, caprylic, capric, lauric, and myristic acid. Butyric and caproic, which are water soluble, and caprylic and capric, which are slightly water soluble, are the acids estimated in the Reichert-Meissl value. Capric, lauric, and myristic comprise the acids estimated in the Polenske value. There is no definite line of demarcation, for the longer the steam distillation is carried on, the more volatile fatty acid is obtained. Nonetheless, if the distillation is carried out under rigidly controlled conditions, consistent results can be obtained. In these methods, the fat or oil is saponified with glycerol soda solution, the fatty acids are liberated from the soaps formed by the addition of acid, and are subsequently isolated by distillation.

The Reichert-Meissl value is the number of milliliters of 0.1 N sodium hydroxide required to neutralize the water soluble fatty acids distilled from 5 g. of the fat or oil under the specific conditions of the method. The Polenske value is the number of milliliters of 0.1 N sodium hydroxide required to neutralize the water insoluble, but alcohol soluble fatty acids distilled from 5 g. of the fat or oil under the specific conditions of the method.

Reichert-Meissl Value.—*Procedure.*[6]—Weigh accurately 5 g. of the sample to be tested into a clean, dry 300-ml. flask; add 20 ml. of the glycerol-soda solution, prepared by adding 20 ml. of (1:1) sodium hydroxide solution to 180 ml. of pure concentrated glycerol. Heat over a flame or asbestos plate until complete saponification occurs, as shown by the mixture becoming perfectly clear. If foaming occurs, shake the flask gently. Add 135 ml. of recently boiled water drop by drop at first to prevent foaming, then add 6 ml. of sulfuric acid (1:4) and a few fragments of pumice stone, previously heated to a white heat, plunged into water and kept there until used. Distill, without previously melting, the fatty acids, using an apparatus of the approximate dimensions illustrated in Fig. 63. Rest the flask on a piece of asbestos board having a hole 5 cm. in diameter in the center, and

---

[6] *Methods of Analysis, A.O.A.C.* Washington, 1945.

so regulate the flame as to collect 110 ml. of the distillate in as near 30 minutes as possible and to allow the distillate to drip into the receiving flask at a temperature not higher than 18-20° C.

When the distillation is complete, substitute for the receiving flask a 25-ml. cylinder to collect any drops that may fall after the flame has been removed. Mix without violent shaking, immerse the flask containing the distillate almost completely in water at 15° C. for 15 minutes, filter the 110 ml. of distillate through a dry filter paper 9 cm. in diameter, and titrate 100 ml. with 0.1 N sodium hydroxide solution, using phenolphthalein indicator. The pink color should remain unchanged for 2 or 3 minutes. The Reichert-Meissl value is the number of milliliters of 0.1 N sodium hydroxide solution used times 1.1, after this result is corrected for the figure obtained in a blank determination.

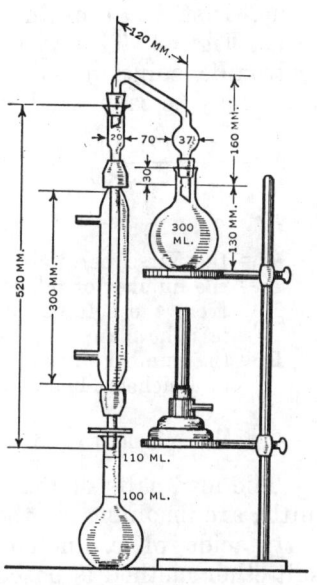

FIG. 63. Reichert-Meissl Polenske Apparatus

Polenske Value.—*Procedure.*—Remove the remainder of the soluble acids from the insoluble acids upon the filter paper by washing with 3 successive 15-ml. portions of water, previously passed through the condenser, the 25-ml. cylinder and the 110-ml. receiving flask. Then dissolve the insoluble acids by passing 3 successive 15-ml. portions of neutral alcohol, 95 per cent by volume, through the filter paper, each portion having previously passed through the condenser, the 25-ml. cylinder and the 110-ml. receiving flask. Titrate the combined alcoholic washings with 0.1 N sodium hydroide solution, using the phenolphthalein solution as indicator. The Polenske value equals the number of milliliters of alkali solution required for the titration.

KIRSCHNER VALUE

The silver salt of butyric acid is soluble in water. By use of this fact the butyric acid portion of the volatile fatty acids may be separated and estimated. The Kirschner value is a number dependent on the amount of soluble silver fatty acids in the Reichert-Meissl distillate. Butterfat gives Kirschner values from 19 to 26; coconut oil gives an average value of 1.9; palm kernel oil gives an average value of 1.0;

the majority of other oils and fats give values varying from 0.1 to 0.2.

Procedure.—To 100 ml. of the Reichert-Meissl distillate, in a 200-ml. Erlenmeyer flask, add 6 drops of phenolphthalein indicator and titrate to a faint pink with 0.1 N barium hydroxide solution. Add 0.3 g. of finely powdered silver sulfate. Allow the mixture to stand for an hour with frequent shaking during that period. Filter and transfer 100 ml. of the filtrate to a 300-ml. flask. Add 10 ml. of sulfuric acid (1:40), 35 ml. of water and several small pieces of pumice stone. Distill 110 ml. in about 20 minutes, using the Polenske apparatus, Fig. 63. Titrate 100 ml. of the distillate with a 0.1 N barium hydroxide solution. Make a blank determination. The Kirschner value may be calculated by use of the following formula:

$$K = \frac{A \times 121(100 + B)}{10,000}$$

where

K = the Kirschner Value
A = the number of ml. of 0.1 N Ba(OH)$_2$ used to neutralize the distillate from the soluble silver salts of the fatty acids (second distillate) less the blank
B = the number of ml. of 0.1 N Ba(OH)$_2$ used to neutralize the 100 ml. of Reichert-Meissl distillate (first distillate).

SATURATED AND UNSATURATED FATTY ACIDS

The lead salts of the solid saturated fatty acids, stearic and palmitic, are insoluble in ether. The lead salts of the liquid unsaturated fatty acids, oleic, linoleic, linolenic, are soluble in ether. The lead-salt-ether method is based on this difference in solubility. The solid unsaturated fatty acids, erucic, iso-oleic, etc., are somewhat soluble in ether, although with difficulty. The liquid saturated acids and the saturated acids having a molecular weight lower than myristic acid give lead salts that are ether soluble. Hence the lead-salt-ether method cannot be applied to fats and oils that contain erucic, elaeostearic, chaulmoogric, hydnocarpic, or similar acids, nor to hydrogenated materials that contain iso-oleic acid, nor to coconut or palm kernel oils that have low molecular weight fatty acids that give ether soluble lead salts. This method is of limited value, time consuming, and laborious; a briefer one than the one customarily employed is the following.

Saturated Acids.—Add 100 ml. of alcoholic potassium hydroxide solution, containing 40 g. of potassium hydroxide per liter, to 20 g. of oil and reflux for 0.5 hour. Add 20 ml. of hydrochloric acid (7 ml.

concentrated hydrochloric acid + 123 ml. of water) and 80 ml. of 95 per cent alcohol and place in an ice bath for 2 hours. Filter the precipitate through a Buchner funnel and wash with three 25-ml. portions of cold 70 per cent alcohol. Combine and save the filtrates.

Transfer the precipitate back to the original flask. Add 80 ml. of water and 20 ml. of concentrated hydrochloric acid, and shake well to decompose the potassium salts. Cool and transfer to a separatory funnel, washing out the flask with successive portions of ether. Shake well to decompose completely the potassium salts, add the ether washings, remove the water layer, and wash the ether layer three times with water. Evaporate the ether solution in a tared dish, dry in an oven at 100° C., and weigh.

Unsaturated Acids.—Evaporate the combined alcoholic filtrates from above to remove the alcohol. Wash into a separatory funnel with water, decompose the potassium salts with 20 ml. of concentrated hydrochloric acid, and extract the liberated fatty acids with ether. Discard the water layer and wash the ether layer 3 times with water. Evaporate the ether solution in a tared dish or flask, dry the unsaturated acids, and weigh.

Calculation.—Determine the iodine numbers of both fractions. The saturated acid fraction has an iodine value which is due to some unsaturated acid present in this fraction.

$$\% \text{ Unsatd. acids in satd. fraction} = \frac{IV \text{ satd. acid fraction}}{IV \text{ unsatd. acid fraction}} \times 100$$

$$\frac{\% \text{ Unsatd. acids in satd. fraction} \times \% \text{ impure satd. acids}}{100} = \frac{\text{correction}}{\text{value}}$$

Subtract correction value from saturated acids found gravimetrically and add to percentage of unsaturated acids found gravimetrically.

FREE FATTY ACIDS OR ACID VALUE

Oils and fats contain more or less fatty acids according to the conditions of manufacture, age, and storage. The glycerides are hydrolyzed to a small degree by enzymes, air, and possibly bacteria. The increase in free fatty acids is generally accompanied by a rancid odor, although the odor itself is not due to the acidity. The acid value is the number of mg. of potassium hydroxide required to neutralize the free fatty acids in 1 g. of the oil or fat.

Procedure.—Transfer a weighed quantity of oil or fat, about 20 g., and add 50 ml. of neutral 95 per cent alcohol. Heat to boiling and shake the flask thoroughly to dissolve the free fatty acids. **Titrate** with aqueous 0.1 *N* potassium hydroxide, shaking thoroughly during the titration until the pink color persists. Calculate the number of milligrams of potassium hydroxide required to neutralize the free fatty acids in 1 g. of the oil or fat from the titration. This is the acid value.

CHOLESTEROL AND SITOSTEROL

All fats and oils contain 0.1 to 2 per cent of complex alcohols known as sterols. The sterols are phenanthrene derivatives, are monohydric alcohols, and are closely related to the bile acids. Cholesterol is the sterol characteristic of animal fats and oils and sitosterol is the sterol characteristic of vegetable oils and fats. Adulteration of animal fats, like butter, with vegetable fat, like coconut oil, may be demonstrated by showing the presence of sitosterol. Similarly, the presence of menhaden or other fish oil in an oil of vegetable origin as linseed can be detected by the presence of cholesterol. One of the methods for this identification, developed by Windaus [7] is based on the precipitation of the sterols with 1 per cent digitonin in alcohol and the subsequent formation of cholesterol and sitosterol acetates. Cholesterol acetate melts at about 114° C., and sitosterol acetate at about 125 to 137° C. Sitosterol is not a pure compound as the older literature reports. Anderson [8] has shown that sitosterol from corn oil is a mixture containing 3 isomeric sterols which differ in their physical properties. The melting point of the sitosterol obtained varies therefore, according to the purity of separation. Mixtures of cholesterol and sitosterol acetate will melt in between these temperatures.

Procedure.—To 50 g. of oil or fat in a separatory funnel, add 20 ml. of a 1 per cent solution of digitonin in 95 per cent alcohol and shake the mixture vigorously for about 15 minutes. Allow the mixture to stand for a time until the emulsion separates. Draw off the lower clear fat layer, being careful not to withdraw any of the bulky, flocculent precipitate which is present in the alcohol layer. Add 100 ml. of ether to the alcohol layer and filter the mixture. Dry the precipitate in air and wash it free from fat with ether. Transfer the precipitate to a tall form 100-ml. beaker and add 2-3 ml. of acetic anhydride. Cover the beaker with a watch glass and boil the mixture gently over a low flame for 30 minutes. Cool and add 30-35 ml. of alcohol, 60 per cent by volume. Stir the contents of the beaker thoroughly. Filter the alcohol solution. Wash the precipitate with more

[7] Windaus, *Chem. Ztg.* **37**, 1001 (1913).
[8] Anderson, *J. Am. Chem. Soc.* **48**, 2976 (1926).

60 per cent alcohol. Dissolve the precipitate on the filter with hot 80 per cent alcohol and place the filtrate in a cool place, 10° C. or below. After the acetates have crystallized, filter them off and recrystallize them from absolute alcohol. When the crystals are dry, determine the melting point. If the melting point is about 114° C. which is confirmed by the melting point of a second recrystallization of the crystals, only cholesterol acetate is present. On the other hand, if the crystals melt between the aforementioned temperature and 125° C., which again is confirmed by the melting point of a second recrystallization, then a mixture of sitosterol and cholesterol acetates is present.

Unsaponifiable Matter

Those substances not soluble in water after saponification of fatty bodies are classed under the general term unsaponifiable matter. The substances composing the greater part of the unsaponifiable matter in pure oils and fats are cholesterol and sitosterol. In some oils—for instance, olive oil—other substances such as squalene may comprise a significant portion of the unsaponifiable matter.

If the unsaponifiable matter exceeds 2 per cent some type of foreign matter is probably present, and adulteration of the fat or oil is indicated. The foreign matter may consist of a mineral or similar hydrocarbon oil, wax or fat, spermaceti, or rosin oil. Estimation of the percentage of unsaponifiable matter will definitely establish such types of adulteration.

The Fat Analysis Committee of the American Chemical Society adopted the following directions for the determination of unsaponifiable matter. Redistill petroleum ether below 75° C. and make a blank determination by evaporating 350 ml. of the reagent with about 0.25 g. of stearin or other hard fat. The blank must not exceed a few milligrams.

Procedure.—Weigh 5 g. (±0.020 g.) of the prepared sample into a 200-ml. Erlenmeyer flask, add 30 ml. of redistilled 95 per cent alcohol and 5 ml. of 50 per cent aqueous potassium hydroxide. Boil the mixture for 1 hour under a reflux condenser. Transfer to the extraction cylinder (glass-stoppered, graduated at 40 ml., 80 ml., and 130 ml. with a diameter of about 1⅜ in. and about 12 in. in height) and wash to the 40-ml. mark with redistilled 95 per cent alcohol. Complete the transfer, first with warm, then with cold, water until the total volume is 80 ml. Rinse the flask with 50 ml. of petroleum ether and add the rinsings to the contents of the cylinder previously cooled to room temperature. Shake as vigorously as possible for 1 minute and allow to settle until both layers are clear, when the volume of the upper layer should be about 40 ml. Draw off the petroleum ether layer as closely as possible by means of a slender glass siphon into a separatory funnel of 500-ml. capacity. Repeat the extraction at least 6 times more, using 50 ml. of petroleum ether for each extraction. Wash the combined extracts in the separatory funnel 3 times with 25-ml. portions of 10 per cent alcohol by volume, shaking vigorously each time. Transfer the petroleum ether layer to a weighed Erlenmeyer flask and distill; or, if desired, evaporate the petroleum ether on a steam bath in a current of air. Heat the flask with the residue until a constant weight is obtained in an oven at a uniform temperature, not less than 100° nor more than 110° C. Deduct the blank from the weight before calculating unsaponifiable matter. Test the final residue for solubility in 50 ml. of petroleum ether at room temperature. Filter, and wash free from the insoluble residue, if any. Evaporate and dry as detailed above.

Ethyl Ether Method.[9]—The oil or fat is saponified by boiling with alcoholic potassium hydroxide solution, and the resulting soap solution is diluted with water and extracted with ethyl ether. The ethereal solution is washed with water, aqueous potassium hydroxide solution, and then finally with water. To insure uniformity, emphasis is laid upon the necessity for attention to detail at every stage.

Procedure.—Weigh accurately a quantity of the oil or fat not exceeding 2.5 g., but not less than 2.0 g. and saponify by boiling for 1 hour, with occasional swirling, under a reflux condenser with 25 ml. of approximately, but not less than 0.5 N alcoholic potassium hydroxide solution. After saponification, during which no loss of alcohol should occur, transfer the alcoholic soap solution to a separatory funnel, washing in with 50 ml. of water in all. Extract the soap solution while still just warm, successively 3 times with 50 ml. of ethyl ether. Use the first quantity of ethyl ether to wash out the saponification flask before adding to the soap solution in the separatory funnel. Make each extraction by shaking the separatory funnel vigorously, allowing the two layers to separate and clarify, running off the aqueous layer at the bottom of the separatory funnel, and pouring the ethereal solution from the top of the separatory funnel into another separatory funnel containing 20 ml. of water. If the ethereal extracts contain solid suspended matter, pass them through a dry fat-free filter into the second separatory funnel, washing the filter subsequently with ethyl ether.

---

[9] Report of Sub-committee on Determination of Unsaponifiable Matter in Oils and Fats, *Analyst* 58, 203 (1933).

The method can be greatly simplified by the use of a Jacobs-Singer separatory flask. Weigh the oil directly into the lower section. Add the 25 ml. of 0.5 N alcoholic potassium hydroxide solution. Attach the section to a reflux condenser through the standard taper joint and saponify for the requisite time. Stopper the lower section with the upper section and add water to the connecting joint. Mix and then extract with 3 successive portions of ethyl ether. Pour off each supernatant ether layer through the mouth of the upper section into a separatory funnel and proceed with the method.

If, for any reason, the presence of metallic soap in the original sample is known or suspected, pour the three ethereal extracts into a second and empty separatory funnel, add 5 drops of hydrochloric acid, and shake vigorously. Wash the combined extracts successively with two quantities of 20 ml. of water, employing vigorous shaking on each occasion, and continue the process beginning with "After one or other of these preliminary treatments, . . ." as detailed below.

If the presence of metallic soaps in the original sample is not known or suspected, rotate the extracts gently without violent shaking with 20 ml. of water and, after allowing to separate, run off the wash water. Then wash the ethereal solution twice with 20 ml. of water, with vigorous shaking on each occasion.

After one or other of these preliminary treatments, wash the ethereal solution 3 times with 20 ml. of 0.5 N aqueous potassium hydroxide solution, shaking vigorously on each occasion, each alkali wash being followed by a wash with 20 ml. of water. After the last 0.5 N aqueous potassium hydroxide treatment, wash with two or more successive quantities of 20 ml. of water until the wash water no longer reacts alkaline to phenolphthalein solution.

Transfer the ethereal extract to a weighed flask, distill off the ethyl ether, and dry the residue to constant weight, not allowing the temperature to exceed 80° C.

In certain unusual cases the unsaponifiable matter appears to suffer a continuous loss during drying, owing to the presence of some material of low volatility, as, for example, residual solvent fractions. In such instances, transfer the washed ethereal extract to a flask containing about 2 g. of a neutral oil, such as peanut, previously brought to constant weight at 80° C. and then proceed as in the ordinary determination. Under these conditions the neutral oil serves to minimize any loss.

The drying may be aided by the use of acetone with very low non-volatile residue. When practically all the ethyl ether is evaporated, add 2 to 3 ml. of acetone. By the aid of a gentle current of air remove the solvent completely from the flask, which is preferably almost entirely immersed, held obliquely, and rotated in a boiling water bath.

After attaining constant weight, dissolve the contents of the flask in 10 ml. of freshly boiled and neutralized 95 per cent alcohol and titrate with 0.1 N alcoholic sodium hydroxide solution, phenolphthalein solution being used as indicator. Provided that, when the determination is carried out in the above described manner, the amount of 0.1 N alcoholic sodium hydroxide solution required does not exceed 0.1 ml., take the unsaponifiable matter as being the amount weighed. If the quantity of 0.1 ml. of 0.1 N alcoholic sodium hydroxide solution is exceeded, repeat the determination from the start, as this limit may correspond with 0.11 per cent of free fatty acid or much larger quantities of acid soap.

If there is any reason to suspect the incomplete separation of saponifiable matter, subject the material, as weighed, to resaponification, re-extraction and washing, under the conditions specified in the method. If on this retreatment the amount of unsaponifiable matter obtained is not the same as that weighed in the first determination, within the limits of manipulative error, ignore the whole test and repeat the determination from the beginning.

Shortened Method.—It is known that with many oils and fats the method described in detail above may be shortened by reducing the number of washes with aqueous potassium hydroxide solution from three to two, and by omitting the intermediate water-washing treatments. Following the three preliminary water washes, in the shortened method wash the ethereal solution twice with 20 ml. of 0.5 N aqueous potassium hydroxide by shaking vigorously on each occasion, and then with two or more successive quantities of 20 ml. of water, until the wash water no longer reacts alkaline to phenolphthalein solution. The directions are then identical with those described in the full method.

RANCIDITY

Deterioration of fats and oils is produced by the auto-oxidation of the unsaturated components. The reaction probably consists in the addition of molecular oxygen to the double bonds of the unsaturated acids with the production of labile peroxides which then further isomerize, or decompose spontaneously into, or react with water to form a complex series of products including aldehydes, ketones, and acids of lower molecular weight.

Kreis Test.—One of these products of oxidation is epihydrinaldehyde, which gives a color with phloroglucinol. It is upon this reaction that the Kreis test depends.

Procedure.—Place 5 ml. of the fatty body into a test tube and add 5 ml. of hydrochloric acid, free from nitrosyl chloride. Stopper the tube with a clean rubber stopper and shake vigorously for 30 seconds. Add 5 ml. of 0.1 per cent ether solution of phloroglucinol, restopper and shake for 30 seconds and then allow to stand for 10 minutes or centrifuge for 2-5 minutes.

If a pink or red color is present in the acid layer, proceed as follows: Make 2 mixtures of the original oil or fat, first, one part of sample and 9 parts of liquid petrolatum and secondly, one part of sample and 19 parts of liquid petrolatum. Test 5-ml. portions of the mixtures as detailed above and note the colors produced. The fatty bodies may then be grouped into 4 classes.

1. No reaction indicates no rancidity.

2. Positive reaction when undiluted indicates no rancidity as far as taste and odor are concerned but that the fat or oil will soon turn rancid.

3. Positive reaction diluted 1 to 10 but none diluted 1 to 20 indicates incipient rancidity, often noticeable by taste and odor.

4. Positive reaction diluted 1 to 20 indicates definite rancidity.

A pink or red color indicates a positive reaction, yellows, oranges and faint pinks should be disregarded. Crude vegetable oils, for example, partially refined cottonseed oil, give a strong Kreis test; hence, care must be taken in the interpretation of this test.

Taufel and Sadler Modification.—The Kreis reaction responds to the formation of epihydrinaldehyde by auto-oxidation of the fatty material. However, fats undergo degradation in other ways as was pointed out and the Kreis reaction may fail. The method described above may be disturbed by the presence of small amounts of allyl alcohol, allylamine, eugenol, etc., which also give a red color with phloroglucinol. Moreover, if the aldehyde content has greatly increased owing to the advanced deterioration of the fatty material, sparingly soluble, colorless phloroglucides may be formed and the Kreis reaction may fail. In order to overcome the foregoing interferences and confirm rancidity, Taufel and Sadler [10] recommend a method based on volatilizing the epihydrinaldehyde.

Procedure.—Mix the oil or melted fat with an equal quantity of

---

[10] Taufel and Sadler, *Z. Untersuch. Lebensm.* **67,** 268 (1934).

ice-cold hydrochloric acid. Insert a cotton-wool plug in the tube at its upper dry part; moisten it with 1 ml. of 0.1 per cent phloroglucinal solution in alcohol and 10 drops of 20 per cent hydrochloric acid. Shake the tube well for 1 to 2 minutes without splashing the cotton and warm if necessary to 40° C. A red coloration at the lower surface of the cotton-wool indicates epihydrinaldehyde.

Lea [11] states that aliphatic aldehydes of medium molecular weight, particularly heptaldehyde and nonaldehyde are compounds mainly responsible for the objectionable odor and flavor of oxidized fats and oils.

Jacobs Modification.—A simple method for the detection of rancidity has been devised by Jacobs.[12]

Measure 5 ml. of the sample of oil or fat, which should be melted into a tube 7¼ x 1 in. Add 5 ml. of concentrated hydrochloric acid and stir for about 30 seconds taking care not to wet the upper portion of the tube. Insert a piece of phloroglucinol paper, prepared by dipping filter paper into a 0.1 per cent solution of phloroglucinol in ether, into the cork which has a slit cut in it to hold the paper and a side channel for an air vent. Then drop 5 or 6 chips of marble or an equivalent amount of calcium carbonate lumps into the tube. Stopper immediately and allow the evolution of gas to proceed, taking care that the resultant foam does not touch the paper. If a pink to lavender or orchid coloration is obtained within 10 to 20 minutes on the paper the sample may be considered rancid. Some measure of the degree of rancidity may be obtained from the depth of color produced.

Modified Kreis Test.[13]—Weigh 3 ml. of melted fat or oil into a test tube, add 1 ml. of a 0.5 per cent solution of phloroglucinal in amyl acetate, and stir the mixture vigorously or bubble air through it for about a minute. Add 2 ml. of a trichloroacetic acid solution, made by dissolving Xg. of the acid in 0.382X ml. of amyl acetate and immerse the tube in a water bath at 45° (±0.1° C.) for 15 minutes with continuous stirring or with air still passing. Remove the tube and dilute the contents immediately with 10 ml. (more if necessary) of an ice-cold solution made by diluting 1 volume of the above trichloroacetic acid solution with 2 volumes of amyl acetate. The purpose of diluting is both to reduce any further reaction while the color is measured, and to bring the color intensity within the region of maximum accuracy of the tintometer. Measure the color at once.

Simultaneously with the test carry a blank out using amyl acetate

[11] Lea, *Ind. Eng. Chem., Anal. Ed.* 6, 241 (1934).
[12] Jacobs, work performed in 1938.
[13] Walters, Juers, and Anderson, *J. Soc. Chem. Ind.* 57, 53 (1938).

in place of the phloroglucinol solution, here stirring only if necessary, instead of aeration. Subtract the blank reading from the observed tintometer reading. It is usually negligible except for highly colored oils. The blank solution is used for the compensating cell in the Zeiss photometer. A small correction may be made for any color produced by interaction of acid and phloroglucinol. A maximum reading (comparing with water) of 0.06-0.67 unit per centimeter on the photometer has been observed.

Peroxide Number.—As mentioned unsaturated acids appear to be able to add oxygen at the double bonds and form peroxides. These peroxides are highly reactive and may be estimated iodometrically. There appears to be a distinct relationship between the peroxide number and rancidity but as noted in the next section on stability tests, the character of the oil is very important for oils with high iodine values will have high peroxide numbers at the start of rancidity and oils with low iodine numbers will have low peroxide values at the initiation of rancidity. In addition, correlation should be established between high peroxide values and organoleptic rancidity before conclusions are drawn.

Procedure.—Dissolve 5 g. of oil in 30 ml. of a solvent mixture consisting of 60 per cent of glacial acetic acid and 40 per cent of chloroform and add 0.5 ml. of a saturated solution of potassium iodide. Shake the flask until clear by giving a rotatory motion to the flask. After exactly 2 minutes from the time of addition of the potassium iodide, add 30 ml. of water and titrate the liberated iodine with 0.1 N or 0.01 N sodium thiosulfate solution, depending upon the amount of iodine liberated. Shake vigorously at the end to remove the last traces of iodine from the chloroform layer.

Make a blank titration on all reagents daily. This should not exceed 0.1 ml. of the standard sodium thiosulfate solution.

The results may be expressed in terms of milliequivalents per 1000 g. of oil or in millimoles per 1000 g. of oil:

$$\text{Milliequivalents per 1000 g.} = \frac{\text{ml.} \times N}{\text{g.}} \times 1000$$

$$\text{Millimoles per 1000 g.} = \frac{0.5 \times \text{ml.} \times N}{\text{g.}} \times 1000$$

where

$$\text{ml.} = \text{ml. of sodium thiosulfate solution}$$
$$N = \text{normality of sodium thiosulfate solution}$$
$$\text{g.} = \text{grams of oil.}$$

The interpretation of these results is discussed in a subsequent section.

STABILITY TESTS

It is a matter of moment as to whether or not a given fat or oil will be stable under conditions of use such as in baked goods, shortenings, and the like. Some measure of this stability or resistance to becoming rancid can be determined by use of stability tests. There are three principal types of tests made to determine the stability of a fat or oil. These are oxygen absorption tests which will be described below, incubation tests which will also be described below, and color reaction tests making use of the Kreis and Schiff color tests, which have been discussed and detailed.

Oxygen Absorption Test.—In these tests the amount of time required by a fat to absorb a certain amount of oxygen, or the amount of oxygen absorbed in a fixed time, is taken as indicative of the keeping qualities of a fat or oil. These tests require rather complicated apparatus.

Thus by aerating a sample of lard in a test tube held at constant temperature in a bath, the aging of the lard is greatly accelerated.[14,15] By starting three or more portions of a sample at intervals of 1 hour apart, the oldest portion or first portion immersed in the bath becomes rancid according to organoleptic tests, while the others which were started later are still in their induction period. Since the personal factor renders organoleptic tests open to some doubt, at times, the results are not based on the organoleptic analysis alone but on the peroxide content of the three portions after treatment.

It is to be noted, however, that the peroxide content at the start of rancidity is dependent on the character of the oil. Thus oils and fats of high iodine number have a high peroxide value at the start of rancidity. Fats and oils which have a low iodine value have a low peroxide value at the start of rancidity, and fats and oils which have a high antioxidant content, either natural or added, have a high peroxide value at the start of rancidity.

Apparatus.—Constant Temperature Bath.—The constant temperature bath consists of a water jacketed vessel in which water is kept at the boiling point with a reflux condenser keeping the volume of water constant. This jacket encloses a mineral oil bath in which a rack accommodates 8 x 1-in. test tubes. This type of bath keeps a temperature of about 208° F. This temperature should be maintained by adding alcohol to lower the temperature, if necessary, or glycerol to raise it.

[14] King, Roschen, and Irwin, *Oil & Soap* 10, 105 (1933).
[15] Committee on Analysis of Commercial Fats and Oils, Am. Chem. Soc., *Ind. Eng. Chem., Anal Ed.* 17, 336 (1945).

Aeration Apparatus.—The aeration apparatus is shown in Fig. 64. It is designed to permit a fixed volume of air, namely 2.33 ml. per second, to be drawn through the apparatus.

The air flow is standardized as follows: Cut capillary tubes about 2 in. long from one piece of thermometer tubing. Standardize each capillary to give the same flow of air at any given pressure. To increase the flow, the capillaries may be ground on a fine emery wheel.

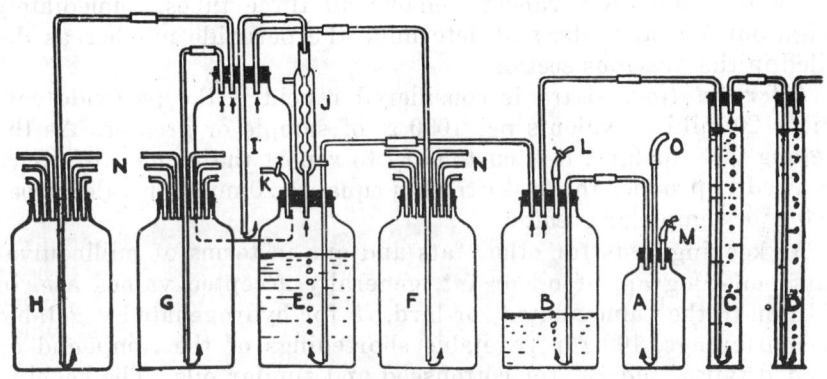

FIG. 64. Diagram of Air-Distributing System [15]

A. Device to control pressure of incoming air
B. Bottle containing water for washing air
C, D. Water columns. Air in space above water in B is kept under constant pressure sufficient to by-pass air through C and D at a steady rate
E. Bottle containing acid dichromate solution. Air from E passes through condenser
J into bottle I, thence into bottles F, G, and H, which distribute air to tubes N, which lead to the aeration tubes
K, L. Pinch cocks to release pressure when shutting off apparatus
M. Screw clamp to regulate flow of air
O. Connection to source of air pressure

Install the capillary tubes in the apparatus. Place 20 ml. of lard in each test tube, and bring up to temperature. Fill bottle B with distilled water to about ⅓ full. Fill bottle E over ½ full with 2 per cent potassium permanganate solution or 2 per cent potassium dichromate solution containing 1 per cent sulfuric acid. Connect the apparatus but place a meter in the air line between bottles B and E.

Adjust the amount of water in cylinder C and D until the amount of air passing is 2.33 ml. per second for each tube. This is approximately 20 in. Then mark B, E, C, and D with a file to indicate the level of the liquid.

Change the permanganate solution in E once a week. Use a screw clamp on bottle A to serve as by-pass.

Procedure for Incubation.—Measure 20 ml. of melted lard into each of three 8 x 1-in. test tubes, calibrated at 20 ml. Heat one of the tubes to approximately 208° F. by immersing it in hot water and place it in the oil bath at that temperature. Start the air flow and record the time. Stopper the second and third tubes and hold them at room temperaure. Exactly 1 hour after starting the first portion, start the second portion in the same way. One hour later start the third portion. At regular 1-hour intervals inspect the first for odor. When it has become definitely rancid, remove all three tubes, immediately weigh out 5-g. aliquots, and determine the peroxide number as detailed in the previous section.

Interpretation.—Lard is considered rancid if the peroxide content is 20 milliequivalents per 1000 g. of sample or greater. As the keeping test for lard, it is customary to report the number of hours required to produce the first titration equal to 20 milliequivalents per 1000 g. of sample or greater.

As keeping tests for other fats and oils in terms of milliequivalents per kilogram of oil or fat, generally accepted values are, in addition to the value quoted for lard, 75 for hydrogenated vegetable oil shortenings, 100 for vegetable shortenings of the compound or blended types, and 125 for cottonseed and similar oils. The keeping qualities of fats and oils in terms of days according to the incubation or Schaal test are approximately equal numerically to keeping qualities in terms of hours by the oxygen absorption method.

Incubation or Schaal Test.—In this test [16] the oils or fats are exposed to the air in a heated oven or incubator and are examined organoleptically at regular intervals, generally in the morning, and the day in which they become rancid is noted. It is clear that this method has a marked disadvantage because of the time element. Another disadvantage is that the test is entirely dependent on personal judgment. The test has been used in the biscuit and cracker industry to test the stability of shortenings.

The use of scrupulously clean glassware is a most important factor in this test. Thorough washing with soap and water, careful rinsing, and drying in an oven are preferable to use of cleaning solution because the latter is not easily removed.

It is well to note that at the end of the induction period, the sample usually darkens and, after the sample becomes rancid, further heating causes the color to become lighter.

Procedure.—Weigh 50 g. of the sample into a 250-ml. Griffin, low-

[16] Joyner and McIntyre, *Oil & Soap* 15, 184 (1938).

form beaker, cover with a 3-in. watch glass and place in an oven at 63.0° C. ± 0.5° C. Smell the sample, preferably in the morning, each day, removing the watch glass only long enough to smell the sample.

If desired the peroxide content may be determined as detailed in a preceding section.

Tristearin in Unhydrogenated Pork Fats

Boemer Number.—This test [17] is applicable to the detection of beef fat in lard. It depends upon the difference in the melting point of the α- and β- forms of monopalmitodistearin. It is the only test based on establishing the presence of a specific glyceride.

Apparatus.—Centrifuge tube, 100-ml., or glass-stoppered cylinder of the same capacity.

Melting point tubes. Capillary glass tubing, inside diameter 1 mm., thin wall, convenient length 5 to 8 cm.

Thermometer. Any convenient thermometer of suitable range but with 0.1° or 0.2° C. subdivisions.

Procedure.—Crystallization of Glycerides.—

1. Weigh 20 g. of the filtered sample into the centrifuge tube or cylinder. Adjust the temperature of the acetone solvent to 30° C. and use at this temperature throughout. Add acetone to the sample to the 100-ml. mark. Shake until a thorough mixture results and allow to stand for about 18 hours at a temperature of 30° ± 2° C.

2. Place the tube in a suitable centrifuge, whirl for 5 minutes, and pour off the supernatant liquid. If a centrifuge is not available a 100-ml. cylinder is used, in which case the supernatant liquid must be siphoned off.

3. Add another 20-ml. portion of acetone to the crystals, shake, centrifuge, and decant or siphon as above.

4. Repeat operation 3, this time mixing well and pouring through a qualitative filter paper. Complete the transfer of the crystals and wash the contents of the filter paper with 5 small portions of acetone.

5. Apply a vacuum to remove as much of the acetone from the crystals as possible. Remove the paper from the funnel, place on a dry smooth surface, and break up any lumps with a spatula. Allow to dry thoroughly. The temperature of the glycerides must not be elevated to the melting point in drying because this will materially

[17] Committee on Analysis of Commercial Fats and Oils, American Chemical Society, *Ind. Eng. Chem., Anal. Ed.* 12, 379 (1940).

influence the final results. After drying, comminute the mass and determine the melting point as directed below.

Preparation of Fatty Acids.—

1. Remove a sufficient amount of the glycerides for the melting point determination and transfer the remainder into a 500-ml. Erlenmeyer flask. Add 100 ml. of 0.5 N alcoholic potassium hydroxide. Place a small funnel in the neck of the flask to prevent loss on boiling and saponify by boiling for 1 hour.

2. Add 100 ml. of water to the soap solution and evaporate on the steam bath to remove as much of the alcohol as possible. Transfer to a 500-ml. separatory funnel. Add a sufficient amount of water to bring the total quantity used to about 250 ml. Neutralize with the hydrochloric acid (1:1) to separate the fatty acids, using a slight excess. Add 75 ml. of ethyl ether and shake.

3. Draw off the aqueous layer and wash the ether layer at least 3 times with water or until the washings are neutral to methyl orange. Withdraw the ethyl ether extract, filter, evaporate the ether on the steam bath, and dry the fatty acids at 100° C. for a few minutes. Protect the fatty acids at all times from ammonia fumes.

Determination of Melting Point.—

1. Seal the glyceride melting point tubes at one end before introducing the crystals. Insert the crystals through the open end and force down into the closed end with a small glass rod or wire.

2. Prepare the fatty acid melting point tubes by dipping the open tubes into the melted acids so that the sample stands about 1 cm. high in the tube. Seal this end of the tube in a gas flame.

3. Allow the tubes containing the fatty acids to stand for 0.5 hour in ice water or hold in a refrigerator overnight (4° to 10° C.).

4. Fasten the melting point tubes containing the fatty acids and the glyceride crystals to the thermometer by a rubber band or any other convenient means. Adjust them so that the sections of the tubes containing the samples are adjacent to the bulb of the thermometer. Suspend the thermometer in a beaker of water (suitably agitated), so that the bottom of the bulb of the thermometer is about 3 cm. below the level of the water. The temperature of the water at this time must be at least 10° C. below the melting point of the sample. Heat th' water at such a rate that the temperature will increase at about 0 C. per minute. The melting point is that point at which the samples become clear and liquid. Determine the melting points of the glycerides and fatty acids at the same time.

*Calculations.—*If the melting point of the glycerides, plus twice the difference between the melting point of the glycerides and the melting point of the fatty acids, is less than 73° C. the lard is regarded as adulterated.

$$B.N. = A + 2(A - B)$$

where

$B.N.$ = Boemer number
A = melting point of glycerides
B = melting point of fatty acids

.*Notes.*—If the quantity of crystals obtained from 20 g. is insufficient, this amount may be increased, providing the acetone is increased proportionally.

The committee's investigation has indicated that 10 per cent beef fat can be detected with certainty and many times amounts smaller than this, even down to 5 per cent, can be found.

Tolman and Robinson have pointed out that this method is not applicable to hydrogenated pork fats.

The results on cooperative samples have indicated that if the melting point of the glycerides alone is used as a criterion, some pure samples of lard may be reported as adulterated. Therefore, use of this value alone is not recommended.

PEANUT OIL

Peanut oil, arachis oil, earthnut oil is the edible oil obtained from the peanut. It contains about 5 per cent arachidic acid as the glyceride. Arachidic acid is insoluble in cold alcohol in comparison to stearic and palmitic acids. The Bellier[18] and Renard[19] tests are based on this fact.

Evers-Bellier Test.—The Evers-Bellier[20] test for peanut oil in other oils uses the following technique. Saponify 1 ml. of the oil with 5 ml. of 1.5 N alcoholic potassium hydroxide solution by heating on a water bath for 5 minutes, avoiding loss of alcohol. Add 50 ml. of 70 per cent alcohol and then 0.8 ml. of hydrochloric acid, specific gravity 1.16. After heating to dissolve any precipitate that may be formed, cool the solution in water, stirring continuously with a thermometer, so that the temperature falls at the rate of about 1° C. per minute. If a turbidity appears before the temperature reaches 9° C., it is best to check the presence of arachis oil by isolating the acid and determining the melting point. If the liquid remains clear at this temperature, arachis oil may be regarded as absent.

[18] Bellier, *Ann. chim. anal.* 4, 4 (1899).
[19] Renard, *Compt. rend.* 73, 1330 (1871).
[20] Evers, *Analyst* 62, 96 (1937).

It is essential that the stirring should be continuous, since local cooling will cause the premature formation of a turbidity. For this reason the cooling water should not rise above the level of the liquid in the flask. The turbidity is best observed by looking through the liquid against a good light and noting the temperature at which a definite precipitate first appears. The point is sharp and the personal error should not be more than ±0.25° C. Occasionally after acidification an oil gives a slight opalescence which is unaffected by warming. This may be disregarded, as it does not affect the true turbidity temperature.

A convenient variation is to transfer the dissolved material after the addition of alcohol and hydrochloric acid to a 100-ml. Nessler tube and then continue with the method. This assists in viewing the onset of turbidity while cooling as required.

COTTONSEED OIL

Cottonseed oil is the edible oil obtained from the seed of the cotton plant, or from the seed of other species of cotton plant. Cottonseed oil may be identified by use of the Halphen [21] reagent.

Halphen Test.—The reagent consists of a 1 per cent solution of sulfur in carbon bisulfide to which an equal volume of amyl alcohol is added.

Procedure.—To 10 ml. of sample in a test tube, add 10 ml. of reagent and some powdered pumice to prevent bumping. Heat in a salt water bath on a hot plate for at least 1 hour. A deep red coloration shows the presence of cottonseed oil. Cottonseed oil subjected to heat circa 250° C. for 10 minutes fails to give this reaction. Kapok oil also gives the red coloration with the Halphen reagent.

This test is often used to detect adulteration of butter and lard with cottonseed stearin. However, butter and lard obtained from cows and hogs fed on cottonseed cakes or meal also give a positive reaction. Hence care must be taken in the interpretation of results. It is well to run controls containing 1 to 10 per cent of cottonseed oil in olive oil or other oil or fat on which the test is being performed. These controls give some measure of the production of light color standards by small amounts of cottonseed oil. They are necessary, for some pure olive oils, especially Tunisian oils, give a faint coloration, which, however, is not comparable to the color obtained from the controls.

---

[21] Halphen, *J. pharm. chim.* [6] **6**, 390 (1899).

SESAME OIL

Villavecchia Test.—Furfural reacts with the sterols peculiar to sesame oil to give a red coloration. This fact is used as the basis for a test showing the presence of sesame oil in other oils.

Reagent.—Prepare a 2 per cent solution of furfural in alcohol by adding 2 ml. of furfural to 100 ml. of 95 per cent alcohol.

Procedure.—To 5 ml. of oil in a test tube, add 2-3' drops of furfural solution and 5 ml. of hydrochloric acid. Shake vigorously for 30 seconds and allow the mixture to separate. A deep red coloration in the acid layer indicates the presence of sesame oil. Confirm by addition of 5 ml. of water and shaking again. If the color is due to sesame oil, the color will remain in the acid layer, whereas that due to other oils will disappear. If the oil or fat be colored with an aniline dye the hydrochloric acid will extract this coloring matter and yield a pink coloration. However, this coloring matter will generally go back into the fat layer on dilution with water. As little as 2 per cent of sesame oil will give a much deeper red color than is usually obtained from the small amount of dye.

The same precautions must be observed in interpretation of results obtained with this test as that with the Halphen test, for cows and hogs fed on sesame cakes yield butter and lard giving a positive Villavecchia or Baudouin reaction.

Modified Villavecchia Test.—In this modification [22] the reagents noted above are used.

Procedure.—Mix 10 ml. of the sample with an equal volume of the hydrochloric acid, specific gravity 1.19.

Add to this mixture 0.1 ml. of the Villavecchia reagent and shake well for 15 seconds.

Note the color of the lower layer as soon as possible after the emulsion has broken. If no pink to crimson color appears, the test may be reported negative at that point. If any color is observed in the lower layer, add 10 ml. of distilled water, shake again, and observe the color as soon as separation has taken place. If the color persists, report the test as positive. If the color disappears, sesame oil is not present.

Notes.—Furfural gives a violet color with hydrochloric acid; therefore, it is necessary to use the dilute solution specified. It is advisable to read the color as soon as possible, so that the pink color, if present, may be observed before it is masked by the development of other noncharacteristic colors.

---

[22] Committee on Analysis of Commercial Fats and Oils, American Chemical Society, *Ind. Eng. Chem., Anal. Ed.* **12,** 379 (1940).

It is advisable, with the Villavecchia test as with others of a similar nature, to run control samples using as standards oils of known
composition.

The test is applicable to hydrogenated as well as unhydrogenated
sesame oil although not with the same degree of sensitiveness. The
committee has found that as little as 0.25 per cent of sesame oil can be
detected, but is of the opinion that this limit should be accepted with
reservations. It is the considered judgment of the committee that
there is every assurance that at least 0.5 per cent of sesame oil is
detectable and that the lower limit with respect to the fully hydrogenated oil is 1 per cent.

The sensitivity of the Villavecchia test to small quantities of
sesame oil may be improved by increasing the amount of Villavecchia
reagent up to 1 ml. However, doing this hastens the rate of development as well as the amount of noncharacteristic colors that are
formed. Therefore, if greater amounts of the reagent are used, relatively greater care must be taken in the observation of the final
color.

NITRIC ACID TEST

A test often applied to olive oil to detect adulterants is the nitric
acid test. This reaction must be interpreted cautiously and at best is
only indicative of adulteration. However, gross adulterations will
respond to this simple sorting test.

Procedure.—To 5 ml. of oil in a test tube add an equal volume of
nitric acid (9:1). Shake thoroughly for 2 minutes and allow to stand.
Pure olive oil remains unchanged, whereas a brown coloration indicates the presence of a foreign oil. Some olive oils give a light brown
color. This reaction was formerly used often as a means of indicating the presence of cottonseed oil in olive oil.

TEA SEED OIL

Fitelson [23] Test.—The chemical composition of olive oil and tea
seed oil as far as the glycerides are concerned is very similar. They
differ somewhat in the character of the sterols present in the unsaponifiable matter of the oils. Chloroform, acetic anhydride, and
sulfuric acid added to tea seed oil produce a deep fluorescent color,
green by reflected light and brown by transmitted light which changes
to an intense red color on the addition of *anhydrous* ethyl ether and
this color finally fades to a light brown. Olive oil and other common
edible oils show a green color but none of these oils show the red

---

[23] Fitelson, *J. Assoc. Official Agr. Chem.* **19**, 493 (1936).

color on the addition of the ether, although some olive oils show a faint pink before fading to the final light brown.

Procedure.—Measure into a test tube exactly 0.8 ml. of acetic anhydride, 1.5 ml. of chloroform and 0.2 ml. of concentrated sulfuric acid. Mix and cool to room temperature. Add 7 drops of the oil to be tested directly to the reagents, mix, and cool again. To measure the 7 drops of oil use glass tubing, 4 mm. outside diameter and approximately 2 mm. inside diameter. These 7 drops should weigh approximately 0.22 g. If the solution of oil in the reagents is cloudy after mixing and cooling, add acetic anhydride dropwise, shaking after each addition until a clear solution is suddenly formed. Appreciable deviations from these quantities, particularly in the sulfuric acid, cause distinct variations in color intensities. Since the mixed reagent deteriorates slowly, do not mix in advance of testing.

After the test tube and contents have remained at room temperature for 5 minutes, note the color produced. Tea seed oil will exhibit a deep green by reflected light and brown by transmitted light. Olive oil will show a green color by reflected and transmitted light, occasionally exhibiting a faint fluorescence. Add 10 ml. of anhydrous ethyl ether from a graduated cylinder and mix immediately by inverting once. Tea seed oil will show a brown color changing to an intense red within a minute or so. This red color reaches a maximum and then fades slowly within a period of a few minutes. Olive oil forms an initial green color on addition of the ether. This color fades slowly to a brown-gray, occasionally passing through a faint pink stage. Mixtures of olive oil and tea seed oil show the characteristic tea seed oil colors which are proportional in intensity to the quantity of tea seed oil present.

For approximately quantitative estimations, drop the oil into the reagents as described above and allow to remain at room temperature for 5 minutes. In the meantime, cool a 10-ml. portion of anhydrous ether in ice water. At the end of the 5-minute period place the test tube containing the oil and reagents in the ice water for 1 minute, add the cold ether, taking care that no water falls into the test tube, and mix. Return the tube to the ice water bath and allow the colors to reach a maximum within 5 minutes. This maximum intensity will remain stable for 5-10 minutes before beginning to fade. Use the deepest colors for comparison. Standards containing known quantities of tea seed oil in an olive oil that gives no pink color with this test, should be run simultaneously with the sample.

In the Fitelson test for tea seed oil, after the addition of the reagents, acetic anhydride, chloroform, sulfuric acid and oil, shake well. After 5-10 seconds a blue to violet color indicates the presence of corn oil. A light green color after 2 minutes indicates the presence of soya oil. These color tests must be confirmed by determinations of the refractive index, the iodine number, and the squalene value.

SQUALENE CONTENT

Nearly all oils contain squalene, an aliphatic hydrocarbon, $C_{30}H_{50}$, containing 6 double bonds and having a theoretical iodine value of 371. Olive oil, however, has a higher content of squalene than most other oils. Fitelson [24] devised a method in which the squalene is concentrated in a fraction obtained by the selective adsorption treatment of the unsaponifiable matter. The unadsorbed residues from olive oil consist almost completely of squalene. The unsaturated component in the much smaller unadsorbed residue from other oils also consists of squalene. On the assumption that the unsaturation of these residues is due to squalene, the concentration of squalene can be calculated from the total absorption of halogen.

TABLE 32. SQUALENE CONTENT OF VARIOUS FATS [a]

Fat	No. Samples	Squalene (Mg./100 G. Fat)	Fat	No. Samples	Squalene (Mg./100 G. Fat)
Olive	44	136-708	Grapeseed	1	7
Cottonseed	12	4- 12	Almond	1	21
Corn	9	19- 36	Cocoa	1	none
Peanut	11	13- 49	Coconut	1	2
Sunflower	3	8- 19	Linseed	1	4
Soyabean	9	7- 17	Butter	1	7
Tea seed	3	8- 16	Cod liver	1	31
Sesame	1	3	Seal	1	35
Rape	1	28	Chicken	1	4
Mustard	1	7	Lard	1	3
Patua	2	2- 5	Beef	1	10
Rice bran	1	332			

[a] Fitelson, *J. Assoc. Official Agr. Chem.* **26**, 509 (1943).

Reagents.—(a) Concentrated Potassium Hydroxide Solution.—Dissolve 60 g. of KOH in 40 ml. of water.

(b) Dilute Potassium Hydroxide Solution.—Dissolve 28 g. of KOH in water and dilute to 1 liter.

(c) Petroleum Benzine.—Skellysolve B (b.p. 63-70° C.) or equivalent.

(d) Aluminum Oxide Adsorbent, 80-200 Mesh.—Adsorption alumina for chromatographic analysis, Fisher Scientific Company,

[24] Fitelson, *J. Assoc. Official Agr. Chem.* **26**, 499 (1943).

Pittsburgh, Pa., or equivalent. Keep in tightly closed container, away from moisture.

(e) Pyridine Sulfate Bromide Reagent, 0.1 N.—Dissolve 8 g. of bromine in 20 ml. of glacial acetic acid (99.5 per cent). Prepare another solution by adding gradually, with cooling, 5.45 ml. of sulfuric acid to a mixture of 20 ml. of glacial acetic acid and 8.15 ml. of pyridine. Mix the two solutions, cool, and dilute to 1 liter with glacial acetic acid.

(f) Sodium Thiosulfate Solution, 0.05 N.—Dissolve 13 g. of $Na_2S_2O_3 \cdot 5H_2O$ in carbon dioxide-free water containing 1 per cent of amyl alcohol. Standardize against an exactly 0.05 N solution of potassium iodate, KIO_3 (1.7835 g./liter) as follows: To a glass-stoppered 125-ml. Erlenmeyer flask, add 10 ml. of 10 per cent potassium iodide solution, 5 ml. of water, 2 g. of sodium bicarbonate, and slowly 5 ml. of about 6 N hydrochloric acid. Mix, add 25 ml. of the potassium iodate solution, wash down the sides of the flask with water, and titrate at once with the thiosulfate solution, using starch indicator toward the end of the titration.

(g) Potassium Iodide Solution.—10 per cent.

(h) Starch Indicator.—1 per cent solution of soluble starch.

Apparatus.—Adsorption Column.—Prepare immediately before use. Place a small wad of cotton in the constricted end of a glass tube, 0.8 cm. inside diameter and 30 cm. long. Add the adsorption alumina in about 10 small portions until a column about 10 cm. high is obtained. Apply gentle suction and tamp each portion of the alumina lightly with the flattened end of a heavy glass rod. Place a small wad of cotton on top of the column and tamp lightly. Wash the column with about 15 ml. of petroleum benzine, remove the suction, and keep the top of the column covered with a shallow layer of petroleum benzine until ready for use.

Procedure.—Weigh accurately (±20 mg.) about 5 g. of sample into a 125-ml. Erlenmeyer flask, add 3 ml. of the concentrated potassium hydroxide solution and 20 ml. of 95 per cent ethyl alcohol, and boil the mixture under an air condenser for 30 minutes. Cool somewhat and, while still warm, add 50 ml. of petroleum benzine; mix, and transfer to a separatory flask. Rinse the flask with 20 ml. of 95 per cent ethyl alcohol and then with 40 ml. of water, adding the rinsings to the solution in the separator. Shake vigorously, allow the two layers to separate completely, and slowly draw off the soap solution. Pour the petroleum benzine extract from the top of the separatory funnel into another separatory funnel containing 20 ml. of water. Repeat the extraction of the soap solution with 50 ml. of petroleum benzine. Rotate the combined extracts gently with the 20 ml. of water and, after allowing the layers to separate, discard the wash water. Repeat the washing by shaking vigorously with 20 ml. of water and again discard the lower layer after separation. Wash the petroleum benzine solution with 20 ml. of the dilute potassium hydroxide solution and then with successive 20-ml. portions of water until the wash liquid is free from alkali, shaking vigorously on each occasion. After the final washing, draw off the last drops of water brought down by swirling the separator. Pour the petroleum benzine solution from the top of the separator into a lipped conical beaker. Rinse the separatory funnel with petroleum benzine and add rinsings to the beaker contents. Add a few pieces of broken porcelain and evaporate almost all of the solvent on a steam bath. Remove the last traces of solvent in a current of carbon dioxide, while warming the beaker.

Dissolve the unsaponifiable matter in 5 ml. of petroleum benzine and transfer to the adsorption tube prepared as described above. (The filtrate, which is caught in a 250-ml. glass-stoppered iodine absorption flask, should emerge dropwise, at a rate of about 1 ml. per minute, gentle suction being used if necessary.) When the solution has been nearly drawn into the column, add about 5 ml. of the petroleum benzine that has been used to rinse the beaker. Continue the addition of the solvent in 5-10-ml. portions, always keeping the surface of the column covered with the liquid, until a total volume of 50 ml. has passed through the adsorption tube. Evaporate most of the solvent in the flask, after adding a few pieces of broken porcelain, and remove the last traces of solvent in an atmosphere of carbon dioxide.

Dissolve the unadsorbed residue in 5 ml. of chloroform and add a quantity of pyridine sulfate bromide reagent sufficient to provide at least 50 per cent excess, 10 ml. will usually suffice. Allow the mixture to remain in the dark for 5 minutes and then add 5 ml. of 10 per cent potassium iodide solution, together with 40 ml. of water. Mix thoroughly, wash down any free iodine on the stopper, and titrate with the 0.05 N sodium thiosulfate solution. Toward the end of the titration add starch indicator solution, shake the flask vigorously, and continue the titration to the disappearance of the blue color. Conduct a blank determination on the pyridine sulfate bromide reagent in the same manner and calculate the milliliters of 0.05 N sodium thiosulfate equivalent to the absorbed halogen. One ml. of 0.05 N sodium thiosulfate solution is equivalent to 1.71 mg. of squalene. Report results as milligrams of squalene/100 g. of sample.

Great care must be exercised in running the squalene number, because many batches of Skellysolve B give a high blank which in turn increases the squalene number. Distillation of the Skellysolve does not eliminate this difficulty. The blank must be run and then subtracted from the result of the test sample.

VALENTA TEST

The Valenta test modified by Fryer and Weston [25] determines the temperature at which glycerides are soluble in acetic acid or in a mixture of amyl alcohol and ethyl alcohol. The temperature at which a turbidity first appears on cooling the mixture is ascertained. Rape oil has the highest Valenta value and coconut oil has the lowest Valenta value of the common edible oils and fats.

TRIACETIN

Fincke [26] gives a method for the detection of the addition of synthetic triacetin to butter or butter substitutes by use of the fact that it is soluble in dilute alcohol.

Procedure.—Determine the Reichert-Meissl value of a portion of the sample in the usual manner. Transfer about 30 g. of the fat to a suitable flask, add 150 ml. of water, 150 ml. of 95 per cent alcohol, and some pumice stone. Reflux for an hour. Transfer the mixture to a separatory funnel. Allow the layer to separate and cool. Draw off the water layer. Transfer the fat to an evaporating dish and evaporate off the alcohol. Dry the fat. Determine the Reichert-Meissl number of the treated fat. If this value of the Reichert-Meissl is lower than the value previously obtained, triacetin, or other material soluble in alcohol and contributing to the Reichert-Meissl value is present in the original sample. If the values obtained are substantially the same, no such material is present in the original sample.

INTERPRETATION OF RESULTS

As was stressed previously, many of these determinations and color tests are used as a means for ascertaining whether butter and olive oil are pure. Deviation of one constant alone from the normal is not sufficient to substantiate belief of adulteration. Thus, if the Reichert-Meissl of a butter was 17 and the other factors were normal, this might well be an abnormal variation of a pure butter.

---

25 Fryer and Weston, *Analyst* **43**, 4 (1918).
26 Fincke, *Z. Nahr. Genussm.* **11**, 666 (1908).

Divers adulterants produce different variations in the characteristic constants of butter and olive oil. Table 33 shows the increase or decrease in the constants of butter fat by the substitution in whole or in part by coconut oil, palm kernel oil, beef tallow, lard, and hydrogenated cottonseed.

Oleomargarine or margarine is the plastic food prepared with one or more ingredients such as rendered fat, oil, or stearin from cattle, sheep, swine, or goats and any vegetable food fat, oil, or stearin into which is mixed a small amount of sweet cream, milk, or nonfat dry milk.[26a] The mixture is cultured for flavor development and artificial flavor such as biacetyl or bacterial starter distillates may be used.[26b]

In general, the smallest adulteration that can be estimated with a fair degree of accuracy is approximately 10 per cent. Therefore small increases or decreases as outlined in Table 33 must be interpreted with caution. Exact estimation of the percentage is not possible; close approximation may be given by calculation on the basis of normal average values of the constants of the oils involved.

Table 34 shows the increase or decrease in the constants of olive oil by substitution in whole or in part by the following adulterants: cottonseed, corn, peanut, sesame, soya, sunflower, rape, and tea seed.

TABLE 33. INCREASE OR DECREASE IN CONSTANTS OF ADULTERATED
BUTTERFAT

Name	Refractive Index	Reichert-Meissl Value	Polenske Value
Coconut	—	—	+
Palm kernel	—	—	+
Beef tallow	+	—	—
Lard	+	—	—
Hydrogenated cottonseed	+	—	—

Name	Iodine Value	Saponification Value	Hehner Value
Coconut	—	+	little change
Palm kernel	—	+	little change
Beef tallow	+	—	+
Lard	+	—	+
Hydrogenated cottonseed	+	—	+

[a] Can be controlled to any iodine value depending on the degree of hydrogenation.

26a U. S. Food Drug Admin., S.R.A., F.D.&C. 2, Part 45, February, 1955.
26b Jacobs, Am. Perfumer Aromatics 69, No. 2, 59 (1957).

TABLE 34. INCREASE OR DECREASE IN CONSTANTS OF ADULTERATED
OLIVE OIL

Name	Refractive Index	Specific Gravity	Iodine Value
Cottonseed....................	+	+	+
Peanut........................	+	no change	+
Sesame........................	+	+	+
Rape..........................	+	slight change	+
Tea seed......................	no change	no change	no change
Soya..........................	+	+	+
Sunflower.....................	+	+	+
Corn..........................	+	+	+

Name	Saponification Value	Specific Tests
Cottonseed................................	no change	Halphen
Peanut....................................	no change	Bellier, Renard
Sesame....................................	no change	Villavecchia
Rape.....................................	decrease	Valenta
Tea seed..................................	no change	Fitelson
Soya.....................................	no change	
Sunflower.................................	no change	
Corn.....................................	no change	odor

TABLE 35. SPECIAL CHARACTERISTICS FOR THE DETECTION OF
ADULTERATION IN BUTTERFAT [a]

Fat	Reichert-Meissl Value	Polenske Value	Kirschner Value
Butterfat	22-32	2-4	20-25
Coconut oil	6-8	14-16	1-2
Palm-kernel oil (African).....	5-7	10-12	0.5-1
Babassu oil	5-7	10-12	0.5-1
Cohune oil	6-8	14-16	1-2
Other common oils	<1	<1	<0.5

[a] Bailey in Jacobs, ed., *Chemistry and Technology of Food and Food Products.* Interscience, New York, 1944.

ANALYTICAL METHODS FOR FINISHED FAT PRODUCTS

In the preceding sections of this chapter, the methods discussed dealt principally with the assay and identification of various oils and fats. Many of these tests are of little value in establishing the usefulness of finished fat products such as salad oils, margarines, and shortenings for the purpose for which they were produced. The tests described in the succeeding portion of this chapter deal with this phase of the subject.

Smoke, Flash, and Fire Points[27] Applicable to Animal and Vegetable Oils and Fats

*Smoke.—Apparatus.—*1. Cleveland Flash Cup, A. S. T. M. Designation D92-33.—

The Cleveland open cup is made of brass and conforms to the dimensional requirements prescribed in Table 36. The beveled edge of the cup shall be at an angle of approximately 45°. There may be a fillet approximately 0.397 cm. in radius inside the bottom of the cup.

2. Heating Plate.—A metal plate, 0.635 cm. thick and 15.24 cm. wide for supporting the flash cup. The plate shall be of brass, cast iron, wrought iron, or steel. In the center of the plate there shall be a plane depression 0.079 cm. deep, and of just sufficient diameter to fit the cup. There shall be a circular opening 5.50 cm. in diameter,

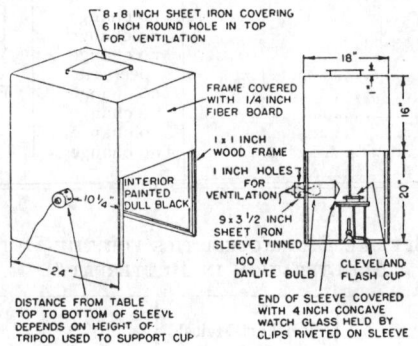

Fig. 65. Apparatus for Determining Smoke Point [27]

cut through the plate, centering with the center of the above-mentioned depression. The plate shall be covered with a sheet of hard asbestos board 0.635 cm. (0.25 in.) thick, and of the same shape as the metal plate. In the center of the asbestos board is cut a circular hole just fitting the cup.

Heat may be supplied from any convenient source. The use of a gas burner, electric heater, or alcohol lamp is permitted, but under no circumstances are products of combustion or free flame allowed to come up around the cup. The source of heat shall be centered under the opening in the plate and shall be of a type that will not produce local superheating. If a flame heater is used, it may be protected from drafts or excessive radiation by any suitable type of shield, that does not project above the level of the upper surface of the asbestos board.

[27] Committee on Analysis of Commercial Fats and Oils, American Chemical Society, *Ind. Eng. Chem., Anal. Ed.* 17, 336 (1945).

3. Thermometer, A. S. T. M. Open Flash.—

	E1 (11C-39)	E1 (11F-39)
Liquid	Mercury	Mercury
Filling above liquid	Nitrogen gas	Nitrogen gas
Temperature range	—6° to +400° C.	+20° to +760° F.
Subdivisions	2° C.	5° F.
Total length	303 to 307 mm.	
Stem diameter	6.0 to 7.0 mm.	
Bulb diameter	Not greater than stem	
Bulb length	Not over 13 mm.	
Bottom of bulb to graduation line at.......	—6° C.	+20° F.
Distance	40 to 50 mm.	
Top of thermometer to graduation line at...	+400° C.	+760° F.
Distance	30 to 45 mm.	
Top finish	Red glass ring	Red glass ring
Longer graduation lines at each	10° C.	10° F.
Graduations numbered at each multiple of...	10° C.	20° F.
Immersion	25 mm.	1 inch
Special marking on thermometer..........	25-mm. imm.	1-inch imm.
	A. S. T. M.	A. S. T. M.
	Open Flash	Open Flash
Scale error at any point up to.............	372° C.	700° F.
When standardized shall not exceed......	1° C.	2.5° F.
Test for permanency of range.	Subject to 360° to	Subject to 680° to
	370° C. for 24 hours	700° F. for 24 hours
Marking on case	A. S. T. M. Open	A. S. T. M. Open
	Flash	Flash
	—6° to +400° C.	+20° to +760° F.
Standardization	Standardize thermometer at ice point and at intervals of approximately 50° C. or 100° F. for 25-mm. or 1-inch immersion and for following temperatures of emergent mercury column.	

Thermometer Reading ° C.	Average Temperature of Emergent Mercury Column ° C.	Thermometer Reading ° F.	Average Temperature of Emergent Mercury Column ° F.
100	44	200	110
150	54	300	129
200	64	400	150
250	77	500	175
300	91	600	205
350	108	700	240

4. Cabinet.—This shall be constructed of the materials and in accordance with the dimensions indicated in Figure 65.

Procedure.—Fill the cup with the sample so that the top of the meniscus is exactly at the filling line of the cup. Adjust the position of the apparatus so that the beam of light is directed across the center of the cup. Suspend the thermometer in the center of the dish with the bottom of the bulb approximately 0.635 cm. (0.25 in.) from the bottom of the cup.

Heat the sample rapidly to within approximately 75° F. of the smoke point. Thereafter regulate the flame so that the temperature of the sample increases at a rate of not less than 9° or more than 11° F. per minute. The smoke point is taken as the temperature at which the sample gives off a thin bluish smoke continuously.

Notes.—In some cases a slight puff of smoke appears before the sample begins to smoke continuously. This is to be disregarded.

It is essential to keep the cup entirely clean and free from any substances which might cause smoke to appear ahead of the true smoke point.

Flash and Fire.—*Apparatus.*—The apparatus including the thermometer is identical with that used for the smoke point except that the cabinet is not used.

TABLE 36. DIMENSIONAL REQUIREMENTS FOR CLEVELAND OPEN FLASH CUP [27]

	Minimum In.	Normal In.	Maximum In.	Minimum Cm.	Normal Cm.	Maximum Cm.
Inside diameter immediately below filling mark......	$2^{15}/_{32}$	$2^{1}/_{2}$	$2^{17}/_{32}$	6.27	6.35	6.43
Outside diameter below flange	$2^{21}/_{32}$	$2^{11}/_{16}$	$2^{23}/_{32}$	6.75	6.83	6.91
Inside height from center of bottom to rim	$1^{9}/_{32}$	$1^{5}/_{16}$	$1^{11}/_{32}$	3.25	3.33	3.41
Thickness of bottom	$7/_{64}$	$1/_{8}$	$9/_{64}$	0.28	0.32	0.36
Distance from rim to filling mark...	$23/_{64}$	$3/_{8}$	$25/_{64}$	0.91	0.95	0.99
Distance from lower surface flange to bottom of cup....	$1^{7}/_{32}$	$1^{1}/_{4}$	$1^{9}/_{32}$	3.10	3.18	3.26
Vertical distance from upper surface flange to rim	$7/_{64}$	$1/_{8}$	$9/_{64}$	0.28	0.32	0.36
Thickness of rim...	$5/_{64}$	$3/_{32}$	$7/_{64}$	0.20	0.24	0.28
Width of lower surface of flange....	$9/_{16}$	$19/_{32}$	$5/_{8}$	1.43	1.51	1.59

Procedure.—1. Suspend the thermometer or hold in a vertical po sition by any suitable device, so that the bottom of the bulb is approxi mately 0.635 cm. (0.25 in.) from the bottom of the cup, and above a point halfway between the center and back of the cup.

2. Fill the cup with the sample to be tested in such a manner that the top of the meniscus is exactly at the filling line at room temperature.

3. The test flame should be approximately 0.397 cm. (0.125 in.) in diameter. Apply the test flame as the temperature read on the thermometer reaches each successive 5° F. mark, so that the flame passes in a straight line (or on the circumference of a circle having a radius of at least 15 cm.) across the center of the cup and at right angles to the diameter passing through the thermometer. The test flame shall, while passing across the surface of the sample, be in the plane of the upper edge of the cup. The time for the passage of the test flame across the cup shall be approximately 1 second.

4. Heat the sample at a rate not exceeding 30° F. rise per minute until a point is reached approximately 100° F. below the probable flash point of the sample. Thereafter decrease the rate of heating; for at least the last 50° F. before the flash point is reached, the rate shall be not less than 9° F. nor more than 11° F. rise per minute.

5. Flash Point.—Take as the flash point the temperature read on the thermometer when a flash appears at any point on the surface of the sample. The true flash must not be confused with a bluish halo that sometimes surrounds the test flame.

6. Fire Point.—After determining the flash point, continue the heating at the specified rate of 9 to 11° F. per minute, and apply the test flame at the specified intervals until the oil ignites and continues to burn for at least 5 seconds. The method of application of the flame shall be the same as for the flash point. The temperature read at the time of the flame application which causes burning for a period of 5 seconds or more is recorded as the fire point.

Notes.—The flash point and fire point tests should be made in a room or compartment free from air drafts. The operator should avoid breathing over the surface of the sample. The room or compartment should be darkened sufficiently so that the flash may be readily discernible.

Cold Test.—The cold test is performed to determine if an oil has been suitably wintered.

Procedure.—Fill a 4-oz. bottle or flask with a winterized oil and adjust the temperature to 2° C. Stopper with a cork and seal by dipping into melted paraffin several times. Submerge the flask or bottle completely in a vessel containing finely cracked ice and add cold water to fill completely the holding vessel. Remove the bottle after a holding period of 5.5 hours and examine the oil. It should be clear and limpid.

The cold test may be used as a rapid sorting test for olive oil.

Procedure.—Place the sample, contained in a 4-oz. bottle, in a refrigerator and hold overnight at about 40-45° F. If the sample gels or freezes, it probably contains about 50 per cent of peanut oil.

CONSISTENCY OF SHORTENINGS

Plasticity [28] as applied to shortening is the degree of pliability or the capability of a shortening to be molded or worked. A synonym which might be used is *ductility*. Plasticity must be distinguished from texture. The latter refers to the physical structure of the component particles comprising the body of the fat or its surface. Thus, when a shortening is termed stiff, reference is made to its plasticity. On the other hand, when a shortening is said to have a high gloss or to be grainy, reference is being made to its texture.

Grease Penetrometer Test.—With the aid of the A. S. T. M. grease penetrometer, plasticity can be expressed in terms of the depth to which a metal cone sinks into the surface of a sample under prescribed conditions.

Apparatus.—The instrument has a cone of standard size and shape attached to a shaft, the two being released by a stop located directly in front of the gauge. This latter measures the distance the cone and shaft penetrate the sample which is directly under it. The weight of the cone and shaft is 150 g. and it is adjustable to a wide

TABLE 37. CHARACTERISTICS OF VARIOUS TYPES OF SHORTENING [a]

	Blended Shortening	All-hydrogenated Shortening	Prime Steam Lard
Iodine value	90	62	67
Stability: Swift method (hours)	20	75	10
Free fatty acids (% as oleic)	0.12	0.02	0.30
Color (Lovibond scale)	20Y-2.5R	10Y-1.2R	5Y-0.8R
Flavor and odor	Slight	None	Characteristic
Plasticity: (by the Freyer micropenetration method)			
at 50° F.	30	20	50
at 70° F.	75	60	110
at 80° F.	140	115	160
at 90° F.	185	215	250
at 95° F.	230	300	400

[28] Rich, *Oil & Soap* 19, 54 (1942).

TABLE 38. CHARACTERISTICS OF CERTAIN COMPLETELY HYDROGENATED
EDIBLE OILS [a]

Oil or Fat	Refractive Index at 60° C.	Melting Point, ° C.	Titer, ° C.
Coconut oil	1.4400	44.8	42.0
Corn oil	1.4470	69.0	65.0
Cottonseed oil	1.4462	62.0	61.5
Sardine oil	1.4464	60.0	55.0
Sesame oil	1.4466	68.0	66.0
Soybean oil	1.4464	68.0	65.5
Sunflowerseed oil	1.4472	69.5	65.0
Tallow, beef	1.4456	61.0	60.5

TABLE 39. MISCELLANEOUS PHYSICAL PROPERTIES COMMON TO VARIOUS
FATS AND OILS [a]

(Refined oils composed largely of C_{18} acids, as cottonseed oil, soybean oil, lard, etc.)

Smoke point, ° F. (0.02% free fatty acid)........................		440-450
Flash point, ° F. (0.02% free fatty acid)........................		615-625
Fire point, ° F. (0.02% free fatty acid)........................		675-685
Specific heat (average, at processing temperatures)................		ca. 0.5
Viscosity (centipoises)	at 20° C. (68° F.)	ca. 48
	at 60° C. (140° F.)	ca. 18
	at 100° C. (212° F.)	ca. 7.5
Solubility of oxygen[b]	at 20° C. (68° F.)	0.125
	at 60° C. (140° F.)	0.142
	at 150° C. (302° F.)	0.182
Solubility of nitrogen[b]	at 20° C.	0.067
	at 60° C.	0.083
	at 150° C.	0.119
Solubility of hydrogen[b]	at 20° C.	0.039
	at 60° C.	0.059
	at 150° C.	0.104
Solubility of water[c]	at 0° C. (32° F.)	0.00074
	at 20° C. (68° F.)	0.00115
	at 35° C. (95° F.)	0.00145
Change in density with temperature...........................		0.000660 per ° C.
		0.000367 per ° F.
Change of refractive index with temperature...................		0.00038 per ° C.

[a] Bailey in Jacobs, ed., *Chemistry and Technology of Food and Food Products*. Interscience, New York, 1944.
[b] Volumes of gas which are soluble in a unit volume of oil.
[c] Weight of water dissolved by a unit weight of oil.

Procedure.—Make the test on a level surface of shortening in cans not over 4 lbs. in size. Set the cone tip exactly at the surface and release it to sink freely into the fat for 5 seconds. Report the penetration as the distance in tenths of a millimeter at 70° F.

Plastic Index Range.—The index is based on the assumption that, if a shortening has a penetration at 70° F. within the limits of good plasticity—that is, in the range of 140-180—the plastic range is better the closer the penetration at 98° F. is to that at 70° F.

Run a penetration test at 98° F. as well as at 70° F., then

$$\text{P. R. I.} = \frac{\text{Penetration at } 98° \text{ F.} - \text{Penetration at } 70° \text{ F.}}{10}$$

AIR INCORPORATION

Bailey and McKinney [29] point out that the only ingredient in a cake mix, exclusive of the foam-type cakes, sponge and angel food, that is active in entrapping air is the shortening. This implies that the logical basis for the calculation of incorporated air is not on the weight of the total mix but on that of the shortening alone.

In making the calculations of percentage of air incorporated, the densities of the major ingredients may be assumed to be

```
Sugar ........................................1.59
Shortening ..................................0.91
Cake flour ..................................1.42
Eggs ........................................1.03
Milk ........................................1.03
```

In mixtures containing eggs and milk, the sugar will be dissolved. Hence, in order to simplify the calculations and make all volumes as well as weights additive, the density of the sugar should be taken as 1.46. This value is the apparent density of sugar dissolved in an equal weight of water.

Procedure.—Pack the mix into a small cup of known capacity and weight and weigh the cup and its contents to the nearest 0.5 g. Calculate the density of the mix without incorporation of air and also the volume percentage of fat from the baking formula.

Then the percentage of air on the basis of the shortening is given by

$$A = \frac{\left(1 - \dfrac{D_2}{D_1}\right)(10000)}{\dfrac{VD_2}{D_1}}$$

---

[29] Bailey and McKinney, *Oil & Soap* **18**, 120 (1941).

where

A = percentage of air.
D_2 = density of mix, as determined
D_1 = calculated density of mix without air
V = % fat by volume in mix

SELECTED REFERENCES

Allen's Commercial Organic Analysis, 5th Ed., Vol. II. Blakiston, Philadelphia, 1924.

Committee on Analysis of Commercial Fats and Oils, Am. Chem. Soc., *Ind. Eng. Chem., Anal. Ed.,* 8, 233 (1936); 17, 336 (1945).

Cox, *Analysis of Foods,* Blakiston, Philadelphia, 1926.

Elsdon, *Edible Fats and Oils.* Van Nostrand, New York, 1926.

Fieser, *Chemistry of Natural Products Related to Phenanthrene.* Reinhold, New York, 1936.

Fryer and Weston, *Oils, Fats and Waxes.* Cambridge, 1920.

Jacobs, *Chemistry and Technology of Food and Food Products.* Interscience, New York, 1944.

Jamieson, *Vegetable Oils and Fats.* Reinhold, New York, 1943.

Lewkowitsch-Warburton, *Oils, Fats and Waxes.* Macmillan, New York. 1922.

Methods of Analysis, A.O.A.C., 8th Ed. Washington, 1955.

U. S. Food Drug Admin., S.R.A., F.D.&C. 2, Part 45, February, 1955.

CHAPTER X

SUGAR FOODS AND CARBOHYDRATES

THE interest of the food analyst in the analysis of carbohydrates stems not only from a desire or need to analyze carbohydrates, such as sugar and starches, or major carbohydrate foods, such as maple sugar, maple syrup, honey, commercial glucose, and other syrups, for identity and assay but also from his desire or need to find out the sugar content of such different food materials as fruits, fruit juices, vegetables, vegetable juices, jams, preserves, jellies, confectionery, milk, preserved meats, and many other products.

Carbohydrates contain carbon, hydrogen, and oxygen, the latter two elements being present in the ratio of 2:1, their ratio in water. In some carbohydrates, however, this ratio does not exist as, for instance, in rhamnose, $CH_3 \cdot C_5H_7O_5$, which is a methyl pentose and in desoxyribose, $C_5H_{10}O_4$. Some organic substances containing carbon, hydrogen, and oxygen, the latter two in the ratio of 2:1, are not commonly considered carbohydrates, for instance, lactic acid, inositol, and acetic acid.

Because of their relative ease of separation, starch, cellulose, sucrose, lactose, dextrose, and levulose were among the first organic substances to be isolated. Their empirical composition was found to be $C_n(H_2O)_x$. When structural organic chemistry was in its infancy, these substances were looked upon as compounds of carbon and water and were thus given the name of *hydrates de carbon* or carbohydrates.

Pigman and Goepp [1] define carbohydrate chemistry as the chemistry of the homologous or isomeric series of polyhydroxy compounds and their derivatives. The term carbohydrates is often used for the unsubstituted members of any of the series of polyhydroxy compounds of biological importance and especially for those occurring naturally. In the field outside of biochemistry, the historical and narrower definition of carbohydrates, in which the term is limited to the saccharides, is employed. In general, the carbohydrates may be considered as the aldehyde or ketone derivatives of polyhydroxy alcohols or their condensation or polymerization products.

Classification.—The carbohydrates from the point of view of or-

---

[1] Pigman and Goepp, Jr., *Chemistry of the Carbohydrates.* Academic, New York, 1948.

ganic chemistry are classified according to length of chain, functional groups, and complexity. Thus the simplest of the carbohydrates is glycolaldehyde, $HO \cdot CH_2 \cdot CHO$. They are, as mentioned, in some reactions aldehydes or ketones, and as to complexity they are considered as mono-, di-, tri-, and polysaccharides.

I. Monosaccharides
  1. *Pentoses*
     a. arabinose
     b. xylose
     c. ribose
     d. desoxyribose
     e. rhamnose
  2. *Hexoses*
     a. dextrose
     b. levulose
     c. galactose
     d. mannose
     e. sorbose
II. Disaccharides
     a. sucrose
     b. lactose
     c. maltose
     d. melibiose
III. Polysaccharides
  1. *Trisaccharides*
     a. raffinose
     b. melezitose
  2. *Dextrins*
     a. amylodextrins
     b. maltodextrins
  3. *Starches and Glycogen*
     a. starch
     b. inulin
     c. glycogen
  4. *Miscellaneous*
     a. gums
     b. pectins ·
     c. hemicelluloses
     d. cellulose

The problem of estimating the sugar content of a food product resolves itself, no matter what the food product may be, into obtaining a water solution of the sugar or mixture of sugars free of interfering substances, upon which solution the identification tests and quantitative test may be performed. This may be done by the use of a group of substances termed clarifiers.

CLARIFIERS

The function of these clarifiers is to precipitate from solution, a process called defecation, substances that might interfere in determining the optical rotation, reducing power or other physical properties or chemical reactions of sugars. The preparation of a number of these clarifiers is detailed below.

Alumina Cream.—Prepare a cold saturated solution of ammonia alum in water. Add ammonium hydroxide with constant stirring until the solution is alkaline to litmus, allow the precipitate to settle, and wash by decantation with water until the wash water gives only a slight test for sulfates with barium chloride solution. Pour off the excess of water and store the residual cream in a stoppered bottle.

Alumina cream is not a good clarifying agent for very dark solutions but is very useful as an additional clarifier and as an aid in filtering. Generally 1 to 2 ml. are used in addition to the other agent.

Basic Lead Acetate Solution.—Boil 430 g. of neutral lead acetate, 130 g. of litharge, and 1 liter of water for 30 minutes. Allow the mixture to cool and settle and then dilute the supernatant liquid to a specific gravity of 1.25 with recently boiled water. Solid basic lead acetate may be substituted for the normal salt and litharge in the preparation of the solution.

This clarifier is used mainly for dark colored solutions of sucrose to be estimated polarimetrically. From 1.5 to 10 ml. of the agent is used according to the depth of color of the material being clarified. It is not to be used for lactose determinations nor for sucrose estimated by chemical methods.

Dry Basic Lead Acetate, Horne's Method.—Add a pinch of the dry salt to the sugar solution after completion to volume and then shake the mixture. Add more salt and shake again, repeating the addition until the precipitation is complete, but avoid any excess.

This clarifier is used for solutions similar to those in which basic lead acetate solution is used and about ⅓ g. of the dry salt is taken as equivalent to 1 ml. of the basic lead acetate solution.

Neutral Lead Acetate Solution.—Add to a solution of the sample to volume. This clarifier must be used for polariscope determinations whenever reducing sugars are to be estimated in the solution for polarizing. It is sometimes used for lactose determinations. Generally 3 to 4 ml. of the reagent is sufficient.

Basic Lead Nitrate, Herle's Method.—This clarifier consists of two solutions. (1) Dissolve 250 g. of lead nitrate in water and make to a volume of 500 ml. (2) Dissolve 25 g. of sodium hydroxide in water and make up to 500 ml. Add equal volumes of both solutions to the solution to be clarified, shake, and add more if clarification is not complete. This clarifier is used in place of basic lead acetate solution, and about 3 ml. of each reagent is generally taken for this purpose.

Phosphotungstic Acid.—Sometimes the dry acid is used as described above for dry basic lead acetate but more often a 20 per cent aqueous solution is used. If much reducing substances are present, as in the case of meats where creatinine and like substances are present, a 1:1 aqueous solution may be found adequate.

Copper Sulfate.—Lactose solutions obtained from milk and milk products are generally clarified by the use of a solution of copper sulfate, 34.639 g. of $CuSO_4 \cdot 5H_2O$ dissolved in water, diluted to 500 ml. and filtered through an asbestos mat; and 0.5 N sodium hydroxide solution . They are added in the proportion of 10 ml. of copper sulfate solution to 8.8 ml. of the alkali solution.

The above-mentioned clarifying agents are the ones generally used in food analysis. The use of these clarifiers produces three types of errors. Those attributable to the volume of the precipitate produced, those resulting from precipitation of sugars from solution, and those due to change in specific rotation. To keep these errors to a minimum, a minimum amount of clarifying agent should be used.

The amount of clarifier to be used is governed by the depth of color and quantity of organic material other than sugars that are present and that must be precipitated in order not to interfere with the estimation of the sugar. The general procedure in the use of these agents is to add a small quantity, say 1 to 2 ml. of the solution of the clarifier, or a pinch or spatula-tipful of a dry defecating agent to the solution being analyzed. Mix and allow the precipitate formed to settle. Add a few drops or a few grains of the agent to see if precipitation is complete. If not, the process is continued to complete precipitation. The procedure is in reality one of trial and error. Experience soon reveals the proper agent to use and the proper quantity of agent to be added.

It may be noted that 10 per cent sodium tungstate solution, 10 g. $Na_2WO_4 \cdot 2H_2O$ dissolved in water and made up to 100 ml. and $\frac{2}{3}$ N sulfuric acid are used to clarify blood on which determinations of sugar are made. Equal volumes of the reagents are used.

Calcined charcoal and allied adsorbing carbons should not be used as clarifying agents because they may adsorb some sugar and produce serious errors in the determinations.

In biochemical determinations the use of clarifiers has reached a far more advanced state than in food chemistry. The author cannot completely recommend these clarifiers because insufficient work has been done in using them in food analyses. They should, however, be fully investigated because they are based on the principle of mutual precipitation of the added salts, leaving the solution to be tested not only free of the materials for whose elimination the clarifier was added but also free of the clarifier itself.

Steiner, Urban, and West [2] recommend ferric sulfate or thorium sulfate and barium carbonate:

$$Fe_2(SO_4)_3 + 3BaCO_3 \rightarrow Fe_2(CO_3)_3 \downarrow + 3BaSO_4 \downarrow$$
$$Th(SO_4)_2 + 2BaCO_3 \rightarrow Th(CO_3)_2 \downarrow + 2BaSO_4 \downarrow$$

West, Scharles, and Peterson [3] recommend mercuric sulfate and barium carbonate:

$$HgSO_4 + BaCO_3 \rightarrow HgCO_3 \downarrow + BaSO_4 \downarrow$$

Kleiner and Tauber [4] suggest the use of copper sulfate and barium hydroxide:

$$CuSO_4 + Ba(OH)_2 \rightarrow Cu(OH)_2 \downarrow + BaSO_4 \downarrow$$

These substances must be added in equivalent solutions in order that the mutual precipitation be virtually complete. Another clarifier is that used by Moir and Hinks [5] and consists of zinc acetate and potassium ferrocyanide. Its use is described in the section, "Total Alkaloids," Chapter XV. In Chapter VII, in connection with the method concerning the pasteurization of milk, Sander and Sager [6] utilize a zinc sulfate-copper sulfate-barium hydroxide protein precipitant. Metaphosphoric acid is widely utilized as a deproteinizing agent in methods for the determination of ascorbic acid in animal

[2] Steiner, Urban, and West, *J. Biol. Chem.* 98, 289 (1932).
[3] West, Scharles, and Peterson, *J. Biol. Chem.* 82, 137 (1929).
[4] Kleiner and Tauber, *J. Biol. Chem.* 100, 749 (1933).
[5] Moir and Hinks, *Analyst* 60, 439 (1935).
[6] Sander and Sager, *J. Dairy Sci.* 30, 909 (1947).

and vegetable foods. Horvath[7] discusses its utilization for other purposes.

MONOSACCHARIDES

Dextrose.—Dextrose is a hexose sugar whose classification name is D-glucose. It is widely distributed being present uncombined in the blood of animals and in the juices of plants. The structure and chemistry of dextrose is very complex for it may form amylene (pyran), butylene (furan), propylene and ethylene oxide rings which in turn

β- D⁺ - Glucose   D⁺- Glucose   α- D⁺- Glucose
                     (Dextrose)
                     (aldehyde form)

can exist in an α- and β-configuration. In freshly prepared solutions of α-glucose the α-form of a predominant ring exists alone and the sugar exhibits a specific rotation of +113.4°. As the solution stands, the rotatory power changes until it becomes constant at +52.3°. This phenomenon of slow change of rotatory power of sugars until an equilibrium is reached is termed *mutarotation*. It is due to the fact that when the sugar is dissolved, the form which predominates α-glucopyranose slowly changes to the other form in part. At equilibrium, although the change is still going on, as many molecules of α-form change into the β-form as the β-form change into the α-form, therefore the rotation remains the same.

Dextrose is one of the end products of the hydrolysis of many polysaccharides. Starch, dextrins, raffinose, sucrose, maltose, and lactose are composed wholly or partially of dextrose residues. It may

---

[7] Horvath, *Ind. Eng. Chem., Anal. Ed.* **18**, 229 (1946).

also be obtained from a group of substances called glucosides—for example, salicin and amygdalin. It is a reducing sugar and hence may be identified by tests for reducing sugars.

Dextrose must not be confused with the product termed commercial glucose. Commercial glucose, or mixing glucose, is a thick syrupy, colorless product made by incompletely hydrolyzing starch or a starch containing product which is decolorized and evaporated.

Formerly it was considered that the more stable form of α- or β-glucose contained the butylene oxide or furan ring structure and the less stable γ dextrose contained the propylene oxide structure. The work of Haworth [8] and his collaborators, however, shows that the more stable dextrose contains an amylene oxide or pyran structure and the less stable glucose contains the butylene oxide structure formerly ascribed to stable dextrose. Haworth terms these forms of dextrose as α- and β-glucopyranose and α- and β-glucofuranose, thus indicating the relationship between these sugars and the pyran and the furan rings. The corresponding isomers of levulose are called fructopyranose and fructofuranose. Sucrose is probably a compound of α-glucopyranose and β-fructofuranose.

α-D-Glucopyranose   β-D-Glucopyranose

Glucofuranose

Levulose.—Levulose is a ketose hexose sugar whose classification name is D-fructose. This sugar is levorotatory and has a specific rotation of $-93°$. It is found in many plant juices, probably as a hydrolysis product of sucrose. It is formed along with dextrose by

[8] Haworth, *The Constitution of Sugars*. Arnold, London, 1929.

the inversion of cane sugar. It is the main hydrolytic product of inulin and comprises about 40 per cent of honey. It is the only impor-

$$
\begin{array}{ccc}
\text{CH}_2\text{OH} & \text{CH}_2\text{OH} & \text{CH}_2\text{OH} \\
| & | & | \\
\text{HO—C} & \text{C}=\text{O} & \text{C—OH} \\
| & | & | \\
\text{HO—C—H} & \text{HO—C—H} & \text{HO—C—H} \\
\qquad\;\;\text{O} & & \qquad\;\;\text{O} \\
\text{H—C—OH} & \text{H—C—OH} & \text{H—C—OH} \\
| & | & | \\
\text{H—C} & \text{H—C—OH} & \text{H--C} \\
| & | & | \\
\text{CH}_2\text{OH} & \text{CH}_2\text{OH} & \text{CH}_2\text{OH} \\
\beta\text{-D-Fructopyranose} & \text{D-Frustose} & \alpha\text{-D-Fructofuranose} \\
 & \text{(Levulose)} & \\
 & \text{(ketone form)} &
\end{array}
$$

tant ketose sugar met with in foods. In the Haworth terminology its forms are fructofuranose and fructopyranose.

α-Fructopyranose   Fructofuranose

Galactose.—Galactose is an aldose hexose sugar that is obtained by the hydrolysis of milk sugar or lactose, galactans, and gums. Its classification name is D-galactose and has a specific rotation of +81.5°. It forms on oxidation under proper conditions, a characteristic acid, namely, mucic acid, which is not formed by other sugars or carbo hydrates that do not contain galactose residues. It forms, as do the aforementioned sugars, galactopyranose and galactofuranose.

Galactopyranose

Pentoses.—Arabinose, xylose, ribose, lyxose, and rhamnose are examples of pentose sugars and are the hydrolytic products of pentosans, nucleoproteins, gums, and similar materials. Of these arabinose and xylose are most important to the food analyst. They have strong reducing power and are nonfermentable. Ribose, as will be detailed in the chapter on vitamins, forms a portion of vitamin B_2, riboflavin. These sugars yield furfural on prolonged boiling with hydrochloric acid. Their pyran ring structures show their similarity to dextrose and the hexose sugars in general.

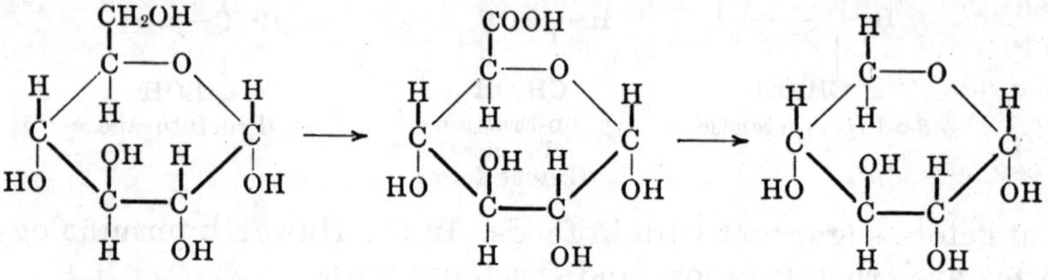

α-Glucopyranose
D-glucose

D-Glycuronic acid

Xylose

DISACCHARIDES

Sucrose.—Sucrose is a sugar characteristic of plant materials. It may be obtained from sugar cane, sugar beet, sugar maple, and other plants. On hydrolysis it yields dextrose and levulose in equimolecular quantities. The syrup formed by the hydrolysis of raw cane sugars is known as invert sugar. Sucrose is dextrorotatory and is a

β-fructofuranose

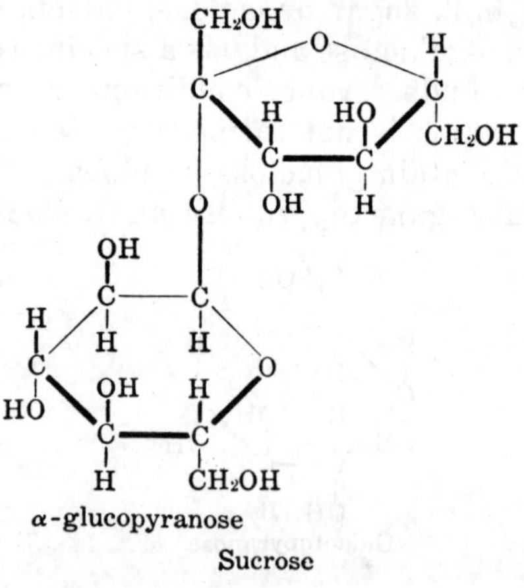

α-glucopyranose

Sucrose

nonreducing sugar. The invert sugar formed by hydrolysis is levorotatory because levulose is more strongly levorotatory than dextrose
is dextrorotatory. Haworth states that sucrose is a compound of
α-glucopyranose and β-fructofuranose. In the hydrolysis the fructofuranose changes to fructopyranose.

Lactose.—Lactose is a sugar that is produced by animals and is
obtained from the milk of mammals. On hydrolysis, it yields equimolecular quantities of dextrose and galactose. It is not hydrolyzed
as easily as sucrose. It is a reducing sugar but is not fermentable by
baker's yeast and thus may be separated by fermentation from dextrose and similar fermentable sugars. Since one of its hydrolysis
products is galactose, lactose also yields mucic acid on oxidation with
nitric acid. Lactose is dextrorotatory and has a specific rotation of
+52.5°. According to Haworth it is β-glucopyranose-β-galactopy-
ranose.

β-Galactopyranose   β-Glucopyranose
                   Lactose

Maltose.—Maltose generally does not occur free in nature. It is
formed by the action of enzymes of plants and animals on starch. It
may also be formed by the action of acids on starch. The further
hydrolysis of maltose yields dextrose. Maltose is present in commercial glucose, beer, malted milk powder, and like products. Maltose is strongly dextrorotatory having a specific rotation of +137°.
It is also a reducing sugar. Maltose is fermentable by yeast. It is
an α-glycoside, more specifically, α-glucopyranose-α-glucopyranose.

Maltose

POLYSACCHARIDES

Raffinose.—Raffinose is a trisaccharide and yields on acid hydrolysis a molecule each of dextrose, levulose, and galactose. It is dextrorotatory. It is a nonreducing sugar. On hydrolysis with enzymes it yields a melibiose residue and fructose with diastase as the enzyme and galactose and a sucrose residue, when emulsin is the enzyme. Hence, it gives the diazouracil reaction and, because melibiose contains galactose, the mucic acid reaction. It is obtained in the manufacture of sugar from sugar beet.

Melibiose residue

Raffinose

Sucrose residue

Dextrin.—Dextrin is a product of the conversion of starch. This conversion may be made by acids or enzymes. Amylodextrin is the first stage of this conversion and is known as soluble starch. This stage still gives a blue color with iodine solution. A further stage of hydrolysis is termed erythrodextrin and yields a red color with iodine solution. Maltodextrin is a stage of conversion at which no further color is obtained with iodine solution.

Starch.—Starch is a polysaccharide whose exact molecular formula is unknown. It is composed of carbon, hydrogen, and oxygen in the proportion $C_6H_{10}O_5$. It is found in the roots, tubers, grains, and seeds of plants. The small white starch granules have a characteristic form for each type of plant and these structures may be identified microscopically. It is insoluble in cold water, alcohol, and ether but swells up and forms a paste with hot water. It is hydrolyzed by enzymes and acids to dextrins, maltose, and dextrose. Final hydrolysis yields dextrose. Starch gives the well-known characteristic blue color with iodine.

Inulin.—Inulin is another polysaccharide of plant extraction with a high molecular weight $(C_6H_{10}O_5)_x$. It is obtained from the tubers of the artichoke and dahlia and from the roots of chicory and dandelion. It is slightly soluble in cold water and easily soluble in hot water. It does not give a red or blue color with iodine and on hydrolysis yields levulose in contradistinction to starch. Inulin is levorotatory.

Glycogen.—Glycogen is a carbohydrate of animal extraction and of high molecular weight $(C_6H_{10}O_5)_x$. It is found in the liver and in small amounts in other parts of the body. It is a white amorphous powder, which is soluble in water yielding a dextrorotatory solution. It is nonreducing and gives a red color with iodine. It may be hydrolyzed in a manner similar to starch by enzymes such as diastase forming dextrins, maltose, and dextrose. The muscle of horse meat contains more glycogen than is found in the muscle of other animals, and hence the presence of glycogen in meat products is an indication of horse meat.

CHEMICAL REACTIONS

Before proceeding with the quantitative determination of sugars and carbohydrates in general, it is best to ascertain the kind of carbohydrate present qualitatively. The food product itself often gives the clue unless it be grossly adulterated. Thus milk and milk products except those sweetened with sucrose, as for example, sweetened condensed milk and ice cream or with maltose in the case of malted milk, will contain lactose and a determination of the sugar reducing power may be interpreted as being due to a definite quantity of lactose. In a like manner, the reducing sugars of malt beverages such as beer, ale, etc., are calculated as maltose.

Molisch Reaction.—Sugars and carbohydrates give color reactions with many phenols. Thymol, α-naphthol, resorcinol, and phloroglucinol are some of the phenols used. The one most often used is α-naphthol, and this reaction is known as the Molisch reaction. In this reaction the furfural or hydroxymethylfurfural that is formed from the carbohydrate by the sulfuric acid condenses with the phenol to give the characteristic color.

Procedure.—Place about 5 ml. of concentrated sulfuric acid into a test tube. Add 2 drops of a 15 per cent alcoholic solution of α-naphthol to 5 ml. of the solution to be tested. Carefully overlay the sulfuric acid with the 5 ml. of solution to be tested containing the Molisch reagent. A reddish-violet ring produced at the interface of the two liquids shows the presence of a carbohydrate. This test is not specific for it is given by all the members of the carbohydrate group capable of yielding furfural in traces.

Thymol may be used instead of α-naphthol. It has the advantage that its solutions are more stable. Use 3 to 4 drops of a 5 per cent alcoholic solution of thymol for the test.

Anthrone Reaction. — Anthrone, 9,10-dihydro-9-ketoanthracene reacts with many carbohydrate materials to give a green color.[9] Furfural also gives a green color but may be differentiated by the fact that the test is rapidly obscured by a brown precipitate when the sample is diluted with 50 per cent sulfuric acid or glacial acetic acid.

Procedure.—Place 1 ml. of an aqueous solution of the sample containing approximately 1 mg. of test material. Add 2 ml. of a 0.2 per cent solution of anthrone in concentrated sulfuric acid. The final sulfuric acid concentration in the test solution should always be greater than 50 per cent. This prevents anthrone from precipitating and giving a milky suspension. The heat produced by the dilution of the sulfuric acid is a necessary part of the test. In the presence of carbohydrate material a clear green color will appear and will rapidly increase in intensity until a dark blue-green solution results. Dilute the test solution with 50 per cent sulfuric acid or glacial acetic acid for comparison. In the absence of carbohydrate material but in the presence of other organic material a brown color is often produced by the action of the sulfuric acid.

Reducing Power.—The sugars are classified chemically into two groups, the reducing sugars and the nonreducing sugars. Dextrose, levulose, galactose, lactose, and maltose are reducing sugars. Sucrose and raffinose are nonreducing sugars. The chemical properties of these reducing sugars are in the case of some sugars, like dextrose, associated with the first carbon atom of the chain so that these sugars are related to the aldehydes. In other sugars, like levulose, the reducing group appears to be associated with the second carbon so that such sugars are related to the ketones.

$$\begin{array}{ccc}
\mathrm{H-C=O} & & \mathrm{H_2C-OH} \\
| & & | \\
\mathrm{H-C-OH} & \text{or} & \mathrm{C=O} \\
| & & | \\
\text{aldoses} & & \mathrm{HO-C-H} \\
& & | \\
& & \text{ketoses}
\end{array}$$

If the terminal groups react as aldehydes, the sugars are known as aldoses; and if the terminal groups react as ketones, the sugars are known as ketoses. Dextrose is an aldose and levulose is a ketose.

However, although the sugars do give reactions associated with aldehydic and ketonic structures, they do not give certain common

---

[9] Dreywood, *Ind. Eng. Chem., Anal. Ed.* **18**, 499 (1946).

reactions like the production of colors with Schiff's reagent given by most aldehydes and ketones. This lack of reactivity is attributable to the formation of the ring structures by the reaction between the carbonyl group and hydroxyl groups noted above.

This reducing power is evidenced by the action of solutions of sugars·on alkaline solutions of copper, silver, mercury, and bismuth. The alkaline solutions of copper have had the most widespread use, and of these the reagents of Fehling, Benedict, Barfoed, Tauber, and Fischl are taken as examples.

Fehling Test.—This test is based on the reduction of an alkaline copper tartrate solution with the formation of red to yellow cuprous oxide by sugar solutions.

Reagent.—The Fehling reagent is composed of two solutions which are mixed immediately prior to the making of the test.

Reagent.—(1) Copper Sulfate Solution.—Dissolve 34.65 g. of $CuSO_4 \cdot 5H_2O$ in water, make up to 500 ml., and filter. (2) Alkaline Tartrate Solution.—Dissolve 125 g. of potassium hydroxide and 173 g. of sodium potassium tartrate (Rochelle salts) in water, make up to 500 ml., and filter through asbestos. Both solutions are kept in separate bottles.

Procedure.—Add to 1 ml. of the mixed Fehling reagent in a test tube, 4 ml. of water. Boil or, better, heat in a boiling water bath and see if a precipitate forms. If one does form, the solutions must be discarded and fresh solutions must be made. To the warm 5 ml., if no precipitate forms, add a few drops of the sugar solution which has previously been clarified with neutral lead acetate and deleaded with potassium oxalate as will be explained in the quantitative methods. Boil or, preferably, heat in a boiling water bath and if necessary add a few more drops of the sugar solution. Heat after each addition and do not add a total of more than 10-12 drops. A yellow to brick red precipitate is produced in the presence of reducing substances. This test is not specific and will be given by any of the reducing sugars.

Benedict Test.—This test is based on the reduction of an alkaline copper citrate solution with the formation of red, yellow, or green precipitates of cuprous oxide. This reagent has the marked advantage that only one solution is needed and that creatinine and uric acid do not interfere markedly.

Reagent.—Dissolve 173 g. of sodium citrate and 100 g. of anhydrous sodium carbonate in 800 ml. of water with the aid of heat. Filter. Dissolve 17.3 g. of copper sulfate, $CuSO_4 \cdot 5H_2O$, in about 100 ml. of water and filter. Make up the citrate-carbonate solution to 850 ml. and place the solution in a beaker. Add the copper sulfate solution slowly and with constant stirring. Make the mixture up to 1 liter. This reagent does not deteriorate on long standing.

Procedure.—To 5 ml. of the Benedict reagent in a test tube add 8 drops of the solution to be tested and boil. Continue boiling for 2 minutes and note the formation of a colloidal colored precipitate. If no precipitate forms, reducing sugars are absent. The sugar solution should be prepared by clarifying with neutral lead acetate and the excess lead should be removed by potassium oxalate. Small quantities of sugar such as dextrose will give a precipitate with this reagent. This test is not specific and will be given by any of the reducing sugars.

Barfoed Test.—This test is based on the reduction of a slightly acid solution of copper acetate.

Reagent.—Dissolve 13.3 g. of neutral crystallized copper acetate in 200 ml. of water. Add 1 ml. of glacial acetic acid and allow the mixture to stand. Use the clear supernatant liquid.

Procedure.—To 5 ml. of the reagent, add 1 ml. of the solution to be tested. Place in a boiling water bath for 3 minutes. At the end of this period note the formation of any red precipitate on the sides or bottom of the tube. If no precipitate has formed, place aside for a few minutes and examine again. A red precipitate shows the presence of a monosaccharide. Lactose and maltose will produce a precipitate only after prolonged boiling. Hence this reagent may be used to distinguish between monosaccharide and disaccharide reducing sugars. If reduction is obtained with either the Fehling or Benedict reagent and none with the Barfoed reagent, a disaccharide is indicated. This is the basis of a qualitative test for the detection of milk products in foods as described in Chapter VIII.

If chlorides are present in the solution to be tested, a green precipitate or a whitish precipitate may form. This is not to be mistaken for a positive sugar reduction. Phosphotungstic acid cannot be used as a clarifying agent if this test is to be performed, for potassium chloride is used to precipitate the excess phosphotungstic acid.

Osazone Reaction.—The reducing sugars have the property of reacting with phenylhydrazine under definite conditions to form difficultly soluble substances called osazones. These compounds can be easily purified and melting points can be obtained. Hence, they are useful in the identification of sugars.

Procedure.—To a mixture of 1 g. of phenylhydrazine hydrochloride and 1.5 g. of sodium acetate add 5 ml. of the sugar solution and place the tube containing the mixture in a boiling water bath for 45 minutes. Allow the tube to cool slowly. The osazones of the monosaccharides crystallize from hot solution, whereas those of the disaccharides crystallize on cooling, consequently the crystals may be separated by filtration.

Although the osazones can be recrystallized and a melting point obtained, the melting points lie so closely together that the osazones are best identified by their characteristic crystalline structure with the aid of a microscope. If the phenylhydrazine solution becomes too concentrated, the osazone may not crystallize out until it is diluted with water. Dextrose and levulose yield the same osazone because they have the same structure in the part of the molecule unaffected by the osazone reaction. The reaction goes in the following manner:

$$\begin{array}{l} \text{R} \\ | \\ \text{CHOH} + \\ | \\ \text{H}-\text{C}=\boxed{\text{O}} \quad \text{H}_2\text{N}\cdot\text{N}\cdot\text{C}_6\text{H}_5 \end{array} \rightarrow \begin{array}{l} \text{R} \\ | \\ \text{CHOH} \\ | \\ \text{H}-\text{C}=\text{NNHC}_6\text{H}_5 \\ \text{H} \end{array} + \text{H}_2\text{N}\cdot\text{N}\cdot\text{C}_6\text{H}_5 \rightarrow \begin{array}{l} \text{R} \\ | \\ \text{C}=\text{O} \\ | \\ \text{H}-\text{C}=\text{NNHC}_6\text{H}_5 \end{array} \begin{array}{l} \text{C}_6\text{H}_5\text{NH}_2 \\ + \text{Aniline} \\ + \text{NH}_3 \end{array}$$

hydrazone

$$\begin{array}{l} \text{R} \\ | \\ \text{C}=\boxed{\text{O}} \quad \text{H}_2\text{N}\cdot\text{N}\cdot\text{C}_6\text{H}_5 \\ | \\ \text{H}-\text{C}=\text{NNHC}_6\text{H}_5 \end{array} \rightarrow \begin{array}{l} \text{R} \\ | \\ \text{C}=\text{NNHC}_6\text{H}_5 \\ | \\ \text{H}-\text{C}=\text{NNHC}_6\text{H}_5 \end{array}$$

osazone

Seliwanoff Reaction.—This reaction is based on the formation of a red color by a ketose sugar in the presence of resorcinol and hydrochloric acid.

Reagent.—Prepare the reagent by dissolving 0.05 g. of resorcinol in 100 ml. of hydrochloric acid (1:2).

Procedure.—To 5 ml. of the reagent in a test tube add 1 ml. of the sugar solution and heat to boiling. A striking red color produced in less than half a minute of boiling indicates the presence of a ketose sugar. This reaction is generally considered to indicate levulose; however, sucrose will give this reaction because it is hydrolyzed rapidly in the boiling acid and yields levulose which in turn gives the red color. The red color develops into a brown-red precipitate which is soluble in alcohol.

Diazouracil Reaction.—This reaction is based on the formation of a blue-green color resulting from the formation of a condensation product of diazouracil and sucrose.

Procedure.—Shake 5 ml. of a 0.05 N sodium hydroxide solution (10° C.) containing 40-50 mg. of sucrose in a corked test tube with 7-10 mg. of diazouracil until the reagent dissolves. A blue-green color develops within a few minutes in the cold. Raffinose is the only other sugar giving this reaction. Other sugars and polysaccharides give yellow to brown-red colors.

Mucic Acid Reaction.—Galactose and any sugar or polysaccharide yielding galactose on hydrolysis—as, for example, lactose, raffinose, and some gums—give mucic acid on oxidation with nitric acid. This reaction has been discussed in some detail in Chapter VIII in connection with the detection of milk powder or milk products in other foods.

Procedure.—To 100 ml. of the sugar solution in a beaker with a graduation at 20 ml., add 20 ml. of nitric acid and evaporate slowly on a hot plate regulated to 80° C. until the volume of the mixture is reduced to 20 ml. Allow to stand for 24 hours; note the formation of a fine white precipitate of mucic acid. If no precipitate is formed or if quantitative estimation of galactans is being made, dilute to 100 ml. with water and add an additional 20 ml. of nitric acid. Again evaporate slowly to 20 ml. on a hot plate or steam bath. A fine white precipitate, insoluble in water but readily soluble in alkali or ammonium carbonate solution and reprecipitated on the addition of nitric acid, is mucic acid. The weight of mucic acid multiplied by the factor 1.33 gives the weight of galactose.

Phloroglucinol Reaction.—This reaction is based on the formation of a red to amber color by heating pentoses, glycuronic acid, galactose, or galactans with hydrochloric acid and a few crystals of phloroglucinol.

Procedure.—To 10 ml. of solution of the carbohydrate add 10 ml. of hydrochloric acid and a few crystals. Heat rapidly to boiling. The production of a red to amber color within 2 minutes indicates the presence of the above-mentioned carbohydrates.

Bial Orcinol Reaction.—The reagent is made by dissolving 1.5 g. of orcinol in 500 ml. of hydrochloric acid (specific gravity 1.15 − 30 per cent) to which 20 drops of 10 per cent ferric chloride solution is added.

Procedure.—Heat 5 ml. of the reagent to boiling, remove from the flame, and add a few drops but not over 1 ml. of the solution to be tested. If pentoses are present, a vivid green color will be developed almost immediately. This reaction will be given by carbohydrates capable of producing pentoses. It will not be given by glycuronic acid, hence it is characteristic of pentoses.

QUANTITATIVE METHODS

The quantitative methods for the estimation of sugars and carbohydrates depend on the properties of reduction and optical rotation that sugars have. The reduction methods resolve themselves to weighing the amount of cuprous oxide precipitated by a given quantity of sugar solution, that is, a gravimetric method; or to estimating the amount of copper in the cuprous oxide titrimetrically; or to measuring the volume of sugar solution necessary to reduce completely a given volume of copper solution. As representative of the gravimetric methods, the Munson and Walker method will be detailed; and as representative of the volumetric method, the Lane-Eynon method will be described. These methods are empirical and therefore the directions must be followed exactly in order to obtain correct results.

Munson and Walker Method.—The Soxhlet modification of the Fehling reagent is used as the copper reagent in this method.

Reagents.—(1) Dissolve 34.639 g. of pure copper sulfate, $CuSO_4 \cdot 5H_2O$, in water, dilute to 500 ml. and filter through asbestos washed successively with hydrochloric acid (1:3), 10 per cent sodium hydroxide solution, and alkaline tartrate solution and then washed thoroughly with water after each treatment. This type of asbestos is termed treated asbestos. (2) Dissolve 173 g. of potassium sodium tartrate (Rochelle salt) and 50 g. of sodium hydroxide in water, dilute to 500 ml., allow to stand 2 days, and filter.

Preparation of Solution.—The sugar solution is generally prepared for estimation by this method, by clarification with neutral lead acetate and alumina cream and the removal of excess lead with a minimum amount of anhydrous potassium oxalate.

Procedure.—Transfer 25 ml. of each of the copper sulfate and alkaline tartrate solutions to a 400-ml. beaker and add 50 ml. of the reducing sugar solution, or, if a smaller volume of sugar solution is used, add water to make the final volume 100 ml. Keep the beaker covered with a watch glass and heat the beaker on an asbestos gauze in such a manner that boiling begins in exactly 4 minutes. Continue the heating for exactly 2 minutes more. Some practice with 50 ml. of the mixed reagents and 50 ml. of water will enable the analyst to regulate the flame properly. Filter the hot solution at once through a prepared Gooch crucible having some of the treated asbestos for a mat. Wash the precipitate thoroughly with hot water (60° C.), then with 10 ml. of alcohol, and finally with 10 ml. of ether. Place the crucible in a constant temperature oven at 100-105° C. for 30 minutes to dry. Cool in a desiccator and weigh as cuprous oxide. Determine the amount of sugar by reference to Table 1, appendix.

If the precipitate be contaminated in any way, it is best to determine the amount of copper by dissolving the cuprous oxide in nitric acid and then proceed with the iodide-thiosulfate method.

Iodide-Thiosulfate Method.—Copper in the cupric form, as noted in Chapter V, reacts with potassium iodide with the formation of cuprous ion and free iodine.

$$Cu^{++} + 2I^- \rightleftharpoons Cu^+ + I^- + \tfrac{1}{2}I_2$$

This is a reversible reaction and can be made to run in either direction by an adequate adjustment of the conditions. Shaffer and Hartmann [10] studied these conditions and concluded that it is essential that the concentration of the potassium iodide in the solution be carefully regulated. They suggested that, if the test solution contains sufficient potassium, iodide should be added to give a concentration at the end of the titration of about 0.25 M. This is achieved by the addition of 4.2 to 5 g. of potassium iodide per 100 ml. of total solution. If greater quantities of copper are present, the potassium iodide should be added as the aqueous solution from a burette with constant agitation in proportionately greater amounts.

The cuprous iodide formed is slightly discolored because of adsorption of iodine. Foote and Vance [11] recommend that ammonium thiocyanate be added near the end of the titration. Since cuprous thiocyanate is more insoluble than cuprous iodide, the surface particles of the latter are changed to the former, thus releasing any adsorbed iodine and in addition the end point is very sharp because the precipitate is entirely white.

Reagents.—Standard Thiosulfate Solution.—Prepare a solution containing 39 g. of pure $Na_2S_2O_3 \cdot 5H_2O$ in 1 liter. Weigh accurately 0.2-0.4 g. of pure copper and transfer to a 250-ml. flask roughly graduated by marks at 20-ml. intervals. Dissolve the copper in 5 ml. of a mixture of equal volumes of nitric acid and water and dilute to 20 or 30 ml.

Boil to expel the red fumes, add a slight excess of strong bromine water, and boil until the bromine is completely driven off. Cool, and add 25 per cent sodium hydroxide solution with agitation until a faint turbidity of cupric hydroxide appears. This requires about 7 ml. of the 25 per cent sodium hydroxide solution. Discharge the turbidity with a few drops of acetic acid and add 2 drops in excess. Prepare a potassium iodide solution, 42 g. of potassium iodide in 100 ml. of solution, made very slightly alkaline to avoid formation of hydriodic acid and its oxidation products.

Observe the volume of the copper solution and add 1 ml. of potassium iodide solution for each 10 ml. of the solution undergoing titration. Titrate at once with the thiosulfate solution until the brown

10 Shaffer and Hartmann, *J. Biol. Chem.* **45**, 349, 365 (1921).
11 Foote and Vance, *J. Am. Chem. Soc.* **57**, 845 (1935).

color becomes very faint. Again observe the volume and add an additional volume of the potassium iodide solution to make the required concentration, noting from the volume of the thiosulfate the approximate copper content of the solution. Add sufficient starch indicator solution to produce a marked blue coloration. Continue the titration cautiously until the color changes toward the end to a faint lilac. Add 2 g. of ammonium thiocyanate. As the end point is approached, add the thiosulfate in fractions of drops, allowing the precipitate to settle slightly after each addition. One ml. of the thiosulfate is equivalent to about 10 mg. of copper.

Knowing the exact amount of copper weighed and the amount of thiosulfate used in the titration the exact strength of the thiosulfate solution may be calculated.

Procedure.—Wash the precipitated cuprous oxide in the Gooch crucible of the Munson and Walker method. Place the crucible in the mouth of a 250-ml. flask roughly graduated at 20-ml. intervals. Cover the crucible with a small watch glass and dissolve the oxide by means of 5 ml. of nitric acid (1:1) directed under the watch glass with a pipette. Wash the crucible with small portions of water, using a total of 25 ml., washing the watch glass with the water. Proceed from this point as directed above, beginning from "boil to expel the red fumes." Calculate the amount of copper present from the thiosulfate titration and ascertain the amount of sugar by reference to Table 1, appendix.

Lane-Eynon Titrimetric Method.—The Lane-Eynon method [12, 13] is a short and rapid method and often the most accurate method for the estimation of reducing sugars. It is based on a determination of the volume of a test solution required to reduce completely a known volume of alkaline copper reagent. The end point is indicated by use of an internal indicator, methylene blue. This thiazine dye is reduced to methylene white by a slight excess of reducing sugar over that necessary to reduce all the copper. However, since methylene white is rapidly oxidized back to methylene blue by air, it is necessary to exclude air. This is accomplished by performing the titration while the solution being analyzed is boiling. The steam produced keeps out the air.

Preparation of Sample.—Dissolve 12.5 g. of the sample in water and add 25 ml. of 10 per cent neutral lead acetate solution. Add some alumina cream and make up to 250 ml. in a volumetric flask. Shake thoroughly and filter. To 100 ml. of the filtrate, add 10 ml. of a 10 per

[12] Lane and Eynon, *J. Soc. Chem. Ind.* **42**, 32T (1923).
[13] Lane and Eynon, *J. Soc. Chem. Ind.* **42**, 463T (1923).

cent solution of potassium oxalate solution. Make to 500 ml., shake, and filter.

Procedure.—Place 10 ml. of the mixed Fehling reagent, prepared as directed in the Munson and Walker Method, into a 250-ml. Erlenmeyer flask. Transfer the sugar solution to a burette, and suspend the burette over the Erlenmeyer flask which has been placed on an asbestos wire gauze. Add 15 ml. of sugar solution to the flask and heat to boiling. Boil about 15 seconds and add rapidly further portions of the sugar solution until only the faintest perceptible blue color remains. Then add 2-5 drops of a 1 per cent aqueous solution of methylene blue and continue the heating and addition of sugar solution dropwise until the titration is complete which is shown by the reduction of the dye. Shake the flask with the aid of a flask holder during the titration and heat may be conveniently applied by a suitable hot plate.

The amount of sugar may be calculated by use of the formula

$$\frac{\text{Factor} \times 100}{\text{Titer}} = \text{mg. of sugar in 100 ml.}$$

The factor is ascertained by reference to Table 2, appendix, in which the factor for each titration from 15 to 50 ml. is given. The tables give the milligrams of sugar corresponding to each factor. If the reduction is complete on the addition of 15 ml. or less of solution, the 100-ml. aliquot must be further diluted. If the reduction is not complete on the addition of 50 ml., then more of the original sample must be taken for the analysis. If an accuracy greater than 1 per cent is desired, then the Fehling copper reagent should be standardized by performing a similar titration against a known amount of pure anhydrous dextrose or against a known amount of pure sucrose, inverted, as described in the subsequent section, "Sucrose by polarization before and after inversion with hydrochloric acid." If any correction is necessary it should be applied to the factors in the table. The final result is then corrected for dilution.

The sugar sample may also be prepared for this determination in the usual way with neutral lead acetate and then deleading the solution with anhydrous potassium oxalate. The analysis is then begun at the point . . . "Place 10 ml. of the mixed Fehling solution . . ."

These methods merely determine the amount of reducing sugar and in no way indicate the kind of sugar present; hence the analyst must decide which of the factors in the table to employ. Furthermore, if sucrose is present other corrections and factors must be applied.

The reader is referred to texts on sugar analysis for full explanations as to these variations.

Tauber Method.—This method [14] may be used for the detection of monoses, that is monosaccharides, in the presence of reducing bioses, or disaccharides. The reagent used is similar to Barfoed's solution, except that the volatile acetic acid is replaced by the nonvolatile lactic acid. The cuprous oxide formed in the test through the oxidation of the monoses is treated with a molybdate solution, the blue color developed being proportional to the amount of monose present. The reagent will keep unaltered, is superior to the Barfoed reagent, and can be used in quantitative work.

Reagents.—(1) Copper Monose Reagent.—Dissolve 24 g. of copper acetate in 450 ml. of boiling water. Any precipitate formed should not be filtered off. To the hot solution add 25 ml. of an 8.5 per cent solution of lactic acid. On shaking and boiling for a short time, the precipitate is mainly dissolved. When cold, filter the mixture and dilute to 500 ml.

(2) Benedict's [15] Molybdate Solution.—Transfer 150 g. of molybdic acid free from ammonia to a large flask and add 75 g. of anhydrous sodium carbonate. Add water in small portions, with shaking until about 500 ml. has been added. Shake thoroughly and heat the mixture to boiling or until nearly all of the molybdic acid is dissolved. Filter and wash the residue with hot water until the total volume of the filtrate is about 600 ml. Add 300 ml. of 85 per cent phosphoric acid to the filtrate. Cool and dilute to one liter.

Procedure.[16] Transfer 0.5 ml. of the neutral monose-biose test solution to a test tube with the aid of a pipette. This solution may be made by the usual clarification with neutral lead acetate and deleading with oxalate or sulfate. Kleiner and Tauber [17] use 4 per cent copper sulfate solution and 1 per cent barium hydroxide solution. The sugar solution should be diluted so that the total reduction is not more than 1.25 mg. per ml. or less than 0.05 mg. per ml. in dextrose equivalents, supposing the solution to contain dextrose and lactose. If the solution contains dextrose and maltose, the total reduction must not exceed 0.5 mg. per ml. Sucrose may be present up to 2.5 mg. per ml. without interfering with the test. In a second test tube place 0.5 ml. of water for a blank test. It is best to run a control on a known sample of the material being tested. Thus if this test is being used to distinguish between added dextrose or lactose

[14] Tauber, *Mikrochemie* **14**, 176 (1933-4).
[15] Benedict, *J. Biol. Chem.* **92**, 141 (1931).
[16] Tauber and Kleiner, *J. Biol Chem.* **99**, 249 (1932).
[17] Kleiner and Tauber, *J. Biol. Chem.* **100**, 749 (1933).

to broken out eggs, a known sample of broken out eggs free of sugars run as a control aids in the interpretation of results. Add a 0.5-ml. portion of the copper-monose reagent to all the tubes and then place them for 8 minutes in a boiling water bath. Cool for 2 minutes. Add a 0.5-ml. portion of the molybdate solution and, after 2 minutes, add 5 ml. of water. A blue color indicates monoses and is proportional to the amount present. Chlorides and some other inorganic salts interfere but as much as 1.5 mg. per ml. of sodium chloride will not.

Fischl Ketose Method.—This method [18] may be used for detecting and determining levulose in the presence of dextrose, other aldoses and sucrose. When sugar solutions are treated under definite conditions with a faintly alkaline copper and Rochelle salt solution containing phosphate, the reduction effected by sugars other than levulose is negligible in amount. As little as 1 mg. of levulose in the presence of 49 mg. of dextrose is thus detectable and determinable.

Reagents.—(1) Copper Ketose Reagent.—Dissolve 15 g. of anhydrous sodium carbonate, 5 g. of crystallized copper sulfate, 300 g. of Rochelle salt, and 100 g. of disodium phosphate ($+12H_2O$) in water in a 1-liter measuring flask to form a volume of about 900 ml. Carry out the solution as far as possible at room temperature and complete on a water bath. Leave the flask in the bath for an hour and, when cold, make up to 1 liter. Mix with two teaspoonsful of active charcoal, and filter. It should be stored in a dark glass bottle with a glass or rubber stopper and should not be kept too long.

Procedure.—The levulose solution to be tested should be neutralized and diluted to contain about 0.5 per cent sugar. If it has been cleared with lead salts, the lead must be removed from the filtrate by means of anhydrous sodium sulfate and not by alkali or sodium phosphate.

Place 10 ml. of the sugar solution and 30 ml. of the copper solution in a 150-ml. flask, which is fitted with a cork carrying an accurate thermometer and channeled at the edge to prevent development of pressure during the heating. Place the flask in a water bath at 70° C. and swirl its contents gently until the temperature is 65° C. Maintain this temperature and the swirling for exactly 5 minutes, after which cool the liquid to room temperature as rapidly as possible in cold water. The appearance of a turbidity or a precipitate of cuprous oxide shows the presence of levulose.

For the determination, wash the thermometer with 20 ml. of water and catch the washings in the flask. Add about 15 ml. of N hydrochloric acid and then a few milliliters of 0.01 N iodine solution by

---

[18] Fischl, *Chem. Ztg.* 57, 393 (1933).

pouring carefully down the wall of the flask, without mixing. Mix the whole batch and add a further quantity of 0.01 N iodine solution so that an excess of iodine solution is present. Thirty to 60 ml. are usually required depending upon the amount of cuprous oxide formed. Stopper the flask and allow the iodine to act for exactly 2 minutes, with occasional shaking. After the addition of starch solution back-titrate the mixture with 0.01 N sodium thiosulfate solution. Unless more than 0.5 ml. of iodine solution has been used up, the presence of levulose cannot be assumed, since 50 mg. of dextrose requires about 0.3 ml. of the iodine solution. This method permits the determination of levulose not only in mixtures of sugars but also in sweet wines, fruit juices, honey, and the like and is suitable for physiological chemical purposes.

Micro-modification of the Fischl Method.—Place 0.5 ml. of the clarified and lead-free sugar filtrate in a small test tube and add 1 ml. of the copper ketose reagent. Place the test tube in a water bath and keep in this bath for 5 to 6 minutes at 65° to 70° with occasional swirling. At the end of this period cool the tube as rapidly as possible to room temperature. Add 1 ml. of Benedict's arsenotungstic reagent. Shake thoroughly. Wait a minute or two for any gas evolved to escape, dilute with 5 ml. of water, and compare against a standard sugar solution treated exactly the same way, in a colorimeter.

Reagent.—Benedict's [19] Arsenotungstic Complex Reagent.—Dissolve 10 g. of sodium tungstate in 60 ml. of water. Add 5 g. of pure arsenic pentoxide and 2.5 ml. of 85 per cent phosphoric acid. Then add 2 ml. of hydrochloric acid. Boil for 20 minutes, adding water occasionally to make up for the loss of water due to evaporation. Cool, add 6 ml. of formalin, 4.5 ml. of hydrochloric acid and 4 g. of sodium chloride. Dissolve all the materials and dilute to 100 ml.

Clinical Colorimetric Methods.—While colorimetric methods for the determination of reducing sugars have been widely employed in clinical analysis, their use for food products has been limited. In a study by Edson and Poe, the 2,4-dinitrophenylate, Lewis-Benedict, Folin-Wu, Benedict, Folin-Sumner, and Kingsbury methods were compared with the Munsun and Walker method. Concentrations of dextrose ranging from 0.5 to 0.3 per cent were compared. The Sumner, Kingsbury, and Poe-Edson methods gave readings which were proportional to the amount of dextrose. The other methods gave results which were not proportional for the higher and lower ranges of dextrose. These methods were compared for syrups, fruit juices, jellies and jam, flavoring syrups and on syrups after inversion. Ed-

[19] Benedict, *J. Biol. Chem.* **68,** 759 (1926).

son and Poe [20] found that the standard colorimetric methods are satisfactory for the determination of reducing sugars in foods. Common clarifying agents, except mercuric nitrate, do not interfere.

SUCROSE

CHEMICAL METHODS

Determine the reducing sugars, clarification having been effected with neutral lead acetate, never with basic lead acetate, as directed under the Munson and Walker method and calculate to invert sugar from Table 1, appendix. Invert the solution as directed in the section, "Sucrose by polarization before and after inversion with hydrochloric acid"; exactly neutralize the acid; and again determine the reducing sugars, but calculate them to invert sugar from the table referred to above, using the invert sugar column alone. Deduct the percentage of invert sugar obtained before the inversion from that obtained after inversion and multiply the difference by 0.95 to obtain the percentage of sucrose. The solutions should be diluted in both determinations so that no more than 240 mg. of invert sugar is present in the quantity taken for reduction. It is important that all the lead be removed from the solution with anhydrous powdered potassium oxalate or sodium carbonate before reduction.

OPTICAL METHODS

All sugars and carbohydrates have asymmetric carbon atoms and therefore exhibit the phenomenon of optical rotation. The polarimeter which was discussed in Chapter II is an instrument which measures the optical rotation of a solution. It was explained that the specific rotation is dependent on the concentration of the active substance in grams per 100 ml., the length of the tube, and the observed rotation. Hence knowing the specific rotation of the sugar, with the same tube and an observed rotation, the concentration of an unknown sugar solution may be calculated.

This procedure is greatly simplified by the use of a saccharimeter which is a polarimeter having an arbitrary scale reading per cent sugar directly, provided a factor weight, such as 26 g./100 ml. is used. Concentration and temperature have a marked effect on the specific rotation of the various sugars. All of the sugars commonly met with in food analysis are dextrorotatory except levulose and invert sugar.

---

[20] Edson and Poe, *J. Assoc. Official Agr. Chem.* **31**, 769-76 (1948).

INTERNATIONAL SUGAR COMMISSION RULES

The rules of the International Commission for Uniform Methods of Sugar Analysis have been generally adopted. A résumé of the work of this commission in its nine sessions held at various times from 1897 to 1936 is given by Bates.[21] The principal rules adopted concerning the analyses discussed in this text are the following.[22]

"In general, all sugar tests shall be made at 20° C.

"The adjustment of the saccharimeter shall be made at 20° C. One dissolves (for instruments arranged for the German normal weight) 26.00 g. of pure sugar in a 100-metric cubic centimeters flask (100-ml. flask), weighing to be made in air, with brass weights, and polarizes the solution in a room, the temperature of which is also 20° C. Under these conditions the instrument must indicate exactly 100.00.

"The temperature of all sugar solutions to be tested is always to be kept at 20° C. while they are being prepared and while they are being polarized.

"However, for those countries the temperature of which is generally higher, it is permissible that the saccharimeters be adjusted at 30° C. (or at any other suitable temperature), under the conditions specified above, and providing that the analysis of sugar be made at that same temperature."

"To make a polarization (sucrose), the whole normal weight for 100 ml. is to be used, or a multiple thereof for any corresponding volume.

"As clarifying and decolorizing reagents there may be used: Subacetate of lead prepared from (three parts by weight of acetate of lead, one part by weight of oxide of lead, ten parts by weight of water) Scheibler's alumina cream concentrated solution of alum. (The A. O. A. C. recommends that whenever reducing sugars are determined in the solution for polarizing, only neutral lead acetate for clarification be used, as basic lead acetate causes precipitation of some of the reducing sugars). Bone black and decolorizing powders are to be absolutely excluded.

"After bringing the solution exactly to the mark and after wiping out the neck of the flask with filter paper, all of the well-shaken, clarified sugar solution is poured upon a dry, rapidly filtering filter. The first portions of the filtrate are to be thrown away and the balance, which must be perfectly clear, is to be used for polarization."

[21] Bates, *Polarimetry, Saccharimetry, and the Sugars Nat. Bur. Standards (U. S.),* Circ. C440, 1942.

[22] International Commission for Uniform Methods of Sugar Analysis, Third Session, Paris, July, 1900.

"Wherever white light is used in polarimetric determinations, the same must be filtered through a solution of potassium dichromate of such a concentration that the percentage content of the solution multiplied by the length of the column of the solution in centimeters is equal to nine." [23] The A. O. A. C.[24] recommends that this concentration be doubled in polarizing carbohydrate materials of high rotation dispersion, such as commercial glucose.

TABLE 40. CONVERSION AND COMPARISON FACTORS OF SACCHARIMETER SCALES

Conversion Factors

Scale	Factor	Equivalent
1° International Sugar Scale[a]	0.34620°	Angular rotation D[b]
1° Angular rotation D	2.8885°	International Sugar Scale
1° French Sugar Scale	0.21667°	Angular rotation D
1° Angular rotation D	4.6153°	French Sugar Scale
1° Wild Sugar Scale	0.13284°	Angular rotation D
1° Angular rotation D	7.52814°	Wild Sugar Scale
1° Ventzke Sugar Scale	0.34657°	Angular rotation D
1° Angular rotation D	2.88542°	Ventzke Sugar Scale
1° Bidecimal Scale	0.26622°	Angular rotation D
1° Angular rotation D	3.75629°	Bidecimal Scale

Comparison Factors

Scale	Factor	Equivalent
1 International Sugar Scale	1.59782°	French Sugar Scale
1° French Sugar Scale	0.62585°	International Sugar Scale
1° International Sugar Scale	2.60614°	Wild Sugar Scale
1° Wild Sugar Scale	0.38371°	International Sugar Scale
1° Wild Sugar Scale	0.61310°	French Sugar Scale
1° French Sugar Scale	1.63106°	Wild Sugar Scale

Normal Weights

International Sugar Scale[a]	26.000 g.	
French Sugar Scale	16.269 g.	
Wild Sugar Scale	10.000 g.	
Ventzke Sugar Scale	26.026 g.	
Bidecimal Scale	20.000 g.	

[a] Bureau of Standards Scale is equivalent to International Sugar Scale.
[b] The designation D refers to sodium light 5893° Angstrøm units.

In the eighth session held at Amsterdam, 1932, the following recommendations were adopted: "(a) That the Commission adopt

[23] International Commission for Uniform Methods of Sugar Analysis, Seventh Session, New York, September, 1912.
[24] *Methods of Analysis*, A.O.A.C. Washington, 1945.

a standard scale for the saccarimeter and that this scale be known as the 'International Sugar Scale.' Rotations expressed in this scale shall be designated as degrees sugar (°S).''

''(b) That the polarization of the normal solution (26.000 g. of pure sucrose dissolved in 100 ml. of solution) and polarized at 20° C. in a 200-mm. tube, using white light and the dichromate filter, as defined by the Commission, be accepted as the basis óf calibration of the 100° (S) point on the International Sugar Scale.''

Sucrose by Polarization Before and After Inversion with Hydrochloric Acid

Weigh double the normal weight of the material to be examined, that is 52 g. exactly, and as rapidly as possible transfer the material to a 200-ml. sugar volumetric flask with water. Add a sufficient amount of clarifying agent avoiding any excess. Shake, dilute to the mark with water, mix thoroughly, and filter. Keep the funnel covered during the filtration and reject the first 25 ml. of the filtrate. If a lead clarifying agent was used, remove the lead from the filtrate by the addition of small portions of anhydrous potassium oxalate, anhydrous sodium carbonate, and sometimes anhydrous potassium sulfate. Avoid any excess of the deleading agent, mix thoroughly, and filter through a dry filter rejecting the first 25 ml. of the filtrate. Pipette 50 ml. of the lead-free filtrate into a 100-ml. volumetric flask, dilute, with water to the mark, and polarize in a 200-mm. tube. The result multiplied by 2 is the direct reading.

Pipette another 50 ml. portion of the lead-free filtrate into a 100-ml. volumetric flask and add 25 ml. of water. Add, in small portions while rotating the flask, 10 ml. of hydrochloric acid (sp. gr. 1.1029). Adjust a constant temperature bath at 70° C. Place the flask containing a thermometer in the bath and, when the thermometer reads 67° C., leave the flask in the bath for exactly 5 minutes longer. At the end of this period remove the flask and place it immediately in a bath at 20° C. When the contents of the flask reach 35° C., remove the thermometer from the flask, rinse it and fill almost to the mark. Allow the flask to remain in the bath for another half hour and then make to volume. Mix well and polarize the solution in a 200-mm. tube at 20° C. This reading must also be multiplied by 2 to give the invert reading. Calculate sucrose by use of the following formula.

$$S = \frac{100(P - I)}{143 + 0.0676 \ (m - 13) - t/2} \qquad [1]$$

in which

S = the percentage of sucrose

P = direct reading, normal solution

I = invert reading, normal solution

t = temperature at which readings are made

m = g. of total solids in 100 ml. of the invert solution read in the polariscope.

The total solids are determined by observing the refractometer reading of the invert solution read in the polariscope and referring to Table 3, appendix, of sugar solids. The percentage total solids is multiplied by the density corresponding to this percentage in the same table. The result is the grams of total solids in 100 ml. of the invert solution.

The inversion may also be made at room temperature by the following procedures. (1) To 50 ml. of the lead-free filtrate in a flask add 10 ml. of the above-mentioned hydrochloric acid and set aside for 24 hours at a temperature not below 20° C.; or (2) if the temperature is above 25° C., set aside for 10 hours. Make up to volume of 100 ml. at 20° C. and polarize as directed above. The above formula must be corrected by changing the factor 143 to 143.2.

Sucrose by the Double Dilution Method

As was mentioned previously in the chapter, one of the errors in determining sugars, if clarifying agents are used, is the volume of the precipitate formed by the clarifier. If the volume of the precipitate is large, more than 1 ml. from 26 g. of material, then the Clerget or the double dilution method may be used to correct this error.

Weigh a half-normal weight of the sample and transfer it with the aid of water to a 100-ml. sugar volumetric flask. Clarify with the appropriate agent and make to volume. At the same time weigh a normal weight of the sample, transfer this material to another 100-ml. sugar volumetric flask, add the same clarifier, and dilute to volume. Filter, and obtain the direct readings as directed above. Invert each solution by one of the methods detailed above and obtain the invert readings.

The true direct polarization of the sample equals 4 times the direct polarization of the diluted solution less the direct polarization of the undiluted solution. The true invert polarization equals 4 times the invert polarization of the diluted solution less the invert polarization of the undiluted solution. Calculate the sucrose from the true polarization thus obtained, using the formula given above.

Sucrose—Low Concentrations.—In sugar processing plants it is necessary to determine sucrose concentrations of the order of 10 to 255 p.p.m. This cannot be done by polarimetric methods. Morse[25] adapted the anthrone test for the quantitative estimation of low concentrations of sucrose.

Procedure.—Transfer, with the aid of a pipette, 2.0 ml. of the test solution to a test tube and underlay with 3.0 ml. of 0.05 per cent anthrone dissolved in concentrated sulfuric acid. Shake the tube to mix completely. Measure the transmittancy of the solution for white light in a photoelectric colorimeter, using water as the standard, with its transmittancy taken as 1.000. Obtain the concentration of sucrose from a previously prepared calibration curve.

Sugar Mixtures

Zerban and Sattler Method for Sugar Mixtures.—In food analysis it is often necessary to analyze complex mixtures containing sucrose, dextrose, levulose, maltose, and lactose. Sucrose can be estimated in the presence of the other four by means of the enzyme invertase. Adequate results for estimation of sugar mixtures have been obtained by microbiological methods utilizing fermentation with specific organisms but such pure cultures are not readily available and many chemical laboratories are not equipped to perform such procedures for they require facilities for the preparation and sterilization of media and glassware, the propagation of the cultures, and other bacteriological techniques. Chemical methods for the determination of dextrose, levulose, maltose, and lactose in sugar mixtures have been developed by Zerban and Sattler.[26]

Principal of the Method.—The determination of four sugars by combined methods requires four equations. Lactose is the only sugar among the four that can be determined independently. It may either be oxidized to mucic acid and weighed in this form, or else the other three sugars may be removed by fermentation, and the residual lactose estimated by any suitable method. The dextrose and levulose may be determined by combining the method of Jackson and Mathews for the selective determination of levulose with the method of Steinhoff for the selective determination of monosaccharides in the pres-

25 Morse, *Anal. Chem.* 19, 1012 (1947).
26 Zerban and Sattler, *Ind. Eng. Chem.*, *Anal. Ed.* 10, 669 (1938).

ence of disaccharides. The total reducing sugars are found by means of Fehling's solution. If

G = mg. dextrose (glucose),
F = mg. levulose (fructose),
M = mg. maltose hydrate,
L = mg. lactose hydrate,
R_1 = mg. apparent levulose by the method of Jackson and Mathews,
R_2 = mg. dextrose plus levulose, expressed as levulose, by the copper acetate method of Steinhoff, and
R_3 = mg. total reducing sugars, expressed as dextrose, by Fehling solution, then

$$L \text{ is determined separately} \qquad (1)$$
$$R_1 = 0.0806\,G + F \qquad (2)$$
$$R_2 = aG + F \qquad (3)$$
$$R_3 = G + bF + cM + dL \qquad (4)$$

The factor 0.0806 in equation (2) is the reducing ratio of dextrose to levulose (12.4 mg. dextrose have the same reducing power as 1 mg. levulose). The factor a is the reducing ratio of dextrose to levulose in Steinhoff's acetate method, and b, c, and d, are the reducing ratios of levulose, maltose, and lactose, respectively, to dextrose for Fehling solution. The values of a, b, c, and d, vary with the concentration and are found from tables.

By solving equations (2) and (3) for G and F, we find

$$G = \frac{R_2 - R_1}{a - 0.0806}, \text{ and}$$

$$F = R_2 - aG.$$

L, G, and F being known, equation (4) gives

$$M = \frac{R_3 - (G + bF + dL)}{c}$$

It has been found that both maltose and lactose have a slight reducing effect on the Jackson and Mathews reagent, as well as on the Steinhoff copper acetate reagent. The procedure for applying the necessary corrections will be discussed under their respective headings.

Lactose.—The method detailed is a modified form of the fermentation procedure of Hoffman, Schweitzer and Dalby.[27] See Chapter VIII.

Procedure.—Place the sample in a 500-ml. volumetric flask, and add a thin suspension of a mixture containing 35 g. of compressed bakers' yeast, 0.5 g. ammonium sulfate, and 0.2 g. sodium bisulfite

---

[27] Hoffman, Schweitzer, and Dalby, *Ind. Eng. Chem., Anal. Ed.* **8**, 298 (1936).

in water. Dilute the mixture in the flask further with water to a volume of about 400 ml. Close the flask with a stopper provided with a delivery tube the outer end of which is immersed 1 cm. below the surface of water in a beaker. Place the flask in a water bath or thermostat kept at 30° C., and shake from time to time.

After the flask has stood a minimum of 4 hours, add 15 ml. of a 20 per cent solution of neutral lead acetate, and the volume made up to the mark at 20° C. Add 1 g. of Filter Cel, shake the contents of the flask thoroughly, and filter through a folded quantitative filter paper. Discard the first, turbid portion of the filtrate, and collect exactly 200 ml. of the clear filtrate in a dry volumetric flask calibrated for 200 and 220 ml. contents. To the 200 ml. of test solution, add 15 ml. of a phosphate-oxalate solution, prepared by dissolving 7 g. of disodium phosphate ($Na_2HPO_4 \cdot 12H_2O$) and 3 g. potassium oxalate ($K_2C_2O_4 \cdot H_2O$) to 100 ml. Make up to the 220-ml. mark at 20° C., add 0.5 g. of Filter Cel, and shake the contents vigorously. Filter the solution through a quantitative filter paper, and collect the clear filtrate for the sugar determination. Determine lactose by the Munson and Walker method detailed in this chapter.

The fermentation method should be carried out with a reliable brand of yeast, and its fermenting power should be checked by appropriate tests. Blank tests run with a good yeast usually yield very low reduction values, and the results are not trustworthy for corrections. It is a much better practice to use pure lactose for the purpose. It is therefore recommended to run parallel determinations with lactose alone, or with a known sugar mixture approximating that of the sample.

Apparent Levulose—Jackson and Mathews [28] Method.—It was found by Zerban and Sattler that both maltose and lactose reduce the copper carbonate reagent. The reducing effect is smaller than that of dextrose, but it is much larger than that of sucrose. It varies not only with the concentration of these sugars alone, being relatively greater for higher concentrations of them, but also with the concentration of dextrose and levulose present in mixtures, the reducing power also increasing with the total sugar concentration. The variations are small, however, and for practical purposes average figures may be used to correct for the reducing effect of maltose and lactose. An average of 26.0 mg. of maltose hydrate, and 25.6 mg. of hydrate lactose were found to be equivalent to 1 mg. of levulose.

Reagents.—Potassium Dichromate Solution.—Prepare a 0.1573 N solution by dissolving 7.7135 g. of potassium dichromate, $K_2Cr_2O_7$,

---

[28] Jackson and Mathews, *Bur. Standards J. Research* **8**, 403 (1932).

in water and diluting to 1 liter. One ml. of this solution is equivalent to 10.00 mg. of copper.

Ferrous Ammonium Sulfate Solution.—Dissolve 61.8 g. of ferrous ammonium sulfate hexahydrate in water, add 5 ml. of concentrated sulfuric acid, and complete to 1 liter.

Ost Solution.—Dissolve 250 g. of anhydrous potassium carbonate in about 700 ml. of hot water and add 100 g. of potassium bicarbonate. Stir until completely dissolved. Cool and add, while stirring vigorously, a solution of 25.3 g. of copper sulfate, $CuSO_4 \cdot 5H_2O$, in 100 to 150 ml. of water. Adjust to room temperature, make to 1 liter, and filter.

Procedure.—Transfer 50 ml. of Ost reagent to a 150-ml. Erlenmeyer flask and add with the aid of a pipette a known aliquot of the solution to be analyzed. This should contain not more than 92 mg. of levulose or its equivalent of a levulose-dextrose mixture, on the basis that dextrose has about $\frac{1}{12}$ the reducing power of levulose. Add sufficient water to make the total volume 70 ml. Immerse the flask in a water bath regulated at 55° C. ± 0.1° C. Digest for exactly 75 minutes, swirling with a rotatory motion at intervals of 10 or 15 minutes. At the end of this time, filter the precipitated copper on a closely packed Gooch crucible and wash the flask and filter thoroughly without attempting to transfer the precipitate quantitatively. Remove the asbestos mat with the aid of a glass rod and transfer to a 400-ml. beaker. Add 5 or 10 ml. of water and disintegrate the asbestos mat. Add a measured volume of 0.1573 N potassium dichromate solution in excess of the volume necessary to oxidize the cuprous oxide. In routine analyses, the amount of copper precipitated will be known roughly and a 3- to 4-ml. excess will be adequate. Add 1 ml. of the dichromate solution to the original flask to dissolve any of the residual copper. Add to the Erlenmeyer flask 50 ml. of hydrochloric acid (1:1) and pour this acidified solution slowly into the 400-ml. beaker with constant stirring. Wash the flask with the aid of a wash bottle and add the washings to the beaker. Immerse the crucible into the acid solution to dissolve any cuprous oxide. Remove the crucible with a glass rod and wash it thoroughly, catching the washings in the beaker. Dilute the solution to about 250 ml. and titrate the excess dichromate with the ferrous sulfate solution electrometrically.

Copper may also be determined by any other suitable method.

Dextrose Plus Levulose — Steinhoff Method.—The Steinhoff method [29] was combined by Zerban and Sattler with the iodometric

[29] Steinhoff, Z. Spiritusind. 56, 64 (1933).

method of Shaffer and Hartmann [30] by modifying it to suit the particular conditions for the analysis of mixtures.

Reagents.—(1) Sodium Acetate Solution.—Dissolve 500 g. of the crystallized salt ($CH_3COONa \cdot 3H_2O$) in about 800 ml. of hot water, cooling, and make up to 1 liter.

(2) Sulfuric Acid.—Dilute 57 ml. of concentrated acid to 1 liter. This is about 2 N.

(3) Potassium Iodide-Iodate Solution.—Dissolve 5.4 g. potassium iodate and 60 g. potassium iodide in water. Make the solution alkaline by adding 0.25 g. of sodium hydroxide dissolved in a little water and dilute to 1 liter.

(4) Saturated Solution of Potassium Oxalate.—Dissolve 165 g. of the hydrated salt ($K_2C_2O_4 \cdot H_2O$) in 500 ml. of hot water, and cool to room temperature.

(5) Sodium Thiosulfate Solution.—Standardize a 0.1 N solution by iodometric determination with standard potassium dichromate solution.

(6) Soxhlet Copper Sulfate Solution (Fehling 1).—Prepare as directed on page 431.

Procedure.—Transfer 10 ml. of the copper sulfate solution and 20 ml. of the sodium acetate solution with the aid of pipettes into a 250-ml. wide-mouth Erlenmeyer flask, and add a measured amount of the sugar solution. The quantity of sugar solution must be such that the thiosulfate solution corresponding to the copper reduced is within the limits of Table 41, and that the thiosulfate solution corresponding to the copper reduced from Fehling solution by the *same* quantity of sugar solution is within the limits of Table 41. A few preliminary experiments are usually necessary to establish the optimum amount for both determinations.

After the sugar solution has been added, complete the volume to a total of 50 ml. by the addition of water. Mix thoroughly, close the Erlenmeyer with a rubber stopper provided with a Bunsen valve to prevent reoxidation of the reduced copper. Place the solution in a briskly boiling water bath, start a stop watch, and remove the flask from the bath after exactly 20 minutes. Cool quickly to room temperature under the water tap. During this time the Bunsen valve must be vented from time to time to prevent boiling caused by the vacuum. After cooling, add 25 ml. of the iodide-iodate solution carefully from a pipette and mix with the solution by gentle shaking. Run 40 ml. of the 2 N sulfuric acid in rapidly from a cylinder, rotating the flask to wash down the inside wall. Follow this by the addition of 20 ml. of the potassium oxalate solution from a cylinder. Mix

[30] Shaffer and Hartmann, *J. Biol. Chem.* 45, 349 (1921).

the contents of the flask well until the precipitate is completely dissolved, and titrate the excess iodine with the standard thiosulfate solution.

Run a blank with water instead of sugar solution. The difference between the thiosulfate titer of the blank and that of the sample is a direct measure of the cuprous oxide precipitated. Results of duplicate determinations usually check within 0.1 to 0.2 ml. thiosulfate solution.

TABLE 41. COPPER ACETATE REAGENT

(Milligrams of dextrose or levulose corresponding to varying volumes of 0.1 N thiosulfate solution; reducing ratios for varying proportions between levulose and dextrose)

Thio-sulfate	R_2 Dex-trose	Levu-lose	100 levulose 0 dextrose	Reducing Ratios, a 75 levulose 25 dextrose	50 levulose 50 dextrose	25 levulose 75 dextrose
Ml.	Mg.	Mg.				
1	5.0	3.2	0.640	0.556	0.829	0.852
2	7.0	5.8	0.829	0.803	0.933	0.956
3	9.0	8.3	0.922	0.952	0.976	1.000
4	11.1	10.7	0.964	1.000	0.981	1.013
5	13.3	12.9	0.970	0.980	0.955	1.010
6	15.5	14.9	0.961	0.953	0.930	0.984
7	17.8	16.8	0.944	0.914	0.905	0.957
8	20.3	18.6	0.916	0.897	0.880	0.931
9	22.9	20.3	0.886	0.870	0.855	0.906
10	25.7	22.1	0.860	0.845	0.830	0.880
11	28.7	23.9	0.833	0.819	0.805	0.853
12	31.8	25.7	0.808	0.793	0.780	0.825
13	35.4	27.7	0.782	0.767	0.755	0.797
14	39.0	29.4	0.754	0.747	0.730	0.769
15	43.3	31.4	0.726	0.715	0.706	0.740
16	47.9	33.6	0.701	0.688	0.681	0.711
17	53.3	35.8	0.672	0.666	0.656	0.681
18	59.5	38.4	0.645	0.640	0.631	0.651
19	66.6	40.9	0.612	0.595	0.605	0.621
20	74.6	43.8	0.587	0.583	0.581	0.593
21	84.1	46.9	0.558	0.555	0.555	0.564
22	95.0	50.5	0.532	0.530	0.529	0.536

The reducing effect of dextrose and levulose, obtained from the National Bureau of Standards, and of mixtures of the two, on the copper acetate reagent were determined by Zerban and Sattler.[31] The results are shown in Table 41, which gives the milligrams dextrose or levulose equivalent to the number of milliliters thiosulfate solution found, and also the values of the factor a in equation 3, for varying proportions between the two sugars.

[31] Zerban and Sattler, Ind. Eng. Chem., Anal. Ed. 10, 669 (1938).

Total Reducing Sugars.—Mix 10 ml. of Soxhlet copper sulfate solution (Fehling I) and 10 ml. of alkaline tartrate solution (Fehling II), page 431 in a 250-ml., wide-mouth Erlenmeyer flask. Add the same quantity of sugar solution as was used in the determination of dextrose plus levulose, make the volume up to a total of 50 ml., and carry out the analysis exactly as described for that determination, except that only 25 ml. of the 2 N sulfuric acid is used instead of 40 ml.

TABLE 42. COPPER TARTRATE REAGENT [a]

(Milligrams Dextrose corresponding to varying volumes of thiosulfate solution, and reducing ratios of levulose, maltose, and lactose, with respect to dextrose.)

Thiosulfate ml.	R_3, Dextrose mg.	Reducing Ratios		
		Levulose, b	Maltose Hydrate, c	Lactose Hydrate, d
1	2.2	0.648	0.394	0.499
2	5.3	0.815	0.481	0.616
3	8.5	0.868	0.508	0.653
4	11.6	0.893	0.521	0.669
5	14.8	0.907	0.528	0.678
6	18.0	0.914	0.532	0.683
7	21.2	0.918	0.535	0.685
8	24.4	0.921	0.537	0.686
9	27.6	0.922	0.538	0.689
10	30.8	0.922	0.538	0.685
11	34.0	0.921	0.538	0.684
12	37.2	0.920	0.538	0.683
13	40.5	0.918	0.537	0.681
14	43.7	0.917	0.537	0.679
15	47.0	0.914	0.536	0.677
16	50.2	0.912	0.536	0.675
17	53.5	0.910	0.535	0.672
18	56.8	0.907	0.534	0.670
19	60.0	0.904	0.533	0.668
20	63.3	0.895	0.531	0.665
21	67.1	0.897	0.534	0.668
22	71.3	0.901	0.540	0.673
23	75.5	0.905	0.545	0.678
24	80.0	0.910	0.551	0.685
25	85.1	0.915	0.561	0.695

[a] Zerban and Sattler, *Ind. Eng. Chem., Anal. Ed.* 10, 669 (1938).

The reducing effect on the copper tartrate reagent has been measured for dextrose, levulose, maltose, and lactose by Zerban and Sattler. The results are shown in Table 42, which gives the milligrams of dextrose corresponding to varying milliliters of 0.1 N thiosulfate solution, and the values of the factors b, c, and d (reducing ratios of levulose, maltose, hydrate, and lactose hydrate with respect to dextrose) in equation 4. These reducing ratios are very close to those for the Munson and Walker Method, as would be expected.

TABLE 43. CORRECTIONS [a]

(To be applied to milliliters of thiosulfate found, for varying quantities of maltose or lactose present in addition to dextrose or levulose)

Dex-trose $Mg.$	Maltose Hydrate and Dextrose			Levu-lose $Mg.$	Maltose Hydrate and Levulose		
	200 mg. maltose	100 mg. maltose	50 mg. maltose		200 mg. maltose	100 mg. maltose	50 mg. maltose
	Correction, Per Cent of Ml. Thiosulfate				Correction, Per Cent of Ml. Thiosulfate		
0	100.0	100.0	100.0	0	100.0	100.0	100.0
5	79.7	72.5	63.0	5	75.3	60.4	41.3
10	61.5	47.0	34.2	10	56.5	39.0	24.1
15	46.5	31.1	20.5	15	41.2	25.0	13.7
20	34.5	20.8	11.0	20	28.6	17.8	8.8
25	25.8	14.4	6.0	25	20.4	12.0	5.9
30	19.6	10.3	3.4	30	13.8	8.3	3.5
35	15.4	7.9	2.3	35	9.2	5.2	2.4
40	12.9	6.9	2.2	40	6.5	3.2	1.6
45	11.2	6.3	2.0	45	5.0	2.4	1.5
50	9.8	5.7	1.9	50	4.5	2.2	1.4
55	8.8	5.1	1.7	55	4.1	2.1	1.3
60	7.8	4.6	1.6	60	3.7	1.9	1.2
65	6.8	4.0	1.4	65	3.3	1.8	1.1
70	5.7	3.5	1.3	70	2.9	1.6	1.0
75	4.7	3.1	1.1	75	2.5	1.5	0.9
80	3.7	2.4	1.0	80	2.1	1.3	0.8

	Lactose Hydrate and Dextrose				Lactose Hydrate and Levulose		
	200 mg. lactose	100 mg. lactose	50 mg. lactose		200 mg. lactose	100 mg. lactose	50 mg. lactose
0	100.0	100.0	100.0	0	100.0	100.0	100.0
5	71.5	62.6	54.5	5	60.7	34.4	19.0
10	47.3	30.5	22.6	10	39.8	19.3	11.7
15	29.5	16.0	9.7	15	26.4	11.2	6.6
20	19.0	7.6	3.4	20	15.8	6.6	2.1
25	13.4	4.3	1.5	25	10.3	3.8	1.0
30	9.4	2.8	1.4	30	5.6	2.0	0.9
35	7.2	2.5	1.3	35	2.4	0.9	0.8
40	6.4	2.4	1.2	40	1.0	0.8	0.8
45	6.1	2.2	1.1	45	0.9	0.8	0.7
50	5.7	2.1	1.1	50	0.8	0.7	0.6
55	5.4	1.9	1.0	55	0.7	0.6	0.5
60	5.0	1.7	0.9	60	0.6	0.5	0.5
65	4.7	1.5	0.8	65	0.6	0.4	0.4
70	4.4	1.4	0.7	70	0.5	0.4	0.3
75	4.0	1.2	0.6	75	0.4	0.3	0.2
80	3.7	1.0	0.5	80	0.3	0.2	0.1

[a] Zerban and Sattler, *Ind. Eng. Chem., Anal. Ed.* 10, 669 (1938).

Calculation.—The results of the analyses are calculated by a series of approximations which are continued until two successive calculations give practically identical results. An example follows. Assume the test solution contains in each 100 ml.

> 30 mg. dextrose,
> 200 mg. levulose,
> 690 mg. maltose hydrate,
> 90 mg. lactose hydrate.

(1) The first check analysis by the mucic acid method or the fermentation method for the determination of the lactose, 100 ml. portions of the solution being taken, gives an average of 14.6 mg. mucic acid, corresponding, according to van der Haar's table, to 38.75 mg. galactose, which multiplied by 2.2, is equivalent to 85.2 mg. lactose in 100 ml., or 8.5 mg. in 10 ml.

(2) Ten ml. of solution gave 0.0728 g. Cu with the Jackson and Mathews reagent; this corresponds to 23.3 mg. apparent levulose (R_1).

(3) Ten ml. solution gave a titration value of 11.6 ml. 0.1 N thiosulfate solution with Steinhoff's copper acetate reagent, equivalent to 25.0 mg. levulose (R_2); and of 20.4 ml. thiosulfate solution with Steinhoff's copper tartrate reagent, equivalent to 64.8 mg. dextrose (R_3).

Hence,

> $L = 8.5$ mg.
> $R_1 = 23.3$ mg.
> $R_2 = 25.0$ mg.; $a = 0.819$
> $R_3 = 64.8$ mg.; $b = 0.896, c = 0.532, d = 0.666$

First approximation:

$$G = \frac{25.0 - 23.3}{0.819 - 0.081} = 2.3 \text{ mg.}$$

$$F = 25.0 - (2.3 \times 0.819) = 23.1 \text{ mg.}$$

Percentage ratio of G to F is as 9 to 91; $a = 0.813$

$$G = \frac{25.0 - 23.3}{0.813 - 0.081} = 2.3 \text{ mg.}$$

$$F = 25.0 - (2.3 \times 0.813) = 23.1 \text{ mg.}$$

$$M = \frac{64.8 - [2.3 + (23.1 \times 0.896) + (8.5 \times 0.666)]}{0.532} = 67.9$$

Result of first approximation, for 10 ml. of solution:

$$G = 2.3 \text{ mg.}$$
$$F = 23.1 \text{ mg.}$$
$$M = 67.9 \text{ mg.}$$
$$L = 8.5 \text{ mg.}$$

Second approximation, correction to R_1:

67.9 mg. M equivalent to 67.9:26	= 2.61 mg. F
8.5 mg. L equivalent to 8.5:25.6	= 0.33 mg. F
Total correction	2.94 mg. F
Corrected R_1 = 23.3 − 2.9	= 20.4 mg.

Correction to R_2:

Total dextrose plus levulose.	= 25.4 mg.
Total maltose plus lactose	= 76.4 mg.

	Correction (Table 43) %
25 F + 76 M	9
25 F + 76 L	2.5
25 G + 76 M	10
25 G + 76 L	2.5
25 F + 68 M gives 9 × 68:76	8.05
25 F + 8 L gives 2.5 × 9:76	0.30
25 F + 68 M + 9 L	8.35
25 G + 68 M gives 10 × 68:76	8.95
25 G + 8 L gives 2.5 × 9:76	0.30
25 G + 78 M + 9 L	9.25
23 F + 68 M + 9 L gives 8.35 × 23:25	7.68
2 G + 68 M + 9 L gives 9.25 × 2:25	0.74
23 F + 2 G + 68 M + 9 L	8.42

Corrected thiosulfate titer, Steinhoff acetate reagent, is 11.6 − (11.6 × 0.0842) = 10.6 ml.

Equivalent corrected R_2 = 23.2; a = 0.844

$$G = \frac{23.2 - 20.4}{0.844 - 0.081} = 3.7 \text{ mg.}$$

$$\jmath = 23.2 - (3.7 \times 0.844) = 20.1 \text{ mg.}$$

Percentage ratio of G to F is as 15.5 to 84.5; $a = 0.835$

$$G = \frac{23.2 - 20.4}{0.835 - 0.081} = 3.7 \text{ mg.}$$

$$F = 23.2 - (3.7 \times 0.835) = 20.1 \text{ mg.}$$

$$M = \frac{64.8 - [3.7 + (20.1 \times 0.896) + (8.5 \times 0.666)]}{0.532} = 70.3 \text{ mg.}$$

Result of second approximation, in 10 ml. of solution:

$$G = 3.7 \text{ mg.}$$
$$F = 20.1 \text{ mg.}$$
$$M = 70.3 \text{ mg.}$$
$$L = 8.5 \text{ mg.}$$

Third approximation, correction to R_1:

70.3 mg. M equivalent to 70.3:26	=	2.70 mg. F
8.5 mg. L equivalent to 8.5:25.6	=	0.33 mg. F
Total correction:		3.03 mg. F

Corrected $R_1 = 23.3 - 3.0$ = 20.3 mg.

Correction to R_2:

Total dextrose plus levulose = 23.8 mg.
Total maltose plus lactose = 78.8 mg.

	Correction %
24 F + 79 M	10.5
24 F + 79 L	3.5
24 G + 79 M	12.0
24 G + 79 L	3.5
24 F + 70 M gives 10.5 × 70:79	9.3
24 F + 9 L gives 3.5 × 9:79	0.4
24 F + 70 M + 9 L	9.7
24 G + 70 M gives 12.0 × 70:79	10.6
24 G + 9 L gives 3.5 × 9:79	0.4
24 G + 70 M + 9 L	11.0
20 F + 70 M + 9 L gives 9.7 × 20:24	8.1
4 G + 70 M + 9 L gives 11.0 × 4:24	1.8
20 F + 4 G + 70 M + 9 L	9.9

The corrected thiosulfate titer, Steinhoff acetate reagent, is 11.6 − (11.6 × 0.099) = 10.45 ml.

Equivalent corrected $R_2 = 22.9$; $a = 0.848$

$$G = \frac{22.9 - 20.3}{0.848 - 0.081} = 3.4 \text{ mg.}$$

$$F = 22.9 - (3.4 \times 0.848) = 20.0 \text{ mg.}$$

Percentage ratio of G to F is as 14.5 to 85.5; $a = 0.839$

$$G = \frac{22.9 - 20.3}{0.839 - 0.081} = 3.4 \text{ mg.}$$

$$F = 22.9 - (3.4 \times 0.839) = 20.0 \text{ mg.}$$

$$M = \frac{64.8 - [3.4 + (20.0 \times 0.896) + (8.5 \times 0.666)]}{0.532} = 71.1 \text{ mg.}$$

Result of third approximation:

$$G = \ \ 3.4 \text{ mg.}$$
$$F = 20.0 \text{ mg.}$$
$$M = 71.1 \text{ mg.}$$
$$L = \ \ 8.5 \text{ mg.}$$

These values agree so closely with those obtained in the second approximation that further calculation is unnecessary. The final results are therefore:

	Taken Mg.	Found Mg.
Dextrose	3.0	3.4
Levulose	20.0	20.0
Maltose hydrate	69.0	71.1
Lactose hydrate	9.0	8.5

COMMERCIAL GLUCOSE

Commercial glucose or corn syrup is the product made from cornstarch by incomplete hydrolysis. Other terms used for the thick, syrupy, colorless products made by incompletely hydrolyzing starch or a starch-containing substance, and decolorizing and evaporating to a proper consistency are glucose, mixing glucose, and confectioner's glucose. Materials other than cornstarch are used for the preparation of such syrups, but, since the greater portion of this type of syrup is made from corn, it is common to use the name corn syrup.

Commercial glucose is not a product of definite composition. It is a mixture of dextrins, including higher dextrins such as erythrodextrin, and the sugars dextrose and maltose. Much of the commercial glucose sold is not water-white, but is colored by the addition of other syrups and materials such as refiners' syrup, sorgo syrup, and molasses.

The terminology and nomenclature given to this and allied products are confusing. There is the chemical substance known as D-glucose, discussed in a prior section of this chapter, and commonly termed dextrose. The definite chemical substance D-glucose should be clearly differentiated in the mind of the analyst from the indefinite material called commercial glucose.

Fiehe Test.—Commercial glucose may be detected in food products by the following qualitative test developed by Fiehe.[32] Dilute or dissolve 10 g. of the sample with 10 ml. of water, warming if necessary to effect solution. Add 10 ml. of a saturated solution of ammonium oxalate and boil. Add animal charcoal and boil again. Filter and place 2 ml. of the clear filtrate in a test tube, add 2 drops of hydrochloric acid and 20 ml. of 95 per cent alcohol and mix. In the presence of commercial glucose or dextrins a marked white turbidity will be formed. This turbidity is due to the precipitated dextrins.

Another test for commercial glucose and better than the Fiehe test for products containing much protein material is the test devised by Jacobs and detailed in Chapter VIII in the malted milk section.

Determination.—The specific rotation of levulose decreases with increasing temperature until at 87° C. it just equals that of the dextrorotatory dextrose, hence in a solution of sugars which has been inverted containing commercial glucose, the commercial glucose will be the only polarizing sugar. The temperature at which exact neutralization occurs varies somewhat with the concentration of the mixture but 87° C. is the accepted temperature. The A. O. A. C. gives the following two methods[33] for the estimation of commercial glucose.

Method 1 (for substances containing little or no invert sugar).— Commercial glucose cannot be determined accurately owing to the varying quantities of dextrin, maltose, and dextrose present in the product. However in syrups in which the quantity of invert sugar is so small as not to affect appreciably the result, commercial glucose may be estimated approximately by the following formula:

$$G = \frac{(a - S)100}{211}$$

[32] Fiehe, *Z. Nahr. Genussm.* **8**, 30 (1909).
[33] *Methods of Analysis*, A.O.A.C. Washington, 1945.

in which

G = percentage of commercial glucose solids
a = direct polarization, normal solution
S = percentage of sucrose.

Express the results in terms of commercial glucose solids polarizing +211° V.

Method 2 (for substances containing invert sugar).—Prepare an inverted half-normal solution of the substance as directed in the section "Sucrose by Polarization before and after Inversion with Hydrochloric Acid," by the rapid method, except cool the solution after inversion, make neutral to phenolphthalein with sodium hydroxide solution, slightly acidify with hydrochloric acid (1:5), and treat with 5-10 ml. of alumina cream before making up to the mark. Filter and polarize at 87° C. in a 200-mm. jacketed metal tube, preferably silver. Multiply the reading by 200 and divide by the factor 196 to obtain the quantity of commercial glucose solids polarizing +211° V.

HONEY

Honey is the nectar and saccharine exudations of plants gathered, modified and stored in the comb by honeybees. Honey is levorotatory and should not contain more than 25 per cent water, 0.25 per cent ash, and 8 per cent of sucrose. The essential components of honey are dextrose, levulose, and sucrose in small amounts together with lesser quantities of mineral matter, proteins, wax, pollen, and sometimes mannitol and dextrins. The composition varies with the feeding of the bees. The presence of more than 8 per cent of sucrose in honey indicates that the bees have been fed on cane sugar and that the honey is not matured or that it is adulterated. In general the invert sugar content is high, circa 60 per cent and, as the definition indicates, honey is levorotatory before and after inversion. The composition of typical honey is tabulated in Tables 44 and 45.

Preparation of Sample

If the sample is clear and appears to be homogeneous, mix thoroughly and it is ready for analysis. If part of the material has crystallized out, the entire sample should be warmed to about 40° C. and stirred until the crystals dissolve and, if parts of the comb or other foreign materials are present, the mixture should be strained through cheese cloth or strained by some similar method.

Total solids, ash, soluble ash, alkalinity of the ash, color, acidity, metals and preservatives may be determined and detected in the usual manner.

TABLE 44. HONEY, LEVOROTATORY [a] (UNITED STATES)

	Maximum	Minimum	Average
Polarizations (°V) Direct			
Immediate	—21.90	+ 3.70	—11.24
Constant	—24.80	— 0.30	—14.73
Birotation	11.60	1.40	3.49
at 87°C	+23.70	+ 0.50	+10.15
Invert			
at 20°C	—29.26	— 1.32	—19.16
at 87°C	+23.21	+ 0.66	+ 7.91
Difference	33.55	23.32	27.07
Complete Analysis			
Water %	26.88	12.42	17.70
Invert Sugar %	83.36	62.23	74.98
Sucrose %	10.01	0.00	1.90
Ash %	0.90	0.03	0.18
Dextrin %	7.58	0.04	1.51
Undetermined %	7.45	0.04	3.73
Free Acid as Formic %	0.25	0.04	0.08

[a] Bryan, *U. S. Dept. Agr., Chem. Bull.* **154** (1912).

TABLE 45. HONEY [a] (CUBAN, MEXICAN, HAITIAN)

	Maximum	Minimum	Average
Polarizations (°V) Direct			
Immediate	—22.90	— 6.05	—13.34
Constant	—24.15	— 8.50	—14.52
Birotation	3.55	0.00	1.18
at 87°C	+17.00	+ 3.20	+10.31
Invert			
at 20°C	—26.07	— 8.86	—16.22
at 87°C	+15.40	+ 2.86	+ 9.08
Difference	28.93	22.77	25.30
Complete Analysis			
Water %	27.00	16.05	21.26
Invert Sugar %	77.56	68.09	72.38
Sucrose %	3.98	0.00	0.80
Ash %	0.58	0.06	0.21
Dextrin %	3.96	0.26	1.24
Undetermined %	8.07	0.66	4.11
Free Acid as Formic %	0.43	0.00	0.15

[a] Bryan, *U. S. Dept. Agr., Bur. Chem. Bull.* **154** (1912).

POLARIZATION

Weigh 26 g. of the sample, the normal weight, and transfer to a 100-ml. sugar volumetric flask with water. Add 5 ml. of alumina cream for clarification purposes, make to the mark at 20° C. and filter. Discard the first portion and polarize immediately in a 200-mm. tube at the customary temperature. This is called the immediate direct polarization. Allow to stand overnight or add anhydrous sodium carbonate until distinctly alkaline or by adding a few drops of ammonium hydroxide solution before making to volume, and again polarize. This is termed the constant direct polarization. The difference between the immediate direct polarization and constant direct polarization is the measure of the mutarotation. Polarize again at 87° C. to obtain the direct polarization at 87°.

Invert 50 ml. of the filtrate as directed in the section on inversion of sucrose and polarize at 20° and at 87° C. to obtain the corresponding invert polarizations.

REDUCING SUGARS

Reducing sugars may be determined before inversion by either the Lane-Eynon method or by the Munson-Walker method. Dilute 10 ml. of the filtrate from the direct polarization determination to 250 ml. and 25 ml. of this solution may be taken for the estimation, in the case of the Munson and Walker method. The result is calculated to percentage invert sugar with Table 1, appendix.

Reducing sugars after inversion may be determined by diluting 10 ml. of the solution used for the invert polarization with water, neutralizing with sodium carbonate, and then diluting to 250 ml. Fifty ml. of this solution is used for the determination of reducing sugars by the Munson and Walker method.

CALCULATION OF SUGAR PERCENTAGES

Sucrose may be calculated from the polarization results by means of formula [1], page 445. Rather better results are obtained by calculating sucrose from the reducing sugar determinations because of the error attributable to the change in optical rotation of levulose before and after inversion.

% invert sugar after inversion − % invert sugar before inversion × 0.95 = % sucrose.

Because of the large amount of levulose and its unequality to the amount of dextrose present, the polarization is not null as would be true in the case of inverted sucrose. Furthermore because of this reason, commercial glucose cannot be detected by the 87° C. polarization method. The percentage levulose may be calculated as follows:

$$L = \frac{a - 1.0315b}{2.3919}$$

where

a = constant direct polarization at 20° C.
b = direct polarization at 87° C.
L = g. of levulose in normal weight of honey (26 g.)
and $L/26 \times 100$ = % levulose.

Levulose may also be estimated by the Fischl method [34] and by the Jackson and Mathews [35] method detailed in this chapter.

DEXTROSE

The percentage of dextrose is equivalent to the percentage of invert sugar less the percentage of levulose.

The percentage of dextrose can also be obtained by use of the formula

$$D = R - (L \times 0.915)$$

in which

R = reducing sugars calculated as dextrose
L = percentage of levulose
D = percentage of dextrose.

Multiplication of the percentage of levulose by the factor 0.915 gives its dextrose equivalent in copper-reducing power. The sum of the dextrose calculated in this way and the levulose determined will be greater than the quantity of invert sugar estimated as reducing sugar.

COMMERCIAL GLUCOSE

Since pure honey contains dextrins, no conclusion can be derived from the mere presence of dextrins. However, in the case of honey the dextrins have been hydrolyzed to such a degree that erythrodextrin is not a normal component. On the other hand, in the case of commercial glucose, the hydrolysis seldom goes so far that all of the erythrodextrin is gone. This is the basis for the detection of commercial glucose in honey.

Erythrodextrin Test.—Dilute some honey with an equal portion of water and add a few milliliters of a solution of 1 g. of iodine and 3 g. of potassium iodide in 50 ml. of water. Run a control with a known sample of honey, adding the same amount of water and reagent. The production of a red or violet color indicates the presence of commercial glucose.

If the honey is colored very deeply so that the red or violet color is obscured, precipitate the dextrins by the addition of sufficient alcohol (several volumes). Allow to stand, decant the supernatant solution, dissolve the precipitate in hot water, cool, and add the iodine solution. The absence of color does not conclusively prove the absence of commercial glucose.

---

[34] Fischl, *Chem. Ztg.* 57, 393 (1933).
[35] Jackson and Mathews, *Bur. Standards J. Research* 8, 403 (1932).

Quantitative Estimation.—The amount of commercial glucose in honey can be calculated by the following formula.

$$\left[\left(\frac{\text{Polarization of invert solution at 20° C.} - \text{Polarization of invert solution at 87° C.}}{\%\text{ Invert sugar after inversion}} \times 77\right)\right]\frac{100}{26.7} = \%\text{ honey}$$

$$100 - \%\text{ honey} = \%\text{ commercial glucose}$$

DEXTRIN

Transfer an accurately weighed portion of from 4-8 g. of honey, depending on the depth of color, with the aid of a Mojonnier pipette and carriage to a 100-ml. volumetric flask. Add 4 ml. of water, 1 ml. at a time, and shake so that the honey dissolves, and then fill to the mark with absolute alcohol, shaking vigorously during the addition of the alcohol. Allow the precipitate to settle completely and decant the supernatant solution through a filter. Wash the precipitate with 10 ml. of 95 per cent alcohol and pour the washings on the same filter. Dissolve the dextrins in the flask with several portions of hot water, passing the water through the filter and catching the filtrate in a tared dish containing dry sand. Evaporate to a small volume and dry to constant weight *in vacuo*. Determine the reducing sugars before and after inversion. Calculate the weight of the invert sugar before inversion and the sucrose and subtract from the weight obtained by drying the filtrate *in vacuo*. This gives the weight of dextrin and dividing by the weight of the original sample taken gives the percentage dextrin. This is an approximate determination.

COMMERCIAL INVERT SUGAR

In the manufacture of invert sugar commercially the invert sugar produced is contaminated with a small amount of furfural or its derivatives. The detection of these substances form the basis for the test for presence of commercial invert sugar in honey.

Fiehe Test.—Transfer 5 ml. of honey to a test tube and add an equal amount of water. Add 5 ml. of ether and shake gently. Allow the layers to separate. Transfer 2 ml. of the clear ether layer to another test tube and then add 1 drop of a recently prepared solution of 1 g. resorcinol in 100 ml. of hydrochloric acid. The production of an orange red to dark red color indicates the presence of commercial invert sugar.[36] Yellow to salmon pink colors are to be disregarded.

Aniline Test.—Another test may be made as follows: Mix 5 ml. of honey with 2.5 ml. of a mixture of 10 ml. of aniline and 3 ml. of

[36] Fiehe, *Z. angew. Chem.* 21, 2315 (1908).

25 per cent hydrochloric acid. The immediate production of a bright red color indicates the presence of commercial invert sugar.[37]

A negative result with these tests for commercial invert sugar does not prove that commercial invert sugar is absent.

DIASTASE

Pure honey if not heated to a temperature high enough to destroy it, contains the enzyme, diastase.

A. O. A. C. Test.—Mix 1 part of honey with 2 parts of sterile water. Treat 10 ml. of this solution with 1 ml. of 1 per cent soluble starch solution and digest at 45° C. for an hour. At the end of this time test the mixture with 1 ml. of iodine solution (1 g. of iodine, 2 g. of potassium iodide, 300 ml. of water). Treat another 10-ml. portion of the honey solution, mixed with 1 ml. of the soluble starch solution without heating to 45° with the reagent and compare the colors produced. If the original honey has not been heated sufficiently to destroy the diastase, an olive-green or brown coloration will be produced in the mixture that has been heated at 45°. Heated or artificial honey becomes blue.

MAPLE PRODUCTS

Maple syrup is the product obtained by the evaporation of the sap obtained by tapping the hard or rock maple. It may also be made by the solution of maple sugar. It should not contain more than 35 per cent water. Maple sugar is the solid product resulting from the evaporation of maple sap or maple syrup. The syrup should be concentrated until it weighs 11 lbs. to the gallon. This weight corresponds to a specific gravity of 1.325.

The principal component of maple syrup is sucrose. The syrup and the solution made from the sugar contain mineral matter, proteins, organic acids, and flavoring materials which give it its characteristic flavor and odor. As in the case of fats and oils and as in jams and jellies, maple products have certain constants that are characteristic. The constants which are most indicative of the purity of these materials are according to Snell and Scott[38] refractometer reading, conductivity value, total ash, alkalinity of the ash, Winton lead number, and the Canadian lead number. Ratios set up on the basis of these constants aid in the interpretation of the results.

---

[37] Browne, *U. S. Dept. Agr., Bur. Chem. Bull.* 110, Rev. (1916). *Methods of Analysis,* A.O.A.C. Washington, 1945.

[38] Snell and Scott, *Ind. Eng. Chem.* 6, 219 (1914).

Preparation of Sample

Because of the nature of the determinations, the samples of maple products must be prepared according to rigid specifications. The A. O. A. C.[39] gives these official details.

Maple Syrup.—(1) *For Solids Determination.*—If the sample contains no sugar crystals or suspended matter, decant sufficient of the clear syrup for use in the determination. If sugar crystals are present, redissolve them by heating. If suspended matter is present, filter the sample through cotton wool.

(2) *For Other Determinations.*—If sugar crystals are present, redissolve them by heating. If other sediment is present, distribute it evenly through the syrup by shaking. Transfer approximately 100 ml. of the syrup, with its suspended sediment, to a casserole or beaker, add ¼ the volume of water, and evaporate over a flame. When the temperature of the boiling syrup approaches 104° C., draw a small quantity into a thin-walled pipette of about 1-ml. capacity and cool to room temperature, in running water. Wipe the outside of the pipette, allow the possibly diluted syrup in the point to escape and make a refractometer measurement of the solids content of the cooled syrup. Repeat the procedure from time to time until a reading is obtained corresponding to 64.5 per cent solids ($n_{20} = 1.4521$), or to such other value as in the experience of the analyst will give a filtered syrup of 65.0 per cent solids. Filter the syrup through a filter which will allow the 100 ml. to pass within 5 minutes and adjust the filtrate to 65.0 ± 0.5 per cent solids (refractometric) by thorough mixing with the appropriate quantity of water.

Maple Sugar and Other Solid or Semi-solid Products.—(1) *For Moisture and Solids Determination.*—Grind in a mortar, if necessary, and mix thoroughly.

(2) *For Other Determinations.*—Prepare a syrup by dissolving approximately 100 g. of the sample in 150 ml. of hot water, boil until the temperature approaches 104° C. and complete the preparation of the resulting syrup as directed above, commencing at "draw a small quantity into a thin-walled pipette."

Using the sample prepared as directed the usual determinations may be made for moisture, solids, ash, soluble and insoluble ash, alkalinity of the ash, polarization, and sugars before and after inversion. Commercial glucose may be detected as previously described.

---

[39] *Methods of Analysis,* A.O.A.C. Washington, 1945.

LEAD NUMBER

When basic lead acetate is added to a maple syrup or to a solution of maple sugar, a precipitate is formed due to the insoluble salts of malic and other organic acids and also of sulfates, chlorides, and proteins. On the other hand, solutions of pure cane sugar do not yield a similar type of precipitate. The amount of lead remaining in solution after the addition of basic lead acetate solution to a maple syrup or solution of maple sugar is a measure of the amount of precipitable material. This is the basis of the following test for the estimation of the Winton lead number. The Canadian lead number is based on the weight of the lead precipitate. Since they are empirical methods, the official directions of the A. O. A. C.[40] are given. They should be followed in detail.

Canadian Lead Number (Fowler Modification).—*Reagent.*—Standard Basic Lead Acetate Solution.—Activate litharge by heating it to 650-670° C. for 2.5-3 hours in a muffle. The cooled product should be a lemon color. In a 500-ml. Erlenmeyer flask provided with a return condenser boil 80 g. of normal lead acetate crystals and 40 g. of the freshly activated litharge with 250 g. of water for 45 minutes. Cool, filter off any residue, and dilute with recently boiled water to a density of 1.25 at 20° C.

Procedure.—Weigh the quantity of syrup containing 25 g. of dry matter, transfer to a 100-ml. flask, and make up to mark at 20° C. Pipette 20 ml. into a large test tube, add 2 ml. of the standard basic lead acetate solution, cork, and allow to stand 2 hours.

Filter with suction on a 25-ml. tared Gooch, having an asbestos mat at least 3-mm. thick. When nearly all the liquid has run through, fill the crucible with cold water. Repeat to a total of 4 washings, taking care to prevent formation of fissures in the precipitate by keeping it covered with water and avoiding too great suction. Dry at 100° C., weigh, and multiply the weight by 20.

Winton Lead Number.—*Reagent.*—Standard Basic Lead Acetate Solution.—To a measured volume of the reagent prepared for the determination of the Canadian lead number, add 4 volumes of water and filter. A blank should be run with each set of determinations.

Determination of Lead in the Blank.—Transfer 25 ml. of the standard basic lead acetate solution to a 100-ml. flask, add a few drops of glacial acetic acid, and make up to the mark with water. Shake and determine lead sulfate in 10 ml. of the solution as directed immediately below. The use of acetic acid is imperative in order to retain all the lead in solution when the reagent is diluted with water.

---

[40] *Methods of Analysis,* A.O.A.C. Washington, 1945.

Procedure.—Transfer 25 g. of the sample to a 100-ml. flask by means of water. Add 25 ml. of the standard basic lead acetate solution and shake. Fill to the mark, shake, and allow to stand for at least 3 hours before filtering. Pipette 10 ml. of the clear filtrate into a 250-ml. beaker, add 40 ml. of water and 1 ml. of sulfuric acid, shake, and add 100 ml. of 95 per cent alcohol. Allow to stand overnight, filter on a weighed Gooch crucible, wash with 95 per cent alcohol, dry in a water oven, and ignite in a muffle or over a Bunsen burner, applying the heat gradually at first and avoiding a reducing flame. Cool, and weigh. Subtract the weight of lead sulfate so found from the weight of lead sulfate found in the blank, and multiply by the factor 27.33. The use of this factor gives the lead number directly without the various calculations otherwise required.

Sy Lead Number.—A simple method based on the volume of the precipitate produced on the addition of the standard basic lead acetate solution is that developed by Sy.[41]

Procedure.—In a 25-ml. graduated cylinder introduce 5 ml. of syrup or 5 g. of sugar which is afterward dissolved in a little water. Add water to the 15-ml. mark and 2 ml. of the basic lead acetate solution. Shake thoroughly and allow the mixture to stand for 20 hours. Then read the volume of the precipitate, which for pure maple products should be at least 3 ml. and is usually over 5 ml. This method is not as accurate as the preceding methods.

Cowles [42] Malic Acid Value

Weigh 6.7 g. of the sample into a 250-ml. beaker; add 5 ml. of water, 2 ml. of a 10 per cent calcium acetate solution, and stir. Add, gradually and with constant agitation, 100 ml. of 95 per cent alcohol and stir the solution until the precipitate settles, or allow it to stand until the supernatant liquid is clear. Filter off the precipitate and wash with 75 ml. of alcohol, 85 per cent by volume. Dry the filter paper and ignite in a platinum dish. Add 10 ml. of 0.1 N hydrochloric acid, and warm gently until all the lime dissolves. Cool, and back titrate with 0.1 N sodium hydroxide solution, using methyl orange indicator solution. The difference in milliliters divided by 10 represents the malic acid value of the sample. Previous to use, the reagents should be tested by a blank determination and any necessary corrections applied.

Rapid Conductivity Method

The detection of adulterated maple syrup may be rapidly and

---

[41] Sy, *J. Am. Chem. Soc.* 30, 1430 (1908).
[42] Cowles, *J. Am. Chem. Soc.* 30, 1285 (1908).

easily performed by a determination of the conductivity of the syrup.[43] Because of the varying nature of maple products, it is impossible to define the two limits between which the conductivity values will fall. Conlin [44] on the basis of over 7,200 samples has adopted for the following method, primarily designed to detect adulteration with white sugar, a minimum value of 100 and values may range up to 200.

The apparatus consists of a Leeds and Northrup sugar ash bridge and conductivity cell, Fig. 66. No constant temperature bath is needed for a knob on the bridge provides for variations in tempera-

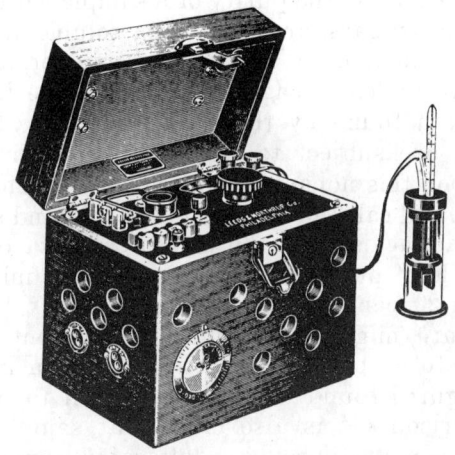

Fig. 66. Sugar Ash Bridge. (Courtesy of Leeds and Northrup.)

ture from 10-35° C. and variations in cell constant from 0.14 to 0.16 and all results are automatically corrected to 20° C.

Procedure.—Place 75 ml. of water in a 100-ml. graduated cylinder, add 25 ml. of syrup, and mix thoroughly. Rinse off the electrodes with approximately 40 ml. of this solution, pour the remainder into the testing cylinder, and insert the electrodes and thermometer. Adjust the temperature compensating knob on the bridge to correspond with the cell reading, place the resistance plug in the proper setting (usually the 1000-ohm block), press the alternating current button, and turn the main slide-wire knob until the galvanometer balances. Take the dial reading. The product of this value and the plug setting is the specific conductance multiplied by 10^6 and reduced to 20° C.

[43] Snell, *Ind. Eng. Chem.* **5**, 740 (1913).
[44] Conlin, *Ind. Eng. Chem.*, *Anal. Ed.* **7**, 426 (1935).

Since the conductivity value is defined by 10^5, the product is 10 times the desired figure.

Measurements at 25° C., which is the temperature required for the official A. O. A. C. method, average about 13 per cent higher than those made at 20° C. The average conductivity value of Canadian syrups is higher than that of American syrups. It is best to confirm doubtful values by the other methods given.

Interpretation of Results, Maple Products

Of all the various analyses for maple products, those used most frequently in determining the purity of a sample with respect to adulteration with other sugars, are: total ash, soluble and insoluble ash, malic acid value, Canadian or Winton lead number, and conductivity.

There is more or less disagreement about the interpretation of the analytical data found by the above methods. Maple products, being unrefined, are subject to wide variations in their nonsugar content. It is upon this nonsugar content, which consists principally of the potassium and calcium salts of malic acid and smaller amounts of similar organic acids, that the analytical data depend. For example, one sample of pure maple sugar or syrup might have a total ash content of 1.20 per cent (calculated to a dry basis), while another, equally pure, might have a total ash content of only 0.80 per cent. It is therefore obvious that, in judging an unknown sample, the minimum figures found in samples known to be pure must be used for comparison. It is also clear that samples showing high analytical data can at times be adulterated considerably without detection.

In Table 46 are tabulated the analyses of typical maple sugars and syrups. For the purpose of the discussion the minimum Vermont figures established by Jones [45] are given: Total Ash = 0.77 per cent, Insoluble Ash = 0.23 per cent, Malic Acid Value = 0.61. These figures are calculated on a moisture free basis. No definite minimum figures have been agreed on for lead numbers and none have been adopted for conductivity value, which is relatively new. (See conductivity section.)

In general, Canadian minimum figures are lower than those of Vermont, and light colored, mild flavored syrups will give lower analytical data than the darker strong syrups. There are, of course, frequent exceptions, but this is always true when composite samples representing very large amounts of syrup are used. Below is a repre-

---

[45] Jones, *Vt. Agr. Exptl. Sta., 17th and 18th Annual Repts.* (1904-5).

sentative analysis of maple sugar made from a mixture of all grades of syrup.[46] It is an average analysis representing 18,000 gal. of syrup. The data are calculated to a moisture-free basis:

Total Ash	0.935%
Soluble Ash	0.572%
Insoluble Ash	0.363%
Alkalinity of Sol. Ash	7.57 ml.
Alkalinity of Insol. Ash	9.77 ml.
Malic Acid Value	0.883
Canadian Lead Number	3.71
Conductivity at 20° C.	124.

TABLE 46. MAXIMA, MINIMA, AND AVERAGE OF ANALYSES OF UNITED STATES AND CANADA MAPLE SUGAR AND SYRUPS [a,b]

	Average (363)	Maximum	Minimum
Sucrose, Clerget %	90.69	98.62	57.04
Invert sugar %	6.19	35.26	0.00
Undetermined %	2.14	8.18	0.00
Total ash %	0.98	1.70	0.76
Soluble ash %	0.62	1.14	0.30
Insoluble ash %	0.36	1.00	0.21
Soluble ash/insoluble ash	1.69	4.07	0.43
Lead number Winton	3.50	5.90	2.20
Malic acid value	0.93	1.72	0.51
Alkalinity			
Soluble ash, ml.	75	140	42
Insoluble ash	87	190	31
Soluble ash/insoluble ash	0.86	2.29	0.37

[a] Bryan, *U. S. Dept. Agr., Bull.* 466 (1917).
[b] Moisture free basis.

CONFECTIONERY

Candy [47] is a food composed of one or more of the following: Sugar (including all acceptable grades and types), syrup of cane juice, honey, maple syrup, molasses, refiners' syrup, starches, corn syrup, other starch syrups, dextrose, lactose, levulose, maltose, cacao products, animal and vegetable products, such as edible oils and fats, eggs and egg products, fruit and fruit products, gelatin, milk and milk products, nuts and nut products, and other wholesome, nutritive substances; harmless acids, harmless coloring, flavoring,

---

[46] Conlin, Private Communication (1937).
[47] Jordan and Langwill, *Confectionery Analysis and Composition.* Manufacturing Confectioner, Chicago, 1946.

preservatives, natural gums and pectins; and harmless nonnutritive substances, incidental to good manufacturing practice that include resinous glaze, mineral oil, petroleum, and waxes, provided that the total weight of such substance or substances does not exceed 0.4 of 1 per cent of the confection weight in which it is found, dry basis; and further provided that less than 0.5 of the 1 per cent of the alcohol by weight, derived solely from the use of flavoring extracts, is a permissible ingredient.

Candy is divided into the following general classes:

Hard candy
Sugar creams (fondants)
Sugar lozenges
Fudge
Caramels including toffee
Marshmallow including whips
Nougat
Jellies and gums
Bon bons (coconut and plain)
Coated candies (chocolate and otherwise)

The many products that fall into this classification have to be analyzed according to the materials making the product. Thus chocolate confectionery and candy should be analyzed according to the methods for cocoa products. Sugared fruits may be treated as marmalade and jams. If the confectionery is composed of different portions, these portions may be separated and analyzed by those methods applicable to those portions. In general, mix or grind the product so as to obtain a homogeneous mass. On this well-mixed sample the usual determinations of moisture or total solids, ash, soluble ash, alkalinity of the ash, metals, color, titratable acidity and pH, preservatives, and protein nitrogen may be performed as described in other sections of the text. Sucrose may be estimated by polarimetric methods on a normal weight of the sample or by determining reducing sugars before and after inversion. Starches and gums may be estimated or detected as detailed in the next chapter. The usual test for commercial glucose may be made. Fat or total ether extract may be determined by the introduction of 4-5 g. of the sample into a Mojonnier extraction tube or Jacobs-Singer separatory flask and then proceeding with the Roese-Gottlieb method exactly as described for the determination of fat in ice cream.

Mineral fillers such as talc, chalk, calcium carbonate may be detected by the methods detailed in Chapter XVIII.

MOISTURE

A method especially designed for the determination of moisture in fondant, hard candy, and fudge is described by Jordan and Langwill. Acetone is used to assist in liberating the moisture.[48]

Procedure.—Weigh approximately 5 g. or 10 g. when the moisture is less than 3 per cent into a 100-ml. beaker after having obtained the tare of the empty beaker along with a short, flat-bottom stirring rod. Add 50 ml. of anhydrous acetone and mash up the sample with the rod. Hard candy must be powdered or broken into fine pieces before sampling. After thorough stirring and disintegration, pour off the clear acetone extract into a second tared beaker and evaporate the extract on a water bath, being careful to prevent bumping.

Rewash the original sample with 4 successive portions of anhydrous acetone, using 50 ml. each for the first two washings and 25 ml. for the last two. Decant the successive extracts, as made, into the second beaker and evaporate concurrently with subsequent extractions of the original sample. After the final extraction, place the first beaker on the water bath to expel the acetone and then dry in an air oven at 105° C. to remove any residual moisture. Dry the second beaker in the oven at the same time and in the same way. Two hours is generally sufficient time to complete drying.

The combined percentages of the acetone residue in the first beaker and the extracted substance in the second beaker subtracted from 100 per cent give the loss attributable to moisture, provided appreciable quantities of other volative materials such as flavors are not present. If such materials are present, allowance may be made by use of a modification of the methods detailed in Chapter XIV.

When large quantities of corn or malt syrup are used in the preparation of the candy, fondant, or fudge, this method does not work too well.

DEXTROSE

For determining dextrose directly in the presence of sucrose and invert sugar, the following method may prove satisfactory.

Procedure.—Invert and polarizé a 0.5 normal solution of the sample at 87° C. as detailed in this chapter. Multiply the polaroscopic reading obtained by the factor 2.71 to get the per cent added dextrose on an anhydrous basis.

---

[48] Jordan and Langwill, *Confectionery Analysis and Composition.* Manufacturing Confectioner, Chicago, 1946.

The factor is derived as follows. The specific rotations for 0.5 normal solutions are:

$$[a]_D^{20} \text{ sucrose } = 66.51$$

$$[a]_D^{20} \text{ dextrose } = 52.83$$

the ratio then is

$$\frac{\text{Sucrose}}{\text{Dextrose}} = \frac{66.51}{52.83} = 1.259$$

The estimated lowering of polarization attributable to the action of acid and to volume expansion at 87° C. can be corrected by dividing the observed polarization by 0.929. Hence, if P equals the observed polarization of a 0.5 normal inverted solution at 87° C., then

$$\frac{2P}{0.929} \times 1.259 = 2.71P$$

PARAFFIN

Saponify the extract in the fat flask obtained in the fat determination by the Roese-Gottlieb method by the addition of 10 ml. of 95 per cent alcohol and 2 ml. of sodium hydroxide solution (1:1). Heat on a steam bath under a reflux condenser until saponification is complete. Remove the condenser and evaporate off the alcohol. Transfer the material in the flask to a separatory funnel with the aid of hot water and extract when cool with 4 successive portions of petroleum ether. Collect the ether layers in a tared flask and evaporate to dryness. Place in an oven and dry to constant weight. An error is introduced due to the inclusion of the normal unsaponifiable matter present in the oil or fat used to make the confectionery, but this would be so small as to be negligible. Any appreciable unsaponifiable matter would indicate the presence of paraffin.

ALCOHOL

Break the candy in such a manner as to collect the syrup and weigh out 30-50 g. of the syrup. Transfer with the aid of 40 ml. water to a distillation flask or better a Florence flask connected with a vertical condenser by means of a Polenske trap. Distill over nearly 50 ml., catching the distillate in a 50-ml. volumetric flask. Make to volume and determine the specific gravity with a pyknometer. Estimate the percentage of alcohol by reference to alcohol tables.

SHELLAC

A simple qualitative test for the presence of shellac on candy is the following. Steep 25 g. of the candy in 50 ml. of 95 per cent alcohol overnight or warm on a steam bath for an hour or so. Decant the alcoholic extract into another beaker and evaporate down to a volume of about 5 ml. Bring 100 ml. of water to boiling in another beaker and pour the remaining alcoholic extract into the boiling water. If shellac is present, it will clump and rise to the surface of the boiling water. This is not done by other resins that may be extracted from candy such as chocolate.

SELECTED REFERENCES

Allen's Commercial Organic Analysis, 5th Ed., Vol. I. Blakiston, Philadelphia, 1923.

Armstrong and Armstrong, *Carbohydrates.* Longmans, London, 1934.

Bates and Associates, *Polarimetry, Saccharimetry and the Sugars; Nat. Bur. Standards, U. S. Circ.* **C440** (1942).

Browne and Zerban, *Sugar Analysis,* 3rd Ed. Wiley, New York, 1941.

Haworth, *Constitution of Sugars.* Arnold, London, 1929.

Hawk, Oser, and Summerson, *Practical Physiological Chemistry.* Blakiston, Philadelphia, 1947.

Jacobs, ed. *Chemistry and Technology of Food and Food Products.* Interscience, New York, 1944.

Jordan and Langwill, *Confectionery Analysis and Composition.* Manufacturing Confectioner, Chicago, 1946.

Leach-Winton, *Food Inspection and Analysis.* Wiley, New York, 1920.

Methods of Analysis, A.O.A.C., 8th Ed. Washington, 1955.

Woodman, *Food Analysis,* McGraw-Hill, New York, 1941.

CHAPTER XI

GUMS, CEREALS, STARCH, OTHER POLYSACCHARIDES, FLOUR, AND BREAD

GUMS

It was mentioned in Chapter VIII that the use of binders and fillers has increased in recent years. One group of these substances is the class of gums. These are amorphous, transparent or translucent substances of wide distribution among plants, which form sticky masses with water and are insoluble in alcohol, acetone, ether, petroleum ether, and, in general, organic solvents. They are usually odorless and tasteless. Some yield clear "solutions" with water and are completely filterable, whereas others swell in water and are only partially soluble and filterable. The first are called real gums and the second are termed vegetable mucilages. Gums are composed of hydrogen, oxygen, and carbon and yield carbohydrate hydrolytic products. Hence, they are generally classified with the carbohydrates.

The gums may be grouped according to their origin as follows:

Plant and Tree Exudates
> Acacia (arabic)
> Ghatti (British Indian gum)
> Karaya (Sterculia Indian gum)
> Tragacanth (bassorin)

Marine Plant Extracts
> Agar agar
> Irish moss (chondrus crispus)
> Algin

Seed Extracts
> Locust kernel and Locust bean
> Guar
> Quince seed

Fruit and Vegetable Extracts
> Pectins

Processed "Gums"
> Dextrins
> British gums
> Cellulose ethers

Acacia, ghatti, Irish moss, quince seed, and the various pectins are readily "soluble" in water and are filterable. Karaya, tragacanth, agar, locust kernel and locust bean are the type that swell in water and are only partially soluble. Agar swells with difficulty in the cold but disperses in boiling water and on cooling sets to a gel. Locust kernel is purified locust bean.

ACACIA

Acacia, arabic is the dried gummy exudation from the stems and branches of leguminous trees of the genus *Acacia* and particularly of species like *Acacia senegal* Willdenow, *Acacia arabica, Acacia verek,* or of some other species of acacia. It forms tears of various sizes, whitish, yellowish white or light amber in color. Gum arabic is composed essentially, according to Hirst,[1] of a mixture of the calcium, magnesium, and potassium salts of a complex molecule containing glycuronic acid, galactose, arabinose, and rhamnose residues. Norman[2] showed the presence of arabinose and galactose as hydrolysis products and concluded that gum arabic has no definite formula, but probably consists of a nucleus acid made up of galactose and a uronic acid, probably, galacturonic acid to which is linked arabinose by glucosidic bonds, so that arabinose is more easily split off than the other components. Arabic contains an oxidase. The composition of acacia, according to Norman[2] is the following:

Ash %	0.24
Furfural yield (ash free) %	13.93
Carbon dioxide yield (ash free) %	4.39
Uronic acid anhydride	17.56
Furfural due to uronic acid %	2.91
Arabinose %	23.52
Galactose %	68.80

According to U. S. P. XV specifications, acacia should contain not more than 4 per cent total ash, not more than 0.5 per cent of acid-insoluble ash, and not more than 15 per cent of moisture. Gum arabic is soluble in twice its weight of water at room temperature. The resulting mixture is acid to litmus and flows readily. It is insoluble in alcohol.

In its broadest sense, however, gum arabic is taken to mean any gum resembling acacia which dissolves in water to form a mucilage.

[1] Hirst, *J. Chem. Soc.* 1942, 70.
[2] Norman, *Biochem. J.* 25, 200 (1931).

GHATTI

Ghatti gum, which is also known as British Indian gum is a gummy exudation from the stem of *Anogeissus latifolia* Wall. It comes from India and Ceylon. Because it is called Indian gum sometimes, it must be carefully differentiated from karaya or sterculia gum. It forms yellowish white tears that are completely soluble in water and, since this property is very similar to that of acacia, it is used sometimes to adulterate that gum. Tests for its distinction from arabic are detailed in the text.

Gum ghatti is 90 per cent soluble and the soluble portion [3] contains the calcium salt of ghattic acid. Free ghattic acid can be isolated by precipitation with alcohol. On hydrolysis with sulfuric acid, the specific rotation of gum ghatti is changed from $-42°$ to $+58°$ and from the hydrolyzed mixture *l*-arabinose may be isolated together with an aldobionic acid. Ghattic acid contains about 50 per cent pentosans and 12 per cent galactose or galacturonic acid.

KARAYA

Karaya, also known as Indian gum, Indian tragacanth gum, or sterculia gum, is an exudate of the trees belonging to the genus *Sterculia urens* Roxb., found in India. Because of its similarity to tragacanth in swelling in water instead of dissolving, it is sometimes used to adulterate tragacanth. Karaya has a high volatile acidity number, varying from 13.4 to 21.3, expressed in terms of acetic acid.[4] Thrun [5] reports somewhat higher values, 17.4-22.7. On long standing the gum develops a distinct odor of acetic acid.

The galactan gelose is a principal constituent of karaya. Thrun and Fuller [6] found it to contain a volatile base also.

TRAGACANTH

Tragacanth is the dried gummy exudation from *Astragalus gummifer* Labillardiere, or other species of Astragalus. The soluble portion of this gum often termed tragacanthin yields arabinose, galactose, and geddic acid on hydrolysis.[7] The insoluble portion which comprises 60-70 per cent is called bassorin. Tragacanth generally contains a small amount of starch and consequently gives a blue color on the addition of iodine solution.

---

[3] Hanna and Shaw, *Proc. S. Dakota Acad. Sci.* 21, 78 (1941).
[4] Tschirch and Fluck, *Pharm. Acta Helv.* 3, 151 (1928).
[5] Thrun, *Ind. Eng. Chem.* 27, 1218 (1935).
[6] Thrun and Fuller, *Ind. Eng. Chem.* 27, 1215 (1935).
[7] O'Sullivan, *J. Chem. Soc.* 79, 1164 (1901).

Norman [8] states that the soluble portion, which he calls traga-canthin, consists solely of uronic acid and arabinose. He was unable to verify the work of O'Sullivan as to the presence of galactose. Van der Haar, however, also found galactose present. The composi-tion of tragacanth according to van der Haar is

	%
Galacturonic acid	32.8
Pentosans (arabinose and xylose)	29.5
Methylpentosan	18.7
Galactose	3.8
Glucose	2.2
Noncarbohydrate material	1.1
Ash	2.6
Water	2.0

Gum tragacanth yields some volatile acid on steam distillation but the acid value of 2-2.4 per cent is very much lower than that of karaya.

AGAR

Agar-agar is the dried mucilaginous substance extracted from marine algae or seaweeds, particularly, *Gelidium corneum* (Hudson) Lamouroux and other species of *Gelidium* (fam. *Gelidiaceae*) and closely related algae, class *Rhodophyceae*. According to Tseng [9] agar is the dried, amorphous, nonnitrogenous extract from *Gelidium* and other red algae being the sulfuric acid ester of a linear galactan.

It generally is marketed in thin transparent membranes which swell with difficulty in cold water but disperse readily in hot water. A dilute neutral, 1 to 2 per cent "solution" sets on cooling to a firm gel at 35-50° C. and becomes fluid again when the temperature is raised to 90-100° C.

According to the specifications of the U. S. P. XIII, agar should contain not more than 1 per cent of foreign organic matter, not more than 1 per cent of acid-insoluble ash, and not more than 20 per cent of water.

Agar contains diatoms which may be identified microscopically. The sulfate radical appears to be an integral part of the complex agar agar molecule and can only be freed by hydrolysis with acid.[10] Agar is used extensively for making culture media in bacteriology.

[8] Norman, *Biochem. J.* 25, 200 (1931).
[9] Tseng, *Sci. Monthly* 58, 24 (1944).
[10] Parkes, *Analyst* 46, 239 (1921).

Irish Moss

Irish moss is the dried, sunbleached plant of *Chondrus Crispus* (Linne) Stackhouse. It is found on the beaches of Ireland, Great Britain, northern European countries, Nova Scotia, Massachusetts, and other New England States. The gum, termed carrageenin, is extracted from the sea plant by boiling in water and is precipitated from the resultant mucilage by alcohol or some other precipitant.

The composition of the dried plant is given by McCance, Widdowson, and Shackleton [11] as follows:

```
Water, g. per 100 g. ...........................13.9
Unavailable carbohydrate (roughage), g. per 100 g.....71.3
Titratable acidity, ml. 0.1 N alkali per 100 g. ......... 8
Reducing sugar, g. per 100 g. ...................... 0.0
Sucrose, g. per 100 g. ............................ 0.4
Starch, as glucose ................................ 0.0
Total available carbohydrate, g. per 100 g., as glucose... 0.4
Total nitrogen, g. per 100 g. ..................... 1.08
Protein, g. per 100 g. ............................ 6.8
```

Irish moss appears to contain two calcium salt sulfuric acid ester carbohydrate complexes,[12] both galactans,[13] for cold water extracts about 47 per cent of soluble matter [14] from the marine plant whereas hot water extracts 70-75 per cent. These complex polysaccharides have different gelling properties. Galactose residues comprise 33 per cent of the hot water extract and 31 per cent of the cold water extract.[15]

Algin

Algin or alginic acid is a gum derived from several species of kelp principally *Macrocystis pyrifera* on the West Coast and from *Laminaria digitata*, the horsetail kelp, and *Laminaria saccharina*, the sugar or broad-leaf kelp, on the American and European coasts of the Atlantic Ocean. Alginic acid has also been obtained from other marine plants of various families, *Undaria pinnatfida*, *Fucus vesiculosus*, *Ascophyllum nodosum*, and *Laminaria flexicaulus*.

There is a considerable literature concerning alginic acid. It is probably a complex anhydride of an aldose sugar acid in which all

---

[11] McCance, Widdowson, and Shackleton, *Med. Research Council, (Brit.) Special Rept. Ser.*, 213 (1936).

[12] Haas, *Biochem. J.* 15, 469 (1921).

[13] Haas and Russell-Wells, *Biochem. J.* 23, 425 (1929).

[14] Dillion and O'Colla, *Nature* 145, 749 (1940).

[15] Buchanan, Percival, and Percival, *J. Chem. Soc.* 1943, 51.

the carboxyl groups are free and all the aldehyde functional groups are conjugated.

Alginic acid is only slightly soluble in water. It reacts, however, with alkali so that it may be titrated. The neutralization equivalent is considered to be of the order of 176-184 and, on boiling with hydrochloric acid as in a uronic acid determination, it yields about 24-25 per cent of carbon dioxide.[16]

Algin may be differentiated from other gums because it is precipitated by trichloroacetic acid.[17] This reaction can be used as a means of separating it from other gums.

Algin and alginates are used in food industries in the preparation of ice cream and other frozen desserts, in confectionery, and in baked goods such as cakes and icings.

Locust Kernel

Locust kernel gum is a gum which has been produced on large commercial basis in recent years. It is the dried mucilaginous extract of locust bean, otherwise known as the carob-bean, *Ceratonia siliqua* L., or St. John's bread. Locust kernel is differentiated from locust bean in that locust bean contains starch and is less pure, from the gum standard, than locust kernel. Knight and Dowsell[18] give as the composition of the commercial gum, the following:

Galactan %	29.18
Mannan %	58.42
Pentosans %	2.75
Protein %	5.29
Nitrogen %	0.83
Cellular tissue %	3.64
Ash %	0.82

The analysis shows that the commercial product is not a pure carbohydrate.

Lew and Gortner[19] found that there were 3 to 4 molecules of mannose to 1 molecule of galactose. One great difference between carob-bean gum and other gums is that no uronic acid or pentoses are found to be present in locust kernel gum.

Quince Seed

This gum may be obtained by extraction from quince seeds,

[16] Lunde, Heen, and Oy, *Kolloid-Z.* 83, 196 (1938).
[17] Hart, *Am. J. Pub. Health* 33, 599 (1943).
[18] Knight and Dowsett, *Pharm. J.* 136, 35 (1936).
[19] Lew and Gortner, *Arch. Biochem.* 1, 325 (1943).

Cydonia vulgaris. With warm water, the seeds swell up and form a mucilaginous mass. The mucilage is used in bandoline and cosmetic hair lotions. The mucilage is sometimes evaporated to dryness and the residue powdered. The powder may then be used to remake the mucilage. Quince seed is representative of a large group of seeds that yield mucilaginous extracts with water.

Hadary[20] discusses the use of quince seed extract as a stabilizer in the manufacture of ice cream.

Renfrew and Cretcher[21] found that arabinose, a mixture of methylated and unmethylated aldobionic acids, and a cellulosic fraction were liberated in the hydrolysis of quince seed gum. Xylose was also identified on further hydrolysis of the aldobionic acids.

GUAR GUM

Guar gum is derived from the endosperm of the seeds of the leguminous plant *Cyamopsis tetragonaloba*. It is obtained from southwestern United States, India, and Pakistan. It is a straight-chain mannan with a galactose branching unit on alternate mannose units. A typical analysis[21a] shows galactomannan 80, protein 5, ash 0.8, and ether extract 0.6 per cent. This gum hydrates in cold water, forming colloidal solutions of high viscosity (distinction from locust kernel); and being a galactomannan like locust kernel, it gives a similar reaction with borax, forming a gel. A solution of the commercial product yields a yellow color when heated with sodium hydroxide solution. Guar is used as a stabilizer in foods like ice cream and also in dog food.

PSYLLIUM SEED

A gum can be isolated from psyllium seed, *Plantago psyllium*, which, as noted, resembles that of quince seed. Some interest has been shown in its use as a food stabilizer. Hanshe and Still[22] give as the composition of the dried aqueous extract of the hulls of psyllium seed the following:

Ash % 2.53		Total phosphorus % .. 0.14	
Total nitrogen % 0.13		Pentose % 83.05	

Again, it can be seen that the mucilage is not a pure carbohydrate.

[20] Hadary, *Food Industries* **15**, No. 2, 76 (1943).
[21] Renfrew and Cretcher, *J. Biol. Chem.* **97**, 503 (1932).
[21a] "Jaguar." Stein, Hall & Co., Inc., New York, N. Y.
[22] Hanshe and Still, *Am. J. Pharm.* **105**, 433 (1933).

Flaxseed Gum

A gum resembling that from quince seed and psyllium seed can be obtained from flaxseed, *Linum usitatissimum* L. When hydrolyzed this gum appears to consist of D-galacturonic acid and L-rhamnose.[23] Its use in foods is small in comparison with other gums.

Pectin

Pectins are a group of carbohydrate materials which occur in fruits and vegetables and which have in common the property of enabling fruit juices to form jellies. A committee of the Agricultural and Food Chemistry Division of the American Chemical Society recommended the following nomenclature for the pectic substances.[24]

Pectic Substances.—"Pectic substances" is a group designation for those complex, colloidal carbohydrate derivatives which occur in or are prepared from plants and contain a large proportion of anhydrogalacturonic acid units which are thought to exist in a chainlike combination. The carboxyl groups of the polygalacturonic acids may be partly esterified by methyl groups and partly or completely neutralized by one or more bases.

Protopectin.—The term "protopectin" is applied to the water-insoluble parent pectic substance which occurs in plants and which, upon restricted hydrolysis, yields pectin or pectinic acids.

Pectinic Acids.—The term "pectinic acids" is used for colloidal polygalacturonic acids containing more than a negligible proportion of methyl ester groups. Pectinic acids, under suitable conditions are capable of forming gels with sugar and acid or, if suitably low in methoxyl content, with certain metallic ions. The salts of pectinic acids are either normal or acid pectinates.

Pectin.—The general term "pectin" (or pectins) designates those water-soluble pectinic acids of varying methyl ester content and degree of neutralization which are capable of forming gels with sugar and acid under suitable conditions.

Pectic Acids.—The term "pectic acids" is applied to pectic substances principally composed of colloidal polygalacturonic acids and essentially free from methyl ester groups. The salts of pectic acids are either normal or acid pectates.

Pectin is then considered to be, essentially, a long-chain, partially methylated, polygalacturonic acid molecule having a very large mo-

[23] Anderson and Crowder, *J. Am. Chem. Soc.* **52**, 3711 (1930).
[24] Kertesz, *Chem. Eng. News* **22**, 105 (1944).

lecular weight. A natural pectin may have a molecular weight of over 200,000. This implies that over 1000 units may be strung together to form each molecule.[25]

Galacturonic Acid Reaction.—Ehrlich [27] describes a typical reaction given by D-galacturonic acid and hydrolyzed pectin, with basic lead acetate solution. If an aqueous solution of alpha or beta

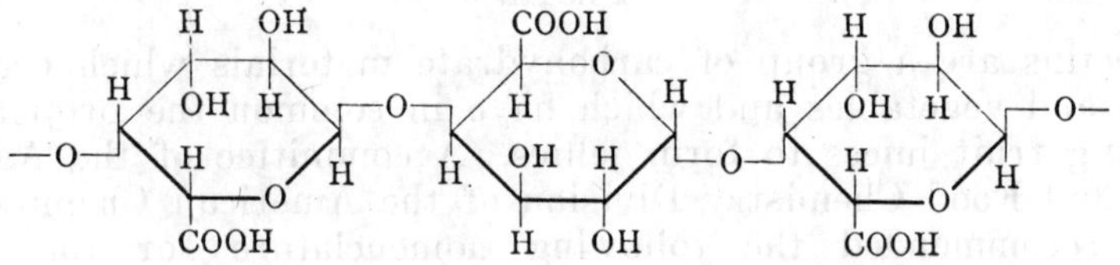

Formula of polygalacturonic acid from pectin
showing three units of the molecule.[26]

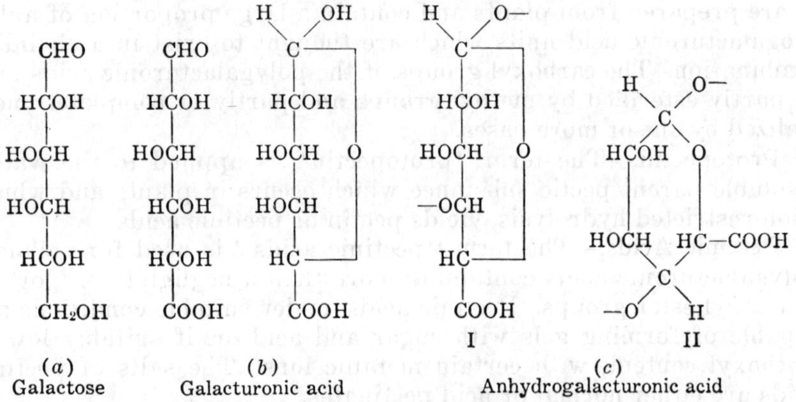

(a)	(b)	(c)
Galactose	Galacturonic acid	Anhydrogalacturonic acid

FIG. 67. Galacturonic Acid Relationships

D-galacturonic acid is treated with a little freshly filtered basic lead acetate solution, a colorless flocculent precipitate is formed which redissolves as more of the reagent is added. The resulting clear solution, when heated in a water bath, becomes turbid and pink in a few seconds, the pink color of the rapidly increasing turbidity visibly changes to a deep red and after about 1 minute a dark blood-red to brick-red precipitate separates in thick flocks and the supernatant liquid becomes colorless or, in the presence of a large excess of lead

[25] Kaufman, Fehlberg, and Olsen, *Food Industries* 14, No. 12, 57 (1942).
[26] Meyer, *Natural and Synthetic High Polymers.* Interscience, New York, 1942.
[27] Ehrlich, *Ber.* 65B, 352 (1932).

acetate solution, yellowish. This amorphous red salt, which is formed by the decomposition of the original basic lead salt of D-galacturonic acid under the influence of the excess of lead acetate in hot solution, is also formed when only enough lead acetate is added to form a heavy white precipitate and the mixture, which should be distinctly alkaline to litmus, is heated 1 minute. In this way 5 mg. of D-galacturonic acid gives a distinct red precipitate and even 1 mg. can be faintly detected after the heated mixture has stood some hours. The soluble salts of D-galacturonic acid behave in a similar manner.

REACTIONS OF THE GUMS

The composition of the gums has never been thoroughly investigated. In fact, some of the previously mentioned citations show that the commercial products are not pure carbohydrates. Furthermore, since many of the gums give common hydrolyzates, little purpose is gained by this type of analysis. Moreover, they are not crystalline bodies nor do they have the characteristic granular structure that starch grains have and possess, at times, even after incorporation into foods; therefore, they cannot easily be distinguished by microscopic means.[28] Jacobs and Jaffe [29] made an extensive survey of the reactions of the gums and showed that the gums differed in their reaction with various inorganic reagents. Weinberger and Jacobs [30] differentiated gums by noting differences in the alcohol precipitate. Hepburn and Laughlin [31] studied the effect of different reagents on psyllium seed mucilage.

Because of the scarcity of material on the subject of gums and the increasing use of these substances in foods, the reactions of the gums are detailed in Table 47. Table 48 gives the characteristic reactions for the identification of each gum.

These reactions are given by the gums in water solutions prepared from the gum as it is sold commercially and depend very often on the concentration of the gum. Thus 5 ml. of 0.5 per cent solution of locust kernel gum is jellied by a few drops of 4 per cent borax solution but an 0.2 per cent solution is not jellied by this reagent. Table 47 gives the reaction of solutions of the gums that vary in concentration from 0.5 per cent to 1 per cent except for agar, for which 0.1 to 0.2 per cent solutions are used. In order to make use of the table properly, care must be taken to notice the distinguishing characteristics of the precipitate formed. That is, whether it is volu-

[28] Wildman, *J. Assoc. Official Agr. Chem.* **18**, 637 (1935).
[29] Jacobs and Jaffe, *Ind. Eng. Chem., Anal Ed.* **3**, 210 (1931).
[30] Weinberger and Jacobs, *J. Am. Pharm. Assoc.* **18**, 34 (1929).
[31] Hepburn and Laughlin, *Am. J. Pharm.* **102**, 565 (1930).

TABLE 47. RESULTS OF TESTS IN VARIOUS GUMS [a]

Gum	Stokes Acid Mercuric Nitrate Reagent	Neutral Lead Acetate 20% Soln.	Basic Lead Acetate (A.O.A.C.)	Potassium Hydroxide 10% Soln.	Neutral Ferric Chloride 5% Soln.	Alcohol[b] Precipitate	Borax 4% Soln.	Millon's Reagent	+2 Vols. Acetic Acid
Arabic	White, fine, opaque ppt, sol. in excess of reagent	No ppt.	White, curdy ppt., insol. in excess	Faint yellow tinge	Ppt. sol. in excess	Very fine floc., non-adherent 40 ml. pt. of definite pptn.	Negative	Yellow, fine ppt.	Negative
Ghatti	No ppt.	No ppt.	Translucent floc. ppt.	Negative	Negative	Fine floc. ppt., non-adherent	Negative	Fine ppt.	Negative
Tragacanth	Voluminous flocculent translucent ppt.	Voluminous flocculent ppt., gels	Voluminous ppt., gels	Bright yellow, stringy ppt.	Gelatinizes	Coag. long and stringy, adherent, 10 ml.	Negative	Vol. floc. ppt.	Negative
Agar Agar	Gelatinizes	Flocculent ppt., gels	Voluminous ppt.	Clarifies soln.	Gelatinizes, heat → excess → ppt.	Heavy floc., adherent to beaker, 20 ml.	Negative	Fine ppt.	Ppt.
Karaya	White, curdy, ppt, settles rapidly	Negative	Stringy ppt., settles rapidly	Negative	Ppt. coag. on heating	Fine filamentous particles, non-adherent, 15-20 ml.	Negative	Fine ppt.	Negative
Irish moss	Gelatinizes	Flocculent ppt., gels	Voluminous flocculent ppt., gels	Gels	Vol., stringy ppt., gels	Coagulated, translucent, stringy, adherent, 20 ml.	Negative	Fine ppt.	Negative
Quince seed	Voluminous flocculent yellowish	Yellowish flocculent ppt., gels	Yellowish vol., flocculent ppt.. gels	Stringy ppt.	Stringy ppt.	Coag., short, stringy, non-adherent, 25 ml.		Floc. ppt.	....
Psyllium seed[c]	....	Flocculent, slimy pale yellow-green ppt.	Curdy ppt.	Yellow froth	Marked darkening	Faint turbidity then floc. ppt.			....
Locust kernel	Gelatinizes	Voluminous ppt., gels	Gels	Slight floc., ppt.	Stringy ppt.	Stringy, clotty, opaque, non-ad., start, 2 ml., complete, 15 ml.	Gels	Fine ppt.	Ppt.
Pectin	Floc. ppt., gelatinizes	Floc. ppt., gelatinizes	Floc. ppt., gelatinizes	Ppt.	Ppt.	Floc. ppt. translucent	Negative	Floc. ppt.	Ppt.
Starch	Negative	Negative	Negative	Negative	Negative	White turbidity ppt.	Negative	Fine ppt.	White ppt.
Dextrin	Negative	Negative	Negative	Negative	Negative	White turbidity	Negative	Fine ppt.	Negative

[a] Jacobs and Jaffe, Ind. Eng. Chem., Anal. Ed., 3, 210 (1931).
[b] Weinberger and Jacobs, J. Am. Pharm. Assoc., 18, 34 (1929).
[c] Hepburn and Laughlin, Am. J. Pharm., 102, 565 (1930).

REACTIONS OF THE GUMS

minous, flocculent, small flocculent, stringy, powdery, curdy, filamentous, etc. With certain reagents there is no apparent change, hence the use of blanks and controls is essential. The word "gel" in the table indicates an actual jellied condition. The word "gelatinizes" implies that the mixture has been thickened. These are clear distinctions and are readily noticeable in making the tests.

TABLE 48. CHARACTERISTIC REACTIONS OF GUMS

Gum	Reaction
Arabic	Very soluble. It is not precipitated by neutral lead acetate solution. Gives a precipitate with 1 drop of 10% ferric chloride which is soluble in excess of ferric chloride.
Agar agar	An aqueous 0.2% solution forms a gel. Five ml. of an 0.1% solution + 1 drop sulfuric acid + 1 drop of congo red and centrifuged yields a flocculent precipitate.
Ghatti	Very soluble. Fresh solutions do not yield a curdy white precipitate with basic lead acetate, distinction from arabic.
Irish moss	Has a distinct odor of seaweed. It is precipitated by a saturated solution of barium chloride.
Karaya	Has a very high volatile acid number and gives a pink color when boiled with phosphoric, hydrochloric, or trichloroacetic acid.
Locust kernel	The addition of 1 to 2 drops of 4% borax solution to 0.5% gum solution yields a gel. Tannic acid gives a precipitate.
Guar	Borates yield gel.
Quince seed	Yields a precipitate with 1 or 2 drops of saturated zinc chloride solution or with ammonium molybdate reagent. Also gives a precipitate with saturated barium chloride solution but Irish moss does not give precipitates with the former reagents. Quince seed also gives a flocculent precipitate when treated as described under agar, but agar does not give the zinc chloride or ammonium molybdate test.
Tragacanth	Boiled with 10% potassium hydroxide solution yields a yellow stringy precipitate and a solution tinged with yellow.
Pectin	Gives a precipitate when treated as described in the chapter on jams and jellies, Chapter XII. Quince seed may also give a precipitate but pectin does not give the zinc chloride or ammonium molybdate reactions.

Reagents.—Stoke's Reagent.—Dissolve metallic mercury in twice its weight of nitric acid and dilute this solution to 25 times its volume with water.

Millon's Reagent.—Dissolve metallic mercury in an equal weight of nitric acid and dilute the solution with an equal volume of water.

Neutral Lead Acetate.—Dissolve 20 g. of lead acetate, $Pb(C_2H_3O_2)_2 \cdot 3H_2O$ in water and make up to 100 ml.

Basic lead Acetate.—Boil 430 g. of neutral lead acetate, 130 g. of litharge, and 1 liter of water for 30 minutes. Allow the mixture to cool and settle and then dilute the supernatant liquid to a specific gravity of 1.25 with recently boiled water. Solid basic lead acetate may be substituted for the normal salt and litharge in the preparation of the solution. The solid salt is dissolved in water and diluted to the proper specific gravity.

Ammonium Molybdate Solution.—Dissolve 100 g. of molybdic acid, MoO_3, in a mixture of 144 ml. of ammonium hydroxide and 271 ml. of water. Pour this solution slowly and with constant stirring into a mixture of 489 ml. of nitric acid and 1148 ml. of water. Keep the final mixture in a warm place for several days or until a portion heated to 40° C. deposits no yellow precipitate of ammonium phosphomolybdate. Decant from any sediment and preserve in glass-stoppered vessels.

Procedure.—Add 2 to 3 drops of the reagent to 5 ml. of the test solution, note the result and then add an excess of the reagent. In the case of some of the reagents namely, potassium hydroxide, phosphoric acid, hydrochloric acid, and ferric chloride, a few drops of the reagent are added to the test solution and the mixture boiled. Then an excess of the reagent is added and the mixture is boiled again.

For noting the type of alcohol precipitate, the technique of Weinberger and Jacobs [32] is recommended. To 20 ml. of the test solution, add 70 ml. of 95 per cent alcohol drop by drop, with constant stirring until precipitation is complete. Note the texture, quality, and the characteristics of the precipitate, as well as the point at which definite precipitation begins.

Microscopic Examination

A system of identification of the common gums by means of certain reagents and examination with the aid of the microscope is detailed by the A. O. A. C. [33]

Reagents.—Chlorozinciodide Solution.—To 100 ml. of a solution of zinc chloride, sp. gr. 1.8, add a solution of 10 g. of potassium iodide and 0.15 g. of iodine in 10 ml. of water. Keep a few crystals of iodine in the solution.

Ruthenium Red Solution.—To a few ml. of 10 per cent lead acetate solution add enough ruthenium red to produce a wine red color.

Preparation of Sample.—Lotions and Jellies.—Mix the sample thoroughly, adding water if necessary to obtain a fluid mixture.

---

[32] Weinberger and Jacobs, *J. Am. Pharm. Assoc.* **18**, 34 (1929).
[33] *Methods of Analysis, A.O.A.C.*, 8th Ed. Washington, 1955.

Transfer 5 or 10 ml. to a centrifuge tube, add 4 volumes of 95 per cent alcohol, mix and centrifuge. If fat or oily material is present wash the precipitated gum with ether,-discard the washings, redissolve in water, and reprecipitate.

TABLE 49.

GROUP I. REACTIONS WITH CHLOROZINCIODIDE [a]

Gum	Alcohol ppt.	Group reaction	Confirmatory Test	Notes
Tragacanth	Stringy, bluish translucent	Blue	Warm with 10% NaOH on steam bath: yellow bright	Irish moss may give dull yellow; arabic faint yellow
Starch	White, compact	Blue black	0.1 N iodine blue	Tragacanth may give faint blue
Quince	Stringy, translucent	Blue	Above tests: negative	Quince is distinguished from tragacanth and starch by negative reactions
Irish moss	Stringy	Brown with small blue particles	Characteristic nodular structures with group reagent	Old prepns. of this gum may fail to show characteristic structures

GROUP II. REACTIONS WITH U.S.P. TINCTURE OF IODINE

Agar	White, opaque	Opaque, blue black	Stains with ruthenium red	Does not dissolve or lose shape when covered with water
Irish moss	Stringy	Brown or lilac	Characteristic blue stain with 1% alc. methylene blue soln.	These reactions are given by both old and new prepns.

GROUP III. REACTIONS WITH RUTHENIUM RED

Karaya	Fine flocculent compact mass on centrifuging	Swells considerably; stained strongly pink granular mass	Heat with concd. HCl: pink	1% aq. methylene blue soln. gives characteristic blue stain

[a] Cannon, *J. Assoc. Official Agr. Chem.* 22, 726 (1939).

TABLE 49—*Continued*

GROUP IV. REACTIONS WITH CONCENTRATED SULFURIC ACID

Gum	Alcohol ppt.	Group reaction	Confirmatory Test	Notes
Arabic		Greenish brown	Ppt. completely sol. in water	Soly. of arabic differentiates it from most other gums
Carob bean	Stringy	Pink or red brown	None	Alc. ppt. resembles that of tragacanth

Gums arabic and agar, and possibly other gums, may not be precipitated promptly. Precipitation can be aided by the addition of a few drops of saturated salt solution.

Controls.—Wet 1 g. of control gum with alcohol. Add 100 ml. of water while stirring steadily and bring the mixture to a boil. Precipitate the gum from 5 or 10 ml. of the control solution as directed above.

Procedure.[34]—Press a small lump of the alcohol precipitate against a slide to form a mat 4 to 8 mm. in diameter. Note the form of the resulting mat as an index to the type of gum; thus quince seed gum and Irish moss form thin translucent films, while arabic, agar, and starch form opaque, white films. Cover the mat of gum with a drop of chlorozinciodide solution and observe with and without magnification. For the direct examination place the slide upon a white surface. For examination with the aid of a microscope use a magnification of about 90 diameters. If no characteristic color is produced within 1 to 2 minutes, examine a fresh mat, using the reagents noted in Table 49. For Group II, add a drop of U.S.P. tincture of iodine and allow it to dry on the mat. Flush off with alcohol, and wash with water and then observe.

VOLATILE ACIDITY

In various parts of this chapter it was indicated that karaya had a high acid number and hence could be differentiated from other gums. The following method may be used for estimating the volatile acidity of gums. This value varies from 13.4 to 21.3 for sterculia gum and from 2 to 2.4 for gum tragacanth. The gum is placed in solution with phosphoric acid and is then steam distilled. The amount

---

[34] Cannon, *J. Assoc. Official Agr. Chem.* 22, 726 (1939).

of acid is subsequently estimated in the distillate by titration with standard alkali.

Procedure.—Transfer about 1 g. of the sample accurately weighed to an 800-ml. Kjeldahl flask and allow the gum to soak in a mixture of 100 ml. of water and 5 ml. of syrupy phosphoric acid overnight or for several hours. Reflux cautiously for 2 hours. Attach the Kjeldahl flask to a steam distillation system having an adequate scrubber trap and steam distill until 600 ml. of distillate have been obtained. Use a small boosting flame under the Kjeldahl flask and a trap or scrubber between the flask and the condenser. Do not allow the contents of the Kjeldahl flask to become so concentrated that the material chars. Titrate the distillate with 0.1 N sodium hydroxide solution using 10 drops of phenolphthalein indicator solution. Correct the determination by running a blank and express the result as the number of milliliters of 0.1 N sodium hydroxide solution required to neutralize the volatile acidity obtained per g. The volatile acid is assumed to be acetic acid.

DETECTION OF GUMS IN FOOD PRODUCTS

In general, it is necessary for the food analyst to determine the presence of gums as a class rather than identify the gum itself. The following method lends itself to the detection and possible identification of gums in various food products.

Procedure.—Weigh 50 g. of the food material into a beaker and, if necessary, remove fatty materials by adding petroleum ether, stirring, and after allowing to settle, pouring off the supernatant petroleum ether layer. Repeat until practically all of the fat has been removed. Three or four washings with petroleum ether suffice. Evaporate off any residual petroleum ether on a steam bath. Add 50 ml. of water to the petroleum ether washed residue and stir vigorously until a paste is formed. Add 10 ml. of 40 per cent trichloroacetic acid and again stir vigorously. Allow to stand for 5 minutes. Transfer the pasty mixture to a centrifuge bottle and centrifuge for 5 to 10 minutes. Filter. Test a portion of the clear filtrate for complete precipitation of proteins with more trichloroacetic acid. If a precipitate develops, add more trichloroacetic acid and recentrifuge and filter. Add 5 volumes of 95 per cent alcohol to the filtrate with constant stirring and allow to stand for 5 minutes. Add ammonium hydroxide solution drop by drop until the mixture is alkaline and again allow to stand for 5 minutes. Any split-proteins not precipitated by the trichloroacetic acid will appear at this point as well as the gums. Add 10 drops of hydrochloric acid and stir vigorously. A flocculent or stringy precipitate at this time indicates the presence of gums, for the split-proteins or pseudo-gelatins will generally dissolve in the hydrochloric acid. Starches and dextrins behave as do the gums and do not dissolve.

With respect to sodium alginate refer to the section on stabilizers in frozen desserts below.

Allow the mixture to stand for at least a hour or preferably overnight to permit the precipitate to settle and agglomerate. Filter the precipitate through a Gooch crucible equipped with a thin pad of asbestos. Wash the residue thoroughly in the crucible with alcohol. Dry by allowing to stand overnight or by placing in an oven for a short time. Place the Gooch crucible and its contents into a 50-ml. beaker and add 30 ml. of water. Boil for 10 minutes, stirring up the asbestos pad. Filter immediately through a fast filter and evaporate the filtrate to 10 ml. To 1 ml. of the filtrate in a small test tube, add 1 to 2 drops of Millon's or Stokes acid mercuric nitrate reagent or basic lead acetate solution. A precipitate having the characteristics of those listed in Table 47 confirms the presence of gums.

To 3 ml. of the filtrate add 12 ml. of 95 per cent alcohol, mix and add 3 ml. of a mixture of 5 ml. of hydrochloric acid and 95 ml. of 95 per cent alcohol. Mix and add 3 ml. of water, a flocculent or stringy precipitate indicates the presence of gums. To 5 ml. of the filtrate add 5 ml. of hydrochloric acid and a few crystals of phloroglucinol and boil for 2 minutes. The production of a pink to red color, fading to amber and the subsequent formation of a brown precipitate confirms, in another manner, the presence of gums.

This method may be applied to such divers food products as cheese, mayonnaise, ice cream, smoked or canned meats, and cake coatings. The concentration of the gum is again a factor and, unless definite flocculent or stringly precipitates are obtained by the alcohol precipitation, gums in quantities over 0.05 per cent are not indicated. Mere turbidities are to be disregarded.

It is well to note that the confirmatory tests eliminate the interference of starch and dextrin, because these substances will not give precipitates with the Stokes acid mercuric nitrate or with the basic lead acetate reagents.

Separation of Gums

The problem of identifying gums in the presence of one another is rather difficult and, at times, impractical. It becomes important when it is necessary to establish the purity of a product during a commercial transaction, in order to ascertain if a cheaper gum has replaced in whole or in part a more costly one.

The presence of one gum in another has been customarily established by measuring the viscosity of a solution containing a weighed amount of the sample.[35] If the viscosity is altered much, the presence of another gum is indicated. Some chemical tests have been developed as, for example, the detection of oxidase [36] as a test for the presence of arabic in tragacanth, or the estimation of the volatile acid number, calculated as acetic acid, for distinguishing between karaya and other gums. However, the acid value of gums varies so much that a large adulteration could at times be made without detection by this test alone. These methods do not tell the type of gum present

Very complex mixtures of gums will not ordinarily be encountered, and if mixtures are made, they will most likely consist of two or three gums. It is possible to effect a more or less definite separation of simple mixtures by making use of the characteristic reactions of the gums. Thus, assuming the mixture to be arabic and tragacanth, precipitate the tragacanth by the addition of neutral lead acetate and filter. The filtrate contains the arabic which may subsequently be precipitated by alcohol. Dissolve the residue from the neutral lead acetate precipitation, which contains the tragacanth, in saturated ammonium acetate solution and recover the tragacanth by the addition of alcohol, which precipitates the gum alone. Filter the alcoholic precipitates, wash the precipitates on the filter paper or Gooch crucible with alcohol, dissolve the gums in hot water as described in the preceding section, and then apply the tests for the characteristic reactions.

In a similar manner by precipitating locust kernel, quince seed, pectin, or Irish moss by neutral lead acetate solution and filtering, separations can be made of the following combinations: arabic and locust kernel, arabic and quince seed, arabic and pectin, arabic and Irish moss. The residue contains one of the aforementioned gums and the filtrate again contains the arabic. By dissolving the lead acetate precipitate in saturated ammonium acetate and by the subsequent addition of alcohol, the gums may be recovered.

Entirely analogous types of separation may be made in: (1) Combinations of quince seed gum and other gums by precipitation of the quince seed by zinc chloride or ammonium molybdate reagent and subsequent filtration. The filtrate contains the other gum and the residue contains the quince seed gum.

(2) Combinations of Irish moss or quince seed gums and other gums by precipitation of the Irish moss and quince seed with sat-

---

35 Middleton, *Quart. J. Pharm.* 9, 506 (1936).
36 Ritsema, *Pharm. Weekblad* 72, 105 (1935).

urated barium chloride solution and subsequent filtration. The filtrate contains the other gum and the residue the Irish moss or quince seed.

(3) Combinations of pectin and other gums by saponification and hydrolysis of the pectin, as described in the chapter on jams and jellies, Chapter XII, with the formation of digalacturonic acid and separation by filtration. The precipitate is digalacturonic acid and the filtrate contains the other gum.

(4) Combinations of locust kernel and other gums by precipitation of the locust kernel with tannic acid solution and separation by filtration. The residue consists of the precipitate and the filtrate contains the other gum.

These separations are, very likely, not quantitative, because of the great similarity of the gums, both chemically and physically. It is possible, in these precipitations, that some of a gum, presumably soluble, is adsorbed and will appear with the precipitate and that, on the other hand, some of the gum, presumably precipitable, remains in solution or suspension. Consequently, only after recovery of the gum by precipitation by alcohol or acetone, which frees the gum from interfering materials, by re-solution of the gum in water, and by the application of the characteristic reactions to the gum can conclusions be drawn as to the nature and type of gums composing the original mixture.

STABILIZERS IN FROZEN DESSERTS

Frozen desserts may contain gelatin, gums, sodium alginate, or a mixture of any of these stabilizers. Sherbets or ices may contain pectin, or any of the foregoing products.

Sodium alginate differs from the true gums in that a precipitate of alginic acid is formed by the addition of trichloroacetic acid. Trichloroacetic acid cannot be used to separate sodium alginate from proteins because it precipitates both proteins and alginic acid. Tannic acid precipitates only the proteins, including gelatin, leaving the gums, including the alginates, in solution.

A method proposed by Hart,[37,38] involves: (1) Differential treatment with trichloroacetic acid to remove proteins and sodium alginate and treatment with tannic acid to remove proteins; (2) addition of alcohol to the filtrates to precipitate gums and any remaining alginates; (3) confirmation with special reagents.

This method does not detect pectin in sherbets and it does not

---

[37] Hart, *Am. J. Pub. Health* 33, 559 (1943).
[38] Hart, *J. Assoc. Official Agr. Chem.* 22, 608 (1939).

indicate definitely whether gelatin is present. Methods for the detection of gelatin in milk which can be used on frozen desserts with little modification have been detailed in Chapter VII.

Apparatus.—Pipette Guard.—The problem of pipetting the lower liquid portion of a centrifuged mixture without contamination from the top layer is facilitated by the use of a pipette guard, called a Vohale tube.[39]

This device consists of a wide-bore glass tube, flared slightly at the upper end and fitted at the lower end with a ground-glass cap, over-all length about ½ in. less than that of the centrifuge bottle. The tube is supported in the neck of the bottle by a slotted rubber stopper. After the material has been centrifuged, the pipette is inserted through the tube, the glass cap being pushed off, and the aliquot part desired is removed without disturbing the supernatant layer.

In lieu of this, a piece of glass tubing, 8 mm. inside diameter and about 7½ in. long, may be used. The tube should be flared at the upper end, and a short cork stopper fitted onto the lower end. This stopper should be of sufficient size so that it does not rise up into the tube when the bottles are centrifuged. The tube is supported in the neck of the bottle by a slotted rubber stopper. After the material in the bottle is centrifuged the cork stopper is pushed through by means of a solid glass rod and the lower layer is removed by means of a pipette inserted through the pipette guard.

Reagent.—Benedict's Reagent.—Dissolve 17.3 g. sodium citrate and 10 g. anhydrous sodium carbonate in about 80 ml. hot water; dissolve 1.73 g. crystalline copper sulfate in 10 ml. water. Filter the alkaline citrate solution, add the copper sulfate solution slowly, with constant stirring, and make up to 100 ml.

Procedure.—Weigh 100 g. ice cream or ice-cream mix into each of two 500-ml. casseroles. Add 100 ml. of hot water to each casserole. Heat to about 80° C., stirring constantly. Transfer the contents of each casserole to an 8-oz. bottle. Add glacial acetic acid drop by drop, shaking after each addition, until separation of casein occurs. Insert a pipette guard. Centrifuge until the casein has collected in the upper part of the bottle (15-20 minutes). Pipette as much of the aqueous layer as possible into a glass or platinum evaporating dish. Evaporate over steam down to about 40 ml. Transfer contents of each dish to a 50-ml. conical centrifuge tube.

To one portion ("Portion A") add 10 ml. of a 50 per cent trichloroacetic acid solution to precipitate proteins and to remove the

[39] Hart, *J. Assoc. Official Agr. Chem.* **20**, 529 (1937); **22**, 605 (1939).

sodium alginate as alginic acid. To the other portion ("Portion B"), add 10 ml. of a 20 per cent tannic acid solution to precipitate proteins.

From this point on treat "Portion A" and "Portion B" identically. Allow to stand 15 minutes. Centrifuge until precipitate has settled compactly (10-15 minutes): Decant through rapid filter paper (Whatman 41H) receiving the filtrate into an 8-oz. bottle. Add 4-5 volumes of 95 per cent alcohol. Allow to stand overnight or until the precipitate begins to settle.

Centrifuge 15-20 minutes. Decant the supernatant alcoholic layer as completely as possible. Wash twice with 80 per cent alcohol, centrifuging, and decanting after each washing. Care must be exercised to wash thoroughly the separated gums (or the gums and proteins) in order to assure complete freedom of the precipitate from the sucrose and/or reducing sugars originally present in the frozen dessert, because these substances, if present in the test portion, will give a positive Benedict's test.

Dissolve residual precipitate in the bottle in 40 ml. hot water. Add 1 ml. glacial acetic acid. Without filtering, add 4-5 volumes of 95 per cent alcohol. Allow the mixture to stand overnight or until the precipitate settles. Centrifuge and decant the supernatant solution. Add 30 ml. hot water to the residue. Transfer to a 150-ml. beaker. Add 5 ml. concentrated hydrochloric acid. Boil gently for 2 minutes. Test both "Portion A" and "Portion B" by Benedict's test and by Tollen's test as follows:

Benedict's Test.—Transfer 1 ml. of the hydrolyzed gum solution to a test tube. Neutralize with approximately 2 N sodium hydroxide solution, using litmus paper as an indicator. Remove the litmus paper, as an indicator. Remove the litmus paper. Add 5 ml. Benedict's qualitative reagent. Boil vigorously 1-2 minutes. Allow to cool spontaneously. A voluminous precipitate, which may be green, yellow, or red, indicates reducing sugars.

Tollen's Test.—Heat the remainder of the hydrolyzed gum solution to boiling and drop in a few crystals of phloroglucinol. A red or deep amber color indicates pentoses. Certain other sugars (as galactose) also give a positive reaction to Tollen's test.

Interpretation.—Sodium alginate is precipitated as alginic acid by trichloroacetic acid and is removed from "Portion A" along with the proteins. Hence when sodium alginate is present as a stabilizer, scanty precipitation or none at all occurs later with alcohol. Alginates are not affected by tannic acid and remain in "Portion B" to be precipitated later by alcohol along with gums. Some vegetable gums will appear in the filtrate after treatment with either protein precipitant. However, locust bean gum and agar-agar gum may be precipitated with tannic acid, and thus be removed wholly or partly with the proteins.[40]

These findings by alcohol precipitation may be summarized thus:

Portion A (Trichloroacetic Acid)	Portion B (Tannic Acid)	Interpretation
+	+	Gums with or without alginates
+	—	Gums, possibly locust bean or agar, but no alginates
—	+	Alginates, but no gums
—	—	No gums or alginates
		(+ and — signs indicate positive and negative tests for sugars on hydrolyzates.)

Voluminous precipitates in both "Portion A" and "Portion B" after addition of alcohol indicate the presence of gums with or without alginates.

A voluminuous precipitate in "Portion A" after addition of alcohol, responding positively to Benedict's test and to Tollen's test, indicates the presence of vegetable gums, but not sodium alginate. A voluminous precipitate in "Portion B," similarly treated with alcohol, responding positively to Benedict's test and to Tollen's test, with scanty precipitate or none at all in "Portion A," indicates sodium alginate.

CEREAL

The proximate analysis of cereals for total solids, ash, protein, fat or ether extract, crude fiber, etc., may be made using methods described in other sections of the text. Often the percentage of carbohydrate is estimated by difference, that is, the amount of water, ash, fat, and protein are determined, these percentages are added, and the sum is subtracted from 100. The remainder is assumed to be carbohydrate. A method for starch is detailed.

Grain and stock feeds may also be estimated by these methods. The methods for pentoses and galactans are detailed.

Tables 50, 51 give the proximate analysis of cereals and flour. In Table 51 the composition and comparison of various flours with that of soy flour is given. It is to be noted, as Table 52 shows, that, although soy flour contains comparatively little starch, it does contain other carbohydrates. It also contains a very large amount of protein in comparison with other flours.

[40] Hart, *J. Assoc. Official Agr. Chem.* 20, 527 (1937).

TABLE 50. ANALYSES OF CEREALS [a] (UNITED STATES)

Type	Water %	Ash %	Protein %	Fat %	Crude Fiber %	Carbohydrate %
Barley						
Maximum........	12.96	2.95	13.83	2.42	5.62	73.47
Minimum........	8.92	1.65	8.32	1.89	1.57	66.75
Average........	10.80	2.44	10.69	2.13	4.05	69.89
Buckwheat						
Maximum........	13.00	2.23	11.90	2.43	12.45	64.14
Minimum........	11.75	1.63	9.19	1.74	9.57	61.01
Average........	12.15	1.89	10.75	2.11	10.75	62.33
Corn						
Maximum........	12.32	1.55	11.55	5.06	2	75.07
Minimum........	9.58	1.19	8.58	2.94	1	68.97
Average........	10.98	1.71	9.88	4.17	1.71	71.95
Sweet Corn [b]	8.44	1.97	11.48	8.57	2.82	66.72
Rye						
Maximum........	11.45	2.41	18.99	2.30	2.50	75.36
Minimum........	9.54	1.71	8.40	1.16	1.65	63.61
Average........	10.62	1.92	12.43	1.65	2.09	71.37
Wheat						
Maximum........	14.53	2.35	17.15	2.50	3.72	76.05
Minimum........	7.11	1.40	8.58	0.28	1.70	66.67
Average........	10.62	1.82	12.23	1.77	2.36	71.18

[a] Wiley, U. S. Dept. Agr., Bur. Chem. Bull. 45 (1895).
[b] Wiley, U. S. Dept. Agr., Bur. Chem. Bull. 50 (1897).

TABLE 51. COMPOSITION AND COMPARISON OF FLOURS [a]

Flour	Water %	Protein %	Fat %	Carbohydrate %	Fiber %	Ash %
Soybean "Aguma"....	8.90	49.01	8.55	26.28	1.47	5.79
Soybean "Aguman"..	10.58	55.90	2.32	24.26	1.77	5.17
Soybean "Soyama"...	....	42.00	18.00	24.00	....	6.00
Soybean "Berczeller"	....	45.50	22.38	....	....	4.81
Soybean Bollman.....	10.7	51.2	0.1	28.8	3.1	6.1
Soybean Deming.....	5.26	44.64	19.43	24.12	2.35	4.20
Wheat.............	12.00	11.0	1.00	77.35	0.20	0.45
Whole wheat........	10.90	12.00	2.00	73.05	1.00	1.05
Rye..............	9.00	12.00	1.50	75.85	0.65	1.10
Corn.............	10.00	8.50	2.70	77.10	0.80	0.90

[a] Horvath, The Soy Bean as Human Food. Union Medical College, Peking, 1925.

STARCH

To obtain satisfactory results with the following A. O. A. C. [41] tentative method for starch in flour, the directions must be followed carefully in every detail. As the steps are timed it is essential to learn the procedure so that no time will be lost in following the

[41] Methods of Analysis, A.O.A.C. Washington, 1945.

method. Arrange everything needed in the determination before the
hydrochloric acid is added to the sample.

Reagent.—To prepare the hydrochloric acid reagent, mix approximately equal volumes of hydrochloric acid and water and adjust
by titration so that 100 ml. of this solution contains 20.5-21.0 g. of
hydrogen chloride.

TABLE 52. COMPOSITION OF SOYBEAN FLOUR

Moisture	7.65[a]	
Protein	40.65	40.8[b]
Fat	20.38	18.3
Starch	traces	2.6
Dextrin	0.00	
Sucrose	5.26	
Reducing soluble carbohydrates	0.00	
Unusable carbohydrate		18.0
Stachyose	5.66	
Araban	4.83	
Galactan	6.18	
Cellulose	1.63	
Ash	6.08	
Phosphatides calculated as lecithin P_2O_4 x 11	3.08	

[a] Kupelwieser, *Veredeltes Soyamehl. Das Oesterreichische Gesundheitswesen.* p. 198,
1932.
[b] *Soybean Flour.* Battle Creek Food Co., 1936.

Procedure.—Weigh accurately a sufficient quantity of finely
ground sample, which should readily pass through a 20-mesh sieve,
to represent 0.5-1.0 g. of starch. Transfer to a funnel fitted with a
9-cm. Schleicher and Schüll No. 589 or Whatman No. 40 filter paper
and extract by nearly filling the filter 4 times with ethyl ether; likewise extract with alcohol, 70 per cent by volume, and with water.
Allow to drain 1 hour uncovered. Transfer the drained filter and
contents to a 50-ml. beaker. In the next step use a stirring rod having
a flattened button-like end 15 mm. in diameter, and, which is very
important, tamp with a twisting motion during the time specified in
order to get the filter paper completely disintegrated and thus insure
the complete suspension of the starch in the hydrochloric acid solution but not to hydrolyze any of it. Complete the maceration while
there is a small amount of hydrochloric acid present and the whole
contents is a rather thick paste. If this optimum condition is obtained, practically duplicate results will follow. Add the hydrochloric
acid reagent at 15° C. to the beaker containing the sample, using a
fast delivering 10-ml. Mohr pipette with 1 ml. marked off at the lower
end with heavy pencil marks. Keep the acid supply on the bench,
but do not allow it to get above 18° C.

Add the hydrochloric acid in the quantities given. Add 1 ml., tamp 1 minute; add 1 ml., tamp 2 minutes; add 1 ml., tamp 2 minutes; add 1 ml., tamp 1 minute; add 1 ml., tamp 1 minute; add 1 ml., tamp 1 minute; add 1 ml., tamp 1 minute.

Fill the beaker half full with the acid and stir for 30 seconds. Fill the beaker ¾ full and stir for 30 seconds. In the 10 minutes during this treatment the paper should be completely disintegrated and in a smooth state of suspension, the tamping should be continued vigorously during this time, and as little time as possible should be spent adding the acid. Immediately transfer to a 100-ml. wide-mouthed volumetric flask, rinsing out the beaker with the hydrochloric acid; carefully make to volume with the hydrochloric acid reagent and add 0.5 ml. for the volume of filter paper. This step should require 2 minutes. Shake the stoppered flask vigorously for 5 minutes, and allow to stand for 5 minutes in a beaker of water at 20° C. Shake twice and filter immediately into a 250-ml. suction flask through a small Büchner funnel, 41 mm. in diameter, fitted with a thin layer of asbestos and filled half full with dry, fluffy asbestos. The filtration requires 1 minute only. Immediately pipette 50 ml. of the filtrate into a tall form 200-ml. beaker containing 115 ml. of 95 per cent alcohol. The quantity of starch finally weighed will then vary from 0.25-0.5 g. The time consumed from the initial addition of the acid is 24 minutes. Allow the pipette to drain completely and then stir with a whipping motion for 1 minute to flocculate the precipitated starch. Wash down the sides of the beaker with 70 per cent alcohol. Allow to stand 3-4 minutes, until nearly all the precipitate has settled, and then carefully decant the supernatant liquid, which is somewhat turbid, so that little or no precipitate passes into the weighed Gooch crucible, which has been fitted with a thin pad of ignited asbestos and is half filled with fluffy ignited asbestos. Wash the precipitate, and filter by decantation, using successively two 40-ml. portions of 70 per cent alcohol, then 4 times, using 30-ml. portions of 95 per cent alcohol, each time breaking up the precipitate by rapid stirring and allowing the precipitate to settle before decantation. After each stirring rinse the sides of the beaker with a small stream of alcohol to prevent the starch from drying and sticking. Finally transfer the starch completely by means of a jet of 95 per cent alcohol and wash the sides of the Gooch and precipitate with a little of the alcohol. All these filtrations are very fast. Dry the crucible and contents uncovered for 2 hours at 130° C.; cover the crucible immediately and place in a desiccator charged with phosphorus pentoxide, fresh sulfuric acid, or freshly ignited calcium oxide; cool 10 minutes and weigh. Multiply the result by 2 and report as starch.

Pentosans

The determination of pentosans, for practical purposes, may be said to depend upon their conversion to furfural or to methylfurfural by distilling with a mineral acid of the proper concentration and subsequent determination of the furfural by oxidation, precipitation by various reagents or by colorimetric means. The A. O. A. C.[42] recommends precipitation with phloroglucinol. The precipitation of furfural with phloroglucinol is not quantitative, and if hydroxymethylfurfural which is of hexose origin is present a small amount of this substance is precipitated along with the furfural. Furthermore, the phloroglucide of furfural is not a compound of definite composition, hence this method has serious errors.

Thiobarbituric Acid Method.—Barbituric acid also precipitates furfural and is recommended by Jager and Unger [43] as better than phloroglucinol since no hydroxymethylfurfural is precipitated. Bailey [44] suggests a method based on the conversion of pentosans to furfural and methylfurfural, their separation by steam distillation and their subsequent precipitation as furfuralmalonylthiourea by means of thiobarbituric acid. Thiobarbituric acid precipitates furfural and methylfurfural quantitatively forming definite compounds of uniform composition such as furfuralmalonylthiourea and, although it reacts with hydroxymethylfurfural, the resulting product is soluble, coloring the solution but not interfering with the determination of pentoses.

Procedure.—Place from 0.1 to 3 g. of material according to the pentosan content into a 125-ml. distillation flask and add 50 ml. of 12 per cent hydrochloric acid. Distill with steam, passing in a stream of steam at a moderate, constant rate throughout the distillation. The temperature, as measured by a thermometer in the vapor in the neck of the flask, should be maintained between 103 to 105° C. by boosting the distillation flask with a burner. Continue distillation until a small sample of the distillate in thiobarbituric acid solution gives no precipitate or turbidity after standing 5 minutes. Aniline acetate paper is worthless for any hydroxymethylfurfural formed from hexoses present will give a positive reaction with this reagent. Precipitate the furfural from the distillate by adding a slight excess of thiobarbituric acid in 12 per cent hydrochloric acid at room temperature and allow to stand overnight. Filter the lemon-yellow com-

---

[42] *Methods of Analysis,* A.O.A.C. Washington, 1945.
[43] Jager and Unger, *Ber.* **36**, 1222 (1903).
[44] Bailey, *Ind. Eng. Chem., Anal. Ed.* **8**, 389 (1936).

pound formed on a tared Gooch crucible, dry at 105° C., and weigh as furfuralmalonylthiourea.

$$CH\!\!-\!\!CH \quad CO\!\!-\!\!NH \quad CH\!\!-\!\!CH \quad CO\!\!-\!\!NH$$

furfural   thiobarbituric   furfuralmalonylthiourea
                  acid

Values obtained in the presence of methylfurfural are only very slightly in error, because the molecular weight of the residue is between furfural and methylfurfural. Calculations may be based on the following relationships:

$$\text{Pentosan} \rightarrow \text{furfural} + 2H_2O$$

$$\text{Weight of pentosan} = \frac{\text{furfural} + 2H_2O}{\text{furfuralmalonylthiourea}} = 59.6\%$$

that is:

Weight of pentosan = 0.596 × weight of furfuralmalonylthiourea.

Other methods for the determination of pentosans are presented by Hughes and Acree [45] and by Bates. [46]

ARABANS

The relative insolubility of arabinose diphenylhydrazone has been utilized by Neuberg and Wohlgemuth [47] as the basis of estimating arabinose in the presence of other monosaccharides. Mannose and fucose may interfere.

Procedure.—After hydrolysis, concentrate 100 ml. of the neutral solution to 30 ml. heat, on a water bath for 0.5 hour with 6 g. of diphenylhydrazine in 50 ml. of 96 per cent alcohol under a reflux condenser. Collect the precipitated hydrazone on a tared Gooch crucible, wash with 50 ml. of 50 per cent alcohol, dry, and weigh. The arabinose factor is 0.4747.

GALACTAN

Galactans yield galactose on hydrolysis and galactose may be converted to mucic acid by oxidation with nitric acid under the proper

[45] Hughes and Acree, *Ind. Eng. Chem., Anal. Ed.* **6**, 123 (1934); **9**, 318 (1937).
[46] Bates, *Polarimetry, Saccharimetry, and the Sugars. Nat. Bur. Standards (U. S.) Circ.* **C**440 (1942).
[47] Neuberg and Wohlgemuth, *Z. physiol. Chem.* **35**, 31 (1902).

conditions. These facts are the basis of Tollen's method for the estimation of galactan for which the A. O. A. C.[48] gives the following details.

Extract a convenient quantity of the sample, representing 2.5-3 g. of the dry material, on a hardened filter with 5 successive portions of 10 ml. of ether; place the extracted residue in a beaker, about 5.5 cm. in diameter and 7 cm. deep; add 60 ml. of nitric acid (sp. gr. 1.15); and evaporate on a steam bath to a volume of 20 ml. Let stand 24 hours, then add 10 ml. of water and allow to stand another 24 hours. Pass through a filter and wash the impure mucic acid crystals with 30 ml. of water to remove as much of the nitric acid as possible, and return the filter and contents to the original beaker. Add 30 ml. of ammonium carbonate solution (consisting of 1 part ammonium carbonate, 19 parts water, and 1 part of ammonium hydroxide) and heat the mixture in a water bath, at 80° C., for 15 minutes, with constant stirring. The ammonium carbonate combines with the mucic acid, forming soluble ammonium mucate. Wash the filter paper and contents several times with hot water by decantation, passing the washings through a filter paper, to which finally transfer the residue, and wash thoroughly. Evaporate the filtrate to dryness on a water bath, avoiding unnecessary heating which causes decomposition; add 5 ml. nitric acid (sp. gr. 1.15); stir the mixture thoroughly; allow to stand for 30 minutes. Collect the precipitated mucic acid on a weighed Gooch crucible or other filter; wash with 10-15 ml. of water, then with 60 ml. of 95 per cent alcohol, and then a number of times with ether; dry at the temperature of boiling water for 3 hours; and weigh. Multiply the weight of the mucic acid by 1.33 to convert to galactose and by 1.20 to convert to galactan.

These factors are empirical and are based on the investigational work of Tollens.[49]

MANNANS

All reducing sugars form hydrazones, but the hydrazone of mannan is particularly insoluble and thus can be utilized for the quantitative estimation of mannans [50] and mannose.

Procedure.—After hydrolysis of the mannan, add to the solution containing about 1 g. of mannose in 16.6 ml. a solution of 1.2 ml. of phenylhydrazine and 1.2 ml. of glacial acetic acid made up to 6 ml. with water. Allow to stand for 8 hours at a temperature not above 10° C. Collect the hydrazone on a Gooch crucible and wash with

---

[48] *Methods of Analysis*, A.O.A.C. Washington, 1945.
[49] Browne, *Sugar Analysis*. Wiley, New York, 1912.
[50] Bourquelot and Hérissey, *Compt. rend.* 129, 339 (1899).

15 ml. of ice water, 10 ml. of absolute alcohol, and 10 ml. of ether. Dry the precipitate in a vacuum over sulfuric acid.

Theoretically 1 g. of mannose yields 1.5 g. of phenylhydrazone, but, since 40 mg. of the hydrazone is soluble in 100 ml. of solution, correction for this solubility increases the precision of the analysis.

Uronic Acids

As noted in the preceding sections of this chapter a number of the gums and the pectins contain galacturonic and glycuronic acid residues. Sugar-cane juice contains from 0.1 to 0.6 per cent (based on ash free solids) of uronic acids, and these are concentrated to about 2 per cent in cane molasses.[51] When uronic acids are heated with 3.290 N (12 per cent) hydrochloric acid, they are decomposed with the formation of furfural and carbon dioxide. The yield of carbon dioxide is quantitative but that of furfural is less than theoretical, hence estimation of the former can be made the basis of the determination of uronic acids.[52,53] The carbon dioxide evolved can be estimated by measuring the amount absorbed in barium hydroxide solution with subsequent determination of the excess barium hydroxide or, gravimetrically, by adsorption on Ascarite in weighed tubes.

FLOUR

Definitions and standards of identity have been promulgated by the Food and Drug Administration for wheat flour and related products and for corn flour and related products.[53a]

Flour may be analyzed by many of the methods detailed in other sections of the text. Thus moisture, ash, protein, may be determined by the methods detailed in Chapter I. Crude fiber may be estimated as detailed in the next chapter. Fat may be determined by a modification of the acid hydrolysis method detailed for cheese, and metals and inorganic components by methods given in Chapter V and in the chapter on inorganic determinations. There are a number of practical tests applied to flour which will be detailed.

Moisture

Weigh about 2 g. accurately into a tared dish, place in an oven thermostatically controlled at 100° C. for 5 hours, remove, place in a desiccator to cool, and weigh. Replace in oven and heat, cool, and weigh, repeating until constant.

[51] Browne and Phillips, *Int. Sugar J.* **41**, 430 (1939).
[52] Dickson, Otterson, and Link, *J. Am. Chem. Soc.* **52**, 774 (1930).
[53] Whistler, Martin, and Harris, *J. Research Nat. Bur. Standards* **24**, 13 (1940).
[53a] *U. S. Food Drug Admin.*, *S.R.A.*, *F.D.&C.* 2, Part 15, March, 1953, as amended 1955.

Tag-Happenstahl Method.—This method [54] is based on the measurement of the change in electrical conductivity of a grain attributable to the change in resistance of the grain with difference in moisture content.

The apparatus consists of two electrodes in the form of corrugated rolls. These are rotated toward each other at a definite speed. The grain being analyzed is fed in a layer one kernel deep between the revolving rollers. The resistance is measured and is the resultant of the resistance for all the individual grains passing between the electrodes. A graph is prepared by plotting readings obtained with samples of known moisture content, and from this graph the moisture content of test samples can be obtained.

GLUTEN

Mix 10 g. of the sample with sufficient water to make a stiff dough. Allow to stand for 1 hour. Knead the mass in a piece of linen under running water until the washings are clear. The fresh gluten obtained by this treatment should have a faint yellow tinge, its consistency should permit it to be pulled out into threads, and it should be tough. Red and gray residues are considered inferior samples. Wheat flour should contain over 7 per cent.

The color of the gluten is sometimes indicative. Thus the gluten from wheat and rye flours is dark and viscous and is nonhomogeneous. Wheat and barley flour gluten is dark, dirty reddish brown, and nonviscous. Wheat and corn flour gluten is yellowish and nonelastic. The gluten from wheat and oat mixtures is dark yellow, and that from wheat and leguminous flours is colored from grayish red to green.

ABSORPTION

Weigh 25 g. of flour into a round-bottom dish. Add 10-12 ml. of water from a burette. Stir the flour to form a smooth dough. Add more water drop by drop. Knead the dough ball with the fingers, adding water until no more dough adheres to the hand. This is considered the point of highest adsorption. A 60 per cent absorption is rated fair.

BLEACHED FLOUR

Shake 0.5 oz. of flour with 50 ml. of petroleum ether or high-grade gasoline. Allow the mixture to settle. Unbleached flour will impart a yellow color to the supernatant liquid. Bleached flour gives no color to the supernatant solvent.

[54] Coleman, *Cereal Chem.* **8**, 328 (1931).

STARCH

Starch may be analyzed by many of the methods detailed. A rapid method for total starch is the acid conversion method.

Procedure.—Treat 3 g. of sample with 50 ml. of cold water for 1 hour stirring frequently. Collect the residue on a filter and wash with sufficient water to make a total volume of 250 ml. The filtrate contains the soluble carbohydrates.

Heat the undissolved residue for 2.5 hours with 2.5 per cent hydrochloric acid, prepared by adding 20 ml. of hydrochloric acid, sp. gr. 1.12 to 200 ml. of water, in a flask equipped with a reflux condenser. Cool, neutralize with sodium hydroxide solution, make up to 250 ml., filter, and determine the dextrose by the Munson and Walker method. Multiply the weight of the dextrose by 0.90 to obtain the weight of starch.

RHEOLOGICAL METHODS

Wheat flour contains two principal protein components, gliadin and glutenin, commonly considered as one factor under the name of gluten. As Amos [55] points out, in a properly made dough the gluten takes the form of an interwoven network of fibrils which constitute the skeletal structure of the mass of dough. The nature of this network, and consequently the number and nature of the individual gluten fibrils, must be such that the dough can be inflated with gas to a suitable degree without serious distortion of the moulded shape and can continue to exhibit a fine and even vesiculation internally.

The dough testing instruments used are based on empirical principles that place them into two main categories. First there are the group in which the resistance of the dough to a given mechanical mixing operation is recorded graphically through a continuous mixing period of 5 to 20 minutes. The second type consists of devices in which the dough is stretched until it breaks, generally a matter of seconds, and in which the stretching force and the corresponding amount of stretch are also recorded in a graphical form.

Farinograph.—The Brabender Farinograph consists of a twin-bladed, water-jacketed mixer directly coupled to an electric motor which is supported so that it is free to rotate in a vertical plane and to the housing of which is connected a pen that can write on a clock-work driven band of ruled paper. When a dough is being mixed in the mixer of this instrument, the resistance that it offers to the movement of the mixer blades causes a rotation of the suspended motor and a consequent movement of the recording pen.

[55] Amos, *Analyst* 74, 392 (1949).

The mean height of the band that the pen produces is directly related in the early stages of the mixing operation to the proportion of water that has been added to the flour. Consequently if the height of the band which corresponds to the dough consistency desired by the baker is known, the instrument can be used to estimate the water absorption of a flour. As the mixing operation continues the height of the band produced decreases because of the continued mechanical abuse of the dough. Flours that exhibit a prolonged dough development time are usually "strong" flours.

In the Farinograph method, the doughs are made up to the same consistency. A preliminary test is performed in order to estimate the proportion of water that needs to be added to the flour to get the standard of consistency that has been adopted.

Mixograph.—Another device of this type is the Mixograph. The mixing head carries four vertical pins which move among four fixed vertical pins in the mixing bowl with a planetary motion when the head is lowered into position. The tendency for the bowl to rotate as a consequence of the pressure exerted on its fixed points is impeded by a standardized spring in such a manner that the torque can be recorded.

Alveographe.—In this device, a disk of non-yeasted dough of fixed dimensions is clamped across an air inlet and then is blown at constant temperature into a bubble until it bursts; the varying pressure within the bubble is depicted throughout the test by a recording manometer.

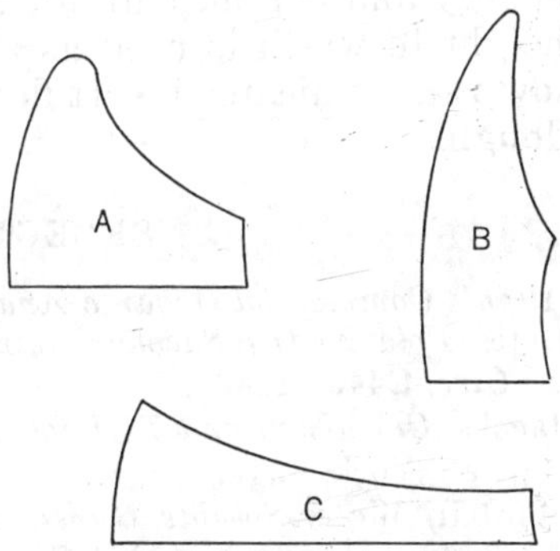

FIG. 68. Alveographe Curves of Equal Area

A. Strength 90
   Stability 105
   Extensibility 66
B. Strength 90
   Stability 165
   Extensibility 42
C. Strength 90
   Stability 61
   Extensibility 93

The doughs used in the Alveographe vary in consistency but they are prepared in such a manner that the ratio of total water—that is, the amount of water in the flour itself plus the added water—to the dry solids is a constant.

Curves of the type of Fig. 68 are obtained with this device. The height of the peak is measured as an index of the stability of the dough, the length of the base of the curve is a measure of its extensibility, and the area enclosed by the curve is a measure of the "strength" of the dough. Curve B shows excessive stability and poor distensibility, a clay-like dough. Curve C indicates high distensibility but poor stability, that is, it is a soft, flowy dough. Curve A shows a dough of good breadmaking qualities.

Extensograph.—The Extensograph is an instrument in which a cylinder of dough supported at each end has a force applied mechanically to the middle so that it is stretched downward into the form of an elongated U until it breaks. The applied force and the degree of extension are recorded on a moving band of ruled paper. As in the Farinograph, the doughs are tested at constant consistency and consequently contain different amounts of water. The maximum height of the curve is termed the resistance; the width of the curve represents the extensibility; and the area enclosed is referred to as the energy and is a measure of the strength. The ratio of the maximum height to width is used as an index of the balance of properties; a low ratio indicates a soft flowy dough and a high ratio a stiff, tough dough.

SELECTED REFERENCES

Allen's Commercial Organic Analysis, 5th Ed. Blakiston, Philadelphia, 1923.

Bates, *Polarimetry, Saccharimetry, and the Sugars, Nat. Bur. Standards, U. S. Circ.* **C440** (1942).

Jacobs, *Chemistry and Technology of Food and Food Products*. Interscience, New York, 1944.

Mantell, *Water-Soluble Gums*. Reinhold, New York, 1947.

Methods of Analysis, A.O.A.C., 8th Ed. Washington, 1955.

Osol and Farrar, *Dispensatory of the United States of America*, 24th Ed. Lippincott, Philadelphia, 1947.

U. S. Food Drug Admin., S.R.A., F.D.&C. 2, Part 15, March, 1953; as amended 1955; Part 17, June, 1952.

CHAPTER XII

JAMS, JELLIES, AND FRUITS

JAMS, preserves, jellies and fruit butters may on the one hand be classed with sugar foods and on the other with fruits and vegetables. Their characteristic properties may be more easily determined from their relationships with fruits rather than with sugars. The methods for the analysis of syrups, maple products, honey, and confectionery have received a great deal of attention and are well adapted for their purpose. On the other hand, those developed for jams, preserves, jellies and fruit butters, although simple, are time consuming and require much interpretation. This is because of the wide variation in the types and kinds of fruit and fruit combinations making the jam or jelly product.

The chemists of the Food and Drug Administration have done considerable work on the determination of the composition of fruits used in the preparation of preserves, jams, and jellies. On the basis of this work and on hearings held at various times definitions have been drawn for these products. These definitions are detailed below.

JAMS AND PRESERVES

Preserves, Jams.—*Identity.*—(a) The preserves or jams for which definitions and standards of identity are prescribed by the United States Food and Drug Administration [1, 2] are the viscous or semisolid foods each of which is made from a mixture composed of not less than 45 parts by weight (see (c) below) of one of the fruit ingredients specified in (b) below to each 55 parts by weight (see (e) (1) below) of one of the optional saccharine ingredients specified in (d) below. Such mixture may also contain one or more of the following optional ingredients:

(1) Spice.

(2) A vinegar, lemon juice, lime juice, citric acid, lactic acid, malic acid, tartaric acid, or any combination of two or more of these, in a quantity which reasonably compensates for deficiency, if any, of the natural acidity of the fruit ingredient.

---

[1] *Federal Register* 5, 3554 (1940).
[2] *U. S. Food Drug Admin., S.R.A., F.D.&C.* 2, Part 29, August, 1954. as amended October, 1955.

509

(3) Pectin, in a quantity which reasonably compensates for deficiency, if any, of the natural pectin content of the fruit ingredient.

(4) Sodium citrate, sodium potassium tartrate, or any combination of these, in a quantity the proportion of which is not more than 3 oz. avoirdupois to each 100 lbs. of the saccharine ingredient used.

(5) Sodium benzoate or benzoic acid or any combination of these, in a quantity reasonably necessary as a preservative.

Such mixture, with or without added water, is concentrated by heat to such point that the soluble solids content of the finished preserve is not less than 68 per cent if the fruit ingredient is specified in group I of (b) below, and not less than 65 per cent if the fruit ingredient is specified in group II of (b) below. The soluble solids content is determined by a prescribed method (equivalent to the method detailed for soluble solids in this chapter) except that no correction is made for water-insoluble solids.

(b) The fruit ingredients referred to in (a) above are the following mature, properly prepared fruits which are fresh, frozen, and/or canned:

GROUP I

Blackberry (other than dewberry)	Huckleberry
Black Raspberry	Loganberry
Blueberry	Orange
Boysenberry	Pineapple
Cherry	Raspberry, Red Raspberry
Crabapple	Rhubarb
Dewberry (other than boysenberry, loganberry, and youngberry)	Strawberry
	Tangerine
Elderberry	Tomato
Grape	Yellow Tomato
Grapefruit	Youngberry

Any combination of two, three, four, or five of such fruits in which the weight of each is not less than one-fifth of the weight of the combination; except that the weight of pineapple may be not less than one-tenth of the weight of the combination.

GROUP II

Apricot	Peach
Cranberry	Pear
Damson, damson plum	Plum (other than greengage plum and damson plum)
Fig	
Gooseberry	Quince
Greengage, greengage plum	Red Currant, Currant (other than black currant)
Guava	
Nectarine	

Any combination of two, three, four, or five of such fruits, or one or more of such fruits with one or more of the individual fruits specified in group I, in which the weight of each is not less than one-fifth of the weight of the combination; except that the weight of pineapple may be not less than one-tenth of the weight of the combination.

Any combination of apple and one, two, three, or four of the individual fruits specified in this group or group I, in which the weight of each is not less than one-fifth, and the weight of apple is not more than one-half, of the weight of the combination; except that the weight of pineapple may be not less than one-tenth of the weight of the combination.

In any such combination of two, three, four, or five fruits, each such fruit is an optional ingredient. For the purposes of this section, the word "fruit" includes the vegetables specified in this section.

(c) Any requirement of this section with respect to the weight of any fruit, combination of fruits, or fruit ingredient means—

(1) The weight of fruit exclusive of the weight of any sugar, water, or other substance added for any processing or packing or canning, or otherwise added to such fruit;

(2) In the case of fruit prepared by the removal, in whole or in part, of pits, seeds, skins, cores, or other parts, the weight of such fruit exclusive of the weight of all such substances removed therefrom; and

(3) In the cases of apricots, cherries, grapes, nectarines, peaches, and all varieties of plums, whether or not pits and seeds are removed therefrom, the weight of such fruit exclusive of the weight of such pits and seeds.

(d) The optional saccharine ingredients referred to in (a) are—

(1) Sugar.

(2) Invert sugar syrup.

(3) Any combination composed of optional saccharine ingredients (1) and (2).

(4) Any combination composed of corn sugar or dextrose and optional saccharine ingredient (1), (2), or (3).

(5) Any combination composed of corn syrup, dried corn syrup, glucose syrup, dried glucose syrup, or any two or more of the foregoing, with optional saccharine ingredient (1), (2), (3), or (4), in which the weight of the solids of corn syrup, dried corn syrup, glucose syrup, and dried glucose syrup—in case two or more of these are used—does not exceed one-fourth of the total weight of the solids of the combined saccharine ingredients.

(6) Honey.

(7) Any combination composed of honey and optional saccharine ingredient (1), (2), or (3), in which the weight of the solids of each component except honey is not less than one-tenth of the weight of the solids of such combination and the weight of honey solids is not less than two-fifths of the weight of the solids of such combination.

(e) For the purposes of this definition—

(1) The weight of any optional saccharine ingredient means the weight of the solids of such ingredient.

(2) The term "sugar" means refined sugar (sucrose).

(3) The term "invert sugar syrup" means a syrup made by inverting or partly inverting sugar or partly refined sugar; its ash content is not more than 0.3 per cent of its solids content, but if it is made from partly refined sugar, color and flavor other than sweetness are removed.

(4) The term "corn syrup" means a clarified, concentrated aqueous solution of the products obtained by incomplete hydrolysis of cornstarch. The solids of corn syrup contain not less than 40 per cent by weight of reducing sugars calculated as anhydrous dextrose.

(5) The term "glucose syrup" means a clarified, concentrated aqueous solution of the products obtained by the incomplete hydrolysis of any edible starch. The solids of glucose syrup contain not less than 40 per cent by weight of reducing sugars calculated as anhydrous dextrose. "Dried glucose syrup" means the product obtained by drying "glucose syrup."

(6) The term "dextrose" means refined anhydrous or hydrated dextrose made from any starch.

JELLY

Fruit Jelly.—*Identity.*—(a) The jellies for which definitions and standards of identity are prescribed by the United States Food and Drug Administration [3][4] are the jelled foods each of which is made from a mixture composed of not less than 45 parts by weight (as determined by the method prescribed in (b) below) of one or any combination of two, three, four, or five of the fruit juice ingredients specified in (c) below to each 55 parts by weight (see section (e) above) of one of the optional saccharine ingredients specified in (d). Such mixture may also contain one or more of the optional ingre-

---

[3] *Federal Register* **5**, 3558 (1940).

[4] *U. S. Food Drug Admin., S.R.A., F.D.&C.* 2, Part 29, August, 1954, as amended October, 1955.

dients listed on page 509 *et seq.* and mint flavoring and harmless artificial green coloring, in case the fruit juice ingredient or combination of fruit juice ingredients is extracted from apple, crabapple, pineapple, or two or all of such fruits.

Such mixture is concentrated by heat to such point that the soluble solids content of the finished jelly is not less than 65·per cent, as determined by means of a refractometer.

(b) Any requirement of this section with respect to the weight of any fruit juice ingredient, whether concentrated, unconcentrated, or diluted, means the weight determined by the following method: Determine the percentage of soluble solids in such fruit juice ingredient by means of a refractometer; multiply the percentage found by the weight of such fruit juice ingredient; divide the result by 100; subtract from the quotient the weight of any added sugar or other added solids; and multiply the remainder by the factor for such fruit juice ingredient prescribed in (c) below. The result is the weight of the fruit juice ingredient.

(c) Each of the fruit juice ingredients referred to in (a) above is the filtered or strained liquid extracted with or without the application of heat and with or without the addition of water, from one of the following mature, properly prepared fruits which are fresh, frozen and/or canned:

Name of Fruit	Factor Referred to in Paragraph (b)	Name of Fruit	Factor Referred to in Paragraph (b)
Apple	7.5	Greengage, Greengage Plum	7.0
Apricot	7.0	Guava	13.0
Blackberry (other than dewberry)	10.0	Loganberry	9.5
		Orange	8.0
Black Raspberry	9.0	Peach	8.5
Cherry	7.0	Pineapple	7.0
Crabapple	6.5	Plum (other than damson,	
Cranberry	9.5	greengage, and prune)	7.0
Damson, damson plum	7.0	Pomegranate	5.5
Dewberry (other than		Quince	7.5
boysenberry, loganberry,		Raspberry, Red Raspberry	9.5
and youngberry)	10.0	Red Currant, Currant (other	
Fig	5.5	than black currant)	9.5
Gooseberry	12.0	Strawberry	12.5
Grape	7.0	Youngberry	10.0
Grapefruit	11.0		

In any combination of two, three, four, or five of such fruit juice ingredients the weight of each is not less than one-fifth the weight of the combination. Each such fruit juice ingredient in any such combination is an optional ingredient.

(d) The optional saccharine ingredients referred to in paragraph (a) above are detailed in paragraphs (d) and (e) on pages 511 and 512.

FRUIT BUTTERS

Fruit Butter.—*Identity.*—(a) The fruit butters for which definitions and standards of identity are prescribed by the United States Food and Drug Administration[5] are the smooth, semisolid foods each of which is made from a mixture composed of not less than five parts by weight (as determined by the method prescribed in (b) (1)) of one or any combination of two, three, four, or five of the optional fruit ingredients specified in (c) to each two parts by weight (see (e) (1)) of one of the optional saccharine ingredients specified in (d), except that the use of such saccharine ingredient is not required when optional ingredient (5) is used. Such mixture may be seasoned with one or more of the following optional ingredients:

(1) Spice.
(2) Flavoring (other than artificial flavoring).
(3) Salt.
(4) A vinegar, lemon juice, lime juice, citric acid, lactic acid, malic acid, tartaric acid, or any combination of two or more of these.

Such mixture may also contain the optional ingredient:

(5) Fruit juice or diluted fruit juice or concentrated fruit juice in a quantity not less than one-half the weight of the optional fruit ingredient.

Such mixture is concentrated by heat to such point that the soluble solids content of the finished fruit butter is not less than 43 per cent as determined by a prescribed method (equivalent to the method detailed for soluble solids in this chapter), except that no correction is made for water-insoluble solids.

(b) (1) Any requirement of this definition with respect to the weight of any optional fruit ingredient, whether concentrated, unconcentrated, or diluted, means the weight determined by the following method:

Determine the percentage of soluble solids in the optional fruit ingredient by the method prescribed for determining soluble solids in (a) above; multiply the percentage so found by the weight of such

---

[5] *Federal Register* **5**, 3561 (1940); *U. S. Food Drug Admin., S.R.A., F.D.&C.* **2**, Part 29, August, 1955, as amended October, 1955.

ingredient; divide the result by 100; subtract from the quotient the weight of any added sugar or any other added solids; and multiply the remainder by the factor for such ingredient prescribed in (c) below. The result is the weight of the optional fruit ingredient.

(2) For the purposes of this section, the weight of fruit juice or diluted fruit juice or concentrated fruit juice (optional ingredient (5)) from a fruit specified in (c) below is the weight of such juice as determined by the method prescribed in (b) (1) above, except that the percentage of soluble solids is determined by means of a refractometer; the weight of diluted or concentrated juice from any other fruit is the original weight of the juice before it was diluted or concentrated.

(c) Each of the optional fruit ingredients referred to in (a) above is prepared by cooking one of the following fresh, frozen, canned, and/or dried (evaporated) mature fruits, with or without added water, and screening out skins, seeds, pits, and cores:

Name of Fruit	Factor Referred to in Paragraph(b)(1)	Name of Fruit	Factor Referred to in Paragraph(b)(1)
Apple	7.5	Plum (other than prune)	7.0
Apricot	7.0	Prune	7.0
Grape	7.0	Quince	7.5
Peach	8.5		
Pear	6.5		

In any combination of two, three, four, or five fruit ingredients, the weight of each is not less than one-fifth the weight of the combination.

(d) The optional saccharine ingredients referred to in (a) above are:

(1) Sugar.

(2) Invert sugar syrup.

(3) Brown sugar.

(4) Invert brown sugar syrup.

(5) Honey.

(6) Any combination of two or more of optional saccharine ingredients (1), (2), (3), and (4).

(7) Any combination of dextrose and optional saccharine ingredient (1), (2), (3), (4), or (6).

(8) Any combination composed of corn syrup, dried corn syrup, glucose syrup, dried glucose syrup, or any two or more of the foregoing with optional saccharine ingredient (1), (2), (3), (4), (6), or (7), in which the weight of the solids of corn syrup, dried corn syrup, glucose syrup, dried glucose syrup or

the sum of the weights of the solids of these in case two or more of them are used, does not exceed one-fourth the total weight of the solids of the combined saccharine ingredients.

(9) Any combination of honey and optional saccharine ingredient (1), (2), (3), (4), (6), or (7), in which the weight of the solids of each component except honey is not less than one-tenth of the weight of the solids of such combination, and the weight of honey solids is not less than two-fifths of the weight of the solids of such combination.

(e) For the purposes of this definition

(1) The weight of any optional saccharine ingredient means the weight of the solids of such ingredient.

(2) The term "sugar" means refined sugar (sucrose).

(3) The term "invert sugar syrup" means a syrup made by inverting or partly inverting sugar or partly refined sugar; its ash content is not more than 0.3 per cent of its solid content, but if it is made from partly refined sugar, color and flavor other than sweetness are removed.

(4) The term "invert brown sugar syrup" means a syrup made by inverting or partly inverting brown sugar.

(5) The term "corn syrup" is defined on page 512.

(6) The term "glucose syrup" is defined on page 512.

(7) The term "dextrose" means refined anhydrous or hydrated dextrose made from any starch.

Jams, preserves, and jellies may be processed in a number of ways, some of which are considered adulterations. Thus one is the addition of thickening agents such as added pectin in excess of that needed to compensate for a deficiency, if any, of the natural pectin content of the fruit ingredient, gums, gelatin, dextrins, and like products. Secondly, waste or rather exhausted material may be added. Thirdly, foreign substances may be added such as color, preservatives, saccharin, fruit acids or lactic acid in excess of that which reasonably compensates for deficiency, if any, of the natural acidity of the fruit ingredient, mineral salts or acids, and many other substances. Fourthly, there may be a fruit shortage or conversely a sugar increase. For the detection of gums, gelatin, dextrins, color, preservatives, saccharin and mineral salts the methods previously detailed in the book may be applied. The other types of adulteration may be detected by determining insoluble solids, soluble solids by refractometer, alcohol precipitate, pectic acid, ash, and the phosphorus pentoxide content especially in grape products.

Preparation of Solution

Weigh 300 g. of the jelly or jam, both of which have been thoroughly mixed by passing the jam or preserve 3 times through a food chopper (avoiding the cracking of seeds in berry preserves as much as possible), and transfer the weighed portion with the aid of 800 ml. of water to a 2-liter beaker. Boil or heat on a steam bath for an hour or two, replacing the water lost by evaporation. Do not boil if sucrose is to be determined. Transfer the mixture to a 2-liter volumetric flask, allow to cool, dilute to volume, and filter through a fluted filter.

Total Solids

Total solids may be determined in the usual manner by weighing about 20 g. of the well-mixed original sample into a tared dish and drying *in vacuo*, at 70° C. However, for all practical purposes the refractive index of the jam or jelly may be taken and the percentage sugar equivalent to this refractive index as given by the sugar table, Table 3 appendix, may be considered as equivalent to the total solids, with correction in some cases for the insoluble solids. The percentage of total solids may be obtained directly refractometrically if the refractometer has a calibrated sugar scale. Obtaining the total solids in this manner is equivalent to obtaining soluble solids.

Soluble Solids.—The percentage soluble solids may also be obtained by the refractometer method and correction for insoluble solids may be made by the following formula when necessary.

$$S = \frac{a(100 - b)}{100}$$

where

     S = per cent soluble solids
     a = per cent soluble solids as determined by refractometer
     b = per cent insoluble solids.

This method is at least as accurate as methods based on the evaporation of an aliquot of the sample and drying *in vacuo* at 70° C.

Insoluble Solids.—Weigh 25 g. of the original well-mixed sample (a blendor is useful) into a 400-ml. beaker and add 200-250 ml. of water. Boil for 30 minutes and replace the water lost from time to time. Prepare a low flat-bottomed dish with a cover and a sheet of 15-cm. filter paper by drying to constant weight in an oven at 100-105° C. Filter the hot mixture through the weighed filter **paper and wash**

thoroughly with hot water. The washing may be considered finished when no positive Molisch reaction is obtained or when the wash water no longer reacts acid to litmus paper. Replace the filter paper and its contents in the dish, replace the dish and contents in the oven, and dry to constant weight. The increase in weight may be calculated as per cent insoluble solids. Run in duplicate.

Use 300-g. portions of samples of damson plum preserves, grape jam, and other preserves that contain pits or large seeds. Weigh (4 g.) a piece of absorbent cotton, 5 sq. in. square, and about half as thick as the customary pound roll of absorbent cotton. Tear off one corner and plug the neck of a funnel lightly. Arrange the absorbent cotton in the funnel. A weighed filter paper may also be used as above. Filter the sample after treatment as above through the cotton making certain that all the pulp remains on the cotton. Pour the hot wash water on the cotton in such a way that all the pulp is loosened from the cotton at each washing. Collect about 800 ml. of filtrate. Fold the cotton carefully to enclose all the particles of pulp while still in the funnel and squeeze gently to press out excess water. Transfer to a dish, including the cotton plug, dry and weigh together with the pits or large seeds to constant weight at 100° C. Pick out the pits and seeds, soak thoroughly in water, free from any adhering pulp and dry at 100° C. Subtract the weight of pits and seeds and the weight of the cotton pad or filter paper from the gross weight to obtain the weight of insoluble solids in the 300-g. portion of the sample.

This determination is not usually performed on jellies because these products are presumed to be prepared from the strained juice and therefore should contain practically no insoluble solids.

Seeds in Berry Fruits.[6]—Prepare the sample by thorough mixing, using a blendor. Take 50 g.±.01 g. of the sample, transfer with 500 ml. of hot water to the mixing chamber of the blendor and mix for 1-2 minutes. Transfer the mixture to a 20-mesh screen and use additional hot water to transfer and wash the bare seeds. Hot tap water is suitable for use in this procedure. Transfer the seeds on the screen to a 70-mm. aluminum dish, previously weighed, with a close-fitting cover. This is readily accomplished by transfer to a 7-cm. Whatman No. 4 circle of filter paper in a Coors 2A Büchner funnel with suction. The paper is previously dried and weighed with the aluminum dish. Dry at 100° C. in a forced draft oven for 30 minutes and weigh. To determine the average weight of one seed, count out and weigh

[6] *J. Assoc. Official Agr. Chem.* 32, 95 (1949).

separately several 100-unit lots. Report average weight of one seed in milligrams, number of seeds per 100 g. of sample, and after determination of the water-insoluble solids content of the sample, calculate and report the per cent of the total that is due to bare seeds and the percent that is due to nonseed water-insoluble solids.

Fruit Juice Ingredient [7]

In any fruit juice product prepared for use as the fruit juice ingredient of a jelly, the weight of the original fruit juice is to the weight of such product as the percentage of the soluble fruit solids in such product is to the percentage of the soluble fruit solids in such original fruit juice.

In using this ratio for computing the weight of fruit juice in a fruit juice ingredient, it is customary to assume that the percentage of soluble fruit solids in the juice of any given kind of fruit is the average percentage of soluble fruit solids of the juice of that kind of fruit. Computations based on such assumption are reasonably accurate and are the most practicable method of arriving at the weight of fruit juice used in making jelly.

The fruits from which the fruit juice ingredients of the jellies, for which definitions and standards of identity are prescribed by the United States Food and Drug Administration are obtained; the percentages by weight of the average soluble fruit solids of such respective fruits; and their respective factors, which are the reciprocals of such percentages multiplied by 100 and rounded out to whole or half numbers are given in Table 53.

The following is a convenient and reasonably accurate method for determining the weight of a fruit juice ingredient.

Procedure.—Determine the percentage of soluble solids in the prepared fruit juice ingredients; multiply the percentage so found by the weight of such prepared fruit juice ingredient; divide the result by 100; subtract from the quotient the weight of any added sugar or other added solids; and multiply the remainder by the factor for the fruit from which the fruit juice was obtained. The result is the weight of the fruit juice present in such prepared fruit juice ingredient.

Fruit Substance in Fruit Butters [8]

In any fruit product screened for use in fruit butter, the weight of the original fruit substance, that is, the fruit less its skins, seeds

---

[7] *Federal Register* 5, 3558 (1940).
[8] *Federal Register* 5, 3562 (1940).

pits, and cores, is to the weight of such product as the percentage of soluble fruit solids in such product is to the percentage of soluble fruit solids in such original fruit substance.

TABLE 53. AVERAGE PERCENTAGE OF SOLUBLE FRUIT SOLIDS [a]

Name of Fruit	%	Factor
Apple	13.7	7.5
Apricot	14.4	7.0
Blackberry (other than dewberry)	10.0	10.0
Black Raspberry	11.2	9.0
Cherry	13.9	7.0
Crabapple	15.4	6.5
Cranberry	10.6	9.5
Damson, Damson plum	14.8	7.0
Dewberry (other than boysenberry, loganberry, and youngberry)	10.0	10.0
Fig	19.0	5.5
Gooseberry	8.2	12.0
Grape	14.1	7.0
Grapefruit	9.1	11.0
Greengage, Greengage Plum	14.8	7.0
Guava	7.6	13.0
Loganberry	10.6	9.5
Orange	12.7	8.0
Peach	11.8	8.5
Pineapple	14.6	7.0
Plum (other than Damson, Greengage, and prune)	14.8	7.0
Pomegranate	17.6	5.5
Quince	13.2	7.5
Raspberry, Red Raspberry	10.5	9.5
Red Currant, Currant (other than black currant)	10.6	9.5
Strawberry	8.0	12.5
Youngberry	10.0	10.0

[a] *Federal Register* 5, 3558 (1940).

In using this ratio for computing the weight of original fruit substance, it is customary to assume that the percentage of soluble fruit solids in any given kind of fruit is the average percentage of soluble fruit solids in that kind of fruit. Computations based on such assumption are reasonably accurate and are the most practicable method of arriving at the weight of fruit substance used in making fruit butter.

The fruits from which the fruit ingredients of the fruit butters, for which definitions and standards of identity are prescribed by the United States Food and Drug Administration are obtained; the percentages by weight of the average soluble fruit solids in the fruit substance of such respective fruits; and their respective factors which are the reciprocals of such percentages multiplied by 100 and rounded out to whole or half numbers, are as follows:

Name of Fruit	Average Percentage of Soluble Fruit Solids	Factor
Apple	13.7	7.5
Apricot	14.4	7.0
Grape	14.1	7.0
Peach	11.8	8.5
Pear	15.5	6.5
Plum (other than prune)	14.8	7.0
Prune	14.8	7.0
Quince	13.2	7.5

The following is a convenient and reasonably accurate method for determining the weight of original fruit substance used in making fruit butter.

Procedure.—Determine the percentage of soluble solids in the screened fruit; multiply the percentage so found by the weight of the screened fruit; divide the result by 100; subtract from the quotient the weight of any added sugar or other added solids; and multiply the remainder by the factor for the fruit from which such screened fruit was obtained. The result is the weight of the original fruit substance in such screened fruit.

ALCOHOL PRECIPITATE

In this determination the amount of combustible alcohol insoluble material present in the preserve or jelly is calculated.

Procedure.—Transfer 100 ml. of the prepared solution by means of a pipette to a tall form 300-ml. beaker and evaporate the solution to 20 ml. If any insoluble matter appears, add 1 or 2 lumps of cube sugar or 4-8 g. of granulated sugar and continue the evaporation to 20 ml. Place 200 ml. of 95 per cent alcohol in a separatory funnel or similar apparatus and add the alcohol to the mixture in a thin stream, with constant stirring. Allow the mixture to stand until the flocculent precipitate settles or overnight. Filter through a smooth 15-cm. filter paper. Do not allow the precipitate to dry before transfer to the filter paper. Wash the precipitate with alcohol. Wash the precipitate back into the original beaker with a stream of hot water and wash the filter paper thoroughly with hot water catching the washings in the beaker. Again evaporate the solution to 20 ml. Cool and add 5 ml. of hydrochloric acid (1:2.5). Reprecipitate with 200 ml. of 95 per cent alcohol as directed above. Allow to stand until the flocculent precipitate settles at least 1 hour or overnight and filter through a tared Gooch crucible which has a thin pad of asbestos, if the precipitate is small. Wash thoroughly with 95 per cent alcohol until free of hydrochloric acid. If the precipitate is large enough to clog the Gooch crucible, filter the precipitate on filter paper. Wash well with alcohol and then wash the precipitate into a tared platinum dish with a stream of hot water. Wash the filter paper well with hot water and catch the washings in the platinum dish. Evaporate the precipitate and washings in the platinum dish to dryness and then dry the dish to constant weight. If a Gooch crucible was used, dry to constant weight also. Ignite the dish or crucible and after cooling in a desiccator weigh again. The loss in weight is the alcohol precipitate.

The alcohol precipitate is sometimes almost invisible but rarely so when the jam or jelly contains dextrins or sometimes starch. The precipitate will be copious, gummy, and sticky because of the dextrins if commercial glucose is present. Therefore care must be taken in the various transfers of the precipitate. If the precipitate does not flocculate properly, the addition of sodium chloride will aid in the flocculation. This determination should be made in duplicate.

PECTIC ACID

In this determination the amount of digalacturonic acid or similar acids obtained from the alcohol precipitate is estimated.

Procedure.—Transfer 200 ml. of the prepared solution to a 300-ml. tall form beaker and evaporate to 25 ml. If a precipitate forms during the evaporation, add 2-4 lumps of cube sugar or 8 to 16 g. of granulated sugar. Continue the evaporation to 25 ml. Precipitate with 200 ml. of 95 per cent alcohol as directed in the alcohol precipitate determination and allow the precipitate to flocculate and settle or to stand overnight. Filter through a 15-cm. smooth filter paper. Wash well with alcohol. Return the precipitate to the original beaker by washing with a stream of hot water. Wash the filter well with hot water and catch the washings in the beaker. Evaporate to 40 ml. Cool to room temperature. If any insoluble matter appears, add 1 ml. of 10 per cent hydrochloric acid, warm if necessary to effect solution and cool again. Add 2 to 5 ml. of 10 per cent sodium hydroxide according to the volume of the precipitate. Dilute to 50 ml. with water and allow to stand for 15 minutes. At the end of this time, add 40 ml. of water and 10 ml. of hydrochloric acid (1:25) and boil for 5 minutes. Filter on a smooth 15-cm. filter paper and wash thoroughly with hot water. Wash the precipitate back into the beaker with a stream of hot water and adjust the volume to 40 ml. either by the addition of water or by evaporation to that volume. Allow to cool and repeat the saponification and precipitation detailed above. Filter again through the same filter paper and wash thoroughly with hot water, until free of acid. The filtrations should proceed rapidly and the filtrates should be clear. Wash the filtrate into a tared platinum dish with a stream of hot water. Evaporate to dryness and then dry to constant weight in an oven at 100°-105° C. Allow to cool and weigh. Ignite and weigh again. The loss in weight is equivalent to the pectic acid. If the precipitate is not too voluminous, it may be filtered the second time through an ignited and tared Gooch crucible with a thin pad of asbestos.

This determination should be made in duplicate. If the saponification goes improperly because of insufficient alkali and colloidal precipitates are formed so that the filtration is not rapid and the filtrates are not clear, the determinations should be repeated.

TOTAL ACIDITY

This determination may be made in the usual way by titrating 25 ml. or a larger aliquot of the prepared solution, diluted with 200 to 250 ml. of recently boiled and cooled water, with 0.1 N sodium hydroxide solution using phenolphthalein solution as an indicator. If despite the high dilution the solution to be titrated is still colored deeply enough to obscure the end point, the titration may be made using phenolphthalein powder as an outside indicator, or the titration may be performed with the aid of a potentiometric device. The result is generally calculated as milliliters of 0.1 N sodium hydroxide solution per 100 g. of the original material.

ASH

Evaporate to dryness in a tared porcelain dish or a silica dish 100 ml. of the prepared solution. Ash the residue in the usual manner at not more than dull red heat. This determination gives total soluble ash. If the total ash is desired, 10 to 25 g. of the original well mixed sample should be weighed into a porcelain or silica dish and then ashed as usual. Platinum dishes should not be used especially where an ash high in phosphoric acid is suspected. Add 2-3 drops of ashless olive oil to avoid excessive swelling or foaming during ashing.

Alkalinity of Ash.—The alkalinity number of an ash is defined as the number of milliliters of N acid required to neutralize 1 g. of ash.

To the ash as determined in the foregoing section an excess of 0.1 N hydrochloric acid is added and the mixture is warmed. Cool and back titrate with 0.1 N sodium hydroxide solution using methyl orange as the indicator. Calculate the alkalinity as indicated in the definition. This determination is valueless if added mineral acid is present.

The sulfur content and the phosphorus pentoxide content may be determined as detailed in Chapter XVIII on inorganic determinations.

INTERPRETATION OF RESULTS

As in the determination of the so-called constants of the oils and fats, the estimation of insoluble solids, alcohol precipitate, pectic acid, ash and acidity serve as the constants of jams, preserves, and jellies. Fruits and vegetables as other natural products have compositions that vary within certain limits and the products made from these natural foods according to definite specifications, such as preserves, will also have compositions or constants that vary within limits. In the case of jams and jellies, if these estimations are set up as ratios, the interpretation to be derived from them is far more clear.

As Sale [9] has pointed out, when a quantity of fruit and a quantity of granulated sugar are concentrated by cooking, the resulting mixture will contain the same quantities of water-insoluble solids (seeds and fibrous material), ash (mineral matter), potash, and phosphate that were present in the amount of fruit taken, since granulated sugar will not contribute to the mixture appreciable amounts of any of the components in question and none of these components will be lost in the process of cooking. Analyses of experimental batches of preserves made in the Food and Drug Administration and the Bureau of Chemistry have shown that the percentages of these components in the mixture are in strict proportion to the amount of fruit used. Furthermore, in making up experimental batches in preserve factories and in the laboratory it has been found that the content of fruit in concentrated mixtures of fruit and sugar will be closely approximated when the average values of two or more of the diagnostic components mentioned above are used as a basis of comparison (Table 54).

The total acidity and pectic acid of a market sample are not dependable indices of the fruit content when the ratios of certain components to each other show that the sample contains added acid and added pectin, as explained below.

---

[9] Sale, *J. Assoc. Official Agr. Chem.* 21, 502 (1938).

TABLE 54.—AVERAGE OF RESULTS ON FRUITS OF KNOWN ORIGIN,[a] AUTHENTIC FRUITS COLLECTED AND ANALYZED BY U. S. FOOD AND DRUG ADMINISTRATION

Kind of Fruit	Total Sugars as Invert %	Soluble Solids %	Insoluble Solids %	Ash of Sample Solution %	Phosphate [P₂O₅] mg./100 g.	Phosphate Per Cent in Ash
Apples	11.1 [29]	13.7 [38]	2.34 [25]	0.32 [38]	24 [37]	7 [37]
Apricots	9.8 [40]	14.4 [40]	1.66 [40]	.80 [42]	56 [41]	7. [41]
Blackberries	7.0 [54]	10.0 [54]	6.24 [54]	.46 [55]	43 [53]	9 [53]
Cherries	9.3 [38]	13.9 [37]	1.41 [45]	.48 [52]	44 [45]	9 [45]
Crabapples	11.2 [30]	15.4 [30]	2.98 [18]	.41 [30]	36 [30]	9 [30]
Currants	6.0 [35]	10.6 [35]	5.71 [25]	.58 [35]	48 [31]	8 [31]
Figs	15.8 [16]	19.0 [16]	2.76 [16]	.55 [16]	47 [14]	8 [14]
Gooseberries	3.9 [8]	8.2 [8]	3.12 [8]	.43 [8]	29 [8]	7 [8]
Guavas	4.3 [24]	7.6 [24]	10.24 [24]	.62 [24]	39 [24]	6 [24]
Grapes	11.6 [31]	14.1 [31]	1.47 [19]	.52 [32]	30 [31]	6 [31]
Loganberries	6.3 [35]	10.6 [35]	5.75 [35]	.50 [35]	44 [35]	9 [35]
Peaches	8.8 [33]	11.8 [33]	1.40 [33]	.49 [34]	44 [34]	9 [34]
Pineapples	12.8 [40]	14.6 [40]	1.12 [40]	.43 [40]	14 [40]	3 [40]
Plums	7.4 [40]	14.8 [40]	1.52 [40]	.57 [40]	38 [40]	7 [40]
Quinces	7.6 [19]	13.2 [21]	4.58 [11]	.44 [21]	37 [21]	8 [21]
Raspberries [Red]	7.2 [58]	10.5 [57]	5.89 [57]	.45 [60]	47 [59]	11 [59]
Raspberries [Black]	7.1 [15]	11.2 [15]	8.85 [15]	.56 [17]	53 [17]	10 [17]
Strawberries	5.3 [172]	8.0 [164]	2.69 [196]	.46 [195]	44 [182]	9 [182]

Kind of Fruit	Potash [K₂O] mg./100 g.	Potash % in Ash	Alcohol Precipitate %	Pectic Acid %	Total Acidity 0.1 N/100 g. ml.	Total Acidity Calculated Per Cent as-
Apples	166 [12]	45 [12]	0.71 [38]	0.38 [38]	77 [38]	0.52 [38] malic
Apricots	454 [20]	54 [20]	.81 [40]	.46 [37]	168 [40]	1.13 [40] malic
Blackberries	235 [15]	49 [15]	.58 [55]	.32 [55]	170 [52]	1.09 [52] isocitric
Cherries	248 [20]	51 [20]	.20 [47]	.09 [47]	203 [46]	1.36 [46] malic
Crabapples			1.09 [30]	.59 [30]	159 [30]	1.07 [30] malic
Currants	290		.61 [35]	.39 [34]	334 [34]	2.14 [34] citric
Figs			.89 [16]	.49 [16]	31 [15]	.20 [15] citric
Gooseberries			.82 [8]	.52 [8]	371 [8]	2.37 [8] citric
Guavas			.90 [24]	.50 [24]	136 [24]	.87 [24] citric
Grapes	260		.41 [32]	.26 [32]	159 [31]	1.19 [31] tartaric
Loganberries			.69 [35]	.37 [35]	342 [35]	2.19 [35] citric
Peaches	246 [14]	51 [14]	.82 [32]	.40 [32]	93 [33]	.62 [33] malic
Pineapples	197 [21]	45 [21]	.12 [40]	.04 [40]	122 [39]	.78 [39] citric
Plums			1.16 [40]	.70 [40]	330 [40]	2.21 [40] malic
Quinces	220		1.03 [21]	.61 [21]	146 [21]	.98 [21] malic
Raspberries [Red]	193 [29]	44 [29]	.63 [58]	.30 [58]	210 [58]	1.35 [58] citric
Raspberries [Black]			.65 [17]	.35 [17]	162 [17]	1.04 [17] citric
Strawberries	209 [25]	50 [25]	.53 [193]	.33 [193]	174 [197]	1.11 [197] citric

[a] Predecessor—Bureau of Chemistry. Numbers in parentheses denote the number of samples analyzed. Sale, *J. Assoc. Official Agr. Chem.* 21, 502 (1938).

For the customary pure granulated sugar, some manufacturers of preserves and jams substitute a partially refined sugar syrup, sometimes called "liquid sugar." This product contains varying amounts of mineral matter (ash), depending upon the degree of refinement. When it is used in the preparation of a "preserve," the ash in the finished product will be obtained from both the fruit and the "liquid sugar." It is also a common practice in the preserving industry for manufacturers to add commercial pectin in the manufacture of preserves to thicken them, and this substance will contribute some mineral matter to them. Other variations in the practice of the manufacture of preserves that must be considered are the use of glucose, acid phosphate, and artificial color. Evidence of the presence of these substances will be found in the ratio of various components in the product to each other.

Ratios

Ratio	Description
$\dfrac{\text{Alcohol precipitate}}{\text{Ash}}$	A high ratio indicates the addition of commercial glucose, or gums, or added pectin. A low ratio indicates the addition of mineral matter or mineral acid or the use of over-ripe or decomposed fruit.
$\dfrac{\text{Pectic acid}}{\text{Ash}}$	A high ratio indicates added pectin and gum. A low ratio, the same as above:
$\dfrac{\text{Insoluble solids}}{\text{Ash}}$	A high ratio indicates the use of exhausted fruit, marc, or pomace. A low ratio indicates the same as above or the use of fruit juice instead of fruit.
$\dfrac{\text{Insoluble solids}}{K_2O}$	Abnormally high ratios indicates the use of fruit from which the juice has been drained off or the addition of fruit pomace.
$\dfrac{\text{Ash}}{K_2O}$	Abnormally high ratio indicates presence of substances other than fruit.
$\dfrac{\text{Insoluble solids}}{\text{Alcohol precipitate}}$	A high ratio indicates the use of exhausted fruit, marc or pomace. A low ratio indicates the addition of pectin, gum or commercial glucose or the use of fruit juice instead of fruit.
$\dfrac{\text{Insoluble solids}}{\text{Pectic acid}}$	The same as above, but not influenced by commercial glucose.
$\dfrac{\text{Insoluble solids}}{\text{Total acid}}$	A high ratio indicates the use of exhausted fruit, marc or pomace. A low ratio indicates added acid.
$\dfrac{\text{Total acid}}{\text{Ash}}$	A high ratio indicates added acid. A low ratio indicates added mineral matter.
$\dfrac{\text{Alcohol precipitate}}{\text{Pectic acid}}$	A high ratio indicates the addition of commercial glucose and starch, and of some gums. Other gums are hydrolyzed in a manner similar to pectin and consequently do not influence the ratio.

In the case of jellies the ratios involving the determination of insoluble solids have little meaning, for in the methods of manufacture all the insoluble solids are presumed to be strained out. On the other hand, these ratios are very important in jams and preserves because of the indication of added exhausted fruit or marc or pomace.

In Tables 54, 63, 64, 65, 66, 67, 68, and 69 the normal range of composition of fruits is tabulated. Since the proportion of fruit to sugar should be at least 45 lbs. to 55 lbs. and preferably more fruit to sugar, the constants of the fruit are about twice as great as the constants of the jams and preserves. However the ratios of these constants remain practically unaffected. Tables 55, 56, 57, 58, 59 and 60 give the composition of authentic jams and jellies.

TABLE 55.—RESULTS ON SUPERVISED FACTORY PACKS OF STANDARD AND SUBSTANDARD PRESERVES [a,b]

Kind of Fruit	Total Sugars as Invert %	Insoluble Solids %	Ash of Sample Solution %	Phosphate [P_2O_5]		Potash [K_2O]	
				Mg./100 g.	Per Cent in Ash	Mg./100 g.	Per Cent in Ash
Apricot	69.2	0.97	0.41	28.2	6.9	225.9	55.1
Apricot	70.7	0.96	0.38	27.9	7.3	215.2	56.6
Peach	68.0	0.23	0.17	10.7	6.5	60.0	36.4
Peach	75.9	0.30	0.14	10.0	7.1	60.1	42.8
Pineapple	68.1	0.49	0.20	6.8	3.4	69.0	34.5
Pineapple	70.8	0.49	0.17	7.0	4.1	60.7	35.7
Raspberry	70.8	2.92	0.26	17.1	6.6	118.3	45.5
Raspberry	67.8	3.15	0.24	19.2	8.0	118.8	49.5
Strawberry	70.1	1.02	0.22	14.0	6.4	92.5	42.0
Strawberry	71.5	0.99	0.18	13.2	7.3	83.1	46.2
Blackberry	69.0	2.32	0.26	12.9	5.0	103.1	39.7
Blackberry	71.3	2.46	0.23	13.3	5.8	99.4	43.2

[a] Odd-numbered samples were made from partially refined commercial sugar syrups. Even-numbered samples were made from granulated sugar. Apricots and peaches were from No. 10 tins labeled "Solid pack." Pineapples were from No. 10 tins labeled "Crushed pineapple in juice." Raspberries, strawberries, and blackberries were in crates out of cold storage. Most samples contained added citric acid and citrus pectin (100 grade).
[b] Sale, J. Assoc. Official Agr. Chem. 21, 506 (1928).

The sophistication of jams and jellies may reach such a high point through the addition of exhausted material to raise the water insoluble solids content and the addition of mineral salts to raise the ash and at the same time allow the ratio to remain normal, that reference to the ratios alone may be insufficient and the normal range of composition of the fruits themselves must be taken into consideration. Jams and jellies containing little or no potassium indicate the lack of fruit juice. The absence of the characteristic acid of the fruit shows the absence of that fruit.

The British [10] place some reliance on the nonsugar solids, that is, total solids minus sugar solids and more reliance on minimum insoluble solids as a true measure of the amount of fruit present.

---

10 Hughes and Maunsell, Analyst 59, 231 (1934).

Added Water.—Sale [11] points out that preserves that are made from fruit and sugar in the normal proportions and that are concentrated by cooking in the usual way will not contain in their finished form any added water. Actually, the weight of the fruit and sugar mixture will be reduced 10 per cent or more by the removal of original fruit moisture in the cooking step.

TABLE 56. ANALYSES OF AUTHENTIC JAMS [a,b]

Type	Insoluble Solids %	Soluble Solids %	Total Solids %	Alcohol Precipitate %	Total Sugars %	Sugar-free Solids %	Ash %
Apricot	1.72	. . . .	70.15	1.14	64.96	5.19	0.35
Currant. . . .	6.32	. . . .	66.32	0.92	54.09	12.23	0.84
Orange.	0.69	. . . .	65.44	0.54	59.33	3.55	0.14
	0.89	. . . .	72.76	1.76	68.26	4.50	0.21
	1.32	. . . .	67.99	1.42	63.11	4.88	0.29
Peach.	1.08	. . . .	67.33	1.49	61.86	5.47	0.28
Plum.	0.96	. . . .	70.19	1.09	63.25	6.94	0.26
	2.57	. . . .	64.78	1.63	57.66	7.12	0.32
(stone free)	0.79	68.36	69.15	2.6	68.4		
(stones) . . .	7.00	46.06	53.06	1.5	44.2	. . . .	. . . .
Raspberry. .	4.24	50.52	54.76	2.80	51.0	. . . .	. . . .
	5.85	63.00	68.85	1.04	58.2	. . . .	. . . .
Strawberry	1.90	. . . .	69.16	0.9	61.07	8.09	0.34
	2.21	71.08	73.29	1.4	73.6	. . . .	. . . .

[a] Munson, *U. S. Dept. Agr., Bur. Chem. Bull.* **66** (1902).
[b] *Canada Inland Revenue Dept. Bull.* **96** (1904).

The proportion of added water in a presumed preserve can be calculated. Determine the number of pounds of fruit used to each 55 pounds of sugar by the procedure described in a preceding section and multiply by the average percentage of sugar as invert in the corresponding kind of fruit. Add the result obtained, pounds of fruit sugar, to the pounds of added sugar expressed as "invert sugar," namely 57.9, and the sum is pounds of total sugar, expressed as invert sugar, in the original mixture. Divide this sum by the percentage of total sugars, expressed as invert sugar, in the finished preserve, and multiply the quotient by 100 to obtain the pounds of finished preserve secured from the original fruit and sugar mixture. The difference between this figure and the weight in pounds of the

[11] Sale, *J. Assoc. Official Agr. Chem.* **21**, 502 (1938).

fruit and sugar used in the original mixture will be the amount of added water in the finished preserve.

TABLE 57. ANALYSES OF AUTHENTIC JELLIES [a,b]

Type	Total Solids	Acidity as H_2SO_4	Total Sugars	Alcohol Precipitate	Sugar-free Solids	Ash	Alkalinity as K_2CO_3
Apple......	62.67	0.24	59.53	0.81	3.11	0.10	0.15
	66.06	0.61	63.05	1.13	2.90	0.26	0.25
	60.86	0.69	55.63	1.30	5.23	0.35	0.32
	60.90	0.13	56.4	3.40	4.50	0.56	0.47
Crab apple .	61.12	0.045	59.2	0.95	1.92	0.30	0.20
Cranberry..	54.76	0.054	51.4	1.50	3.36	0.16	0.12
Currant....	64.21	1.36	59.00	3.36	5.14	0.44	0.42
	62.46	1.06	59.90	1.39	2.52	0.27	0.26
	66.22	0.92	65.70	1.18	0.52	0.22	0.22
	70.31	0.93	65.79	2.11	4.43	0.33	0.31
	75.03	1.15	69.25	2.87	5.78	0.47	0.44
	68.77	1.08	64.49	1.70	4.28	0.41	0.39
Red currant	66.78	0.098	65.7	1.25	1.08	0.44	0.32
	65.76	0.031	67.4	0.65	....	0.17	0.11
Grape......	70.66	0.31	64.13	1.25	6.53	0.21	0.28
	72.22	0.83	67.74	2.49	4.48	0.35	0.23
Guava.....	79.97	0.47	74.89	1.36	5.08	0.45	0.41
	79.14	0.47	75.91	1.59	4.23	0.38	0.35
Lemon.....	64.62	0.46	61.49	0.73	3.13	0.21	0.19
Raspberry..	70.07	0.33	63.25	1.45	6.82	0.24	0.26
	74.53	0.83	70.59	0.77	3.94	0.31	0.24
Strawberry	68.11	0.29	63.07	0.89	5.04	0.21	0.21
	67.50	0.29	63.56	1.47	5.44	0.21	0.31

[a] Munson, *U. S. Dept. Agr., Bur. Chem.* **66** (1902).
[b] *Canada Inland Revenue Dept. Bull.* **96** (1904).

For example, if the proportion of fruit to sugar as found by the procedure previously described is 14 lbs. of fruit to each 55 lbs. of sugar and the average content of sugar of the fruit is 6 per cent, then the pounds of sugar in the fruit will be 0.84. In the process of cooking, the 55 lbs. of added sugar in the original mixture will be inverted to a greater or lesser extent. Despite the degree of inversion, the conclusions will be accurate if both the sugar of the original mixture and the sugar in the preserve are compared on the basis of complete inversion. When 55 lbs. of sugar is completely inverted, it will weigh 57.9 lbs. The sum of the sugar in the fruit and the added sugar is 58.76 lbs. If the total sugars after inversion in the preserve are 67 per cent, as determined by analysis, the weight of the finished preserve will be 87.7 lbs.

$$\left(\frac{58.74}{67} \times 100 \right)$$

TABLE 58. MAXIMA, MINIMA, AND AVERAGE COMPOSITION OF AUTHENTIC JAMS AND JELLIES [a]

	Jellies			Jams		
	Average	Maximum	Minimum	Average	Maximum	Minimum
Total solids %	67.13	80.28	45.56	65.98	82.46	50.43
Insoluble solids %				1.92	6.32	0.09
Protein %	0.21	0.42	0.07	0.43	1.41	0.18
Acidity % H_2SO_4	0.63	1.57	0.17	0.54	1.36	0.16
Sugars						
Reducing %	37.07	65.52	3.95	36.41	61.02	13.20
Sucrose %	25.96	65.22	3.47	22.15	54.23	0.30
Alcohol precipitate %	2.53	3.36	0.73	1.12	1.76	0.09
Sugar-free solids %	4.05	6.82	0.52	7.71	14.58	3.55
Ash %	0.34	0.73	0.10	0.32	0.84	0.14
Alkalinity of ash as % K_2CO_3	0.30	0.51	0.15	0.26	0.60	0.10

TABLE 59. RANGE OF ALCOHOL PRECIPITATE CONTENT OF JELLY [a]

	Minimum %	Maximum %	Average %
Apple	0.81	1.30	1.08
Currant	1.18	3.36	2.10
Grape	1.25	2.49	1.87
Guava	1.36	1.59	1.47
Lemon			0.73
Strawberry	0.89	1.47	1.18
Raspberry	0.77	1.45	1.11

TABLE 60. RANGE OF ALCOHOL PRECIPITATE CONTENT FOR JAM [a]

	Minimum %	Maximum %	Average %
Apricot			1.14
Currant			0.92
Orange	0.54	1.76	1.24
Peach			1.49
Plum	1.09	1.63	1.36
Strawberry			0.90

[a] Munson, U. S. Dept. Agr., Bur. Chem. Bull. 66 (1902).

The difference between the weight of the finished preserve, 87.7, and the weight of the original fruit and sugar mixture, 69 (14 plus 55) will be 18.7, which is the pounds of added water present in 87.7 lbs. of preserve or 21 per cent of added water.

EXAMINATION OF FRUITS AND JAMS BY LEAD PRECIPITATION

The quantity of lead precipitate formed, when a lead acetate solution is added to a solution of a jam, affords some indication of the amount of fruit in the sample. The method will not, by itself, indicate the percentage of fruit in a jam with any more certainty than the other methods previously described. It may, however, be particularly useful in the analysis of jams made from a mixture of fruits, the acids of which differ, for example, one fruit containing citric acid and the other malic acid. It is sometimes useful in indicating whether a fruit juice or a commercial pectin has been used in the jam. By use of the method it is possible to place the various fruits in three classes, i. e. those containing mainly (1) citric acid, (2) malic acid, (3) lactic acid, or an acid having similar lead precipitating properties.

Hinton [12] found that the theoretical amount of 2 per cent lead acetate solution, equivalent to each 0.1 g. of acid, as citric was 13.55 ml. He found that malic acid was not precipitated by lead acetate solution but that, in the presence of citric acid, lead acetate solution precipitated all of the citric acid and 43 per cent of the malic acid. Tartaric acid is also completely precipitated by lead acetate solution but lactic acid is not precipitated and, if present, would make the apparent malic acid content high. However, the lead salts of citric, malic, and tartaric acids are only very slightly soluble in 50 per cent acetone, whereas that of lactic acid is soluble, consequently the amount of lactic acid may be estimated by calculation.

Preparation of Extract.—Weigh 250 g. of the sample into a beaker, and add 250 ml. of water. Mix well to break up the jam, then heat to boiling with continual stirring, and boil gently for an hour, with occasional stirring, keeping the beaker covered, and maintaining the volume by adding water if necessary. Cool, transfer to a 500-ml. volumetric flask, make up to volume, shake well and filter through a coarse filter.

Preparation of Pectin-free Filtrate.—Transfer 250 ml. of the aqueous extract to a 500-ml. flask, and add acetone, while swirling round without incorporating too much air, to the mark. Mix well, and

---

[12] Hinton, *Analyst* 59, 248 (1934).

filter through a large dry filter, with precautions to avoid loss of acetone by evaporation.

Titration of Free Acid.—Pipette 40 ml. of the pectin-free filtrate into a large beaker, add about 500 ml. of boiled and cooled water, and titrate with 0.1 N sodium hydroxide, using phenolphthalein as indicator. Carry out a blank titration of 500 ml. of water similarly, and deduct this from the jam titration The difference, multiplied by 0.07, gives the percentage of free acid in the jam, expressed as hydrated citric acid.

Ash of Jam.—Evaporate 50 to 100 ml. of the pectin-free filtrate to dryness in a platinum dish on a water bath, char and ash at a dull red heat, preferably in a muffle. Cool and if desired weigh.

Titration of Combined Fruit Acid.—Dissolve the ash in a measured 15-ml. portion of 0.1 N hydrochloric acid, filter into a 175-ml. conical flask, and wash through thoroughly. Boil for a few minutes. Cool, add a drop of methyl orange solution, titrate to yellow with 0.1 N sodium hydroxide solution, and then to the neutral tint with 0.1 N hydrochloric acid. Calculate the amount of acid consumed by the ash to the number of milliliters of 0.1 N acid per 100 g. of jam. This is the methyl orange alkalinity of the ash.[13] Next acidify the titrated solution with about 2 ml. of 0.1 N hydrochloric acid, and evaporate to about 15 ml. It is advisable to do so on a sand bath to minimize bumping. Cool, and neutralize carefully to methyl orange with 0.1 N sodium hydroxide solution. Add a few drops of phenolphthalein solution and 10 ml. of a strong neutral calcium chloride solution. Boil again for a few minutes, and titrate to the phenolphthalein end point with 0.1 N sodium hydroxide solution. Calculate the number of milliliters of 0.1 N sodium hydroxide solution per 100 g. of jam, and multiply by 3/2. This gives the phosphoric acid in the ash as its equivalent of 0.1 N sodium hydroxide solution. Then find the "total alkalinity" of the ash, equivalent to all the alkali and alkaline earth metals present, by adding to the "methyl orange alkalinity" ⅓ of the phosphoric acid equivalent. Finally multiply this total alkalinity by 0.007 to obtain its value as percentage of combined fruit acid as citric acid.

Total Fruit Acid.—The total fruit acid, including any phosphates present in the extract, is the sum of the free acid and combined acid found as above, and is expressed as the percentage of total citric acid, hydrated.

[13] Pfyl, *Z. Nahr. Genussm.* **43**, 313 (1922).

Aqueous Lead Number.—Take an amount of pectin-free filtrate, to the nearest 5 to 10 ml. containing approximate amounts of total fruit acids according to the following scheme:

1. Gooseberry, apricot, or blackberry jams, 0.50 g.
2. Strawberry, raspberry, red currant, or black currant jams, 0.35 g.
3. Apple, cherry, plum, greengage, or damson jams, 0.65 g.

Procedure.—Remove the acetone by distillation, transfer the residue to a 250-ml. volumetric flask and cool. Then add, in the case of groups (1) and (2), 3.0 ml. of 10 per cent malic acid solution, or in the case of group (3), 3.0 ml. of 5 per cent citric acid solution, pipetted accurately. The strength of the acid used should be correct to within 1 per cent of the total. Dilute to about 200 ml. Make sure that the temperature of the solution is at about 16 to 20° C., then add from a pipette, while rotating the flask, 20 ml. of lead acetate solution, containing 100 g. of normal lead acetate crystals, and 12.5 g. of acetic acid per liter, and make up to volume with water. Shake well and filter without delaying more than a few minutes. Titrate 50 ml. of the filtrate, diluted with 50 ml. of water, at or near the boiling point, with ammonium molybdate solution, 9.3 g. of ammonium molybdate per liter, using a 0.5 per cent solution of tannic acid as an outside indicator by spotting on a tile. The first appearance of a distinct yellow color in the test drop, or a definite increase in a slight existing yellowish color marks the end point.

For a blank titration, dilute 20 ml. of the lead acetate solution to 250 ml. and titrate 50 ml. of this plus 50 ml. of water in the same way. Correct the difference between the two titrations for any lack of correct strength in the lead or molybdate solutions. If the difference is not in the range 11 to 14 ml. repeat the determination on a larger or smaller quantity, as the case may require.

From the corrected titration difference deduct 3.6 ml., when malic acid was initially added, or 4.1 ml. when citric acid was added. Calculate the remainder back to the number of ml. of 2 per cent lead acetate solution, which would be completely precipitated by 10 g. of the original sample, or 40 ml. of the pectin-free filtrate. This is the lead number, aqueous, of the jam. That is, if the amount of pectin-free filtrate taken for the test be P ml. and the corrected titration difference D, then,

$$\text{Aqueous lead number} = L = \frac{200D}{P}$$

The "lead number per 0.1 g. of acid" is then found by simply dividing L by the percentage of total fruit acid (a) in the sample

$$l = \frac{L}{a}$$

Acetone (50 Per Cent) Lead Number.—Take an amount of pectin-free filtrate, to the nearest 5 to 10 ml., containing approximate amounts of total fruit acid as follows:

1. Gooseberry, strawberry, raspberry, red currant, black currant, apricot, or blackberry jams, 0.50 g.

2. Apple cherry, plum, greengage, or damson jams, 0.40 g.

If more than 200 ml. would be required, the amount must be restricted to this figure.

Place the required amount in a 250-ml. volumetric flask, and in case of jams of group (2) add 3.0 ml. of 5 per cent citric acid solution, measured accurately. Should it have been necessary to limit the amount of pectin-free filtrate taken to 200 ml., the deficiency of fruit acid may be made up by a suitable addition of 5 per cent citric acid solution, its effect being allowed for later.

Add acetone, while rotating the flask, to make up the total amount of acetone present to 125 ml. Then add from a pipette, 20 ml. of 10 per cent lead acetate solution, as in the aqueous test, and make up to volume with water. Mix and filter, taking precautions to avoid loss of acetone by evaporation. Titrate 50 ml. of the filtrate, diluted with 50 ml. of water, as in the aqueous test. Correct the difference between the titration and the blank for any factors of the lead of molybdate solutions. If the difference so corrected is not approximately 14-15 ml. make a further correction as follows:

Lead Titration Difference ml.	Correction ml.
9.0–11	−0.3
11.1–13	−0.2
13.1–14	−0.1
14.1–15	0
15.1–15.7	+0.1
15.8–16.2	+0.2
16.3–16.5	+0.3
16.6–16.8	+0.4
16.9–17.1	+0.5

From the corrected titration difference deduct 1.5 ml. for each 1 ml. of added 5 per cent citric acid solution, if any.

Calculate the remainder, as before, back to the number of milliliters of 2 per cent lead acetate solution completely precipitated by 10 g. of the sample. This is the lead number, acetone, L^1 of the jam. The lead number per 0.1 g. of the acid is then obtained as before.

$$l^1 = \frac{L^1}{a}$$

Interpretation of Results.—If l^1 is appreciably less than 15, and plums or greengages are not present, lactic acid from a commercial pectin is probably present, and its amount can be approximately found from

$$k = a - \frac{L^1}{15}$$

where k is the percentage of lactic acid in the sample. A corrected value for l is then obtained by deducting the lactic acid from the total acid:

$$l \text{ (corrected)} = \frac{L}{a - k}$$

The value for l, corrected or not as required, and the total acidity, a, with any lactic acid deducted, are then used in some of the following formulas to find the fruit content of the sample.

The lead number is due to the acid components of the fruits. Furthermore the acid components themselves can, by means of the lead precipitation be separated into groups containing a preponderance of citric or malic acid. Thus it is possible, within certain rough limits, to discover whether the acid components of a jam are normal to the class of fruit used, provided that any foreign fruit or fruit juice added belongs to a different class. In particular, the addition of apple pulp or juice, or pomace extract, to strawberry, raspberry, etc., jam, should make itself evident by disturbing the normal citric acid preponderance of these fruits.

For this purpose it is not necessary to calculate from the lead number the actual quantities of citric and malic acids present. The lead numbers themselves can be used and compared with established data for the several kinds of fruits concerned, with due allowance for the natural variations. It may be pointed out, what is evident from Tables 61 and 62, that it is not the lead number itself that is specially characteristic of a particular fruit. Thus, a lead number of 6.5 found for a jam might be given by 40 per cent of strawberries alone, or by 25 per cent of strawberries and 26 per cent of apple pulp, or by 60 per cent of apple alone. The characteristic property which makes it possible to gain some idea of the proportions of the components in such mixtures is the lead number relative to the acid content, or, according to the empirical method for expressing this property, the "lead number per 0.1 g. of acid." It should be remembered that total acid is always meant here, as obtained from the free acidity and ash titrations. Hence, from a consideration of the average values for this figure for the fruits concerned, their respective

TABLE 61. LEAD NUMBERS OF JAM FRUITS (AQUEOUS PRECIPITATION [a])

Fruit	Lead Number ml.		Total Acidity as Citric Acid %	Lead Number per 0.1 g. Acid ml.
Gooseberries (17)	max.	29.9		11.7
	min.	17.3		9.5
	ave.	25.5	2.44	10.4
Strawberries (15)	max.	22.7		13.3
	min.	9.8		10.5
	ave.	16.3	1.31	12.4
Raspberries (11)	max.	34.2		14.3
	min.	19.0		12.1
	ave.	26.8	2.02	13.3
Red currants (7)	max.	43.1		14.3
	min.	34.8		12.0
	ave.	37.6	2.81	13.4
Black currants (10)	max.	64.8		14.4
	min.	39.5		12.2
	ave.	52.7	3.92	13.5
Apples (4)	max.	10.9		7.4
	min.	6.9		6.3
	ave.	9.4	1.33	7.1
Plums (5)	max.	18.0		7.6
	min.	6.9		2.5
	ave.	10.2	2.14	4.8
Greengages (2)	1	12.0		5.9
	2	8.1		5.5
	ave.	10.1	1.77	5.7
Damsons (4)	max.	19.4		7.0
	min.	9.4		3.0
	ave.	13.0	2.77	4.7
Blackberries (12)	max.	22.5		10.2
	min.	7.2		8.1
	ave.	13.9	1.51	9.2
Apricots (2)	1	20.8		11.0
	2	14.7		9.6
	ave.	17.8	1.72	10.3

[a] Hinton, *Analyst* **59**, 248 (1934).

proportions in an unknown mixture, or rather, the proportions of their acids, can be approximately determined.

This can be made clear by the following example. Suppose the lead number of a strawberry and apple jam be 6.4, and the percentage of total acid in the jam be 0.61, then the acid in the 10 g. of jam equivalent to the 6.4 ml. of 2 per cent lead acetate solution is

TABLE 62. LEAD NUMBERS OF JAM FRUITS (50 PER CENT ACETONE PRECIPITATION [a])

Fruit	Lead Number ml.		Total Acidity as Citric Acid %	Lead Number per 0.1 g. Acid ml.
Gooseberries (2)	1	40.9		15.4
	2	39.8		15.4
	ave.	40.4	2.62	15.4
Strawberries (3)	max.	27.0		15.6
	min.	15.8		15.3
	ave.	21.4	1.39	15.4
Raspberries (3)	max.	37.4		1
	min.	35.0		
	ave.	36.6	2.38	15.4
Red currants (2)	1	51.8		17.1
	2	50.7		16.7
	ave.	51.3	3.04	16.9
Black currants (2)	1	78.2		17.3
	2	74.1		16.3
	ave.	76.2	4.53	16.8
Apples (4)	max.	22.2		15.1
	min.	16.5		14.9
	ave.	19.9	1.33	15.0
Plums (3)	max.	34.4		14.6
	min.	19.0		12.3
	ave.	26.1	1.88	13.9
Greengages (3)	max.	28.6		13.0
	min.	15.0		10.3
	ave.	20.3	1.68	12.1
Damsons (2)	1	43.3		15.6
	2	34.9		14.6
	ave.	39.1	2.59	15.1
Blackberries (2)	1	34.4		15.6
	2	22.2		15.2
	ave.	28.3	1.84	15.4
Apricots (2)	1	25.9		15.3
	2	23.5		13.6
	ave.	24.7	1.72	14.4

[a] Hinton, *Analyst* 59, 248 (1934).

0.061 g. and the "lead number per 0.1 g. of acid" is 6.4/0.61 = 10.5. The averages for strawberries and apples are, respectively, 12.4 and 7.1, Table 61. Hence the proportion of the acids attributable to strawberries is $\dfrac{10.5 - 7.1}{12.4 - 7.1} \times 100$ per cent of the total = 64 per cent. Thus, the percentage of acid in the jam due to strawberries, is 0.61 × 64/100 = 0.39 per cent; and that due to apples will, therefore, be 0.61 − 0.39 = 0.22 per cent. Taking the average total acid contents of strawberries and apples as 1.31 per cent and 1.33 per cent respectively, Table 61, the amounts of the two fruits in the jam are:

Strawberry: $\dfrac{0.39}{1.31} \times 100 = 30$ per cent

Apple: $\dfrac{0.22}{1.33} \times 100 = 17$ per cent

The same method can be applied to other mixtures, provided the characteristic figures for the fruits concerned, "lead number per 0.1 g. of acid" are sufficiently far apart. Thus the method breaks down for such mixtures as raspberry and red currant, plum and apple, etc., and is of doubtful value for blackberry and apple, for strawberry and gooseberry, as Table 61 indicates.

The calculation can be expressed in the form of a simple formula.

Let F_1 and F_2 be the respective percentages of two fruits in a mixed product,

L_1 and L_2 be the average values of the "lead number per 0.1 g. acid," Table 61,

A_1 and A_2 be the average values for total acid in the two fruits, Table 61,

l be the actual "lead number per 0.1 g. of acid" found in the sample and

a be the actual per cent of total acidity found,

Then

$$F_1 = \frac{100a(l - L_2)}{A_1(L_1 - L_2)} \tag{1}$$

and

$$F_2 = \frac{100a(L_1 - l)}{A_2(L_1 - L_2)}$$

Many jams on the market contain added pectin, which is used either in the form of a direct extract from pomace, or a specially prepared proprietary "fruit pectin," usually prepared from apple residues. The latter preparations, which are more or less pure pectin, should introduce no lead precipitating acids but themselves into the jam and, as they are removed along with the natural fruit pectins by the preliminary precipitation with acetone, they should cause no complications.

The case is different with pomace extracts, etc. These, when they are aqueous extracts, may be considered as apple extracts deprived of a portion of their natural acids. The remaining acids, however, will still have the same lead-precipitating properties as the acids of the whole fruit, apart from the pectinous components, which can be removed before the lead test is made. Thus the formulas immediately above can still be applied, although the apparent percentage of apple juice indicated will be low because of the removal of part of its acid. There will be no interference with the calculation of the amount of the main fruit component, F_1.

Some commercial pectins, however, appear to have been prepared by an extraction of pomace with lactic acid. The calculation of the fruit-content from the aqueous lead number by formula (1) would be erroneous in jams containing such pectin preparations. The effect of the lactic acid would be, as previously noted, to depress the "lead number per 0.1 g. of acid" making the proportion of apple acids appear too high. The solution of the difficulty is afforded by the second, or acetone lead number. This gives a value for the total acids excluding the lactic acid, so that a corrected value can be obtained from the "lead number per 0.1 g. of acid" which refers only to the malic and citric types of acid. Formula (1) can then be applied to these corrected values for a and l.

This calculation is of use only for those fruits which themselves show no appreciable amount of the lactic type of acid, namely, the soft fruits and damsons. But as plums and greengages have an aqueous lead number so similar to that of apples as to preclude its use in calculating fruit content in their case, the restriction is not a material one.

It is desirable to note in connection with the calculation of fruit-content in mixtures containing commercial pectins, experience in the analysis of a large number of jam samples of various origin has shown the figure of 7.1 for "lead number per 0.1 g. of acid" of apples, Table 61, to be rather high for general application. A figure giving results more in accordance with other analytical indications is 6.5. This, too, is about the figure given by commercial pectins when due allowance is made for the extraneous lactic acid.

Hence, for the fruit-content of jams with added pectin, from apples, formula (1) may be simplified to

$$F_1 = \frac{100a(l - 6.5)}{A_1(L_1 - 6.5)}$$

The a and l of this formula should be suitably corrected for any lactic acid shown to be present by the acetone lead number. Values of A_1 and L_1 appropriate to the various fruits may be obtained from Table 61.

FRUITS

Fruits were very likely the first plant food of man. Later tubers and cereals were cultivated. The water content of fruits varies from 60 to 90 per cent. The pulpy nature of the fruit protects the seeds until the spring when conditions are favorable for growth.

Fruits undergo a series of progressive changes from the unripe stage, to the ripe stage, to the rotten and fermentative stages. The ripening point of a fruit is considered to be that point at which the sugar content is a maximum. In oranges and grapefruit where many means are used to make the fruit appear ripe, maturity is based on a minimum invert sugar-citric acid ratio of eight to one. Apples are considered ripe when the last trace of starch disappears.

The browning of fruit flesh on exposure to air is due to the oxidation of the tannin in the fruit by an oxidase in the presence of the oxygen of the air. In drying fruits this effect is countered by sulfuring, that is drying by exposing to the fumes of burning sulfur. By the use of this process fruits are obtained that are not really dry and which will subsequently lose or take on moisture according to whether the fruit is stored under dry or humid circumstances.

Fruit juice is the product obtained by pressure from fruit. Fruit juices are sold as such and those more commonly encountered are apple juice or sweet cider, grape juice, orange juice, pineapple juice, grapefruit juice, and tomato juice. Some of these are frequently concentrated and sold as syrups for fountain use in sodas. When allowed to ferment, these juices form the well-known products, hard cider from apple juice, wine from grape juice, and perry from pear cider.

Only occasionally does the food analyst analyze fruits, as such. His interest lies in the analysis of fruit products. However, it is necessary to perform many investigational determinations in order to establish more definitely the normal constants of fruits. The determinations may be made in a manner entirely similar to those previously detailed for jams and jellies. Added color in oranges and similar citrus fruits may be detected by the method detailed in coloring matters in foods, Chapter III. The vitamin C content of citrus fruits may be estimated by one of the methods outlined in the chapter on vitamins, Chapter XVII. Inorganic constituents may be esti-

TABLE 63. ANALYSES OF FRESH FRUITS [a]

Fruit		Total Solids %	Total Sugars % as Invert Sugar	Pectin %	Insoluble Solids %
Cooking apples.........	max.	15.61	8.72	1.60	2.47
	min.	10.25	3.64	0.84	1.95
	av.	13.04	•7.11	1.29	2.17
Eating apples...........	max.	17.98	12.58	0.93	1.91
	min.	12.29	3.16	0.71	1.51
	av.	15.12	9.72	0.82	1.70
Cherries (without stones)	max.	24.70	15.30	0.54	3.10
	min.	14.74	8.28	0.24	1.29
	av.	18.64	11.47	0.35	2.05
Apricots (without stones)	max.	14.30	7.61	1.32	2.49
	min.	10.13	1.57	0.71	1.57
	av.	12.97	5.19	1.03	2.00
Blackberries...........	max.	18.67	4.36	1.19	10.00
	min.	13.62	2.59	0.68	6.45
	av.	16.24	3.48	0.94	8.13
Black currants.........	max.	24.43	7.44	1.79	6.18
	min.	15.93	3.66	1.37	4.78
	av.	19.44	5.50	1.52	5.51
Gooseberries...........	max.	13.90	6.54	1.20	2.76
	min.	7.93	2.00	0.95	1.66
	av.	11.38	3.98	1.08	2.26
Greengages............ (without stones)	max.	18.27	9.77	1.32	1.99
	min.	11.01	4.68	0.95	1.40
	av.	14.10	6.45	1.14	1.56
Plums (without stones)...	max.	15.18	8.76	1.48	1.75
	min.	9.65	2.28	0.75	1.00
	av.	12.87	6.31	0.96	1.22
Raspberries...........	max.	24.82	8.67	0.86	6.22
	min.	12.38	2.54	0.58	4.23
	av.	16.78	4.80	0.71	5.50
Red currants..........	max.	20.72	7.88	1.50	5.65
	min.	12.70	2.95	0.91	3.99
	av.	16.12	5.38	1.16	4.77
Strawberries...........	max.	13.04	7.07	0.73	2.13
	min.	8.95	3.37	0.60	1.70
	av.	10.80	5.56	0.68	1.90
Loganberries..........	max.	17.11	5.92	0.68	7.25
	min.	16.69	2.66	0.62	7.13
	av.	16.92	4.04	0.65	7.19

[a] Lampitt and Hughes, *Analyst* 53, 32 (1928).

TABLE 64. ANALYSES OF FRESH FRUITS [a,b]

Fruit	Total Solids %		Total Sugar as Invert %	Non-sugar Solids %	Insoluble Solids %	Acidity No. ml. 0.1 N per 100 g.	Crude Calcium Pectate %	Refractometer Reading of Juice Sugar Scale %
Gooseberries..	max.	14.0	7.7	9.0	2.8	415	1.2	10.1
	min.	7.9(51)°	2.0(51)	4.4(51)	1.7(7)	176(36)	0.3(9)	5.2(9)
	ave.	11.2	4.4	6.8	2.3	235	0.8	7.0
Strawberries..	max.	13.2	8.2	7.4	2.4	200	0.7	10.2
	min.	8.2(145)	3.4(145)	2.9(145)	1.5(13)	90(125)	0.2(16)	6.1(74)
	ave.	10.2	5.4	4.8	1.9	145	0.5	6.6
Raspberries...	max.	21.3	8.7	17.7	6.2	390	0.9	12.3
	min.	11.0(107)	3.2(107)	7.4(107)	4.2(13)	106(90)	0.6(13)	5.3(57)
	ave.	14.4	4.8	9.7	5.4	203	0.7	7.9
Red currants .	max.	20.7	7.9	17.2	7.8	495	1.5	....
	min.	12.7(26)	2.2(26)	7.4(26)	4.0(5)	275(13)	0.9(5)	....
	ave.	16.0	5.0	10.9	5.5	375	1.1	....
Black currants	max.	24.4	10.2	17.7	6.3	622	1.8	14.5
	min.	13.7(37)	1.6(37)	10.8(37)	4.8(5)	121(23)	1.4(5)	8.0(5)
	ave.	19.0	5.0	14.1	5.7	449	1.6	11.0
Cherries...... (stone-free)	max.	24.7	15.3	9.8	3.1	145	0.5	18.3
	min.	10.9(41)	6.4(41)	3.3(41)	1.3(7)	96(27)	0.2(5)	10.0(10)
	ave.	16.3	10.2	6.1	1.9	107	0.3	13.9
Plums various (stone-free)	max.	21.9	13.3	11.7	2.0	386	1.5	22.4
	min.	8.1(91)	2.3(91)	3.4(91)	1.0(14)	25(70)	0.7(13)	10.0(36)
	ave.	14.0	7.8	6.2	1.4	213	1.2	14.1
Greengages (stone-free)	max.	21.5	13.9	11.8	2.0	435	1.4	19.7
	min.	11.0(49)	4.1(49)	5.1(49)	1.4(9)	88(39)	1.0(7)	10.2(14)
	ave.	15.6	7.9	7.8	1.5	189	1.2	16.2
Blackberries..	max.	21.2	10.4	16.0	10.5	206	1.2	11.4
	min.	14.1(29)	1.7(29)	8.4(29)	6.3(9)	90(18)	0.6(9)	6.5(12)
	ave.	16.8	4.0	12.8	8.4	135	0.8	8.5
Apricots...... (stone-free)	max.	18.4	11.8	10.4	2.5	349	1.3	18.5
	min.	8.6(55)	3.0(55)	4.1(55)	1.2(10)	123(43)	0.7(11)	8.0(24)
	ave.	12.4	5.6	6.8	1.7	235	1.0	12.9
Loganberries..	max.	23.3	7.3	22.2	7.3	420	0.7	....
	min.	13.2(19)	1.1(19)	7.3(19)	7.1(2)	151(10)	0.6(2)	....
	ave.	16.6	4.5	12.1	....	315	....	....
Apples, whole.	max.	19.5	13.5	9.8	3.4	410	1.6	17.0
	min.	10.3(147)	3.2(147)	1.1(147)	1.4(12)	25(115)	0.5(16)	9.8(39)
	ave.	15.1	10.3	4.9	2.2	162	0.8	13.4
Apples, edible portion	max.	19.0	14.2	6.9	2.3	450	1.0	17.3
	min.	11.5(80)	6.2(80)	1.4(80)	1.5(5)	20(75)	0.4(9)	9.8(38)
	ave.	15.2	10.8	4.4	2.0	93	0.6	13.5
Pears, whole..	max.	21.9	12.6	9.9	....	47	....	18.6
	min.	14.6(22)	7.3(22)	5.6(22)	....	10(23)	....	12.0(18)
	ave.	17.9	10.3	7.6	....	23	....	16.0
Pears, edible portion	max.	20.2	12.8	9.7	1.8	42	0.7	19.2
	min.	13.5(23)	7.8(23)	5.2(23)	1.7(3)	13(25)	0.3(3)	12.0(18)
	ave.	17.1	10.5	6.6	1.8	26	0.0	13.2

[a] Hughes and Maunsell, *Analyst* 59, 231 (1934).
[b] Results are very similar to those of Macara, *Analyst* 56, 39 (1931).
[c] Number of samples in parentheses.

TABLE 65. ANALYSES OF FRESH FRUITS [a]

Fruit		Total Solids %	Protein %	Ash %	Sugars % as Invert	Acid % as Malic (M) Citric (C)
Avocado[b]	max.	39.5	4.4	1.93	1.6	....
	min.	15.7	0.8	0.54	0.3	....
	ave.	.25.9(129)[c]	2.0(112)	1.28(80)	0.7(23)	....
Banana	max.	34.6	2.0	1.4	25.7	0.55 M
	min.	16.6	0.8	0.5	14.5	0.26
	ave.	25.2(69)	1.2(59)	0.84(62)	19.2(36)	0.39(21)
Figs	max.	45.0	2.4	1.05	20.5	0.38 C
	min.	12.0	0.8	0.26	3.5	0.02
	ave.	22.0(53)	1.4(59)	0.64(68)	16.2(68)	0.17(44)
Grapefruit	max.	14.0	0.6	0.54	8.5	1.58 C
	min.	6.9	0.3	0.30	4.6	0.9
	ave.	11.2(61)	0.5(10)	0.42(8)	6.5(47)	1.17(47)
Grapes	max.	28.0	2.2	0.6	14.4	1.67 M
	min.	14.1	0.7	0.3	7.0	0.86
	ave.	18.1(28)	1.4(10)	0.45(10)	11.5(30)	1.21(14)
Guavas	max.	24.2	1.5	1.00	10.0	0.88 C
	min.	15.4	0.3	0.46	3.0	0.34
	ave.	19.4(17)	1.0(13)	0.70(17)	6.1(12)	0.62(10)
Lemons	max.	11.9	1.1	0.71	....	....
	min.	9.5	0.6	0.5	....	....
	ave.	10.7(6)	0.9(5)	0.54(6)	2.2	5.07 C
Limes	max.	14.6	0.9	1.0	0.6	7.2 C
	min.	12.4	0.6	0.7	0.3	4.2
	ave.	14.0(3)	0.8(2)	0.8(3)	0.5(3)	5.9(3)
Muskmelons	max.	12.5	1.2	1.02	11.3	....
	min.	3.5	0.2	0.2	2.4	....
	ave.	7.2(70)	0.6(11)	0.57(45)	5.4(60)	....
Peaches	max.	18.1	1.0	0.63	13.1	1.5 M
	min.	10.0	0.2	0.32	5.76	0.35
	ave.	13.1(154)	0.5(31)	0.47(31)	8.8(157)	0.64(165)
Pineapples	max.	18.9	0.6	0.7	15.3	1.10 C
	min.	9.9	0.2	0.3	8.2	0.39
	ave.	14.7(131)	0.4(46)	0.42(46)	11.9(34)	0.72(30)

[a] Chatfield and McLaughlin, *U. S. Dept. Agr.*, *Circ.* **50** (1928); see also Chatfield and Adams, *U. S. Dept. Agr.*, *Circ.* 549 (1940).

[b] The fat content of fruits is small and varies from 0.0 to 1.5% except for avocados, which have max. 28.8, min. 7.1, and av. 17.2%.

[c] Number of samples in parentheses.

mated on the ash of the fruit by methods detailed in Chapter XVIII. Spray residue determinations have been fully detailed in the chapter on metals in foods, Chapter V. Estimations of characteristic acids may be made by the subsequent methods.

TABLE 66. RANGE OF ALCOHOL PRECIPITATE CONTENT OF FRUITS [a]

	Minimum	Maximum	Average
Strawberry............................	0.48	0.56	0.54
Blackberry............................	0.61	0.74	0.68
Cherry................................	....	....	0.67
Currant...............................	....	....	0.80
Red raspberry.........................	0.70	0.78	0.74

[a] Munson, *U. S. Dept. Agr., Bur. of Chem., Bull.* **66** (1902).

TABLE 67. RANGE OF ALCOHOL PRECIPITATE CONTENT OF FRUIT JUICES [a]

	Minimum	Maximum	Average
Black raspberry.......................	1.00	1.62	1.36
Red raspberry.........................	0.65	0.73	0.69
Strawberry............................	0.48	0.69	0.56

[a] Munson, *U. S. Dept. Agr., Bur. of Chem. Bull.* **66** (1902).

MATURITY TEST FOR ORANGES AND GRAPEFRUIT

Total Solids—Acid Ratio.—Since the color of grapefruit and oranges is inadequate as a measure of maturity, as mentioned in Chapter III, it is customary to use a more objective standard. This is based on the ratio of total solids to acid present in the juice. Since the major part of the total solids is sugar, this is often termed the sugar:acid ratio. In California and in Israel [13a] the minimum ratio is 8:1 for oranges. In California a ratio of 7:1 is applicable to grapefruit. In South Africa the ratios are 5:1 for seedling oranges, 5.5:1 for Valencias, and 6:1 for navels but such standards are considered too low.

[13a] Braverman, *Citrus Products.* Interscience, New York, 1949.

Procedure.—Obtain the percentage of total solids of the juice composite by means of a Brix hydrometer as described in Chapter I, or by means of a refractometer as detailed in Chapter II. Determine the acidity by titration with 0.1 N sodium hydroxide solution and express the acidity in terms of percentage of citric acid. The maturity ratio is then the percentage of total solids divided by the percentage of citric acid.

Oranges, grapefruit, and tangerines are commonly analyzed by this method.

MATURITY TESTS FOR APPLES

The maturity of apples has been considered from several points of view but no completely satisfactory test has as yet been devised. Thus the firmness of the fruit, the starch content, the soluble solids, the rate of respiration, the development of red color, the number of days from full bloom, and the color of the seed have all been considered but none is fully accepted.

Starch Test.—As apples mature, the amount of starch decreases and in some varieties it is negligible. This test is not reliable for varieties grown in the United States.

Procedure.—Dip slices of apple into a solution of iodine in potassium iodide for 1 minute. Use the depth of the blue staining as an index of the amount of starch present.

Soluble Solids.—As apples mature, the soluble solids content increases. However, this increase is small and is often less than the variation in solids content of apples from tree to tree and from season to season, hence the results are difficult to evaluate.

Procedure.—Catch a few drops of juice from a number of apples as pressure tests are made in an appropriate vessel. Determine the soluble solids of the composite by means of a refractometer.

Color of Seeds.—As apples mature, the seeds turn brown to black in color. In some varieties, like Northern Spy, the seeds turn dark in color before maturation, hence the test has serious limitations.

CITRIC ACID

This method, in contradistinction to oxidation to acetone and pentabromoacetone, consists in precipitating citric acid with barium acetate in dilute alcoholic solution to separate it from interfering substances. The precipitated citrate, which is soluble in acid is then oxidized by permanganate in the presence of mercuric sulfate, yielding a very insoluble precipitate which may be determined gravimetrically. It was developed by Bruce.[14]

---

[14] Bruce, *Ind. Eng. Chem., Anal. Ed.* **6**, 283 (1934).

In order to make the method suitable for food products, which might contain interfering substances such as sugars, the citric acid is first separated from the sugars by precipitation with calcium hydroxide. The precipitated calcium citrate is recovered and subsequently analyzed.[15]

Preparation of Sample.—Make the sample alkaline with lime water and add an equal volume of alcohol. Centrifuge. Wash the precipitate with 50 per cent alcohol, heat with 85 per cent phosphoric acid and then analyze the solution for citrate according to the method detailed above. This treatment avoids any possible interference from sugars or nonacidic substances.

Reagents and Apparatus.—(a) A fresh solution of analyzed anhydrous citric acid containing 2 mg. of citric acid per ml.

(b) A solution of mercuric sulfate, made by dissolving a suspension of 50 g. of mercuric oxide in 500 ml. of water by the gradual addition of 200 ml. of 96 per cent sulfuric acid and diluting to 1 liter.

(c) Heavy walled Pyrex glass centrifuge tubes of 15-ml. capacity, numbered, weighed, and marked at 10 ml. A copper stand holds the centrifuge tubes in a water bath and a special stirrer is used when the permanganate is added to each in turn. The stirrer is made from a 25 cm. length of 3 mm. glass rod, fitted with a glass bearing and rubber stopper as holder. The upper end is bent to a small hook and weighted. The lower end is bent to a short spiral small enough to enter the centrifuge tubes, with a 2 cm. tip to prevent the stirrer from sticking in the cone. This stirrer when mounted is operated by a string in the hand of the analyst. Graduated 1-ml. pipettes are used to measure the mercuric sulfate and phosphoric acid solutions. Potassium permanganate is added from a 10-ml. burette having a rubber connection with a glass pearl and a bent tip. This type of burette permits more accurate control of the rate of addition of the reagent than the type with a glass stopcock. A silver or platinum wire is used for stirring the precipitate in the wash liquid.

Procedure.—For approximately known quantities of citric acid, six centrifuge tubes containing 1 to 5 ml. of solution containing 2 mg. per ml. make a convenient series. Add to each 1 ml. of 10 per cent barium acetate solution and 4 drops of saturated barium hydroxide. Complete the precipitation of the barium salt by the addition of two volumes of 95 per cent alcohol. After standing 10 minutes, centrifuge the precipitate for 5 minutes and wash 3 times by centrifuging with 3 ml. of 50 per cent alcohol containing 1 per cent barium acetate. After the precipitate is well drained (10 minutes at 30° C.), dissolve

---

[15] Bruce, personal communication (1937).

it in a mixture of 3 ml. of water and 0.16 ml. of 85 per cent phosphoric acid. Place this solution in a boiling water bath for 5 minutes to insure complete elimination of the alcohol.

TABLE 68. MAXIMA, MINIMA AND AVERAGE OF CONSTANTS OF SMALL FLORIDA ORANGES [a]

	Ave.	Max.	Min.	Composite	Grand Ave.
Wt. of orange in g.	204	229	184	....	....
Wt. of peel	54	64	41	....	....
Per cent of peel	26.5	31.8	20.6	....	....
Wt. of juice	86	99	64	....	....
Per cent of juice	41.1	48.7	31.3	....	....
Wt. of pulp	65	93	51	....	....
Per cent of pulp	32.5	44.7	25.4	....	....
Specific gravity @ 20/20	1.048	1.050	1.048	1.047	1.047
Per cent solids by sp. gr.	11.5	12.4	10.7	11.7	11.6
Refractive index @ 20° C.	1.3500	1.3519	1.3484	1.3503	1.3500
Per cent solids by refractive index	11.5	12.6	10.4	11.6	11.5
Per cent solids gravimetrically	11.94	12.6	10.6	....	....
Average per cent solids	11.3	12.4	10.4	11.6	11.4
Per cent ash	.43	.45	.41	.42	.42
Alkalinity of ash ml. N acid per g. ash	10.4	11.0	9.4	10.4	10.4
Total acidity ml. N/10 alk. per 100 ml.	104	122	81	11.5	107.7
Citric acid g. per 100 ml.	.67	.78	.52	.71	.69
Ratio solids to acid	17.7	22.9	15.0	16.3	17.0
Per cent nitrogen	.12	.13	.12	.12	.12
Per cent protein	.75	.80	.72	.74	.74
Mg. vitamin C per ml. of juice	.544	.625	.440	.549	.546

[a] Analyses of author made in 1936 on 100 oranges. First published in first edition of this text, 1938.

TABLE 69. ANALYSES OF STRAWBERRIES [a]

	Maximum	Minimum	Average
Ash %	0.62	0.53	0.55
Alkalinity of ash, ml. N acid per 1 g. of ash	14.2	9.6	13.0
Water insoluble solids %	4.10	3.00	3.46
Alcohol precipitate %	1.07	0.70	0.82
Pectic acid %	0.69	0.37	0.52

[a] Wichmann, *J. Assoc. Official Agr. Chem.* **8**, 123 (1924).

To the hot solution, add 1 ml. of the mercuric sulfate reagent. After the precipitate has settled, dilute the solution with water to 10 ml., centrifuge, and decant through a filter into a small beaker.

Transfer 8 ml. of the filtrate to a weighed 15-ml. centrifuge tube, and add 0.2 ml. of mercuric sulfate reagent and 1 ml. of water. Place the tube for 1 minute in a water bath at 85° C., add a drop of 3 per cent hydrogen peroxide and, with continuous stirring, add 1 per cent potassium permanganate at a rate not exceeding 1 drop in 10 seconds until a faint pink color persists for 10 seconds. Add 1 drop of hydrogen peroxide. After 1 minute rinse off the stirrer, cool the tube, and centrifuge for 5 minutes. Upon decanting the supernatant liquid, stir the precipitate with 3 ml. of 50 per cent alcohol and centrifuge again. Repeat the washing 3 times. During the process a slight scum occasionally escapes from precipitation; but the combined scum from eight such tubes weighs less than 0.2 mg. After draining for 5 minutes, wipe the tube and dry at 100° C. for an hour, or in a vacuum oven for half an hour.

The relation between the mercuric sulfate complex and citric acid may be taken from Table 70.

TABLE 70. CITRIC ACID—MERCURIC SULFATE COMPLEX

Citric acid, anhydrous, mg.	Mercuric sulfate complex, mg.
2	4
3	8
4	12
5	16
6	20
7	24
8	28

It is apparent that there is a straight-line relationship between the milligram of anhydrous citric acid and milligram of mercuric sulfate complex. However, since this relationship is empirical the directions for the determination must be rigidly adhered to, in order to obtain proper results. The weight of the mercuric sulfate complex varies to 1 mg. above the figures quoted above 12 and to 1 mg. below the figures quoted below 12.

No interference is encountered from formic, acetic, succinic, malic, lactic, or tartaric acids present in amounts comparable with the citric acid present. Where much larger amounts of substances precipitable by barium acetate are present, it is necessary to use larger amounts of the reagent. The supernatant liquid from the precipitation of an unknown should be tested with more reagent before discarding. Aconitic acid interferes in this method.

TARTARIC ACID

King Method.—This method developed by Kling [16] and modified by King [17] depends on the insolubility of calcium racemate in dilute acetic and hydrochloric acid.

Preparation of Sample.—In jams, jellies, and other type products that contain much sugar, pectin, or gelatin or' other alcohol insoluble material, add sufficient alcohol and a few drops sulfuric acid to a known portion of the sample to precipitate those materials. Allow the mixture to stand until the precipitate has settled and filter off an aliquot portion through a coarse fluted paper or pad of cotton wool. Where esters of tartaric acid may have been formed it may be necessary to saponify before proceeding with the determination. Other samples such as cider may be treated directly.

Reagents.—(a) Diammonium Citrate Solution.—Dissolve 29 g. of citric acid in about 200 ml. of water, carefully neutralize to methyl red with ammonia, add 14.5 g. of citric acid, and make up to 1 liter with water. This solution contains 50 g. per liter.

(b) Ammonium L-Tartrate Solution.—Dissolve 3.2 g. of the salt, entirely free from D-tartrate, in water, add 1 ml. of commercial formalin as a preservative, and dilute to 200 ml.

(c) Calcium Acetate Solution.—Dissolve 16 g. of calcium carbonate in 120 ml. of glacial acetic acid diluted with sufficient water, make up to 1 liter and filter.

(d) Hydrochloric Acid —Dilute 34 ml. of pure acid to 1 liter.

(e) Calcium and Sodium Acetate Solution.—Dissolve 5 g. of calcium carbonate in 20 g. of acetic acid and sufficient water, add 100 g. of sodium acetate, make up the solution to 1 liter and filter.

(f) Potassium Permanganate Solution.—Prepare a solution in water containing 6.974 g. in a liter. Standardize this reagent against pure tartaric acid, employing the complete precipitation process as for the sample taken. One ml. of the potassium permanganate solution equals nearly 2.5 mg. of the D-tartaric acid originally present or nearly 5 mg. of racemic acid.

(g) Oxalic Acid Solution.—Prepare a solution containing 13.879 g. per liter and titrate against the permanganate solution.

Procedure.—Weigh or measure such a portion of the sample as will contain not more than 0.2 g. of tartaric acid in the final aliquot portion, adjust to 35 ml. by dilution or concentration, add 3 ml. of N sulfuric acid, pour into a 250-ml. flask, rinse with 15 ml. of warm water and then with 95 per cent alcohol, and make up to the mark

---

[16] Kling, *Ann. fals.* **14**, 185 (1911).
[17] King, *Analyst* **58**, 135 (1933).

with the 95 per cent alcohol. Shake the mixture, and allow it to stand for half an hour, filtering if necessary. Transfer a convenient aliquot portion of the clear alcoholic solution so obtained into a centrifuge tube, and add a slight excess of neutral lead acetate solution. Shake vigorously for 2 minutes, and centrifuge for 15 minutes at about 1000 r.p.m. Drain off the supernatant liquid thoroughly, and wash once with alcohol, centrifuging and draining as before. Transfer the lead salts to a beaker with warm water, and pass in a rapid stream of hydrogen sulfide until the reaction is complete. Filter, wash thoroughly, boil the filtrate until it is free from hydrogen sulfide, and adjust the volume to 150 ml. Up to this point it has been assumed that the difficulties mentioned in the preparation of the sample have been encountered; if they have not, take a portion of the sample, which will contain not more than 0.2 g. tartaric acid, and dilute to 150 ml.

To the 150 ml. obtained by either process, add 15 ml. of diammonium citrate reagent, 25 ml. of ammonium L-tartrate reagent, and 20 ml. of calcium acetate reagent, stir vigorously until calcium racemate begins to precipitate, and allow the mixture to stand overnight at room temperature. Filter by decantation on to a thin, lightly tamped pad of asbestos, and transfer the precipitate to the crucible with a portion of the filtrate. Wash the contents of the crucible 5 times with water, filling the crucible about half full and sucking dry each time. Treat the precipitate and mat after removal from the Gooch with 20 ml. of hydrochloric acid reagent and wash the crucible thoroughly. Adjust the volume of solution to 150 ml. with water. Bring 50 ml. of the calcium and sodium acetate reagent to the boiling point and pour it through the Gooch crucible into the 150 ml. mentioned above, then bringing the temperature of the whole to 80° C.; cool, stir vigorously, and leave for at least 4 hours, stirring occasionally. Filter and wash as described in the first operation. Transfer the pad and precipitate to a beaker with 150 ml. of water, add 50 ml. of sulfuric acid, 10 per cent by volume, and heat to 80° C. Immediately add standardized potassium permanganate solution until an excess is indicated. Again heat to 80° C., add an additional 5 ml. of the potassium permanganate solution and allow the beaker to stand for about 1 minute. After re-heating to 80° C. immediately add 10 ml. of the standardized oxalic acid solution, and titrate back with the potassium permanganate solution. One ml. of the potassium permanganate solution equals 2.5 mg. of D-tartaric acid.

Fenton Method.—The well-known Fenton [18] reaction in which a

---

[18] Fenton, *Chem. News* **33, 190 (1876)**.

violet color is produced by a tartrate in the presence of ferrous sulfate, hydrogen peroxide, and sodium hydroxide may be made the basis for a colorimetric method. This color reaction appears to be specific for tartaric acid. Fenton found that citric acid, succinic, malic, and oxalic acid and sugars do not give the test. The following method, developed by Anderson, Rouse, and Letonoff [19] may be used to determine tartaric acid in tartrate baking powders with or without aluminum.

Reagents.—(1) One per cent Ferrous Sulfate Solution.—Dissolve 1 g. of ferrous sulfate in 80 ml. of water, heating gently and stirring to aid solution. Cool, transfer to a 100-ml. volumetric flask, and make up to volume.

(2) Transfer 16 g. of dry D-tartaric acid to a 100-ml. volumetric flask, dissolve in water and make up to volume.

(3) Working Tartaric Acid Standard Solution.—Transfer 5 ml. of solution (2) to a 100-ml. volumetric flask, add 10.66 ml. of N sodium hydroxide solution and make up to volume. This solution contains 0.80 g. of tartaric acid per 100 ml. and has a pH of 6.2.

Procedure.—Transfer to a small beaker a 2 g. sample of baking powder. Add water drop by drop, until carbon dioxide ceases to be evolved. Next add 45 ml. of water, and stir thoroughly to dissolve the tartrates present. To remove the starch, filter into a 100-ml. volumetric flask and wash the residue 3 times with 15 ml. water at each washing. Make up to volume with water. This solution should have a pH of approximately 6.2. If the pH varies from 6.2 by more than ±0.5, another sample should be prepared and the pH adjusted before making up to volume. The pH of the solution may be determined colorimetrically, using chlorophenol red as an indicator. As a rule tartrate baking powders require no adjustment.

Transfer 10 ml. of the above solution to a 25-ml. volumetric flask. Add 0.2 ml. of 1 per cent ferrous sulfate solution and 0.2 ml. of commercial 3 per cent hydrogen peroxide solution and mix thoroughly. Upon the addition of the hydrogen peroxide, the solution will turn yellow. Allow the solution to stand until it becomes brownish in color and then place it in an ice bath until the brown color disappears and the color becomes definitely lavender. Add immediately 5 ml. of N sodium hydroxide solution. Stopper the flask, mix by inversion twice, and place the flask in the ice bath for 10 minutes. At the end of this time remove the flask from the ice bath, mix by inversion twice, and compare in a colorimeter with a standard prepared simultaneously. For the standard, 10 ml. of the working tar-

19 Anderson, Rouse, and Letonoff, *Ind. Eng. Chem.*, *Anal Ed.* 5, 19 (1933).

taric acid standard solution containing 0.08 g. of tartaric acid is used. The results may be calculated:

$$\frac{\text{Reading of standard} \times 0.08 \times 10 \times 100}{\text{Reading of unknown} \times \text{weight of sample}} = \% \text{ tartaric acid}$$

Glyoxylic Acid Method.—On occasion it becomes necessary to prove the presence of tartaric acid in a given product, as for instance in a fruit juice other than grape juice. This may be done as pointed out by Mathers [19a] by use of the selective oxidizing action of periodic acid on a-, β-glycols to convert tartaric acid into glyoxylic acid. The glyoxylic acid may be converted into a p-nitrophenylhydrazone and the rearrangement of this compound in alkaline solution gives a compound colored intensely red.

$$\begin{array}{l} CHOH \cdot COOH \\ | \\ CHOH \cdot COOH \end{array} + HIO_4 \rightarrow 2CHO \cdot COOH + HIO_3 + H_2O$$

$$CHO \cdot COOH + p\text{-}NO_2C_6H_4NHNH_2 \rightarrow p\text{-}NO_2C_6H_4NHN : CH \cdot COOH + H_2O$$

$$p\text{-}NO_2C_6H_4NHN : CH \cdot COOH + NaOH \rightarrow \text{Red Compound}$$

Adsorption Tube.—Any small chromatographic column is satisfactory. Pack the column to a height of 40 to 50 mm. with glass powder held in place by asbestos fibers. The tube may be made by sealing about 1″ of 6 mm. o.d. tubing to the bottom of a 16 × 150 mm. test tube.

Reagent.—p-Nitrophenylhydrazine Reagent.—Dissolve 1 g. of p-nitrophenylhydrazine, or one of its salts, in 7.5 ml. of concentrated sulfuric acid and dilute to 75 ml. with ethyl alcohol (95%).

Procedure.—Dilute 10 ml. of sample to 50 ml. with water; add approximately 0.5 g. of activated carbon, warm on steam bath about 10 minutes and filter into a centrifuge bottle. To the filtrate add 5 ml. of 5 per cent neutral lead acetate solution, centrifuge, and decant the supernatant liquid. Wash the precipitate with 40 ml. of water, again centrifuge and decant. Dissolve the precipitate in 5 ml. of 20 per cent ammonium acetate in glacial acetic acid solution, warming if necessary to effect solution. If a clear solution is not obtained because of coagulated material or inorganic salts, filter into a second centrifuge bottle, washing the filter paper with 5 ml. of glacial acetic acid. Add 5 ml. of 3 per cent lead acetate solution in glacial acetic acid, plus 25 ml. of anhydrous ethyl alcohol. Centrifuge and decant the supernatant liquid. To the precipitate add 4 ml. of 10 per cent sodium bisulfate solution, 10 ml. of 1 per cent periodic acid solution,

[19a] Mathers, *J. Assoc. Official Agr. Chem.* 32, 418 (1949).

and 10 ml. of water. Allow oxidation to take place at room temperature for 20 minutes. Destroy the excess oxidizing agent by addition of powdered sodium bisulfite. Add 4 ml. of p-nitrophenylhydrazine reagent and place the centrifuge bottle in boiling water for 10 minutes. Filter the solution into a separatory funnel, cool, and extract the aqueous layer with 50 ml. of ethyl ether. Wash the ether extract with 5 ml. of water and pass the washed extract through the packed adsorption column. Wash the column with 30 ml. of ethyl ether and discard the washings. Elute with 20 ml. of anhydrous ethyl alcohol. Dissolve about 0.1 g. of solid sodium hydroxide in the eluate. A brilliant pink to red coloration is obtained if tartrates were present in the original sample. The color is stable upon dilution with an equal quantity of water.

LACTIC ACID

Methods for the determination of lactic acid are detailed in Chapters IV and XX.

HARDNESS TESTER

The principle of the tester, Fig. 70, devised by Ross[20] is the measurement of that gas pressure necessary to force the blunt end of a piston a very small but fixed distance into the test material.

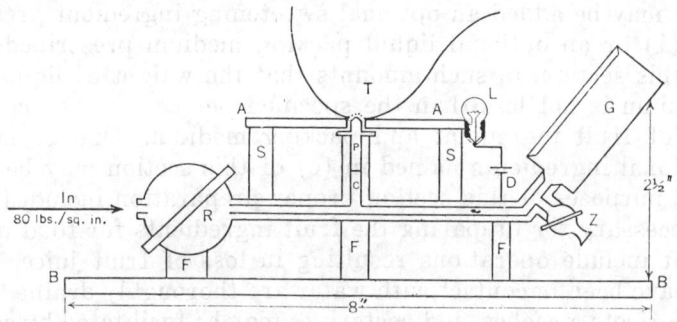

FIG.. 70. Hardness Tester.

A—top plate; B—base plate; C—cylinder; D—dry cell battery; F—metal frame; G—pressure gauge; L—indicator lamp P—piston; R—regulator valve; S—insulated support; T—test fruit; Z—release petcock

This tester forces a rounded brass tip, $5\!/\!32$ in. in diameter $1\!/\!32$ in. into the fruit or food product. Ross used pears. A top plate serves both as a stop, restricting penetration to $1\!/\!32$ in. and as an electrical contact to complete a circuit which lights an indicator lamp when maximum penetration is reached.

[20] Ross, *Science* 109, 204 (1949).

Pressures required to effect this penetration into normal green pears varied from 50 to 65 lbs./sq. in. Abnormally hard pears were found to test above 65 lbs./sq. in. Tips of other sizes and penetrations of different depths may be used for other foods. Compressed air or nitrogen may be used as the source of gas pressure. The pears or other fruits are held firmly against the top plate during the test which results in only a barely visible indentation.

Other testers have been devised for apples and for vegetables. See Chapter XIX.

Frozen Fruits and Fruit Products

The Food and Drug Administration set a hearing [21] with respect to definitions and standards of identity and fill of container for frozen fruits. In 1949 these had not yet been adopted.

Frozen Fruits.—Identity.—(a) The frozen fruits for which definitions and standards of identity are prescribed by the Food and Drug Administration are the frozen foods each of which is made by freezing a properly prepared fruit ingredient or mixture of fruit ingredients specified in fill of container section below, but do not include frozen pureed fruits. To such fruit ingredient or mixture of fruit ingredients (other than unstemmed currants and unstemmed grapes) may be added an optional sweetening ingredient prescribed in (c) (1) or an optional liquid packing medium prescribed in (c) (2) of this section, in such amounts that the weight of liquid packing medium is not less than the specified per cent of the combined weight of fruit ingredient and packing medium. One or more of the optional ingredients named in (e) of this section may be added. For the purposes of this section proper preparation includes operations necessary for preparing the fruit ingredients for food use, but does not include operations resulting in loss of fruit juice. Fruits which have been in contact with water are thoroughly drained. Peeling of apricots, peaches, and nectarines may be facilitated by scalding with steam, hot water, or lye solution, with removal of the lye solution when it is used. The fruit ingredients, before or after addition of the optional sweetening ingredient or the optional liquid packing medium, may be cooked.

(b) (1) See fill of container section below.

(2) Any mixture of properly prepared fruit ingredients, other than pureed fruits, which contains one or more of the fruit ingredi-

---

[21] *Federal Register* **13**, 1456 (1948).

ents listed below. In such mixtures the per cent by weight of the fruit ingredients present in smaller amounts, computed on the basis of total weight of fruit ingredients present, is not less than:

Number of Fruits in Mixture	Minimum Per Cent by Weight of Fruit
2	Neither less than 25%
3	None less than 20%
4	One not less than 5%, each of others not less than 15%
5 or more	Each of 2 not less than 5%, each of others not less than 10%

(3) **Number of Pits.**—For the purposes of this section, pitted cherries are cherries containing not more than 1 pit in each 20 oz. of the frozen food. Partially pitted cherries are incompletely pitted cherries containing more than 1 pit in each 20 oz. of the frozen food. The number of pits is determined as follows:

Procedure.—Collect at random at least 24 lbs. of sample from two or more containers. Count the pits and pieces of pit shell in the weighed sample. Count a piece of pit shell equal to or smaller than one-half pit shell as one-half pit, and a piece of pit shell larger than one-half pit shell as one pit; but when two or more pieces of pit shell are within or attached to a single cherry, count such pieces as one-half pit if their combined size is equivalent to that of one-half pit shell or less, and as one pit if their combined size is equivalent to that of more than one-half pit shell. From the total number of pits so counted and the combined weight of the contents of all the containers, calculate the number of pits present in each 20 oz. of frozen food.

(c) (1) The optional sweetening ingredients referred to in (a) are:

(i) Sugar (ii) Any mixture of sugar with dextrose or corn syrup solids or both, which contains not less than 66⅔ per cent by weight of sugar.

(2) The optional liquid packing media referred to in (a) are:

    (i) Heavy syrup.
    (ii) Medium syrup.
    (iii) Light syrup.
    (iv) Corn syrup.

Each of the liquid packing media named in subdivisions (i), (ii), and (iii) of this subparagraph is an aqueous syrup made with one or any combination of the following saccharine ingredients, within the limitations prescribed herein:

> Sugar
> Invert sugar syrup
> Dextrose
> Corn syrup solids
> Corn syrup
> Glucose syrup

The total solids of each of the liquid packing media, subdivisions (i), (ii), and (iii) of this definition, shall contain not less than 66⅔ per cent by weight of sugar or invert sugar or any mixture of these.

The per cent of solids, as determined by the Brix hydrometer (see Chapter I), of heavy syrup, is not less than 60; of the medium syrup less than 60 but not less than 50; and of light syrup less than 50 but not less than 40. When corn syrup is used as the sole packing medium it contains not less than 75 per cent by weight of solids.

(3) Light syrup and medium syrup shall not be used with sliced strawberries or with red sour pitted cherries.

(d) The terms "sugar," (2) "invert sugar syrup," (3) "dextrose," and (4) "corn syrup" have been previously defined.

(5) The term "corn syrup solids" means dried corn syrup.

(6) The term "glucose syrup" means a syrup which conforms to the definition for corn syrup, except that it is made from starch other than cornstarch.

(e) The optional ingredients referred to in (a) are ascorbic acid. citric acid, and salt.

Fill of Container.—(a) The standards of fill of container for frozen fruits with liquid packing medium are:

The maximum percentages by weight of a liquid packing medium which shall be used with specified frozen fruits, stated as per cent by weight of the combined weights of fruit ingredient and packing medium are given in Table 71.

TABLE 71. MAXIMUM PER CENT BY WEIGHT OF LIQUID PACKING MEDIUM IN COMBINED WEIGHTS OF FRUIT INGREDIENT AND PACKING MEDIUM

Kinds and Forms of Frozen Fruits	%
Apricots, peeled dice	25
Apricots, unpeeled dice	25
Apricots, peeled slices	30
Apricots, unpeeled slices	30
Apricots, peeled quarters	35
Apricots, unpeeled quarters	35

TABLE 71—*Continued*

Apricots, peeled halves	37
Apricots, unpeeled halves	37
Apricots, peeled whole	37
Apricots, unpeeled whole	37
Apricots. peeled pitted	30
Apricots, unpeeled pitted	30
Apricots, pieces of irregular sizes and shapes	30
Blackberries	25
Blueberries	25
Boysenberries	25
Cherries, pitted dark sweet	25
Cherries, partially pitted dark sweet	25
Cherries, unpitted dark sweet	25
Cherries, pitted light sweet	25
Cherries, partially pitted light sweet	25
Cherries, unpitted light sweet	25
Cherries, pitted red sour (or tart)	25
Cherries, partially pitted red sour (or tart)	25
Cherries, unpitted red sour (or tart)	25
Currants, stemmed	25
Currants, unstemmed	0
Gooseberries	25
Grapes, stemmed	25
Grapes, unstemmed	0
Huckleberries	25
Loganberries	25
Nectarines, unpeeled dice	25
Nectarines, unpeeled slices	30
Nectarines, unpeeled quarters	35
Nectarines, unpeeled halves	35
Nectarines, unpeeled whole	37
Nectarines, unpeeled pitted	37
Nectarines, pieces of irregular sizes and shapes	30
Freestone peaches, peeled dice	25
Freestone peaches, peeled slices	31
Freestone peaches, peeled quarters	35
Freestone peaches, peeled halves	35
Freestone peaches, peeled whole	37
Freestone peaches, pieces of irregular sizes and shapes	30
Cling peaches, peeled dice	25
Cling peaches, peeled slices	31
Cling peaches, peeled quarters	35
Cling peaches, peeled halves	35
Cling peaches, peeled whole	37
Cling peaches, pieces of irregular sizes and shapes	30
Plums, slices	30
Plums, halves	30
Plums, pitted	30
Plums, unpitted	35

TABLE 71—*Continued*

Raspberries, red 30
Raspberries, black 30
Rhubarb, cuts 30
Rhubarb, stalks 30
Strawberries, slices 25
Strawberries, whole 30
Youngberries 25
Mixed fruits 30

SELECTED REFERENCES

Allen's Commercial Organic Analysis, 5th Ed., Vol. X, Blakiston, Philadelphia, 1933.

Braverman, *Citrus Products.* Interscience, New York, 1949.

Brooks, *Critical Studies in the Legal Chemistry of Foods.* Chemical Catalog, New York, 1927.

Cox, *Analysis of Foods.* Blakiston, Philadelphia, 1926.

Federal Register 5, 3554 (1940).

Hinton, *Analyst* 59, 248 (1934).

Jacobs, ed., *Chemistry and Technology of Food and Food Products,* 2nd Ed. Interscience, New York, 1951.

Joslyn, *Methods in Food Analysis Applied to Plant Products.* Academic, New York, 1950.

McCance, Widdowson, and Shackleton, *The Nutritive Value of Fruits, Vegetables, and Nuts. Med. Research Council, Special Rept. Ser.,* 213 (1936).

Methods of Analysis, A.O.A.C., 8th Ed. Washington, 1955.

Sale, *J. Assoc. Official Agr. Chem.* 21, 502 (1938).

U.S. Food Drug Admin., S.R.A., F.D.&C. 2, Part 29, August, 1954, as amended October, 1955.

CHAPTER XIII

VEGETABLE PRODUCTS

THE analysis of these products, that is, the estimation of moisture, ash, metallic constituents, sugars, acids, volatile acids, preservatives, and coloring matters are made in a manner entirely similar to other foods and as described in various sections of the text. Chlorides may be determined by the method described under the section, "Fish," Chapter XVI. The specific gravity of tomato products may be ascertained by the specific gravity centrifuge bottle method detailed in Chapter I. Organoleptic examination should be made as directed in Chapter XIX.

TABLE 72. PROXIMATE COMPOSITION OF VEGETABLES [c]

Type	Moisture	Protein	Fat	Ash	Carbo-hydrate
Artichokes[a]	79.5	2.6	0.2	1.0	16.7
Asparagus[a]	94.0	1.8	0.2	0.7	3.3
Beans[a], dried	12.6	22.5	1.8	3.5	59.6
String beans[b]	89.2	2.3	0.3	0.8	7.4
Beets[b]	87.5	1.6	0.1	1.1	9.7
Cabbage[b]	91.5	1.6	0.3	1.0	5.6
Carrots[b]	92.3	1.8	0.5	0.7	4.7
Celery[b]	94.5	1.1	0.1	1.0	3.3
Corn[b]	75.4	3.1	1.1	0.7	19.7
Cucumber[b]	95.4	0.8	0.2	0.5	3.1
Lettuce[b]	94.7	1.2	0.3	0.9	2.9
Mushrooms[a]	88.1	3.5	0.4	1.2	6.8
Onions[b]	87.6	1.6	0.3	0.6	9.9
Parsnip[b]	83.0	1.6	0.5	1.4	13.5
Peas, dried[a]	9.5	24.6	1.0	2.9	62.0
Peas, green[b]	74.6	7.0	0.5	1.0	16.9
Potatoes[b]	78.3	2.2	0.1	1.0	18.4
Sweet potatoes[b]	69.0	1.8	0.7	1.1	27.4
Radishes[b]	91.8	1.3	0.1	1.0	5.8
Spinach[a]	92.3	2.1	0.3	2.1	3.2
Squash[b]	88.3	1.4	0.5	0.8	9.0
Tomatoes[a]	94.3	0.9	0.4	0.5	3.9
Turnips[b]	89.6	1.3	0.2	0.8	8.1

[a] As purchased.
[b] Edible portion.
[c] Atwater and Bryant, U. S. Dept. Agr., Bull. 28 (1906).

TABLE 73. PROXIMATE COMPOSITION OF FRESH VEGETABLES, EDIBLE PORTION [a]

Type		Water %	Protein %	Fat %	Ash %	Total Carbo-hydrate [b]	Fiber	Sugars	Starch
Asparagus	ave.	93.0	2.2	0.2	0.67	3.9	0.7	1.34	0.4
	max.	94.4	3.4	0.3	0.97	....	0.9	2.96	0.7
	min.	90.8	1.1	0.0	0.49	....	0.7	0.59	0.1
Beets	ave.	87.6	1.6	0.1	1.11	9.6	0.9	......	......
	max.	94.1	2.2	0.3	2.0	....	1.7	......	......
	min.	82.3	0.9	0.0	0.7	....	0.6	......	......
Cabbage	ave.	92.4	1.4	0.2	0.75	5.3	1.0	3.5	......
	max.	94.8	3.1	0.5	1.07	....	1.4	4.8	......
	min.	88.4	0.8	0.1	0.34	....	0.5	2.9	......
Carrots	ave.	88.2	1.2	0.3	1.02	9.3	1.1	7.5	......
	max.	91.1	2.3	0.7	1.55	....	2.3	8.7	......
	min.	83.1	0.7	0.0	0.62	....	0.7	6.2	......
Sweet Corn...	ave.	73.9	3.7	1.2	0.66	20.5	0.8	4.29	14.6
	max.	86.1	4.9	2.1	0.84	....	1.4	7.56	26.2
	min.	61.3	2.8	0.5	0.4	....	0.5	1.58	3.4
Jerusalem Atichoke ...	ave.	79.5	2.2	0.1	1.17	17.0 [c]	0.8	......	......
	max.	84.2	3.1	0.2	2.0	....	1.4	......	......
	min.	74.2	1.1	0.0	0.87	....	0.6	......	......
Lettuce	ave.	94.8	1.2	0.2	0.91	2.9	0.6	1.6	......
	max.	97.4	1.9	0.6	1.41	....	1.1	2.2	......
	min.	91.5	0.5	0.0	0.5	....	0.3	0.9	......
Mushrooms ...	ave.	91.1	(0.57) [d]	0.3	1.14	....	0.9	......	......
	max.	94.7	(0.98)	0.8	1.86	....	1.3	......	......
	min.	87.9	(0.27)	0.1	0.58	....	0.2	......	......
Onions	ave.	87.5	1.4	0.2	0.58	10.3	0.8	6.7	0.5
	max.	95.2	2.7	0.8	1.20	....	1.8	8.4	......
	min.	70.2	0.4	0.1	0.17	....	0.4	3.7	......
Parsnips	ave.	78.6	1.5	0.5	1.15	18.2	2.2	9.5	2.4
	max.	89.2	2.1	0.8	1.9	....	3.0	14.2	8.0
	min.	72.6	1.1	0.2	0.7	....	1.4	4.5	0.0
Peas	ave.	74.3	6.7	0.4	0.92	17.7	2.2	3.2	8.2
	max.	84.1	9.9	0.6	1.2	....	2.9	6.9	15.9
	min.	56.7	3.5	0.1	0.55	....	1.3	0.4	1.8
Potatoes	ave.	77.9	2.0	0.1	0.99	19.1	0.4	0.87	14.7
	max.	85.2	3.9	0.3	1.9	....	0.9	1.5	16.4
	min.	66.0	0.9	0.0	0.5	....	0.2	0.21	12.1
Spinach	ave.	92.7	2.3	0.3	1.53	3.2	0.6	0.3	......
	max.	95.0	3.4	0.6	2.0	....	0.7	0.4	......
	min.	89.4	1.9	0.1	1.06	....	0.5	0.2	......
Sweet Potato	ave.	68.5	1.8	0.7	1.07	27.9	1.0	5.35	20.2
	max.	82.7	4.4	2.5	1.85	....	1.8	11.9	29.8
	min.	58.5	0.5	0.2	0.4	....	0.6	1.15	8.8
Tomatoes	ave.	94.1	1.0	0.3	0.57	4.0	0.6	3.37	......
	max.	96.7	1.8	0.5	1.0	....	1.2	4.06	......
	min.	90.6	0.7	0.1	0.34	....	0.2	2.3	......
Turnips	ave.	90.9	1.1	0.2	0.73	7.1	1.1	4.6	......
	max.	95.7	2.1	0.4	1.0	....	1.4	......	......
	min.	85.6	0.7	0.1	0.5	....	0.6	......	......

[a] Chatfield and Adams, U. S. Dept. Agr., Circ. 146 (1931); Circ. 549 (1940).
[b] Total carbohydrate, by difference.
[c] Mostly inulin.
[d] As nitrogen, protein cannot be calculated from nitrogen because of much non-protein nitrogen.

The composition, by means of proximate analysis, of some vegetables is given in Tables 72 and 73. An extended survey of the chemical composition not only of vegetables but also of fruits and nuts is given by McCance, Widdowson, and Shackleton.[1]

BLANCHING ADEQUACY

Catalase Method.—A rapid method for the determination of catalase and thus estimating its inactivation after blanching and before dehydration and freezing was devised by Thompson.[2] The amount of oxygen liberated from hydrogen peroxide is measured.

Procedure. — Grind an accurately weighed sample with 0.6 g. of calcium carbonate and 1.0 g. of fine sand. Add 10 ml. of water and continue grinding for about 2 minutes. Transfer 1 ml. of this mixture with the aid of a pipette into one-half of the specially divided flask, Fig. 71, and place 2 ml. of hydrogen peroxide (Dioxygen) in the other half. Attach the flask to the manometer and suspend the entire apparatus in a water bath controlled thermostatically at 20° C. When the apparatus reaches the temperature of the bath, close the stopcock (water level in U-tube set at 0.0 ml.) and shake the apparatus for 2 minutes. Read the pipette to determine the amount of oxygen liberated. Report the catalase as milliliters of oxygen liberated by 0.1 g. in 2 minutes.

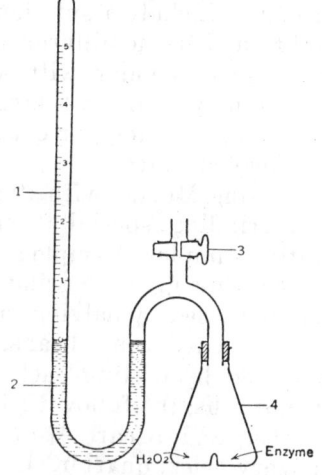

FIG. 71. Catalase Apparatus

1. 5-ml. pipette in 0.1 ml. fused to U-piece of tubing of same diameter
2. Water in tube, at 0 ml. level of pipette at start of reaction
3. Stopcock to adjust level of water after flask is attachel
4. 50-ml. Erlenmeyer flask with divided bottom

Peroxidase Method.—A test for the adequacy of blanching of frozen vegetables has been devised by the Western Regional Research Laboratory[3] of the United States Department of Agriculture. It is based upon the determination of peroxidase activity by a method which has given good correlation with the keeping quality of certain frozen vegetables held in freezing storage at −5° F. for a period of

[1] McCance, Widdowson, and Shackleton, *Med. Research Council, Special Rept. Ser.*, 213 (1936).

[2] Thompson, *Ind. Eng. Chem., Anal. Ed.* 14, 585 (1942).

[3] *Western Regional Research Laboratory*, AIC-34, Rev. May, 1947.

four years. This test is applicable to frozen peas, snap beans, Lima beans, asparagus, and cut corn. It is probably applicable to other frozen vegetables.

Sampling.—To insure adequate blanching of all units of the product, select the largest pieces or parts, since these are the most likely to be underblanched. At least, select a sample with a large proportion of large pieces. With some vegetables, like broccoli, for example, a more complete knowledge of adequacy of blanch can be obtained by testing stalk and floral parts separately. With such products it is especially important to use a large sample, or more than one sample, so as to include a sufficient number of units to overcome size variability and tissue differentiation.

Using Blendor with 100- to 200-g. Samples.—With vegetables which may be in very large pieces, like broccoli, either test different tissues separately, or cut so as to secure normal proportions from a number of units.

Using Mortar with 10-g. Samples.—Because of the limits of mortar grinding, special attention should be given to securing representative samples. Sample size can be increased to 20 or 30 g. with additional time given to grinding, or the number of units represented can be increased by halving or quartering if they are symmetrical. With large-seeded Lima beans, for example, the number of units can be doubled by cutting each seed in half across the cotyledons. With asparagus, the following procedure is recommended:

Cut and discard ¾ in. from butt end. Split spears lengthwise or, if very thick, quarter. Use alternate ½-in. cuts from half or quarter of each spear, discarding every other cut. In case of doubt, test more than one sample.

Reagents.—Guaiacol Solution.—Dissolve 0.5 g. of guaiacol in 50 per cent alcohol and make up to 100 ml.

Hydrogen Peroxide Solution.—Dilute 2.8 ml. of 30 per cent hydrogen peroxide with water to 1 liter. This is a 0.08 per cent solution. Store in a dark bottle and keep in a refrigerator. Prepare fresh each week.

Procedure with Blendor.—Weigh out a representative 100- to 200-g. sample. Place in a blendor with 3 ml. of water for each gram of sample. Blend for 1 minute at moderate or high speed. Filter through a cotton milk filter. Add 2 ml. of filtrate to 20 ml. of water in a test tube.

Prepare a blank by adding 2 ml. of filtrate to 22 ml. of water in a second test tube, mix, and use as a color comparison tube. Do not add any guaiacol or peroxide to this tube.

Add 1 ml. of 0.5 per cent guaiacol solution to first tube, without mixing. Add 1 ml. of 0.08 per cent hydrogen peroxide solution to the same tube, without mixing. Mix contents thoroughly by inverting and watch for *development of any color differing from blank, regardless of hue,* but of *sufficient intensity* to show an obvious contrast to blank. This is a *positive* test, and indicates *inadequate blanching.* If no such color contrast develops in 3½ minutes, consider the test *negative* and the product *adequately blanched.* If color develops after 3½ minutes, it is to be disregarded, and the test still considered negative.[4]

Procedure Using Mortar.—Cut tissue to be tested into small pieces and weigh out representative 10-g. sample with aid of balance sensitive to 0.1 g. Place 30 ml. water in graduated cylinder. Place sample in mortar with a little clean sand, add minimum amount of water from graduated cylinder to give best consistency for thorough maceration, and grind for 3 minutes. Add remainder of water from graduate, mix, and proceed with the method as above.

ALCOHOL-INSOLUBLE SOLIDS

The percentage of alcohol-insoluble solids is used as a criterion of the quality of canned peas, of the maturity of frozen Lima beans, and of other vegetables. In the method [5] detailed, sugars are extracted from frozen vegetables with the aid of a Waring blendor and 85 per cent alcohol. The alcohol-insoluble solids content is determined by filtering the macerate and drying the residue.

As a means of preserving the samples until a convenient time for analysis, the vegetables are frozen and stored at $-23°$ C. This also affords an opportunity for obtaining a representative sample for analysis by grinding one or more pounds of the frozen material twice in a food chopper at $-23°$ C.

Procedure.—Weigh 20 g. of the finely ground vegetable into a 100-ml. beaker in the cold room. At room temperature wash the ground sample into a blendor cup (500-ml. capacity, having a rubber gasket under the screw top) with 150 ml. of 85 per cent ethyl alcohol (specific gravity 0.850). After 5 minutes of maceration, wash the contents of the blendor cup into a 600-ml. beaker with 85 per cent ethyl alcohol from a wash bottle. Allow the solids to settle and pour the contents of the beaker onto a weighed 5.5-cm. No. 40 Whatman filter paper in a Büchner funnel inserted through a two-holed rubber stopper placed in the mouth of a 500-ml. Kohlrausch sugar flask

---

[4] Masure and Campbell, *Fruit Products J.* 23, 369 (1944).
[5] Moyer and Holgate, *Anal. Chem.* 20, 472 (1948).

Where heavy-walled or Kohlrausch flasks are not available, a 500-ml. Pyrex volumetric flask can be adapted by sealing onto the neck a piece of tubing, 37 mm. in outside diameter and 6 cm. long. The use of a Kohlrausch flask and the danger of possible collapse of the flask by the vacuum can be avoided by collecting the filtrate in a 500-ml. volumetric flask under the high-form glass cover of a Fisher Filtrator.

Apply suction to the interior of the Kohlrausch flask through a small piece of glass tubing bent at right angles and inserted into the other hole of the rubber stopper. By decanting off most of the alcoholic solution before adding the solids to the funnel, the filtration is greatly accelerated. Rinse the beaker out with more 85 per cent ethyl alcohol which is poured into the Büchner funnel. Wash the residue with alcohol three or four times. After each washing allow the residue to become partially dry but take care that the precipitate is not too thoroughly dried or the filter paper will pull away from the sides of the funnel and the next addition of alcohol will wash the solids into the flask. When the volume of the filtrate approaches the 500-ml. mark, allow the residue to dry and carefully remove to a weighing dish for complete drying at 95° C. overnight. The weight of the dried residue represents the alcohol-insoluble solids content of the sample. Make the volume of the filtrate up to the 500-ml. mark with the 85 per cent alcohol and mix the contents of the flask thoroughly.

SUGAR [6]

The alcohol in a small aliquot of the filtrate from the alcohol-insoluble solids determination is evaporated and an aqueous solution of the residue is clarified by Somogyi's barium hydroxide-zinc sulfate procedure. For estimation of the total sugar content, an aliquot of the clarified extract is inverted with invertase. The sugar content before and after inversion is determined by Nelson's colorimetric method,[7] using Somogyi's new copper reagent.[8]

Clarification.—Somogyi Method.—Reagents.—Barium Hydroxide Solution.—Mix 56 g. of barium hydroxide with 2 liters of hot, boiled water and filter into a storage bottle through a small Büchner funnel. Equip the stopper in the bottle with a soda-lime tube and a syphon tube to a 10-ml. burette having a three-way stopcock.

Zinc Sulfate Solution.—Dissolve 100 g. of zinc sulfate heptahydrate in 2 liters of water. Adjust the final concentration of zinc sulfate so that a 10-ml. aliquot mixed with 50 ml. of water requires 9.5 ml.

---

[6] Moyer and Holgate, *Anal. Chem.* 20, 472 (1948).

[7] Nelson, *J. Biol Chem.* 153, 375 (1944).

[8] Somogyi, *J. Biol. Chem.* 160, 61 (1945).

of barium hydroxide, added dropwise, to give a faint pink end point with phenolphthalein that is stable for one minute.

Procedure.—Transfer 10 ml. of an alcoholic extract with a pipette to a 100-ml. beaker and evaporate the contents to near dryness on a steam bath. Wash the walls of the beaker down with approximately 5 ml. of water and add 2 ml. of barium hydroxide solution, followed by 2 ml. of zinc sulfate solution with constant agitation during the addition of each reagent. Wash the contents of the beaker into a funnel 50 mm. in diameter and collect the filtrate in a graduated test tube. Wash the precipitate with a fine stream of water until a filtrate of 35-ml. volume is attained. For the determination of the reducing sugar content, use a 2-ml. aliquot of this filtrate in the colorimetric procedure. When determining the total sugar content, use a 5-ml. aliquot of the filtrate for inversion.

Inversion.—Of the two methods commonly used to hydrolyze sucrose, Moyer and Holgate have found enzymatic action better suited for use with the colorimetric procedure. With acid hydrolysis the concentration of acid necessary for inversion required very careful neutralization and did not lend itself to routine analyses.

Reagents.—Sodium Acetate Buffer Solution.—To prepare the sodium acetate buffer, dissolve 13.6 g. of sodium acetate trihydrate in water, add 8 ml. of glacial acetic acid, and dilute the mixture to 500 ml.

Invertase Solution.—For the invertase solution, dissolve 200 mg. of Wallerstein Laboratories Blue Label invertase scales in 100 ml. of water and keep the solution in the ice box under a layer of toluene. This solution contains an excess of invertase activity for the amounts of sugar encountered in the analysis. Information supplied by the manufacturer indicates that the solution will have a k value in the neighborhood of 0.02.

Procedure.—Pipette a 5-ml. aliquot of the clarified extract into a graduated test tube. Add 2 drops of the sodium acetate buffer and 5 drops of the invertase solution to the test tube before incubation overnight at 35° C. Dilute the contents of the test tube to 35 ml. with water before a 2-ml. aliquot is used for color development. The overnight incubation period has been used for convenience. A much shorter time at a higher temperature would very likely accomplish the same degree of inversion.

Nelson-Somogyi Method.—*Reagents.*—Somogyi's Copper Solution.—Add 56 g. of anhydrous disodium phosphate slowly with stirring to 1400 ml. of water. Then, with continued stirring, add 80 g. of Rochelle salts, followed by the slow addition of 200 ml. of 1 N sodium hydroxide solution. Prepare a cupric sulfate solution by dissolving

16 g. in 160 ml. of water, and add this solution to the phosphate-tartrate mixture. Finally, add 360 g. of anhydrous sodium sulfate slowly with stirring. Dilute the mixture with water to the 2000-ml. mark and allow to stand for two days before filtration.

Nelson's Arsenomolybdate Solution.—Dissolve 100 g. of ammonium molybdate with stirring in 1800 ml. of water. Add 84 ml. of concentrated sulfuric acid slowly with continued agitation and finally 12 g. of sodium arsenate heptahydrate. When the arsenate is dissolved, dilute the solution to 2000 ml. with water and store at 37° C. for 48 hours. At the end of this period, filter the solution and store in a brown bottle.

Procedure.—Transfer a 2-ml. aliquot of the clarified extract (for reducing sugars) or of the inverted extract (for total sugars) in a Folin-Wu blood sugar tube with an Ostwald pipette. Add 2 ml. of Somogyi's copper reagent from a 25-ml. burette and place the tube in boiling water for 20 minutes. (When a number of samples are being analyzed, sufficient space must be provided around each tube to permit adequate circulation of the boiling water. Wire test tube supports serve this purpose very well.) Cool the sugar tubes in water at room temperature and add 2 ml. of Nelson's arsenomolybdate solution from a 25-ml. burette. Because the copper reagent has a high specific gravity, the solutions are most effectively mixed by moderate vertical agitation with a small knob on the end of a glass rod. Wash this rod and dilute the contents of the sugar tube to the 25-ml. mark with water before shaking to insure thorough mixing. It has been found advantageous to allow the tubes to stand for 15 minutes to permit maximum color development before the solutions are read at 600 mμ in a photoelectric colorimeter or spectrophotometer which has been adjusted to give 100 per cent transmittance with water. With each series of unknown samples, treat tubes containing 2 ml. of water for a blank and 2 ml. of standard solutions having 0.10 and 0.20 mg. of glucose in a similar manner to obtain a standard reference graph.

MATURITY IN FROZEN VEGETABLES

This method [9, 10] depends upon the determination of the specific gravity of the sample by means of the difference in weight in air and in a liquid of known specific gravity.

The equipment used is a suitable balance sensitive to 0.1 g. which can be supported on a stand or shelf and to which a basket can be attached by a hook under the pan. A basket made of 16-mesh brass

[9] Lee, *Ind. Eng. Chem., Anal. Ed.* **13**, 38 (1941).
[10] Lee, DeFelice, and Jenkins, *Ind. Eng. Chem., Anal. Ed.* **14**, 240 (1942).

screen 8.1 cm. (3¼ in.) high and 5.6 cm. (2¼ in.) in diameter was found suitable. A basket of these dimensions conveniently holds a 100-g. sample. Equipment for larger samples can be used if desired.

Peas.—Thaw peas and drain for 2 minutes. Weigh the peas in air. Immerse the peas in the solvent. Shake to free from air bubbles and weigh. Obtain the weight of the peas in the reference solvent by subtracting the weight of the basket in the solvent from that of the peas and the basket in the same solvent. The weight in air minus the weight in the solvent gives the difference of weight. Calculate the specific gravity by use of the formula:

$$\text{Specific gravity} = \frac{\text{wt. in air} \times \text{sp. gr. of liquid}}{\text{difference of wt. in liquid and in air}}$$

With water containing sufficient salt to adjust the specific gravity to 1.000, Lee [11] suggests the following standards:

```
Fancy ..........................1.072 and lower
Standard .......................1.073 to 1.084
Substandard ....................1.085 and higher
```

These standards can be revised if and when an extra standard grade is generally packed.

With a mixture of xylene and carbon tetrachloride whose specific gravity has been adjusted to 1.000 the suggested tentative standards for frozen peas, based upon the comparison of the specific gravity values with the organoleptic tests and upon a knowledge of the samples themselves, might be set as follows:

```
Fancy ..........................1.084 and lower
Standard .......................1.085 to 1.094
```

Samples having a specific gravity of 1.095 and higher should be considered substandard.

Whole Kernel Corn.—Seal the samples of corn [12] individually in pliofilm bags and immerse in water at 150° F. in order to thaw; follow by draining the contents for 2 minutes on an 8-mesh circular sieve, 8 in. in diameter, before starting the work. The samples can be thawed by exposure to room temperature overnight if desired.

Procedure.—Proceed as detailed above.

Tentative standards using brine for the determination were found as follows:

```
Fancy ..........................1.080 to 1.118
Reject, immature ...............1.079 and lower
Reject, overmature .............1.119 and higher
```

[11] Lee, *Ind. Eng. Chem., Anal. Ed.* 14, 241 (1942).
[12] Lee and DeFelice, *The Canner*, May (194?).

These standards could be revised if and when other grades are generally packed, or if ideas of the maturity to be classed as fancy change.

Lima Beans.—The maturity of Lima beans was ascertained by Lee [13] by determinations of specific gravity, alcohol-insoluble solids and of total solids. He suggested the values given in Table 74 as standards.

TABLE 74. GRADING OF FROZEN LIMA BEANS [13]

Grade	Henderson's Bush	Clark's Bush (an all-green type)
Specific Gravity		
Fancy	Up to 1.104	Up to 1.120
Extra Standard	1.105-1.122	1.121-1.145
Reject	1.123 and higher	1.146 and higher
Alcohol-Insoluble Solids		
Fancy	Up to 26.5%	Up to 30.0%
Extra Standard	26.6-30.0%	30.1%-35.0%
Reject	30.1% and higher	35.1% and higher
Total Solids		
Fancy	Up to 30.0%	Up to 34.0%
Extra Standard	30.1%-34.0%	34.1-39.0%
Reject	34.1% and higher	39.1% and higher

Prunes.—Nichols and Reed [14] used a specific gravity method for determining the quality of dried prunes.

In 1949 standards of identity were published by the Food and Drug Administration for canned peas and beans. Standards of quality were also set and methods were detailed for evaluating conformance with such standards. These are given in the ensuing text.

CANNED PEAS [15]

Identity.—(a) Canned peas is the food prepared from one of the following optional pea ingredients:

---

[13] Lee, *N. Y. State Agr. Exp. Sta.*, *Bull.* **729** (1948).
[14] Nichols and Reed, *Hilgardia* **6**, 561 (1932).
[15] *U. S. Food Drug Admin.*, *S.R.A.*, *F.D.&C.* **2**, Parts 51, 52, 53, March, 1954.

(1) Shelled, succulent peas (*Pisum sativum*) of Alaska or other smooth skin varieties.

(2) Shelled, succulent peas (*Pisum sativum*) of sweet, wrinkled varieties.

(3) Shelled, dried peas (*Pisum sativum*) of Alaska or other smooth skin varieties.

(4) Shelled, dried peas (*Pisum sativum*) of sweet, wrinkled varieties.

(b) To one such optional pea ingredient water is added.

(c) The following optional ingredients may be used:

(1) Salt.

(2) Sugar.

(3) Dextrose.

(4) Spice.

(5) Flavoring.

(6) Artificial coloring.

and in case optional pea ingredient (1) or (2) is used,

(7) Sodium carbonate, sodium bicarbonate, sodium hydroxide, calcium hydroxide, magnesium hydroxide, magnesium oxide, or magnesium carbonate or any mixture or combination of them in such quantity that the pH of the finished canned peas is not more than 8, as determined by the glass electrode method for the hydrogen-ion concentration.

(d) The food may be seasoned with one or more of the following optional seasonings:

(1) Green peppers.

(2) Mint leaves.

(3) Onions.

(4) Garlic.

(5) Horseradish.

(e) The food is sealed in a container and so processed by heat as to prevent spoilage.

Quality.—(a) The standard of quality for canned peas is as follows:

(1) Not more than 4 per cent by count of the peas in the container are spotted or otherwise discolored;

(2) Standard canned peas are normally colored, not artificially colored;

(3) The combined weight of pea pods and other harmless extraneous vegetable material is not more than one-half of 1 per cent of the drained weight of peas in the container;

(4) The weight of pieces of peas is not more than 10 per cent of the drained weight of peas in the container;

(5) The skins of not more than 25 per cent by count of the peas in the container are ruptured to a width of ⅟₁₆ inch or more;

(6) Not less than 90 per cent by count of the peas in the container are crushed by a weight of not more than 907.2 g. (2 lbs); and

(7) The alcohol-insoluble solids of Alaska or other smooth skin varieties of peas in the container, is not more than 23.5 per cent, and of sweet, wrinkled varieties, not more than 21 per cent.

(b) Canned peas are tested by the following methods to determine whether or not they meet the requirements of (a) above.

(1) **Drained Weight.**—After determining the fill of the container as prescribed in Chapter I, distribute the contents of the container over the meshes of a circular sieve made with No. 8 woven-wire cloth which complies with the specifications for such cloth set forth on page 3 of "Standard Specifications for Sieves," published October 25, 1938, by U. S. Department of Commerce, National Bureau of Standards. The diameter of the sieve used is 8 in. if the quantity of the contents of the container is less than 3 lbs., or 12 in. if such quantity is 3 lbs. or more. Without shifting the peas, so incline the sieve as to facilitate drainage. Two minutes from the time drainage begins, remove the peas from the sieve and weigh them. Such weight is considered to be the drained weight of the peas.

(2) **Extraneous Vegetable Material.**—From the drained peas obtained in (1) above, promptly segregate and weigh the pea pods and other harmless extraneous vegetable material, and the pieces of peas.

(3) **Spotted and Ruptured Number.**—From the drained peas obtained in (1) above take at random a subdivision of 100 to 150 peas, and count them. Immediately cover these peas with a portion of the liquid obtained in (1) above, and add the remaining liquid to the drained peas from which the subdivision was taken. Count those peas in the subdivision which are spotted or otherwise discolored, and also those peas the skins of which are ruptured to a width of ⅟₁₆ inch or more.

(4) **Resistance to Crushing.**—Immediately after each pea is examined by the method prescribed in (3), test it by removing its skin, placing one of its cotyledons, with flat surface down, on the approximate center of the level, smooth surface of a rigid plate, lowering a horizontal disk to the highest point of the cotyledon, and measuring the height of the cotyledon. The disk is of rigid material and is affixed to a rod held vertically by a support through which the rod can freely move upward or downward. The lower face of the disk is a smooth, plane surface horizontal to the vertical axis of the rod. A device to which weight may be added is affixed to the upper end of

the rod. Before lowering the disk to the cotyledon, adjust the combined weight of disk rod, and device to 100 g. After measuring the height of the cotyledon, and shifting the plate, if necessary, so that the cotyledon is under the approximate center of the disk, add weight to the device at a uniform, continuous rate of 12 g. per second until the cotyledon is pressed to one-fourth its previously measured height, or until the combined weight of disk, rod, and device is 907.2 g. (2 lbs.). A pea so tested shall be considered to be crushed when its cotyledon is pressed to one-fourth its original height.

(5) **Alcohol-Insoluble Solids.**—Drain the liquid from the peas which remained after taking the subdivision as prescribed in (3). Transfer the peas to a pan, and rinse them with a volume of water equal to twice the capacity of the container from which such peas were drained in (1). Immediately drain the peas again by the method prescribed in (1). After the 2 minutes' draining, wipe the moisture from the bottom of the sieve. Comminute the peas thus drained, stir them to a uniform mixture, and weigh 20 g. of such mixture into a 600-ml. beaker. Add 300 ml. of 80 per cent alcohol (by volume), stir, cover beaker, and bring to a boil. Simmer slowly for 30 minutes. Fit a Büchner funnel with a previously prepared filter paper of such size that its edges extend ½ in. or more up the vertical sides of the funnel. The previous preparation of the filter paper consists of drying it in a flat-bottomed dish for 2 hours at 100° C., covering the dish with a tight-fitting cover, cooling it in a desiccator, and promptly weighing. After the filter paper is fitted to the funnel, apply suction and transfer the contents of the beaker to the funnel. Do not allow any of the material to run over the edge of the paper. Wash the material on the filter with 80 per cent alcohol until the washings are clear and colorless. Transfer the filter paper with the material retained thereon to the dish used in preparing the filter paper. Dry the material in a ventilated oven, without covering the dish, for 2 hours at 100° C. Place the cover on the dish, cool it in a desiccator, and promptly weigh. From this weight, subtract the weight of the dish, cover, and paper, as previously found. The weight in grams thus obtained, multiplied by 5, shall be considered to be the per cent of alcohol-insoluble solids.

Fill of Container.—(a) The standard of fill of container for canned peas is a fill such that, when the peas and liquid are removed from the container and returned thereto, the leveled peas (irrespective of the quantity of the liquid), 15 seconds after they are so returned completely fill the container. A container with lid attached by double seam shall be considered to be completely filled when it is filled to the level ³⁄₁₆ in. vertical distance below the top of the double seam; and a glass container shall be considered to be completely filled when it is filled to the level of ½ in. vertical distance below the top of the container.

(b) If canned peas fall below the standard of fill of container prescribed they are considered substandard fill.

CANNED GREEN BEANS [16]

Identity.—(a) Canned green beans is the food prepared from stemmed, succulent pods of the green-bean plant and water. It may be seasoned with salt, sugar, or dextrose, or any two or all of these. The pods are prepared in one or more of the following forms:

(1) Whole pods, including pods which after removal of either or both ends are less than 2¾ in. in length, or transversely cut pods not less than 2¾ in. in length. There may be present such broken pieces of pods as normally occur in the commercial packing of such product.

(2) Pods sliced lengthwise.

(3) Pods cut transversely into pieces less than 2¾ in. in length but not less than ¾ in. in length, with or without shorter end pieces resulting therefrom.

(4) Pieces of pods of which not less than 75 per cent by count are less than ¾ in. in length and not more than 1 per cent by count are more than 1¼ in. in length.

Any such form is an optional ingredient. Mixtures of two or more optional ingredients may be used. The food is sealed in a container and so processed by heat as to prevent spoilage.

Quality.—(a) The standard of quality of canned green beans is as follows:

When tested by the method prescribed in (b) below:

(1) In the case of cut beans and mixtures of two or more of the optional ingredients specified in (a) (1) to (a) (4), above, not more than 60 units per 12 oz. drained weight are less than ½ in. long; provided, that where the number of units per 12 oz. drained weight exceed 240, not more than 25 per cent by count of the total units are less than ½ in. long.

(2) The trimmed pods contain not more than 25 per cent by weight of seed and pieces of seed.

(3) In case there are present pods or pieces of pods 27/64 in. or more in diameter, there are not more than 12 strings per 12 oz. of drained weight which will support ½ lb. for 5 seconds or longer.

---

[16] *U. S. Food Drug Admin., S.R.A., F.D.&C.* 2, Parts 51, 52, 53, March, 1954.

(4) The deseeded pods contain not more than 0.15 per cent by weight of fibrous material.

(5) There are not more than 8 per cent by count of blemished units. A unit is considered blemished when the aggregate blemished area exceeds the area of a circle ⅛ in. in diameter.

(6) There are not more than 6 unstemmed units per 12 oz. of drained weight.

(7) The combined weight of loose seed and pieces of seed is not more than 5 per cent of the drained weight. This provision does not apply in case the green-bean ingredient is pods sliced lengthwise [(a) (2)].

(8) The combined weight of leaves, detached stems, and other extraneous vegetable matter is not more than 0.6 oz. per 60 oz. drained weight.

(b) Canned green beans shall be tested by the following method to determine whether they meet the requirements of paragraph (a) of this section.

(1) **Drained Weight.**—Distribute the contents of the container over the meshes of a circular sieve which has been previously weighed. The diameter of the sieve is 8 in. if the quantity of the contents of the container is less than 3 lbs. and 12 in. if such quantity is 3 lbs. or more. The bottom of the sieve is woven-wire cloth which complies with the specifications for such cloth set forth under "2380 Micron (No. 8)" in Table I of "Standard Specifications for Sieves," published March 1, 1940, in L. C. 584 of the U. S. Department of Commerce, National Bureau of Standards. Without shifting the material on the sieve, so incline the sieve as to facilitate drainage. Two minutes from the time drainage begins, weigh the sieve and the drained material. Record, in ounces, the weight so found, less the weight of the sieve, as the drained weight.

(2) **Units per 12 Ounces.**—Pour the drained material from the sieve into a flat tray and spread it in a layer of fairly uniform thickness. Count the total number of units. For the purpose of this count, loose seed, pieces of seed, loose stems, and extraneous material are not to be included. Divide the number of units by the drained weight recorded in (b) (1) and multiply by 12 to obtain the number of units per 12 oz. drained weight.

(3) **Blemished Units.**—Examine the drained material in the tray, counting and recording the number of blemished units, number of unstemmed units, and, in case the material consists of the optional ingredient specified in (a) (3) of this section or a mixture of two or more of the optional ingredients specified in (a) (1) to (4), inclusive, count and record the number of units which are less than ½ in. long.

If the number of units per 12 oz. is 240 or less, divide the number of units which are less than ½ in. long by the drained weight recorded in (b) (1) and multiply by 12 to obtain the number of such units per 12 oz. drained weight. If the number of units per 12 oz. exceeds 240, divide the number of units less than ½ in. long by the total number of units and multiply by 100 to determine the percentage by count of the total units which are less than ½ in. long.

Divide the number of blemished units by the total number of units in the container and multiply by 100 to obtain the percentage by count of blemished units in the container.

Divide the number of unstemmed units by the drained weight recorded in subparagraph (1) of this paragraph and multiply by 12 to obtain the number of unstemmed units per 12 oz. of drained weight.

(4) **Loose Seed Percentage.**—Except in the case of pods sliced lengthwise, remove the loose seed and pieces of seed, weigh and record weight and return to tray. Divide the weight of loose seed and pieces of seed by the drained weight recorded in (b) (1) and multiply by 100 to obtain the percentage by weight of loose seed and pieces of seed in the drained material.

(5) **Extraneous Vegetable Material.**—Remove from the tray the extraneous vegetable material, weigh, record weight, and return to tray.

(6) Remove from the tray one or more representative samples of 3½ to 4 oz., covering each sample as taken to prevent evaporation. If the tray includes pods or pieces of pods $^{27}\!/_{64}$ in. or more in diameter, weigh and record weight in ounces of each representative sample.

(7) **Weight of Seed in Trimmed Pods.**—From each representative sample selected in (b) (6) discard any loose seed and extraneous vegetable material and detach and discard any attached stems. Except with optional ingredient specified in (a) (2) (pods sliced lengthwise), trim off, as far as the end of the space formerly occupied by the seed, any portion of pods from which seed have become separated. Remove and discard any portions of seed from the trimmings and reserve the trimmings for (b) (9). Weigh and record the weight of the trimmed pods. Deseed the trimmed pods and reserve the deseeded pods for (b) (9). If the original container contained pods $^{27}\!/_{64}$ in. or more in diameter, remove strings from the pods during the deseeding operation. Reserve these strings for testing as prescribed in (b) (8). Collect the seed on a sieve of mesh fine enough to retain them, and so distribute them that any liquid drains away.

Weigh the seed, divide by the weight of the trimmed pods, and multiply by 100 to obtain the percentage by weight of seed in the trimmed pods.

In the case of pods sliced lengthwise remove seed and pieces of seed and reserve the deseeded pods for use as prescribed in (b) (9).

(8) **String Toughness.**—If strings have been removed for testing, as prescribed in (b) (7), test them as follows:

Fasten clamp, weighted to ½ lb., to one end of the string, grasp the other end with the fingers (a cloth may be used to aid in holding the string), and lift gently. Count the string as tough if it supports the ½-lb. weight for at least 5 seconds. If the string breaks before 5 seconds, test such parts into which it breaks as are ½ in. or more in length, and if any such part of the strong supports the ½-lb. weight for at least 5 seconds count the string as tough. Divide the number of tough strings by the weight of the sample recorded in (b) (6) and multiply by 12 to obtain the number of tough strings per 12 oz. of drained weight.

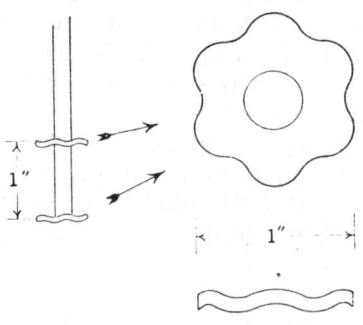

Fig. 72. Two Button Rotor; Scalloped Buttons

(9) **Fibrous Material.**—Combine the deseeded pods with the trimmings reserved in (b) (7), and, if strings were tested as prescribed in (b) (8), add such strings, broken or unbroken. Weigh and record weight of combined material. Transfer to the metal cup of a malted-milk stirrer and mash with a pestle. Wash material adhering to the pestle back into cup with 200 ml. of boiling water. Bring mixture nearly to a boil, add 25 ml. of 50 per cent (by weight) sodium hydroxide solution and bring to a boil. (If foaming is excessive, 1 ml. of capryl alcohol may be added.) Boil for 5 minutes, then stir for 5 minutes with a malted-milk stirrer capable of a no-load speed of at least 7200 r. p. m. Use a rotor with two scalloped buttons shaped as shown in Fig. 72.

Transfer the material from the cup to a previously weighed 30-mesh monel metal screen having a diameter of about 3½ to 4 in. and side walls about 1 in. high, and wash fiber on the screen with a stream of water using a pressure not exceeding a head (vertical distance between upper level of water and outlet of glass tube) of 60 in., delivered through a glass tube 3 in. long and ⅛ in. inside diameter inserted into a rubber tube of ¼ in. inside diameter. Wash the pulpy portion of the material through the screen and continue washing until the remaining fibrous material, moistened with phenolphthalein solution, does not show any red color after standing 5 minutes. Again wash to remove phenolphthalein. Dry the screen containing the fibrous material for 2 hours at 100° C., cool, weigh, and deduct weight of screen. Divide the weight of fibrous material by the weight of combined deseeded pods, trimmings, and strings and multiply by 100 to obtain the percentage of fibrous material.

(10) If the drained weight recorded in (b) (1) was less than 60 oz., open and examine separately for extraneous material, as directed in (b) (5'), additional containers until a total of not less than 60 oz. of drained material is obtained. To determine the combined weight of extraneous vegetable material per 60 oz. of drained weight, total the weights of extraneous vegetable material found in all containers opened, divide this sum by the sum of the drained weights in these containers and multiply by 60.

(c) If the quality of the canned green beans falls below the standard of quality prescribed, they are considered substandard quality.

Canned Wax Beans [17]

Canned wax beans conform to the definition and standard of identity, and are subject to the requirements for label statement of optional ingredients prescribed for canned green beans by (a) and (b), except that they are prepared from stemmed, succulent pods of the wax-bean plant.

The standard of quality for canned wax beans is that prescribed for canned green beans.

Canned Corn

Standards of identity and fill of container for canned corn and canned field corn [18] were proposed in 1949 and were promulgated subsequently.[17] These standards are essentially the following:

Canned Corn, Canned Sweet Corn, and Canned Sugar Corn.—Identity.—(a) Canned corn, canned sweet corn, canned sugar corn is the food prepared from one of the optional corn ingredients specified in (b), and water. It may be seasoned with one or more of the following optional seasonings:

(1) Salt.
(2) Sugar.

---

[17] *U. S. Food Drug Admin., S.R.A., F.D.&C.* 2, Parts 51, 52, 53, March, 1954.
[18] *Federal Register* 14, 489 (1949).

(3) Pieces of red sweet peppers, or green sweet peppers, or a mixture of these.

It is sealed in a container and so processed by heat as to prevent spoilage.

(b) The corn ingredients referred to in (a) of this section consist of succulent sweet corn of the white or yellow color groups, or mixtures of these:

(1) Cut kernels from which the hulls have not been separated.

(2) Pieces of the inner portion of corn kernel substantially free from hull.

(3) Ground kernels from which the hulls have not been separated.

(4) A mixture of the form described in (b) (1) with one or both of the forms described in (b) (2) and (3). When necessary to insure smoothness, starch may be added in a quantity not more than sufficient for that purpose.

(5) Cut and cooked kernels from which most of the moisture has been evaporated.

In preparing each of the foregoing corn ingredients the tip caps are removed.

(c) The name of the food is "corn" or "sweet corn" or "sugar corn," with the name of the color group used—"white," "yellow," or "golden"—or with the names of the color groups used—"white and yellow" or "white and golden"—when the white color group predominates, and "yellow and white" or "golden and white" when the yellow group predominates, and:

(1) By the words "whole kernel" or "whole grain" when the form of kernels specified in (b) (1) is used. When the weight of the drained liquid in a container is not more than 20 per cent of the net weight, as determined by the method prescribed in (d) (2), and the vapor pressure within the container is not more than 15 in. of mercury, the words "vacuum pack" or "vacuum packed" precede or follow the name or the words "whole kernel" or "whole grain."

(2) By the word "fritter" when the form of corn kernels specified in (b) (2) is used.

(3) By the words "crushed" or "ground" when the form of corn kernels specified in (b) (3) is used.

(4) By the words "cream style" when the form of corn kernels specified in (b) (4) is used.

(5) By the word "evaporated" when the corn ingredient specified in (b) (5) is used.

(d) The methods referred to in (b) and (c) (1) are:

(1) **Spreading Test.**—In the forms specified in (b) (2), (3), and (4) allow the container to stand at least 24 hours at a temperature of

68° F. to 85° F. Determine the gross weight, open the container, transfer into a pan, and stir thoroughly in such a manner as not to incorporate air bubbles. If the contents of a single container are less than 18 oz., open, determine the gross weight, and mix the contents of just sufficient containers to obtain at least 18 oz. From the mixed material fill level full a hollow truncated cone so placed on a polished horizontal plate as to prevent leakage. The cone has an inside bottom diameter of 3 in., inside top diameter of 2 in., and height of 4 27/32 in. As soon as the cone is filled, lift it vertically. Determine the average diameter of the sample 30 seconds thereafter. Obtain the net weight.

(2) **Drained Liquid.**—Determine the per cent of drained liquid specified in (c) (1) by the following method: Obtain the gross weight of the container. Open the container and distribute the contents over the meshes of a circular sieve which has been previously weighed. The diameter of the sieve is 8 in. if the quantity of the contents of the container is less than 3 lbs., and 12 in. if such quantity is 3 lbs. or more. The bottom of the sieve is woven wire cloth which complies with the specifications for such cloth set forth under "2380 Micron (No. 8)" in Table 1 of "Standard Specifications for Sieves," published March 1, 1940, in L.C. 584 of the U.. S. Department of Commerce, National Bureau of Standards. Without shifting the material on the sieve, so incline the sieve as to facilitate drainage. Two minutes from the time drainage begins, weigh the drained liquid. Dry and weigh the empty container and subtract this weight from the gross weight previously determined, to obtain weight of the food. Compute the percentage of drained liquid in the food.

Canned Corn.—Quality.—(a) The standard of quality for canned corn is as follows: When tested by the methods prescribed in (b):

(1) Except in the whole kernel form, there is not more than one brown or black discolored piece for each 2 oz.; in the whole kernel form, there is not more than one brown or black discolored kernel for each 2 oz. of drained weight.

(2) In the cream style form, there are not more than two pulled kernels for each 1 oz. of washed drained residue; in the whole kernel form, there are not more than two pulled kernels for each 1 oz. of drained weight.

(3) Except in the whole kernel form, there is not more than 1 ml. of pieces of cob for each 20 oz.; in the whole kernel form, there is not more than 1 ml. of pieces of cob for each 14 oz. of drained weight.

(4) Except in the whole kernel form, there is not more than 1 sq. in. of husk for each 20 oz.; in the whole kernel form, there is not more than 1 sq. in. of husk for each 14 oz. of drained weight.

(5) Except in the whole kernel form, there are not more than 6 in. of silk for each 1 oz.; in the whole kernel form, there are not more than 7 in. of silk for each 1 oz. of drained weight.

(6) In the cream style form and the whole kernel form, the alcohol-insoluble solids in the drained material does not exceed 27%.

(b) Canned corn may be tested by the following methods to determine whether it meets the requirements of paragraph (a) above:

(1) After completing the spreading test described in (d) (1), above, transfer the material from the plate, cone, and original container or containers onto the sieve specified in (d) (2), above. Set the sieve in a pan of appropriate size. Add enough water to bring the level within $\frac{3}{8}$ to $\frac{1}{4}$ in. of the top of the sieve. Wash the material on the sieve by combined up-and-down circular motion for 30 seconds. Pour washings from pan, reserving them, as well as the subsequent washings, for further tests. Repeat washing with a second portion of water. Remove sieve from pan, incline to facilitate drainage and drain for 2 minutes. Record weight of material on sieve as washed drained residue.

Pour both washings through a 20-mesh sieve and discard wash water. Count, but do not remove, the brown or black discolored kernels or pieces and the pulled kernels in the material. (A pulled kernel is an entire, uncut kernel.) Remove pieces of silk more than $\frac{1}{4}$ in. long, husk, and cob. Measure total length of such silk. Spread the husk flat and measure its total area. Measure the cob by placing all pieces of cob under a measured amount of water in a cylinder which is graduated to 0.2 ml. The increase is the volume of the cob.

In the cream style form, comminute the material remaining on the 8-mesh sieve. Weigh 10 g. of the comminuted material into a 600-ml. beaker. Add 300 ml. of 80 per cent alcohol, stir, cover beaker, and bring to a boil. Simmer slowly 30 minutes. Fit into Büchner funnel a filter paper of appropriate size (previously prepared by drying in flat-bottomed dish at temperature of boiling water, covering with a tight-fitting cover, cooling in desiccator, and weighing at once). Apply suction, and transfer contents of beaker to the Büchner funnel in such a manner as to avoid running over edge of paper. Suck dry and wash the alcohol-insoluble solids on filter with 80 per cent alcohol until washings are clear and colorless. Transfer filter paper and alcohol-insoluble solids to dish used in preparation of filter paper, dry uncovered for 2 hours at temperature of boiling water, place cover on dish, cool in desiccator, and weigh at once. From this weight deduct weight of dish, cover, and paper, and calculate the remainder to percentage.

(2) In the whole kernel form, drain as directed in (d) (2) above. Two minutes from the time drainage begins, determine the weight of material retained on the sieve and record as the weight of drained corn. Count, but do not remove, all discolored kernels and all pulled kernels. Also remove and measure pieces of silk more than 1/4 inch long, husk, and cob, as described in (b) (1). Comminute the material from which the silk, husk, and cob have been removed, and determine the alcohol-insoluble solids as described in (b) (1).

Tomato Products [19]

Tomato Juice.—Identity.—Tomato juice is the unconcentrated liquid extracted from mature tomatoes of red or reddish varieties, with or without scalding followed by draining. In the extraction of such liquid, heat may be applied by any method which does not add water thereto. Such liquid is strained free from skins, seeds, and other coarse or hard substances, but carries finely divided insoluble solids from the flesh of the tomato. Such liquid may be homogenized, and may be seasoned with salt. When sealed in a container it is so processed by heat, before or after sealing, as to prevent spoilage.

Yellow Tomato Juice.—Identity.—Yellow tomato juice is the unconcentrated liquid extracted from mature tomatoes of yellow varieties. It conforms, in all other respects, to the definition and standard of identity for tomato juice.

Catsup, Ketchup, Catchup.—Identity.—(a) Catsup, ketchup, catchup, is the food prepared from one or any combination of two or all of the following optional ingredients:

(1) The liquid obtained from mature tomatoes of red or reddish varieties.

(2) The liquid obtained from the residue from preparing such tomatoes for canning, consisting of peelings and cores with or without such tomatoes or pieces thereof.

(3) The liquid obtained from the residue from partial extraction of juice from such tomatoes.

Such liquid is obtained by so straining such tomatoes or residue, with or without heating, as to exclude skins, seeds, and other coarse or hard substances. It is concentrated and is seasoned with salt, a vinegar or vinegars, spices or flavorings or both and is sweetened with sugar or a mixture of sugar and dextrose or a mixture of sugar (or sugar and dextrose) with corn syrup or dried corn syrup or both or with glucose syrup or both, in such quantity that the weight of the solids of the corn syrup or dried corn syrup or both, or glucose

[19] *U. S. Food Drug Admin., S.R.A., F.D.&C.* 2, Parts 51, 52, 53, March, 1954, as amended October, 1955.

syrup or dried glucose syrup or both is not more than one-third of the weight of the solids of such mixture. Glucose syrup is defined on page 512. When sealed in a container it is so processed by heat, before or after sealing, as to prevent spoilage.

Tomato Puree, Tomato Pulp.—Identity.—Tomato puree, tomato pulp, is the food prepared from one or any combination of two or all of the following optional ingredients:

(1) The liquid obtained from mature tomatoes of red or reddish varieties.

(2) The liquid obtained from the residue from preparing such tomatoes for canning, consisting of peelings and cores with or without such tomatoes or pieces thereof.

(3) The liquid obtained from the residue from partial extraction of juice from such tomatoes.

Such liquid is obtained by so straining such tomatoes or residue, with or without heating, as to exclude skins, seeds, and other coarse or hard substances. It is concentrated and may be seasoned with salt. When sealed in a container it is so processed by heat, before or after sealing, as to prevent spoilage. It contains not less than 8.37 per cent, but less than 25.00 per cent, of salt-free tomato solids, as determined by the following method:

Salt-free Solids.—Determine total solids by the method described on page 21 and sodium chloride by the method described on page 676. Subtract the per cent of sodium chloride found from the per cent of total solids found; the difference shall be considered to be the per cent of salt-free tomato solids.

Tomato Paste.—Identity.—Tomato paste is the food prepared from one or any combination of two or all of the following optional ingredients:

(1) The liquid obtained from mature tomatoes of red or reddish varieties.

(2) The liquid obtained from the residue from preparing such tomatoes for canning, consisting of peelings and cores with or without such tomatoes or pieces thereof.

(3) The liquid obtained from the residue from partial extraction of juice from such tomatoes.

Such liquid is obtained by so straining such tomatoes or residue, with or without heating, as to exclude skins, seeds, and other coarse or hard substances. It is concentrated, and may be seasoned with one or more of the optional ingredients:

(4) Salt.

(5) Spice.

(6) Flavoring.

It may contain, in such quantity as neutralizes a part of the tomato acids, the optional ingredient:

(7) Baking soda.

When sealed in a container it is so processed by heat, before or after sealing, as to prevent spoilage. It contains not less than 25.00 per cent of salt-free tomato solids, as determined by the method detailed above.

Canned Tomatoes.—Identity.—Canned tomatoes are mature tomatoes of red or reddish varieties which are peeled and cored and to which may be added one or more of the following optional ingredients:

(1) The liquid draining from such tomatoes during or after peeling and coring.

(2) The liquid strained from the residue from preparing such tomatoes for canning, consisting of peelings and cores with or without such tomatoes or pieces thereof.

(3) The liquid strained from mature tomatoes of such varieties.

(4) Purified calcium chloride, calcium sulfate, calcium citrate, monocalcium phosphate, or any two or more of these calcium salts, in a quantity reasonably necessary to firm the tomatoes, but in no case such that the amount of the calcium contained in such salts is more than 0.026 per cent of the weight of the finished canned tomatoes.

It may be seasoned with one or more of the optional ingredients:

(5) Salt.

(6) Spices.

(7) Flavoring.

It is sealed in a container and so processed by heat as to prevent spoilage.

Quality.—(a) The standard of quality for canned tomatoes is as follows:

(1) The drained weight, as determined by the method described in (b) (1) below, is not less than 50 per cent of the weight of water required to fill the container, as determined by the general method for water capacity of containers described in Chapter I.

(2) The strength and redness of color, as determined by the method described in (b) (2), is not less than that of the blended color of any combination of the color disks described in such method, in which one-third the area of disk 1, and not more than one-third the area of disk 2, is exposed;

(3) Peel, per pound of canned tomatoes in the container, covers an area of not more than 1 sq. in.; and

(4) Blemishes, per pound of canned tomatoes in the container, cover an area of not more than 0.25 sq. in.

(b) Canned tomatoes shall be tested by the following method to determine whether or not they meet the requirements of (a) (1) and (2), above.

(1) **Drained Weight.**—Remove lid from container, but in the case of a container with lid attached by double seam, do not remove or alter the height of the double seam. Tilt the opened container so as to distribute the contents over the meshes of a circular sieve which has previously been weighed. The diameter of the sieve used is 8 in. if the quantity of the contents of the container is less than 3 lbs., or 12 in. if such quantity is 3 lbs. or more. The meshes of such sieve are made by so weaving wire of 0.054-in. diameter as to form square openings 0.446 in. by 0.446 in. Without shifting the tomatoes, so incline the sieve as to facilitate drainage of the liquid. Two minutes from the time drainage begins, weigh the sieve and drained tomatoes. The weight so found, less the weight of the sieve, shall be considered to be the drained weight.

(2) **Color.**—Remove from the sieve the drained tomatoes obtained in (b) (1). Cut out and segregate successively those portions of least redness until 50 per cent of the drained weight, as determined under (b) (1), has been so segregated. Comminute the segregated portions to a uniform mixture without removing or breaking the seeds. Fill the mixture into a black container to a depth of at least 1 in. Free the mixture from air bubbles, and skim off or press below the surface all visible seeds. Compare the color of the mixture, in full diffused daylight or its equivalent, with the blended color of combinations of the following concentric Munsell color disks of equal diameter, or the color equivalents of such disks:

(i) Red—Munsell 5 R 2.6/13 (glossy finish).

(ii) Yellow—Munsell 2.5 YR 5/12 (glossy finish).

(iii) Black—Munsell N 1/ (glossy finish).

(iv) Grey—Munsell N 4 (mat finish).

(c) If the quality of canned tomatoes falls below the standard prescribed in (a) they are considered substandard.

Fill of Container.—The standard of fill of container for canned tomatoes is a fill of not less than 90 per cent of the total capacity of the container, as determined by the general method for fill of containers described in Chapter I.

If canned tomatoes fall below the standard of fill of container prescribed, they are considered substandard fill.

NUTS AND NUT PRODUCTS

Nuts and nut products may be analyzed by methods detailed in other sections of this text. The A. O. A. C.[20] suggests the samples be preserved in glass top fruit jars or analogous containers that can be tightly closed. These samples should be stored at 5 to 10° C. Preparation of samples for analysis details the following procedure.

Preparation of Sample.—(a) Nuts in shell.—Remove meats from shells, being careful to remove all particles of shell from meats. Prepare separated meats as in following.

(b) *Nut meats, shredded coconut, or small pieces.*—Grind not less than 250 g. twice through an Enterprise No. 5 food chopper, equipped with a revolving knife blade, and plate with holes about ⅛ inch in diameter. Other types of food choppers, graters, or comminuting devices that give a smooth homogeneous paste without loss of oil may be used. Mix sample well and store in air-tight glass container.

(c) *Nut butters and pastes.*—Transfer the sample to container of convenient size and shapes, warming semisolid products, and mix carefully with stiff bladed spatula or knife. Electric powered mixers or stirrers may be used instead of a spatula or knife if the material is of the right consistency to give uniform mixture. Store sample in air-tight glass container.

Protein nitrogen and ash may be estimated as detailed in Chapter I, coloring matters in Chapter III, preservatives in Chapter IV, metals in Chapter V, sucrose and reducing sugars in Chapter X, crude fiber in Chapter XIV, and chloride as detailed in Chapter XVI under the section on fish, using a 2-g. sample. Moisture may be determined on a 2-g. sample by heating at 95-100° C. under reduced pressure of less than 100 mm. for about 5 hours.

FAT

Total ether extractable material may be estimated by a modification of the Sohxlet method described in Chapter I.

Reagent.—Anhydrous Ether.—Wash ether in a separatory funnel with three portions of water. Transfer the ether to a flask that contains some potassium or sodium hydroxide and allow to stand until all or most of the water has reacted. Decant the ether into another dry flask and add clean pieces of metallic sodium and allow to stand until there is no further production of hydrogen gas. Store the dehydrated ether in a loosely stoppered bottle over metallic sodium.

---

[20] *J. Assoc. Official Agr. Chem.* **32**, 96 (1949).

Procedure.—Since large quantities of soluble carbohydrate may interfere with the extraction of fat, extract a weighed portion of the sample, about 2 g. is adequate, with water, dry as mentioned above, cool, weigh, transfer the dried material to a Sohxlet extractor, and extract with the anhydrous ether for 16 hours. Dry at 100° C. in a constant temperature oven for 1 to 1.5 hours, cool, and weigh.

SELECTED REFERENCES

Jacobs, ed., *Chemistry and Technology of Food and Food Products*, 2nd ed. Interscience, New York, 1951.
Joslyn, *Methods in Food Analysis Applied to Plant Products*, Academic Press, New York, 1950.
McCance, Widdowson, and Shackleton, *The Nutritive Value of Fruits, Vegetables, and Nuts. Med. Research Council, Special Rept. Ser.* 213 (1936).
U. S. Food Drug Admin., S.R.A., F. D. & C. 2, Parts 51, 52, 53, March, 1954, as amended October, 1955.

CHAPTER XIV

SPICES, FLAVORS, AND CONDIMENTS

SPICES are aromatic vegetable substances used for the seasoning of food. They are true to name, and from them no portion of any volatile oil or other flavoring principle has been removed. In Table 75 are tabulated the advisory standards promulgated by the Food and Drug Administration. These advisory standards were promulgated when the Food and Drug Administration was an arm of the U. S. Department of Agriculture. They were not in 1958 in the same category as the definitions and standards promulgated for a number of foods [1] but were rather held to be a guide for the enforcement of the Act.

In general, some of the determinations of spices follow closely those methods we have previously encountered. Thus moisture, ash, nitrogen, copper reducing materials, and starch may be estimated as described in other sections of the text. Care must be taken in grinding the samples so that a uniform mixture results. If possible, the spice should be ground to an impalpable powder. Tables 76 and 77 contain analyses of genuine spices.

EXTRACTS

Volatile and Nonvolatile Extract.—Extract 2 g. of the ground sample in a continuous extractor for 20 hours with anhydrous ether. Evaporate the ether to a small volume. Transfer to a small weighed beaker or micro-beaker, taking care to wash out the original flask with small portions of anhydrous ether. Evaporate at room temperature and allow to stand for 18 hours over sulfuric acid in a desiccator. Weigh and calculate the gain in weight as percentage of total ether extract. Heat the extract gradually and then at 110° C. until successive weighings show only a small loss. The difference in weight equals the volatile ether extract. The residue is the nonvolatile ether extract.

Alcohol Extract.—The alcohol extract is determined by shaking, at intervals, 2 g. of the sample in a 100-ml. volumetric flask with 95 per cent alcohol for 8 hrs. Filter, and evaporate a 50-ml. aliquot to dryness in a tared dish. Dry at 100° C. The residue is the alcohol extract.

---

[1] *U. S. Food Drug Admin., S.R.A., F.D.&C.* 2, Rev. 1 (1949).

STANDARDS FOR SPICES 587

TABLE 75. STANDARDS FOR SPICES [a]

Allspice, Pimento.—The dried, nearly ripe fruit of *Pimenta officinalis* Lindl. It contains not less than 8 per cent of quercitannic acid, calculated from the total oxygen absorbed by the aqueous extract, not more than 25 per cent of crude fiber, not more than 6 per cent of total ash, nor more than 0.4 per cent of ash insoluble in hydrochloric acid.

Anise, Aniseed.—The dried fruit of *Pimpinella anisum* L. It contains not more than 9 per cent of total ash, nor more than 1.5 per cent of ash insoluble in hydrochloric acid.

Caraway, Caraway Seed.—The dried fruit of *Carum carvi* L. It contains not more than 8 per cent of total ash, nor more than 1.5 per cent of ash insoluble in hydrochloric acid.

Cardamom Seed.—The dried seed of cardamom, *Elettaria cardmomum* Maton. It contains not more than 8 per cent of total ash, nor more than 3 per cent of ash insoluble in hydrochloric acid.

Red Pepper.—The red, dried, ripe fruit of any species of *Capsicum*. It contains not more than 8 per cent of total ash, nor more than 1 per cent of ash insoluble in hydrochloric acid.

Cayenne Pepper, Cayenne.—The dried, ripe fruit of *Capsicum frutescens* L., *C. baccatum* L., or some other small-fruited species of *Capsicum*. It contains not less than 15 per cent of nonvolatile ether extract, not more than 1.5 per cent starch, not more than 28 per cent crude fiber, not more than 8 per cent of total ash, nor more than 1.25 per cent of ash insoluble in hydrochloric acid.

Paprika.—The dried ripe fruit of *Capsicum annuum* L. It contains not more than 8.5 per cent of total ash, nor more than 1 per cent of ash insoluble in hydrochloric acid. The iodine number of its extracted oil is not less than 125, nor more than 136.

Hungarian Paprika.—The paprika having the pungency and flavor characteristic of that grown in Hungary. (a) Rosenpaprika, rosapaprika, rosé paprika, is Hungarian paprika prepared by grinding specially selected pods of paprika, from which the placentae, stalks, and stems have been removed. It contains no more seeds than in normal pods, not more than 18 per cent of nonvolatile ether extract, not more than 23 per cent of crude fiber, not more than 6 per cent of total ash, nor more than 0.4 per cent of ash insoluble in hydrochloric acid, (b) Koenigspaprika, king's paprika, is Hungarian paprika prepared by grinding whole pods of paprika without selection, and includes the seeds and stems naturally occurring with the pods. It contains not more than 18 per cent nonvolatile ether extract, not more than 23 per cent crude fiber, nor more than 6.5 per cent of total ash, nor more than 0.5 per cent of ash insoluble in hydrochloric acid.

---

[a] *U. S. Food Drug Admin., S.R.A., F.D. 2 (1936).*

TABLE 75.—(*Continued*)

Pimenton, Pimiento, Spanish Paprika.—Paprika having the characteristics of that grown in Spain. It contains not more than 18 per cent of nonvolatile ether extract, not more than 21 per cent of crude fiber, not more than 8.5 per cent of total ash, nor more than 1 per cent of ash insoluble in hydrochloric acid.

Celery Seed.—The dried fruit of *Celeri graveolens* (L.) Britton (*Apium graveolens* L.). It contains not more than 10 per cent of total ash, nor more than 2 per cent of ash insoluble in hydrochloric acid.

Ground Cinnamon, Ground Cassia.—The powder made from cinnamon, the dried bark of cultivated varieties of *Cinnamomum zeylanicum* Nees or of *C. cassia* (L.) Blume. It contains not more than 5 per cent of total ash, nor more than 2 per cent of ash insoluble in hydrochloric acid.

Cloves.—The dried flower buds of *Caryophyllus aromaticus* L. They contain not more than 5 per cent of clove stems, not less than 15 per cent of volatile ether extract, not less than 12 per cent of quercitannic acid, calculated from the total oxygen absorbed by the aqueous extract, not more than 10 per cent of crude fiber, not more than 7 per cent of total ash, nor more than 0.5 per cent of ash insoluble in hydrochloric acid.

Coriander Seed.—The dried fruit of *Coriandrum sativam* L. It contains not more than 7 per cent of total ash, nor more than 1.5 per cent of ash insoluble in hydrochloric acid.

Cumin Seed.—The dried fruit of *Cuminum cyminum* L. It contains not more than 9.5 per cent of total ash nor more than 1.5 per cent of ash insoluble in hydrochloric acid, nor more than 5 per cent of harmless foreign matter.

Dill Seed.—The dried fruit of *Anethum graveolens* L. It contains not more than 10 per cent of total ash, nor more than 3 per cent of ash insoluble in hydrochloric acid.

Fennel Seed.—The dried fruit of cultivated varieties of *Foeniculum vulgare* Hill. It contains not more than 9 per cent of total ash nor more than 2 per cent of ash insoluble in hydrochloric acid.

Ginger.—The washed and dried, or decorticated and dried, rhizome of *Zingiber officinate* Roscoe. It contains not more than 8 per cent of crude fiber, not less than 42 per cent of starch, not more than 1 per cent of lime, calcium oxide, not less than 12 per cent of cold water extract, not more than 7 per cent of total ash, nor more than 2 per cent of ash insoluble in hydrochloric acid, nor less than 2 per cent of ash soluble in cold water.

Mace.—The dried arillus of *Myristica fragrans* Houtt. It contains not less than 20 per cent nor more than 30 per cent of nonvolatile ether extract, not more than 10 per cent of crude fiber, not more than 3 per cent of total ash, nor more than 0.5 per cent ash insoluble in hydrochloric acid.

TABLE 75.—(*Continued*)

Marjoram, Leaf Marjoram.—The dried leaves, with or without a small proportion of the flowering tops, of *Majorana hortensis* Moench. It contains not more than 16 per cent of total ash, not more than 4.5 per cent of ash insoluble in hydrochloric acid, nor more than 10 per cent of stems and harmless foreign material.

Mustard Seed.—The seed of *Sinapis alba* L., white mustard, *Brassica nigra* (L.) Koch, black mustard, *B. juncea* (L.) Cosson, or varieties or closely related species of the types *B. nigra* and *B. juncea*. White mustard contains no appreciable amount of volatile oil. It contains not more than 5 per cent of total ash nor more than 1.5 per cent of ash insoluble in hydrochloric acid.

Brassica nigra, black mustard, and *B. juncea* yield 0.6 per cent of volatile mustard oil, calculated as allylisothiocyanate. The varieties and species closely related to the types of *B. nigra* and *B. juncea* yield not less than 0.6 per cent of volatile mustard oil, similar in character and composition to the volatile oils yielded by *B. nigra* and *B. juncea*. These mustard seeds contain not more than 5 per cent of total ash, nor more than 1.5 per cent of ash insoluble in hydrochloric acid.

Nutmeg.—The dried seed of *Myristica fragrans* Houtt deprived of its testa, with or without a thin coating of lime, calcium oxide. It contains not less than 25 per cent of nonvolatile ether extract, not more than 10 per cent of crude fiber, not more than 5 per cent of total ash, nor more than 0.5 per cent of ash insoluble in hydrochloric acid.

Black Pepper.—The dried immature berry of *Piper nigrum* L. It contains not less than 6.75 per cent of nonvolatile ether extract, not less than 30 per cent starch, not more than 7 per cent of total ash, not more than 1.5 per cent of ash insoluble in hydrochloric acid.

White Pepper.—The dried mature berry of *Piper nigrum* L. from which the outer coating, or the outer and inner coatings, have been removed. It contains not less than 52 per cent of starch, not less than 7 per cent of nonvolatile ether extract, not more than 5 per cent of crude fiber, not more than 3.5 per cent of total ash, nor more than 0.3 per cent of ash insoluble in hydrochloric acid.

Saffron.—The dried stigma of *Crocus sativas* L. It contains not more than 10 per cent of yellow styles and other foreign matter, not more than 14 per cent of volatile matter when dried at 100° C., not more than 7.5 per cent of total ash, nor more than 1 per cent of ash insoluble in hydrochloric acid.

Thyme.—The dried leaves and flowering tops of *Thymus vulgaris* L. It contains not more than 14 per cent of total ash, nor more than 4 per cent of ash insoluble in hydrochloric acid.

Star Aniseed.—The dried fruit of *Illicium verum* Hook. It contains not more than 5 per cent of total ash.

VOLATILE OIL

The volatile oil is that oil that can be obtained from a spice by distillation.[2] Since such oils may have densities lower or higher than

TABLE 76. ANALYSES OF GENUINE PEPPER [a]

Type		Moisture %	Nonvolatile ether extract %	Crude fiber %	Dextrose %	Alcohol extract %	Nitrogen %
Lampong	max.	12.38	10.74	13.70	51.32	12.26	2.28
	min.	8.73	7.30	11.76	44.28	10.22	1.84
	ave.	10.06	9.29	12.50	47.20	11.20	2.09
Alleppi	max.	10.86	10.46	13.02	51.76	14.34	2.22
	min.	9.80	7.97	10.85	47.48	10.00	1.95
	ave.	10.13	8.95	11.66	50.36	11.15	2.08
Tellicherry	max.	10.27	9.37	14.36	53.36	11.20	2.16
	min.	8.20	7.25	11.56	51.20	9.60	1.97
	ave.	9.34	8.31	13.08	51.90	10.35	2.06
White and decorticated	max.	11.65	9.70	4.86	76.56	9.36	2.02
	min.	10.34	6.18	1.03	64.88	7.38	1.56
	ave.	11.05	7.79	3.64	68.64	8.50	1.89

ASH

Type		Total %	Water soluble %	Water insoluble %	Acid insoluble %	Alkalinity of soluble[b]	Alkalinity of insoluble[b]
Lampong	max.	6.29	2.97	3.88	1.02	2.9	4.8
	min.	4.39	1.98	1.99	0.11	2.3	2.5
	ave.	5.05	2.33	2.72	0.41	2.6	4.1
Alleppi	max.	5.83	3.61	2.66	0.36	3.0	3.9
	min.	4.18	2.50	1.70	0.02	2.4	2.3
	ave.	4.74	2.88	1.86	0.11	2.6	3.3
Tellicherry	max.	4.75	2.87	1.99	0.11	2.6	3.5
	min.	4.41	2.54	1.71	0.07	2.4	2.4
	ave.	4.55	2.71	1.84	0.08	2.5	3.0
White and decorticated	max.	4.84	0.50	4.34	1.28	0.3	2.2
	min.	0.83	0.07	0.76	0.05	0.1	0.9
	ave.	2.01	0.23	1.78	0.37	0.2	1.5

[a] Smith, Alfend, and Mitchell, *J. Assoc. Official Agr. Chem.* 9, 340 (1926).
[b] Alkalinity of water soluble and insoluble ash in ml. 0.1 *N* acid per 1 g.

that of water, it is necessary to use collection traps that are suitable for each type of oil recovered.

[2] Clevenger, *Am. Perfumer Essential Oil Review* 23, 467 (1928).

Apparatus.—The apparatus (Figs. 73 and 74) consists essentially of three parts: (a) A round-bottomed flask, in which are placed the material containing the volatile oil and a given quantity of water; (b) a separator, in which the oil is automatically separated from the distillate in a graduated tube, thereby permitting a direct reading of the quantity of the oil; and (c) a convenient condenser. The size of the flask may vary from 100 ml. to approximately 2 liters, depending upon the nature of the material and the percentage of volatile oil

TABLE 77. ANALYSES OF PURE SPICES [a]

		Moisture %	Total Ash %	Water Soluble Ash %	Ash In- soluble in. HCl %	Volatile Ether Extract %	Nonvol- atile Ether Extract %	Alcohol Extract %
Black pepper ..	max.	12.95	6.52	3.20	1.19	1.60	10.37	11.86
	min.	10.63	3.09	1.75	0.00	0.65	6.86	8.47
	ave.	11.96	4.76	2.54	0.47	1.14	8.42	9.62
White pepper	max.	14.47	2.96	0.80	0.20	0.95	7.94	8.55
	min.	12.72	1.03	0.28	0.00	0.49	6.26	7.19
	ave.	13.47	1.77	0.47	0.10	0.73	6.91	7.66
Cayenne	max.	7.08	5.96	4.93	0.23	2.57	21.81	27.61
pepper	min.	3.67	5.08	3.30	0.05	0.73	17.17	21.52
	ave.	5.73	5.43	3.98	0.15	1.35	20.15	24.35
Ginger	max.	11.72	9.35	4.09	2.29	3.09	5.42	6.58
	min.	8.71	3.61	1.73	0.02	0.96	2.82	3.63
	ave.	10.44	5.27	2.71	0.44	1.97	4.10	5.18
Cinnamon	max.	10.48	5.99	2.71	0.58	1.62	1.68	13.60
	min.	7.79	4.16	1.40	0.02	0.72	1.35	9.97
	ave.	8.63	4.82	1.87	0.13	1.39	1.44	12.21
Cassia	max.	11.91	6.20	2.52	2.42	5.15	4.13	16.74
	min.	6.53	3.01	0.71	0.02	0.93	1.32	4.57
	ave.	9.24	4.73	1.68	0.56	2.61	2.12	8.29
Cloves	max.	8.26	6.22	3.75	0.13	20.53	6.67	15.58
	min.	7.03	5.28	3.25	0.00	17.82	6.24	13.99
	ave.	7.81	5.92	3.58	0.06	19.18	6.49	14.87
Allspice or	max.	10.14	4.76	2.69	0.06	5.21	7.72	14.27
pimento	min.	9.45	4.15	2.29	0.00	3.38	4.35	7.39
	ave.	9.78	4.47	2.47	0.03	4.05	5.84	11.79
Nutmegs	max.	10.83	3.26	1.46	0.01	6.94	36.94	17.38
	min.	5.79	2.13	0.82	0.00	2.56	28.73	10.42

[a] Winton, Ogden, and Mitchell, *Conn. Agr. Exp. Sta. Ann. Rept.* (1898).

TABLE 77—Continued

		Reducing matter [b] %	Starch by Diastase %	Crude Fiber %	Protein %	Total Nitrogen %	Cold Water Extract %	Lime %
Black pepper..	max.	43.47	39.66	18.25	13.81[e]	2.53	....	....
	min.	28.15	22.05	10.75	10.50	2.03	....	....
	ave.	38.63	34.15	13.06	12.05	2.26	....	....
White pepper..	max.	64.92	63.60	4.25	11.19[e]	2.13	....	....
	min.	56.43	53.11	0.54	10.44	1.95	....	....
	ave.	59.17	56.47	3.14	10.89	2.04	....	....
Cayenne pepper	max.	9.31	1.46	24.91	14.63	2.34	....	....
	min.	7.15	0.80	20.35	13.31	2.13	....	....
	ave.	8.47	1.01	22.35	13.67	2.18	....	....
Ginger	max.	62.42	60.31	5.50	9.75	1.45	17.55	3.53
	min.	53.43	49.05	2.37	4.81	0.77	10.92	0.20
	ave.	57.45	54.53	3.91	7.74	1.23	13.42	0.80
Cinnamon	max.	22.00	....	38.48	4.06	0.65		Querci-
	min.	16.65	....	34.38	3.25	0.52		tannic
	ave.	19.30	....	36.20	3.70	0.59	Oxygen	acid
							absorbed	equiva-
Cassia	max.	32.04	....	28.80	5.44	0.87	by	lent to
	min.	16.65	....	17.03	3.31	0.53	aqueous	oxygen
	ave.	23.32	....	22.96	4.34	0.69	extract	absorbed
Cloves	max.	9.63	3.15	9.02	7.06	1.13	2.63	20.54
	min.	8.19	2.08	7.06	5.88	0.94	2.08	16.25
	ave.	8.99	2.74	8.10	6.18	0.99	2.33	18.19
Allspice or pimento ...	max.	20.65	3.76	23.98	6.37	1.02	1.59	12.48
	min.	16.56	1.82	20.46	5.19	0.83	1.03	8.06
	ave.	18.03	3.04	22.39	5.75	0.92	1.24	9.71
Nutmegs	max.	25.60	24.20	3.72	7.00	1.12	....	....
	min.	17.19	14.62	2.38	6.56	1.05	....	....

[b] By direct inversion, calculated as starch.
[e] Total nitrogen less nitrogen in nonvolatile ether extract × 6.25.

present. The quantity of the material taken should be such as to obtain, if possible, from 1 to 3 ml. of volatile oil. The flask should not be heated by a direct flame on account of the danger of charring the material containing the volatile oil, thereby giving erroneous results. An oil-bath heated electrically or by a suitable gas flame has been found satisfactory for this purpose.

Procedure.—(a) For Plant Products Containing Little or No Starch or Mucilage.—If the material consists of roots or thick products, place a suitable quantity of the coarsely comminuted material in a flask of suitable size and add water until the flask is half full. Leaf-like material may be placed directly in the flask without first being comminuted or ground. Set up the apparatus as indicated in Fig. 74. Boil the contents for approximately two hours, or until all of the volatile oil has been driven off. The steam carrying the volatile oil condenses and falls into the graduated tube of the separator. The water is separated from the oil by gravity and automatically

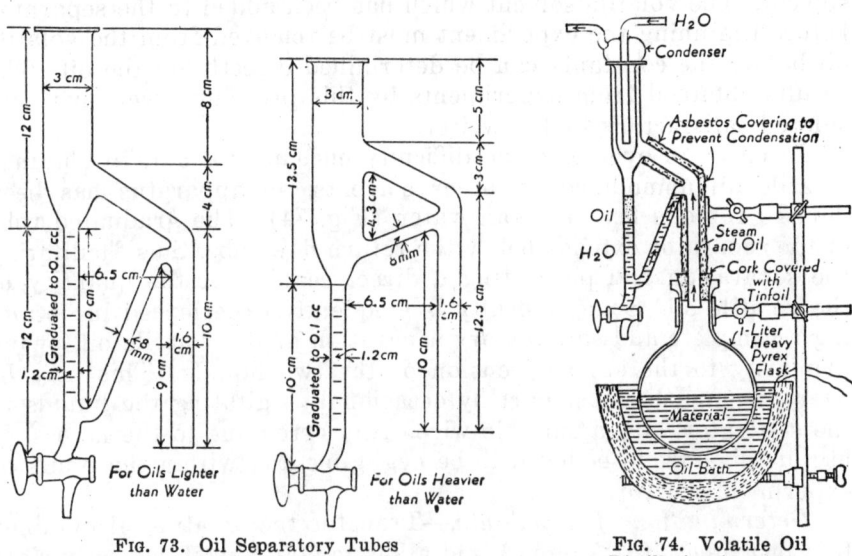

FIG. 73. Oil Separatory Tubes

FIG. 74. Volatile Oil
Distillation Apparatus

flows back into the distillation flask. It has been found that a small amount of paraffin added to the flask prevents excessive foaming which occurs especially in case of powdered products. (Care must be taken that the distillation be conducted at a rate sufficiently slow to prevent the escape of vapors around the condenser, thus insuring against loss of volatile oil.)

(b) *For Crude Material Containing Considerable Starch or Mucilage.*—Exhaust a weighed amount of material with a suitable solvent (alcohol or ether) in an automatic extractor. Transfer the extract to a suitable flask and evaporate the solvent, using a current of air, on a slowly simmering steam-bath until the odor of the solvent is no longer detected. Proceed as outlined in (a).

(c) *For Fluidextracts Containing Volatile Oil.*—Transfer a given quantity or extract to a suitable flask and evaporate as outlined in (b) and, after evaporation, proceed as outlined in (a).

A number of volatile oils having a specific gravity slightly greater than that of water, for example, clove oil, will slowly settle in water. Such oils cannot be determined directly by the use of the separator for oils lighter than water (Fig. 73). Attempts have been made to render the same apparatus serviceable for oils heavier than water by adding a given quantity of a volatile solvent lighter than water and immiscible with it directly to the separatory tube before beginning the determination and making appropriate corrections for the added solvent. The volatile solvent which has been added to the separator before beginning the experiment must be removed from the volatile oil before the constants can be determined directly on the oil. The results obtained from experiments by this procedure were not considered satisfactory by the writer.

In order to overcome the difficulty encountered and to obtain a volatile oil unmodified by other solvents, an apparatus has been devised for oils heavier than water (Fig. 74). The graduated tube of the separator extends below the return flow tube, thus facilitating the separation and permitting a direct reading of the quantity of the volatile oil. Some difficulty is frequently experienced in obtaining a complete and satisfactory separation of the volatile oil, owing primarily to the surface tension of the two liquids. This may be overcome for the most part by occasionally agitating the liquids in the separator with a suitable wire. Any error due to the necessary use of this wire is believed to be negligible and within the limits of experimental error.

Determination of Constants.—Transfer the volatile oil obtained to a test tube (10 × 75 mm.) and allow to stand until perfectly clear (overnight is usually sufficient) or dry at once with a suitable dehydrating agent, such as anhydrous sodium sulfate, and filter. The index of refraction may be determined by the usual method. In spite of the small amounts of volatile oil usually obtained (0.75 to 1.5 ml.), the determination of the optical rotation may be made in a 50-mm. micropolarizing tube and the specific gravity may be determined in a Sprengel specific gravity bottle of approximately 1-ml. capacity.

Minimum Boiling Point.—The minimum boiling point may be determined by the method described by Smith and Menzies.[2] In this method the volatile oil is introduced into a small glass bulb, approximately 5 mm. in diameter, from which a bent capillary tube, approximately 6 cm. long, extends. The volatile oil is introduced into the bulb by immersing the end of the capillary tube in the oil in question and applying a small flame to the bulb for a short time. This causes the air in the bulb to expand and escape from the end of the capillary tube. Upon cooling the bulb, the volatile oil enters. The bulb thus approximately half filled with oil is attached to a thermometer and immersed in a molten paraffin-bath. The paraffin-bath is gently heated until bubbles of the volatile oil freely flow from the end of the capillary tube. The heat is then removed and the bath is allowed to cool until the level of the paraffin in the capillary tube becomes level with the volatile oil in the bulb. This temperature is noted and regarded as the minimum boiling point for the volatile oil.

Volatile Oil in Mustard Seed.—Place 5 g. of the ground seed (No. 20 powder) in a 200-ml. flask, add 100 ml. water, stopper tightly, and macerate for 2 hours at about 37° C. Then add 20 ml. of 95 per cent alcohol and distill about 60 ml. into a 100-ml. volumetric flask containing 10 ml. of ammonium hydroxide solution (1:2), taking care that the tip of the condenser dips below the surface of the solution. Add 20 ml. of 0.1 N silver nitrate solution to the distillate, set aside overnight, heat to boiling on a water bath in order to agglomerate the silver sulfide, cool, make up to 100 ml. with water, and filter. Acidify 50 ml. of the filtrate with about 5 ml. nitric acid and titrate with 0.1 N ammonium thiocyanate, using 5 ml. of 10 per cent ferric ammonium sulfate solution as an indicator. One ml. of 0.1 N silver nitrate solution is equivalent to 0.004956 g. of allylisothiocyanate.

CRUDE FIBER

By the term "crude fiber" is meant in food analysis the combustible residue that is left after the other carbohydrates and the proteins have been removed by successive treatments with boiling acid and alkali. This residue is largely cellulose and consists of carbohydrates not assimilable by humans. The method devised by Henneberg has been rigidly standardized by the A. O. A. C.[3]

Reagents.—(a) Sulfuric Acid Solution.—Prepare a solution containing 1.25 g. of H_2SO_4 per 100 ml.

(b) Sodium Hydroxide Solution.—Prepare a solution containing 1.25 g. of NaOH per 100 ml. free, or nearly free from sodium carbonate. A concentrated solution of sodium hydroxide nearly free from sodium carbonate can be prepared as follows: Dissolve 50 g. of sodium hydroxide in 50 g. of water. Transfer to a tall Pyrex test tube, stopper with a rubber stopper, and allow to stand until all the carbonate has settled. Withdraw sufficient of the supernatant solution to give the required amount of sodium hydroxide. Prepare solutions using a weighed portion of the supernatant solution.

---

[3] *Methods of Analysis*, *A.O.A.C.*, 8th ed. Washington, 1955.

The strength of these solutions must be accurately checked by titration. Adjust the concentration by dilution or addition of more concentrated reagent as necessary.

(c) Asbestos.—Digest on a steam bath or at an equivalent temperature for at least 8 hours with an approximately 5 per cent sodium hydroxide solution and thoroughly wash with hot water; then digest in a similar manner for 8 hours with hydrochloric acid (1:3) and again wash thoroughly with hot water. Dry, and ignite at bright red heat.

Apparatus.—(a) Condenser.—Use a condenser that will maintain a constant volume of solution throughout the process of digestion.

(b) Digestion Flasks.—Use digestion flasks of such size and shape that the solution will be not less than 1 in. nor more than 1.5 in. in depth. A 700-750-ml. Erlenmeyer flask is adequate.

(c) Filtering Cloth.—Use filtering cloth of such character that no appreciable solid matter passes through when filtering is rapid. Butchers linen or dress linen with about 45 threads to the inch or No. 40 filtering cloth made by the National Filter Cloth and Weaving Co., or its equivalent, may be used.

Procedure.—Extract 2 g. of the dry material with ordinary ether, or use the residue from the ether extract determination and transfer the residue, together with about 0.5 g. of asbestos, to the digestion flask. If the residue from the ether extract is used and the proper quantity of asbestos has already been added, further addition is unnecessary. Add 200 ml. of the boiling sulfuric acid solution, immediately connect with the condenser, and heat. It is essential that the contents of the flask come to boiling within 1 minute and that the boiling continue briskly for exactly 30 minutes. Rotate the flask about every 5 minutes in order to mix the charge thoroughly. Take care to keep the material from remaining on the sides of the flask out of contact with the solution. A blast of air conducted into the flask will serve to reduce frothing of the liquid. At the end of 30 minutes remove the flask, immediately filter through linen in a fluted funnel, and wash with boiling water until the washings are no longer acid. Bring a quantity of the sodium hydroxide solution to boiling and keep at this temperature under a reflux condenser until used. Wash the charge and the asbestos back into the flask with 200 ml. of the boiling sodium hydroxide solution, using a wash bottle marked to deliver 200 ml. The boiling sodium hydroxide solution is conveniently transferred to the 200-ml. wash bottle by means of a bent tube through which the liquid is forced by blowing into a tube connected with the top of the reflux condenser attached to the sodium hydroxide flask. Then connect the flask with the reflux condenser and boil for exactly 30 minutes. The boiling with the alkali should be timed so that the contents of the different flasks will reach boiling point approximately 3 minutes apart, which permits sufficient time for filtration. At the end of 30 minutes, remove the flask and immediately filter through a Gooch crucible prepared with an asbestos mat, through an alundum crucible, or through the filtering cloth in a fluted funnel. If the filter cloth is used, thoroughly wash the residue with boiling water and then transfer it to a Gooch crucible prepared with a thin but close layer of ignited asbestos. After thorough washing with boiling water, wash with about 15 ml. of 95 per cent alcohol. Dry the crucible and contents at 110° C. to constant weight. Cool in desiccator and weigh. Ignite the contents of the crucible in an electric muffle or over a Meker burner at a dull red heat, until the carbonaceous matter has been consumed, which will take about 20 minutes. Cool in a desiccator and weigh. Report the loss in weight as crude fiber.

Alternative Method.—Weigh 2 g. of sample and transfer to a beaker. Add 200 ml. of 5 per cent hydrochloric acid. Heat on a steam bath at 90-95° C. for 2 hours. Filter on linen cloth, wash back into beaker with 200 ml. of 5 per cent sodium hydroxide solution, and heat on steam bath again for 2 hours. Filter through a Gooch crucible equipped with an asbestos pad, wash thoroughly with hot water, alcohol, and ether, and dry at 120° C. Weigh, ignite in a muffle, cool in a desiccator, and weigh again. The loss in weight is crude fiber.

STARCH

Qualitative Test.—Spices such as mustard, cayenne pepper, and cloves contain no starch. Hence a qualitative test for starch will indicate adulteration with flour or analogous carbohydrate material.

Procedure.—Add 10 ml. of water to 2 g. of spice in a test tube. Boil for 10 minutes, cool, add a few drops of starch. A blue color indicates flour or starch.

Quantitative Method.—Macerate 3 g. of sample with 50 ml. of cold water for 1 hour with frequent stirring. Collect the residue on a filter and wash with sufficient water to yield a total volume of 250 ml. The filtrate contains the soluble carbohydrates. Heat the undissolved residue with 2.5 per cent hydrochloric acid, prepared by mixing 20 ml. of hydrochloric acid, specific gravity 1.12 and 200 ml. of water, in a flask equipped with a reflux condenser. Cool, neutralize with sodium carbonate solution, transfer to a 250-ml. volumetric flask, and make up to volume. Filter, and determine the dextrose in an aliquot portion of the filtrate as detailed in Chapter XI. The

598 SPICES, FLAVORS, AND CONDIMENTS

weight of dextrose multiplied by 0.9 is equivalent to the weight of the starch.

MICROSCOPIC EXAMINATION

A valuable means and, in some cases, a better means, of detecting adulteration in spices is microscopic examination. The reader is referred to more specialized texts, such as Winton [4] for detailed methods of microscopic examination of spices.

MAYONNAISE AND SALAD DRESSINGS

Definitions for mayonnaise and salad dressings were promulgated and reissued in 1952.[5]

Mayonnaise, Mayonnaise Dressing.—Identity.—(a) Mayonnaise, mayonnaise dressing, is the emulsified semisolid food prepared from edible vegetable oil, one or both of the acidifying ingredients specified in (b) below, and one or more of the egg-yolk-containing ingredients specified in (c). It may be seasoned or flavored with one or more of the following ingredients:

(1) Salt.

(2) Sugar, dextrose, corn syrup, invert sugar syrup, nondiastatic maltose syrup, glucose syrup, honey. The foregoing sweetening agents may be used in the syrup or dried form.

(3) Mustard, paprika, other spice, or any spice oil or spice extract, except that no turmeric or saffron is used and no spice oil or spice extract is used which imparts to the mayonnaise a color simulating the color imparted by egg yolk.

(4) Monosodium glutamate.

(5) Any suitable, harmless food seasoning or flavoring (other than imitations), provided it does not impart to the mayonnaise a color simulating the color imparted by egg yolk.

Mayonnaise may be mixed and packed in an atmosphere in which air is replaced in whole or in part by carbon dioxide or nitrogen. Mayonnaise contains not less than 65 per cent by weight of vegetable oil.

(b) The acidifying ingredients referred to in (a) are:

(1) any vinegar or any vinegar diluted with water to an acidity, calculated as acetic acid, of not less than 2.5 per cent of weight, or any such vinegar or diluted vinegar mixed with the additional optional acidifying ingredient citric acid, but in any such mixture the weight of citric acid is not greater than 25 per cent of the weight

---

[5] *U. S. Food Drug Admin.*, *S.R.A.*, *F.D.&C.* 2, Part 25, June, 1952.

of the acids of the vinegar or diluted vinegar calculated as acetic acid. For the purpose of this definition any blend of two or more vinegars is considered to be a vinegar.

(2) Lemon juice or lime juice or both or any such juice in frozen, canned, concentrated, or dried form, or any one or more of these diluted with water to an acidity, calculated as citric acid, of not less than 2.5 per cent by weight.

(c) The egg-yolk-containing ingredients referred to in (a) are liquid egg yolks, frozen egg yolks, liquid whole eggs, frozen whole eggs, or any one or more of the foregoing with liquid egg white or frozen egg white.

French Dressing.—Identity.—(a) French dressing is the separable liquid food or the emulsified viscous fluid food prepared from edible vegetable oil and one or both of the acidifying ingredients specified below. It may be seasoned or flavored with one or more of the following ingredients: (a) (1), (2), and (4) of the mayonnaise definition; mustard, paprika, other spice, or spice oil or spice extract; any suitable harmless food seasoning or flavoring other than imitations; and tomato paste, tomato puree, catsup, sherry wine.

French dressing may be emulsified with certain optional emulsifying agents. It may be mixed and packed in an atmosphere in which air is replaced in whole or in part by carbon dioxide or nitrogen.

The acidifying ingredients are those mentioned for mayonnaise except that there is no requirement that the acid ingredient contain 2.5 per cent acid by weight. The emulsifying agents are gum acacia or arabic, carob bean or locust bean gum, guar gum, karaya, tragacanth, extract of Irish moss, pectin, propylene glycol ester of alginic acid, sodium carboxymethyl cellulose, or any mixture of two or more of these, or the egg ingredients mentioned in mayonnaise. The quantity of any such emulsifying ingredient or mixture used should not amount to more than 0.75 per cent by weight of the finished French dressing.

Salad Dressing.—Identity.—(a) Salad dressing is the emulsified semisolid food prepared from edible vegetable oil, one or more acidifying ingredients, one or more egg-yolk-containing ingredients, and a cooked or partly cooked starchy paste prepared with a food starch, tapioca flour, rye flour, or any two or more of these. In the preparation of such starchy paste water may be added. Salad dressing may be seasoned or flavored with ingredients (a) (1), (2), (3), (4), and (5) of the mayonnaise definition. It may contain an emulsifying agent and be packed with carbon dioxide or nitrogen. Salad dressing contains not less than 30 per cent by weight of vegetable oil and not less egg-yolk-containing ingredient than is equivalent in egg-yolk

solids to 4 per cent by weight of liquid egg yolks. The optional acidifying ingredients and emulsifying agents and quantities are those
mentioned for French dressing.

Mayonnaise and salad dressings may in general, be analyzed by
methods detailed previously. The presence of gums, starch, or other
thickening agents or emulsifying agents may be shown by methods
completely detailed in the Chapters on Milk, Milk Products, and
Gums. Coloring matter may be detected in the usual manner, although a much simpler and more rapid method is to follow the procedure as outlined in the method for gums, by using trichloroacetic
acid to break the emulsion. The water soluble colors will be in the
clear filtrate and may be taken up directly on wool, and subsequently
redyed. The oil soluble colors are in the petroleum ether washings
and may be extracted in the usual manner.

The egg content determination is based on the lipide-phosphorus
pentoxide content, assuming that all of the lipide-phosphorus pentoxide comes from eggs. This is not necessarily the case and will be
discussed in a subsequent section. Lipide-phosphorus pentoxide content may be estimated according to the method detailed in the section
"Egg and Egg-products," Chapter XVI.

FAT

Rapid Method.—Kaufman [6] devised a rapid method for the estimation of fat in mayonnaise and salad dressings in which the emulsion is broken by use of alcohol and the amount of fat is estimated
centrifugally with Babcock cream flasks. The addition of alcohol in
small portions avoids clumping of starch if any is present and assists
in the production of a clear fat layer.

Procedure for Mayonnaise.—Weigh into a 50-ml. beaker a 5.00-g.
sample on a torsion balance. Add 0.5 ml. of 95 per cent alcohol and
stir into the sample until it is homogeneous. Add another 0.5-ml.
portion and stir thoroughly. Finally add 1.0 ml. of 95 per cent alcohol
and stir again. It is important to make the test sample homogeneous
and free from lumps by this process. Wash the contents of the
beaker into a Babcock 9-g. cream bottle, using a minimum quantity
of ethyl alcohol, 50 per cent by volume. Centrifuge for 10 minutes at
approximately 1200 r.p.m. Add more 50 per cent alcohol to bring
the layer of fat just below the neck of the flask and again centrifuge
for 5 minutes. Add sufficient 50 per cent alcohol to bring the column
of fat well within the neck of the flask and again centrifuge for 5
minutes. Remove the flask from the centrifuge and read the percentage of fat directly. Readings should be made from the bottom of

6 Kaufman, *Chemist-Analyst* 37, 38 (1948).

the lower meniscus to the bottom of the upper meniscus at room temperature. Multiply the reading by 1.8 to obtain the percentage of fat in the sample, since the bottle is calibrated for a 9-g. sample.

Procedure for Salad Dressing.—Weigh a 9.00-g. sample as above. Add respectively 1.0, 1.0, and 2.0 ml. of 95 per cent alcohol. Continue with the method as detailed above. Since the Babcock flask is calibrated for 9-g. samples, it is not necessary to correct the reading of the percentage of fat.

ACIDITY

Total acidity may be estimated by diluting a weighed sample with a considerable amount of water and titrating with 0.1 N sodium hydroxide solution using phenolphthalein as the indicator.

SUGARS

Sugars before inversion, after inversion, and sucrose may be estimated as detailed in Chapter X, but the sample must be prepared by removing fat and protein.

Preparation of Sample.—Removing Fat.—Weigh 20 g. of the sample into a 125-ml. centrifuge bottle, add 60-80 ml. of petroleum ether, stir well, and centrifuge. Draw off the ether layer with the aid of a siphon. Repeat the extraction 4 times or until all the fat has been removed as can be seen by the color of the supernatant liquid. Free the residue from petroleum ether with the aid of air.

If starch is present, defat the sample in a manner similar to that used for the Roese-Gottleib method for ice cream, using analogous ratios of ammonium hydroxide, alcohol, and the mixed ethers. The material can then be transferred to a volumetric flask with 50 per cent alcohol and made up to volume with that solvent. The method described below for mineral oil detection may also be used to break starch emulsions.

Removal of Protein.—Reagent.—Dissolve 5 g. of transparent lumps of glacial phosphoric acid, HPO_3, first washing off any white coating with water, in water and dilute to 100 ml.

Transfer the residue from the defatting process to a 100-ml. volumetric flask. Add 5-10 ml. of the freshly prepared metaphosphoric acid reagent, mix, and complete to volume. Filter and transfer 80 ml. of the filtrate to another 100-ml. volumetric flask, neutralize with strong sodium hydroxide solution (1:1) using phenolphthalein solution as indicator, make to volume and proceed as detailed in Chapter X.

Mineral Oil in Mayonnaise and Salad Dressing

A mayonnaise, according to definition, should contain 50 per cent of edible vegetable oil. Some manufacturers have stressed the use of mayonnaise in so-called nonfattening diets. Indeed in certain cases, the edible oil has been replaced in whole or in part by mineral oil in order to substantiate such claims.

Procedure.—Place 50 g. of mayonnaise or salad dressing in a beaker and add 25 ml. of alcohol. Warm on a steam bath until the emulsion is broken and the oil separates sufficiently to float on top. Transfer to a separatory funnel and add 200 ml. of warm water. When the layers separate, run off the water and transfer the oil to a cylinder. After allowing to stand the clear oil may be poured off.

Saponify 2 g. of the clear oil with 15 ml. of an alcoholic solution of potassium hydroxide, 40 g. potassium hydroxide per liter of alcohol, and evaporate off the alcohol on the steam bath. Add 50 ml. of water and dissolve the soaps by heat, if necessary. If mineral oil is present, in comparatively large amounts, a cloudy emulsion will be formed with globules of oil floating on the surface. Normally a slight amount of unsaponifiable matter is present in an oil or fat and a slight cloud may result which may be neglected. If doubtful cases arise, the presence of much unsaponifiable matter may be demonstrated by the quantitative method described in the chapter on oils and fats, Chapter IX.

Calculation of Composition

The A. O. A. C.[7] gives the following formulas for the calculation of the composition of mayonnaise salad dressing, and other salad dressings.

Major Components:

When $P = \%$ total P_2O_5 and $N = \%$ total nitrogen, then
$$\% \text{ yolk} = 75.69\ P - 1.802\ N;$$
$$\% \text{ white} = 60.80\ N - 114.59\ P;$$
$$\% \text{ total egg} = \% \text{ yolk} + \% \text{ white};$$
$$\% \text{ white in egg component} = \frac{\% \text{ white}}{\% \text{ total egg}} \times 100$$
$$\text{Vegetable oil} = \text{total fat} - (\text{yolk} \times 0.3188);$$
$$\text{Vinegar (4\% acid strength)} = \text{total acidity as acetic} \times 25$$

Minor Components:

$$\text{(sugar, salt, spices, stabilizers)} = \text{total solids} - (\text{yolk} \times 0.5047)$$
$$- (\text{white} \times 0.1221)$$
$$- \text{vegetable oil}$$

---

[7] *Methods of Analysis*, A.O.A.C. Washington, 1945.

$$\text{Added water} = 100\% - \text{total egg} - \text{vegetable oil}$$
$$- \text{vinegar}$$
$$- \text{minor components}$$

The total fat may be estimated by the acid hydrolysis modification of the Roese-Gottlieb method as detailed under "Cheese," Chapter VIII or as detailed in a previous section. The oil may be identified by the methods outlined in the Chapter on Oils and Fats, Chapter IX.

MUSTARD FOR MAYONNAISE [8]

Moisture Holding Ability.—In order to be useful for mayonnaise and salad dressings, mustard flour must have the ability to absorb and hold adequate amounts of water—that is, an amount of water equal to twice its weight. In addition it should be able to form a thick paste with sufficient plasticity to mold into a ball and retain that shape.

Procedure.—Mix 10 g. of mustard flour and 20 ml. of water into a paste. Allow to stand for 1 hour and examine. An adequate mustard flour on dilution to 120 ml. will form a suspension which will not separate readily.

Keeping Quality.—Mix 14.17 g. of mustard flour and sufficient boiled and cooled water to make a paste. Dilute to 240 ml. with boiled and cooled water. Place in jar. No mold growth should be in evidence on standing 1 month at room temperature.

Cupping Test.—To test the mustard blending ability, mix 14.17 g. of mustard with 30 ml. of water in a jar. Note the odor immediately after mixing and 1 hour later to see if there is any change. Cap the jar after diluting to 240 ml. and allow to stand for 1 week. A good mustard should retain its aroma for this period.

FLAVORING EXTRACTS

A flavoring extract is a solution in ethyl alcohol of proper strength of the sapid and odorous principles derived from an aromatic plant, or parts of the plant, with or without its coloring matter, conforming in name to the plant used in its preparation. They should contain no ingredients that may render them harmful to health.

Pending the establishment of standards for specific flavoring extracts and oils, the Food and Drug Administration has issued the descriptions for flavoring extracts given in Table 78.

Distinction between "Extract" and "Flavor."—The vehicle or menstruum of a flavoring extract is ethyl alcohol of proper strength. The terms "extract" and "flavor" are not synonymous. The term

---

[8] Kilgare, *Glass Packer*, February (1934).

TABLE 78 [a]

Almond Extract.—The flavoring extract prepared from oil of bitter almonds, free from hydrocyanic acid. It contains not less than 1 per cent by volume of oil of bitter almonds.

Oil of Bitter Almonds, Commercial.—The volatile oil obtained from the seed of the bitter almond (*Amygdalus communis* L.), the apricot (*Prunus armeniaca* L.), or the peach (*Amygdalus persica* L.).

Anise Extract.—The flavoring extract prepared from oil of anise. It contains not less than 3 per cent by volume of oil of anise.

Oil of Anise.—The volatile oil obtained from aniseed.

Celery Seed Extract.—The flavoring extract prepared from celery seed or the oil of celery seed, or both. It contains not less than 0.3 per cent by volume of oil of celery seed.

Oil of Celery Seed.—The volatile oil obtained from celery seed.

Cinnamon Extract, Cassia Extract, Cassia Cinnamon Extract.—The flavoring product prepared from oil of cinnamon. It contains not less than 2 per cent by volume of oil of cinnamon.

Oil of Cinnamon, Oil of Cassia, Oil of Cassia Cinnamon.—The lead-free volatile oil obtained from the leaves or bark of *Cinnamomum cassia* (L.) Blume. It contains not less than 80 per cent by volume of cinnamic aldehyde.

Ceylon Cinnamon Extract.—The flavoring extract prepared from oil of Ceylon cinnamon. It contains not less than 2 per cent by volume of oil of Ceylon cinnamon.

Oil of Ceylon Cinnamon.—The lead-free volatile oil obtained from the bark of the Ceylon cinnamon (*Cinnamomum zeylanicum* Nees). It contains not less than 65 per cent by weight of cinnamic aldehyde and not more than 10 per cent by weight of eugenol.

Clove Extract.—The flavoring extract prepared from oil of cloves. It contains not less than 2 per cent by volume of oil of cloves.

Oil of Cloves.—The lead-free volatile oil obtained from cloves.

Ginger Extract.—The flavoring extract prepared from ginger. It contains in each 100 ml. the alcohol-soluble matters from not less than 20 g. of ginger.

Lemon Extract.—The flavoring extract prepared from oil of lemon, or from lemon peel, or both. It contains not less than 5 per cent by volume of oil of lemon.

Oil of Lemon.—The volatile oil expressed, without the aid of heat, from the fresh peel of the lemon (*Citrus limonia* Osbeck), with or without previous separation of the pulp and peel.

Terpeneless Extract of Lemon.—The flavoring extract prepared by shaking oil of lemon with dilute alcohol or by dissolving terpeneless oil of lemon in dilute alcohol. It contains not less than 0.2 per cent by weight of citral derived from oil of lemon.

---

[a] *U. S. Food Drug Admin.*, "Flavoring Extracts under the Federal Food, Drug, and Cosmetic Act." March, 1946.

TABLE 78—(*Continued*)

Terpeneless Oil of Lemon.—Oil of lemon from which all or nearly all of the terpenes have been removed.

Nutmeg Extract.—The flavoring extract prepared from oil of nutmeg. It contains not less than 2 per cent by volume of oil of nutmeg.

Oil of Nutmeg.—The volatile oil obtained from nutmegs.

Orange Extract.—The flavoring extract prepared from oil of orange or from orange peel, or both. It contains not less than 5 per cent by volume of oil of orange.

Oil of Orange.—The volatile oil obtained, by expression or alcoholic solution, from the fresh peel of the orange (*Citrus aurantium* L.). It has an optical rotation (25° C.) of not less than +95° in a 100-mm. tube.

Terpeneless Extract of Orange.—The flavoring extract prepared by shaking oil of orange with dilute alcohol or by dissolving terpeneless oil of orange in dilute alcohol. It corresponds in flavoring strength to orange extract.

Terpeneless Oil of Orange.—Oil of orange from which all or nearly all of the terpenes have been removed.

Peppermint Extract.—The flavoring extract prepared from oil of peppermint, or from peppermint, or both. It contains not less than 3 per cent by volume of oil of peppermint.

Peppermint.—The leaves and flowering tops of *Mentha piperita* L.

Oil of Peppermint.—The volatile oil obtained from peppermint. It contains not less than 50 per cent by weight of menthol.

Rose Extract.—The flavoring extract prepared from attar of roses, with or without red rose petals. It contains not less than 0.4 per cent by volume of attar of roses.

Attar of Roses.—The volatile oil obtained from the petals of *Rosa Damascena* Mill., *R. centifolia* L., or *R. moschata* Herrm.

Savory Extract.—The flavoring extract prepared from oil of savory, or from savory, or both. It contains not less than 0.35 per cent by volume of oil of savory.

Oil of Savory.—The volatile oil obtained from savory.

Spearmint Extract.—The flavoring extract prepared from oil of spearmint, or from spearmint, or both. It contains not less than 3 per cent by volume of oil of spearmint.

Spearmint.—The leaves and flowering tops of *Mentha spicata* L.

Oil of Spearmint.—The volatile oil obtained from spearmint.

Star Anise Extract.—The flavoring extract prepared from oil of star anise. It contains not less than 3 per cent by volume of oil of star anise.

Oil of Star Anise.—The volatile oil distilled from the fruit of the star anise (*Illicium verum* Hook.).

Sweet Basil Extract.—The flavoring extract prepared from oil of sweet basil, or from sweet basil, or both. It contains not less than 0.1 per cent by volume of oil of sweet basil.

TABLE 78—(Continued)

Sweet Basil, Basil.—The leaves and tops of *Ocimum basilicum* L.
Oil of Sweet Basil.—The volatile oil obtained from basil.
Sweet Marjoram Extract, Marjoram Extract.—The flavoring extract prepared from the oil of marjoram, or from marjoram, or both. It contains not less than 1 per cent by volume of oil of marjoram.
Oil of Marjoram.—The volatile oil obtained from marjoram.
Thyme Extract—The flavoring extract prepared from oil of thyme, or from thyme, or both. It contains not less than 0.2 per cent by volume of oil of thyme.
Oil of Thyme.—The volatile oil obtained from thyme.
Vanilla Extract.—The flavoring extract prepared from vanilla bean, with or without one or more of the following: Sugar, dextrose, glycerol. It contains in 100 ml. the soluble matters from not less than 10 g. of the vanilla bean.
Vanilla Bean.—The dried, cured fruit of *Vanilla fragrans* (Salisb. Ames (*V. planifolia* Andr.)).
Wintergreen Extract.—The flavoring extract prepared from oil of wintergreen. It contains not less than 3 per cent by volume of oil of wintergreen.
Oil of Wintergreen.—The volatile oil distilled from the leaves of *Gaultheria procumbens* L.

"extract" implies an alcoholic product. Flavoring products prepared with vehicles other than alcohol should be labeled with the term "flavor." Articles labeled "lemon flavor," "orange flavor," etc., should contain the same kinds and proportions of flavoring ingredients as are contained in lemon (or orange, etc.) extract. The term "flavor" as used below will include both extracts and non-alcoholic flavors.

Shortly after the enactment of the Federal Food, Drug, and Cosmetic Act, certain foods—including orange extract, lemon extract, and vanilla extract—were listed as exempted from the labeling requirements as to ingredient declaration. In the *Federal Register* of Sept. 17, 1957, a notice was published announcing that effective in one year (Sept. 17, 1958) the exemption as to the listed foods, including lemon extract and orange extract, would end.

VANILLA EXTRACT

Vanilla extract should have the composition noted above. The composition of vanilla extract; of vanilla and vanillin flavor; and of vanilla, vanillin, and other flavors has been described by the Food and Drug Administration as given below. Synthetic and artificial vanilla flavors are discussed in detail by Jacobs.[9,10]

---

[9] Jacobs, *Am. Perfumer Essential Oil Review* **48**, No. 2, 59 (1946); **48**, No. 3, 56 (1946).
[10] Jacobs, *Synthetic Food Adjuncts.* Van Nostrand, New York, 1947.

Composition of Vanilla.—One U. S. gallon of vanilla extract should contain the soluble matter from not less than 13.35 oz. (avoirdupois) of vanilla beans. Some manufacturers use only 12.8 oz. of vanilla beans per gallon, in the mistaken belief that this quantity is proper. The finished flavor should contain at least 35 per cent of alcohol by volume to keep this soluble matter in solution.

"Vanilla and Vanillin Flavor."—The name "Vanilla and Vanillin Flavor" implies that approximately as much of the total flavor of the product is due to true vanilla as to vanillin. Such a name should not be applied to an article that owes its flavor chiefly to vanillin. Experts of the Food and Drug Administration have found that a standard vanilla extract is equivalent in flavoring strength, although not necessarily in flavoring quality, to 0.7 per cent vanillin solution. Expressed in another way, 1 lb. of vanilla beans has a flavoring strength equivalent to about 1⅛ oz. of vanillin. They also established that 1 part of coumarin is equivalent in flavoring strength to 3 parts of vanillin, that 1 part of heliotropine or piperonal is equivalent in flavoring strength to 2 parts of vanillin, and that a standard tonka extract is equivalent in flavoring strength to a 0.3 per cent vanillin solution.

In the light of the foregoing results, it is a simple matter to construct formulas for flavoring products of different strength that can be legitimately designated "Vanilla and Vanillin Flavor." For example, if a manufacturer desires to make a product of this type equal in flavoring strength to vanilla extract, he should use 0.5 gal. of vanilla extract (13.35 oz. of beans per gallon) and 0.5 gal. of 0.7 per cent vanillin solution (0.93 oz. of vanillin per gallon). For a double-strength extract of this type he should dissolve 0.93 oz. of vanillin in 1 gal. of standard vanilla extract. For higher concentration it is necessary to use a concentrated vanilla extract or a vanilla oleoresin in order to obtain a proper proportion of true vanilla.

In March of 1946, the Food and Drug Administration issued a mimeographed release entitled, "Flavoring Extracts under the Federal Food, Drug, and Cosmetic Act." This release contained information concerning "vanillin and coumarin flavor," and "vanilla, vanillin, and coumarin flavor." In 1953, a commercial toxicology laboratory running routine toxicity tests on a new flavoring product found that this material was apparently toxic. This led to a more complete testing of the individual ingredients of the product and these more complete tests revealed that the toxicity was attributable to coumarin. On the basis of these results the major manufacturers of synthetic coumarin decided voluntarily to withdraw it from sale for food use and submitted their experimental results to the Food and Drug Administration for its consideration.

Jacobs[10a] reviewed the literature concerning the toxicity of coumarin and noted that the physiological response of this substance had been studied for about 100 years, that is, for a period longer than the period which has elapsed since it was synthesized by Perkin in 1867. Thus Ellinger pointed out in 1908 that Buchheim and his student Walewski had studied the physiological response of coumarin as early as 1855.

It is to be noted that coumarin is widely distributed in nature. A principal source is the tonka bean, that is the seeds of *Dipteryx odorata* (Aubl.) Willd. Fam. Leguminosae, formerly classified as *Coumarouna odorata* Aubl. This is a large tree indigenous to Guiana and this type is known as Dutch tonka. Another source is *Dipteryx oppositifolia*. This is a tree native to Brazil and yields the type of seeds known as English tonka beans. Coumarin is found in several species of clover and in many other plants.

As a result of the new information concerning coumarin the Food and Drug Administration issued a statement of policy and interpretation which was published under the title, "Status of foods containing added coumarin," in 1954. In brief this statement announced that food containing added coumarin will be regarded as adulterated.

SUMMARY OF RELATIVE FLAVORING STRENGTHS

The relative flavoring strengths of the ordinary constituents of imitation vanilla flavors have been determined organoleptically in the laboratories of the Food and Drug Administration and found to be as follows:

One part of vanilla beans is equivalent to 0.07 part of vanillin.

A standard vanilla extract is equivalent to a 0.7 per cent vanillin solution.

One part of heliotropine or piperonal is equivalent to 2 parts of vanillin.

One part bourbonal (ethylvanillin) is equivalent to 3-4 parts of vanillin.

Glycerol, total solids, ash, sucrose and other sugars, coloring matters, and caramel may be detected and estimated as detailed in the respective chapters covering these determinations. Alcohol and alcohol denaturants may be estimated as detailed by Herstein and Jacobs.[11]

[10a] Jacobs, *Am. Perfumer Essent. Oil Rev.* **62,** 53 (1953).

[11] Herstein and Jacobs, *Chemistry and Technology of Wines and Liquors.* Van Nostrand, New York, 1948.

Hess and Prescott Method.—A gravimetric method was devised by Hess and Prescott [12] for the determination of vanillin and coumarin in various preparations.

TABLE 79. COMPOSITION OF VANILLA EXTRACT [a]

	Maximum	Minimum	Average
Vanillin g. 100 ml.	0.31	0.11	0.19
Lead number	0.74	0.40	0.54
Acidity of extract, ml. 0.1 N alkali per 100 ml.			
Total	52	30	42
Equivalent to Vanillin	20	7	12
Other than Vanillin	42	14	30
Ash g./100 ml.			
Total	0.432	0.220	0.319
Water soluble	0.357	0.179	0.265
Water insoluble	0.081	0.027	0.054
Alkalinity of ash, ml. 0.1 N acid per 100 ml.			
Total	54	30	42
Water soluble	40	22	30
Water insoluble	42	30	12

[a] Winton, Albright, and Berry, *Ind. Eng. Chem.* 7, 516 (1915).

Preparation of Sample.—Transfer by means of a pipette 50 ml. of the extract to a 250-ml beaker with marks at 80 ml. and 50 ml. Dilute the extract to 80 ml. and evaporate to 50 ml. on a water bath kept at 70° C. Dilute again with water to 80 ml. and again evaporate to 50 ml. Transfer to a 100-ml. flask, rinsing the beaker with hot water, add 25 ml. of 8 per cent neutral lead acetate solution, make up to volume with water, shake, and allow to stand overnight at 37° C. in an incubator, or water bath thermostatically controlled or on the outside top of an oven whose inside temperature is 100° C. Filter through a small dry filter, reserving the filtrate for the determination of vanillin and coumarin, and lead number.

Vanillin.—Transfer a 50-ml. aliquot by means of a pipette to a Jacobs-Singer separatory flask and extract with four 15-ml. portions of ether washed twice to remove the alcohol. Pour the clear successive ether layers from the orifice of the separatory flask into a separatory funnel. Wash the combined ether layers with 4 or 5 portions of ammonium hydroxide solution (1:11), using 10 ml. the first time and 5 ml. each time after the first. Save the ether layer for the coumarin determination. Slightly acidify the combined ammoniacal solutions with hydrochloric acid (1:2), cool, and transfer to a Jacobs-Singer separatory flask. Extract with 4 portions of washed ether, using the same procedure as detailed above with a total of about

[12] Hess and Prescott, *J. Am. Chem. Soc.* 21, 256 (1899).

40 ml. of ether. Evaporate the combined ether extractions in a 50-ml. tared beaker at room temperature, dry over sulfuric acid, and weigh. The residue, if it is pure white and has the characteristic odor of vanillin, may be considered vanillin.

If the residue is discolored or gummy or has not crystallized after standing in the desiccator overnight, extract with 15 successive small portions of boiling petroleum ether. Combine the petroleum ether extracts and evaporate in a small, tared beaker and dry the residue, which should be vanillin, over sulfuric acid and weigh. A series of separatory funnels can be used instead of the separatory flask. Vanillin melts at approximately 80° C. Dissolve a small quantity of vanillin in hydrochloric acid. Add a few crystals of resorcinol. A red to pink coloration should be produced.

A more rapid, although somewhat less accurate method of estimating the amount of vanillin is to heat the dry residue at 105° C. for 2 hours, cool, reweigh, and regard the loss as vanillin. Prior to heating the vanillin may be sublimed and the sublimate subjected to the characteristic test for vanillin.[13]

Coumarin.—Evaporate at room temperature, in a small tared beaker, the original ether extract obtained from the sample from which the vanillin has been removed by means of the ammonia washings. Dry over sulfuric acid and weigh. If the residue is pure white, identification tests may be made on it directly; if not, the residue may be purified by recrystallization from petroleum ether in the manner described under vanillin. Pure coumarin melts at approximately 67° C. and has a characteristic odor resembling "sweet grass" and tonka beans.

The presence of coumarin may be confirmed by applying the Leach [14] test. Dissolve some of the residue in a few drops of water by warming, and transfer to a small porcelain dish or to a spot plate. Add a few drops of 0.1 N iodine solution. Coumarin yields a brown precipitate which gathers into green flecks, leaving a clear brown solution.

Wichmann [15] Rapid Method for Coumarin.—If coumarin, alone, is to be determined, transfer 50 ml. of the vanilla extract to a 100-ml. volumetric flask, add lead acetate in slight excess, make up to volume and filter. Remove the excess lead by means of anhydrous potassium oxalate and filter. Extract 50 ml. of the deleaded filtrate with three or four portions of washed ether. Add to the ether layers a few drops of phenolphthalein indicator and a slight excess of alcoholic

[13] Hiltner, *U. S. Dept. Agr., Bur. Chem., Bull.* **152** (1911).
[14] Leach-Winton, *Food Inspection and Analysis.* Wiley, New York, 1920.
[15] Wichmann, *Ind. Eng. Chem.* **10**, 537 (1918).

potassium hydroxide solution. Remove the vanillin salt by washing with several 10-ml. portions of water until the disappearance of the pink phenolphthalein color in the wash water shows the absence of alkali. Evaporate the washed ether solution in a small tared dish, dry, and weigh the coumarin.

Folin and Denis Colorimetric Method for Vanillin.—This method [16] is based on the production of an intense blue color due to the reduction of the complex phosphotungsticmolybdic reagent by vanillin which is a phenolic compound.

Reagent.—Phosphotungstic-phosphomolybdic acid.—To 100 g. of sodium tungstate and 20 g. of phosphomolybdic acid or its equivalent of molybdic acid, add 100 g. of 85 per cent phosphoric acid and 700 ml. of water. Boil over a free flame for 24 hours, making up the loss by evaporation by the addition of water. Cool, filter, and make up to 1 liter.

Procedure.—Transfer to a 100-ml. volumetric flask 5 ml. of the vanilla extract or less if the sample contains more than 8-12 mg. of vanillin per 5 ml. Add 75 ml. of water, 4 ml. of lead acetate solution, containing 5 per cent basic lead acetate and 5 per cent neutral lead acetate, make up to volume and mix. Filter rapidly through a dry filter and transfer 5 ml. of the clear filtrate by means of a pipette to a 50-ml. volumetric flask. Into another 50-ml. volumetric flask, pipette 5 ml. of standard vanillin solution (1 ml. = 0.1 mg. of vanillin). To each of these flasks add by means of a pipette 5 ml. of the phosphotungstic-phosphomolybdic reagent, allowing it to flow down the neck of the flask in such a way as to wash down the vanillin solution that may be on the sides of the flask. Mix the contents of the flasks by rotating and, after 5 minutes, dilute the contents to 50 ml. with saturated sodium carbonate solution. Mix thoroughly by inverting the flasks several times and allow to stand for at least 10 minutes, so that the precipitate that forms may separate completely. Filter the solutions through a dry filter and compare the blue colors produced in a colorimeter. Calculate in the usual manner and report as grams of vanillin per 100 ml. of extract.

Wichmann Test for Coumarin.—This test [17] is based on the conversion of coumarin to potassium salicylate when it is fused with potassium hydroxide, whereas vanillin under the same conditions is converted to potassium protocatechuate.

Procedure.—To 10 ml. of the vanilla extract add 10 per cent sodium hydroxide, sufficient to make the solution alkaline. Dilute with 15 ml. of water to reduce the alcoholic strength and extract with

[16] Folin and Denis, *Ind. Eng. Chem.* 4, 670 (1912).
[17] Wichmann, *Ind. Eng. Chem.* 10, 536 (1918).

20 ml. of ether in a separatory funnel. The ether layer will be slightly colored when the brown aqueous layer has been drawn off. Add a few milliliters of strong alcoholic potassium hydroxide solution, shake the mixture, and wash with 10 ml. of water. The ether layer should then be colorless. This procedure removes all organic acids, vanillin, coloring matter or saccharin that may be present. Place 1 ml. of 50 per cent potassium hydroxide solution in a test tube. Overlay the potassium hydroxide in the test tube with the ether layer, shake thoroughly, and rapidly evaporate the ether on a steam or water bath. Heat the tube over a free flame, evaporating the water. Fuse the potassium hydroxide. If coumarin is present in any amount a change of color will be noticed as the evaporation of the water proceeds and fusion begins. Even very small quantities of coumarin in strong, hot potassium hydroxide solution will show a greenish-yellow color that suddenly disappears as the heating is continued. The disappearance of the color shows that the coumarin has been converted into the salicylate and heating should be discontinued. Take up the melt with a few milliliters of water, acidify with sulfuric acid and extract in a small separatory funnel with 5-10 ml. of benzene. Benzene is preferable as the solvent, because it will dissolve little of the mineral acid and the protocatechuic acid formed from vanillin that may have been carried over with the ether. Remove the acid layer from the funnel, wash the benzene into a test tube, and test for salicylic acid with 1 ml. of water containing a few drops of ferric chloride solution. If no color develops on shaking, neutralize any trace of mineral acid that may be present and prevent the development of the purple color, by the addition of 1 or 2 drops of 0.1 N sodium hydroxide.

The change of color on fusion indicates its own end point, and gives, together with the purple salicylate color, a double test for coumarin. Coumarin is changed to the salt of coumaric acid by the hot potassium hydroxide. The development of the yellow color shows this phase. The sudden disappearance of the color indicates the conversion into the colorless salicylate.

LEAD NUMBER

When a lead salt is added to vanilla extract, a precipitate is produced, in a manner similar to that in maple products, which is attributable to materials whose lead salts are insoluble. In the case of vanilla extract, however, only neutral lead acetate may be used, for basic lead acetate yields a precipitate with vanillin. The residual lead, that is, the amount of lead not precipitated, is determined by either the Winton [18] or the Wichmann [19] method and is known as the Winton or Wichmann lead number. Since these lead numbers are empirical, the details of the methods must be followed minutely in order to obtain results that have meaning. The directions of the A. O. A. C.[20] follow.

Winton Method.—Determine lead as sulfate or chromate in 10 ml. of the filtrate from the lead acetate precipitate of the vanillin and coumarin gravimetric determination, and in the filtrate from a blank determination, using water and 5 drops of glacial acetic acid in place of the sample as directed under the Wichmann number. Calculate the lead number and report as "Lead Number—Winton."

Wichmann Method.—Place 175 ml. of boiled water in a round-bottomed flask of 1 liter capacity. Add by means of a pipette 25 ml. of clear lead acetate solution, 8 g. per 100 ml., and 50 ml. of sample. Place the flask in a hole in an asbestos board that is large enough to prevent the heating of the upper portion of the flask. When the contents of the flask are reduced to 50 ml. of liquid, the level of the liquid should be even with the top of the board, or slightly above it. Connect the flask to a condenser, and with a moderate flame distill 200 ml. into a volumetric flask, reserving the distillate for the determination of alcohol, if desired. Transfer the residual solution to a 100-ml. volumetric flask by means of carbon dioxide-free water and a bent glass rod provided with a rubber tip. When cool, dilute to 100 ml. with carbon dioxide-free water, mix, and filter through a dry filter (Solution A). Conduct a blank determination, using 5 drops of glacial acetic acid in place of the sample and distilling 150 ml. instead of 200 ml. Determine lead as directed below, either as sulfate or chromate, and calculate the lead number and report as "Lead Number—Wichmann."

Determination of Lead.—*As Sulfate.*—Pipette 10 ml. of Solution A into a 250-ml. beaker, add 25 ml. of water, 2 ml. of sulfuric acid (1:1) and 100 ml. of 95 per cent alcohol, stir, and allow to settle overnight. Filter on a Gooch crucible, wash with 95 per cent alcohol, ignite at low redness, cool in a desiccator, and weigh. The difference between the weight of lead sulfate obtained from the blank and that obtained from the sample $\times$ 13.66 = the lead number of the extract.

As Chromate.—Pipette 10 ml. of solution A into a 400-ml. beaker and add 2 ml. of glacial acetic acid, 25 ml. of water, and 25 ml. of approximately 0.1 N potassium dichromate solution. Heat the beaker and contents immediately with a moderate flame until the precipitate

[18] Winton, *U. S. Dept. Agr., Bur. Chem., Bull.* **95** (1912).
[19] Wichmann, *Ind. Eng. Chem.* **13**, 414 (1921).
[20] *Methods of Analysis*, A.O.A.C. Washington, 1945.

changes in color from yellow to orange. Filter on a Gooch crucible; wash thoroughly with hot water and then with a few milliliters each of alcohol and ether. Dry at 100° C., cool in a desiccator, and weigh. The difference between the weight of lead chromate obtained from the blank and that obtained from the sample × 12.82 = the lead number.

Bourbonal [21] **in Vanilla Extracts.**—Dealcoholize 50 ml. of the sample, add lead acetate solution, and extract with ether as described in the Hess and Prescott method for vanillin detailed in this chapter. Place the ether extract in a small beaker and allow the residue to remain overnight in a desiccator. Then add 1 ml. of hydrochloric acid (2 + 1). Place the beaker in a water bath at 55° C. until the residue is dissolved, then pour the solution into a medium-sized test tube. Use no wash water, as a quantitative transfer is not required. Add 1 ml. of 3 per cent hydrogen peroxide solution to the test tube and shake frequently, while the color changes to yellow, brown, then red. Finally a deep purple color appears and a blue precipitate forms.

After standing 15 minutes, add 5 ml. of benzene and place the test tube in the water bath at 55° C. Shake frequently and allow the test tube to remain in the water bath until the lower aqueous layer becomes a dirty yellowish brown (about 15-20 minutes). Then remove it from the water bath and carefully pour or pipette a major part of the benzene layer into a small dry test tube. If the benzene is colored violet, ethylvanillin is present in the original sample. In the absence of bourbonal, the benzene is colored a light or dirty yellow.

VINEGAR

The use of vinegar as a condiment is probably as old as the use of wine for it was very likely first obtained from subsequent acetous fermentation of wine.

The standards for vinegar [22] were promulgated when the Food and Drug Administration was an arm of the U. S. Department of Agriculture. They were not in 1958 in the same category as the definitions and standards promulgated for a number of foods [23] but were rather held to be a guide for the enforcement of the Act.

1. *Vinegar, Cider Vinegar, Apple Vinegar* is the product made by the alcoholic and subsequent acetous fermentations of the juice of apples. It contains, in 100 ml. (20° C.), not less than 4 g. of acetic acid.

---

[21] Chenoweth, *Ind. Eng. Chem., Anal. Ed.* 12, 98 (1940).
[22] *U. S. Dept. Agr., Food Drug Admin., S.R.A., F.D.* 2 (1936).
[23] *U. S. Food Drug Admin., S.R.A., F.D.& C.* 2.

2. *Wine Vinegar, Grape Vinegar* is the product made by the alcoholic and subsequent acetous fermentations of the juice of grapes. It contains, in 100 ml. (20° C.), not less than 4 g. of acetic acid.

3. *Malt Vinegar* is the product made by the alcoholic and subsequent acetous fermentations, without distillation, of an infusion of barley malt or cereals whose starch has been converted by malt. It contains, in 100 ml. (20° C.), not less than 4 g. of acetic acid.

4. *Sugar Vinegar* is the product made by the alcoholic and subsequent acetous fermentations of sugar syrup, molasses, or refiners syrup. It contains, in 100 ml. (20° C.), not less than 4 g. of acetic acid.

5. *Glucose Vinegar* is the product made by the alcoholic and subsequent acetous fermentations of a solution of glucose. It is dextrorotatory and contains, in 100 ml. (20° C.), not less than 4 g. of acetic acid.

6. *Spirit Vinegar, Distilled Vinegar, Grain Vinegar* is the product made by the acetous fermentation of dilute distilled alcohol. It contains, in 100 ml. (20° C.), not less than 4 g. of acetic acid.

Because vinegars may be made in a variety of ways, although all by means of acetous fermentation, and furthermore because of the liberality of governmental definitions, it is more or less easy to reconstitute a vinegar to simulate a natural product. It is well to remember that even though vinegar is essentially a dilute solution of acetic acid, it is a product of fermentation and consequently contains many characteristic substances that are not present in an imitation vinegar. The determination of the presence of these other fermentation products provides an index of its genuineness. In Tables 80, 81, the analyses of a number of typical vinegars are given. As in other food materials, analysis of a vinegar and comparison with the normals will at times indicate whether or not it is standard or substandard. More skillful sophistication may be detected by other means.

In general the analysis of vinegar proceeds along lines that have been previously detailed. Specific gravity, solids, ash, soluble ash, insoluble ash, alkalinity of ash, and acidity may be estimated as usual. Fixed acids, which in the case of vinegar is largely lactic acid and not malic acid, volatile acids and alcohol are determined as detailed by Herstein and Jacobs.[24] Phosphates and other inorganic determinations may be made as detailed in the chapters on inorganic determinations and metals, Chapters XVIII and V. Sugars and reducing substances are estimated as outlined in Chapter X. Alcohol precipitate, dextrin, pentosans, characteristic organic acids in a man-

---

[24] Herstein and Jacobs, *Chemistry and Technology of Wines and Liquors.* Van Nostrand, New York, 1948.

ner similar to the methods given in the chapters on jams and jellies and gums, Chapters XII and XI. The methods for glycerol, for the detection of mineral acids and for the differentiation between natural and artificial vinegars will be discussed more fully and particularly.

TABLE 80. COMPOSITION OF VINEGAR [a,b]

	Average	Maximum	Minimum
Total acid as acetic acid %	4.94	7.96	3.29
Total solids %	2.54	4.52	1.37
Non-sugar solids %	1.90	2.89	1.26
Reducing sugars in solids %	19.6	45.0	5.0
Total ash %	0.367	0.52	0.20
Alkalinity of water soluble ash, ml.	35.7	56.0	21.5
Ash in nonsugar solids %	18.8	26.5	11.2
Soluble phosphoric acid (mg. P_2O_5)	17.3	39.9	6.7
Insoluble phosphoric acid (mg. P_2O_5)	12.0	32.0	4.3
Total phosphoric acid (mg. P_2O_5)	29.3	64.2	15.1
Polarization (direct) V°	—1.46	—3.6	—0.2
Polarization (invert) V°	—1.69	—3.1	±0.0

[a] Balcom, *U. S. Dept. Agr., Bull.* 132 (1909).
[b] Of about 100 samples, in g. per 100 ml.

TABLE 81. CHEMICAL COMPOSITION OF CIDER VINEGARS [a]

	Filtered [b]	Cleared [b]
Sp. gr. $\frac{15° C.}{15° C.}$	1.0193	1.0153
Alcohol % by vol.	0.04	0.055
Glycerol	0.235	0.245
Solids	1.315	1.205
Sugars as invert		
Before inversion	0.45	0.45
After inversion	0.12	0.11
Non-sugar solids	1.19	1.10
Volatile reducing substances	0.33	0.34
Total acid as acetic	6.47	6.63
Volatile acid as acetic	6.44	6.58
Fixed acid as malic	0.03	0.055
Volatile esters as ethyl acetate	0.80	0.88
Pentosans	0.083	0.076
Formic acid	0.0004	0.0004
Ash	0.30	0.31
Alkalinity of ash ml. 0.1 N acid per 100 ml.		
Soluble ash	34.5	34.6
Insoluble ash	5.8	5.9

[a] Hartman and Tolman, *Ind. Eng. Chem.* 9, 759 (1917).
[b] In g. per 100 ml. unless otherwise stated.

GLYCEROL

The official A. O. A. C. method [25] is based on the separation of the glycerol in vinegars from other oxidizable materials, the oxidation of the glycerol by strong potassium dichromate solution and the subsequent calculation of the amount of glycerol from a control oxidation of ferrous ammonium sulfate.

Reagents.—(a) Strong Potassium Dichromate Solution.—Dissolve 74.55 g. of dry recrystallized potassium dichromate $K_2Cr_2O_7$ in water; add 150 ml. of sulfuric acid, cool, and dilute with water to 1 liter at 20° C. One ml. of this solution is equivalent to 0.01 g. of glycerol. Because of the high coefficient of expansion of this strong solution, it is necessary to make all volumetric measurements of the solution at the same temperature (20° C.) as that at which it was diluted to volume.

(b) Dilute Potassium Dichromate Solution.—Transfer 25 ml. of the strong potassium dichromate solution at 20° C. to a 500-ml. volumetric flask and dilute to the mark with water at room temperature. Twenty ml. of this solution is equivalent to 1 ml. of (a).

(c) Ferrous Ammonium Sulfate Solution.—Dissolve 30 g. of crystallized ferrous ammonium sulfate $Fe(NH_4)_2(SO_4)_2 \cdot 6H_2O$, in water, add 50 ml. of sulfuric acid, cool, and dilute with water to 1 liter at room temperature. One ml. of this solution is equivalent to approximately 1 ml. of (b). As its value changes slightly from day to day, it must be standardized against (b) whenever used.

(d) Diphenylamine Indicator.—Dissolve 1 g. of diphenylamine in 100 ml. of sulfuric acid.

(e) Retarder.—Dilute 150 ml. of syrupy phosphoric acid with 600 ml. of water and 250 ml. of sulfuric acid.

(f) Milk of Lime.—Introduce 150 g. of calcium oxide selected from clean hard lumps, prepared preferably from marble, into a large porcelain or iron dish; slake with water, cool, and add sufficient water to make 1 liter.

(g) Silver Carbonate.—Dissolve 0.1 g. of silver sulfate in about 50 ml. of water, add an excess of sodium carbonate solution, allow the precipitate to settle, and wash with water several times by decantation until the washings are practically neutral. This reagent must be freshly prepared immediately before use.

Procedure.—Make evaporations on a water bath maintained at a temperature of 85-90° C. The area of the dish exposed to the bath should not be greater in circumference than that covered by the liquid inside.

---

[25] *Methods of Analysis*, A.O.A.C. Washington, 1945.

Evaporate 100 ml. of the vinegar to 5 ml., add 20 ml. of water, and again evaporate to 5 ml. to expel acetic acid. Treat the residue with about 5 g. of 40 mesh sand and 15 ml. of the milk of lime and evaporate almost to dryness, with frequent stirring, avoiding the formation of a dry crust or evaporation to complete dryness. Treat the moist residue with 5 ml. of water; rub into a homogeneous paste; add slowly 45 ml. of absolute alcohol, washing down the sides of the dish to remove adhering paste; and stir thoroughly. Heat the mixture on a water bath, with constant stirring, to incipient boiling; transfer to a suitable vessel; and centrifuge. Decant the clear liquid into a porcelain dish and wash the residue with several small portions of hot alcohol, 90 per cent by volume, by the aid of the centrifuge. If a centrifuge is not available, decant the liquid through a folded filter into a porcelain dish. Wash the residue repeatedly with small portions of hot 90 per cent alcohol, twice by decantation, and then by transferring all the material to the filter. Continue the washing until the filtrate amounts to 150 ml. Evaporate to a syrupy consistency, add 10 ml. of absolute alcohol to dissolve this residue, and transfer to a 50-ml. glass-stoppered cylinder, washing the dish with successive small portions of absolute alcohol until the volume of the solution is 20 ml. Add 3 portions of 10 ml. each of anhydrous ether, shaking thoroughly after each addition. Let stand until clear, pour off through a filter, and wash the cylinder and filter with a mixture of 2 volumes of absolute alcohol and 3 of anhydrous ether. If a heavy precipitate has formed in the cylinder, centrifuge at low speed, decant the clear liquid, and wash 3 times with 20-ml. portions of the alcohol-ether mixture, shaking the mixture thoroughly each time and separating the precipitate by means of the centrifuge. Wash the paper with the alcohol-ether mixture and evaporate the filtrate and washings on the water bath to about 5 ml.; add 20 ml. of water, and again evaporate to 5 ml.; again add 20 ml. of water and evaporate to 5 ml.; finally add 10 ml. of water and evaporate to 5 ml.

These evaporations are necessary to remove all the ether and alcohol, and when conducted at 85-90° C. they result in no loss of glycerol if the concentration of the latter is less than 50 per cent.

Transfer the residue with hot water to a 50-ml. volumetric flask, cool, add the silver carbonate prepared from 0.1 g. of silver sulfate, shake, and allow to stand 10 minutes. Then add 0.5 ml. of basic lead acetate solution, prepared as directed in the section "Clarifiers" in Chapter X; shake occasionally, and allow to stand 10 minutes. Make up to the mark, shake well, filter, rejecting the first portion of the filtrate, and pipette 25 ml. of the clear filtrate into a 250-ml. volumetric flask. Add 1 ml. of sulfuric acid to precipitate the excess of lead and then 30 ml. of reagent (a), the strong potassium dichromate solution. Add carefully 24 ml. of sulfuric acid, rotating the flask gently to mix the contents and avoid violent ebullition, and then place in a boiling water bath for exactly 20 minutes. Remove the flask from the bath, dilute, cool, and make up to the mark at room temperature. The quantity of strong dichromate solution used must be sufficient to leave an excess of about 12.5 ml. at the end of the oxidation, the quantity given above, namely 30 ml., being sufficient for ordinary vinegar containing about 0.35 g. or less of glycerol per 100 ml.

Standardize the ferrous ammonium sulfate solution by pipetting 20 ml. into a 250-ml. beaker, adding 20 ml. of the retarder, 4 drops of the indicator, and about 100 ml. of water. Titrate with the dilute potassium dichromate solution until the liquid assumes a dark green color, then add the dichromate slowly dropwise, stirring continuously, until the color changes from a blue gray to a deep violet. Designate the milliliters of dilute dichromate solution used as (a). In place of the dilute dichromate solution, substitute a burette containing the oxidized glycerol with an excess of the strong dichromate solution and titrate 20 ml. of the ferrous ammonium sulfate solution as before, designating the milliliters used as (b).

Calculation.—From the figures obtained calculate the glycerol, g. per 100 ml. of vinegar by the following formula

$$G = \left[D - \frac{250(a)}{20(b)} \right] 0.02$$

in which

G = g. of glycerol per 100 ml. of vinegar

and

D = ml. of the strong potassium dichromate solution used to oxidize the glycerol.

Free Mineral Acids

Many tests used to detect free mineral acids in vinegars depend on the fact that, in general, the hydrogen-ion concentration will be much greater for a certain percentage (total titratable acidity) of mineral acid than for a corresponding equal percentage (total titratable acidity) of organic acid. This greater hydrogen-ion concentration may easily be demonstrated by use of an indicator which will be changed to a different color if the pH falls below a certain value.

Allen's [26] **Test, Logwood Method.**—Prepare an extract of logwood by pouring 100 ml. of boiling water upon 2 g. of fresh logwood chips. Allow the infusion to stand for a few hours, and filter. Place several drops of the liquid on a porcelain surface or spot plate and dry on a water bath. Add to one of the spots a drop of the vinegar to be tested and evaporate to dryness. A yellow tint remains if free mineral acids are absent, and a red tint if they are present. The method may be made more sensitive for colored vinegar samples by adsorbing the color on activated carbon, filtering, and then applying the test.

Methyl Violet Method.—Add 5-10 ml. of water to 5 ml. of vinegar and after mixing add 4 or 5 drops of methyl violet solution, 1 part of methyl violet 2B in 10,000 parts of water. A blue or green color indicates the presence of a free mineral acid.

Topfer's Reagent.—Topfer's reagent, 0.5 g. dimethylaminobenzene in 100 ml. 95 per cent alcohol may be used in a manner similar to that of the methyl violet method. In the presence of mineral acid the color of the indicator changes from yellow or salmon pink to red.

Quantitative [27] **Method.**—To a measured quantity of the sample add a measured excess of standard alkali, evaporate to dryness, incinerate at low red heat, and titrate the ash with standard acid, using methyl orange indicator. The difference between the number of milliliters of alkali first added and the number of milliliters of acid needed to titrate the ash represents the free mineral acid present.

DETECTION OF ARTIFICIAL VINEGAR

The problem of detecting artificial vinegars and distinguishing between spirit vinegar and other fermentation vinegars is not always easy. The definitions of the Food and Drug Administration are so drawn as to exclude substances not mentioned, and, since color is not designated, the presence of color in an unknown sample of vinegar would be indicative of artificiality. However, a far better means of distinction is to determine the presence or absence of characteristic substances normally present in natural vinegars.

Acetyl Methyl Carbinol Method.—One of these characteristic substances is acetyl methyl carbinol. This substance is also present in very small amounts in butter. Methods for the differentiation of fermentation vinegar from artificial vinegar by estimating acetyl methyl carbinol have been developed.

---

[26] Allen's *Commercial Organic Analysis*, 5th Ed., Vol. I. Blakiston, Philadelphia, 1923.
[27] *Methods of Analysis*, A.O.A.C. Washington, 1945.

One of these methods is that noted by Dingemans [28] based on the distillation of the acetyl methyl carbinol in the presence of ferric chloride with the formation of diacetyl which may then be estimated as nickel dimethylglyoxime as detailed under the section, diacetyl, in Chapter VIII. A more rapid method depends upon the reduction of cold Fehling's solution.[29] The cuprous oxide so formed is transferred to a weighed, narrow centrifuge tube, in which it is washed by cencentrifuging with water, alcohol and ether, dried at 30° to 40° C., and weighed.

Formic Acid Method.—Genuine fermentation vinegar contains only very little formic acid,[30] but artificial vinegar made from acetic acid manufactured from acetylene, even when controlled during the process of manufacture by means of the permanganate test, may contain relatively large amounts of formic acid. This may be estimated as described in the chapter on preservatives, Chapter IV. This determination is of no value as a means of detection when pure commercial acetic acid is used.

Permanganate Oxidation Value.—There are a number of other methods for the differentiation of, not only fermentation vinegar from artificial vinegar, but also for the differentiation of wine vinegar from spirit vinegar, as well as from artificial vinegar. These methods are based on the oxidation of the reducing substances normally present in fermentation vinegar and absent in artificial vinegar and spirit vinegar. Thus Schmidt [31] defines the "oxidation value" of a vinegar as the number of milliliters of 0.1 N potassium permanganate solution required to give a permanent pink color to 50 ml. of vinegar containing 3 per cent acetic acid, in the presence of sulfuric acid. If the vinegar contains more than 3 per cent acetic acid, it must be diluted to that value, and if it contains caramel color, it must be decolorized with activated carbon by treatment for 2 minutes at room temperature. The oxidation value should be determined before and after treatment with the active carbon.

Procedure.[32]—Take a quantity of vinegar sufficient to yield a solution of 3 per cent acidity when diluted to 100 ml. To this quantity add 1 g. of activated carbon, dilute to a volume of 100 ml., shake for 2 minutes, and filter. Take 50 ml. of the filtrate, add 2 ml. of sulfuric acid (1:1), heat just to boiling and titrate with 0.1 N potassium permanganate, until the pink color developed lasts for half a minute.

28 Dingemans, *Ann. fals.* 26, 346 (1933).
29 Arbenz and Pritzker, *Mitt. Lebensm.* 22, 354 (1931).
30 Kreutz and Buchner, *Z. Untersuch. Lebensm.* 52, 295 (1926).
31 Schmidt, *Z. Untersuch. Lebensm.* 73, 441 (1937).
32 *Conn. Agr. Exp. Sta. Rept., Analyst* 62, 550 (1937).

The volume in milliliters of 0.1 N potassium permanganate consumed, corrected for a blank determination made with water carried through the same procedure, is the oxygen or oxidation value of the vinegar.

The end point is fugitive, except with uncolored dilute acetic acid, and cannot be determined with the same degree of accuracy as in inorganic permanganate determinations.

Interpretation.—For wine, the value is 50 or more; for wine vinegar of acidity 9.2 to 9.4 per cent, the value is about 30, and for wine vinegar of 3 per cent strength, it varies from 8 to 12. Distilled vinegar has values about 5; artificial vinegar, generally not more than 1 to 1.5 and dilute acetic acid has a value of zero. The method cannot be applied easily to cider vinegars because the end point cannot be read to even approximate accuracy owing to the large amount of manganese dioxide formed. They give oxygen values well over 100.

Permanganate Number.—The permanganate method of the Connecticut Agricultural Experiment Station has been modified by using excess permanganate to liberate iodine and estimating the latter titrimetrically.[33]

Procedure.—Adjust the acidity of the vinegar to 4 g. per 100 ml. expressed as acetic acid. Distill in an all-glass apparatus 50 ml. with steam, maintaining the volume so as to have a residue of 45 ml. for 50 ml. of distillate, catching the distillate in a 250-ml. glass-stoppered bottle. Keep the distillate and reagents at 25° C. Add 10 ml. of sulfuric acid (1:1) and 25 ml. of N potassium permanganate solution. Hold at 25° C. for exactly 1 hour. Add immediately thereafter 20 ml. of potassium iodide solution, prepared by dissolving 30 g. of KI in 100 ml. of water and filtering, and mix thoroughly. Titrate the liberated iodine with 0.5 N sodium thiosulfate solution.

Run a blank on an acetic acid solution, prepared by dissolving 4 g. of acetic acid in 100 ml. of water. This acetic acid solution should have a negligible permanganate oxidation value.

The permanganate number may also be determined on 50 ml. of the adjusted vinegar without distillation and on the residue in the distillation flask after adjusting the volume to 50 ml.

Calculation.—Divide the milliliters of thiosulfate solution used by 2 and subtract from 25 to obtain the permanganate oxidation value.

If this value is greater than 15, repeat the determination taking half the original quantity of vinegar. Repeat this reduction by half until the milliliters of N potassium permanganate solution used is less than 15. Calculate the value to the basis of a 50-ml. sample.

33 O'Neill and Henry, *J. Assoc. Official Agr. Chem.* 27, 263 (1944).

Interpretation.—The study of O'Neill and Henry and that of Michael [34] indicate that samples of cider vinegar and white distilled vinegar whose permanganate numbers fall below 3.0 are generally to be viewed with suspicion. However, Michael found that certain vinegars made by the Frings process had permanganate numbers as low as 2.26. In a survey of 91 open market samples the following results were obtained:

PERMANGANATE No. ABOVE 3.0

Ave. Permanganate No.

61 Samples of cider vinegar 5.08
8 Samples of white vinegar 5.48
Total average 5.13

PERMANGANATE No. BELOW 3.0

14 Samples of cider vinegar 2.02
8 Samples of white vinegar 2.23
Total average 2.10

TABLE 82. VINEGAR PERMANGANATE OXIDATION NUMBERS [a]

Product	Before Distillation	Distillate	Undistilled Residue
Destructively distilled glacial acetic acid dil...	0.0	0.1	0.0
Above, colored with caramel..............	0.8	0.1	0.8
Synthetic glacial acetic acid diluted.........	0.0	0.1	0.0
Above, colored with caramel..............	0.9	0.1	0.7
Above, colored by soaking with charred oak chips	13.6	0.3	12.6
Distilled vinegar (a)	7.9	6.2	1.0
Distilled vinegar (b)	5.1	3.4	0.9
Distilled vinegar (b) colored with caramel....	6.0	3.3	1.9
Molasses vinegar	15.+	3.3	15.+
Cider vinegar	15.+	3.5	15.+

[a] O'Neill and Henry, *J. Assoc. Official Agr. Chem.* 27, 263 (1944).

Iodine Oxidation.—Another means of determining the amount of oxidizable material is to use iodine solution.

Procedure.—Add excess 0.01 *N* iodine solution, measured accurately, to 25 ml. of the vinegar, which has been made alkaline. Th. excess iodine must be at least 25 per cent. Allow the mixture to stand

[34] Michael, *Assoc. Food Drug Officials U. S., Quarterly Bull.* 13, 93 (1949).

for 15 minutes, then acidify with dilute hydrochloric acid, and back-titrate the iodine with 0.01 *N* sodium thiosulfate solution.

Interpretation.—The estimation is made both before and after treatment with active carbon for 2 minutes at room temperature, and samples containing more than 3 per cent of acetic acid are diluted to that concentration. Spirit vinegar gives values from 30 to 60, wine vinegars from 90 to 300, according to strength, and artificial vinegars from 4 to 7. If the value is below 20 the sample may be regarded with suspicion.

Schmidt suggests as minimum values for wine vinegars corresponding with 3 per cent of acetic acid: Oxidation value, 8 ml. of 0.1 *N* potassium permanganate solution per 50 ml. of 3 per cent vinegar solution; iodine value, 90 ml. of 0.01 *N* iodine solution per 25 ml. per 3 per cent vinegar solution.

Schiff's Reagent Method.—A method using Schiff's reagent, prepared as directed under formaldehyde, Chapter IV, is the following and may be applied in the presence of caramel.

Procedure.—To 10 ml. of the vinegar, add 1 ml. of phosphoric acid and 1 ml. of 3 per cent potassium permanganate solution and allow the mixture to stand for 10 minutes. Rapid decolorization takes place with wine vinegars; the decolorization is slower with spirit vinegars and is negligible with artificial vinegars. On the further addition of 1. ml. of saturated oxalic acid solution, 1 ml. of 4 *N* sulfuric acid, followed after a short interval by 1 ml. of Schiff's reagent, the rapid formation of an intense red-violet color is characteristic of wine vinegars, a pale violet of spirit vinegars, and the absence of color of artificial vinegars. The formation, after some time, of a pale brown color is due to the presence of caramel.

SELECTED REFERENCES

Cox, *Analysis of Foods*, Blakiston, London, 1926.
Brooks, *Critical Studies in the Legal Chemistry of Foods*. Chemical Catalog, New York, 1927.
Leach-Winton, *Food Inspection and Analysis*. Wiley, New York, 1920.
Methods of Analysis, A.O.A.C., 8th ed. Washington, 1955.
Thurston, *Pharmaceutical and Food Analysis*. Van Nostrand, New York, 1922.
Winton, *Structure and Composition of Foods*. Wiley, New York, 1932, 1935, 1937.

CHAPTER XV

NONALCOHOLIC BEVERAGES AND ALLIED PRODUCTS

ALTHOUGH coffee, tea, and other beverages have little food value in themselves, except for that given by added foodstuffs such as milk, cream, and sugar, the flavor, stimulation of appetite, relaxation, and satisfaction that they give and produce during and at the end of meals make them valuable food adjuncts. Cacao products, such as chocolate, are concentrated foods and have actual food value as well as flavoring and stimulating properties.

The exhilarating effects of coffee, tea, cocoa and some carbonated beverages are due to purine derivatives such as caffeine, theobromine, and xanthine which are closely related to uric acid.

COFFEE

The definitions and standards for food products, when the Food and Drug Administration was an arm of the U. S. Department of Agriculture defined coffee as follows:

Coffee is the seed of cultivated varieties of *Coffea arabica, C. liberica,* and *C. robusta.*

(a) Green coffee, raw coffee, unroasted coffee, is coffee freed from all but a small portion of its spermodern, and conforms in variety and in place of production to the name it bears.

(b) Roasted coffee, "coffee" is properly cleaned green coffee which by the action of heat, roasting, has become brown and has developed its characteristic aroma.

The beverage, coffee, is the liquid made from the roasted bean by a number of processes commonly known as percolation, decoction, and infusion.

The most important constituents of raw coffee as can be seen from Table 83 are oil, cellulose, water, and reducing sugar. The process of roasting caramelizes the sugar, reduces moisture content and at the same time develops the characteristic flavors of coffee. Only a small amount of caffeine is lost in the roasting process. Comparative compositions are given in Table 84.

Coffee may be analyzed by methods previously described for moisture, ether extract, protein crude fiber, reducing sugars, ash, etc. Certain types of adulteration may best be detected microscopically, as for example in the case of chicory. However, it is well to note that chicory is one of the substances that contains inulin, which, on hydrolysis, yields levulose. A high levulose content in a coffee would be indicative of the presence of chicory.

TABLE 83. COMPOSITION OF COFFEE [a]

	Raw Coffee		Roasted Coffee	
	Minimum	Maximum	Minimum	Maximum
Water %	8.0	12.0	0.4	4.0
Caffeine %	0.8	1.8	0.8	1.8
Fat %	11.4	14.2	10.5	16.5
Reducing sugar %	5.8	7.8	0.0	1.1
Cellulose %	16.6	42.3	26.3	51.0
Total nitrogen %	1.1	2.2	1.3	2.7
Ash %	3.5	4.0	4.0	5.0

[a] König, *Chemie der menschlichen Nahrungs und Genussmittel.* Springer, Berlin, 1889.

TABLE 84. COMPOSITION OF COFFEE BEFORE AND AFTER ROASTING. [a]

	Raw Coffee	Roasted Coffee
Water %	11.23	1.15
Caffeine %	1.21	1.24
Fat %	12.27	14.48
Sugar %	8.55	0.66
Cellulose %	18.17	10.89
Nitrogenous substances %	12.07	13.98
Other non-nitrogenous matter %	32.58	45.09
Ash %	3.92	4.75

[a] König, *Chemie der menschlichen Nahrungs und Genussmittel.* Springer, Berlin, 1889.

TABLE 85. COMPOSITION OF COFFEE BEFORE AND AFTER ROASTING [b]

	Raw %	Roasted %
Water	10.73	2.16
Sugar	8.62	0.75
Caffeine	1.07	1.20
Crude fiber	24.00	13.03
Ether extract	11.08	13.75 (petroleum ether cold extract)
Aqueous extract	30.35	12.62
Ash	3.00	4.03
Nitrogenous substances	12.64	2.27 (total nitrogen)
Other nitrogen free extractives	19.30	....
Dextrin	0.86	....
Tannic acid	9.02	....

[b] Bailey, *Food Products and Their Sources,* 3rd Ed. Blakiston, Philadelphia, 1928.

CAFFEINE

Power-Chesnut Method.—In this method [1] caffeine is extracted with alcohol, liberated from its combination in the food material by use of magnesia and subsequently isolated by means of chloroform.

Procedure.—Moisten with 95 per cent alcohol 10 g. of the sample, prepared by grinding to 30 mesh. Transfer to a Soxhlet or similar extractor, and extract with 95 per cent-alcohol for 8 hours or as much longer as is necessary to complete the extraction. Transfer the alcohol extract with the aid of hot water to a porcelain dish containing 10 g. of heavy magnesium oxide in suspension, in 100 ml. of water. Evaporate slowly on a steam bath with frequent stirring to a dry, powdery mass. Rub the residue with a pestle into a paste with boiling water and transfer with hot water to a smooth filter, cleaning the dish with a rubber policeman. Collect the filtrate in a liter flask marked at 250 ml. and wash with boiling water until the filtrate reaches the mark. Add 20 ml. sulfuric acid (1:9) and boil gently for 30 minutes with a funnel in the neck of the flask. Cool, filter through a moistened double paper into a separatory funnel, and wash with small portions of sulfuric acid (1:199). Extract with 6 successive 25-ml. portions of chloroform. Wash the combined chloroform extracts in a separatory funnel with 5 ml. of 1 per cent potassium hydroxide solution. Filter the chloroform into an Erlenmeyer flask. Wash the potassium hydroxide solution with two 10-ml. portions of chloroform, adding them to the Erlenmeyer flask. Evaporate or distill on a steam bath to a volume of 10-15 ml., transfer with chloroform to a small tared beaker, evaporate carefully with the aid of air, dry for 30 minutes at 100° C., and weigh. Test the purity of the residue, which is considered to be caffeine, by determining nitrogen by the usual method and multiply by the factor 3.464.

Fendler-Stüber Rapid Method.—Transfer 10 g. of the sample ground to pass through a 30-mesh sieve to a glass-stoppered bottle. Add 10 ml. of ammonium hydroxide (1:2) and 200 g. of chloroform. Place in a shaking machine for 30 minutes. Chill in a refrigerator. Pour the entire contents of the bottle on a 24-cm. fluted filter, covering immediately with a watch glass, catching the filtrate in a tared flask resting in an ice bath. Stopper the flask as soon as the solvent ceases to run in a continuous stream and weigh. Evaporate on a steam bath with the aid of a current of air. Digest the residue with 80 ml. of hot water for 10 minutes on the steam bath, shaking frequently, and allow to cool. Treat the solution with 1 per cent potas-

[1] Power and Chesnut, *J. Am. Chem. Soc.* 41, 1298 (1919).
[2] Fendler and Stüber, *Z. Nahr. Genussm.* 28, 9 (1914).

sium permanganate solution using 20 ml. for roasted coffee and 10 ml. for green coffee and let stand for 15 minutes at room temperature, shaking occasionally. Add 2 ml. of a solution consisting of 100 ml. of 3 per cent hydrogen peroxide, acetanilid-free, and 1 ml. of glacial acetic acid. If the mixture is still red or reddish, add the hydrogen peroxide solution, 1 ml. at a time until the excess potassium permanganate is destroyed. Place the flask on a steam bath for 15 minutes and add 0.5-ml. portions of the hydrogen peroxide until the mixture ceases to become lighter. Cool, and filter by suction through a Gooch crucible, washing with cold water. Transfer the filtrate to a separatory funnel and extract 6 times with 25-ml. portions of chloroform. Evaporate the combined chloroform extracts to a small volume, transfer to a small tared beaker, finish the evaporation, dry at 100° C., generally 30 minutes is sufficient, and weigh the residue as caffeine. The weight of caffeine, multiplied by 2000, divided by the weight of the chloroform aliquot obtained from the first filtration, equals the percentage of caffeine in the 10 g. sample.

Spectrophotometric Method.—The characteristic absorption of caffeine at 272 mμ is utilized by Ishler, Finucane and Borker[3] to measure quantitatively its presence in coffees and crude caffeine. Interfering impurities found in these samples are removed by treatment with heavy magnesium oxide and zinc ferrocyanide, plus in some cases permanganate oxidation. Rapidity and specificity for caffeine are outstanding characteristics of the method.

Reagent.—Buffered Zinc Acetate Solution, 1.0 molar.—Dissolve 438.0 g. of zinc acetate dihydrate in water, add 60 ml. of glacial acetic acid, and make to 2000 ml.

Crude Caffeine and Green Coffee.—Preparation of Sample.— Crude Caffeine.—Weigh 0.1 g. of sample, transfer to a 500-ml. volumetric flask, add 400 ml. of hot water, and shake thoroughly. Cool the flask to room temperature and make to the mark with water. Pipette a convenient aliquot containing approximately 2 mg. of caffeine to a 100-ml. volumetric flask. Add water to a total volume of approximately 50 ml.

Green Coffee.—Weigh 2.0 g. of sample and transfer to a tared 1-liter Erlenmeyer flask. Add 50 ml.,of 0.1 N sulfuric acid and 450 ml. of water, heat to boiling, and boil 30 minutes. Cool flask to room temperature, make to weight (tare plus 502.0 g.), and filter. Pipette a 50-ml. aliquot into a 100-ml. volumetric flask.

Ferrocyanide and Magnesium Oxide Clarification.—To the volumetric flask add 7.0 ml. of zinc acetate solution and swirl vigorously.

---

[3] Ishler, Finucane, and Borker, *Anal. Chem.* **20**, 1162 (1948).

Dropwise add 6.0 ml. of 0.25 M potassium ferrocyanide solution with constant swirling of the flask. (An excess of zinc acetate must be present to prevent interference by the ferrocyanide.) Make to the mark with water. Mix thoroughly, filter through a No. 41 (15-cm.) Whatman filter paper, and discard first 10-ml. portion. Pipette a 50-ml. portion into a tared 250-ml. Erlenmeyer flask containing 5.0 g. of heavy magnesium oxide, add 50 ml. of water, and boil for 20 minutes. Cool to room temperature and make to weight (tare plus 105 g.). Filter and discard first portion (10 ml.). Read a portion of the filtrate in the spectrophotometer at 272 mμ. Density reading at 272 mμ divided by the factor previously determined equals milligrams of caffeine per 100 ml. of filtrate.

Green, Roasted, and Soluble Coffee.—Preparation of Sample and Magnesium Oxide Clarification.—Green and Roasted Coffee.—Weigh 1.0 g. of roasted coffee or 1.2 g. of green coffee and transfer to a tared 1-liter Erlenmeyer flask containing 50 ml. of 0.1 N sulfuric acid. Add 250 ml. of water and boil for 20 minutes. Then add 50 g. of heavy magnesium oxide and continue boiling for 20 minutes. Cool and make to weight (tare plus 350 g. plus weight of sample). Filter, pipette a 25-ml. aliquot into a 100-ml. volumetric flask, and add 25 ml. of water.

Soluble Coffee.—Take a sample weight containing approximately 100 mg. of caffeine and transfer to a 500-ml. volumetric flask. Make to volume, mix thoroughly, and pipette a 25-ml. aliquot into a tared Erlenmeyer flask containing 25 ml. of 0.1 N sulfuric acid. Add 200 ml. of water and boil for 20 minutes. Then add 25 g. of heavy magnesium oxide and continue boiling for 20 minutes. Cool and make to weight (tare plus 275 g.). Filter and pipette a 50-ml. aliquot into a 100-ml. volumetric flask.

Permanganate and Ferrocyanide Clarification.—In the volumetric flask place 10 ml. of 1 per cent potassium permanganate solution. After 10 minutes, add 3.0 ml. of 5 per cent sodium sulfite solution, then add 0.5 ml. of glacial acetic acid, and titrate with sulfite solution to the disappearance of the manganese dioxide precipitate. Add 7.0 ml. of zinc acetate solution and swirl vigorously. Dropwise add 6.0 ml. of 0.25 M potassium ferrocyanide solution with constant swirling of the flask. Make to volume with water. Mix thoroughly and filter through No. 42 Whatman filter paper, discarding the first portion. Read a portion of the filtrate in the spectrophotometer. From the density reading, calculate the per cent caffeine in the sample.

Ishler, Finucane, and Borker found that the factor that fit best for their data and instrument was 0.510 density unit per milligram of caffeine per 100 ml. of solution. Each analyst will need to establish his own factor for density in terms of caffeine concentration to account for differences in the instrument and in the cells.

CHICORY

Boil 10 g. of the ground coffee sample with 250 ml. of water. Strain and add an excess of basic lead acetate to the mixture. Allow the precipitate to settle. A colorless supernatant indicates the absence of chicory. A colored supernatant indicates the presence of chicory.

TEA

Tea is prepared and cured by recognized methods of manufacture from the tender leaves, leaf buds, and tender internodes of different varieties of *Thea sinensis L.* It conforms in variety and place of production to the name it bears; contains not less than 4 per cent nor more than 7 per cent of ash; and meets the provisions of the act of Congress approved March 2, 1897, as amended, regulating the importation and inspection of tea. The beverage is a water infusion and is also called tea.

Tea may be analyzed by methods similar to those that have been described in other sections of the book. Thus moisture, ash, petroleum ether extract, inorganic materials, protein, crude fiber, and volatile oil may be estimated by the methods detailed under those headings. Facing attributable to minerals may be detected in the ash. Caffeine may be estimated by the Power-Chesnut method or by the one subsequently detailed. Tables 86, 87, and 88 give the composition of tea.

WATER EXTRACT

To 2 g. of the ground sample in a 500-ml. flask, add 200 ml. of water and boil over a low flame for 1 hour, rotating occasionally. Close the flask with a rubber stopper through which passes a glass tube 30 in. long for a condenser. Boil very slowly so that no steam escapes from the top of the air condenser. Cool, dilute to volume, mix thoroughly, and filter through a dry filter paper. Transfer an aliquot of 50 ml. to a weighed dish and evaporate to dryness on a steam bath. Place in the oven, heat at 100° C. for 1 hour, cool, and weigh.

CAFFEINE

Bailey-Andrew[4] **Method.**—To 5 g. of the sample, prepared by

---

[4] Andrew, *J. Assoc. Official Agr. Chem.* **6**, 107 (1922).

TABLE 86. ANALYSES OF TEA [a]

Type	Moisture %		Total Ash %	Water Soluble Ash %	Water Insoluble Ash %	Hot Water Extract %
Green tea	max.		8.39	4.01	5.59	40.08
(26)	min.		6.13	2.80	2.25	30.82
	ave.	5.18	7.16	3.61	3.54	37.41
Black tea	max.		7.42	4.75	3.82	44.92
(53)	min.		5.57	2.87	1.94	28.48
	ave.	5.94	6.27	3.58	2.69	36.23
Mixed tea						
(10)	ave.	6.16	6.72	3.72	3.00	34.35

[a] Winton, Ogden, and Mitchell, *Conn. Agr. Exp. Sta. Rept.* (1898). Number of samples in parentheses.

TABLE 87. COMPOSITION OF GREEN TEA [a]

	Minimum %	Maximum %
Moisture	6.42	9.83
Extract	45.4	56.6
Total ash	5.5	10.85
Water soluble ash	2.3	4.3
Water insoluble ash	1.8	7.75
Ash insoluble in HCl	0.3	3.73
Silica	0.1	2.7
Tannin	3.98	13.35
Caffeine	1.77	3.83

[a] McLachlan and Stern, *Analyst* **59**, 385 (1934).

TABLE 88. COMPOSITION OF TEA [a]

	Maximum %	Minimum %	Average %
Water	16.20	3.93	9.51
Protein	36.61	15.91	24.50
Caffeine	4.70	1.00	3.58
Tannin	26.13	8.16	15.65
Ash	8.37	3.82	5.65
Essential oil	0.89	0.54	0.68

[a] König, *Chemie der menschlichen Nahrungs und Genussmittel*, p. 1011, 3rd ed. Springer, Berlin, 1889.

grinding so that it will pass through a 30-mesh sieve, in a 500-ml. volumetric flask, add 10 g. of heavy magnesium oxide and 200 ml. of water. Boil gently over a low flame for 2 hours, using a small bore glass tube 30 in. long as a condenser. Cool, dilute to volume, and filter through a dry paper. Transfer an aliquot portion of 300 ml., equivalent to 3 g. of original material, to a flask of 1-liter capacity; add 10 ml. of sulfuric acid (1:9); and boil until the volume is reduced to about 100 ml. Filter into a separatory funnel, washing the flask with small portions of sulfuric acid (1:99), and shake 6 times with chloroform, using 25, 20, 15, 10, 10, 10-ml. portions. Treat the combined extracts with 5 ml. of a 1 per cent solution of potassium hydroxide and, when the liquids have completely separated, draw off the chloroform layer into a suitable flask or beaker. Wash the alkaline solution in the separatory funnel with 2 portions of chloroform of 10 ml. each, and unite the washings with the main bulk of extract. Evaporate or distill off the chloroform to a small volume, transfer to a small weighed beaker with the aid of chloroform, evaporate to dryness, and complete drying in an oven at 100° C. Test the purity of the residue by determining nitrogen and multiplying by the factor 3.464. This gives the value for anhydrous caffeine.

TANNIN

Lowenthal-Procter Method.—Tannin may be estimated by determining its oxidizability by potassium permanganate solutions.[5,6]

Reagents.—(a) Oxalic Acid Solution.—Prepare a 0.1 N solution of oxalic acid. One ml. if equivalent to 0.00416 g. of tannin.

(b) Standard Potassium Permanganate Solution.—Dissolve 1.333 g. of potassium permanganate in 1 liter of water and standardize against (a) above.

(c) Indigo Solution.—Dissolve 6 g. of sodium indigotindisulfonate in 500 ml. of water by heating; cool, add 50 ml. of sulfuric acid, make up to 1 liter, and filter.

(d) Gelatin Solution.—Soak 25 g. of gelatin for an hour in saturated sodium chloride solution, heat until the gelatin is dissolved, cool, and dilute with a saturated sodium chloride solution to 1 liter.

(e) Acid Sodium Chloride Solution.—Acidify 975 ml. of saturated sodium chloride solution with 25 ml. of sulfuric acid.

Procedure.—Boil 5 g. of the tea for 30 minutes with 400 ml. of water, cool, transfer to a 500-ml. volumetric flask, and dilute to the mark. To 10 ml. of the infusion, filtered, if not clear, add 25 ml. of the indigo carmine solution and about 750 ml. of water. Add the

[5] Lowenthal, *Z. anal. Chem.* **16**, 33 (1877).
[6] Procter, *Chem. News* **37**, 256 (1878).

potassium permanganate solution from a burette, a little at a time while stirring, until the color becomes light green, then dropwise until the color changes to bright yellow or to a faint pink at the rim. Designate the number of milliliters of potassium permanganate used as a. Mix 100 ml. of the clear infusion of tea with 50 ml. of the gelatin solution, 100 ml. of the acid sodium chloride solution, and 10 g. of powdered kaolin, and shake several minutes in a stoppered bottle. After allowing the mixture to settle, decant through a filter. Mix 25 ml. of the filtrate with 25 ml. of the indigo carmine solution and about 750 ml. of water and titrate with potassium permanganate solution as before. The number of milliliters of potassium permanganate solution used subtracted from that obtained above, a, gives the quantity of potassium permanganate solution required to oxidize the tannin. One ml. of 0.1 N oxalic acid equals 0.00416 g. of tannin (gallotannic acid).[7]

ASH

Total ash, water insoluble ash, and ash insoluble in hydrochloric acid may be estimated as detailed in Chapter I.

Interpretation.—In many teas the ash is between 5 and 6 per cent. In faced teas the proportion of ash may be as high or higher than 10 per cent.[8] In "lie-tea," the proportion may reach 30 per cent. In spent tea, however, the ash drops below 3 per cent and the ash, itself, is high in calcium. In genuine teas the water insoluble ash seldom exceeds 3 per cent. In genuine tea about 50 per cent of the ash is water soluble ash, that is, it is in the range of 3-3.5 per cent, whereas in exhausted teas it drops to about 0.5 per cent.

The following formula [8] may be used to calculate the percentage of spent tea:

$$E = (6-2S)$$

in which

E = spent tea
S = per cent of water soluble ash.

In genuine tea the ash insoluble in dilute hydrochloric acid ranges from 0.3 to 0.8 per cent. On the other hand, in faced tea the proportion may reach 2 to 5 per cent.

CACAO PRODUCTS

The analysis of cacao or cocoa products follows methods described in the foregoing text. Thus moisture, ash, soluble and insolu-

---

[7] Mitchell, *Analyst* **61**, 295 (1936).

[8] Fienc and Blumenthal, *Handbook of Food Manufacture*. Chemical Publishing, New York, 1938.

ble ash, alkalinity of ash, and protein content may be estimated as previously detailed. Crude fiber in cocoa products not containing milk products is determined on the dried residue after exhaustively washing by centrifuging and decanting successive portions of ether, water and then alcohol, and water. Crude fiber in cocoa products containing milk products is determined on the dried residue after exhaustively washing by centrifuging and decanting successive portions of ether, 1 per cent sodium oxalate solution, in which solution the sample is allowed to steep for 30 minutes, water, and then alcohol, and water. The fat in cocoa products may be extracted by use of ether and recovered by evaporation of the ether. The fat may then be subjected to the examination detailed in Chapter IX. The qualitative presence of lactose may be established by defatting the cocoa product and then proceding along lines given in Chapter VIII. Quantitatively the sucrose and lactose content are ascertained by a combination of the polariscopic and copper reducing methods outlined in Chapter X, with special formulas and corrections for the sucrose content in the lactose determinations.

The Federal Government definitions for these products follow and the composition of some cocoa products are given in Table 89.

TABLE 89. ANALYSES OF COCOA PRODUCTS [a]

	Plain Chocolate	Sweet Chocolate	Cocoa	Cocoa Shells
Moisture %	3.78	2.17	6.23	4.87
Ash %	3.15	1.40	5.49	10.43
Theobromine %	0.78	0.35	1.15	0.49
Caffeine %	0.13	0.08	0.16	0.16
Protein %	12.36	4.58	18.34	14.46
Crude fiber %	2.86	0.95	4.48	16.55
Sugar %	0.00	56.44	0.00	0.00
Starch %	8.11	2.88	11.14	4.13
Other nitrogen-free material %	16.64	7.64	26.32	46.15
Fat %	52.19	23.51	26.69	2.76

[a] Winton, Bailey, and Silverman, *Conn. Agr. Exp. Sta. Ann. Rept.* (1903).

DEFINITIONS OF CACAO PRODUCTS

The following definitions [9] are given by the Food and Drug Administration:

Cacao Nibs, Cocoa Nibs, Cracked Cocoa.—Identity.—(a) Cacao nibs, cocoa nibs, cracked cocoa is the food prepared by heating and

[9] *U. S. Food Drug Admin., S.R.A., F.D.&C.* 2, Part 14, August, 1952, as amended, August, 1956.

cracking dried or cured and cleaned cacao beans and removing shell therefrom. Cacao nibs or the cacao beans from which they are prepared may be processed by heating with one or more of the following optional alkali ingredients, added as such or in aqueous solution: Bicarbonate, carbonate, or hydroxide of sodium, ammonium, or potassium; or carbonate or oxide of magnesium; but for each 100 parts by weight of cacao nibs used, as such or before shelling from the cacao beans, the total quantity of such alkalis used is not greater in neutralizing value (calculated from the respective combining weights of such alkalis used) than the neutralizing value of 3 parts by weight of anhydrous potassium carbonate. The cacao shell content of cacao nibs is not more than 1.75 per cent by weight (calculated to an alkali-free basis if they or the cacao beans from which they were prepared have been processed with alkali), as determined by the method prescribed under "Shell in Cacao Nibs—Tentative" beginning on p. 223 of "Official and Tentative Methods of Analysis of the Association of Official Agricultural Chemists," 6th Ed., 1945.

Chocolate Liquor, Chocolate, Baking Chocolate, Bitter Chocolate, Cooking Chocolate, Chocolate Coating, Bitter Chocolate Coating.— Identity.—(a) Chocolate liquor, chocolate, baking chocolate, bitter chocolate, cooking chocolate, chocolate coating, bitter chocolate coating is the solid or semiplastic food prepared by finely grinding cacao nibs. To such ground cacao nibs, cacao fat or a cocoa or both may be added in quantities needed to adjust the cacao fat content of the finished chocolate liquor. (For the purposes of this definition the term "cocoa" means breakfast cocoa, cocoa, low-fat cocoa, or any mixture of two or more of these.) Chocolate liquor may be spiced, flavored, or otherwise seasoned with one or more of the following optional ingredients, other than any such ingredient or combination of ingredients specified in (1), (2), or (3) below, which imparts a flavor that imitates the flavor of chocolate, milk, or butter:

(1) Ground spice.

(2) Ground vanilla beans; any natural food flavoring oil, oleoresin, or extract.

(3) Vanillin, bourbonal (ethylvanillin), or other artificial food flavoring.

(4) Butter, milk fat, dried malted cereal extract, ground coffee, ground nut meats.

(5) Salt.

Any optional ingredient used with the cacao beans or cacao nibs from which such chocolate liquor is prepared, or used with any cocoa added in preparing such chocolate liquor, shall be considered to be an optional ingredient used with such chocolate liquor. The optional alkali ingredients specified for use with cacao nibs (a) may be used as optional ingredients with chocolate liquor; but for each 100 parts by weight of cacao nibs used in preparing the chocolate liquor, the total quantity of such alkalis used is not greater in neutralizing value (calculated from the respective combining weights of such alkalis used) than 3 parts by weight of anhydrous potassium carbonate. The finished chocolate liquor contains not less than 50 per cent and not more than 58 per cent by weight of cacao fat. Unless the chocolate liquor is seasoned with butter, milk fat, or ground nut meats, the percentage of cacao fat is determined by a prescribed method.

Breakfast Cocoa, High Fat Cocoa.—Identity.—(a) Breakfast cocoa, high fat cocoa is the food prepared by pulverizing the residual material remaining after part of the cacao fat has been removed from ground cacao nibs. It may be spiced, flavored, or otherwise seasoned with one or more of the following optional ingredients, other than any such ingredient or combination of ingredients which imparts a flavor that imitates the flavor of chocolate, milk, or butter:

(1) Ground spice.

(2) Ground vanilla beans; any natural food flavoring oil, oleoresin, or extract.

(3) Vanillin, bourbonal (ethylvanillin), or other artificial food flavoring.

(4) Salt.

Any optional ingredient used with the cacao beans, cacao nibs, or ground cacao nibs from which such breakfast cocoa is prepared shall be considered to be an optional ingredient used with such breakfast cocoa. The optional alkali ingredients specified for use with cacao nibs (a) may be used as optional ingredients with breakfast cocoa; but for each 100 parts by weight of cacao nibs used in preparing the breakfast cocoa, the total quantity of such alkalis used is not greater in neutralizing value (calculated from the respective combining weights of such alkalis used) than 3 parts by weight of anhydrous potassium carbonate. The finished breakfast cocoa contains not less than 22 per cent of cacao fat.

Cocoa, Medium Fat Cocoa.—Identity.—Cocoa, medium fat cocoa conforms to the definition and standard of identity prescribed for breakfast cocoa, except that it contains less than 22 per cent but not less than 10 per cent of cacao fat.

Low-fat Cocoa.—Identity.—Low-fat cocoa conforms to the definition and standard of identity prescribed for breakfast cocoa, except that it contains less than 10 per cent of cacao fat.

Sweet Chocolate, Sweet Chocolate Coating.—Identity.—(a) Sweet chocolate, sweet chocolate coating is the solid or semiplastic food the ingredients of which are intimately mixed and ground, prepared from chocolate liquor (with or without the addition of cacao fat) sweetened with one of the optional saccharine ingredients specified in (b) below. It may be spiced, flavored, or otherwise seasoned with one or more of the following optional ingredients, other than any such ingredient or combination of ingredients which imparts a flavor that imitates the flavor of chocolate, milk, or butter:

(1) Ground spice, ground vanilla beans, any natural food flavoring oil or oleoresin or extract, ground coffee, ground nut meats, honey, molasses, brown sugar, maple sugar, dried malted cereal extract, salt.

(2) Vanillin, bourbonal (ethylvanillin), or other artificial food flavoring.

One or a mixture of both of the following optional emulsifying ingredients may be added in a total quantity not more than 0.5 per cent of the weight of the finished food (such ingredient or mixture may be added in combination with a vegetable food fat carrier, such combination containing not less than 60 per cent by weight of the emulsifying ingredient or mixture):

(3) Lecithin, with or without related natural phosphatides.

(4) Monoglycerides and diglycerides of fat-forming fatty acids in combination with monosodium phosphate derivatives thereof.

One or any mixture of two or more of the following optional dairy ingredients may be used in such quantity that the finished sweet chocolate contains less than 12 per cent by weight of milk constituent solids:

(5) Butter, milk fat, cream, milk, concentrated milk, evaporated milk, sweetened condensed milk, dried milk, skim milk, concentrated skim milk, evaporated skim milk, sweetened condensed skim milk, nonfat dry milk solids, concentrated buttermilk, dried buttermilk, malted milk.

If chocolate liquor with any optional ingredient specified in that definition is used, such ingredient shall be considered to be an optional ingredient used with the sweet chocolate. The finished sweet chocolate contains not less than 15 per cent by weight of chocolate liquor, calculated by subtracting from the weight of chocolate liquor used the weight of cacao fat therein and the weights therein of alkali and seasoning ingredients, if any, multiplying the remainder by 2.2, dividing the result by the weight of the finished sweet chocolate, and multiplying the quotient by 100. Bittersweet chocolate is sweet chocolate which contains not less than 35 per cent by weight of chocolate liquor, calculated in the same manner.

(b) The optional saccharine ingredients referred to in paragraph (a) of this section are:

(1) Sugar, or partly refined cane sugar, or both.

(2) Any mixture of dextrose and sugar or partly refined cane sugar or both in which the weight of the solids of the dextrose used is not more than one-third of the total weight of the solids of all the saccharine ingredients used.

(3) Any mixture of dried corn or glucose syrup and sugar or partly refined cane sugar or both in which the weight of the solids of the dried corn or glucose syrup used is not more than one-fourth of the total weight of the solids of all the saccharine ingredients used.

(4) Any mixture of dextrose, dried corn or glucose syrup and sugar or partly refined cane sugar or both, in which three times the weight of the solids of the dextrose used plus four times the weight of the solids of the dried corn or glucose syrup used is not more than the total weight of the solids of all the saccharine ingredients used.

(c) For the purpose of this section:

(1) The term "dextrose" means the anhydrous refined monosaccharide obtained from hydrolyzed starch.

(2) The term "dried corn syrup" means the product obtained by drying incompletely hydrolyzed cornstarch; its solids contain not less than 58 per cent by weight of reducing sugars. The term "dried glucose syrup" is defined on page 512.

(d) "Semisweet chocolate," "bittersweet chocolate," "semisweet chocolate coating," and "bittersweet chocolate coating" are alternate names for sweet chocolate that contains not less than the minimum quantity of chocolate liquor prescribed for bittersweet chocolate by (a) of this section.

Milk Chocolate, Sweet Milk Chocolate, Milk Chocolate Coating, Sweet Milk Chocolate Coating.—Identity.—(a) Milk chocolate, sweet milk chocolate, milk chocolate coating, sweet milk chocolate coating is the solid or semiplastic food the ingredients of which are intimately mixed and ground, prepared from chocolate liquor (with or without the addition of cacao fat) and one or more of the optional dairy ingredients specified in (b) of this section, sweetened with one of the optional saccharine ingredients specified above. It may be spiced, flavored, or otherwise seasoned with one or more of the following optional ingredients, other than any such ingredient or combination of ingredients which imparts a flavor that imitates the flavor of chocolate, milk, or butter:

(1) Ground spice, ground vanilla beans, any natural food flavoring oil or oleoresin or extract, ground coffee, ground nut meats, honey, molasses, brown sugar, maple sugar, dried malted cereal extract, salt.

(2) Vanillin, bourbonal (ethylvanillin), or other artificial food flavoring.

One or a mixture of both of the following optional emulsifying ingredients may be added in a total quantity not more than 0.5 per cent of the weight of the finished food (such ingredient or mixture may be added in combination with a vegetable food fat carrier, such combination containing not less than 60 per cent by weight of the emulsifying ingredient or mixture):

(3) Lecithin, with or without related natural phosphatides.

(4) Monoglycerides and diglycerides of fat-forming fatty acids in combination with monosodium phosphate derivatives thereof.

If chocolate liquor with any optional ingredient specified in that definition is used, such ingredient shall be considered to be an optional ingredient used with the milk chocolate. The finished milk chocolate contains not less than 3.66 per cent by weight of milk fat, not less than 12 per cent by weight of milk solids, and not less than 10 per cent by weight of chocolate liquor as calculated by subtracting from the weight of chocolate liquor used the weight of cacao fat therein and the weights therein of alkali and seasoning ingredients, if any, multiplying the remainder by 2.2, dividing the result by the weight of the finished milk chocolate, and multiplying the quotient by 100.

(b) The optional dairy ingredients referred to in (a) of this section are milk, concentrated milk, evaporated milk, sweetened condensed milk, dried milk, butter, milk fat, cream, skim milk, concentrated skim milk, evaporated skim milk, sweetened condensed skim milk, and nonfat dry milk solids; but in any such ingredient or combination of two or more of such ingredients used, the weight of nonfat milk solids is not more than 2.43 times and not less than 1.20 times the weight of milk fat therein.

Skim Milk Chocolate, Sweet Skim Milk Chocolate, Skim Milk Chocolate Coating, Sweet Skim Milk Chocolate Coating.—Identity.— Skim milk chocolate, sweet skim milk chocolate, skim milk chocolate coating, sweet skim milk chocolate coating conforms to the definition and standard of identity, and is subject to the requirements for label statement of optional ingredients, prescribed for milk chocolate, except that:

(1) The dairy ingredients used are limited to skim milk, concentrated skim milk, evaporated skim milk, sweetened condensed skim milk, nonfat dry milk solids, and any combination of two or more of these.

(2) The finished skim milk chocolate contains less than 3.66 per cent by weight of milk fat and, instead of milk solids, it contains not less than 12 per cent by weight of skim milk solids.

Buttermilk Chocolate, Buttermilk Chocolate Coating.—Identity.— Buttermilk chocolate, buttermilk chocolate coating conforms to the definition and standard of identity prescribed for milk chocolate, except that:

(1) The dairy ingredients used are limited to sweet cream buttermilk, concentrated sweet cream buttermilk, dried sweet cream buttermilk, or any combination of two or all of these.

(2) The finished buttermilk chocolate contains less than 3.66 per cent by weight of milk fat and, instead of milk solids, it contains not less than 12 per cent by weight of sweet cream buttermilk solids.

Mixed Dairy Product Chocolates, Mixed Dairy Product Chocolate Coatings.—Identity.— (a) The articles for which definitions and standards of identity are prescribed by this section are the foods each of which conforms to the definition and standard of identity prescribed for milk chocolate, except that:

(1) The dairy ingredient used in each such article is a mixture of two or more of the following four components:

(i) Any dairy ingredient or combination of such ingredients specified in milk chocolate (b) which is within the limits of the ratios specified therein for nonfat milk solids to milk fat.

(ii) One or more of the five skim milk ingredients specified in skim milk chocolate definition.

(iii) One or more of the three sweet cream buttermilk ingredients specified in the buttermilk chocolate definition.

(iv) Malted milk..

(2) Each of the finished articles may contain less than 3.66 per cent by weight of milk fat and, instead of milk solids, it contains not less than 12 per cent by weight of milk solids of the ingredients used. The quantity of each ingredient used in any such mixture is such that no ingredient contributes less than one-third of the weight of milk component solids contributed by that ingredient used in largest proportion. When any such mixture is of ingredient (i) and (ii) of (1), the quantity of nonfat milk solids in such mixture is more than 2.43 times the quantity of milk fat therein.

Sweet Chocolate and Vegetable Fat (Other Than Cacao Fat) Coating.—Identity.— (a) Sweet chocolate and vegetable fat (other than cacao fat) coating conforms to the definition and standard of identity prescribed for sweet chocolate, except that:

(1) In its preparation is added one or any combination of two or more vegetable food oils or vegetable food fats, other than cacao fat, which oil, fat, or combination may be hydrogenated and which has a melting point lower than that of cacao fat.

(2) The requirement of the sweet chocolate definition that the milk constituent solids be less than 12 per cent by weight does not apply.

(b) The provisions of this definition shall not be construed as applicable to any article by reason of the addition thereto of a vegetable food fat other than cacao fat as a carrier of emulsifying·ingredients, as authorized and within the limit prescribed by the sweet chocolate definition.

Sweet Cocoa and Vegetable Fat (Other Than Cacao Fat) Coating.—Identity.—Sweet cocoa and vegetable fat (other than cacao fat) coating conforms to the definition and standard of identity prescribed for sweet chocolate, except that:

(1) In its preparation cocoa is used, instead of chocolate liquor, in such quantity that the finished food contains not less than 6.8 per cent by weight of the nonfat cacao portion of such cocoa, calculated by subtracting from the weight of cocoa used the weight of cacao fat therein and the weight therein of alkali and seasoning ingredients, if any, dividing the remainder by the weight of the finished food, and multiplying the quotient by 100. (For the purposes of this section the term "cocoa" means breakfast cocoa, cocoa, low-fat cocoa, or any mixture of two or more of these.)

(2) In its preparation is added one or any combination of two or more vegetable food oils, vegetable food fats, or vegetable food stearins, other than cacao fat, which oil, fat, stearin, or combination has a melting point higher than that of cacao fat. Any such oil or fat may be hydrogenated.

(3) The requirement of the sweet chocolate definition (a) that the milk constituent solids be less than 12 per cent by weight does not apply.

FAT

Weigh accurately about 1 g. of chocolate, ground to pass a 30-mesh sieve, or cocoa powder into the bottom section of a Jacobs-Singer separatory flask. Add 10 ml. of water and stir by shaking into a paste. Add 2 ml. of ammonia and again shake vigorously. Boil gently for 5 minutes or else immerse in a boiling water bath for 5 minutes.[10] Cool, stopper with the upper section of the flask. Add 11 ml. of alcohol and proceed with the Roese-Gottlieb method as detailed in Chapter VII, making certain to add additional alcohol before each succeeding extraction with the mixed ethers.

The chocolate may be weighed into a 100-ml. tall form beaker, processed as detailed above and then transferred to a Mojonnier flask

---

10 Wiseman, *Analyst* 55, 685 (1930).

or similar apparatus as described in the section "cheese," Chapter VIII.

Refractive Index Method for Total Fat in Cacao Products [11]

Tricresyl phosphate, diethyl phthalate, and dibuty phthalate may be used satisfactorily as solvents for a rapid refractometric method for determining total fat in cacao products. The tricresyl phosphate is used for the wide range refractometer, that is, an Abbé refractometer, and the phthalates for the narrow range butyro-refractometer. When equal weights of sample and solvent are taken and the refractive index of the filtrate obtained, the percentage of fat may be read directly from Table 90.

Table 90. Refractometric Determination of Fat in Chocolate

Cocoa Butter per 10 g. of Solvent	Refraction		
	35°C.	40°C.	45 C.
0.000	1.55039	1.54841	1.54659
0.250	1.54772	1.54563	1.54380
0.500	1.54498	1.54283	1.54115
0.750	1.54243	1.54031	1.53823
1.000	1.53892	1.53774	1.53593
1.250	1.53765	1.53533	1.53343
1.500	1.53547	1.53336	1.53167
1.750	1.53348	1.53150	1.52940
2.000	1.53146	1.52950	1.52700
2.250	1.52943	1.52738	1.52548
2.500	1.52774	1.52560	1.52400
2.750	1.52613	1.52445	1.52213
3.000	1.52434	1.52223	1.52049
3.250	1.52283	1.52076	1.51900
3.500	1.52148	1.51953	1.51778
3.750	1.52026	1.51814	1.51649
4.000	1.51856	1.51695	1.51524
4.250	1.51740	1.51540	1.51380
4.500	1.51649	1.51444	1.51253
4.750	1.51515	1.51307	1.51130
5.000	1.51358	1.51177	1.50997
5.250	1.51261	1.51081	1.50921
5.500	1.51185	1.50960	1.50819
5.750	1.51049	1.50906	1.50723
6.000	1.50967	1.50775	1.50619
6.500	1.50825	1.50615	1.50448
7.500	1.50495	1.50309	1.50110
8.500	1.50196	1.50014	1.49828
9.500	1.49950	1.49772	1.49570
10.000	1.49804	1.49627	1.49450

---

[11] Stanley, *Ind. Eng. Chem., Anal. Ed.* 9, 132 (1937).

Procedure.—Weigh 20 g. of sample accurately to 0.05 g. and add 20 g. of tricresyl phosphate, also accurately weighed to 0.05 g. and melt in an oven at 100° C. Mix thoroughly. Filter through a 15 cm. No. 1 Whatman fluted filter, by placing the paper in a 100-ml. beaker without using a funnel. Preferably centrifuge in order to obtain the ½ ml. Let the filtration take place in the oven. When about ½ ml. has filtered, read the refractive index between 35° C. and 45° C., noting the exact temperature. Obtain the percentage fat from Table 90.

The concentration of cocoa butter is expressed as grams of fat per 10.000 g. of solvent.

The zero point for the solvent is obtained by making a blank test without the fat, and the difference between this figure and that given in the Table 90 is the correction to be used. It seldom exceeds a few units in the fourth decimal place.

In the case of samples running over 60 per cent fat, it is advisable to double the amount of solvent, and then to double the per cent read in the table. It is preferable to use tricresyl phosphate as the solvent and the wide range refractometer, in order to obtain more accurate results, because of the greater difference in index of refraction between this solvent and cocoa butter.

Butterfat and lecithin do not interfere, but coconut oil does.

FAT IN MILK CHOCOLATE

A straight ether extraction of milk chocolate as in the A. O. A. C. methods leaves unextracted a small amount of fat which is principally milk fat. By use of dioxane [12] as an additional solvent, this discrepancy can be overcome.

Procedure.—Place 40 g. of milk chocolate in a centrifuge bottle, add 100 ml. of ether, and shake vigorously until the chocolate is disintegrated. Centrifuge and decant or syphon off the ether layer. Make 2 more extractions with 100 ml. of ether. Combine the extracts and evaporate off most of the ether. Add to the residue in the centrifuge bottle 25 ml. of 1,4-dioxane, and heat for 20 minutes in a water bath at approximately 90° C. with frequent stirring. Cool, add 100 ml. of ether, stopper, shake thoroughly, centrifuge, and decant. Make 2 more extractions with 50 ml. of ether, combine the extracts, evaporate the ether, and remove the dioxane completely by means of a current of air on the steam bath. Dissolve the fat in ether, filter into the flask containing the preliminary ether extract, evaporate the ether, dry at 100° C., and weigh. If any insoluble

---
[12] Ferris, *J. Assoc. Official Agr. Chem.* 31, 728 (1948).

material appears in the dry fat, filter in a hot oven before weighing out the 5-g. sample for the Reichert-Meissl number, detailed in Chapter IX.

MILK PROTEINS

Casein in contradistinction to other proteins is soluble in sodium oxalate solution. Upon this distinction, the following method is based.

Procedure.—Weigh exactly 10 g. of the finely grated chocolate into a suitable 8-oz. centrifuge bottle. Add two 100-ml. portions of ether, centrifuge, and decant the supernatant liquor after each addition. Dry the residue in an oven at about 100° C., and powder the residue in the bottle with a flattened glass rod. Add 200 ml. of 3 per cent sodium oxalate solution and let stand 4 hours, shaking frequently. Centrifuge and filter through a small folded filter. Discard the first 5-10 ml. of the filtrate and determine nitrogen in 50 ml. of this filtrate. Pipette 100 ml. of the filtrate into a 200-ml. volumetric flask and dilute almost to the mark with water. Precipitate the proteins by the addition of 2 ml. of glacial acetic acid. Make to volume, shake, filter, and determine nitrogen in 100 ml. of the filtrate. The difference between the two nitrogen figures obtained is the nitrogen of the casein contained in 2.5 g. of the sample. This figure $\times$ 4 $\times$ 6.38 = the total casein contained in the 10 g. taken for the analysis. Casein $\times$ 1.25 = total milk protein.

ESTIMATION OF BUTTERFAT AND MILK PRODUCT CONTENT

The quantity of butterfat in cocoa products may be estimated by the following formula based on the assumption of a Reichert-Meissl value of 0.5 for cocoa butter.

$$M.F. = \frac{5R - 2.5}{23.5}$$

in which

$$M.F. = \text{g. of butterfat in 5 g. of mixed fat}$$

and

$$R = \text{Reichert-Meissl value of the extracted fat.}$$

Using the above formula for the calculation of butterfat content and the foregoing method for the estimation of milk protein, the approximate total milk solids content may be calculated from the following formula: [13]

$$T.M.S. = M.F. + L + M.P. + 0.05 \ (M.F. + L + M.P.) \ \text{or more simply}$$
$$= 1.05 \ [M.F. + L + M.P.]$$

---

[13] Offutt, *J. Assoc. Official Agr. Chem.* **18**, 424 (1935).

in which

T.M.S. = per cent total milk solids
M.F. = per cent butterfat
L = per cent lactose
M.P. = per cent milk protein

The lactose content may be approximately calculated from the milk protein content by multiplying the milk protein content by the factor, 1.4. This is an average factor and actually varies about 10 per cent either way but serves for rapid and approximate calculation. The factor 0.05 represents the material to be added to account for milk ash. Janssen and Dehut [14] use the factor 0.0674.

TOTAL ALKALOIDS [15]

The term alkaloids, as far as cocoa is concerned, means theobromine and caffeine. The cocoa is digested repeatedly with 80 per cent alcohol by volume and a little magnesia and filtered. The filtrate is evaporated, and water is added to replace the alcohol lost. The mixture is clarified with zinc ferrocyanide and again filtered. The filtrate is evaporated to a small volume and extracted with chloroform. Nitrogen is determined in the extract and the total alkaloids are calculated by use of an appropriate factor.

Reagents.—Zinc Acetate Solution.—Dissolve 21.9 g. of crystallized zinc acetate, $Zn(C_2H_3O_2)_2 2H_2O$, and 3 ml. of glacial acetic acid in water and make up to 100 ml.

Potassium Ferrocyanide Solution.—Dissolve 10.6 g. of crystallized potassium ferrocyanide in water and make up to 100 ml.

Procedure.—Grind 2 g. of prepared cocoa or 2 or more g. of cocoanib or chocolate to a smooth paste with a little 80 per cent alcohol, and transfer to a 200-ml. flask with more alcohol of the same strength, sufficient to produce a total volume of about 100 ml. Add 1 g. of freshly-ignited magnesium oxide and digest in a boiling water bath for 1½ hours under a reflux air condenser, with occasional shaking. Filter while hot through a small Büchner funnel, return the residue to the flask, and re-digest for half an hour with 50 ml. of the alcohol. Filter and repeat the digestion with a further 50-ml. portion of 80 per cent alcohol. Evaporate the combined extract on the water bath, adding hot water from time to time to replace the alcohol lost. When all the alcohol is removed, finally evaporate to about 100 ml., add 2 to 3 drops of 10 per cent hydrochloric acid and transfer the liquid to a 150-ml. volumetric flask. Cool, add 5 ml. of the zinc ace-

[14] Janssen and Dehut, *Analyst* 61, 45 (1936).
[15] Moir and Hinks, *Analyst* 60, 439 (1935).

tate solution and mix, and then add 5 ml. of the potassium ferrocyanide solution. Make up to the mark with water and mix
thoroughly by shaking. Allow the flask to stand for a few minutes,
and filter through a dry filter paper. Discard the first portion.

Evaporate a measured quantity, 120 ml., of the filtrate to about
10 ml., transfer to a separatory funnel and extract by vigorous shaking with five successive 30-ml. portions of chloroform. Run off the
chloroform after each extraction into a second separatory funnel,
and wash the combined extracts with 3 to 5 ml. of water. Repeat the
above process of extraction with five successive quantities of chloroform, wash the second chloroform extracts with the same wash water,
combine the whole of the extracts, and remove the chloroform by
distillation. Dissolve the residue in a little hot water, transfer to a
Kjeldahl flask, add 0.2 g. of sucrose and 10 ml. of sulfuric acid. Heat
over a small flame until frothing ceases, add 0.02 g. of selenium, and
digest until colorless. Heat for 1 hour longer, and determine the
ammonia in the usual way by distillation into 0.1 N acid. Use the
factor 3.26 for the conversion of nitrogen into alkaloid.

THEOBROMINE

In the method devised by Moores and Campbell [16] the theobromine and caffeine are extracted from the plant materials by percolation with hot water in the presence of magnesium oxide, and the
extract is clarified by treatment with zinc acetate-potassium ferrocyanide reagents. The alkaloids in the clarified extract are adsorbed
on a column of English XL fuller's earth and the theobromine is
selectively eluted with dilute sodium hydroxide. The alkaline solution is adjusted to pH 6.4 with sulfuric acid and treated with an
excess of silver nitrate, and the nitric acid formed by the reaction
between theobromine and silver nitrate is titrated electrometrically
with sodium hydroxide.

Apparatus.—Extraction Tubes.—Attach 100-mm. lengths of
8-mm. glass tubing to the bottom of 25 × 200 mm. test tubes.

Absorption Tubes.—Attach 80-mm. lengths of 6-mm. glass tubing
to the bottom of 18 × 150 mm. test tubes.

Salt Bridge for Electrometric Titration.—Seal a thread of asbestos into the end of an 18 × 100 mm. test tube.

Reagents.—Zinc Acetate Solution.—Dissolve 219 g. of crystal-
lized zinc acetate, $Zn(C_2H_3O_2)_2 \cdot 2H_2O$, and 30 ml. of glacial acetic acid
in water and make up to 1 liter.

Potassium Ferrocyanide Solution.—Dissolve 106 g. of crystal-

---

[16] Moores and Campbell, *Anal. Chem.* **20**, 40 (1948).

lized potassium ferrocyanide, $K_4Fe(CN)_6 \cdot 3H_2O$, in water and make up to 1 liter.

Silver Nitrate Solution, 1 N and 0.1 N.—Dissolve 170 g. (1 N) or 17 g. (0.1 N) of crystalline silver nitrate in water and make up to 1 liter.

Sodium Hydroxide Solution, 0.1 N and 0.025 N.—Prepare from carbonate-free sodium hydroxide as described on page 38 from 50 per cent sodium hydroxide solution after standing 10 days or longer.

Potassium Acid Phthalate Buffer, 0.05 molar.

Sodium Nitrate Solution.—Add about 100 g. of sodium nitrate to 100 ml. of water.

Mixed Indicator (0.625 g. of methyl red and 0.412 g. of methylene blue). Dissolve methyl red in 450 ml. of 95 per cent ethyl alcohol. Filter through asbestos. Dissolve methylene blue in 50 ml. of water. Combine methyl red and methylene blue solutions, make to 500 ml., and mix. pH range, 5.34 (acid) to 5.65 (alkaline). Reddish purple to green.

Fuller's Earth-Celite Absorption Mixture.—Mix thoroughly equal quantities by weight of English Superfine XL fuller's earth and Celite 535 or 545. The fuller's earth should be tested for its capacity to adsorb theobromine. The following test is recommended: Stir 2 g. of the fuller's earth for 10 minutes with 200 ml. of a neutral water solution containing 100 mg. of theobromine. Centrifuge and determine the theobromine in a 100-ml. aliquot of the supernatant solution by the silver nitrate titration procedure described below. The fuller's earth should absorb at least 75 mg. of the theobromine.

For the best accuracy, sample weights should be adjusted so that the materials going through the clarification and adsorption steps contain about 40 mg. of theobromine.

Extraction and Clarification.—Place a weighed portion of the prepared sample (usually 2.0 to 3.0 g.) and 1 g. of heavy magnesium oxide in a 200-ml. casserole. Add 4 g. of Celite 545 and mix thoroughly with sufficient hot water (about 10 ml.) to make a smooth paste. Dilute to about 50 ml. with hot water, mix thoroughly, and transfer to a 25 × 200 mm. extraction tube containing a glass wool plug and Celite 545 filter bed (about 1 g.) placed in a 1-liter filter flask. Connect the extraction tube to a supply of boiling water and attach a vacuum line to the filter flask. Regulate the vacuum on the filter flask to percolate boiling water through the bed at such a rate that 500 ml. of extract will be collected in 30 to 40 minutes. The tube should not be allowed to go dry at any time during the percolation. The water on top of the sample should not be more than 2.5 cm.

(1 in.) deep in order to avoid cooling. The temperature of the sample bed should be 80° to 90° C.

Neutralize the extract to about pH 6 with 1 N sulfuric acid, using universal indicator paper. Transfer the extract to a 1-liter beaker and concentrate to about 150 ml. Transfer the concentrate to a 200-ml volumetric flask, using a small amount of wash water and keeping the total volume to about 170 ml. Add 7 ml. of zinc acetate reagent and mix, then immediately add with swirling 7 ml. of potassium ferrocyanide reagent. Make to volume with water and mix thoroughly by shaking. After a minimum of 3 or a maximum of 5 minutes' standing, filter the solution through a dry paper (Green's No. 488½). Discard the first 5 to 8 ml. of the filtrate and collect the remainder. Place 150 ml. of the theobromine solution in a separatory funnel equipped with a rubber stopper to fit the top of the adsorption tube.

Adsorption of Theobromine by Fuller's Earth.—Place a plug of glass wool in the bottom of the adsorption tube and press it firmly and evenly in place by use of a glass rod. Then introduce about 0.5 g. of Celite 545 to make a bed about 10 mm. thick. On top of the Celite bed, place about 6 g. of the 1 to 1 fuller's earth-Celite mixture. To obtain a compact uniform column that will not channel, tap the tube to distribute the packing, then compress the bed by drawing air through the dry fuller's earth column. This is conveniently carried out by placing the adsorption tube on a 500-ml. suction flask and drawing a vacuum on the flask. Test the column for channeling by adding water to the tube while the vacuum is released. Turn the vacuum on slowly. If no channeling is observed, immediately connect to the adsorption tube the separatory funnel holding the clarified solution containing preferably 40 mg., and not more than 60 mg., of theobromine. Draw the theobromine solution through the fuller's earth column. The filtration time for the adsorption step should be 20 to 30 minutes. As soon as the theobromine solution is drawn through, all except 4 to 5 ml. on top of the clay bed, wash the fuller's earth column with 50 ml. of water. It is desirable to keep some liquid on top of the adsorption column at all times to avoid channeling.

Elution of Theobromine from Fuller's Earth Column.—Remove the suction flask containing the alkaloid-free filtrate and wash water and replace it with a clean 250-ml. suction flask. Place 75 ml. of 0.1 N sodium hydroxide in the separatory funnel and elute the theobromine by connecting the funnel to the adsorption tube and drawing the alkali through the column by applying vacuum to the suction flask. The time for the elution step should be 10 to 20 minutes, making the total time for the absorption, washing, and elution about 40 to 60 minutes.

Titration of Theobromine.—Add 2 drops of mixed indicator to facilitate the rough adjustment of the pH, then neutralize the sodium hydroxide eluate with 1 N sulfuric acid to a faint red end point. Transfer the neutralized sodium hydroxide eluate to a 250-ml. beaker, using water to rinse the flask. In the beaker place a glass electrode and a calomel electrode, making contact with the solution by means of a salt bridge filled with a saturated solution of sodium nitrate. Adjust the solution to pH 6.40 ± 0.05, using 0.025 N sodium hydroxide for the final adjustment. The total volume should be about 125 ml. Add 25 ml. of 0.1 N silver nitrate solution from a graduate or automatic pipette and stir thoroughly. With thorough stirring titrate to pH 6.40 ± 0.05. For best results, this titration should be made without interruptions and should not require more than 5 minutes for its completion. Record the amount of 0.025 N sodium hydroxide required for the titration.

Blanks and Controls.—Carry a sample of water through the zinc acetate-potassium ferrocyanide treatment, adsorption, elution with sodium hydroxide, neutralization, and titration in the presence of silver nitrate. Perform each operation in the exact manner prescribed for unknown samples. This blank titration should require not more than 0.20 ml. of 0.025 N sodium hydroxide.

Carry a sample containing 40 mg. of theobromine through the same procedure to establish the quantitative nature of all steps involved.

Calculations.

$$\% \text{ Theobromine} = \frac{100 \text{ (vol. of NaOH for sample} - \text{vol. of NaOH for blank)} \times \text{theobromine factor}}{\text{weight of sample clarified}}$$

$$\text{Theobromine factor} = \text{normality of NaOH} \times 0.180 \times \frac{\text{total volume clarified}}{\text{aliquot analyzed}}$$

(The addition of a 3 per cent correction as recommended in the discussion may be made by substituting 103 for 100 in the above formula.)

For 0.025 N sodium hydroxide and using a 150-ml. aliquot from a total volume of 200 ml. clarified, the factor is 0.006.

Analysis of Pure or Crude Theobromine.—In the analysis of pure or crude theobromine samples the clarification and adsorption steps may be omitted.

Procedure.—Pulverize the pure or crude theobromine samples with a mortar and pestle and weigh accurately about 0.1 g. into a

500-ml. Erlenmeyer flask. Add about 250 ml. of water, place a small funnel in the flask, and boil for about 50 minutes. Filter while still hot through a coarse filter paper into a 400-ml. beaker. Filtration may be omitted for samples containing 90 to 100 per cent theobromine. Wash the filter paper thoroughly with about 50 ml. of hot water. Cool the filtrate and washings to room temperature, adjust to pH 6.40, add 5 ml. of 1 N silver nitrate, and titrate back to pH 6.40 with 0.025 N sodium hydroxide. Correct for water blank, using 250 ml. of boiled water and 5 ml. of 1 N silver nitrate.

CAFFEINE

For determination of caffeine,[17] the clarified solution following the zinc acetate-potassium ferrocyanide treatment is made alkaline with sodium phosphate and the caffeine is selectively extracted by shaking with chloroform. The chloroform is evaporated and the caffeine is measured by a Kjeldahl nitrogen determination.

Reagents.—Sodium Phosphate, 0.5 Molar.—Solution.—Dissolve 190 g. of trisodium phosphate, $Na_3PO_4 \cdot 12H_2O$, in 1 liter of water.

The reagents for the extraction and clarification are the same as those specified under the theobromine procedure.

Extraction and Clarification.—Mix 4 g. of the sample with 2 g. of heavy magnesium oxide and 8 g. of Celite 545 and extract by the same procedure used for theobromine. Neutralize the extract, concentrate to 150 ml., and clarify with zinc ferrocyanide solution. Add 10 ml. of 0.5 M sodium phosphate solution to 150 ml. of filtrate, make to 200 ml., and filter on fluted paper. Collect 150 ml. and transfer to a 500-ml. separatory funnel.

Chloroform Separation.—Shake the clarified water extract five successive times for 1 minute each with 30-ml. portions of chloroform. After each shaking, draw off the chloroform solution into a 250-ml. separatory funnel. After the chloroform extraction is completed, add 5 ml. of 1 N sulfuric acid to the chloroform solution in the 250-ml. separatory funnel. Mix thoroughly, let stand about 10 minutes, and draw off the chloroform solution through a cotton plug in the stem of the separatory funnel into a 650-ml. Kjeldahl flask. Wash the acid with one 30-ml. portion of chloroform and add the washings to the Kjeldahl flask. Recover the major portion of the chloroform by distillation. Do not allow the distillation to proceed too rapidly or go to dryness, but stop when the Kjeldahl flask contains 10 to 15 ml. of solution.

Nitrogen Determination.—Determine the total nitrogen in the

---

[17] Moores and Campbell, *Anal. Chem.* 20, 40 (1948).

chloroform extract by the Kjeldahl-Gunning method as detailed in Chapter I. Run a nitrogen blank on the same reagents used for the determination, starting with 180 ml. of chloroform.

Calculations.

$$\% \text{ Caffeine} = \frac{100 \text{ (vol. of NaOH for blank } - \text{ vol. of NaOH for sample) } \times N \text{ of NaOH } \times 0.0485}{\text{weight of sample analyzed}}$$

CARBONATED BEVERAGES

There are many types of carbonated beverages sold commercially under the name of soda water in the United States. The name has been shortened by usage to "soda," "soda pop," or "pop." The Food and Drug Administration in 1950 continued to use certain definitions and standards for some of these. They are:

Ginger ale is the carbonated beverage prepared from ginger ale flavor, harmless organic acid, water, and a syrup of one or more of the following: sugar, invert sugar, dextrose; with or without the addition of caramel color. Ginger ale flavor or ginger ale concentrate is the beverage flavor in which ginger is the essential constituent, with or without aromatic and pungent ingredients, citrus oils, fruit juices, and caramel color.

Sarsaparilla is the carbonated beverage prepared from sarsaparilla flavor, water, and a syrup of one or more of the following: sugar, invert sugar, dextrose; with or without harmless organic acid, and with or without the addition of caramel color. Sarsaparilla flavor is the beverage flavor prepared from oil of sassafras [18] and methyl salicylate or oil of wintergreen or oil of sweet birch, with or without other aromatic and flavoring substances and caramel color. It derives its characteristic flavor from oil of sassafras and methyl salicylate.

Root beer is the carbonated beverage prepared from root beer flavor, water, and a syrup of one or more of the following: sugar, invert sugar, dextrose; with or without harmless organic acid, and with or without the addition of caramel color. Root beer flavor, or root beer concentrate, is the beverage flavor in which oil of sassafras and methyl salicylate, or oil of wintergreen or oil of sweet birch, are the principal flavoring constituents. It contains other flavoring substances, with or without the addition of caramel color.

Birch beer is the carbonated beverage prepared from birch beer flavor, water, and a syrup of one or more of the following: sugar, invert sugar, dextrose; with or without harmless organic acid, and with or without the addition of caramel color. Birch beer flavor or

---

[18] In 1957, the toxicity of safrole, a component of oil of sassafras, was under review. If it is deemed to be toxic, it may be banned from use in foods and beverages.

birch beer concentrate is the beverage flavor in which methyl salicylate, or oil of sweet birch or oil of wintergreen, and oil of sassafras are the principal flavoring constituents, with or without other flavoring substances, and with or without the addition of caramel color. The flavor of methyl salicylate predominates.

Cream soda water, "cream soda," is the carbonated beverage prepared from cream soda water flavor, water, and a syrup of one or more of the following: sugar, invert sugar, dextrose; with or without harmless organic acid, and with or without the addition of caramel color. Cream soda water flavor or cream soda water concentrate is the beverage flavor prepared from vanilla, vanillin, or bourbonal, singly or in combination, together with other flavoring substances; with or without the addition of caramel color.

No definitions have been promulgated for the so-called fruit beverages both carbonated and not carbonated. Most of the carbonated fruit beverages are in reality imitation fruit beverages, containing little or no fruit extract and consisting in the main of artificial fruit flavor, artificial color, organic and at times mineral acid such as phosphoric acid, sugar, invert sugar or dextrose and carbonated water. Other types of carbonated beverages, such as the "cola" type contain added caffeine and act as mild stimulants.

Carbonated beverages may be analyzed by the methods described in other sections of the book. Thus sugars may be estimated as detailed in Chapter X most easily by use of a refractometer. The estimation of some organic acids and the presence of mineral acid is outlined in the chapters on jams and jellies and on spices and flavors. Chapters XII and XIV. Total acidity may be estimated by customary methods and hydrogen-ion concentration by means of a pH meter. Saccharin and other sweetening agents may be detected as described in Chapter IV and determined as detailed in Chapter XXIV. Artificial coloring matters may be detected as detailed in Chapter III.

A method for the estimation of orange juice content of orange drinks is explained in some detail in the chapter on vitamins, Chapter XVII. A procedure for the detection of malted milk powder in malted milk drinks is discussed in the chapter on milk products, Chapter VIII.

SELECTED REFERENCES

Bailey, *Food Products and Their Sources,* 3rd ed. Blakiston, Philadelphia, 1928.

Jacobs, ed., *Chemistry and Technology of Food and Food Products,* 2nd ed. Interscience, New York, 1951.

Leach, *Food Inspection and Analysis.* Wiley, New York, 1920

Methods of Analysis, A.O.A.C., 8th ed. Washington, 1955.

Woodman, *Food Analysis.* McGraw-Hill, New York, 1941.

CHAPTER XVI

MEAT, MEAT PRODUCTS, FISH, AND EGGS

It is seldom necessary for the food analyst to analyze fresh meat as such except for investigational work. However, other occasions do arise when such analyses must be made. Meat is the properly dressed flesh (the edible part of the striated muscle of an animal) derived from cattle, from swine, from sheep, or from goats, sufficiently mature and in good health at the time of slaughter, but restricted to that part of the striated muscle which is skeletal or that which is found in the tongue, in the diaphragm, in the heart, or in the esophagus, and does not include that found in the lips, in the snout, or in the ears, with or without the accompanying and overlying fat, and the portions of bone, skin, sinew, nerve, and blood vessels which normally accompany the flesh, and which may not have been separated from it in the process of dressing it for sale. *Beef* is the meat derived from cattle nearly one year of age or older. *Veal* is the meat derived from young cattle one year or less of age. Veal generally implies meat from an immature milk-fed bovine animal usually not over three months of age. *Calf,* on the other hand, implies meat from an immature bovine animal which for a considerable time had subsisted in part or entirely on feeds other than milk. *Mutton* is meat derived from sheep nearly one year of age or older. *Lamb* is the meat derived from young sheep one year or less of age. *Pork* is the meat derived from swine.

In general, the food analyst has occasion to analyze prepared meats, for these products are subject often to a wide variety of forms of adulteration or, as some authorities might say, substitution. The excessive addition of water, cereal, starch, and soya flour are examples of adulteration and substitution in such products as smoked meats, sausage, and potted meat.

It is not only instructive, but an aid, in judging the analyses to be made on a sample, to bear in mind the definitions of meat products.

Prepared meat is the product obtained by subjecting meat to a process of comminuting, of drying, of curing, of smoking, of cooking, of seasoning, or of flavoring, or to any combination of such processes.

Cured meat is the product obtained by subjecting meat to a process of salting, by the employment of dry common salt or of brine, with or without the use of one or more of the following: Sodium nitrite, sodium nitrate, potassium nitrate, sugar, dextrose, a syrup, honey, spice.

Dry salt meat is the prepared meat which has been cured by the application of dry common salt, with or without the use of one or more of the following: Sodium nitrite, sodium nitrate, potassium nitrate, sugar, dextrose, a syrup, honey, spice; with or without the injection into it of a solution of common salt to which may have been added one or more of the following: Sodium nitrite, sodium nitrate, potassium nitrate, sugar, dextrose, a syrup, honey.

Corned meat is the prepared meat which has been cured by soaking in with or without injecting into it, a solution of common salt, with or without one or more of the following, each in its proper proportion: Sodium nitrite, sodium nitrate, potassium nitrate, sugar, dextrose, a syrup, honey, and with or without the use of spice.

Sweet pickled meat is the prepared meat which has been cured by soaking in, with or without injecting into it a solution of common salt, with sugar and/or dextrose, a syrup, and/or honey, together with one or more of the following, each in its proper proportion: Sodium nitrite, sodium nitrate, potassium nitrate, and with or without the use of spice.

Dried meat is the product obtained by subjecting fresh meat or cured meat to a process of drying, with or without the aid of artificial heat, until a substantial portion of the water has been removed.

Smoked meat is the product obtained by subjecting fresh meat, dried meat, or cured meat to the direct action of the smoke either of burning wood or of similar burning material.

Canned meat is fresh meat or prepared meat, packed in hermetically sealed containers, with or without subsequent heating for the purpose of sterilization.

Hamburg steak or hamburger steak is comminuted fresh beef, with or without addition of suet and/or of seasoning.

Potted meat or deviled meat is the product obtained by comminuting and cooking fresh meat and/or prepared meat, with or without spice. It is usually packed in hermetically sealed containers.

Sausage meat is fresh meat or prepared meat, or a mixture of fresh meat and prepared meat. It is sometimes comminuted. The term "sausage meat" is sometimes applied to bulk sausage containing no meat by-products.

Meat by-products are any properly dressed edible parts, other than meat, which have been derived from one or more carcasses of cattle, of swine, of sheep or of goats, sufficiently mature and in good health at the time of slaughter.

Meat food products are any articles of food or any articles that enter into the composition of food which are not prepared meats but which are derived .or prepared in whole or in part, by a process of manufacture from any portion of the carcasses of cattle, swine, sheep, or goats, if such manufactured portion be all, or a considerable and definite portion, of the article, except such preparations as are for medicinal purposes only.

Meat loaf is the product consisting of a mixture of comminuted meat with spice and/or with cereals, with or without milk and/or eggs, pressed into the form of a loaf and cooked.

Pork sausage is chopped or ground fresh pork, with or without one or more of the following: Herbs, spice, common salt, sugar, dextrose, a syrup, water.

Brawn is the product made from chopped or ground and cooked edible parts of swine, chiefly from the head, feet, and/or legs, with or without the chopped or ground tongue.

Headcheese, or mock brawn, is the product made from chopped or ground, cooked edible parts of meat or meat by-products.

Souse is the product consisting of meat and/or meat by-products; after cooking, the mixture is commonly packed into containers and covered with vinegar.

Scrapple is the product consisting of meat and/or meat by-products mixed with meal or the flour of grain, and cooked with seasoning materials, after which it is poured into a mold.

Frankfurter and *bologna style sausage* are important meat products. The following definitions, although not those of the Food and Drug Administration, are entirely descriptive. Frankfurter style sausage of the best quality is prepared from meat, usually a mixture of beef and pork trimmings and certain beef cuts.[1] The ground mixture is stuffed in animal or artificial casings and is then smoked and cooked for a short time. Cheaper grades of frankfurter style sausage are made of a mixture of meat and meat by-products. Cereal is frequently added to these grades of sausage.

Bologna style sausage is prepared from ingredients similar to those used in frankfurter style sausage and in a similar manner, except that the former is stuffed in larger casings or containers and is cooked for a longer time. The cheaper grades of this product are made from much the same materials as those used for the same grades of frankfurter style sausage. Kosher bologna style sausage and kosher frankfurter style sausage contain no pork.

---

[1] Hoagland, *U. S. Dept. Agr. Circ.* 230 (1932).

Composition of Meat

The constituents and components of the muscle of any animal vary between certain limits just as the constituents and components of milk, blood, urine, or any body fluid or section vary between certain limits, and as we set the term constants for the percentage of these components, variations in these constants may be used for the detection of adulteration and sophistication.

The muscle matter of animals generally contains about 75 per cent water and 25 per cent total solids. The total solids consist of about 20 per cent nitrogenous and non-nitrogenous material and extractives, and mineral salts or ash. The remaining 80 per cent is protein material. Among the important nitrogen extractives are creatine, creatinine, uric acid, urea, and the xanthine bodies or purine bases, xanthine, guanine, and carnine. Among the important non-nitrogenous extractives are lactic acid, glycogen, and fatty materials. The moisture content of lean meat tends to approach 80 per cent as a maximum. Lean meat consists essentially of muscle tissue, connective tissue, and intramuscular fat cells, the percentage of the latter two components being small.

Tables 91, 92, 93, 94 and 95 give the percentage composition of lean and prepared meats. It can be seen from the preceding paragraph and from Table 96 that in lean normal meat the moisture-

TABLE 91. Composition of Fresh Lean Beef [a,b]

Type	Moisture %	Protein %	Fat %	Ash %
Ox..........................	76.7	20.0	1.5	1.2
Ox..........................	76.71	20.78	1.5	1.18
Cow.........................	76.35	20.54	1.78	1.32
Medium Ox.................	73.7	19.8	5.3	1.2
Chuck.......................	71.3	20.2	8.2	1.0
Chuck.......................	72.84	19.81	7.13	....
Cattle, fat free.............	76.5	21.88	....	1.1
Cattle.......................	76.11	....	....	....
Cattle, emaciated...........	80.09	18.90	0.45	0.99
Cattle, thin.................	78.84	19.65	0.75	1.03
Very thin....................	74.24	20.25	2.78	0.93
Thin.........................	73.52	19.81	4.91	1.00
Very very thin..............	76.37	18.94	1.87	1.05
Thin cow....................	68.70	20.00	9.52	0.98
Chuck, very lean...........	73.8	22.3	3.9	1.0
Chuck rib:.................	75.8	22.2	1.4	1.1
Chuck rib..................	71.3	19.5	8.3	0.8
Chuck.......................	74.1	22.6	2.8	1.1
Flank.......................	70.7	25.9	3.3	1.2

[a] *Allen's Commercial Organic Analysis, Vol. IX, 5th Ed.* Blakiston, Philadelphia, 1932.
[b] Atwater and Bryant, *U. S. Dept. Agr., Bull.* 28 (1906).

protein ratio tends to become 4:1 as a maximum with the more usual
ratio as 3.5:1. As meat is processed, treated, or prepared in any way
whatever, it tends to lose moisture unless this loss is deliberately
counteracted. Hence the tendency of manufacturers and processors
is to attempt to equalize the loss by the incorporation of water. At
times this incorporation is carried to fraudulent lengths. As meat or
meat products stand they also tend to lose moisture. In order to prevent the excessive loss of moisture in such products as sausage, it
has become customary to incorporate not only a small portion of
water, as ice, which aids in the manufacturing process, but also
binders such as cereal, starch, soybean flour, skim milk powder and
other materials. At times the addition of these materials assumes
deceitful dimensions.

TABLE 92. COMPOSITION OF PREPARED BEEF [a,b]

Type	Moisture %	Protein %	Fat %	Ash %
Boiled	51.8	25.5	22.5	1.3
Corned canned	51.8	26.3	18.7	4.0
Dried canned	44.8	39.2	5.4	11.2
Roast canned	58.9	25.9	14.8	1.3
Luncheon canned	52.9	27.6	15.9	4.8
Corned	53.6	15.6	26.2	4.9
Spiced and rolled	30.0	12.0	51.4	6.8
Roast cooked	48.2	22.3	28.6	1.3
Round steak	63.0	27.6	7.7	1.8
Sandwich meat	58.3	28.3	11.0	2.8
Loin steak	54.8	23.5	20.4	1.2
Corned	63.86	26.01	8.06	2.80 (salt)

[a] Atwater and Bryant, *U. S. Dept. Agr., Bull.* 28 (1906).
[b] Eakins, *Military Meat and Dairy Hygiene*, p. 317. Williams and Wilkins, Baltimore, 1924.

TABLE 93. COMPOSITION OF FRESH AND PREPARED BEEF TONGUE [a,b,c]

Type	Moisture %	Protein %	Fat %	Ash %
Fresh	70.8	18.9	9.2	1.0
Canned, ground	49.9	21.4	25.1	4.0
Canned, whole	51.3	19.5	23.2	4.0
Pickled	62.3	12.8	20.5	4.7
Fresh, ox	68.3	18.13	11.46	1.18
Fresh	64.00	15.20	19.15	0.80
Potted	41.52	18.46	32.85	6.7

[a] Atwater and Bryant, *U. S. Dept. Agr., Bull.* 28. (1906).
[b] Wright and Forsyth, *J. Soc. Chem. Ind.* 46, 36T (1927).
[c] König, *Chemie der Menschlichen Nahrungs und Genussmittel.* Springer, Berlin, 1898.

Analysis for the percentage of the constituents or components of a meat or meat product and comparison with the normal analyses shown in Tables 93 to 95 aid in the interpretation of whether a processed, prepared, or treated meat or meat product has been manufactured along normal lines, or whether it has been adulterated.

TABLE 94. COMPOSITION OF FRESH AND PREPARED HAM [a,b]

Type	Moisture %	Protein %	Fat %	Ash %
Fresh lean	60.0	25.0	14.4	1.3
Smoked lean	53.5	19.8	20.8	5.5
Cooked	49.2	22.5	21.0	5.8
Deviled	44.1	19.0	34.1	3.3
Visible fat removed	64.5	19.2	16.2	0.9
Smoked, boiled	51.3	20.2	22.4	6.1
Fresh	51.55	15.29	32.37	0.8
Cured	64.61	21.06	13.6	...

[a] Atwater and Bryant, *U. S. Dept. Agr., Bull.* 28 (1906).
[b] *Allen's Commercial Organic Analysis, Vol. IX, 5th Ed.* Blakiston, Philadelphia, 1932.

TABLE 95. AVERAGE CHEMICAL COMPOSITION OF SAUSAGE AND OTHER MEAT FOOD PRODUCTS [a]

Product	Water %	Ash %	Fat %	Protein %	Starch %
1st grade frankfurter style	60.88	2.64	22.00	13.69	....
2nd grade frankfurter style	61.33	2.95	19.70	14.51	....
Frankfurter style and cereal	64.29	3.12	14.06	15.24	1.20
1st grade bologna style	63.98	2.82	18.11	14.35	....
2nd grade bologna style	64.03	3.16	17.34	14.48	....
Bologna style with cereal	62.38	3.31	15.92	14.78	1.81
Pure pork	41.93	2.09	44.83	10.81	....
Fresh link	44.75	2.48	41.17	11.28	....
Braunschweiger style	56.20	2.67	23.75	15.39	....
Liver sausage	58.97	2.23	20.57	16.69	....
Headcheese	62.04	2.23	20.26	15.04	....
Blood sausage	47.09	2.30	34.64	14.81	....
Meat loaf	63.99	3.52	13.45	16.14	....
Souse	72.87	1.87	12.34	13.18	....
Luncheon roll	56.43	3.39	23.79	15.89	....
Polish style	56.04	3.55	23.06	16.41	....
Country style	51.74	3.92	27.45	16.21	....
Bockwurst	63.53	2.43	21.85	11.70	....

[a] Hoagland, *U. S. Dept. Agr., Circ.* 230 (1932).

PREPARATION OF SAMPLE

In the analysis of meat and meat products, it is usual to remove the skin, bones, and visible fat and to perform the analysis on the lean meat. This portion of the meat is passed through a meat chopper at least 3 times and, if any water separates out, it is reincorporated. The mixed meat is then kept in hermetically sealed containers, until needed for the analysis or the entire sample may be dried, reground, and then kept in containers until the analysis may be performed. If it is desired to make the analysis on the entire sample, the bones are separated and pulverized and then the entire mass, skin, bones, and meat passed through the meat chopper until the mass is uniform in composition. Or the entire sample may be dried and then ground and reground until a homogeneous mass is obtained. In the case of sausages, the outer wrapper is removed and the interior is then prepared as above.

TABLE 96. WATER-PROTEIN RATIO OF LEAN MEAT

Type	Ratio
Pork[a]	3 : 1
Beef[a]	3.26 : 1
Mixed meat[a]	3.13 : 1
Composite lean pork[b] (118)	3.4 : 1
Composite lean beef[b] (34)	3.6 : 1
Fresh pork[c]	3.4 : 1
Fresh beef[c]	3.6 : 1
Mixed meat[c]	3.5 : 1

[a] Stubbs and More, *Analyst* 44, 125 (1919).
[b] Moulton, *Meat Through the Microscope*. Chicago, 1929.
[c] Jackson and Jones, *Analyst* 57, 562 (1932).

MOISTURE

The moisture content of meat and meat products may be estimated simply by the direct heat method.

Procedure.—Weigh accurately approximately 5 g. of the prepared sample into a tared low, flat-bottomed dish and place in an oven thermostatically controlled at 75°-80° C. for 24 hours. At the end of this period, remove the dish, place in a desiccator, allow to cool, and weigh. The loss in weight is calculated as moisture. This determination should be done in triplicate, for some determinations, yielding low results because of the inclusion of unevenly distributed cartilage, should be discarded.

The moisture content may also be determined by the toluene distillation method described in Chapter I. For very accurate results, a weighed portion of the sample should be dried over sulfuric acid.

PROTEIN

At nearly the same time that the moisture determination is performed, if it be found convenient, weigh accurately approximately

2 g. of the sample on a piece of filter paper on a watch glass counterpoised by another watch glass and filter paper. Transfer the filter paper and meat to an 800-ml. Kjeldahl flask and proceed to determine the albuminoid-nitrogen as directed in the Kjeldahl-Gunning-Arnold method in Chapter I. Run a blank as directed and include the filter paper. The protein content is calculated by multiplying the nitrogen content by the factor 6.25.

If the residue from the moisture determination is practically fat free, the residue may be ground in a mortar and the protein determination may be made on a 0.7-g. portion of the dried and ground material. If the original sample was dried and ground, also perform the analysis on a 0.7-g. portion.

These determinations may also be made on the residue from the ether extraction and corrections are mathematically applied for the determinations made on the moisture- and fat-free basis.

ADDED WATER

Added water may be calculated in meat or a meat product in which the accepted moisture-protein ratio is 4:1 by use of the following formula.

$$W - 4P = \text{added water}$$
$$W = \text{moisture content}$$
$$P = \text{protein content.}$$

The determination of added water in meat and meat products is becoming an estimation of increasing importance. For many years it has been customary to inject hams with the pickling solution near the bone to prevent bone and shank sour. This method of intramuscular injection aided in promoting a surer cure. Recently a method of intravenous injection or pumping has been developed and used in the processing of beef tongues. In corned beef, also, the intramuscular method has been applied. The original object of the injection or pump method of cure was to provide a faster and at the same time a surer cure for, without prior injection, the meat had to be soaked for long periods before it was completely cured. This vat method of soaking is known as the "dry" cure method.

The ability of the meat to retain water thus injected into it is very high, and, consequently, unscrupulous manufacturers and dealers have used the pump cure method to increase excessively the weight of the cured meat. This type of adulteration may easily be determined by means of estimation of the moisture and protein content.

Not only have brine and pickling solutions been injected but also solutions of gelatin. This problem and that of the injection of viscolized fats will be discussed subsequently.

In the manufacture of sausage, the meat loses moisture because of heat developed by the friction of the knives and as a result of other causes. This is counteracted by the processor by the inclusion of chopped ice in the comminuting step in the preparation of the sausage. Some authorities feel that it is very difficult to prepare a succulent sausage without the addition of water. At any rate, governmental agencies have held that the addition of more than 3 per cent of moisture to uncooked sausage and of 10 per cent of moisture in smoked or cooked sausage is an adulteration. Added water may be estimated in an analogous manner to that detailed above.

Since binders are permitted in the manufacture of sausage the addition of moisture and its estimation becomes far more difficult than the simple procedure outlined in the case of tongue, corned beef, and ham. Many of the binders permitted are protein bearing materials, as, for example, skim milk powder and soybean flour. In the case of skim milk powder the casein content may be determined with a fair degree of accuracy as detailed in Chapter VIII, section, "The Detection of Milk Powder in Foods." The total protein found should then be corrected for the casein content. Methods for the detection and estimation of soybean flour will be discussed subsequently. Corrections for the amount of soybean flour protein present should be applied in calculating the added water.

Other factors to bear in mind are that the nitrogen thus determined comes not only from the protein material in the meat and meat product but also from the nitrogenous extractives, such as creatine and creatinine. However, the inclusion of nitrogen from such sources acts to increase the apparent protein content.

FAT

Gerber Method.—At times it is of value and necessary to estimate the fat content of meat. A rapid, although only fairly accurate, method that may be used is a modified Gerber Method.[2]

Procedure.—Weigh 2.5 g. of the meat into a scoop described in the Gerber method for soft-curd cheese. Place the scoop and its contents into a cheese butyrometer and add 9 ml. of 4 per cent borax solution and heat the mixture in a water bath until the meat is softened. Add 1 ml. of amyl alcohol and then cautiously add 10 ml. of the acid (to 13 parts water add 87 parts sulfuric acid, specific gravity 1.82), used

---

[2] Jacobs, work performed in 1936.

for ice cream determinations. Stopper the butyrometer and then shake vigorously. When the meat has been thoroughly disintegrated by the acid, centrifuge for the required 10 minutes. Remove the butyrometer from the centrifuge and read in the usual manner. If any small particles of meat have not been dissolved by the acid, place the butyrometer in a boiling water bath and shake at intervals until all the meat is dissolved. Recentrifuge the butyrometer and read the fat after removal from the centrifuge. This method will give results that are accurate within the limits of the readability of the butyrometer.

Modified Babcock Method.—A rapid method for the determination of fat in meat and meat products was devised by Oesting and Kaufman.[3]

Procedure.—Weigh out 25 g. of finely ground sample and transfer to a Waring blendor. Add 100 g. of cracked ice or water at 1-3° C., and 2 g. of household Oakite. Run the blendor for 10 minutes with the cover in place, stopping it occasionally to swirl the contents and dislodge any lumps of meat that may adhere to the side of the container. After thorough mixing, weigh 10 g. of the emulsion to the nearest 0.1 g. into an 18-g., 8 per cent Babcock flask. Add 5 ml. of glacial acetic acid, and mix thoroughly to coagulate the proteins. Add 10 ml. of concentrated sulfuric acid, specific gravity 1.84, a little at a time, swirling until all lumps are dissolved. Add just enough hot water to form a layer above the acid mixture to assist in reducing charring, for the fat rises above the water layer. Add 5 ml. more of sulfuric acid, mix, and centrifuge at approximately 1000 r. p. m. for 5 minutes. Add hot water to the neck of the bottle and centrifuge for 2 minutes. Add hot water to within 1 to 2 cm. of the top of the neck and centrifuge for 1 minute. Immerse the flask in water at 70° C. and read after 2 minutes on a descending fat column. The column will begin to descend when the flask is removed from the water bath. Read the column from the top of the upper meniscus to the bottom of the lower meniscus, applying the calibrations on the bottle for this measurement. Multiply directly by 9.2, in order to convert to per cent fat. A correction must be applied for the difference from exactly 10.0 g. in the weight of the sample used.

Rapid Method.—By use of a kerotolytic agent, the proteins of meat can be dissolved and the fat liberated. By this means a modified Babcock method for fat in meat useful as a screening test particularly for meat products for which fat specifications have been set (pork sausage not to exceed 45 per cent fat, hamburger not more than

---

[3] Oesting and Kaufman, *Ind, Eng. Chem., Anal. Ed.* 17, 125 (1945).

30 per cent fat, deviled ham not more than 35 per cent fat) was devised by Copeland.[4]

Reagent.—Dissolve 64.5 g. of sodium salicylate, 35.5 g. of potassium carbonate, and 16.5 g. of sodium hydroxide in 100 ml. of isopropyl alcohol and 300 ml. of water.

Procedure.—Grind or mash the sausage finely. Weigh 9 g. into a cream Babcock bottle. Add 15 ml. of reagent. Set a water bath on high heat and heat for 15 minutes. Centrifuge for 5 minutes. Add boiling water to the lower part of the neck of the Babcock flask. Centrifuge for 2 minutes. Add boiling water to bring the fat into the graduated portion of the flask. Centrifuge for 1 minute. Read from the top to the bottom of the meniscus for the percentage fat. Run in duplicate.

Ether Extraction Method.—As an alternative method, weigh 1 g. of the properly prepared sample into a Jacobs-Singer separatory flask or into a tall form 100-ml. beaker and add 9 ml. of water and 1 cc. of ammonium hydroxide. Warm on a hot plate, stirring with a glass rod or by shaking until the meat is thoroughly softened. Add ½ ml. of hydrochloric acid and stir. Add 10 ml. more of hydrochloric acid and a pinch of sand. If a beaker was used, cover the beaker with a watch glass on glass hooks and boil the mixture in the flask or beaker gently for 5 minutes or until the meat is completely dissolved. Cool the mixture and transfer if a beaker was used to a Mojonnier extraction tube or to a Jacobs-Singer separatory flask. Rinse the tall form beaker with sufficient water to bring the level of the water up to the middle of the constriction in the Mojonnier extraction tube or the flask and then rinse the beaker, watch glass, and glass hooks with 25 ml. of ethyl ether in small portions and add the washings to the Mojonnier tube. Stopper the tube and shake vigorously. Repeat the washings and the shaking with a 25-ml. portion of petroleum ether. Allow the layers to separate and draw off the ether layer into a tared fat flask. From this point proceed as directed in the Roese-Gottlieb method for the determination of fat.

Continuous Extraction Method.—The percentage fat or ether extract may also be determined and far more accurately by continuous extraction. Transfer the residue from the moisture determination to an extraction shell or alundum thimble. If the moisture determination was done in a soft metal dish or thin metal dish such as lead or aluminum, the dish may be cut and then entire dish and its contents placed in the extraction shell or alundum thimble. Place

---

[4] Copeland, V. C. Laboratory, Memphis Army Service Forces Depot, Memphis, Tennessee, 1944.

the thimble in a Soxhlet or other type continuous extractor and extract for 16-24 hours with petroleum ether. Evaporate the ether in the weighed receiving flask and dry the extract to constant weight in an oven at 100-105° C. In order to make certain that the extraction is complete the extraction may be stopped in the middle of the extraction period, the contents removed from the thimble, ground, replaced in the thimble, and the extraction with the petroleum ether continued to completion. The gain in weight in the receiving flask may be calculated as fat or ether extract. This process not only includes the fat but all ether soluble material.

ASH

Ash determinations may be made in the usual manner. These determinations should be made in porcelain or silica dishes because repeated ashing of meat in platinum dishes ruins the platinum. The ashing should be done below or at a dull red heat. The charred mass may be leached, filtered, the filtrate evaporated in the original dish and then the filter paper and residue ashed in the original dish.

COLD WATER EXTRACT

At times it is necessary to separate the nitrogenous constituents and obtain an idea of their proportion in the meat or meat product. This is especially important in estimating added gelatin and in detecting the addition of material that consists of nitrogenous material with only a small portion water soluble. In general, a weighed portion of meat is exhausted by treating with successive portions of cold water. These extracts are filtered, made to a definite volume, and portions of the filtrate are then analyzed according to the usual methods.

Procedure.—Weigh 7-25 g. of the sample into a 150-ml. beaker. Add 50 ml. of water and stir the mixture during a 15-minute interval. Filter the extract into a 500-ml. volumetric flask. Use three successive 50-ml. portions of water, followed by four 25-ml. portions. Transfer the meat to the filter and wash finally with three 10-ml. portions of water. Make the filtrate to the mark and mix thoroughly. Use portions of the filtrate for the subsequent determinations.

Determine total solids in the cold water extract by evaporating 100 ml. of the extract to dryness and then drying to constant weight in an oven. If the evaporation is performed in a porcelain or silica dish, the ash may be estimated by igniting the residue. Estimate total nitrogen by the Kjeldahl-Gunning-Arnold method on a 50-ml. portion of the extract.

COAGULABLE PROTEINS

Evaporate 150 ml. of the extract in a beaker to 40 ml. on a steam bath. Make neutral to phenolphthalein. Add 1 ml. of 0.1 N acetic acid and boil the mixture for 5 minutes. Filter through quantitative paper and transfer all of the coagulum to the filter with the aid of hot water and a rubber policeman. Wash 3 times more on the filter. Make the filtrate up to definite volume and reserve for the determination of proteose, peptone and gelatin, and creatine. Transfer the filter paper and the coagulum to an 800-ml. Kjeldahl flask and proceed with the nitrogen determination as previously detailed except that the original beaker is washed with the concentrated sulfuric acid before the acid is added to the digestion flask. This is done in order to dissolve any of the coagulated material not transferred to the filter.

PROTEOSES

Evaporate the filtrate from the coagulable protein determination to 30 ml. Cool, add 1 ml. of 50 per cent sulfuric acid, and saturate with zinc sulfate. Heat on a steam bath with stirring until clear, allow to stand for 12 hours, and filter. Wash the precipitate thoroughly with saturated zinc sulfate solution slightly acid with sulfuric acid. Determine nitrogen in the precipitate in the usual manner.

PROTEOSE, PEPTONE, AND GELATIN NITROGEN

Transfer a 50-ml. aliquot of the filtrate obtained from the coagulable nitrogen determination to a 100-ml. volumetric flask, add 15 g. of sodium chloride and 10 ml. of cold water, shake until the sodium chloride has dissolved, and cool to 12° C. ·Add 30 ml. of 24 per cent tannic acid solution cooled to 12° C., dilute to the mark with water previously cooled to 12° C. for 12 hours, or overnight. Filter at 12° C., transfer 50 ml. of the filtrate to a Kjeldahl flask, and add a few drops of sulfuric acid. Place the flask in a steam bath, connect with a vacuum pump and evaporate to dryness. Determine nitrogen in the residue as directed in the Kjeldahl-Gunning-Arnold method. Conduct a blank determination, using the same quantity of reagent, and correct the result accordingly. Multiply the corrected result by 2, and deduct the quantity of nitrogen found from the nitrogen determined in another 50-ml. aliquot of the filtrate from the coagulable nitrogen determination without the tannin-salt treatment; the difference × 6.25 = the percentage of proteose, peptone, and gelatin.

CREATINE, FOLIN METHOD

Procedure.—Evaporate an aliquot or the remaining portion of the filtrate and washings from the coagulable nitrogen to 5-10 ml. Transfer with a minimum quantity of hot water to a 50-ml. volumetric flask, keeping the volume below 30 ml.; add 10 ml. of 2 N hydrochloric acid and mix. Hydrolyze in an autoclave at 117-120° C. for 20 minutes, allow the flask to cool somewhat, remove, and chill under running water. Partially neutralize the excess of acid by adding 7.5 ml. of 10 per cent sodium hydroxide solution, free from carbonates, dilute to the mark, and mix. Make a preliminary reading on 20 ml. with a Duboscq colorimeter to ascertain the volume to use to obtain a reading of approximately 8 mm. Transfer such a volume of the solution to a 500-ml. volumetric flask and add 10 ml. of 10 per cent sodium hydroxide solution and 30 ml. of saturated picric acid solution. Mix, rotate for 30 seconds, and let stand exactly 4.5 minutes. Dilute to the mark at once with water. Shake thoroughly, and read in a Duboscq colorimeter set at 8 mm., comparing the color with 0.5 N potassium dichromate.

If the reading is too high or too low (above 9.5 or below 7 mm.), calculate the quantity necessary to obtain a reading of about 8 mm. The strength of the dichromate solution used must be checked against a standard creatine solution. Divide 81 by the reading and multiply by the volume factor to obtain the milligrams of creatinine. Multiply the value obtained by 1.16 to obtain creatine, which, divided by the weight of the sample and multiplied by 100 gives the percentage of creatine.

Example.—Extract 29 g. of meat with water as directed under cold water extract and dilute the extract to 500 ml. Treat 150 ml. of this latter solution which is equivalent to 6 g. of meat as directed in the section coagulable nitrogen. Evaporate the filtrate thus obtained, hydrolyze as above, and dilute to 50 ml. Treat 25 ml. of this last solution with sodium hydroxide solution and picric acid solution as directed above and dilute to 500 ml. This latter solution gives a Duboscq reading of 9 mm.

$$\frac{81}{9} \times \frac{50}{25} = \text{mg. of creatinine}; \quad \frac{0.018 \times 1.16 \times 100}{6} = 0.35\% \text{ creatine.}$$

MEAT BASES

Deduct from the percentage of total nitrogen, the sum of the percentages of nitrogen obtained in the determination of insoluble nitrogen, coagulable nitrogen and proteose, peptone, and gelatin, to obtai the percentage of nitrogen of the meat bases. Multiply the result by 3.12 to obtain the percentage of meat bases.

NON-PROTEIN NITROGEN

Mezincescu and Szabo [5] suggest a method for the determination of the non-protein nitrogen in tissue that is based on estimation of the nitrogen in a trichloroacetic acid extract.

Procedure.—Pass about 100 g. of sample through a small meat grinder. Thoroughly mix the ground tissue with the addition of water if necessary until it is transformed into a more or less homogeneous paste, and determine its dry weight by heating a sample at 105° C. If water is added to the sample, the water content of the sample is used to calculate the dry weight and the nonprotein nitrogen is calculated from the figure corresponding to the dry weight.

Weigh 3 to 7 g. of the paste into a glass-stoppered cylinder of about 120-ml. capacity. Add 10-20 glass beads and about 50 ml. of water; shake the contents for 10 minutes and then allow to stand for 30 minutes. At the end of this time add 50 ml. of 20 per cent trichloroacetic acid solution and shake the cylinder again for 10 minutes. Leave in a refrigerator for 3 hours and then filter. The filtrate should be perfectly clear, the first portions being refiltered if necessary. Determine nitrogen in an aliquot of the filtrate and recalculate the percentage to the dry weight of the sample, or to the fresh weight of the sample.

AMINO ACIDS

The content of essential amino acids in some foods and proteins has been summarized by Horn, Breese Jones, and Blum.[5a] These investigators used, in the main, microbiological methods. Thus *Lactobacillus arabinosus* 17-5 was used for the determination of methionine; *Leuconostoc mesenteroides* P-60 was employed for lysine, histidine, leucine, phenylalanine, and isoleucine; and *Streptococcus faecalis* No. 9790 was used for arginine, threonine, and valine. These organisms can be obtained from the American Type Culture Collection, Georgetown University School of Medicine, Washington, D. C. Tryptophan was estimated by colorimetric methods.

STARCH

This method for the estimation of starch in meat products depends on the solution of the protein, other nitrogenous, fatty, and

---

[5] Mezincescu and Szabo, *J. Biol. Chem.* 115, 131 (1936).
[5a] Horn, Breese Jones and Blum, *U. S. Dept. Agr., Misc. Pub.* 696 (1950).

salt components of meat in alcoholic potash, in which starch and materials of a like nature are insoluble. These materials are separated by filtration and then hydrolyzed to dextrose by sulfuric acid. The dextrose thus formed is estimated by the usual methods.

Any starch present in meat products generally comes from cereal or flour. Actually not only is the starch estimated but all materials of a carbohydrate nature, such as gums, insoluble in alcohol and capable of being hydrolyzed to dextrose or other reducing sugar are included in the final result. Thus, for example, although soybean flour has practically no starch, its other carbohydrate content yields reducing sugars when subjected to this method.

Procedure.—Weigh 10 g. of the finely divided meat into a 250-ml. beaker. Add 75 ml. of an 8 per cent alcoholic solution of potassium hydroxide and heat on a steam or water bath until all the meat is dissolved. This generally requires 30-45 minutes. Add an equal volume of 95 per cent alcohol, cool, and allow to stand for at least an hour. Filter through a thin layer of asbestos in a Gooch crucible. Wash twice with a warm 4 per cent alcoholic solution of potassium hydroxide, 50 per cent by volume, and then twice with warm 50 per cent alcohol. Discard the washings. Retain as much of the precipitate in the beaker as possible until the last washing. Place the crucible with its contents in the original beaker and add 40 ml. of water and 25 ml. of sulfuric acid. Stir during the addition of the acid and make sure that the acid comes in contact with all the precipitate. Allow to stand about 5 minutes, add 40 ml. of water, heat just to boiling while stirring constantly. Transfer the solution to a 250-ml. volumetric flask, add 2 ml. of 20 per cent phosphotungstic acid solution, allow to cool to room temperature, and make up to the mark with water. Filter through a starch-free filter paper, pipette 100 ml. of the filtrate into a 200-ml. volumetric flask, neutralize with 20 per cent sodium hydroxide solution, make up to volume, and determine the dextrose present in a 50-ml. portion of the filtrate as directed in the Munson and Walker method, detailed in the chapter on sugars, Chapter X. If much dextrose is obtained, the Lane-Eynon volumetric method may be used. The weight of dextrose multiplied by 0.9 equals the weight of starch.

Rapid Method for Sausage and Other Meat Products.—Digest 10 g. of the sample on the steam bath for 30 minutes with 50 ml. of alcoholic potash solution (80 g. stick KOH in 1 liter of 95 per cent alcohol) with occasional stirring to facilitate digestion.

Transfer to a 100-ml. graduated oil tube (A.S.T.M. long form), washing the sediment from the beaker with a stream of 95 per cent alcohol from a wash bottle. Bring the solution in the tube up to the

100-ml. mark. Mix and allow the tube to stand for 1 hour, giving it a gentle rotation from time to time to assist sedimentation. At the end of one hour's standing, read volume of solids.

A volume of solids exceeding 1 ml. indicates an added cereal product. A volume of solids less than 3 ml. indicates that less than 3.5 per cent of cereal has been added to the product. Samples of cereal added products showing a less volume than 3 ml. may be passed as in conformity with regulations.

A volume of solids in excess of 3.5 ml. indicates the presence of added cereal or vegetable flour in excess of 3.5 per cent. Samples found to show a volume of sediment materially in excess of 3.5 ml. may be reported as containing cereal in excess of 3.5 per cent without further test.

It has been shown that 0.5 per cent of soy flour increases the volume of the precipitate 0.2 ml. Hence 1 per cent of soy flour is equivalent to 0.4 ml. and 3.55 per cent of soy flour is equivalent to 1.4 ml. Since 3.5 per cent of wheat flour gives a sediment of 1.9 ml., soy flour gives a sediment ¾ of starchy cereal. One ml. is generally considered a tolerance, hence 2.4 ml. is a limit for soy flour.

Hemoglobin in Meat Scraps and Tankage

According to the definitions of the Association of American Feed Control Officials (1946), meat scrap differs from tankage in that blood is excluded from the former but not from the latter. Hence the presence of blood in meat scraps, except in traces attributable to normal factory practice, is considered an adulteration.[6]

The amount of hemoglobin in meat scraps and tankage can be estimated by use of blood meal solvent of alcohol-sodium hydroxide-pyridine in which hemochromogen is formed by reduction with hydrosulfite.

Procedure.—Transfer 1 g. of finely ground meat scraps or 0.5 g. of tankage to a 300-ml. ground-glass joined Erlenmeyer flask and add 100 ml. of a solution containing 55 ml. of 95 per cent alcohol, 5 ml. of pyridine, 5 g. of sodium hydroxide, and water to volume. Reflux for 30 minutes, cool, and centrifuge. Add a few milligrams of sodium hydrosulfite to 10 to 20 ml. of the clear supernatant solution, and determine the increase in density of the reduced solution over that of the unreduced at 553 mμ.

Calculate the hemoglobin content from the equation

$$Hb = \frac{D \times V \times 100}{e \times L \times S \times 100}$$

---

[6] Reiser, *Anal. Chem.* 19, 114 (1947).

in which Hb equals the percentage of hemoglobin in the sample, e is the extinction coefficient, L is the depth of the solution in centimeters, S is the weight of the sample in grams, and V is the volume of the solution.

HORSE MEAT

The definition of meat as given by the U. S. Department of Agriculture excludes the use of horse meat unless labeled as such. In certain communities the use of horse meat without declaration is considered not only misbranding but also adulteration. Horse meat unmixed with other meats can be detected and identified by microscopic examination of the flesh structure and by a determination of the constants of the intramuscular fat. The iodine value of the intramuscular fat of horse flesh is much higher than that of beef and considerably higher than that of lard. Thus the respective values are:

Lard	47—66.5
Beef tallow	35.4—42.3
Mutton tallow	48—61
Horse fat	75—86

Possibly the most definite method of identifying horse meat in the presence of other animal tissue is the use of serological methods. Antisera to horse meat are obtained from rabbits by inoculating them with a horse meat extract. When an extract of a meat preparation containing horse meat is prepared and titrated by· bacteriological techniques with antisera mentioned above, a precipitin reaction is obtained.

In living animals, the glycogen content of the muscle is intimately connected with the movements of the muscle. In horse muscle, the amount of glycogen is generally greater than that found in other animals and consequently the presence of comparatively large amounts of glycogen in meat or meat products is evidence of horse meat. However, it must be remembered that glycogen is transitory, being broken up into simpler substances, and consequently may not appear in the meat unless examined immediately after slaughtering the animal. Furthermore, glycogen is stored in the liver of animals and, therefore, the liver of all animals is likely to have comparatively large amounts of glycogen. If the meat is comminuted and composed of different species the difficulties of detecting horse flesh are greatly increased.

Glycogen.—Brautigam and Edelmann Test.—This test [7] is based,

---

[7] Thurston, *Pharmaceutical and Food Analysis.* Van Nostrand, New York, 1922.

as is the following test, on the production of a red coloration with iodine solution if glycogen is present.

Procedure.—Boil 50 g. of the finely divided flesh for an hour with four times the volume of water. After cooling, add dilute nitric acid to the resulting broth to precipitate the proteins and decolorize the liquid. Filter, and test a portion of the filtrate with a freshly prepared, saturated aqueous solution of iodine, which is added so as to form a supernatant layer on the surface of the liquid. The presence of glycogen is shown by the production of a wine-red ring at the liquid junction.

If the color does not appear, or is uncertain, it may be confirmed by heating the portion to be tested with sufficient potassium hydroxide solution so as to have 3 per cent potassium hydroxide in the resulting mixture, on a water bath until the muscular fiber is decomposed. Concentrate the broth to half its volume, precipitate the proteins with nitric acid, and add the iodine solution as directed above.

Qualitative Test.—Boil 50 g. of the macerated sample with 50 ml. of water for 15-30 minutes. Filter the broth through moistened filter paper or fine linen. To a portion of the filtrate in a test tube add a few drops of a mixture of 2 parts of iodine, 4 parts of potassium iodide and 100 parts of water. If a considerable quantity of glycogen is present, it produces a dark brown color; this color is destroyed by heating, but it reappears on cooling. If starch is present, it may be precipitated by treating the water extract with two volumes of glacial acetic acid and, after filtering, applying the test for glycogen to the filtrate.

Quantitative Estimation.—This method [8] is based on the separation of glycogen from interfering substances by dissolving those materials in potassium hydroxide solution and alcohol, in which mixture glycogen is insoluble. The glycogen is subsequently estimated either by direct weighing or by hydrolysis to dextrose which is then determined in the usual manner.

Procedure.—Weigh by difference 25 g. of the finely ground and thoroughly mixed sample and transfer to a 400-ml. beaker. Mix with 50 ml. of potassium hydroxide solution (1.5:1). Cover the beaker with a watch glass and digest on a steam bath for 2 hours, stirring occasionally. At the end of the 2 hours, dilute to approximately 200 ml. with cold water. Add to the solution an equal volume of 95 per cent alcohol, cover with a watch glass, and set aside overnight. Decant the supernatant liquid through a large folded filter, allowing the glycogen to remain in the beaker, and wash by decantation 4 times

---

[8] Trowbridge and Francis, *Ind. Eng. Chem.* **2,** 21, 215 (1910).

with a mixture of 2 volumes of 95 per cent alcohol and 1 of water or until the glycogen is white, or nearly so. Transfer the washed precipitate from the beaker to the filter and wash 2 or 3 times more with the solvent mixture. The albuminous substance present retards the filtration if it is permitted to dry on the paper; therefore the funnel should be covered with a watch glass to prevent excessive evaporation.

After the washing is completed, close the bottom of the funnel by a piece of rubber tubing and a pinch cock. Fill the funnel with warm water, cover with the watch glass, and let stand overnight. Open the pinch cock and allow all the solution to pass through the filter into a beaker. Close the funnel with the pinch cock and fill with warm water again. Allow the water to remain in the funnel for 1 hour and then filter as before. At first the glycogen solution appears quite turbid. Continue washing with warm water until the filtrate becomes perfectly clear. To the solution of glycogen in water, add double its volume of 95 per cent alcohol and let stand overnight to complete reprecipitation of the glycogen. Filter and wash as before with the solvent mixture.

The last filtration may be made through a tared Gooch crucible and the weight of the glycogen determined by drying to constant weight. This procedure gives results that are approximately correct. More satisfactory results are obtained by hydrolyzing the glycogen with hydrochloric acid (1:3) and determining the resultant dextrose. Dissolve the glycogen on the filter in warm water as directed above, collecting the filtrate and washings in a 300-ml. volumetric flask and keeping the volume within 225 ml. Add 12.5 ml. hydrochloric acid to the combined filtrate and washings, mix, place in a boiling water bath for 3 hours, cool, and neutralize with sodium hydroxide solution. Cool again, make up to volume with water, and estimate dextrose in an aliquot as directed in the chapter on sugar foods. The corresponding weight of dextrose $\times$ 0.9 = its equivalent of glycogen. Correct this result for dilution to obtain the percentage glycogen in the sample.

DETECTION OF ADDED GELATIN IN SMOKED MEAT PRODUCTS

Pumped meats are sometimes injected with a solution of gelatin. The gelatin solution solidifies during storage and aids both in giving added resistance to knife-cutting of the meat and in the retention of water. In smoked tongues, this addition may be detected organoleptically by cutting the tongue perpendicularly to the veins and then examining the veins. The presence of gelatin will be evidenced by the expulsion of the jelly-like mass from the veins, themselves, by pressure. An aqueous solution of the jelly, after precipitation of proteins, may then be subjected to confirmatory tests for gelatin as described under cream, Chapter VII. *Viscolized fats* will also be present in the veins if injected and may be analyzed as directed in Chapter IX. A modification of the Jacobs and Jaffe method [9] may be used to detect the gelatin in the meat itself.

Procedure.—Weigh 25 g. of the ground sample into a small beaker and add 50 ml. of water. Stir at 5-minute intervals for 20 minutes and then filter through a wire cone or skimmer or through a rapid filter and disregard any cloudiness. Evaporate the filtrate to 10 ml. Cool. Add 3 ml. of lead nitrate solution, 250 g. lead nitrate dissolved in water and made up to 500 ml., and stir. Add 3 ml. of 5 per cent sodium hydroxide solution and stir vigorously. Allow to stand for 5 minutes and filter. To 3 ml. of the filtrate add 2 drops of nitric acid and then a few drops of freshly or recently prepared 5 per cent tannic acid solution. The presence of gelatin is indicated by a white or brownish voluminous precipitate. In the absence of gelatin the solution remains clear. To another portion of the filtrate, add an equal volume of saturated picric acid solution. A marked turbidity indicates the presence of gelatin.

SOYBEAN FLOUR

In a preceding section it was mentioned that soybean flour is used as a binder and sometimes as a substitute for meat in the manufacture of sausages. It is of course desirable to be able to tell when such substitution is made. A rapid method for total cereal product in sausage has been detailed in the section on starch in meat. Kerr [10] recommends the following methods.

Urease Method.—This test depends on the presence of the enzyme urease which normally occurs in soybean flour. This test is not reliable with products which have been heated to a temperature sufficiently high to destroy the urease.

Procedure.—Mix approximately 0.5 g. of the sample with 5 ml. of a solution of urea, 20 g. per liter, in a small test tube or flask containing a strip of red litmus paper partially immersed in the liquid. Stopper the tube or flask and hold at 40° C. for 3 hours. Appearance of a blue color in the litmus paper indicates soybean flour.

Microscopic Examination.—Digest 10 g. of the sample in a 150-ml. beaker, covered with a watch glass, with 50 ml. of alcoholic potash

---

[9] Jacobs and Jaffe, *Ind. Eng. Chem., Anal. Ed.* 4, 418 (1932).
[10] Kerr, *J. Assoc. Official Agr. Chem.* 19, 410 (1936).

solution, 80 g. stick potassium hydroxide in 1 liter of 95 per cent alcohol, with occasional stirring to facilitate digestion. Transfer to a 100-ml. graduated oil tube (A. S. T. M. form) or other similar pointed tube, washing the sediment from the beaker with a stream of 95 per cent alcohol from a wash bottle. Bring the volume to 100 ml. with 95 per cent alcohol. Mix, and allow the tube to stand for an hour or more, giving it a gentle rotation from time to time to assist sedimentation. Syphon off the supernatant alcoholic potash solution and examine the sediment microscopically with a magnification of 120-150 diameters. Use a cover slip to cover the preparation. If the sediment is transferred by means of 15-20 ml. of water, to a small centrifuge tube and centrifuged, a much better preparation is obtained. Instead of centrifuging, allow the oil tube to stand after mixing with the water until sedimentation is complete or filter, if desired. Look for the large hour-glass or I-shaped cells (sometimes called bearer cells) characteristic of soybeans. Examine with polarized light. The cells stand out quite brilliantly in polarized light. A volume of sediment materially exceeding 0.5 ml., due to spices, in a product containing no starch, except spice starch, warrants suspicion that soybean flour has been used. Identification of the characteristic soybean cells in the sediment is proof that soybean flour has been used.

Neither cooking the frankfurter in the casing, as is usually practiced, nor smoking affects the qualitative test for soy, based on the liberation of ammonia from urea by the urease, naturally present. If, however, the soy flour is first made into a stiff paste and heated above 100° C. or under pressure, all of the urease is destroyed.[11] If then, sausage has been heated or processed in such a way as to destroy not only the characteristic enzyme urease but also to destroy the structure of the characteristic cells of soybean, it becomes necessary to examine the meat product for other characteristic components of soybean flour. As noted in Chapter XI, the soybean flour is composed roughly of 40 to 45 per cent protein, 20 per cent oil, 6 to 7 per cent ash, 7 to 10 per cent moisture. The remaining material is carbohydrate in nature.

Examination of the ether extract of the meat product will readily show the presence of an oil not normal to meat if a soybean flour, not defatted, has been used in large amounts. The lecithin content of soybean flour is given by various investigators as 1.65-3.0 per cent of the flour. The carbohydrate of soybean consists largely of indigestible material. A high content of crude fiber in a meat product is evidence of the substitution of soybean for other flours which have only

---

[11] LaWall and Harrison, J. Assoc. Official Agr. Chem. 18, 644 (1935).

a small percentage of crude fiber. A high carbohydrate content, a high crude fiber content, a high lecithin content, a low non-protein nitrogen content and changed constants of the fat are indicative of soybean flour.

FISH

The proximate analysis of fish may be made by methods previously described. Thus moisture or conversely total solids, ash and nitrogen, may be ascertained as described in an appropriate method in Chapter I. Fat or oil may be estimated by the rapid acid hydrolysis method described for meat or by one of the methods detailed below. Metals, preservatives or added color may be detected as detailed in the chapters concerned with those matters.

OIL

Fish oils are difficult to extract quantitatively from fish flesh because they are easily oxidized and polymerized. The fish flesh may be dried by the vacuum desiccator method and then the fat may be estimated. Stansby and Lemon [12] recommend the following methods; one, rapid and depending on the extraction of oil by shaking with an oil solvent and a dehydrating agent but only approximately quantitative, the other a quantitative continuous acetone extraction method.

Rapid Method.—Weigh 20 g. of the finely ground flesh, free of skin and bones, into a shaking bottle equipped with a cork stopper. Add 25 g. of anhydrous sodium sulfate and exactly 100 ml. of ethyl ether. Shake the bottle and its contents for 60 minutes. Allow most of the fine solid particles to settle out and transfer a 20-ml. aliquot of the solution by means of a pipette, through a filter into a small weighed beaker. Wash the filter with three 3- to 5-ml. portions of ether. Evaporate off the ether on a water bath, dry the oil for 1 hour at 100° C., and then weigh.

Acetone Continuous Extraction Method.—Place a 15- to 20-g. sample of the finely ground fish flesh in a continuous extractor and heat for 16 hours, replacing the solvent with fresh acetone after 2 hours. Heat the solutions on a steam bath until all of the acetone and most of the water are removed. Place in a vacuum desiccator over freshly boiled sulfuric acid and evacuate the desiccator. When practically all of the water has been removed, as indicated by the cessation of foaming and bubbling or after about 3 hours, remove the oils and add 35 ml. of ethyl ether. After thorough shaking, pour the solution through a filter, preferably one of the sintered-glass type. Wash the

---

12 Stansby and Lemon, *Ind. Eng. Chem., Anal. Ed.* 9, 341 (1937).

residue with several small portions of ether until the wash liquid is colorless, pouring the solution in each case through the filter. Wash the oil solution into a weighed beaker and remove the ether by means of a carefully regulated blast of air. When the odor of ether is no longer evident, place the beaker in an oven at 100° to 105° C. for 45 minutes. Remove the beaker, place in a desiccator, cool, and weigh.

By these methods, Stansby and Lemon established that the average value of oil in mackerel varied from 12 to 15 per cent rather than the value of 7.1 per cent given in Table 97.

TABLE 97. PROXIMATE COMPOSITION OF FISH [a]

Type	Moisture	Protein	Fat	Ash
Bass	77.7	18.6	2.8	1.2
Cod	82.6	16.5	0.4	1.2
Flounder	84.2	14.2	0.6	1.3
Haddock	81.7	17.2	0.3	1.2
Halibut	75.4	18.6	5.2	1.0
Herring	72.5	19.5	7.1	1.5
Mackerel	73.4	18.7	7.1	1.2
Perch, white	75.7	19.3	4.0	1.2
Perch, yellow	79.3	18.7	0.8	1.2
Pickerel, pike	79.8	18.7	0.5	1.1
Salmon	64.6	22.0	12.8	1.4
Shad	70.6	18.8	9.5	1.3
Smelt	79.2	17.6	1.8	1.7
Sturgeon	78.7	18.1	1.9	1.4
Trout	77.8	19.2	2.1	1.2
Whitefish	69.8	22.9	6.5	1.6

[a] Atwater and Bryant, *U. S. Dept. Agr., Bull.* 28 (1906).

CHLORIDE

Chloride [13] expressed as sodium chloride may be estimated by simple titration with thiocyanate solution with but slight modification.

Procedure.—Place 10 g. of the sample into a flask or beaker, add a known volume of silver nitrate solution which is more than sufficient to precipitate all the chloride as silver chloride and then add 20 ml. of nitric acid. Boil gently on a hot plate, until all solid matter is dissolved except silver chloride. Cool; add 50 ml. of water; 5 ml. of a saturated solution of hydrazine sulfate, to remove any nitrous acid formed; 5 ml. of a saturated solution of ferric ammonium alum as indicator; 1 ml. of nitrobenzene for each 0.05 g. of chloride; [14] and titrate the excess of silver with 0.1 N potassium thiocyanate solution until a permanent light brown color appears. Subtract the milliliters of 0.1 N thiocyanate used from the milliliters of 0.1 N silver nitrate

[13] Grigsby, *J. Assoc. Official Agr. Chem.* 20, 410 (1937).
[14] Caldwell and Moyer, *Ind. Eng. Chem., Anal. Ed.* 7, 38 (1935).

added and calculate the quantity of chlorine as sodium chloride. With a 10-g. sample, each milliliter of 0.1 N silver nitrate solution is equivalent to 0.058 per cent sodium chloride.

If it is desired to make the determination on an ash, add 10 per cent calcium acetate solution as the fixative and ash at 550° C. Dissolve the ash in 25 ml. of nitric acid (1:3) and proceed as directed above.

DETECTION OF PHENOL IN SMOKED FISH [15]

Procedure.—Weigh 50 g. of sample into a 400-ml. beaker. Add 200 ml. of water-alcohol mixture (1:1), macerate for 10 minutes, allow to settle and filter. Chill in refrigerator and then refilter.

To 5 ml. of filtrate add 5 ml. of 0.5 per cent sodium borate solution and mix. Add 1 ml. of 2,6-dibromoquinonechloroimide. Allow to stand for 1 hour at room temperature. Extract with 5 ml. of N butyl alcohol and note the color. A blue color is positive for phenol.

OYSTERS

It is generally sufficient for the chemical determination of "floating" or excessive washing of oysters to estimate the solids, ash, and salt content of oysters. "Floating" is the practice of soaking oysters, fish, and the like, in fresh water for the purpose of making them appear plump and increasing their weight. Since the oysters have a higher salt content than the surrounding fresh water, the water passes into the oysters by the process of osmosis and thus bloats them.

Where oysters are grown on muddy or soft bottoms, the shells often become coated with mud. The most efficient method of washing such oysters and, for that matter, all oysters, is by using a blower with a solution of salt. If the oysters are subjected to a reasonable washing with fresh water, a gain in volume from 3 to 10 per cent may take place.[16]

The volume of free liquor in oysters may be ascertained by draining the meats on an oyster skimmer for 1 minute and measuring the shrinkage in volume of the meats after they are returned to the original measuring vessel. The skimmer is a metal tray with holes ¼ in. in diameter, spaced 1¼ in. in a square pattern and of sufficient area to take the entire volume of oysters drained at one time in a single layer.[17]

---

[15] Tucker, *J. Assoc. Official Agr. Chem.* **25,** 779 (1942).
[16] Hunter and Harrison, *U. S. Dept. Agr., Tech Bull.* 64 (1928).
[17] Grigsby, *J. Assoc. Official Agr. Chem.* **20,** 410 (1937).

Atwater [18] and Bryant give the proximate composition of oysters as: total solids—13.1 per cent; protein—6.2 per cent; fat—1.2 per cent; carbohydrate—3.7 per cent; ash—2.0 per cent. More recent analyses are tabulated in Table 98.

TABLE 98. NEW ENGLAND OYSTERS[16]

	Unwashed			Washed		
	Max.	Min.	Ave.	Max.	Min.	Ave.
Meat.............	95.7	75.6	85.1	98.5	86.1	93.1
Liquor...................	24.4	4.3	14.9	13.9	1.5	6.9
Meats						
Solids.................	24.05	16.94	20.57	21.44	15.29	18.20
Ash...................	3.03	1.84	2.29	1.99	1.20	1.50
Salt...................	1.41	0.67	0.92	0.63	0.24	0.35
Liquor						
Solids.................	6.67	4.60	5.32	7.00	3.62	5.01
Ash...................	2.76	2.40	2.57	1.72	1.08	1.34
Salt...................	2.66	1.81	2.13	1.56	0.49	0.97
Entire Sample						
Solids.................	21.58	15.55	18.29	21.0	14.41	17.28
Ash............... ...	2.70	1.97	2.25	1.77	1.25	1.44
Salt...................	1.53	0.80	1.10	0.68	0.27	0.40

The Food and Drug Administration [19] has set standards and definitions for raw and canned oysters. These are abstracted below.

RAW OYSTERS

Oysters, Raw Oysters, Shucked Oysters.—*Identity.*—(a) Oysters, raw oysters, shucked oysters, are the class of foods each of which is obtained by shucking shell oysters and preparing them in accordance with the procedure described in (b) below. The name of each such food is the name specified in the applicable definition and standard of identity described below.

(b) If water, or salt water containing less than 0.75 per cent salt, is used in any vessel into which the oysters are shucked, the combined volume of oysters and liquid when such oysters are emptied from such vessel is not less than four times the volume of such water or salt water. Any liquid accumulated with the oysters is removed. The oysters are washed, by blowing or otherwise, in water or salt water, or both. The total time that the oysters are in contact with water or salt water after leaving the shucker, including the time of washing, rinsing, and any other contact with water or salt water, is

---

[18] Atwater and Bryant, *U. S. Dept. Agr. Bull.* **28** (1906).

[19] *U. S. Food Drug Admin.*, *S.R.A.*, *F.D.&C.* 2, Part 36, June, 1952, as amended January, 1954.

not more than 30 minutes. In computing the time of contact with water or salt water, the length of time that oysters are in contact with water or salt water that is agitated by blowing or otherwise shall be calculated at twice its actual length. Any period of time that oysters are in contact with salt water, containing not less than 0.75 per cent salt before contact with oysters, shall not be included in computing the time that the oysters are in contact with water or salt water. Before packing into the containers for shipment or other delivery for consumption the oysters are thoroughly drained and are packed without any added substance.

(c) For the purposes of this definition:

(1) "Shell oysters" means live oysters of any of the species, *Ostrea virginica, Ostrea gigas, Ostrea lurida,* in the shell, which after removal from their beds, have not been floated or otherwise held under conditions which result in the addition of water.

(2) **Definition, Thoroughly Drained.**—"Thoroughly drained" means one of the following:

(i) The oysters are drained on a strainer or skimmer which has an area of not less than 300 sq. in. per gal. of oysters, drained, and has perforations of at least 0.25 in. in diameter and not more than 1¼ in. apart, or perforations of equivalent areas and distribution. The oysters are distributed evenly over the draining surface of the skimmer and drained for not less than 5 minutes; or

(ii) The oysters are drained by any method other than that prescribed by (i) above whereby liquid from the oysters is removed so that when the oysters are tested within 15 minutes after packing by draining a representative gallon of oysters on a skimmer of the dimensions and in the manner described in (i) for 2 minutes, not more than 5 per cent of liquid by weight is removed by such draining.

Extra Large Oysters, Oysters Counts (or Plants), Extra Large Raw Oysters, Raw Oysters Count (or Plants), Extra Large Shucked Oysters, Shucked Oysters Counts (or Plants) are of the species *Ostrea virginica* and are of such size that 1 gal. contains not more than 160 oysters and a quart of the smallest oysters selected therefrom contains not more than 44 oysters.

Large oysters, oysters extra selects, large raw oysters, raw oysters extra selects, large shucked oysters, shucked oysters extra selects are of the species *Ostrea virginica* and are of such size that 1 gal. contains more than 160 oysters but not more than 210 oysters; a quart of the smallest oysters selected therefrom contains not more than 58 oysters and a quart of the largest oysters selected therefrom contains more than 36 oysters.

Medium oysters, oysters selects, medium raw oysters, raw oysters selects, medium shucked oysters, shucked oysters selects are of the species *Ostrea virginica* and are of such size that 1 gal. contains more than 210 oysters, but not more than 300 oysters; a quart of the smallest oysters selected therefrom contains not more than 83 oysters, and a quart of the largest oysters selected therefrom contains more than 46 oysters.

Small oysters, oysters standards, small raw oysters, raw oysters standards, small shucked oysters, shucked oysters standards are of the species *Ostrea virginica* and are of such size that 1 gal. contains more than 300 oysters but not more than 500 oysters; a quart of the smallest oysters selected therefrom contains not more than 138 oysters and a quart of the largest oysters selected therefrom contains more than 68 oysters.

Very small oysters, very small raw oysters, very small shucked oysters are of the species *Ostrea virginica* and are of such size that 1 gal. contains more than 500 oysters, and a quart of the largest oysters selected therefrom contains more than 112 oysters.

Olympia oysters, raw Olympia oysters, shucked Olympia oysters are of the species *Ostrea lurida* and conform to the definition and standard of identity described for oysters.

In 1954 the definitions and standards of identity for Pacific oysters, raw Pacific oysters, and shucked Pacific oysters were revised with the 6 former groups being reduced to four.

Large Pacific oysters, large raw Pacific oysters, large shucked Pacific oysters are of the species *Ostrea gigas* and are of such size that 1 gal. contains not more than 64 oysters and the largest oyster in the container is not more than twice the weight of the smallest oyster therein.

Medium Pacific oysters, medium raw Pacific oysters, medium shucked Pacific oysters are of the species *Ostrea gigas* and are of such size that 1 gal. contains more than 64 and not more than 96 oysters, and the largest oyster in the container is not more than twice the weight of the smallest oyster therein.

Small Pacific oysters, small raw Pacific oysters, small shucked Pacific oysters are of the species *Ostrea gigas* and are of such size that 1 gal. contains more than 96 and not more than 144 oysters, and the largest oyster in the container is not more than twice the weight of the smallest oyster therein.

Extra small Pacific oysters, extra small raw Pacific oysters, extra small shucked Pacific oysters are of the species *Ostrea gigas* and are of such size that 1 gal. contains more than 144, and the largest oyster in the container is not more than twice the weight of the smallest oyster therein.

The amended definitions of 1954 compare with the original definitions of 1949 in the following manner: large Pacific oysters are equivalent to the former 5 to 8 per pint size, medium Pacific oysters are equivalent to both the 8 to 10 and 10 to 12 per pint sizes, small Pacific oysters are equivalent to both the 12 to 15 and the 15 to 18 per pint sizes, and extra small Pacific oysters are equivalent to the more than 18 per pint size.

CANNED OYSTERS

Canned Oysters.—Identity.—(a) Canned oysters is the food prepared from one or any mixture of two or all of the forms of oysters specified in (b) below, and a packing medium of water, or the watery liquid draining from oysters before or during processing, or a mixture of such liquid and water. The food may be seasoned with salt. It is sealed in containers and so processed by heat as to prevent spoilage.

(b) The forms of oysters referred to in paragraph (a) of this section are prepared from oysters which have been removed from their shells and washed and which may be steamed while in the shell or steamed or blanched or both after removal therefrom, and are as follows:

(1) Whole oysters with such broken pieces of oysters as normally occur in removing oysters from their shells, washing, and packing.

(2) Pieces of oysters obtained by segregating pieces of oysters broken in shucking, washing, or packing whole oysters.

(3) Cut oysters obtained by cutting whole oysters.

(a) **Fill of Container.**—The standard of fill of container for canned oysters is a fill such that the drained weight of oysters taken from each container is not less than 59 per cent of the water capacity of the container.

(b) **Water Capacity.**—Water capacity of containers is determined by the general method described in Chapter I.

(c) **Drained Weight.**—Drained weight is determined by the following method.

Procedure.—Keep the unopened canned oyster container at a temperature of not less than 68° or more than 95° F. for at least 12 hours immediately preceding the determination. After opening, tilt the container so as to distribute its contents evenly over the meshes of a circular sieve which has been previously weighed. The diameter of the sieve is 8 in. if the quantity of the contents of the container is less than 3 lbs. and 12 in. if such quantity is 3 lbs. or more. The bottom of the sieve is woven-wire cloth which complies with the specifications for such cloth set forth under "2380 Micron (No. 8)," in Table I of "Standard Specifications for Sieves," published March 1, 1940, in L. C. 584 of the U. S. Department of Commerce, National Bureau of Standards. Without shifting the material on the sieve, so incline the sieve as to facilitate drainage. Two minutes from the time drainage begins, weigh the sieve and the drained oysters. The weight so found, less the weight of the sieve, shall be considered to be the drained weight of the oysters.

(d) If canned oysters fall below the standard of fill of container described in· (a) above they are considered substandard fill.

CANNED SHRIMP

Canned Wet Pack Shrimp and Canned Dry Pack Shrimp (in nontransparent containers).—**Fill of Containers.**—(a) The standard of fill of nontransparent containers for canned wet pack shrimp is a fill such that the cut-out weight of shrimp taken from each can is not less than 64 per cent of the water capacity of the container, and, for canned dry pack shrimp (except that packed in the nontransparent cylindrical container which is $2^{11}/_{16}$ in. in diameter and 4 in. in height), is a fill such that the cut-out weight of shrimp taken from each can is not less than 60 per cent of the water capacity of the container. The standard of fill for canned dry pack shrimp packed in the nontransparent cylindrical container which is $2^{11}/_{16}$ in. in diameter and 4 in. in height is a cut-out weight of not less than $6\frac{1}{2}$ avoirdupois oz. of shrimp for each container.

Water Capacity.—Water capacity of containers is determined by the general method provided in Chapter I.

Cut-out Weight.—Cut-out weight is determined by the following method.

Procedure.—Keep the unopened canned shrimp container at a temperature of not less than 68° nor more than 95° F. for at least 12 hours immediately preceding the determination. After opening, tilt the container so as to distribute the shrimp evenly over the meshes of a circular sieve which has been previously weighed. The diameter of the sieve is 8 in. if the quantity of the contents of the container· is less than 3 lbs. and 12 in. if such quantity is 3 lbs. or more. The bottom of the sieve is woven-wire cloth which complies with the specifications for such cloth set forth under "2380 Micron (No. 8)" in Table I of "Standard Specifications for Sieves," published March 1, 1940, in L. C. 584 of the U. S. Department of Commerce, National Bureau of Standards. Without shifting the material on the sieve, so incline the sieve as to facilitate drainage. Two minutes from the time drainage begins, weigh the sieve and the drained shrimp. The weight so found, less the weight of the sieve, shall be considered to be the cut-out weight of the shrimp.

(b) If canned wet pack shrimp or canned dry pack shrimp, in nontransparent containers, falls below the applicable standard of fill of container described in (a) they are considered substandard fill.

EGGS AND EGG PRODUCTS

Eggs are almost as complete a food as milk although not as well balanced.

Definitions.—The Food and Drug Administration [20] gives the following definitions for eggs and egg products.

Eggs.—No regulation has been promulgated, fixing and establishing a reasonable definition and standard of identity for the food commonly known as eggs.

Liquid eggs, mixed eggs, liquid whole eggs, mixed whole eggs are eggs of the domestic hen, broken from the shells, and with yolks and whites in their natural proportions as so broken. They may be mixed, or mixed and strained.

Frozen eggs, frozen whole eggs, frozen mixed eggs are the food prepared by freezing liquid eggs.

Dried eggs, dried whole eggs are the food prepared by drying liquid eggs. They may be powdered. They contain not less than 92 per cent total egg solids.

Egg yolks, liquid egg yolks, yolks, liquid yolks are yolks of eggs of the domestic hen so separated from the whites thereof as to contain not less than 43 per cent total egg solids. They may be mixed, or mixed and strained.

Frozen yolks, frozen egg yolks are the food prepared by freezing egg yolks.

Dried egg yolks, dried yolks are the food prepared by drying egg yolks. They contain not less than 95 per cent total egg solids.

---

[20] *U. S. Food Drug Admin., S.R.A., F.D.&C. 2,* Part 42, June, 1952.

Whole Eggs

Whole eggs are seldom analyzed chemically by the food analyst except for investigational purposes. However, organoleptic examination by "candling," that is, examination of the egg by rotation before a light, is often necessary. Whole eggs are generally classified as follows:

1. *Firsts* and *extras* are freshly laid, sound, whole, clean shell, medium and good-sized eggs. The best eggs of the year are laid during the spring months and these eggs are usually selected for storage for the winter trade.

2. *Seconds* include all eggs not considered firsts except "spots" and "rots." Generally this group includes all the undersized, dirty, checked, cracked shell, or weak eggs. These terms signify:

   a. *Undersized* are eggs which may be classed as first except for size.

   b. *Checks* and *cracks* are eggs, the shells of which have become broken by careless handling but which have shell membranes intact.

   c. *Leakers* are eggs whose shells and shell membranes are sufficiently broken to permit a portion of their contents to escape.

   d. *Dirties* are eggs, the shells of which have become soiled from unclean nests and through other means.

   e. *Weak eggs* are those in which the albumin has become weak or watery due to high or varying temperature changes.

   These seconds are considered suitable for food purposes when handled under proper conditions, except "leakers" which are usually badly contaminated.

3. *Spots* and *rots* consist of all the discards and are considered useful only for manufacturing purposes, such as for tanning leather. They are eggs which show a spot before the candle. These spots may be due to developing embryos, blood rings, or molds, or by the attachment of the yolk to the shell, which may be caused by holding in one position. White rot or sour rot designate eggs whose white and yolk are just beginning to mix. Stale eggs are those whose shell shows an enlarged air space and whose yolk has gained in opacity and fallen to the pointed end of the shell. Spot rot designate eggs in which the yolk becomes adherent to one or both of the shell membranes. Moldy eggs are those which show dense black areas of varying size inside the shell. Light spot eggs are those which has been incubated sufficiently to show a small darkened area.

Analysis of Eggs

The rapid growth of the egg-breaking industry has led to increased need for accurate and reliable laboratory methods of analysis and control, especially because eggs are a product that is highly perishable and cannot be pasteurized as are milk, cheese, and similar dairy products. Preservation is accomplished generally by means of rapid or sharp freezing and any changes in composition of the product packed must be made within a few hours, or before the batch in question is frozen solid. The eggs must be packed in different compositions for different industries. Thus egg whites are used by the baking and candy industries; plain yolk and salted yolk by the mayonnaise and noodle trades; whole eggs, sugar yolk, and glycerol yolk by the baking, ice cream, and confectionery trades. Special types of egg products are packed to specification.

Guthmann and Terre [21] recommend that, in order to overcome discrepancies in sampling, at least 3 samples be taken from each batch churned, one at the top of the churn, one at the middle and one at the bottom. The three samples are then mixed in a quart container such as a mason jar and constitute a composite sample. Frozen eggs in 30-lb. cans may be sampled by the use of a drill in a manner similar to the use of a trier as described in Chapter I.

In general, analyses may be made in a manner similar to that of other food products. Total solids and moisture are most accurately determined by the prior evaporation to dryness of a weighed portion in a tared dish with final drying in a vacuum oven, preferably under 25-in. vacuum at 100° C. for 7 hours. Bailey [22] gives a refractometric method for liquid eggs, which is useful for rapidity although not yet completely satisfactory. Fat may be determined on an appropriate weight of sample by the acid hydrolysis modification of the Roese-Gottlieb method detailed under the section "Cheese." Nitrogen may be determined by the Kjeldahl-Gunning-Arnold method. Ash, phosphates, after making alkaline with 10 per cent sodium carbonate solution, are estimated as usual.

FAT

Rapid Method.—Paley [23] devised a modified Babcock method for fat in dried eggs using the Paley test bottle. Samples are weighed out on a regular cream test scale and hydrochloric acid (4:1) is used for the digestion and hydrolysis of the protein. Sulfuric acid (3.5:1) as in the Pennsylvania method is also used.

Procedure.—Weigh 3 g. of the well-mixed sample into a 9-g., 20 per cent Paley test bottle, using a cream test scale. Add 20 ml. hydro-

21 Guthmann and Terre, *Ind. Eng. Chem., Anal. Ed.* 8, 377 (1936).
22 Bailey, *Ind. Eng. Chem., Anal. Ed.*, 7, 385 (1935).
23 Paley and Rubin, *Food Industries* 18, 1194 (1946).

chloric acid (4:1), insert rubber stopper, and mix well. Place the bottle in a water bath at 70° C., and gradually heat the water to the boiling point. Shake at frequent intervals, using care to prevent foaming that may occur during the early stages of heating. Continue to heat in the boiling water bath until digestion is completed. This is indicated, when lumps are no longer visible and a clear layer of fat is observed on the surface. The digestion time is normally about 30 minutes. Remove from water bath, cool under running tap water, add 3 ml. n-butyl alcohol and shake. Cautiously add 8 ml. sulfuric acid (3.5:1), shake, and place the bottle in a water bath at 60 to 65° C. Temper for 10 minutes with occasional shaking. Place bottle in a heated centrifuge, as used for the Babcock test, with the stopper facing the center of the machine. Whirl for 5 minutes. Add hot water to base of neck. Centrifuge for 2 minutes. Add hot water to bring the fat into the graduated neck. Centrifuge for 1 minute and place in a tempering bath at 54° C. for 5 minutes. Add colored mineral oil and read. Multiply the reading by 3 to obtain the percentage of fat.

Lipoids and Lipoid Phosphoric Acid [24]

Reagent.—Alcoholic Sodium Hydroxide Solution.—Prepare a saturated solution free from carbonates by dissolving 100 g. of sodium hydroxide in 100 ml. of water. Allow the mixture to stand until clear, or filter through a hard filter paper which has been soaked in alcohol. Five ml. of the sodium hydroxide solution contains approximately 4 g. of NaOH. Dissolve 50 ml. of this solution in 900 ml. of 95 per cent alcohol and dilute with 95 per cent alcohol to 1 liter.

Preparation of Solution.—(a) Liquid Eggs.—Weigh accurately by difference with the aid of a Mojonnier pipette and carriage approximately 4 g. of the well-mixed sample into a 100-ml. volumetric flask, add very slowly, dropwise, from a pipette, 25 ml. of a mixed solvent consisting of equal volumes of chloroform and absolute alcohol, shaking constantly until the proteins become coagulated and then thoroughly broken up. Add 60-65 ml. more of the solvent and allow to stand 1 hour, shaking at 5-minute intervals. Fill to the mark with the solvent, shake, and allow the mixture to stand until clear.

(b) Dried Eggs.—Transfer 2 g. of the well-mixed sample into a 100-ml. volumetric flask, add 85-90 ml. of a mixed solvent consisting of equal volumes of chloroform and absolute alcohol, and allow to

---

[24] *Methods of Analysis,* A.O.A.C. Washington, 1945.

stand 1 hour, mixing at 5-minute intervals. Proceed as directed above.

Procedure.—Transfer a 50-ml. aliquot to a 150-ml. beaker and evaporate the extract to dryness on a steam bath. Use an electric fan or a gentle blast of dry air to hasten evaporation. Place the beaker into an oven at 100° C. for 5-10 minutes to remove any remaining moisture. Dissolve the dry extract in 5-10 ml. of chloroform and filter the solution into a weighed 100-ml. Pyrex beaker through a pledget of cotton packed into the stem of a funnel, transferring all soluble extract from the bottom and sides of the beaker by means of chloroform from a wash bottle. Finally wash the funnel and stem tip. The filtrate should be clear. Evaporate the chloroform on a steam bath, and dry the beaker and contents in an oven at 100° C. to minimum weight. This generally requires approximately 90 minutes. Allow the beaker to stand in air until no further change in weight takes place, approximately 30 minutes. Weigh, and report the percentage of lipoids.

Procedure for Lipoid Phosphorus Pentoxide.—Dissolve the dried lipoids in 2-3 ml. of chloroform, add 10-20 ml. of the alcoholic sodium hydroxide solution, evaporate to dryness on a steam bath, using care to avoid spattering, and place the beaker into an oven at 100° C. for 30 minutes to remove any remaining moisture. Transfer the beaker while hot to an electric muffle heated to 500° C., faint redness, and allow it to remain at that temperature for 1 hour. Cool, add a few drops of water, break up the charge with a glass rod, having a flattened end, cover the beaker with a watch glass, add slowly 5 ml. nitric acid (1:3), mix, remove the watch glass, and filter, collecting the filtrate in a 300-ml. or 500-ml. flask. Thoroughly wash the charred material and filter paper with water, by means of a wash bottle. Determine phosphorus pentoxide as directed in Chapter XVIII. Report as percentage lipoid phosphorus pentoxide.

DISTINCTION BETWEEN EGG YOLK AND VEGETABLE LECITHIN

With the increased use of soybean flour and of vegetable lecithin, itself, which is used as an emulsifying agent, the estimation of lecithin-phosphorus pentoxide as a means of determining egg content is dubious unless there is assurance that vegetable lecithin is not present. Differences in the nitrogen-phosphorus ratio are used by some investigators.[25] Kluge [26] uses the lecithin-phosphorus to sterols ratio which is greater in egg yolk than in vegetable licithin. Because this

---

[25] Nottbohm and Mayer, *Chem. Ztg.* **56**, 881 (1932).
[26] Kluge, *Z. Untersuch. Lebensm.* **69**, 9 (1935).

ratio is very much greater in egg yolk than in vegetable lecithin preparations, it may be utilized in conjunction with other tests for the detection of vegetable lecithin in food pastes. If the egg content of the material, as calculated from the lecithin-phosphorus contest, is greater than that calculated from the sterols present, the presence of a vegetable lecithin preparation is indicated.

Weyl's Lutein Test.—A qualitative test is Weyl's [26] lutein test. The dry residue from an acetone extraction is dissolved in ether and treated with solid sodium nitrite and dilute sulfuric acid. With the oil from egg food pastes, the deep yellow etheral solution undergoes almost complete decolorization, whereas with that from vegetable lecithin the yellow color persists virtually unchanged.

Another test is that the iodine value of egg lecithin is lower than that of vegetable lecithin. The former has a value about 70 whereas the latter has a value from 108 to 117.

GLYCEROL

Glycerol yolks are packed chiefly for the baking trade. They contain from 5-7 per cent of glycerol and have the property of keeping the cakes fresh and moist for a longer period of time. The following method is not applicable in the presence of sugar in the yolks. The reagents used are those detailed in the section "Glycerol," Chapter XIV.

Procedure.—Weigh out from 8 to 10 g. of the sample into a 50-ml. beaker, add 1 ml. of 5 per cent acetic acid, and heat over a low flame on a wire gauze with constant stirring until the egg is thoroughly coagulated. The material must not be allowed to char. The lumps of egg should be dry in character and not of a sticky or pasty consistency.

Transfer the coagulated material to a 250-ml. volumetric flask, washing the beaker with 25 to 50 ml. of water. Add about 100 ml. of washed alumina cream and make up the mixture to the mark with water; shake thoroughly, allow to stand for 5 minutes, shake again, and filter through a coarse filter paper. The filtrate should be perfectly clear. Transfer 25 ml. of the filtrate to a 250-ml. volumetric flask and add 25 ml. of sulfuric acid and 2 ml. of silver nitrate, 10 per cent solution. Add 10 ml. of a strong potassium dichromate solution, slowly from a pipette at 20° C., place the volumetric flask on a steam bath, and heat for exactly 30 minutes. Cool and dilute to the mark with water. Place this solution in a burette and titrate into a beaker containing 100 ml. of water, 15 ml. of sulfuric-phosphoric acid retarding mixture and 10 ml. of ferrous ammonium sulfate solution measured accurately with a pipette. Add 3 drops of a solution of 1 g. of diphenylamine in 100 ml. of sulfuric acid, as indicator.

As the titration proceeds the color changes from green to blue-gray to a pure violet at the end point.

Before each determination the ferrous ammonium sulfate solution is standardized against dilute dichromate solution. Measure 25 ml. of strong dichromate solution at 20° C. into a 500-ml. volumetric flask and dilute to the mark with water at room temperature. Twenty ml. of this solution equals 1 ml. of strong dichromate. Place the dilute dichromate in a burette, and titrate into a beaker containing 100 ml. of water, 15 ml. of sulfuric-phosphoric acid mixture, and 10 ml. of the ferrous ammonium sulfate. Three drops of diphenylamine solution are used as an indicator just as in the determination.

The percentage of glycerol is calculated according to the following formula:

When 10 ml. of strong dichromate solution and 10 ml. of ferrous ammonium sulfate solution are used

$$\% \text{ of glycerol} = \frac{10\left[10.0 - \left(\dfrac{12.5 \times \text{ml. of dilute } K_2Cr_2O_7}{\text{ml. of oxidized glycerol solution}}\right)\right]}{\text{sample weight}}$$

There is some oxidizable material present in ordinary egg yolk; hence the determination of glycerol in a prepared glycerol yolk is on an average 0.5 per cent high. Controls with pure egg yolks should be run at the same time or else the above-mentioned figure should be deducted.

GLYCEROL IN PRESENCE OF SUGAR [27]

Proceed as detailed in the glycerol determination, taking a sample so that the combined weight of glycerol and sugar is not more than 3.0 g. If starch is present, it must be removed by the usual alcohol procedure or by some other method. If alcohol is used, the alcohol should be removed by evaporation. Boil for 20 minutes with the sulfuric acid in order to insure complete inversion. Determine the amount of potassium dichromate needed to oxidize both sugar and glycerol. Determine invert sugar as detailed in the next section. Calculate percentage of glycerol after deducting the amount of potassium dichromate required by sugar.

1 ml. $K_2Cr_2O_7$ is equivalent to 0.0100 g. of glycerol.

1 ml. $K_2Cr_2O_7$ is equivalent to 0.01142 g. of invert sugar.

The sugar may be removed with milk of lime and the glycerol may then be estimated as previously described.

---

[27] Hoyt and Pemberton, *Cotton Oil Press* 14, 54 (1922); 14, 340 (1922).

SUGAR

Sugar before and after inversion may be estimated by the Lane-Eynon method.

Procedure.—Wash about 25 g. of the sample into a 200-ml. volumetric flask with 75 ml. of water. Make slightly acid by adding 2 ml. of 5 per cent acetic acid for white or whole egg and 1 ml. for yolk. Mix, and immerse the flask in boiling water until the egg material is thoroughly coagulated. This requires from 15 minutes to 0.5 hour. Cool to room temperature and make up to the mark with washed alumina cream. Shake the sample vigorously for 1 minute, allow it to stand for 5 minutes, and then shake for 1 minute. Filter through a dry folded filter. Dilute 25 ml. of the filtrate to 100 ml. and proceed to determine reducing sugars before inversion by the Lane-Eynon method as detailed in Chapter X. Invert 25 ml. of the filtrate by the hydrochloric acid inversion method, make to volume after neutralization, and proceed with the Lane-Eynon method to determine total reducing sugars.

SALT

Salt yolks are extensively packed for the mayonnaise trade, since the presence of about 10 per cent salt not only acts as a preservative but permits the egg to be thawed as a syrupy liquid, whereas a plain yolk when thawed has a gummy, semi-solid consistency. Since salt is very much cheaper than egg yolks, both the packer and the consumer are vigilant in seeing that the amount of salt added stays very close to the predetermined figure.

Procedure.—Weigh 5 g. of the sample into a 500-ml. volumetric flask containing 200 ml. of water. Make up to the mark and shake thoroughly. Pipette 50 ml. of this solution into an Erlenmeyer flask, and add 10 ml. of 0.1 N silver nitrate solution with shaking. Add 10 ml. strong nitric acid and 5 ml. of a saturated solution of ferric ammonium sulfate, and titrate with 0.1 N ammonium thiocyanate solution to a permanent light brown color. One ml. of 0.1 N silver nitrate equals 0.0058 g. of sodium chloride. Correction factor: Subtract 0.3 per cent of sodium chloride from the result.

QUALITY FACTORS

Fluorescence.—The fluorescence exhibited by spray-dried eggs must not exceed 30 to avoid penalty and 35 with penalty in order to conform to the specifications of Quartermaster Corps, tentative specification C. Q. D. No. 1178 of May, 1945. The method used for determining this value is the following.

Preparation of Quinine Reference Solution.—Stock Solution.—This is to be supplied by the Quartermaster Corps Subsistence Research and Development Laboratory, 1819 West Pershing Road, Chicago 9, Illinois The stock solution contains 50 micrograms quinine sulfate per ml. Keep it in a dark glass container in a refrigerator (35-40° F.). This stock solution, if kept in a dark glass container under refrigeration, is stable for several months.

Reference Solution.—Dilute 1.00 ml. of stock solution to 250.0 ml. with 0.1 N sulfuric acid. This solution contains 0.200 micrograms of quinine sulfate per milliliter and should be stored in a refrigerator (35-40° F.).

Calibration of the Photofluorometer.—Switch on the instrument and permit it to warm up for approximately 15 minutes before attempting *any* adjustment. Balance the photofluorometer, Coleman or equivalent, according to the manufacturer's instructions. Place 15 ml. of reference solution adjusted to 25° C. into a sample tube. Insert this into the instrument and adjust the needle to 50. The instrument is now ready for measuring the fluorescence of the saline extracts.

Purification of Chloroform.—Distill chloroform, using low heat. Wash distilled chloroform twice with equal volumes of 10 per cent of ammonium hydroxide. Wash twice with equal volumes of distilled water. Dry chloroform by adding an excess of anhydrous calcium chloride to it in a flask. Carefully decant the chloroform from the calcium chloride. Redistill, using about 10 g. of phosphorus pentoxide per liter of chloroform in the distilling flask.

Defatting of Sample.—Brush approximately 2.5 of egg powder into a 125-ml. Erlenmeyer flask, and wash down with 25 ml. of purified chloroform. Swirl the mixture by hand for 1 minute. (Cork or rubber stoppers must not be used.) Pour the mixture on to a No. 1 Whatman filter paper. After the powder is reasonably dry, return it to the flask and wash down any adhering powder with the second 25 ml. of chloroform. Extract the powder twice more (4 times in all) using fresh chloroform each time and a fresh filter paper for the final filtration.

Saline Extraction.—Dry the defatted powder (until free of chloroform odor) by spreading it out on the filter paper and place it in a warm (not to exceed 35° C.), well-ventilated place. Weigh out 1.000 g. of the defatted, dry powder and transfer to a clean, dry 250-ml. flask. Add 100.0 ml. of a 10 per cent sodium chloride solution and shake for 30 minutes at 300 strokes per minute in a Kahn or equivalent shaker. Filter the solution through a No. 1 Whatman filter paper and collect the filtrate in a clean, dry graduate. Collect the first 15 ml. of clear filtrate, adjust the temperature of the solution to 25° C. and determine the fluorescence.

Determine the fluorescence on 15 ml. of the 10 per cent salt solution adjusted to 25° C. (used for extracting defatted egg powder). Subtract this value from the fluorescence of the egg extract.

Oxygen.—The oxygen content may be determined by the Orsat method [28] or by an equivalent procedure. The analysis should be performed on an empty can immediately after the sealing operation, processed through the gassing chamber in the same manner as a can filled with the finished product.

Beating Test.—Procedure.—Dissolve 1½ oz. of powdered egg white in 15 oz. of water at 70° F. Place in a 10-qt. bowl and beat in a Hobart electric mixer (or any other equivalent mixer) with a wire whip for 90 seconds in second speed, and 90 seconds in third speed. Remove the whip, level the foam, and measure its depth at a point midway between the center and the side of the bowl.

Off-odors and Flavors.—The presence of off-odors and flavors in powdered whole egg may be determined in accordance with the following procedure.

Mix 24 g. of sample with 100 ml. of water and thoroughly reconstitute in a mechanical mixer. Transfer the sample to a 400-ml. beaker and cook in a water bath at 90 to 95° C. with constant stirring (using glass rods) until the mixture reaches the consistency of scrambled egg. Place hot 10-g. portions in coded Petri dishes and cover. Serve the samples immediately for evaluation against a fresh current-receipt shell egg control.

Evaluate the cooked samples in duplicate in a quiet odor-free room. Not more than twelve determinations should be made by any judge at any scoring session. Judges should not converse or in any way influence each other's decision. Smoking, and cosmetics on the person, should not be permitted during the scoring period. Individuals cooking, preparing, and serving the samples should not be judges.

SELECTED REFERENCES

Allen's Commercial Organic Analysis, 5th Ed., Vol. IX. Blakiston, Philadelphia, 1932.

Hawk, Oser, and Summerson, *Practical Physiological Chemistry*. Blakiston, Philadelphia, 1947.

Jacobs, ed., *Chemistry and Technology of Food and Food Products*, 2nd Ed. Interscience, New York, 1951.

Leach, *Food Inspection and Analysis*. Wiley, New York, 1920.

Methods of Analysis, A.O.A.C., 8th Ed. Washington, 1955.

Moulton, *Meat through the Microscope*. Chicago, 1929.

Thurston, *Pharmaceutical and Food Analysis*. Van Nostrand, New York, 1922.

---

[28] Furman, *Scott's Standard Methods of Chemical Analysis*. Van Nostrand, New York.

CHAPTER XVII

VITAMINS

Definition.—The vitamins and vitagens are organic substances which perform specific and necessary functions in relatively small concentration within the body.

Vitamins are defined [1,2] as organic compounds which are required for the normal growth and maintenance of the lives of animals, including man, which, as a rule, are unable to synthesize these compounds by anabolic processes independent of environment other than air. These compounds are effective in small amounts, do not furnish energy, and are not utilized as building units for the structure of the organism, but are essential for the transformation of energy and for the regulation of the metabolism of structural units.

Vitagens are placed in a different category [1,2] from that of vitamins because of the function of these compounds in the organism. Thus the vitagens are essential not only as transformers of energy or regulators of the metabolism of structural units, but also as suppliers of energy or as structural building units.

It is well known that in the absence of vitamins and vitagens clinical deficiency symptoms will occur or borderline deficiency may result. At any rate one suffers abnormal development or uneasiness, or actual illness because of lack of vitamins in the diet.

Classification.—Over the period of years since Funk in 1912 named these substances, *vitamines,* some 16 vitamins or groups of vitamins have been isolated, synthesized, or otherwise identified as being necessary for proper metabolic functioning of either man or animal. Although these substances have been isolated and identified, not all of them have been synthesized.

We can classify and discuss these substances according to chemical structure or alphabetically. The latter is more convenient for it is in accordance with the common nomenclature used. It is not yet possible to place all the vitamins in groups according to their chemical or physiological function.

---

[1] Rosenberg. *Chemistry and Physiology of the Vitamins.* Interscience, New York, 1942.
[2] Jacobs, ed., *Chemistry and Technology of Food and Food Products,* 2nd ed. Interscience, New York, 1951.

These 16 vitamins or groups of vitamins are:

Vitamin A, its precursors or provitamins A, and vitamin A_2.
The B complex consisting of:
  Thiamine or vitamin B_1;
  Riboflavin or vitamin B_2;
  Niacin and niacin amide (nicotinic acid and nicotinic acid
    amide);
  Pyridoxine or vitamin B_6, pyridoxal, pyridoxamine;
  Pantothenic acid;
  Inositol;
  Biotin or vitamin H;
  p-Aminobenzoic acid;
  Folic acid group, pteroylglutamic acid, and its conjugates;
  Vitamin B_{12}.
Ascorbic acid or vitamin C.
Vitamins D and their provitamins.
The vitamins E.
The vitamins K.
Vitamin P.

Nutritional Value.—The vitamins recognized as most important in human nutrition are vitamin A and its precursors, vitamin B_1 or thiamine, vitamin B_2 or riboflavin, niacin and niacin amide, vitamin C or ascorbic acid, and the vitamins D.

While the quantity of each vitamin necessary for adequate nutrition has not been fully established, the daily allowance recommended by the National Research Council, as given in Table 99, can serve as a guide. Data on the daily allowance for calories, protein, calcium, iron, fat, water, salt, iodine, phosphorus, and copper are included for the sake of completeness.

It should be stressed that vitamins are essential food factors. They are necessary for the adequate nutrition of human beings but they cannot take the place of other essential food factors such as the necessary proteins, fats, carbohydrates, minerals, water, and air. Vitamins should be used with foods or to supplement deficiencies in our diet and not as medicines unless a clinical deficiency exists. The indiscriminate use and sale of vitamins are certainly not to be recommended.

Within recent years the chemical nature of many of the vitamins has been elucidated. Some satisfactory tests have been developed for their detection and estimation. It is necessary that any method

TABLE 99. RECOMMENDED DAILY DIETARY ALLOWANCES[a]

	Age Years	Weight kg. (lb.)	Height cm. (in.)	Calories	Protein g.	Calcium g.	Iron mg.	Vitamin A I.U.	Thiamine mg.	Riboflavin mg.	Niacin mg.	Ascorbic Acid mg.	Vitamin D I.U.
Men......	25	65 (143)	170 (67)	3200[b]	65	0.8	12	5000	1.6	1.6	16	75	
	45	65 (143)	170 (67)	2900	65	0.8	12	5000	1.5	1.6	15	75	
	65	65 (143)	170 (67)	2600	65	0.8	12	5000	1.3	1.6	13	75	
Women......	25	55 (121)	157 (62)	2300[b]	55	0.8	12	5000	1.2	1.4	12	70	
	45	55 (121)	157 (62)	2100	55	0.8	12	5000	1.1	1.4	11	70	
	65	55 (121)	157 (62)	1800	55	0.8	12	5000	1.0	1.4	10	70	
Pregnant (3rd trimester)				Add 400	80	1.5	15	6000	1.5	2.0	15	100	400
Lactating (850 ml. daily)				Add 1000	100	2.0	15	8000	1.5	2.5	15	150	400
Infants[c]......	0-1/12[d]												
	1/12-3/12	6 (13)	60 (24)	kg.x120	kg.x3.5[c]	0.6	6	1500	0.3	0.4	3	30	400
	4/12-9/12	9 (20)	70 (28)	kg.x110	kg.x3.5[c]	0.8	6	1500	0.4	0.7	4	30	400
	10/12-1	10 (22)	75 (30)	kg.x100	kg.x3.5[c]	1.0	6	1500	0.5	0.9	5	30	400
Children......	1-3	12 (27)	87 (34)	1200	40	1.0	7	2000	0.6	1.0	6	35	400
	4-6	18 (40)	109 (43)	1600	50	1.0	8	2500	0.8	1.2	8	50	400
	7-9	27 (59)	129 (51)	2000	60	1.0	10	3500	1.0	1.5	10	60	400
Boys......	10-12	35 (78)	144 (57)	2500	70	1.2	12	4500	1.3	1.8	13	75	400
	13-15	49 (108)	163 (64)	3200	85	1.4	15	5000	1.6	2.1	16	90	400
	16-20	63 (139)	175 (69)	3800	100	1.4	15	5000	1.9	2.5	19	100	400
Girls......	10-12	36 (79)	144 (57)	2300	70	1.2	12	4500	1.2	1.8	12	75	400
	13-15	49 (108)	160 (63)	2500	80	1.3	15	5000	1.3	2.0	13	80	400
	16-20	54 (120)	162 (64)	2400	75	1.3	15	5000	1.2	1.9	12	80	400

[a] Food and Nutrition Board, National Research Council, Recommended Daily Dietary Allowances, Revised 1953, Publication 302, Designed for the Maintenance of Good Nutrition of Healthy Persons in the U.S.A. (allowances are considered to apply to persons normally vigorous and living in temperate climate). In planning practical dietaries, the recommended allowances can be attained with a variety of common foods which will also provide other nutrient requirements less well known; the allowance levels are considered to cover individual variations among normal persons as they live in the United States subjected to ordinary environmental stresses.

[b] These calorie recommendations apply to the degree of activity for the reference man and woman described in the bulletin. For the urban "white-collar" worker they are probably excessive. In any case, the caloric allowance must be adjusted to the actual needs of the individual as required to achieve and maintain his desirable weight.

[c] The recommendations for infants pertain to nutrients derived primarily from cow's milk. If the milk from which the protein is derived is human milk or has been treated to render it more digestible, the allowance may be in the range of 2-3 gms. per kg. There should be no question that human milk is a desirable source of nutrients for infants even though it may not provide the levels recommended for certain nutrients.

[d] During the first month of life, desirable allowances for many nutrients are dependent upon maturation of excretory and endocrine functions. Therefore no specific recommendations are given.

TABLE 99—*Continued*

Fat.—Allowances must be based more on food habits than on physiological requirements. It is desirable that fat be included in the diet to the extent of at least 20 to 25% of the total calories and that fat intake include essential unsaturated fatty acids (such as linoleic and arachidic acids of natural fatty acids) to the extent of at least 1% of the total calories. For a very active person consuming 4500 calories and for children and for adolescent persons, it is desirable that 30 to 35% of the total calories be derived from fat. Since footstuffs such as meat, milk, cheese, nuts, etc., contribute fat to the diet, it is necessary to use separated or "visible" fats such as butter, oleomargarine, lard, or shortenings to supply only ⅓ to ½ the amounts indicated.

Water.—A suitable allowance for adults is 2.5 quarts (2.5 liters) daily. An ordinary standard is 1 ml. for each calory of food. Most of this quantity is contained in prepared foods. At work or in hot weather, requirements may reach 5 to 13 liters daily. Water should be allowed *ad libitum*, since sensations of thirst usually serve as adequate guides to water intake except for infants and sick persons.

Salt.—The needs for salt and for water are closely interrelated. The average normal intake of sodium chloride is 7 to 15 g. daily, an amount which meets the requirements for water intake up to 4 liters daily. When sweating is excessive, one additional gram of salt should be consumed for each liter of water in excess of 4 liters daily. With heavy work or in hot climates 10 to 15 or more g. daily may be consumed. Even then most persons do not need more salt than usually occurs in prepared foods. After acclimatization persons produce sweat that contains only about 0.5 g. per liter in contrast with a content of 2 to 3 g. for sweat of the unacclimatized person; after acclimatization the need for increase of salt beyond that of ordinary food disappears.

Iodine.—The requirement for iodine is probably about 0.002 to 0.004 mg. daily for each kg. of body weight. This need is met by the regular use of iodized salt; its use is especially important in adolescence and pregnancy.

Phosphorus.—The allowances should be at least equal to those for calcium in the diets of children and of women during the latter part of pregnancy and during lactation. In the case of other adults the allowances should be approximately 1.5 times those of calcium. In general, it is safe to assume that if the calcium and protein needs are through common foods, the phosphorus requirement will be covered because the common foods richest in calcium and protein are also the best sources of phosphorus.

Copper.—The requirements for adults is about 1 to 2 mg. daily. Infants and children require approximately 0.05 mg. for each kg. of body weight. A good diet normally will supply sufficient copper.

Vitamin K.—The requirement usually is satisfied by any good diet except for the infant *in utero* and for the first few days after birth. Supplemental vitamin K is recommended during the last month of pregnancy. When it has not been given in this manner, it is recommended for the mother preceding delivery or for the baby immediately after birth.

Folic Acid. (pteroylglutamic acid, vitamin B$_c$, L casei factor, or vitamin M).—The quantitative requirement could not be closely estimated from evidence available in 1953.

based on chemical or physiochemical procedures have a fair degree of correlation with microbiological and biological assays, otherwise the chemical or physical method is valueless as a measure of the metabolic effect. Chemical methods are of great and growing importance, for as the chemistry of the vitamins becomes clearer and as the various vitamin factors are isolated and synthesized, methods for their estimation follow as a matter of course. Even if the chemical properties are ascertained before isolation, methods can be developed. It must be recognized that some of the microbiological and biological assays are subject to error and that even approximate chemical methods are valuable adjuncts to biological procedures. Some of the methods that are applicable in a food laboratory and do have a good correlation with biological assays are detailed.

Because, as mentioned above, a number of substances may display a certain type of vitamin activity, it is not always possible to evaluate the total vitamin activity by chemical methods and it becomes necessary to use biological and microbiological methods of assay.

It must not be forgotten that biological methods of assay are sometimes inconsistent in themselves, and therefore, although it is necessary that some degree of correlation exist between a chemical and biological assay, caution must be observed before assuming that a chemical method is valueless because the degree of correlation with a biological assay is small.

VITAMIN A

Vitamin A is generally considered to be an oxidation product of half the carotene molecule and is an unsaturated alcohol of high molecular weight. Carotene has been assigned the formula $C_{40}H_{56}$, structurally:

β-Carotene

α-Carotene differs from β-carotene in that the positions of the double bonds are changed so that the molecule has an asymmetric carbon atom. Carotene has been shown by numerous investigators to give rise to vitamin A in the body and hence it is considered to be provitamin A.

In addition to α- and β-carotene, K- and γ-carotene, cryptoxanthene, echinenone, myxoxanthin, aphanin, and aphanicin are considered precursors of vitamin A, but it should be noted that these substances vary in biological value.

The necessity of vitamin A for growth was established by McCollum and Davis and by Osborne and Mendel. Vitamin A can be prepared as pale yellow crystals which melt at 63-64° C. and boil under reduced pressure at 120-125° C. at 5×10^{-3} mm. It is soluble in most organic solvents and in fats, a property to which its original name, fat-soluble A, is attributed, and it is insoluble in water. Its absorption maximum occurs at 325-328 mμ and the absorption maximum of the antimony trichloride reaction product occurs at 620 mμ. Vitamin H has a characteristic green fluorescence in ultraviolet light.

Vitamin A is found principally in the form of its esters rather than as the free alcohol.

Vitamin A

Neither vitamin A nor its precursors have been synthesized on a commercial scale, although, because of the preparation some vitamin A concentrates undergo and the processing of vitamin A acetate, the latter may be considered as semisynthetics. It should be noted, however, in addition, that *kitol,* which is a compound that has been isolated from whole liver oil can be decomposed thermally to give a compound having vitamin A activity. It appears to be a divitamin A with two hydroxyl groups.

Vitamin A acetate is made by acetylating the unsaponifiable fraction of some distilled vitamin A ester concentrates and then redistilling the crude vitamin A acetate produced in this manner. It is a light yellow, limpid oil with a very mild odor and with high vitamin potency. All the vitamin A present is in the acetate ester form, which makes it as stable as the natural ester when compared at the same concentration. It is much more stable toward decomposition by oxygen and acids than vitamin A alcohol concentrates or most natural oils. Although the material is soluble in 100 per cent alcohol solutions, it is not recommended in this form since vitamin A esters are not especially stable in polar solvents. Since all of the fatty matter typical of fish liver oils has been removed in this process the vitamin A acetate concentrate will not revert to give the strong fishy flavor characteristic of many other vitamin A products.

Average bioassays indicate that the same conversion factor from spectrographic units to biological units will hold for this material as for distilled vitamin A natural ester concentrates. The potency of one commercial product is 700,000 U.S.P. units per gram which provides an ample margin of potency for general usage.

In common with other fat-soluble vitamins, it has been shown that a group of substances carry vitamin A activity. Thus, in addition to vitamin A itself and its esters, another material having vitamin A activity has been found in fresh-water fish, and it has been designated vitamin A_2. Vitamin A_2 has an absorption maximum at 345-352 mμ, and its antimony trichloride reaction product has an absorption maximum at 691 mμ.

Vitamin A is required for normal growth, and for adequate maintenance of well-being. Deficiency of this vitamin causes a disease of the eye known as xerophthalmia, night blindness, and keratosis of the skin. The minimum daily requirements set by the Food and Drug Administration for this vitamin are: for infants, 1500; for children up to 12, 3000; and for adults, 5000 U.S.P. units. However, the optimum requirement given by the National Research Council for growing children of 10-12 years is 4500 U.S.P. units and for women during pregnancy and lactation 6000-8000 units.

Vitamin A is optional in oleomargarine but if used, it should be added as a fish liver oil or as a concentrate of vitamin A from fish liver oil in such quantity that the finished margarine contains not less than 9000 U.S.P. units of vitamin. It has been used for the fortification of other foods like milk.

Antimony Trichloride Method.—It was noticed a number of years ago that cod liver oils gave a blue or purple color with dehydrating agents such as sulfuric acid. Rosenheim and Drummond found that arsenic trichloride gave an intense blue color that did not fade rapidly. Carr and Price[3] found that although the color with antimony trichloride was less intense, it was more stable and hence should serve as a basis for its estimation. On the development of this color, many tests for the quantitative determination of vitamin A were developed. Although the blue color produced is transient, its optical density is a linear function, within limits, of the vitamin A concentration at 620 mμ.

Procedure.—Weigh 1 g. of oil accurately into a volumetric flask of 10-ml. capacity. Dissolve the oil in chloroform and make to volume with this solvent. Transfer 0.2 ml. of this solution with the aid of a pipette to a glass cell of 1-cm. internal thickness, and run in exactly 2 ml. of a 30 per cent solution of antimony trichloride in chloroform from a burette. During the addition of the reagent shake the solution gently. The blue color produced is matched in a Rosenheim-Schuster colorimeter.[4] The final match is made at the point of maximum intensity of the blue color. The blue color is then correlated to a blue value expressed in blue units based on the assumption that the blue value of pure vitamin A is equal to 80,000.

---

3 Carr and Price, *Biochem. J.* 20, 497 (1926).
4 Griffiths, Hilditch and Raë, *Analyst* 58, 65 (1933).

Many observers have found that the biological potency of cod liver oil and other liver oils is far in excess of the chemical blue value and state that the absorption at 328 mμ gives far better agreement with the biological measure of vitamin A. Some investigators determine the blue value of the unsaponifiable fraction of the oil and determine the vitamin A corresponding to this value. However, it is doubtful whether the increase in blue value compensates for the labor involved in obtaining the unsaponifiable residue.

In the old Carr-Price method, the chloroform solution of the oil was diluted in such a manner that 0.2 ml. of this solution mixed with 2 ml. of a saturated solution of antimony trichloride in dry alcohol-free chloroform produced a color between 4.0 and 6.0 units of Lovibond blue. The result was calculated to the proportionate color for 0.04 g. of the original oil. Either daylight or artificial daylight was used and, at the point of maximum intensity, red or yellow glasses should be used in addition to blue glasses in order to obtain a proper match.

Although the blue value of an oil does not in any of these methods necessarily yield the true vitamin A content of an oil, nevertheless it is indicative of the approximate vitamin A content. These tests are of great value in establishing the presence or absence of appreciable amounts of vitamin A.

Saponification Method.[5, 6]—*Reagent.*—Antimony Trichloride Solution.—Great care must be exercised in preparing this reagent because it is very corrosive. Wash all glassware that comes in contact with the reagent with chloroform, a mixture of alcohol and ether, or dilute or concentrated hydrochloric acid before regular washing because antimony trichloride forms antimonyl chloride which is insoluble in water.

Weigh an unopened bottle of antimony trichloride, generally a 4-oz. bottle. Open it and empty it into a glass-stoppered, wide-mouth, amber-colored bottle containing 100 ml. of chloroform. Reweigh the 4-oz. bottle and obtain the weight of antimony trichloride added to the chloroform by difference. Add sufficient chloroform to maintain a ratio of 100 ml. for each 25 g. of antimony trichloride. Shake or warm to dissolve and filter or decant into a dry, glass-stoppered, amber-colored bottle. Store in the dark at room temperature.

Preparation of Sample.—*Butter and Margarine.*—Clarify the fat by melting at 45-50° C. and filtering through absorbent cotton.

---

[5] Dann and Evelyn, *Biochem. J.* 32, 1008 (1938).
[6] Oser, Melnick, and Pader, *Ind. Eng. Chem., Anal. Ed.* 15, 724 (1943).

Milk, Cream, and Cheese.—Extract 50 to 100 ml. of milk, 20 ml. of cream, and an appropriate amount of cheese depending on its fat content with mixed ethers as in the modified Mojonnier or Roese-Gottlieb methods detailed in Chapters VII and VIII to obtain the fat.

Feed Concentrates.—Extract in a continuous extraction device with ether.

Fish Oils, Oils, and Premixes.—Saponify directly.

Candy and Starch Products.—Dissolve or suspend in water.

Procedure.—Weigh 1 g. of an oil or an amount of fat containing 200 U.S.P. units of other products into a saponification flask equipped with a T neck, add 15 ml. of alcoholic potassium hydroxide solution, prepared by dissolving 12 g. of KOH in 100 ml. of alcohol or sufficient alcoholic potassium hydroxide solution to equal about half the weight of the sample being analyzed, and attach a reflux condenser. Heat for 15 minutes or until saponification is complete. Wash the condenser with 10 ml. of water. Cool, dilute with 50-100 ml. of water, and transfer to a separatory funnel. Wash the saponification flask with 50 ml. of ether or more if necessary to equal twice the volume of alcoholic potassium hydroxide solution used. Add the ether to the separatory funnel and shake carefully to avoid formation of emulsions. Allow the layers to separate and draw off the aqueous layer into another separatory funnel. Rinse the saponification flask with another 50 ml. of ether and extract the aqueous layer with this portion of ether. Allow the layers to separate. Draw off the aqueous layer into another separatory funnel and combine the ether layers. Extract again with 35-50 ml. of ether. Repeat two more times. Combine all five ether extracts in the first separatory funnel and wash by pouring 50-100 ml. of water through the ether without shaking. Draw off the aqueous layer and discard. Wash the ether extract with 50 ml. of 0.5 N sodium hydroxide solution, shaking gently. Allow to separate, draw off the aqueous layer, and discard. Wash with five to eight 50-ml. portions of water until free of alkali, testing with phenolphthalein. Draw off the final aqueous rinse, filter through anhydrous sodium sulfate placed on filter paper in a funnel into a 250-ml. flask. Rinse the separatory funnel with two 25-ml. portions of ether and add the rinses to the 250-ml. flask. Add some glass chips and evaporate the ether to dryness on a water bath, removing the flask from direct heat toward the end of the evaporation. Take up the residue immediately in chloroform adjusting the concentration to 7-15 U.S.P. units per milliliter.

Transfer 2 ml. of chloroform to a colorimeter tube or cuvette and add 9 ml. of antimony trichloride reagent with the aid of a rapid delivery pipette. Set the galvanometer at full deflection with the light beam at 620 mμ. Place 1 ml. of the unknown solution, 1 ml. of chloroform, in a tube or cuvette, set the tube in the colorimeter and add 9 ml. of antimony trichloride solution. Read the galvanometer and obtain the per cent transmission. The reading should be made within 3 to 6 seconds after addition of the antimony trichloride reagent. To a third tube or cuvette, add 1 ml. of the unknown solution, 1 ml. of a known vitamin A solution in chloroform approximately equivalent to that of the unknown, and treat as above.

Calculation.—Convert all galvanometer readings to optical density.

$$\text{Optical density} = 2 - \log G_{620}$$

where

$$G_{620} = \text{galvanometer reading at 620 m}\mu$$

$$\text{U.S.P. units per ml. unknown} = \frac{\text{corrected opt. dens. unknown}}{\text{opt. dens. known} + \text{unknown}} $$
$$- \text{opt. dens. unknown}$$
$$\times \text{concn. of standard (U.S.P. units/ml.}$$

$$\text{Vitamin A/g. sample} = \text{U.S.P. units/ml. unknown} \frac{\text{final volume}}{\text{sample weight}}$$

Subtract the optical density of a solution of 1 ml. of the unknown plus 10 ml. of chloroform from the optical density of the unknown to correct for any color in the final unknown chloroform solution.

Glycerol Dichlorohydrin Method.—This method devised by Sobel and Werbin [7, 8] depends on a colorimetric reaction between vitamin A and practical or activated glycerol dichlorohydrin. It has several advantages over the Carr-Price Method in that the violet color produced is stable from 2 to 10 minutes after the addition of the reagent, the reagent is not affected by traces of moisture and hence can be used on humid days, no film of antimony oxychloride is left on cuvettes, and it is relatively noncorrosive.

Reagent.—Activated Glycerol Chlorohydrin.—Add 2 per cent by weight of antimony chloride dissolved in chloroform to 1 liter of practical glycerol chlorohydrin. Distill the mixture under vacuum at 86-92° C. at 30-40 mm. after discarding the chloroform fraction. The reagent prepared in this manner should give an $L_{1\ cm.}^{1\%}$ value (550 mμ) of from 1150 to 1250 in the Coleman universal spectrophotometer, model 11, and should be colorless. It should be stored in glass-stoppered bottles and is generally stable for 2 months.

---

[7] Sobel and Werbin, *Ind. Eng. Chem.*, *Anal. Ed.* 18, 570 (1946).

[8] Sobel and Werbin, *Anal. Chem.* 19, 107 (1947).

Procedure.—To 4.0 ml. of activated glycerol chlorohydrin add 1.0 ml. of chloroform solution containing between 2 and 5 micrograms of vitamin A. Mix the solution and place in a water bath at 25° C. for about a minute and a half. Pour into a cuvette and read the absorption at 550 mμ 2 minutes from the time of the addition of the vitamin A.

Guaiacol or Catechol Method.—Another method for vitamin A was developed by Rosenthal and Erdélyi.[9] The oil containing vitamin A is diluted with alcohol-free chloroform in the proportion of 0.1 to 0.5 ml. of oil to 1 ml. of chloroform, according to the vitamin content. To 1 ml. of this solution is added 1 ml. of a freshly prepared 0.5 per cent solution of catechol in chloroform and 3 ml. of a cold saturated solution of antimony trichloride in chloroform. The tube containing the mixture is heated immediately to 60° C. for 1 to 2 minutes. The blue color first produced changes to an intense violet red. Quantitative determinations may be made with the violet-red stage by comparison with a 0.01 per cent solution of potassium permanganate. Using guaiacol instead of catechol, the color is produced without heating.

Carotene, lycopin, zeaxanthin, capsanthin, physalin, do not give the reaction. Ergosterol turns pink first and then becomes blue. The color produced with guaiacol and with catechol is more stable than that produced in the original Carr-Price reaction.

Spectrographic Method.—It has been shown that the ultraviolet absorption of vitamin A at 328 mμ (3280 Å) is not due to carotene or to similar coloring matters. Thus carotene (British Drug House) absorbs at about 450 mμ, which is in the visible region, and its absorption in the ultraviolet is negligible. Calciferol or vitamin D absorbs at about 265 to 270 mμ.[10]

The determinations are made by taking a series of pairs of photographs with the aid of a quartz spectrograph fitted with a short-focus rotating sector photometer mounted directly on the instrument. Each pair of photographs of the spectrum consists of one spectrogram taken through a solution of the substance and another taken through the same thickness of the solvent in a quartz cell. By means of the sector photometer, the time of exposure of the individual halves of each pair of spectrograms is varied so that the relative exposures through the solvent are decreased quantitatively. In this way points can be found on each pair of photographs where the density of the two halves is the same, showing that at these points

[9] Rosenthal and Erdélyi, *Biochem. J.* **28**, 41 (1934); **29**, 2112 (1935).
[10] Crews and Cox, *Analyst* **59**, 85 (1934).

the intensities of the two beams of light, multiplied by the relative exposures, are equal. As the ratio of the exposures decreases, the points of equal density of the spectrograms become closer together, until at the head of the band they become coincident. This point is used for the calculation of the extinction coefficient, E, which is the logarithm of the ratio of the intensity of the transmitted light to the incident light.

In the case of vitamin A preparations, it is usual to calculate to a 1 per cent concentration and a 1-cm. cell, and to specify the particular wave length or band at which absorption has taken place. This is usually expressed as $E_{1\%}^{1\,cm.}$, 328 mμ. Solvents used must be specifically purified for spectropho.ographic work; generally absolute alcohol and cyclohexane are used. $E_{1\%}^{1\,cm.}$ 328 mμ for pure vitamin A equals 1600.

Ultraviolet Light Absorption Method.—In this method [11] the absorption of light attributable to the amount of vitamin A in a solution is measured. This absorption is proportional to the amount of vita- A when measured for light of 328 mμ, actually light of 324-330 mμ is used. This method is customarily used for clear fish oils having a vitamin A concentration of over 10,000 U.S.P. units per g. and to vitamin A concentrates.

Procedure.—Weigh 1 g. of fish oil or an equivalent amount of other samples to the nearest milligram, transfer to a 50- or 100-ml. volumetric flask, and complete to volume with isopropyl alcohol. Dilute an aliquot of this solution with additional isopropyl alcohol so that the concentration is in the range of 5 to 15 units per ml. This will depend on the spectrophotometer or spectrograph being used, and instructions are generally given with the apparatus. Fill a matched cuvette with the diluted unknown sample and another cuvette with isopropyl alcohol. Set the cuvettes in the instrument, adjusting the isopropyl alcohol to zero optical density at 328 mμ and take the optical density reading of the unknown solution.

Calculation.—When $E_{1\,cm.}^{1\%}$ 328 mμ is the extinction coefficient at 328 mμ, c is the concentration of the sample in g. per 100 ml., l is the length of the light path through the solution, and d is the optical density of the solution at the wave length noted, then

$$E_{1\,cm.}^{1\%}\ 328\ m\mu = \frac{d}{cl}$$

and

$$\text{Potency in U.S.P. units per g.} = E_{1\,cm.}^{1\%}\ 328\ m\mu \times 2{,}000$$

---

[11] Zscheile and Henry, *Ind. Eng. Chem., Anal. Ed.* **14**, 422 (1942).

THIAMINE—VITAMIN B₁

Vitamin B₁

Thiamine, vitamin B_1, aneurin, oryzanine, is amorphous as the free base, thiamine chloride, but crystallizes as the hemihydrate from alcoholic solution and melts at 248-250° C. with some decomposition. It is very soluble in water for 1 g. of the vitamin dissolves in 1 ml. It is much less readily soluble in alcohol for only 1 g. dissolves in 100 ml. of 95 per cent alcohol. About 5 g. dissolves in 100 ml. of glycerol. It is insoluble in common organic solvents such as ether, chloroform, benzene, acetone, and petroleum ether. Thiamine is stable in relatively strong acid solution for the hydrochloride can be heated to 120° at pH 3.5 without decomposition, but in weak acid solution it decomposes and is no longer biologically active. Vitamin B_1 decomposes readily in alkaline and neutral solutions if heated and is also easily destroyed in solution by both reducing and oxidizing agents. Thiamine has a nutty or yeast-like odor.

The commercial vitamin B_1 is thiamine chloride-hydrochloride. It has a biological activity of 333,000 U.S.P. units or International units per gram; that is, 3 micrograms of the crystalline substance are equivalent to 1 U.S.P. unit.

A marked deficiency of vitamin B_1 is evidenced by illness known as polyneuritis and *beriberi* because of injury sustained by the nervous system. The oxidation of carbohydrate does not proceed properly when there is vitamin B_1 deficiency; thus more thiamine is needed for a diet high in carbohydrates. It has been shown that high fat diets have a sparing action on vitamin B_1, consequently less of this vitamin is needed with high fat diets.

The daily requirements of vitamin B_1 for human beings vary from 0.3 mg. (83 I.U.) for infants, and up to 1.0-1.9 mg. for adults. See Table 99. An increase in vitamin B_1 in the diet assists in producing a feeling of well-being.

Vitamin B_1 is one of the most commonly used vitamins for fortification and restoration of foods. During World War II it was a required additament in the enrichment of flour and bread. According to the standard for enriched flour published in March, 1953, promulgated by the Food and Drug Administration [12] under the Federal Food, Drug, and Cosmetic Act of 1938, enriched flour, enriched bromate flour, and enriched self-rising flour must contain not less than 2.0 nor more than 2.5 mg. of thiamine per pound of flour. The corresponding requirements for the addition of vitamin B_1 to white bakery products is a minimum of 1.1 mg. and maximum of 1.8 mg. of thiamine per pound of bread, rolls, or other bakery product. This requirement became effective for bread, buns, and rolls on October 1, 1943, and for all other bakery products on May 1, 1944 under war food orders. Enriched farina must contain the same amount of thiamine per pound as enriched flour.[12a]

Thiochrome Reaction.—A method for the estimation of vitamin B_1 has been developed on the observation of Barger [13] the vitamin is oxidized to thiochrome, to which has been assigned the formula

Thiochrome

$C_{12}H_{14}ON_4S$, by potassium ferricyanide in alkaline solution. Jansen [14] using this method found good agreement with biological assay.

Procedure.—Treat 10 ml. of a 25 per cent extract of yeast in water at a pH 3 with 20 ml. of methyl alcohol. Remove the precipitate by centrifuging and mix 1 ml. with 0.1 to 0.3 ml. of 1 per cent ferricyanide solution, 1 ml. of methyl alcohol and 1 ml. of 30 per cent sodium hydroxide solution. Extract this mixture with 13 ml. of isobutyl alcohol. Separate the two layers by centrifuging and from the upper layer pipette 10 ml. into the vessel of the fluorometer. Measure the fluorescence attributable to the thiochrome by a galvanometer through which the fluctuating current of the photoelectric cell of the fluorometer passes.

Pure thiamine may be assayed in a similar manner. Place 0.1 ml. containing 1 to 20 micrograms in a 25-ml. glass-stoppered cylinder (rubber stoppers give rise to fluorescent substances). Add 0.1 per cent solution of potassium ferricyanide in the proportion of 0.01 to

[12] *U. S. Food Drug Admin., S.R.A., F.D.&C.* 2, Part 15, March, 1953.
[12a] *U. S. Food Drug Admin., S.R.A., F.D.&C.* 2, Part 15, March, 1953, Amendment No. 1, April, 1955.
[13] Barger, *Ber.* 68, 2257 (1935).
[14] Jansen, *Rec. trav. chim. Pays-Bas* 55, 1046 (1936).

0.1 ml. for 1 microgram, 0.03 to 0.1 ml. for 10 micrograms and 0.1 to 0.2 ml. for 20 micrograms. Mix and add 3 ml. of 10 per cent sodium hydroxide. Mix again thoroughly and after 1 to 2 minutes extract with 13 ml. of isobutyl alcohol. Centrifuge, transfer 10 ml. of the supernatant solvent to the cell of the fluorometer and read.

The International Standard power (adsorbate) may be assayed in a similar manner. Weigh accurately from 10 to 50 mg. of the material and transfer to a 25-ml. glass-stoppered cylinder and add 10 ml. of 10 per cent sodium hydroxide solution. Add the requisite amount of 0.1 per cent potassium ferricyanide solution. For 10 mg. take 0.05 to 0.1 ml., for 30 mg. take 0.1 to 0.2 ml., and for 50 mg. take 0.3 ml. of the ferricyanide solution. Stir the mixture for 5 to 10 minutes after which extract with 13 ml. of isobutyl alcohol. Read the fluorescence of the supernatant liquid as before in the fluorometer.

Standardized Thiochrome Method.—The thiochrome method for the estimation of vitamin B₁ has been standardized for cereal products.[15]

Apparatus.—Exchange Tubes.—The barrel of the exchange tubes is 0.7 × 15 cm. The capillary tube at the lower end which controls the rate of flow is 0.03 × 3.0 cm. The reservoir at the upper end of the barrel has a capacity of 50 ml. These may be set up in any convenient holder. Push a plug of glass wool down to the lower end of the barrel in such a way that the strands lie across the capillary opening. Introduce sufficient prepared zeolite into the tube to a 6-cm. column and wet by drawing through 10 ml. of 3 per cent acetic acid.

Reagents.—Activated Zeolite.—Prepare 60- to 80-mesh activated zeolite, such as Decalso, in bulk using 100 to 500 g. Stir or shake for 15 minutes each with 4 portions of 3 per cent acetic acid. The volume of acid used should be 8 to 10 times that of the zeolite. Between the third and fourth acid washing, treat the zeolite for 20 minutes with 4 volumes of 25 per cent potassium chloride solution. Finally wash the zeolite repeatedly with water, filter on a Büchner funnel with the aid of suction, allow to dry in air, and bottle.

Sodium Acetate Solution.—Prepare a 2.5 M solution of sodium acetate by dissolving 345 g. of $NaC_2H_3O_2 \cdot 3H_2O$ or 205 g. of the anhydrous salt in water and dilute to 1 liter.

Alkaline Potassium Ferricyanide Solution.—Dissolve 1.0 g. of $K_3Fe(CN)_6$ in water and dilute to 100 ml. Dilute 3 ml. of this solution to 100 ml. with cool 15 per cent sodium hydroxide solution. The alkaline potassium ferricyanide solution must be prepared fresh when used.

---

[15] Hennessy, *Cereal Chem. Bull.* 2, No. 2 (1942).

Acid Potassium Chloride Solution.—Dissolve 250 g. of potassium chloride in water and dilute to 1 liter. Dilute 8.5 ml. of concentrated hydrochloric acid to 1 liter with the 25 per cent potassium chloride solution.

Isobutyl Alcohol.—It is necessary that the fluorescence of the isobutyl alcohol should not exceed 10 per cent of the fluorescence of the quinine standard. This is achieved by redistilling in an all-glass apparatus and collecting the fraction boiling in the range 105-108 C.

Enzyme Solution.—Solutions made with Mylase, Polidase, Clarase, or Taka-diastase are generally suitable. Suspend 6 g. of the enzyme in 2.5 M sodium acetate solution by shaking thoroughly and dilute to 100 ml. with the same buffer solution.

Standards.—Thiamine Stock Solution.—Dry thiamine hydrochloride over phosphorus pentoxide in a desiccator for more than 24 hours. Dissolve 100 mg. in 25 per cent ethyl alcohol or in 0.01 N hydrochloric acid and dilute to 1 liter with the same solvent. This stock solution is stable for about five months if kept below 5° C.

Standard Thiamine Solution.—Dilute 5.0 ml. of the thiamine stock solution at room temperature to 100 ml. with water. Transfer 4.0 ml. of this diluted solution with the aid of a pipette to a flask containing 75 ml. of 0.1 N sulfuric acid and 5 ml. of 2.5 M sodium acetate solution and make up to 100 ml. with water. Each milliliter of this dilution contains 0.2 microgram of thiamine. It must be prepared fresh.

Quinine Sulfate Stock Solution.—Dissolve 100 mg. of U.S.P. quinine sulfate in 0.1 N sulfuric acid and dilute to 1 liter with the reagent. Store in an amber or brown bottle.

Quinine Sulfate Reference Solution.—Dilute 3 ml. of quinine sulfate stock solution to 1 liter with 0.1 N sulfuric acid. This solution contains 0.3 microgram per milliliter.

Preparation of Samples.—Finely divide cereal products by grinding. Air dry bread and then divide finely by grinding. Analyze yeasts as such. A determination of the moisture content is necessary if the results are to be recalculated to a dry basis.

Preparation of the Extract.—The weight of the sample chosen for extraction is based on the principle that the most efficient extraction results from the use of the largest ratio between the volume of solvent and the weight of the sample. The limitations imposed by the potency and the sampling error have led to the following tabulation of Hennessy.

Because in flour samples to which thiamine has been added, there may be inhomogeneity, it is advisable to use a larger sample and a larger volume of extracting liquid. Multiplication of all quantities by five will probably give a representative sample, especially if the flour is sieved twice and the samples selected by quartering.

Sample Wt. To Be Extracted by 75 ml. of 0.1 N H_2O_4	Expected Potency Range I.U.B.$_1$/g.
1.0	10
2.0	5–10
3.0	2–5
4.0	1–2
5.0	1

Procedure.—Introduce 25 ml. of 0.1 N sulfuric acid into a 125-ml. Erlenmeyer flask. Add the weighed sample and mix by careful swirling. The introduction of 25 ml. of the extracting liquid into the flask before the sample keeps it from adhering to the bottom of the flask. Add an additional 25 ml. of extracting acid and set the flask in a boiling water bath. Shake the sample frequently during the first 5 minutes and then once every 5 minutes for 0.5 hour. Set the flask in cold water and cool the contents below 50° C. Add 5 ml. of the enzyme suspension, freshly prepared. Stopper the flask and keep at 45-50° C. for 2 hours. Make up to 100 ml. by the addition of water and, after mixing thoroughly, filter. Discard the first 10 ml. of the filtrate and collect the remainder.

Base Exchange Purification.—Introduce 10 to 50 ml. of extract containing 0.5 to 10 micrograms with the preferred range being 3 to 5 micrograms of vitamin B₁ as thiamine chloride into the reservoir of the exchange tube and allow to pass through the zeolite column. Wash the exchanger column by drawing three 10-ml. portions of water through the column.

Remove the vitamin B₁ from the zeolite by treating with acid potassium chloride solution. Place a 15-ml. portion of this solution into the reservoir and allow it to pass through the zeolite. Collect the effluent in a graduate. When the tube has stopped dripping add sufficient acid potassium chloride solution so that the final volume of the effluent will be 23 to 25 ml. Make up the effluent to 25 ml. with acid potassium chloride solution and mix well. Use 5-ml. aliquots for thiochrome production and for the blank determination. Repeat with an aliquot of the standard thiamine solution using 5.0 micrograms of thiamine in place of the unknown.

Production of Thiochrome.—All samples at this stage should be in solution in the 22-25 per cent potassium chloride solution. Effluents from the base exchange treatment are already in such solution. Extracts or solutions which are not put through the base exchange columns must be diluted from 1 to 25 up to 3 to 25 with 25 per cent potassium chloride solution after bringing the concentration of thiamine chloride to 1-10 micrograms per milliliter with water.

Place 5 ml. of the potassium chloride solution containing 0.4-4.0 micrograms of vitamin B_1 as thiamine chloride in each of two separatory centrifuge tubes. To the first, A, add quickly with mixing 3 ml. of 15 per cent sodium hydroxide—0.03 per cent potassium ferricyanide solution; to the second, B, 3 ml. of 15 per cent sodium hydroxide solution. Add 15 ml. of redistilled isobutyl alcohol to each. Insert stoppers and shake the tubes vigorously for 1.5 minutes. Centrifuge the tube at 500-600 r.p.m. for ¾ minute. Remove the stoppers, drain off the lower layer, add approximately 2 g. of anhydrous sodium sulfate to each, shake vigorously for a moment, and, if necessary for complete clarification, centrifuge for 30 seconds.

Fluorometry.—There are various types of fluorometer used. One type measures the fluorescent light intensity as a galvanometer deflection, the other type as a potentiometer setting. In either case, because of the instability of thiochrome in the ultraviolet light, the readings must be made quickly. Fifteen seconds should be the maximum exposure of the thiochrome solution to the ultraviolet beam. As an example of the standardization of the fluorometer and of the fluorometry of an unknown, the procedure as conducted with the Pfaltz and Bauer Fluorophotometer, Model A or B, is described.

(a) Standardization. Put 12 ml. of the clear isobutanol solutions produced from 1.00 micrograms of thiamine chloride in identical cuvettes A and B. Place a third cuvette C, containing 12 ml. of the working standard quinine sulfate solution in the instrument and adjust the iris diaphragm to give a fixed reading (Q_1) on the higher half of the scale of the galvanometer. The concentration of the working standard quinine sulfate is chosen so as to give a reading at the upper end of the galvanometer scale when the instrument is at full sensitivity, that is, with the iris diaphragm fully open. Insert cuvette B and take the reading (Y_1) without changing the diaphragm opening. Likewise for cuvette A, whose reading is (X_1). The value of $X_1 - Y_1$ must be established by repeated checks, each on fresh aliquots of the KCl solution of thiamine chloride. The reading of the quinine solution (Q_1) must be maintained constant at all times by readjustment of the iris diaphragm, if necessary, to give the original value just before putting the cuvettes containing the isobutanol solution in place.

(b) Fluorometry of Unknown. Place two 5-ml. aliquots of the thiamine in potassium chloride solution in each of two separatory centrifuge tubes and treat as described under *Production of Thio-*

chrome. Measure the fluorescence of the resulting isobutanol solutions in cuvettes A and B after verifying the reading of the standard quinine solution as Q_1. Reading of A = X; B = Y. Then $\dfrac{X - Y}{X_1 - Y_1}$ = micrograms of thiamine chloride in the 5 ml. aliquot used. The micrograms of thiamine chloride per gram of original sample is calculated as follows: a 4.00-g. sample is made up to 100 ml. after extraction and incubation, 15 ml. are put through base exchange and 25.0 ml. of effluent collected, 5-ml. aliquots being used for the thiochrome reaction and the blank.

Micrograms of thiamine chloride/g. sample will be equal to

$$\frac{X - Y}{X_1 - Y_1} = \frac{25}{5} \cdot \frac{100}{15} \cdot \frac{1}{4.00}$$

and I.U. B_1/g. sample = ⅓ micrograms thiamine chloride.

Sometimes it is desirable to increase or decrease the galvanometer deflection per microgram of thiamine chloride by changing the intensity of the exciting light, that is, to change the sensitivity of the fluorometer. The deflection per microgram at the new sensitivity is equal to $(X_1 - Y_1)$ times the deflection for the quinine sulfate standard at the new sensitivity divided by Q_1.

Diazo Method.—One of these developed by Prebluda and McCollum [16] depends on the formation of a water-insoluble compound from the condensation of vitamin B_1 and a diazotized compound. These investigators found that a solution of either *p*-aminoacetanilide or methyl-*p*-aminophenylketone (*p*-aminoacetophenone) treated with nitrous acid yields a diazotized compound which, when added to vitamin B_1 under definite conditions, yields a characteristic purple red compound that is stable and highly insoluble in water. The reagents prepared from the aforementioned amines were reacted with samples of wheat germ, rice polishings, and a number of commercial preparations. In each case the same characteristic product previously mentioned may be obtained.

Fermentation Method.—Another method correlates the added amount of carbon dioxide, liberated in the fermentation of dextrose, with the quantity of vitamin B_1 present.[17] Determinations are made of the volume of carbon dioxide liberated in 3 hours by the fermentation of dextrose in the presence of vitamin B_1. With known amounts of natural crystalline vitamin B_1, by means of the following method, the tabulated results were obtained.

---

16 Prebluda and McCollum, *Science* **84**, 448 (1936).
17 Schultz, Atkin, and Frey, *J. Am. Chem. Soc.* **59**, 948, 2457 (1937).

For each test, 1 g. of yeast, 100 ml. of an aqueous solution containing a synthetic salt mixture as a buffer and 3 g. of dextrose is used. The temperature is 30° C. and the oscillations are 100 per minute.

Natural Crystalline Vitamin B_1 mg.	Ml. of Gas in 3 Hours
None	185
0.001	215
0.005	305
0.010	350
0.040	395
0.100	405

This method agrees with estimations of vitamin B_1 made with the rat-growth method.

Formaldehyde-Azo Test.—Vitamin B_1, having a pyrimidine ring, will couple with a diazo compound under controlled conditions. Kennersley and Peters [18] found that the test for the vitamin with diazotized sulfanilic acid requires a definite pH and is more stable in the presence of formaldehyde.

Reagents.—(a) Sulfanilic Acid Solution.—Dissolve 4.5 g. of sulfanilic acid and 45 ml. of hydrochloric acid (sp. gr. 1.19) in water and make up to 500 ml.

(b) Sodium Nitrite Solution.—Dissolve 25 g. of 90 per cent sodium nitrite in water and make up to 500 ml.

Add 1.5 ml. of solution (a) to 1.5 ml. of solution (b) and then the mixture is left on ice for 5 minutes. Add 6 ml. more of (B), placing the mixture again on ice for 5 minutes. Dilute to 50 ml. and replace on ice for 15 minutes before use. This solution is the diazotized sulfanilic acid.

(C) Dissolve 5.76 g. of sodium bicarbonate in water and make up to 100 ml. Add 100 ml. of 1 N sodium hydroxide solution.

Procedure.—Add 0.5 ml. of the diazotized sulfanilic acid to 1.25 ml. of the sodium bicarbonate-sodium hydroxide reagent in a small test tube. After 1 minute, 0.3 ml. of 40 per cent formaldehyde is added. Then 0.1 to 0.3 ml. of the solution of vitamin B_1 of acidity greater than pH 4.0 is added. A pink color develops slowly, and increases in intensity for 30-60 minutes, after which time it is practically constant. Comparison may be made in a colorimeter against standard vitamin treated the same way.

---

[18] Kennersley and Peters, *Biochem. J.* **28**, 667 (1934).

RIBOFLAVIN—VITAMIN B$_2$

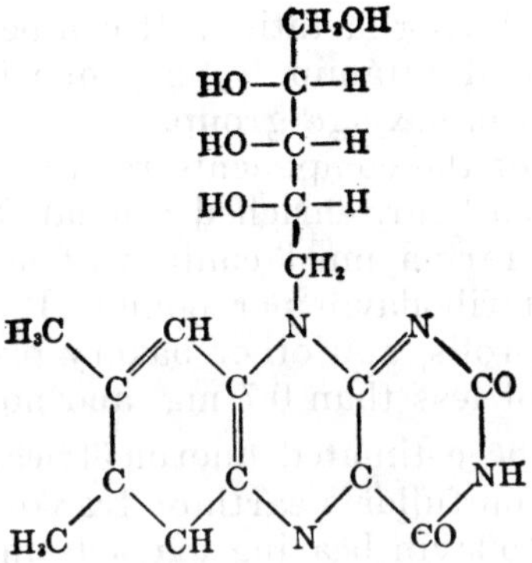

$$CH_2OH$$
$$HO—C—H$$
$$HO—C—H$$
$$HO—C—H$$
$$CH_2$$

Riboflavin, vitamin B$_2$, lactoflavin is a solid crystallizing in fine orange-yellow needles which melt at 282° C. with some decomposition. It is not very soluble in either water or alcohol for only about 11 mg. dissolve in 100 ml. of water and 4.5 mg. dissolve in 100 ml. of alcohol at room temperature. However, it is very soluble in alkaline solutions. Riboflavin is not very soluble in amyl alcohol, amyl acetate, and cyclohexanol, and it is insoluble in acetone and other common organic solvents. Aqueous solutions of riboflavin are greenish-yellow in color and exhibit an intense yellow-green fluorescence with a maximum at pH 6.7-6.8 which is destroyed by acids and bases. While crystalline vitamin B$_2$ is stable at room temperature when protected against light, it is unstable in solution, particularly alkaline solution, and when exposed to either visible or ultraviolet light. It has relatively good thermostability. Its optical rotation in 0.1 N sodium hydroxide solution is $[\alpha]_D^{20} - 114°$.

Riboflavin probably acts physiologically by taking part in the various enzyme systems of the oxidation-reduction type. When the diet contains inadequate amounts of this vitamin there is evident impairment of health and retarded growth. The most obvious signs are shown by changes in the nervous system and the skin. With decrease in the incidence of pellagra, the large amount of illness resulting from lack of riboflavin became evident. It is necessary for the proper metabolism of the carbohydrates.

One of the earliest methods of its synthesis was from 1,2-dimethyl-4-amino-5-d-1'-ribitylaminobenzene and alloxan, which reacts in the lactim form to yield 6,7-dimethyl-9-d-1-ribitylisoalloxazine, that is riboflavin.

The daily allowance of riboflavin recommended by the National Research Council varies from 0.3 mg. for infants under 1 year to 1.5 mg. for a woman during lactation. It can be seen from Table 99 that the requirement of vitamin B_2 is proportional to the caloric requirements of a given sex-age group.

Riboflavin is one of the components required for the enrichment of flour. This enriched flour, enriched bromated flour, and enriched self-rising flour and farina must contain at least 1.2 mg. and not more than 1.5 mg. of riboflavin per pound. It is also used for the enrichment of bread, rolls, and other bakery products which should contain per pound not less than 0.7 mg. and not more than 1.6 mg.

Vitamin B_2 may be estimated fluorometrically.[19] It may be obtained by adsorption on fuller's earth or frankonite from fresh liver extracts or other lactoflavin bearing extracts in neutral or strongly acid solutions. The adsorption is generally complete in 10 minutes. The adsorbates may be eluted with diluted 10-50 per cent aqueous diethylamine or 0.2 per cent sodium hydroxide solution. The amount of riboflavin is then determined by estimation of the fluorescence.

Rapid Method for the Determination of Riboflavin in Milk.—It was observed by Whitnah, Kunerth and Kramer [20] that the trichloroacetic acid serum from milk used for vitamin C titrations often had a greenish color, whereas the mercuric nitrate serum used for the determination of sugar was colorless. Based on this fact these investigators developed the following procedure for the fluorimetric estimation of flavin in milk.

Procedure.—Add 15 ml. of 10 per cent trichloroacetic acid to 10 ml. of milk, let stand 30 to 60 minutes, centrifuge for 5 minutes. Neutralize 10 ml. of the resulting serum, using methyl orange as the indicator, and dilute until the sample can be matched in the light of a lamp,[21] with standard flavin solution containing 0.12 to 0.06 microgram of flavin per ml. Calculate the flavin content on the basis of dilutions made. Dilutions, until the portions read, contain less than 0.12 micrograms per ml., seem essential as the values for stronger solutions are easily underestimated.

Standardized fluorimetric methods based on measuring the fluorescence of riboflavin solutions at 400-500 mμ before and after the chemical destruction of the riboflavin are detailed in several of the references given at the end of this chapter. A number of authorities prefer the use of microbiological methods for the assay of riboflavin

[19] Lepkovsky, Popper and Evans, *J. Biol. Chem.* 108, 257 (1935).
[20] Whitnah, Kunerth and Kramer, *J. Am. Chem. Soc.* 59, 1153 (1937).
[21] Everready Fluoray lamp.

because of the greater applicability of the microbiological method.[22, 23]

NIACIN AND NIACIN AMIDE

Nicotinic Acid   Nicotinamide

Niacin, nicotinic acid, 3-pyridinecarboxylic acid is another member of the vitamin B group. It crystallizes in colorless crystals which melt at 235-237° C. Niacin is soluble in cold water—about 1.5 g. in 100 ml.; it is readily soluble in hot water; and about 1 g. is soluble in 100 ml. of 95 per cent alcohol. The alkali salts of nicotinic acid are soluble in water and niacin is also soluble in hot glycerol. Niacin has an absorption maximum at 385 mμ. While nicotinic acid is often called the vitamin, and although it is probably converted in the body to the vitamin, actually the vitamin itself is nicotinamide.

Niacin amide, nicotinamide, nicotinic acid amide, 3-pyridinecarboxylic acid amide, is a crystalline solid which melts at 129° C. and boils at 150-160° under reduced pressure (10^{-4} mm.). Nicotinamide is very soluble in water and alcohol—1 g. is soluble in 1 ml. of water and 0.66 g. is soluble in 1 ml. of ethyl alcohol. It is very slightly soluble in ether. Niacinamide has an absorption maximum at 212 mμ. Both nicotinic acid and nicotinamide are stable in dry form, in solutions, on exposure to light, and to change in acidity.

Nicotinamide takes part in several of the dehydrogenation enzyme systems, taking on two hydrogens and becoming dihydronicotinamide. The latter substance readily gives up two hydrogens and reverts to niacin amide again. It has been shown that niacin amide is necessary for human beings. It is ordinarily stated that a deficiency of this substance results in the illness known as pellagra.

Actually, however, niacin amide cures and prevents blacktongue in dogs, a deficiency disease which has symptoms corresponding to pellagra in human beings. The administration of niacin amide or niacin does not completely cure a person suffering from pellagra. An adequate diet is also necessary. The suggestion of the National Research Council is that about ten times as much niacin amide is needed as thiamine, that is, about 10 to 18 mg. of this vitamin are

[22] Snell and Strong, *Ind. Eng. Chem., Anal. Ed.* **11**, 346 (1939).
[23] Arnold, *Cereal Chem.* **22**, 455 (1945).

necessary for the adequate nutrition of human beings. In 1958, the Food and Drug Administration set the minimum requirements at 2.5 mg. for infants, 5 mg. for children under 6 years, 7.5 mg. for children of 6 or more years, and 10 mg. for adults.

Niacin amide (niacin) is one of the vitamins that are required for the fortification of flour, and bread and other bakery products. Enriched ·flour, enriched bromated flour, enriched self-rising flour, and enriched farina must contain not less than 16.0 mg. nor more than 20 mg. of niacin amide or niacin per pound. Enriched bread should contain not less than 10 mg. nor more than 15 mg. of this vitamin per pound.

Microbiological methods [24] are preferred to chemical methods for the assay of foodstuffs for niacin content, particularly in product control routines.

PYRIDOXINE—VITAMIN B₆

$$\begin{array}{c} CH_2OH \\ | \\ HO-C \\ / \backslash \\ C \quad C-CH_2OH \\ \| \quad \| \\ C \quad CH \\ / \backslash \quad \backslash / \\ H_3C \quad N \end{array}$$

Pyridoxine, vitamin B_6, adermin, is a colorless crystalline powder in the free base form, melting at 160° C. It has a bitter taste. Pyridoxine is readily soluble in water and alcohol, and it is also soluble in acetone. Since it is an organic base, it forms salts with mineral acids such as hydrochloric acid. Pyridoxine hydrochloride is also soluble in water (about 20 g. per 100 ml.) and in alcohol (1.1 g. in 100 ml.). The hydrochloride is an odorless, white powder with a salty taste, which is characteristic property of nearly all organic base hydrochlorides (see page 60). Pyridoxine is stable to heat, alkali and acid but is affected by light. Various methods are available for the synthesis of this compound and the synthetic product is commercially available.

It is not known what the physiological mechanism of pyridoxine is, nor is the exact amount for human beings known, but the range is probably from 0.1 to 1.0 mg. per day. Pyridoxine appears to be necessary in order to prevent certain symptoms that are associated with pellagra and the need for pyridoxine in human nutrition has been established.[25]

---

[24] Snell and Wright, *J. Biol. Chem.* **139**, 675 (1941).
[25] *Natl. Acad. Sci.—Natl. Research Council, Publication* **302** (1953).

Pyridoxal, pyridoxal phosphate, and pyridoxamine also display vitamin B_6 activity.

Pyridoxine may be estimated by chemical, microbiological, and bioassay methods. Since pyridoxine has a phenolic functional group it can couple to form dyes and this function is the basis of several tests.[26]

Hochberg, Melnick, and Oser Method.—Bound pyridoxine may be hydrolyzed by heating with hydrochloric acid. The free vitamin is adsorbed from solution at pH 3, eluted with sodium hydroxide solution, and the eluate is clarified with isopropyl alcohol. Three aliquots of the supernatant are treated with Gibbs' reagent, one alone, the second in the presence of added pyridoxine acting as an internal standard, and the third in the presence of boric acid which renders the vitamin nonreactive without affecting the ability of other reacting compounds to couple with the 2,6-dichloroquinonechloroimide. It is best in performing the test to avoid unnecessary exposure to light.

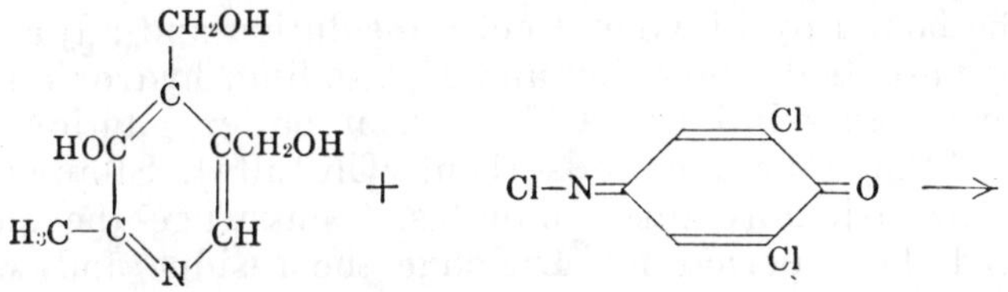

CH₂OH

HOC CCH₂OH

H₃C-C CH
    N

Pyridoxine

+

Cl-N= =O —→
    Cl

2,6-Dichloroquinonechloroimide

CH₂OH

HO CH₂OH

H₃C N

—N= =O +HCl
    Cl
    Cl

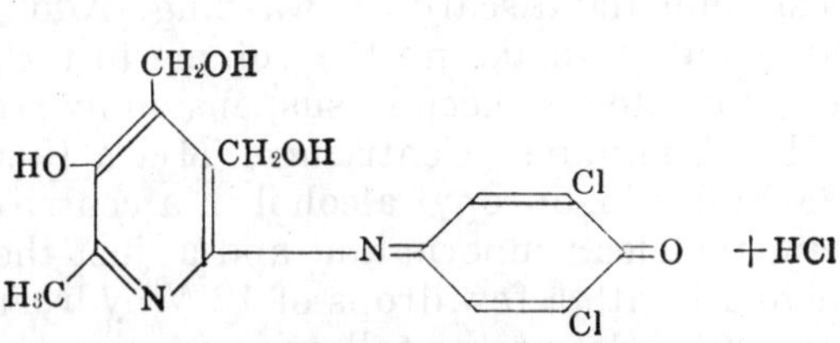

Blue pigment

Reagents.—Gibbs' Reagent.—Dissolve 100 mg. of recrystallized 2,6-dichloroquinonechloroimide in 250 ml. of isopropanol. Store the reagent in an amber, glass-stoppered bottle and keep in the refrigerator, withdrawing portions as needed. Discard after one month or sooner if a pink color develops.

Purify the compound by dissolving 1 g. in 50 ml. of acetone and precipitating by the gradual addition of small amounts of water

²⁶ Hochberg, Melnick, and Oser, *J. Biol. Chem.* 155, 109, 119 (1944).

while stirring. Recover the crystals on a Büchner funnel, air dry by suction, and store in a sealed bottle in a refrigerator.

Ammonia-Ammonium Chloride Solution.—Dissolve 160 g. of ammonium chloride in 700 ml. of water, add 160 ml. of concentrated ammonium hydroxide solution, and dilute to 1 liter with water.

Pyridoxine Hydrochloride Solution.—Dissolve 100 mg. of the crystalline vitamin in 1 liter of 0.1 N hydrochloric acid. This stock solution is stable for three months if kept in a refrigerator. Prepare working standards daily as needed from the stock solution.

Buffer Solution.—Dissolve 73 g. of disodium hydrogen phosphate dihydrate, $Na_2HPO_4 \cdot 2H_2O$, and 167 g. of citric acid in water and dilute to 1 liter. This will give a pH of 3.

Preparation of Test Extract.—Weigh up to 3 g. of sample, containing from 30 to 200 micrograms of pyridoxine, preferably 100 micrograms, into a test tube calibrated at the 20-ml. mark. Add 10 ml. of 4 N hydrochloric acid. Insert a glass stirring rod with a loop at one end to facilitate mixing and adjustment of pH. Immerse the tube in a boiling water bath for 1 hour with occasional stirring to hydrolyze the bound pyridoxine. Cool the solution and adjust to a pH of 3 using an outside indicator and 1 N sodium hydroxide and 1 N hydrochloric acid solutions. Add 3 ml. of buffer solution and then add 2.5 g. of Lloyd's reagent adsorbent (Eli Lilly). Stopper the tube and shake occasionally for 5 minutes. Centrifuge the suspension and discard the supernatant. Dislodge the residue and wash with 15 ml. of 0.001 N hydrochloric acid. Centrifuge the finely divided adsorbent and residuum and discard the washing. Add 5 ml. of 2.0 N sodium hydroxide solution, make up the volume to the 20-ml. mark, and disperse the adsorbate and keep in suspension by frequent inversions for a period of 3 minutes. Centrifuge. Mix a 10-ml. aliquot of the eluate with 500 ml. of isopropyl alcohol in a centrifuge tube and centrifuge. Decant the clear supernatant and adjust the pH with an outside indicator to 5-7 with a few drops of 12 N hydrochloric acid.

Color Development.—Set up the following tubes and add the solutions in the order mentioned to avoid precipitation of salts: Tube 1, 6 ml. of test solution plus 2 ml. of ammonia-ammonium chloride solution, 1 ml. of saturated boric acid solution; Tube 2, 6 ml. of test extract plus 2 ml. of ammonia-ammonium chloride solution plus 1 ml. of water; Tube 3, 6 ml. of test extract plus 2 ml. of ammonia-ammonium chloride solution plus 1 ml. of working pyridoxine standard containing 10 micrograms. For each series, set up a direct reading photoelectric colorimeter with a 620 mμ filter at 100 per cent transmission with Tube 1, the blank, 60 seconds after the addition of 1 ml. of chloroimide reagent. To each of the remaining tubes, add

1 ml. of chloroimide reagent and measure the blue color exactly 60 seconds later. The use of Tube 1 as a blank eliminates simultaneously the necessity for correcting for absorption attributable to the color of the test extract, that of the reagents, and that due to the coupling of compounds other than pyridoxine which yield colored derivatives.

Calculations.—

$$\frac{L_2}{L_3 - L_2} \times \frac{10}{6 \text{ ml.}} \times \frac{60 \text{ ml.}}{10 \text{ ml.}} \times \frac{18.5 \text{ ml.}}{W} = \gamma \text{ pyridoxine/g.}$$

In the equation, L_2 represents the photometric density ($2 - \log G$, where G is the per cent transmission) attributable to the pyridoxine in Tube 2 containing 6 ml. of the final test extract. $L_3 - L_2$ is the increment in photometric density attributable to the added 10 micrograms of pyridoxine. W represents the weight in grams of the test sample, and $60/10 \times 18.5$ is the dilution factor. A correction is made in the calculation for the volume of 1.5 ml. occupied by the 2.5 g. of Lloyd's reagent in a total volume of 20 ml.

Pyridoxine may also be estimated chemically by means of diazotized sulfanilic acid [27] and diazotized *p*-aminoacetophenone.[28]

INOSITOL

CHOH
/ \
HOHC CHOH
| |
HOHC CHOH
\ /
CHOH

Inosital, cyclohexanehexol, is a sweet-tasting, colorless crystalline solid that melts at 225-226° C. and boils at 319° under reduced pressure (15 mm.). Inositol dihydrate melts at 215-216° C. About 4.5 g. are soluble in 100 ml. of water. Inositol is insoluble in alcohol. The need for inositol in human nutrition has not been established, although there has been some suggestion that it is necessary for the maintenance and growth of hair. Inositol has been synthesized by the hydrogenation of hexahydroxybenzene. However, since it occurs in many natural products it will probably be made commercially from one of those sources.

The principal methods for the assay of inositol are microbiological.[29]

[27] Bina, Thomas, and Brown, *J. Biol. Chem.* 148, 111 (1943).
[28] Brown, Bina, and Thomas, *J. Biol Chem.* 158, 455 (1945).
[29] Williams, Stout, Mitchell, and McMahan, *Univ. Texas Publication* 4137, 27 (1941).

PANTOTHENIC ACID

Pantothenic acid, $HOCH_2C(CH_3)_2CHOHCONHCH_2CH_2COOH$, pantothen, is a viscous, colorless to pale yellow syrup that is soluble in water and alcohol. The d-form has an optical rotation of $+37.5°$ at 25° C. This vitamin is generally used in the form of its calcium or sodium salt. Calcium pantothenate is a fine, white, dextrorotatory, crystalline powder which is colorless but has a slightly bitter taste. It is known that pantothenic acid is a part of coenzyme A and participates in acetylation and many other acylation reactions. The need for pantothenic acid in human nutrition has been established.[30] Several authorities quote the amount of 10 mg. per day as adequate for human beings. It has been synthesized by condensation of an ester of β-aminopropionic acid with the lactone of α-γ-dihydroxy-β-dimethylbutyric acid or by direct condensation of the amino acid and the lactone.

The principal methods for the assay of pantothenic acid are microbiological[31] and bioassay.

Colorimetric Method.—Szalkowski, Mader, and Frediani[31a] have detailed a colorimetric method based on hydrolysis of the pantothenate to pantoyl lactone, formation of hydroxamic acid by addition of hydroxylamine hydrochloric acid in alkaline solution and the development of a purple color with ferric chloride.

Hydrolytic cleavage of pantothenic acid in alkaline or acid media results in the formation of β-alanine and α, α-dihydroxy-β, β-dimethylbutyric acid. β-Alanine when oxidized with potassium permanganate in the presence of potassium bromide under properly regulated conditions yields an insoluble hydrazone with 2,4-dinitrophenylhydrazine. The ratio between the amount of hydrazone formed and the hydrolyzed pantothenate oxidized is constant and may be estimated by dissolving the dinitrophenylhydrazone in pyridine, diluting with sodium hydroxide and measuring the resulting blue color spectrophotometrically at 570 mμ. Beer's law is followed over a suitable concentration range, with either acid or alkaline hydrolysis. Thus, for pure calcium pantothenate, as well as in some mixtures, either the acid or alkaline hydrolysis may be used. In certain mixtures as for example, with soya flour, it is necessary to use acid hydrolysis.

The specificity of this reaction with respect to the other vitamins

[30] *Natl. Acad. Sci.—Natl. Research Council, Publication* 302 (1953)
[31] Roberts and Snell, *J. Biol. Chem.* 163, 499 (1946).
[31a] Szalkowski, Mader, and Frediani, *Am. Chem. Soc. Abstracts Paper, 117th Meeting,* Philadelphia, 1950.

and organic acids usually encountered has been studied. The presence of relatively large amounts of such compounds as acetic, lactic, tartaric, glycollic and succinic acids, α-alanine, ethyl alcohol and riboflavin has been found not to affect the pantothenate estimation. Compounds which yield insoluble dinitrophenylhydrazones after oxidation with potassium permanganate, however, do interfere. Compounds such as carbohydrates, ascorbic acid, thiamine hydrochloride, pyridoxine hydrochloride, niacin, niacinamide and soya flour fall into this class. This interference may be eliminated, except for ascorbic acid, by chromatographic isolation of the calcium pantothenate.

Chromatographic Column.—Use a glass column 12 mm. inside diameter, approximately 50 cm. in length and constricted at the lower end. Place a plug of glass wool in the lower end and add slowly 8 g. of special chromatographic grade aluminum oxide, applying gentle suction. Wash the column with 50 ml. of 1:1 hydrochloric acid and then with 100 ml. of water immediately before use.

Preparation of Sample.—Weigh an amount of sample containing about 200 mg. of calcium pantothenate into a 250-ml. volumetric flask and shake with 200 ml. of water for 10 minutes. Dilute to the mark and mix thoroughly. Filter this mixture or centrifuge and use the clear solution.

Chromatography.—Pass an aliquot of the sample containing about 25 mg. of calcium pantothenate through the column and wash with successive portions of water, 100 ml. in all. Elute from the column with 50 ml. of 1:1 sulfuric acid followed by 10 ml. of water. Reflux the combined acid and water washings for one hour, cool to room temperature, transfer to a 250-ml. volumetric flask, and make up to volume with water.

Oxidation and Measurement.—Transfer a 25-ml. aliquot of the solution obtained to a 125-ml. glass-stoppered flask, and add 5 ml. of 1:1 sulfuric acid. Adjust the temperature to 22-25° C. by placing in a water bath, and add 5 ml. of potassium bromide (12/100) and 10 ml. of 5 per cent potassium permanganate. Stopper the flask and allow to stand in the bath for 10 minutes, after which cool in an ice bath for 5 minutes. Add 20 per cent freshly prepared sodium sulfite solution dropwise to decolorize the excess potassium permanganate. To the clear, colorless oxidized solution, add 10 ml. of 2,4-dinitrophenylhydrazine (5 g./l. of 1:4 hydrochloric acid). Mix the solution with the precipitate thoroughly and heat on a steam bath for at least 15 minutes after which cool the flask and contents to room temperature. Transfer the yellow precipitate to a small sintered glass filter, wash with five 5-ml. portions of water and dry for 30 minutes at 100° C. Meanwhile, drain the flask thoroughly, and take up the last traces of precipitate in hot pyridine used in two successive portions of 3 ml. each. Transfer the pyridine washings to a 25-ml. volumetric flask. Fit the filter into a filtrator containing the 25-ml. volumetric flask. Add boiling pyridine in small portions to the filter, and gently triturate the contents with a glass rod, after which apply suction and draw the resultant pyridine solution through each time; 3 or 4 washings are usually sufficient to dissolve and transfer all the precipitate to the volumetric flask. Make up the pyridine solution to exactly 25 ml.

Transfer a 5-ml. aliquot of this pyridine solution to a 100-ml. volumetric flask and add 50 ml. of water, followed by 5 ml. of 5 N sodium hydroxide solution. Dilute the solution to the mark and read the blue color in a Beckman spectrophotometer set at 570 mμ. This color has been found to be stable for about one hour. Any other suitable colorimeter may be used.

AMINOBENZOIC ACID

p-Aminobenzoic acid, $NH_2C_6H_4COOH$, is a yellowish-red crystalline solid melting at 187° C. About 0.4 g. is soluble in 100 ml. of water and about 11 g. is soluble in 100 ml. of alcohol. Although it has been shown that this substance is required by chicks for propor growth, and by rats for proper growth and maintenance of hair, there has been no convincing evidence that it is necessary for adequate human nutrition or that it will have an effect on the color of the hair. There has been some experimental work showing that it will interfere with the action of the sulfonamides.[32]

The principal methods for the assay of p-aminobenzoic acid are microbiological.[33]

BIOTIN

$$\begin{array}{c}
CO \\
\diagup \quad \diagdown \\
HN \qquad NH \\
| \qquad | \\
HC\!-\!-\!-\!-\!CH \\
| \qquad | \\
H_2C \qquad CH\!-\!CH_2\!-\!CH_2\!-\!CH_2\!-\!CH_2\!-\!COOH \\
\diagdown \quad \diagup \\
S
\end{array}$$

Biotin, vitamin H, is often considered as a member of the vitamin B complex. It is an optically active crystalline substance melting at

[32] *U. S. Food Drug Admin.*, *F.D.&C. Act Trade Correspondence* TC-401, April 19, 1943.

[33] Thompson, Isbell, and Mitchell, *J. Biol Chem.* 148, 281 (1943).

230-232° C. with some decomposition. It is soluble in both water and alcohol. Biotin is not readily affected by acids, alkalis, light and air at ordinary temperatures, but high temperatures and oxidizing agents affect it. Biotin has been synthesized. It is known that this vitamin is a growth-promoting factor but its exact role in human nutrition has not been completely elucidated, nor have its requirements been established.

Both bioassay and microbiological assay methods [34] are utilized for the estimation of biotin.

FOLIC ACID

Folic Acid or Pteroylglutamic Acid

For a number of years, folic acid, which is present, as can be gathered from the name, in many green leaves, was considered one of the nonidentified vitamins. That is, its actual nutritional significance in both human and animal nutrition had not been established. Late in 1945, however, substantial evidence was obtained to indicate that it was a nutritional factor necessary for the prevention of anemia. Since it has already been established that folic acid was either a constituent or component of animal tissue, particularly liver and kidney, and since the role of liver in anemia has been very well established, it appears that folic acid is one of the necessary vitamins.

Folic acid is the name applied to a group of substances of vitamin-like character which have an antianemia activity. Vitamin B$_c$ and vitamin M are considered either to be the same material, to be a precursor, or to belong to the same group.

One such substance is pteroylglutamic acid, named systematically according to the *Chemical Abstract* system N-{p-{[(2-amino-4-hydroxypyrimido [4,5-b] pyrazin-6-yl) methyl] amino} benzoyl} glutamic acid and also by the name N[4{[(2-amino-4-hydroxy-6-pteridyl) methyl] amino} benzoyl] glutamic acid. It thus contains a pterin group attached to 1 mole of p-aminobenzoic acid linked to 1 mole of glutamic acid.

Folic acid from different sources may be conjugates of various sizes. They are useful in preventing and treating nutritional macro-

[34] Wright and Skeggs, *Proc. Soc. Exptl. Biol. Med.* 56, 95 (1944).

Lactobacillus casei Factor. Vitamin B$_c$ Has a Total of 7 Glutamic Acid Radicals

cytic anemia. Synthetic pteroyl glutamic acid is a yellow crystalline material that is slightly soluble in water. The sodium salt is more soluble. It is reported to be unstable in alkaline medium and is destroyed when heated with mineral acids. Ultraviolet light causes some decomposition. About 1.4 mg. is considered the daily requirement.

The principal methods of analysis are microbiological and bioassay.

VITAMIN B$_{12}$

Vitamin B$_{12}$ has been isolated from commercial liver extracts [35] and from a mold of the Streptomyces species.[36] It is a crystalline, red, heat-stable cobalt complex containing 3 molecules of phosphorus and some nitrogen with a probable molecular weight of between 1550-1750. This substance appears to be the most potent antianemic material known for Addisonian pernicious anemia. It also appears to be a growth factor.[37]

ASCORBIC ACID—VITAMIN C

Ascorbic acid, L-ascorbic acid vitamin C, is an odorless, white crystalline solid that melts at 190-192° C. It has a slightly sour taste. Ascorbic acid is very soluble in water, 33 g. dissolving in 100 ml., and it is soluble in alcohol (2 g. in 100 ml.) and in methyl alcohol. It is

[35] Rickes, *Science* 107, 396 (1948).
[36] Rickes, *Science* 108, 635 (1948).
[37] Wetzel, Fargo, Smith, and Helikson, *Science* 110, 651 (1949).

insoluble in the usual organic solvents. Vitamin C is optically active, having a specific rotation of $+23°$-$24°$ in water and of $+48°$ in methyl alcohol. It has an absorption maximum at 265 mμ and a small band between 350 and 400 mμ. Although vitamin C is stable in solid form, it is relatively readily oxidized in solution to dehydroascorbic acid which has a similar effective physiological action but is more readily hydrolyzed so that there is complete loss of antiscorbutic power.

$$
\begin{array}{ccc}
\text{O}=\text{C}\underset{\displaystyle|}{} & \text{O}=\text{C}\underset{\displaystyle|}{} & \text{O}=\text{C}\underset{\displaystyle|}{} \\
\text{HO}-\text{C} & \text{O}=\text{C} & \text{HO}-\text{C}-\text{OH} \\
\text{HO}-\text{C} & \text{O}=\text{C} & \text{HO}-\text{C}-\text{OH} \\
\text{H}-\text{C} & \text{H}-\text{C} & \text{H}-\text{C} \\
\text{HO}-\text{C}-\text{H} & \text{HO}-\text{C}-\text{H} & \text{HO}-\text{C}-\text{H} \\
\text{H}_2\text{COH} & \text{H}_2\text{COH} & \text{H}_2\text{COH}
\end{array}
$$

| Vitamin C (ascorbic acid) | Dehydroascorbic acid (ketone structure) | Dehydroascorbic acid (hydrate structure) |

Ascorbic acid, which is a lactone, has been synthesized by 3 principal methods—namely, by lactone formation after formation of 2- and 3-ketohexonic acids; by an ester condensation of α-hydroxy acids, and by condensation of two aldehydes of lower molecular weight. The 2-ketohexoic acid method is used commercially.

The role that ascorbic acid plays in nutrition has been well characterized, although the mechanism has not been entirely established. The name given to the illness caused by a deficiency of this vitamin is *scurvy*.

Vitamin C is necessary in human nutrition because it keeps the cement substance lying between the cells of the tissue of the body in proper condition. As a result of this action it prevents sore and bleeding gums, hemorrhages through tissue, and some forms of anemia. It assists in the proper development of bone structure, normal calcification and muscle tone. The daily allowances recommended by the National Research Council for vitamin C·vary from 30 mg. for infants under one year to 150 mg. for women during lactation. See Table 99.

Ascorbic acid has been used in the fortification of a number of foods and for the restoration of other foods. Thus, for instance, it has been included as a component of orange drinks and has been added to orange juice undergoing concentration in order to standardize the product produced.

Waugh and King [38] isolated the vitamin from lemon juice. They found that this crystalline material was identical with the hexuronic acid of Szent-Gyorgyi.[39] It is a reducing substance and reduces Fehling solution, neutral silver nitrate, and neutral potassium permanganate. It is easily oxidized by iodine and Benedict solution as well as by atmospheric oxygen in alkaline solution. This property is the basis of most of the methods for the estimation of the vitamin. Vitamin C has been synthesized by a number of investigators.

Tillmans [40] noted the reducing power of some unknown substance, later identified as ascorbic acid, vitamin C, in natural foods and used the oxidation-reduction indicator 2,6 dichlorophenolindophenol, sodium 2,6 dichlorobenzenoneindophenol, to distinguish between natural juices and artificial juices and later pointed out that there was a close degree of correlation between the reducing capacity of these juices and their antiscorbutic activity. This correlation was confirmed by many other investigators.

Preparation of Sample.—In substances from which the juice may easily be extracted by mechanical pressure and subsequent centrifuging such as oranges, lemons, limes, and other citrus fruits, the resultant clear juice may be used directly for the determination.

In order to keep oxidation of ascorbic acid at a minimum it is necessary for raw materials that are to be chopped, cut, or otherwise processed to be handled rapidly and stabilized with an acid such as 6 per cent metaphosphoric acid.[41] This acid retards the oxidation of ascorbic acid by inactivating ascorbic acid oxidase and inhibiting the catalytic action of copper. If iron is present, as may be the case in canned goods, 8 per cent acetic acid may be used to minimize interference from ferrous ion. Dehydrated foods should be rehydrated by adding to a known weight of sample 100 ml. of 3 per cent metaphosphoric acid, holding for 15 minutes, blending, and then proceeding with the method. In the case of sulfited dehydrated food, the sulfite may be removed by means of acetone.

Other food materials, such as milk, may be treated as described in Chapter VII, section, "Reconstituted Milk," that is preparing a trichloroacetic acid serum. Solid materials may be ground and mixed with sand and then extracted with trichloroacetic acid solution and

---

[38] Waugh and King, *J. Biol. Chem.* 97, 325 (1932).
[39] Szent-Gyorgyi, *Nature* 129, 943 (1932).
[40] Tillmans, Hirsch, and Reinhagen, *Z. Untersuch. Lebensm.* 56, 272 (1928).
[41] Mapson, *Biochem. J.* 36, 196 (1942).

subsequently diluted so that the final concentration of the trichloro-acetic acid is about 5 per cent.

Iodine Titration Method.—This method is especially applicable to fruit and tomato juices and is a simple and rapid method for the differentiation of true or natural juices and artificial juices. Cut the fruit in half and extract the juice with a hand orange squeezer. Centrifuge and filter through a large size Gooch crucible fitted with a Fisher cotton filter pad.

Procedure.—Pipette 5 ml. of the juice or the prepared solution and transfer to a 125-ml. flask. Add 20 ml. of water and 2 ml. of a 1 per cent soluble starch. Titrate rapidly with an accurately standardized 0.01 N iodine solution containing 16 g. potassium iodide per liter. Each milliliter of iodine is equivalent [42] to 0.88 mg. of ascorbic acid, lactone form. The milligram of vitamin C per milliliter can be calculated from this relationship. The titration must be rapidly performed because other substances such as glutathione and cysteine are oxidized slowly by iodine solution, consequently rapid titration minimizes this possible error.

When juices with a low vitamin C content or suspected artificial juices or fruit drinks with a small proportion of natural fruit juices are analyzed, 25 ml. of the original material may be taken for the analysis and no additional water need be added.

Dye Method.—Weigh accurately 125 mg. of sodium 2,6-dichloro-benzenoneindophenol, dissolve in small portions of warm water and filter. The filtered solution, when cool, is made up to 250 ml. The indicator solution keeps better if prepared in a 7.2 pH phosphate buffer. The strength of the dye solution may be estimated in three ways. (1) Transfer 5 ml. of the dye to a 50-ml. flask and dilute with 20 ml. of water and a few drops of acetic acid. Titrate with lemon juice prepared as directed above, with the aid of a microburette or semimicroburette until the dye is decolorized. The color of the dye changes from blue to red to colorless. The blue is the neutral or alkaline color and the red is the acid color. Then titrate 5 ml. of the lemon juice with the standard 0.01 N iodine solution. Since each milliliter of 0.01 N iodine solution used is equivalent to 0.88 mg. of ascorbic acid, then the number of milliliters of lemon juice multiplied by the ascorbic acid content as derived by the iodine titration and divided by 5 yields the value of ascorbic acid in milligrams per milliliter of dye solution.

(2) Titrate 25 ml. of dye solution with a 0.01 N solution of ascorbic acid made by dissolving 176 mg. ascorbic acid in water and mak-

---

[42] Bessy and King, *J. Biol. Chem.* **103**, 687 (1933).

ing up to 100 ml. volume. The equivalent value of each milliliter of dye solution is thus determined directly.

(3) Pipette 15 ml. of the dye solution into a small flask. Add 0.5 to 1.0 g. of potassium iodide and 0.5 to 1.0 ml. sulfuric acid (1:4). Shake to facilitate oxidation and titrate the liberated iodine with 0.01 N sodium thiosulfate solution using starch solution as indicator. One ml. of 0.01 N thiosulfate solution is equivalent to 0.88 mg. ascorbic acid.[43, 44]

Procedure.—Measure 5 ml. of the dye solution by means of a pipette into a 125-ml. flask and add 20 ml. of water. Titrate with the prepared juice from a microburette or semimicroburette. Five ml. of the dye generally is equivalent to 1.06 mg. of ascorbic acid. Divide this factor or the appropriate factor by the number of milliliters of juice required to decolorize 5 ml. of the dye solution. The result is the number of milligrams vitamin C per milliliter of juice. This method is modified from that of Birch, Harris, and Ray.[45]

Standardized Dye Method.—This method, as in the preceding method, is based upon the reduction of 2,6-dichlorobenzenoneindophenol by vitamin C in acid solution.

Reagents.—Standard 2,6-Dichlorobenzenoneindophenol Solution. —Dissolve 50 mg. of sodium 2,6-dichlorobenzenoneindophenol in 150 ml. of hot water containing 42 mg. of sodium bicarbonate and dilute with water to 200 ml. This is a 0.025 per cent solution. Transfer to an amber, glass-stoppered bottle and store in a refrigerator at 3° C. Prepare fresh each week. Standardized each day, this solution is used against standard ascorbic acid solution as follows:

Dilute a 5-ml. aliquot of ascorbic acid solution containing 1 mg. with 5 ml. of 3 per cent metaphosphoric acid. Titrate with the dye solution until a pink color which persists for 15 seconds is obtained. This volume of dye solution represents 1 mg. of ascorbic acid, hence 1 ml. of dye solution is equivalent in terms of ascorbic acid to 1 divided by the number of milliliters used in the titration.

Metaphosphoric Acid Solution.—Dissolve 60 g. of metaphosphoric acid beads (HPO_3) in 900 ml. of water without heating and dilute to 1 liter. Store in a refrigerator. This is a 6 per cent solution. Metaphosphoric acid hydrolyzes on standing to orthophosphoric acid, H_3PO_4, therefore it must be prepared anew each week. Dilute 500 ml. of the 6 per cent solution to 1 liter with water to give a 3 per cent solution.

---

[43] Buck and Ritchie, *Ind. Eng. Chem., Anal. Ed.* 10, 26 (1938).
[44] Menaker and Guerrant, *Ind. Eng. Chem., Anal. Ed.* 10, 25 (1938).
[45] Birch, Harris, and Ray, *Biochem. J.* 27, 590 (1933).

Standard.—Dissolve 100 mg. of ascorbic acid in 3 per cent metaphosphoric acid and dilute to 500 ml. Five ml. of this solution contains 1 mg. of ascorbic acid.

Procedure.—Transfer equal weights of the sample to be analyzed and 3 per cent'metaphosphoric acid, of the order of 200-300 g. to a blendor and mix to obtain a homogeneous slurry. With the aid of a weight pipette and carriage, transfer 10 to 30 g. of the slurry to a 100-ml. volumetric flask and make to volume with 3 per cent metaphosphoric acid. Filter the mixture, rejecting the first portion of the filtrate. Transfer a 10-ml. aliquot with the aid of a pipette to a 25-ml. or 50-ml. Erlenmeyer flask and titrate immediately with the standard dye solution to a faint pink which persists for 15 seconds. Calculate the concentration of ascorbic acid with the aid of the following equation.

$$\text{Mg. vitamin C per 100 g.} = \frac{W_1 + W_2}{W_1 \times W_3} \times \frac{V_1}{V_2} \times 100(V \times F)$$

in which

W_1 = weight of sample in grams
W_2 = weight of extracting acid in grams
W_3 = weight of slurry taken for analysis
V_1 = volume to which slurry sample is diluted in milliliters
V_2 = volume of filtrate taken for titration in milliliters
V = volume of dye solution used for titration
F = ascorbic acid equivalent of dye in milligrams/milliliters

Photometric Method.—This variation of the dye method is based on the measurement of the amount of decolorization produced in a 2,6-dichlorobenzenoneindophenol solution by the ascorbic acid present in a sample.

Standardization of Dye Solution.—Prepare the dye solution by dissolving sufficient sodium 2,6-dichlorobenzenoneindophenol in water so that a G_1 reading of about 30 is given with an Evelyn photoelectric colorimeter. Generally 13 mg. per liter will give this reading.

Standardize the dye by noting the 10-second reading with a 520 mμ filter (when the instrument is calibrated to 100 with water) given by a tube containing 1 ml. of 1 per cent metaphosphoric acid (or 0.25 per cent oxalic acid when used) and 9 ml. of the dye solution. This value is G_1. Obtain L_1 by reference to the calibration chart provided with the instrument. Substitute L_1 in the equation in the calculation formula below.

Blending Sample.—Blend 25 or 50 g. of fresh or frozen fruit or vegetable tissue with 350 ml. of 1 per cent metaphosphoric acid in a blending machine operated for 5 minutes at high speed. If the material is of high ascorbic acid content, such as leafy vegetables, raspberries, strawberries, or asparagus, use the smaller quantity. Use 50 g. for fresh or frozen foods containing less ascorbic acid, such as potatoes, carrots, sweet potatoes, plums, and apricots. If a very large sample is desired, in order to assure greater uniformity, double these weights, and use 650 ml. of acid extractant. This is about the maximum capacity of the blending cups. Frozen material need not be defrosted prior to blending. When a dehydrated fruit or vegetable is being analyzed, use 5 or 10 g. of sample and 350 ml. of extractant, in accordance with the classification mentioned above. If metaphosphoric acid is not available, the same volume of 0.25 per cent oxalic acid may be substituted as the extractant.[46]

Clarification of Extract.—After blending the extracts clarify, preferably by contrifugation. Moderate turbidities do not interfere, since the instrument is calibrated with proper blanks. Extracts of starchy vegetables that filter slowly (such as the extract of potato) may lose considerable ascorbic acid during filtration but will lose none during centrifugation. Even with centrifugation the extracts should be tested within a few minutes to avoid loss of ascorbic acid.

Procedure.—Pipette 1-ml. portions of the centrifugate into three matched tubes from the Evelyn photoelectric colorimeter. Add 9 ml. of water to one tube and adjust the colorimeter to read 100 with this tube, using filter No. 520. To each of the other tubes add 9 ml. of the previously standardized indophenol dye solution from a calibrated rapid-delivery pipette. Take a reading in the photoelectric colorimeter, using the same filter, 10 seconds after the beginning of the addition of dye.

This reading is G_2, from which the corresponding L_2 value is obtained from the calibration chart provided with the instrument. This value (L_2) and the value for L_1 obtained in the dye standardization are substituted in the following equation which contains the factor 10.8, as determined by Evelyn, Malloy, and Rosen.[47]

$$\text{Ascorbic acid (mg./100 g.)} = 10.8 \ (L_1 - L_2) \frac{\text{ml. acid extractant} + (\% \text{ liquid in sample}) (\text{g. sample})}{\text{g. sample}}$$

[46] Ponting, *Ind. Eng. Chem., Anal. Ed.* 15, 389 (1943).
[47] Evelyn, Malloy, and Rosen, *J. Biol. Chem.* 126, 645 (1938).

For most fresh or frozen fruits the sum of the percentages of soluble solids and of water will be so near 100 that the insoluble material can be neglected. Then the formula becomes:

$$\text{Ascorbic acid (mg./100 g.)} = 10.8\,(L_1 - L_2)\,\frac{\text{ml. acid} + \text{g. sample}}{\text{g. sample}}$$

For dehydrated vegetables the formula becomes:

$$\text{Ascorbic acid (mg./100 g.)} = 10.8\,(L_1 - L_2)\,\frac{\text{ml. acid}}{\text{g. sample}}$$

For material low in ascorbic acid such as potatoes and carrots, the ratio of extract to dye can be altered to give greater sensitivity. The use of 2 ml. of centrifugate plus 9 ml. of dye solution is also satisfactory, since the sensitivity is doubled. The factor 6.25 must then be used instead of 10.8 given in the formulas above. A larger proportion of extract to dye is not recommended, since the fading of the dye will be accelerated.

A reservoir-type of automatic pipette for delivery of the acid and nonreservoir types for delivery of the 1-ml. aliquots of filtrate and 9-ml. portions of the dye solution will be found advantageous. The 9-ml. pipette must extend to near the surface of the liquid in the tube to avoid splashing and must be calibrated to drain uniformly in less than 5 seconds.

It is difficult to calibrate the automatic 1-ml. pipette to deliver exactly 1 ml. It is simpler to adjust the pipette to approximately 1 ml. and then determine the exact factor for each pipette by testing a solution of pure ascorbic acid of known concentration, freshly prepared in acid. This factor is then used instead of the 10.8 mentioned previously.

For most fruits and vegetables the fading of the dye when added to the extracts in a 9-plus-1 ratio will be no greater than the fading of the dye in the acid alone—that is, not over one-fourth scale division in a 10-second interval. Where the tissue extract contains interfering substances which cause greater fading, an approximate correction should be made. This can be accomplished by taking both a 10- and a 20-second reading and subtracting the difference between the two.

Violent splashing and loss of liquid may occur when the blendors are started. Bits of dehydrated material will be thrown out of the liquid and may stick on the upper surfaces of the container, where they do not become macerated and extracted. These losses can be prevented by using auto-transformers in series with the blendor so that they serve as starting switches. The starting impulse can then be applied gradually.

Although the factor 10.8 given by Evelyn has been confirmed many times, one worker has reported that he obtained a factor other than 10.8 by using the suggested amount of ascorbic acid solution and dye, the Evelyn colorimeter, and the proper filter. In case of doubt, therefore, the instrument itself can be checked by determining whether the proper factor is obtained on a pure, freshly prepared solution of ascorbic acid.

Photoelectric colorimeters other than the Evelyn can be used, provided the cells permit very rapid mixing and the instrument can be read rapidly. The factor must be redetermined for the specific cells, filter, and instrument used.

Dehydroascorbic Acid

In the Roe and Kuether [48] method for the determination of total vitamin C, an acid extract of the material is passed through Norit (1 to 3 g. Norit for 50 ml. of filtrate); this converts ascorbic acid to dehydroascorbic acid. The latter is reacted with 2,4-dinitrophenyl-hydrazine to form the hydrazone which, when treated with sulfuric acid, forms a red color, that can be measured photometrically. By the addition of thiourea to stabilize the ascorbic acid and the omission of the Norit step, the amount of dehydroascrobic acid alone can be ascertained.[48a]

Dehydroascorbic Acid Standard Solution.—Treat a solution of 25 mg. of ascorbic acid in 25 ml. of 5 per cent metaphosphoric acid with 1 or 2 drops of bromine, shake until yellow, decant from the excess bromine into a large test tube, and aerate until colorless. Make appropriate standards by diluting with 5 per cent metaphosphoric acid containing 1 per cent of thiourea.

Procedure.—Grind the plant tissue with 20 to 50 parts but never less than 20 parts of a solution containing 5 per cent metaphosphoric acid and 1 per cent thiourea. Filter and place 4 ml. of the extract into each of 3 matched colorimeter tubes one of which is reserved for a blank. To each of the others add 1 ml. of 2 per cent 2,4-dinitro-phenylhydrazine in 9 N sulfuric acid. Hold at 37° C. for 3 hours, cool together with the blank in ice water. To each of the 3 tubes, while in the ice water batch, add 5 ml. of 85 per cent sulfuric acid, a drop at a time during not less than 1 minute. Add 1 ml. of the 2,4-dinitro-phenylhydrazine solution to the blank tube. Shake thoroughly under

---

[48] Roe and Kuether, *J. Biol. Chem.* 147, 399 (1943).
[48a] Roe and Oesterling, *J. Biol Chem.* 152, 511 (1944).

the ice bath, wipe dry, and read in a photoelectric colorimeter with a 540 mμ filter.

Run standards along with the unknowns or prepare a calibration curve using standard solution of dehydroascorbic acid in concentration ranging from 0.25 to 15 micrograms per ml.

Stability of Vitamin C.—It has been the general impression that the vitamin C content of orange juice, other juices, and milk degrades rapidly. The work of various investigators shows that the vitamin content of these food materials when kept under moderate refrigeration decreases slowly. Nelson and Mottern [49] kept orange juice hermetically sealed under atmospheres of oxygen, nitrogen, and air and frozen for a year. They found that there was no appreciable loss of vitamin C content at the end of that time. Other investigations show that heating foodstuffs does not destroy all of the vitamin C content. Thus pasteurized milk has less vitamin C and correspondingly less antiscorbutic activity than raw milk, but the vitamin is not completely destroyed.

The vitamin C present in milk and other materials can be regenerated by the addition of hydrogen sulfide.[50] The hydrogen sulfide may then be removed by passing a stream of nitrogen through the the solution until no test for hydrogen sulfide is obtained in the exhaust gas. The titration for the vitamin may then be performed in the usual manner. This regeneration is of value, for investigators have shown that the body can do it also.

Orange Drink.—There are three main reasons for the purchase of orange drink. First, in order to quench thirst, secondly, to obtain the benefits of a fruit drink containing vitamin C, and thirdly, to build up the alkaline reserve through the ash content of the fruit juice. The second and third are the only reasons when one orders orange drink with a meal, as for example, breakfast or a rapid lunch. The layman has been educated in the importance of having vitamin rich foods. In many cases, it becomes necessary for him to pay particular attention to the quantity of vitamin he ingests in order to combat some disorder. This person may be misled by some flagrantly adulterated and misbranded orange drink.

The quantity of orange juice in an orange drink may be estimated by analyses of total solids, ash, alcohol precipitate, pectin, pectic acid, acid, and protein. However, a much shorter means may be used, if it is borne in mind that the vitamin C content does not degrade rapidly when the orange drink is stored under proper conditions.

[49] Nelson and Mottern, *Ind. Eng. Chem.* **25**, 216 (1933).
[50] Johnson, *Biochem. J.* **27**, 1287 (1933).

This method is one based on the average vitamin C content of oranges. The average vitamin C content or ascorbic acid content of orange juice is 0.5 mg. per milliliter.

Determine the number of milligrams of ascorbic acid per milliliter in the orange drink or orangeade and divide by the factor 0.5 as defined above. The quotient multiplied by 100 yields the per cent of orange drink in the product, based, of course, on the above-mentioned average vitamin C content.

The same procedure may be developed for other fruit drinks and for tomato juice.

VITAMINS D

The work of Windaus, Waddell, Bills, and others has shown that vitamin D is not a single substance. The term vitamin D designates the entire group of substances which are antirachitically active. In 1939 Bills [51] pointed out that eight forms of vitamin D had been artificially prepared and at least two forms were found in fish oil which might be identical with certain of those made synthetically. Since then other substances with antirachitic activity have been isolated or prepared.

Vitamin D_3
Activated 7—dehydrocholesterol

Vitamin D_2
Activated ergosterol

Vitamin D_2, which is known under the names of *viosterol* and *calciferol*, is a solid, crystallizing in white odorless crystals, which melts at 115-117° C. and has an $[a]_D^{20} = +82.6°$ C. in acetone, and which is soluble in oils and fats, in alcohol and propylene glycol, and insoluble in water. It is relatively stable in oil solution but oxidizes readily if the irradiation products of some of the vitamins D and provitamins D are present. This vitamin is affected by light and has poor thermostability.

Vitamin D_3, activated 7-dehydrocholesterol, is also a white, odor-

---

[51] Bills, *J. Am. Med. Assoc.* **108**, 13 (1937).

less, crystalline solid. It melts at 82-83° C. and has a specific rotation $[\alpha]_D^{20} = +83.5°$ in acetone. It is soluble in oils and alcohol and is insoluble in water. The vitamins D have a characteristic absorption spectrum with a maximum at 265 mμ.

The principal physiological action of vitamin D is the control of the mineral balance of the system, particularly in the regulation of calcium and phosphorus metabolism. Inadequate supply of this vitamin causes the illness known under the name of *rickets*. Most vitamin D is used for prevention rather than the cure of rickets.

Vitamin D happens to be one of the vitamins not readily supplied in sufficient quantity in the ordinary diet of human beings, and consequently foods need vitamin D enrichment. This has been done in milk, bread, oleomargarine, and butter principally by the addition of oil solutions of irradiated ergosterol and fish liver concentrates. The principal method of obtaining ergosterol is by the extraction of yeast yielding vitamin D_2 on irradiation. Vitamin D_3 is obtained by the irradiation of 7-dehydrocholesterol which is obtained synthetically from cholesterol.

Vitamins D must be considered as to source for there is a difference in their antirachitic potency. Thus vitamin D_2, or the activated ergosterol type, is considered to be the plant type vitamin D. Vitamin D_3, or the activated 7-dehydrocholesterol type, is considered to be the animal type. Only the latter type is of value to chicks, and for human beings the animal type is about 1.5 times as effective as the plant type vitamin D.

A number of food products have been fortified with vitamin D. Among the more important are milk, flour, bread, breakfast cereals, and margarine. The potency of milk is generally adjusted by producers at 400 U.S.P. units per quart. Enriched flour, enriched bromated flour, and enriched self-rising flour may be fortified with vitamin D, the minimum amount being 250 U.S.P. units and the maximum being 1000 U.S.P. units per pound of flour. Farina may also be enriched with a minimum of 250 U.S.P. units per pound.

For adequate human nutrition, particularly for infants and young children, from 400 to 800 units of vitamin D are needed daily. U.S.P. units for vitamin D are equal to International units for this vitamin.

The principal methods for the estimation of vitamin D content are bioassay methods, because, as explained above, there is a difference in the antirachitic value of vitamins D from various sources.

Antimony Trichloride Reaction of Vitamin D.[52]—Vitamins D_2 and

---

[52] Brockmann and Chen, *Z. physiol. Chem.* **241**, 129 (1936).

D_3 give with antimony trichloride in chloroform [53] an orange yellow color which soon reaches maximum intensity and shows a sharp absorption band at 500 mμ. Tachysterol behaves similarly. None of the other sterols, nor vitamin A, have this sharp absorption band, hence this absorption may be used to estimate the vitamins. Place 0.2 ml. of the solution to be tested in the absorption cell and add 4 ml. of a saturated solution of antimony trichloride in dry chloroform. After 10 to 15 minutes the extinction coefficient is measured, and the vitamin concentration may be read from a calibration curve.

A variation of this method has been developed which is based on the separation of vitamins D from vitamin A and other interfering substances by chromatographic adsorption,[53] and then measurement of the extinction coefficient at 500 mμ of the reaction product in antimony trichloride.

Tzoni Method.—This method [54] is based on the production of a red-violet color when vitamin D, pyrogallol, and aluminum chloride react. If the vitamin is in a pure form and dissolved in absolute alcohol, benzene, petroleum ether, or chloroform or other suitable solvent, it may be estimated directly. Otherwise, for example, if it is dissolved in oil or fat, it must be separated by prior saponification and subsequent extraction. The oil or fat is saponified and the soap is extracted 4 times with petroleum ether. The petroleum ether extract is washed with a solution of 1 part of alcohol and 1 part 10 per cent sodium chloride solution. The wash solution is washed in turn with another portion of petroleum ether. The combined petroleum ether layers are dried with anhydrous sodium sulfate and the ether solution is then distilled in vacuum in an atmosphere of carbon dioxide. The residue is dried in vacuum over sulfuric acid after which it may be dissolved in one of the dry solvents mentioned above.

If the vitamin is mixed with other sterols in moderate proportions, no interference results. However, if large quantities of other sterols are present, they should be eliminated by precipitation with 1 per cent solution of digitonin in alcohol. The precipitate is filtered off and is washed with a solution of 73 parts of acetone, 18 parts of water, and 9 parts of absolute alcohol. The filtrate is mixed with an equal volume of 10 per cent sodium chloride solution, and the vitamin is extracted with petroleum ether as directed above.

Vitamin A, carotene, and other substances that react with the reagents must be eliminated also.

---

[53] Ewing, Kingsley, Brown, and Emmett, *Ind. Eng. Chem., Anal. Ed.* 15, 301 (1943).
[54] Tzoni, *Biochem. Z.* 287, 18 (1936).

Reagents.—(a) Pyrogallol Solution.—Prepare a 0.1 per cent solution of pure pyrogallol in absolute alcohol. This solution is kept in a dropping bottle.

(b) Aluminum Chloride Solution.—Place as rapidly as possible large pieces of anhydrous aluminum chloride in an Erlenmeyer flask and add sufficient absolute alcohol to obtain a 10 per cent solution. Stopper the flask with a stopper fitted with a calcium chloride tube and then shake the flask carefully until all the aluminum chloride is dissolved. Filter the solution rapidly into a dropping bottle and stopper securely.

Procedure.—Measure a portion of the vitamin solution, not to exceed 2 ml., into a dry clean test tube and add 5 drops of the pyrogallol solution. Evaporate to about 0.1 ml. Add 3 drops of the aluminum chloride solution and again heat in a water bath. The presence of vitamin D is indicated by the rapid development of a red-violet crust. On solution in absolute alcohol, a lilac-red to red-violet color develops.

This color may be matched against a mixture of 1 ml. of a solution of 0.01 g. of acid fuchsin in 1 liter of water and 2 ml. of a solution of 0.09 g. of Renol black in 1 liter of water, diluted with 20 ml. of water.

It is necessary, for proper procedure, that all the glassware and the reagents be dry and that the pyrogallol be added prior to the addition of aluminum chloride for aluminum chloride also reacts alone with the vitamin.

Glycerol Dichlorohydrin Reaction.—Vitamins D_2 and D_3 react with glycerol dichlorohydrin or related compounds in the presence of acetyl chloride or other acid halides.[55] With calciferol an immediate yellow color is formed which turns green in 1 minute, reaches a maximum in 15 minutes, and remains stable at 625 mμ for several hours. With ergosterol a weak pink color is observed immediately. This turns to orange in 15 to 20 minutes and later to a fluorescent green. With 7-dehydrocholesterol no color is observed for several minutes. A faint pink is then produced which becomes more intense in 24 hours. No color is produced with cholesterol. Under conditions suitable for quantitative estimation at 625 mμ, ergosterol has less than 4 per cent and 7-dehydrocholesterol less than 0.3 per cent of the absorption attributable to vitamins D_2 and D_3. This reaction serves to distinguish not only vitamins D_2 and D_3 from ergosterol and 7-dehydrocholesterol but also serves to distinguish between ergosterol and 7-dehydrocholesterol.

---

[55] Sobel, Mayer, and Kramer, *Ind. Eng. Chem., Anal. Ed.* **17,** 160 **(1945)**

VITAMINS E

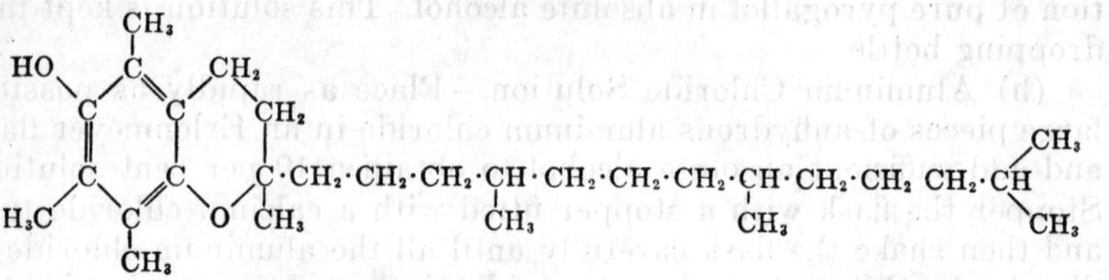

HO
CH_3
CH_2
CH_2
H_3C
$C \cdot CH_2 \cdot CH_2 \cdot CH_2 \cdot CH \cdot CH_2 \cdot CH_2 \cdot CH_2 \cdot CH \cdot CH_2 \cdot CH_2 \quad CH_2 \cdot CH$
CH_3
CH_3
CH_2
CH_3
CH_3
CH_3

α-Tocopherol

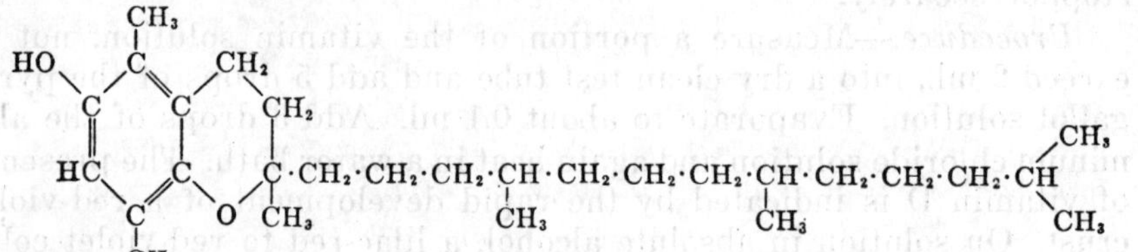

HO
CH_3
CH_2
CH_2
HC
$C \cdot CH_2 \cdot CH_2 \cdot CH_2 \cdot CH \cdot CH_2 \cdot CH_2 \cdot CH_2 \cdot CH \cdot CH_2 \cdot CH_2 \cdot CH_2 \cdot CH$
CH_3
CH_3
CH_3
CH_3
CH_3

β-Tocopherol

HO
CH
CH_2
CH_2
H_3C
$C \cdot CH_2 \cdot CH_2 \cdot CH_2 \cdot CH \cdot CH_2 \cdot CH_2 \cdot CH_2 \cdot CH \cdot CH_2 \cdot CH_2 \cdot CH_2 \cdot CH$
CH_3
CH_3
CH_3
CH_3
CH_3

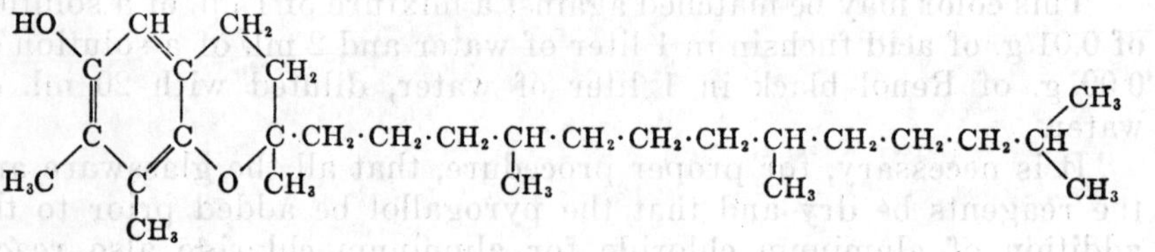

γ-Tocopherol

The group of vitamins E consists of the α-, β-, γ-, and δ-tocopherols and certain of their esters, such as the acetates and allophanates. While much of vitamin E produced is obtained from natural sources, such as wheat-germ oil, rice-germ oil, lettuce oil, dl-tocopherol acetate, a racemic synthetic product is preferred for medicinal use.

The tocopherols are oils which are soluble in lipid solvents but are insoluble in water. The tocopherols are resistant to acids, alkalis, heat, and visible light but are easily oxidized and are affected by ultraviolet light.

The physiological action of this vitamin is to aid in the normal functioning of reproduction. While it has not been proved that this vitamin is necessary for normal human nutrition, it is used extensively for the prevention of spontaneous abortion. About 10 mg. per day is considered adequate. Synthetic α-tocopherol acetate is considered to be the standard for the physiological activity of this group of vitamins for it has been shown that it has the same potency as natural α-tocopherol. One gram of racemic α-tocopherol is equivalent to 1000 International units of this vitamin.

The tocopherols are potent antioxidants, but while α-tocopherol has the greatest vitamin activity with β-tocopherol having less and γ-tocopherol the least, the reverse order of activity exists in so far as antioxidant potency is concerned.

Methods for the estimation of tocopherols are based on the quantitative reduction of ferric ions by the tocopherols and the determination of the amount of ferrous ions thus formed by means of the color reaction with α, α'-dipyridyl.[56-58] Carotene and fats interfere with the determination. By keeping the amount of potassium hydroxide used to saponify the fats at a minimum, the destruction of vitamin attributable to this reagent can be minimized and carotene can be eliminated by adsorption.

Reagent.—Methyl Alcohol Solution of Potassium Hydroxide.—Dissolve 112 g. of potassium hydroxide in methyl alcohol and make up to 1 liter with this solvent. This is a 2 N solution.

Floridin XS Column.—Fill a 12 × 30 mm. tube with the purified adsorbent. To purify digest on a boiling water bath for 1 hour with hydrochloric acid. Repeat with fresh portions of acid at room temperature. Wash with water until free of acid, then with ethyl alcohol, and with benzene. Dry at room temperature.

Procedure.—Saponify 1 g. of the oil in a test tube attached to a reflux condenser with 2 ml. of 2 N methyl alcohol solution of potassium hydroxide for 10 minutes at 72-74° C. in an atmosphere of nitrogen. Dilute with 8 ml. of methyl alcohol, add 10 ml. of water, and extract 3 times with 50 ml. of peroxide-free ether. Wash the combined ether extracts with water, with 2 per cent aqueous potassium hydroxide solution, and again with water until the alkali is removed. Dry the extract which consists of the unsaponifiable matter over anhydrous sodium sulfate and evaporate under vacuum in an atmosphere of carbon dioxide.

Carotene Removal.—Dissolve the residue in 5 ml. of benzene and pass the solution through the Floridin XS column previously wet with benzene. Wash with benzene until the eluate volume is 25 ml. The adsorbent earth is colored a greenish blue by carotenoids and dark blue by vitamin A.

Color Production.—If carotene is not present, dissolve the residue directly in 25 ml. of ethyl alcohol. To 1 ml. or a greater volume of the unsaponifiable matter solution, add 1 ml. of 0.2 per cent solu-

[56] Emmerie and Engel, *Rec. trav. chim. Pays-Bas* 57, 1351 (1938).
[57] Hove and Hove, *J. Biol. Chem.* 156, 601 (1944).
[58] Kaunitz and Beaver, *J. Biol. Chem.* 156, 653 (1944).

tion of ferric chloride in absolute ethyl alcohol, prepared fresh from $FeCl_3 \cdot 6H_2O$, and mix. Add 1 ml. of a 0.5 per cent solution of a, a'-dipyridyl in absolute ethyl alcohol, mix, and make up to a volume of 25 ml. Prepare a blank in a similar manner. Allow to stand for 10 to 15 minutes and compare the colors in a photometer with a standard solution prepared from pure tocopherol treated with the same amount of reagents. Correct the known and unknown for the blank determination.

Kaunitz and Beaver Modification.—Reagents.—Petroleum Ether. —Purify Skellysolve B, by shaking with concentrated sulfuric acid and washing with water, dilute sodium hydroxide solution, and 5 times with water. Dry over anhydrous sodium sulfate and distill.

Devlin and Mattill Reagent.[59]—Dissolve 250 mg. of ferric chloride and 500 mg. of a, a'-dipyridyl in glacial acetic acid and make to volume with the acid.

Procedure.—Prepare petroleum ether solutions containing 10 per cent of the oil or fat. Prepare a second set of petroleum ether solutions containing from 25 to 200 micrograms of a-tocopherol per milliliter. Pipette 2 ml. of the petroleum ether solution into the comparison cell of a Coleman universal spectrophotometer. Add 10 ml. of the iron-dipyridyl reagent. Into an identical cell, pipette 1 ml. of the oil solution and 1 ml. of petroleum ether containing from 0 to 200 micrograms of a-tocopherol. Add 10 ml. of the iron-dipyridyl reagent. Start a stopwatch the instant the reagent is added and record the reading for 10 minutes. Obtain the tocopherol concentration by extrapolation.

Quaife Modification.—A macro [59a] and a micro [59b] method for the determination of tocopherols have been devised in which the proteins are precipitated by ethyl alcohol and the tocopherols are extracted by xylene. An aliquot of the xylene extract is diluted with a, a-dipyridyl in n-propyl alcohol, the light absorption attributable to the carotenoids is measured at 460 mμ, ferric chloride solution is added, and the absorption at 520 mμ is measured.

Tocopherol.—A method [60] for the determination of γ-tocopherol in vegetable oils in the presence of a-tocopherol is detailed by Fisher. It is based on the oxidation of a-tocopherol by nitric acid in the presence of acetic acid followed by photometric estimation of the red color produced. The method assumes β-tocopherol is not present.

No vegetable oil has been reported to contain both β- and γ-toco-

[59] Devlin and Mattill, *J. Biol. Chem.* **146**, 123 (1942).
[59a] Quaife and Harris, *J. Biol. Chem.* **156**, 499 (1944).
[59b] Quaife, Scrimshaw, and Lowry, *J. Biol. Chem.* **180**, 1229 (1949).
[60] Fisher, *Ind. Eng. Chem., Anal. Ed.* **17**, 224 (1945).

pherol and only wheat-germ oil has been shown to contain β-tocopherol. Hence if either α- or γ-tocopherol can be estimated, the concentration of the other can be calculated.

VITAMINS K

Vitamin K₁: R = — CH₂·CH=C·CH₂·[CH₂·CH₂·CH·CH₃]₂·CH₂·CH₂·CH·CH₃ (with CH₃ substituents)

Vitamin K₂: R = — CH₂·[CH=C·CH₂·CH₂]₆·CH=C·CH₃ (with CH₃ substituents)

2-Methyl-1,4-naphthoquinone: R = H

Although the original vitamins K were isolated from natural sources—namely vitamin K₁ (2-methyl-3-phytyl-1,4-naphthoquinone) and vitamin K₂ (2,3-difarnesyl-1,4-naphthoquinone)—intensive investigation demonstrated that there were a large number of substances which had the specific action of the vitamins K, namely, that of exerting an effect on prothrombin deficiencies. The simplest of these was 2-methyl-1,4-naphthoquinone, known as *Menadione*.

Menadione is a yellow crystalline substance which melts at 106° C. It has a faint but characteristic odor. It is slightly soluble in water and is soluble in alcohol and the common organic solvents. Since it is insoluble in water a number of its more soluble derivatives are used, such as sodium 2,3-naphthohydroquinone diphosphate and sodium 2-methyl-1,4-naphthohydroquinone disulfate have been used. These have less potency but are absorbed more readily.

Vitamin K has been shown to be necessary for proper human metabolism. See page 696. Although the mechanism of its action is not known, it appears that it takes part in oxidation-reduction systems through its quinone structure. About 1 to 2 mg. is required daily.

A method [61] for the quantitative estimation of 2-methyl-1,4-naphthoquinone and related substances, applicable to oil solutions or alcoholic extracts depends on the interaction of 2,4-dinitrophenylhydrazine with 2-methyl-1,4-naphthoquinone with the subsequent pro-

---

[61] Menotti, *Ind. Eng. Chem., Anal. Ed.* **14**, 418 (1942).

duction of a blue to blue-green color by the addition of alcoholic ammonia. Another method [62] is based on the catalytic reduction of the quinone in butyl alcohol solution in the presence of phenosafranine as indicator. The resulting vitamin hydroquinone is then treated with an excess of a butyl alcohol solution of sodium 2,6-dichlorobenzene-oneindophenol in the absence of air. The decrease in color is proportional to the quinone originally present.

SELECTED REFERENCES

Assoc. Vitamin Chem., *Methods of Vitamin Assay.* Interscience, New York, 1947.

Food and Nutrition Board, "Recommended Dietary Allowances," *Natl. Acad. Sci.—Natl. Research Council, Publication* 302 (1953).

Jacobs, ed. *Chemistry and Technology of Food and Food Products*, Vol. I, 2nd ed. Interscience, New York, 1951.

Methods of Analysis, A.O.A.C., 8th ed. Washington, 1955.

Rosenberg, *Chemistry and Physiology of the Vitamins.* Interscience, New York, 1942.

U. S. Pharmacopeia XV.

---

[62] Scudi and Buhs, *J. Biol. Chem.* 141. 451 (1941).

CHAPTER XVIII

INORGANIC DETERMINATIONS

THERE is little reason to explain the necessity of this chapter in a book on food analysis. The importance of the inorganic constituents in foods is too well recognized. In this chapter the methods described are generally specific. As the author has mentioned previously, methods which do not have specificity are to be regarded with caution and must be followed in great detail after the elimination of interferences before definite conclusions can be drawn.

FLUORIDE

Methods for the detection of fluorides in food with reference to their use as preservatives and as a cause of food poisoning have been detailed in Chapters IV and VI. See also Chapter XXIII.

Distillation Method.—The determination of fluorine as developed by Willard and Winter [1] is based on the isolation of fluorine accurately and expeditiously from interfering materials by distillation as hydrofluosilicic acid which may subsequently be estimated colorimetrically by the bleaching of a zirconium-alizarin lake or by titration with thorium or cerous nitrate. These estimations are of importance because fluorine causes mottling and discoloration of teeth as well as being a poison.

The estimation of small amounts of fluorine is difficult because, unless the proper fixative is used, the fluorine will be lost in the ashing or will not be completely volatilized when the ash is distilled with perchloric acid. Winter [2] recommends magnesium acetate as the most satisfactory fixative in the following method.

Preparation of Sample.—Place 5-25 g. of material, according to the fluorine content, in a crucible or porcelain dish, add sufficient 5 per cent magnesium acetate solution to moisten completely and no more. Dry in an oven for at least 24 hours, and ash in a muffle at dull redness. Brush the ash into the distillation flask.

Apparatus.—The distillation apparatus, Fig. 75, consists of a Claissen flask with necks 10 cm. long instead of the usual length. The side arm that connects the flask with the condenser is bent upward

---

[1] Willard and Winter, *Ind. Eng. Chem., Anal. Ed.* 5, 7 (1933).
[2] Winter, *J. Assoc. Official Agr. Chem.* 19, 362 (1936).

for about 4 cm. and then downward at two points in order to fit a vertical condenser. Preferably the side arm should fit the condenser by a ground glass joint. More elaborate trapping devices are inadvisable, because the possible adsorption causes fluorine deficiencies. The straight neck of the distilling flask carries a rubber stopper fitted with a thermometer and a dropping funnel whose stem has been drawn to a capillary. Both the thermometer and the dropping funnel extend to within 5 mm. of the flask bottom.

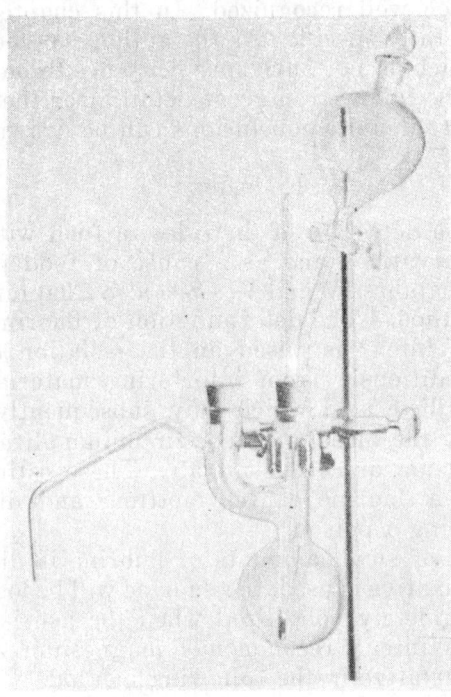

Procedure.—Wash the crucible or dish several times with water and a small amount of sulfuric acid, adding the wash solution to the flask. Connect the apparatus as directed above. Remove the stopper, add sulfuric acid slowly from a pipette until the effervescence ceases, and then add approximately 12 ml. more of sulfuric acid. Replace the stopper, boil at 135-140° C., and collect the distillate in a 100-ml. volumetric flask. When the liquid temperature reaches 135° C. sufficient water is slowly dropped from the funnel to compensate for the water distilling out and to maintain the temperature at 135° C. The distillation requires constant supervision. After the 100-ml. flask is filled (distillate 1) collect another 50 ml. (distillate 2) to be certain that all the fluorine has been volatilized.

Churchill, Bridges, and Rowley [3] point out that phosphates may interfere in this determination, for in some food products the phosphates are possibly reduced to a form which is readily carried over in the distillate. They recommend a double Willard-Winter distillation of the fluorine from the ash of foods. The first should be made with sulfuric acid to eliminate hazard, as some carbonaceous material may be present, and the second distillation should be made with perchloric acid at 135° C. This procedure yields a distillate free from sulfate and phosphate. Great care should be exercised in all distillations and operations in which perchloric acid is employed because of danger from explosion.

The fading of the zirconium-alizarine lake is a measure of the amount of fluorine present, a procedure developed by Smith and Dutcher.[4] If quantities of fluorine fall below the range given, less dye should be used, and if quantities above the range are to be determined more dye should be used.

The individual tubes show a fading in color, which increases as the fluorine content increases, hence a comparison of the fading caused by an aliquot of the unknown with the standards prepared gives the measure of the fluorine content of the sample.

Colorimetric Method.—*Reagent.*—Dissolve 0.87 g. of zirconium nitrate, $Zr(NO_3)_4 \cdot 5H_2O$, in 100 ml. of water, and 0.17 g. of sodium alizarinate in 100 g. of water. Mix equal parts of the two solutions and dilute the mixture (1:4) with water.

Procedure.—Make up a series of standards in Nessler tubes or test tubes of about 80-ml. capacity by placing 0.02, 0.04, 0.06, 0.09, 1.20 mg. of fluorine in each tube, respectively. Add water to make about 50 ml., 10 ml. of hydrochloric acid (1:1), mix thoroughly, add 2 ml. of the dye solution and bring all the tubes to the same level with water. Again mix thoroughly, place the tubes in a steam bath for 30 minutes, and cool. For unknowns take aliquots of the distillates whose fluorine contents fall within the above range of standards. The fluorine is determined from the nearest standard.

Titration Method.[5]—Cerous Nitrate Modification.—Add several drops of 0.04 per cent phenol red solution to the distillate obtained as detailed in the foregoing and neutralize the liquid with dilute sodium hydroxide, avoiding a large excess. Boil the alkaline solu-

[3] Churchill, Bridges, and Rowley, *Ind. Eng. Chem., Anal. Ed.* 9, 222 (1937).
[4] Smith and Dutcher, *Ind. Eng. Chem., Anal. Ed.* 6, 61 (1934).
[5] Scott and Henne, *Ind. Eng. Chem., Anal. Ed.* 7, 299 (1935).

tion and bring back repeatedly to the apparent neutral point with 0.02 N or 0.01 N perchloric acid. During this neutralization, reduce the volume to 5 to 10 ml. When the faint pink color is no longer restored by boiling (carbonate free), cool the solution, transfer quantitatively to a 50-ml. beaker and boil from about 25 ml. down to 2 to 3 ml. Add 2 drops of a saturated alcoholic solution of methyl red and 10 drops of 0.04 per cent bromocresol green solution. (The bromocresol green is weighed out exactly and neutralized with standard sodium hydroxide to yield the monosodium salt. This prevents alteration of the neutrality of the solution when the indicator is added.) Titrate the liquid at 80° C. to the maximum red color with cerous nitrate solution 1 ml. of which is equivalent to 0.5 mg. of fluorine. When the amount of fluorine is less than 0.2 mg., thorium nitrate solution, 1 ml. of which is equivalent to 0.1 mg. of fluorine, may be substituted for the cerous nitrate, using the same mixed indicator.

Thorium Nitrate Modification.—The thorium nitrate solution may be standardized by titration against known volumes of 0.02 N fluoride solution.

Procedure.—Transfer a known aliquot of standard 0.02 N fluoride solution to a flask, add water to bring the volume to 20 ml. and then add an equal volume of ethyl alcohol. Add 6 drops of alizarin red indicator (prepared by dissolving 1 g. of sodium alizarin sulfonate in 100 ml. of ethyl alcohol, filtering off the residue and making up the filtrate to 250 ml. with alcohol) and then only enough dilute hydrochloric acid to destroy the color. Avoid excess acid. Titrate with the thorium nitrate solution, over a white surface in a good light to a faint permanent reappearance of color. Titrate slowly near the end point. Run a blank titration on the indicator by determining the volume of standard 0.02 N fluoride solution necessary to cause disappearance of color in a slightly acid water-alcohol solution of 6 drops of the indicator and compare this with the volume of standard thorium nitrate necessary to discharge the color. Calculate the strength of the thorium nitrate solution by use of the following equation:

$$1.0 \text{ ml. Th}(NO_3)_4 = \frac{\text{ml. of } 0.02 \ N \ F^- \text{ soln.}}{\text{ml. of Th}(NO_3)_4 \text{ soln.}} \times 0.38 = A \text{ mg. of } F^-$$

The unknown distillates may be titrated in a similar manner by making alkaline, neutralizing, and concentrating as described above. Add 6 drops of indicator, dilute acid until the color of the indicator just disappears, and then an equal volume of alcohol. The solution should be only faintly acid. If no fluorides are present, the color will not be discharged. Titrate at once, if fluorides are present with the standardized thorium nitrate solution to the faint reappearance of the pink color.[6]

Williams Modification.—The thorium nitrate titration of fluorides has been simplified by Williams [7] by using a single titration against a permanent color standard.

Reagents.—Acidified Standard Thorium Nitrate Solution.—Stock Solution.—Dissolve 1.27 g. of thorium nitrate, $Th(NO_3)_4 \cdot 4H_2O$, and 72 ml. of N hydrochloric acid in water and make up to 100 ml. Dilute solution. Dilute 5 ml. of the stock solution to 500 ml. with fluorine-free water. One ml. of the dilute solution is equivalent to 5 micrograms of fluoride.

Acid-Indicator Solution.—Dissolve 0.020 g. of sodium alizarin-monosulfonate (alizarin-S) in water, add 100 ml. of the dilute acidified thorium nitrate solution, and 14.3 ml. of N hydrochloric acid and make up to 200 ml. Two ml. of this solution added to 50 ml. of fluorine-free water and 10 ml. of 2 N sodium chloride solution in a Nessler tube should give the correct end point color; if not impurities in the salt or other chemicals may be responsible and the proportion of the dilute standard thorium nitrate solution used should be modified accordingly. The color should be judged when making the acid-indicator solution for it is likely to alter on standing.

2,5-Dinitrophenol Indicator Solution.—Prepare a 0.05 per cent aqueous solution.

Color Standards.—Temporary.—Dilute an aliquot of a standard fluoride solution made from sodium fluoride prepared from the purest sodium carbonate and hydrofluoric acid, containing 100 micrograms of fluorine to 50 ml. with water. Add 10 ml. of 2 N sodium chloride solution, 2 ml. of the acid indicator solution, 20 ml. of the dilute standard thorium nitrate solution, and mix. This color is stable for several hours.

Permanent.—Mix 3 ml. of 10 per cent hydrochloric acid with 50 ml. of a solution containing 1 per cent of cobalt chloride, $CoCl_2$, add 30 ml. of 0.1 per cent potassium chromate solution, and dilute to 100 ml. Dilute 3 ml. of this stock mixture with water to a volume approximately equal to the anticipated volume of the test solution when titrated. The color should be identical with that of the temporary standard. If it is not, possibly because of differences in the thorium nitrate used, the proportions of ingredients should be adjusted until a match is obtained. Alternatively, alizarin-S may be used as a color standard in a buffered solution of suitable pH.

---

[6] Boruff and Abbott, *Ind. Eng. Chem., Anal. Ed.* **5**, 236 (1933).
[7] Williams, *Analyst* **71**, 175 (1947).

Procedure.—Add 3 drops of 2,5-dinitrophenol indicator solution to an appropriate aliquot of the distillate or fluoride solution in a Nessler tube and add 0.05 N sodium hydroxide solution until the solution, when mixed, assumes a faint yellow color. Then add a drop or sufficient 0.01 N hydrochloric acid solution just to discharge the color. It is useful to have a Nessler tube containing water for comparison, because the color becomes very pale near the end point. If the presence of free halogen is suspected, add 1 ml. of 1 per cent hydroxylamine hydrochloride solution just before the neutralization.

Transfer accurately 50 ml. of the neutralized solution, containing between 0.5 and about 150 micrograms of fluorine to another Nessler tube, add 10 ml. of 2 N sodium chloride solution, 2 ml. of the acid-indicator solution, and mix. Titrate with the dilute standard thorium nitrate solution until the color exactly matches that of the standard color solutions. For a 5-g. sample, 1 ml. of dilute thorium nitrate solution is equivalent to 1 part per million, subject to correction blanks.

SELENIUM

Selenium traced to vegetation grown in soil areas that contained selenium caused illness in animals and may very likely cause illness in human beings. Thus Munsell, DeVaney, and Kennedy [8] measured the toxicity of food containing selenium by its effect on rats. Because of the possible danger to human beings, rapid and accurate methods are necessary. The following method developed by Robinson, Dudley, Williams, and Byers [9] is based on the fact that selenium may be separated from all other elements except arsenic and germanium by distillation with concentrated hydrobromic acid. The selenium must be in, or converted into, the sexavalent condition before distillation in order to insure its distillation with the acid will be complete. In most cases the conversion may be accomplished by the use of bromine. The excess of bromine distills at a low temperature and the hydrobromic acid then reduces the selenium to the quadrivalent condition. In this form it readily distills along with the hydrobromic acid. The selenium is subsequently estimated in the distillate by reduction with hydroxylamine hydrochloride and sulfur dioxide.

Preparation of Sample.—Vegetable Matter.—Stir 100 g. of the well-ground and mixed vegetation into a concentrated solution of 25 g. of magnesium nitrate, and add 5 g. of magnesium oxide. Dry

---

[8] Munsell, DeVaney and Kennedy, *U. S. Dept. of Agr., Tech. Bull.* 534 (1936).
[9] Robinson, Dudley, Williams, and Byers, *Ind. Eng. Chem., Anal. Ed.* 6, 274 (1934).

the mass over a water bath and finally in an oven at 105° C. Ignite the dried material slowly in a muffle until the ash is a uniform gray color. After ignition, triturate the ash with 100 ml. of concentrated hydrobromic acid, capable of being completely decolorized with sulfur dioxide, and 2 ml. of bromine, transfer to a distilling flask and estimate as detailed below.

Animal Matter.[10]—Place the material in a suitable state of subdivision in a beaker of 400- to 600-ml. capacity, covered with 150 to 200 ml. of nitric acid, and allow to stand at room temperature for from 2 to 3 hours, during which period stir it vigorously at intervals. Add 50 ml. of hydrogen peroxide, 30 per cent by weight, and allow the mixture to stand overnight. If frothing occurs on the addition of the hydrogen peroxide, foaming over is prevented by vigorous stirring of the foam. After standing overnight, warm the mixture slowly on the steam bath until frothing ceases, after which add 50 ml. more of hydrogen peroxide, together with 20 ml. of sulfuric acid. Evaporate the mixture to essentially complete dryness on the steam bath or hot plate. Treat the cooled black paste with 100 ml. of concentrated hydrobromic acid to which has been added sufficient bromine to make it deep yellow in color. Transfer the material to the distillation flask.

Williams and Lakin [11] recommend the following procedure for preparation of the sample prior to the selenium distillation. To prepare a sample of air-dry vegetation, grind it to pass a 2-mm. mesh sieve, then mix and quarter. Stir a weighed sample, usually 10 g., into a mixture of 50 ml. of sulfuric acid and 100 ml. nitric acid in a 600-ml. Pyrex beaker. Stir the mixture with a thermometer until it becomes homogeneous, after the first few minutes with gentle heating, without allowing the temperature to rise above 100° C. After all frothing has ceased, raise the temperature of the mixture to a maximum of 120° C. until all evolution of nitrogen peroxide has ceased. The end of the operation is marked also by an incipient carbonization of the mixture, although longer heating at 120° C. does little harm. After the mixture is cooled, transfer it to the all-glass distilling flask, described below, add 100 ml. of hydrobromic acid and 1 ml. of bromine and collect 75 ml. of the distillate. Care must be taken that the first portion of the distillate contains a small excess of bromine.

Apparatus.—The apparatus, Fig. 76, consists of a Pyrex 500-ml. round-bottom flask fitted with a ground-glass stopper into which has been sealed a thistle tube with a stem long enough to reach within 5 mm. of the bottom of the flask. The ground-glass stopper also has a

[10] Dudley and Byers, *Ind. Eng. Chem., Anal. Ed.* **7,** 3 (1935).
[11] Williams and Lakin, *Ind. Eng. Chem., Anal. Ed.* **7,** 409 (1935).

side arm with a ground-glass end fitted to a condenser whose end is drawn out into a long adapter, bent, and with a capillary tip so that it may fit easily into a 100-ml. wide-mouth Erlenmeyer flask, which acts as the receiver.

Procedure.—Connect the distillation apparatus described with the adapter just below the surface of 2 to 3 ml. of bromine water in the receiver flask and apply heat gradually. One or 2 g. of bromine should distill over in the first few milliliters of distillate. If insufficient bromine has been added to produce this quantity of bromine, more must be added through the thistle tube. A somewhat greater excess of bromine does no harm, but too great an excess is to be avoided because of the formation of too much sulfuric acid later. Collect 30 to 50 ml. of the distillate by increasing the heat. Make a second, or even third distillation with intervening additions of hydrobromic acid and bromine through the thistle tube, unless it is certain from experience that all the selenium is in the first distillate. Remove the distillate and pass in sulfur dioxide until the yellow color due to bromine is discharged. Add 0.25 to 0.5 g. of hydroxylamine hydrochloride, stopper the flask loosely, put on the steam bath for an hour, and allow to stand overnight at room temperature. If selenium is present, it will appear as a characteristic pink or red precipitate. If much selenium is present it will shortly turn black.

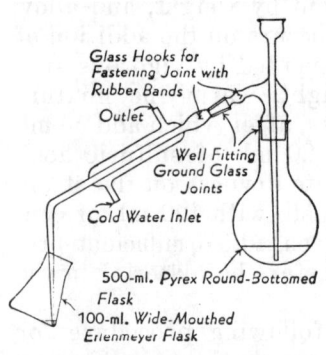

Glass Hooks for Fastening Joint with Rubber Bands

Outlet

Well Fitting Ground Glass Joints

Cold Water Inlet

500-ml. Pyrex Round-Bottomed Flask

100-ml. Wide-Mouthed Erlenmeyer Flask

FIG. 76. Selenium Distillation Apparatus

Collect the precipitated selenium on an asbestos pad in a porcelain crucible, and wash slightly with hydrobromic acid containing a little hydroxylamine hydrochloride. Dissolve the selenium on the pad by passing through 10 to 15 ml. of a solution of 1 ml. of bromine in 10 ml. of hydrobromic acid in small quantities and wash into a 25-ml. measuring flask if the quantity is small and is to be estimated colorimetrically. If over 0.5 mg., filter into a small beaker, precipitate as before, gather on an asbestos pad as before, wash with hydrobromic acid containing a little hydroxylamine hydrochloride, and then with water. Prepare a tare in the same way. Dry at 90° C. for 1 hour, place in a vacuum desiccator, and exhaust the air while the crucibles are still hot. Cool 0.5 hour. Allow the air to enter the desiccator, cool an additional 0.5 hour, and weigh against the tare. Check the weight by drying again.

Colorimetric Procedure.—If the quantity is small and is to be estimated colorimetrically, add 1 ml. of a solution containing 5 per cent gum arabic and precipitate the selenium by sulfur dioxide and hydroxylamine hydrochloride. Prepare comparison solutions containing known quantities of selenium in exactly the same manner and allow them to stand overnight. Shake the standard and test solutions and compare the depth of color in Nessler tubes. This comparison is best carried out in sunlight. It is difficult to match solutions containing more than 0.5 mg. of selenium in 25 ml. and the color comparison is most satisfactory when 0.01 to 0.1 mg. is present.

PHOSPHORUS

In food materials phosphorus is usually determined and estimated as phosphoric acid expressed as P_2O_5. This may be done gravimetrically as magnesium pyrophosphate, $Mg_2P_2O_7$, or titrimetrically, or colorimetrically. In general, the food sample is ashed by one of the number of methods previously described, and then a suitable aliquot of the ash or the entire ash dissolved in sulfuric acid or in nitric acid or in both, or the wet ash is used.

Gravimetric Method.—*Reagents.*—(a) Molybdate Solution.—Dissolve 100 g. of molybdic acid, MoO_3, in a mixture of 144 ml. of ammonium hydroxide and 271 ml. of water. Pour this solution slowly and with constant stirring into a mixture of 489 ml. of nitric acid and 1148 ml. of water. Keep the final mixture in a warm place for several days or until a portion heated to 40° C. deposits no yellow precipitate of ammonium phosphomolybdate. Decant the solution from any sediment and preserve in glass-stoppered bottles.

(b) Magnesia Mixture.—Dissolve 11 g. of magnesium oxide in hydrochloric acid (1:4), avoiding an excess of the acid; add a little magnesium oxide in excess; boil a few minutes to precipitate iron, aluminum, and phosphorus pentoxide, and filter. To the filtrate add 140 g. of ammonium chloride and 130.5 ml. of ammonium hydroxide and dilute to 1 liter. Or dissolve 55 g. of magnesium chloride, $MgCl_2 \cdot 6H_2O$, in water, add 140 g. of ammonium chloride and dilute to 870 ml. Add ammonium hydroxide to each required portion of the solution just before using, in the proportion of 15 ml. to 100 ml. of solution.

(c) Magnesium Nitrate Solution.—Dissolve 150 g. of magnesium oxide in nitric acid (1:1), avoiding an excess of the acid; add a little magnesium oxide in excess, boil, filter from the excess of magnesium oxide, ferric oxide, etc., and dilute to 1 liter.

Procedure.—Transfer the solution containing the ash to a 250-ml. beaker; add ammonium hydroxide in slight excess, and barely dissolve the precipitate formed with a few drops of nitric acid, stirring vigorously. If hydrochloric acid or sulfuric acid had been used as a solvent, add about 15 g. of crystalline ammonium nitrate or a solution containing that quantity. To the hot solution add 70 ml. of the molybdate solution for every decigram of phosphorus pentoxide present. Digest at about 65° C. for 1 hour, and determine whether or not the phosphorus pentoxide has been completely precipitated by adding more molybdate solution to the clear supernatant liquid. Filter, and wash with cold water or preferably with a solution of a 100 g. of ammonium nitrate dissolved in and diluted to a liter of water. Dissolve the precipitate on the filter with ammonium hydroxide (1:1) and hot water and wash into a beaker to a volume of not more than 100 ml. Neutralize with hydrochloric acid, using litmus paper or bromothymol blue as indicator; cool; and from a burette add slowly, at about 1 drop per second, stirring vigorously, 15 ml. of the magnesia mixture for each decigram of phosphorus pentoxide present. After 15 minutes, add 12 ml. of ammonium hydroxide. Let stand until the supernatant liquid is clear (about 2 hours), filter, wash the precipitate with ammonium hydroxide (1:9) until the washings are practically free of chlorides, dry, burn at a low heat, and then ignite in an electric furnace at 950-1000° C., cool in a desiccator and weigh as magnesium pyrophosphate, $Mg_2P_2O_7$. Calculate the result as percentage phosphorus pentoxide.

Titrimetric Method.—Add 5-10 ml. of nitric acid, depending on the manner of solution or add the equivalent in ammonium nitrate. Add ammonium hydroxide solution until the precipitate that forms dissolves but slowly on stirring vigorously, dilute to 75-100 ml., and adjust to a temperature of 25-30° C. Add sufficient molybdate solution to insure complete precipitation. Five ml. of nitric acid must be added to every 100 ml. of molybdate solution which is then filter immediately before use. Place the solution in a shaking or stirring apparatus and shake for 30 minutes at room temperature. Decant at once through a filter, and wash the precipitate twice by decantation with 25- to 30-ml. portions of water, agitating thoroughly and allowing to settle. Transfer the precipitate to the filter and wash with cold water until the filtrate from 2 fillings of the filter yields a pink color upon the addition of phenolphthalein and 1 drop of the standard alkali. The standard alkali is prepared by diluting 328.81 ml. of N alkali to 1 liter. One ml. of this solution is equivalent to 1 mg. of phosphorus pentoxide. Transfer the precipitate and filter to the beaker or precipitating vessel, dissolve the precipitate in a small excess of the standard alkali solution, add a few drops of phenolphthalein indicator, and titrate with standard acid. The standard acid is prepared to be equal to or ½ the normality of the standard alkali solution.

Colorimetric Method.[12]—*Reagents.*—Ammonium Molybdate Solution.—Dissolve 25 g. of ammonium molybdate in 200 ml. of water heated to 60° C. and filter. Cool and dilute with water to 1 liter. This solution then contains 2.5 g. of ammonium molybdate per 100 ml.

Sulfuric Acid Solution.—Dilute 280 ml. of arsenic and phosphorus-free sulfuric acid to a liter with water. This is approximately a 10 N sulfuric acid solution.

Stannous Chloride Solution.—Place 25 g. of stannous chloride, $SnCl_2 \cdot H_2O$, in a solution of 100 ml. of hydrochloric acid diluted to 500 ml. with water and let stand in a warm room until dissolved; then dilute to 1 liter with water. Filter if necessary. This solution may be stored in a bottle with a side opening near the bottom and arranged with a stopcock for delivering the solution in drops. The solution may be protected from the air by floating a layer of white mineral oil about 5 mm. thick on the surface.

Standard Phosphate Solution.—Dissolve 0.2195 g. of recrystallized potassium dihydrogen phosphate, KH_2PO_4, in water and dilute to a liter. This solution contains 50 p.p.m. of phosphorus and is too concentrated to use directly. A second stock solution may be made by taking 50 ml. of the first stock solution and diluting to 500 ml. The standard solution for color comparison is made by diluting 5 ml. of the second stock solution to 91 ml. with water; 4 ml. each of the ammonium molybdate solution and sulfuric acid solution are added and mixed thoroughly by swirling in a 150-ml. Erlenmeyer flask. Six drops of the stannous chloride solution are added and the solution is swirled again. The solution is diluted to 100 ml. and again mixed by shaking in the Erlenmeyer flask. The standard phosphate solution is ready for use although it is necessary to add a drop of stannous chloride solution every 10-12 minutes to obtain full color. One ml. of this standard phosphate solution contains 0.00025 mg. phosphorus per ml.

Procedure.—Dissolve the ash in 1 ml. of 10 N sulfuric acid. Add sufficient water to insure complete solution and transfer the solution to a small beaker. Neutralize with ammonia using phenolphthalein as indicator. Transfer the solution to a 100-ml. volumetric flask and make to volume.

An appropriate aliquot is transferred by means of a pipette to a 100-ml. volumetric flask with a mark at 91 ml., and is diluted with water to that volume. Four ml. each of the ammonium molybdate and 10 N sulfuric acid solutions is added, swirling after each addi-

12 Truog and Meyer, *Ind. Eng. Chem., Anal. Ed.* 1, 136 (1929).

tion. Add 6 drops of stannous chloride solution, shake and make up to volume. Compare in a colorimeter within 10 minutes with a standard prepared as directed above.

Comparison may also be made in Nessler tubes by using varying proportions of the standard phosphorus solution.

Microcolorimetric Method.—Phosphorus in the form of phosphate reacts with ammonium molybdate to form a complex molybdiphosphate, known familiarly as phosphomolybdate. This complex is reduced to form a molybdeum blue by the use of solutions of hydroquinone and sodium sulfite.[13]

Reagents.—Ammonium Molybdate Solution.—Dissolve 25 g. of ammonium molybdate in 300 ml. of water. Dilute 75 ml. of concentrated sulfuric acid to 200 ml. with water and add to the ammonium molybdate solution.

Hydroquinone Solution.—Dissolve 0.5 g. of hydroquinone in 100 ml. of water and add 1 drop of sulfuric acid to retard oxidation.

Sodium Sulfite Solution.—Dissolve 200 g. of sodium sulfite, Na_2SO_3, in water, make up to 1 liter and filter. Keep this solution well stoppered or prepare a fresh equivalent each time it is to be used.

Standards.—Potassium Dihydrogen Phosphate Solution.—Dissolve 0.4394 g. of pure, dry potassium dihydrogen phosphate, KH_2PO_4, in water and make up to 1 liter. Dilute 50 ml. of this solution to 200 ml. Each ml. of the latter solution is equivalent to 0.05 mg. of phosphorus.

Procedure.—Convert the phosphorus and phosphorus compounds to phosphate as detailed above. Transfer to a volumetric flask and make to volume. Transfer a 5-ml. aliquot to a 10-ml. volumetric flask. Add 1 ml. of the ammonium molybdate solution, rotate the flask to mix and allow to stand a few moments. Add 1 ml. of the hydroquinone solution, again rotate the flask; add 1 ml. of the sodium sulfite solution and mix. Make to volume with water. Stopper the flask and shake thoroughly. Allow to stand 30 minutes for development of the blue color and compare immediately thereafter in a colorimeter with 2 ml. of the standard potassium dihydrogen phosphate solution treated at the same time with the same reagents as the test solution and in the same way. With either the test solution or the standard set at 25.0 mm., readings within 10 mm.—that is, with a range of 20 mm.—are accurate. If concentrations of the test solution are outside this range, larger or smaller aliquots of the sample solution should be used.

---

[13] Briggs, *J. Biol. Chem.* 59, 255 (1924).

SULFATES

Titrimetric Method.—In general, sulfates are estimated by precipitation with barium chloride solution in hydrochloric acid solution However, a direct titration method has been developed by Schroeder.[14] This method is based on the use of the specific indicator tetrahydroxyquinone for barium in the titration of sulfate. The tetrahydroxyquinone is used as an internal indicator. Sheen and Kahler [15] recommend the following details.

Reagents.—Standard barium chloride solution, the strength varying from 1 ml. = 1 mg. sulfate to 1 ml. = 50 mg. of sulfate standardized gravimetrically. An indicator composed of disodium tetrahydroxyquinone ground with dried potassium chloride in a 1 to 300 ratio, and passing a 100-mesh screen. Ethyl alcohol or alcohol denatured by formula No. 30 or No. 3-A or isopropyl alcohol. Phenolphthalein indicator and bromocresol green indicator, if phosphates are present.

Procedure A.—Carefully neutralize a 25-ml. sample containing up to approximately 2000 p. p. m. of sulfate with approximately 0.02 N hydrochloric acid until just acid to phenolphthalein. The temperature of the solution should be below 35° C., and it is advisable to work between 20-25° C. Add either 25 ml. of ethyl alcohol or one of the other solvents. Introduce the tetrahydroxyquinone, using 0.1 g. of the indicator for sulfate up to 100 p. p. m. and 0.2 g. for sulfate up to 2000 p. p. m. Swirl the flask to dissolve the indicator; the solution will be colored a deep yellow. Titrate with standard barium chloride solution, the strength to be employed depending on the approximate sulfate content of the sample. Add the standard barium chloride solution at a steady dropping rate with constant swirling of the flask until the yellow color changes to a rose. The rose color is the end point and is due to the appearance of the red barium salt of tetrahydroxyquinone. The rose color should appear throughout the body of the solution and not as spots of color.

Procedure B.—(sulfate range from 200 to 30,000 p. p. m.) Add sodium chloride according to the Table 100. The procedure is the same as in A for neutralization and titration.

Procedure C.—(with phosphate up to 60 p. p. m.) Carefully neutralize a 25-ml. filtered sample with approximately 0.02 N hydrochloric acid until just acid, yellow range, to bromocresol green, approximately pH 4. Follow the procedure as in A or B; no correction will be required for the phosphate ion present.

14 Schroeder, *Ind. Eng. Chem., Anal. Ed.* 5, 403 (1933).
15 Sheen and Kahler, *Ind. Eng. Chem., Anal. Ed.* 8, 127 (1936).

IODIDE

In many countries, where people live in the interior away from sea air, endemic goitre prevails. For these regions iodized salt is recommended. However, the amount should be and is rigidly controlled in certain countries, to not less than 1 part nor more than 2 parts potassium or sodium iodide in 250,000 parts of salt. The amount of potassium or sodium iodide may be easily determined by a method based on the replacement of iodine by chlorine, or bromine. Andrew and Mandeno [16] found that bromine is better. The action goes in the following manner:

$$KI + 3Br_2 + 3H_2O \rightarrow KIO_3 + 6HBr$$

and after the excess bromine is removed, the iodine is liberated by potassium iodide in faintly acid solution:

$$5KI + KIO_3 + 6HCl \rightarrow 6KCl + 3I_2 + 3H_2O$$

TABLE 100. TETRAHYDROXYQUINONE REQUIRED FOR VARIOUS SULFATE CONCENTRATIONS

Sulfate Concentrations p.p.m.	Quantity of THQ Indicator g.	Strength [a] of BaCl₂ Solution	NaCl Required g.
Up to 100 [b]	0.1	1	....
100 to 1,000 [b]	0.2	1	....
1,000 to 2,000	0.2	4	....
2,000 to 4,000	0.4	10	2
4,000 to 10,000	0.4	10	4
10,000 to 20,000	0.6	50	8
20,000 to 30,000	0.8	50	8

[a] 1 ml. = mg. SO₄.
[b] Subtract 0.1 ml. as a blank in titration.

Procedure.—Dissolve 100 g. of salt in water and make up to 500 ml. in a volumetric flask, and filter. Transfer 200 ml. of the filtrate to a 500-ml. Erlenmeyer flask and add 1 ml. of bromine water and 2 ml. of N hydrochloric acid. Add a few pieces of pumice to prevent bumping and boil gently until the salt begins to precipitate. Dissolve the precipitate in water and cool. Neutralize with N sodium hydroxide solution. Make acid to methyl orange, add 0.2 g. of potassium iodide, 1 ml. of starch solution and titrate with 0.1 N sodium thiosulfate solution. Calculate the percentage of iodide.

16 Andrew and Mandeno, *Analyst* 60, 801 (1935).

Alcoholic Potash Method.—Methods for the determination of iodine in food materials have been employed using the following means for placing the iodine in solution: digestion with sulfuric acid and hydrogen peroxide, fusion with potassium hydroxide and nitrate, and combustion tube methods. The alcoholic potash digestion method used by Almquist and Givens [17] for liquid eggs, assures in the case of liquid eggs, a more complete recovery of iodine. This method is undoubtedly applicable to other food materials such as meat and meat products and to milk and milk products. The treatment with alcoholic potash probably converts the organically bound iodine into iodide ion and thus stabilizes it to action of ashing.

Procedure.—Place the liquid contents of a number of eggs in a flask and add an equal volume of 95 per cent alcohol and 10 g. of potassium hydroxide per egg. Boil the mixture under a reflux condenser, gently, for 16 to 24 hours. The boiling takes place without foaming or bumping. The product is a dark brown liquid containing very little solids. Place an amount of this liquid equivalent to 1 egg in a 50-ml. nickel crucible or Pyrex beaker and evaporated to dryness on a hot plate. The crucible or beaker and contents are placed in a muffle furnace and ashed for about 4 hours at about 600° C.

Extract the ash with 50 ml. of hot water, filter off the extract and wash the residue with hot water, running the washings into the filtrate. Acidify the filtrate carefully with 6 N sulfuric acid to the acid point of methyl red. Add 5 more drops of acid. Add saturated bromine water to the solution until it has a permanent, strong yellow color. Boil off the excess bromine and evaporate the solution to about 15 ml. on a steam bath, cool, and transfer to a small separatory funnel. Remove any crystalline matter formed during cooling and evaporate and wash. Add the washings to the solution.

Add a crystal of potassium iodide and extract the iodine formed with five 1-ml. portions of carbon tetrachloride. Determine the iodine in the carbon tetrachloride colorimetrically by comparison with standard solutions of iodine in carbon tetrachloride. Corrections for the iodine content of the reagents used are determined by following the same procedure.

Instead of estimating the iodine colorimetrically the iodine may be estimated as in the previous method using 0.1 N or 0.01 N sodium thiosulfate solution according to the amount of iodine liberated by the potassium iodide.

CHLORIDE

Fajans Method.—The use of dyes which are absorbed near the

---

[17] Almquist and Givens, *Ind. Eng. Chem., Anal. Ed.* 5, 254 (1933).

end point of a titration has been investigated by Fajans [18] and by Kolthoff.[19] Dichlorofluorescein is adequate for dilute chloride solutions and concentrations of the order of 0.0005 N in chloride ion may be titrated with an accuracy of 1 to 2 per cent.

Indicator Solution.—Prepare a 0.1 per cent solution of the dye in 60-70 per cent alcohol or a 0.1 per cent solution of the sodium salt of the dye in water. About 2.5 ml. of 0.1 N sodium hydroxide solution are required to neutralize 100 mg. of the indicator.

Procedure.—Add 2-4 drops of indicator solution to 50 ml. of the test solution. As the end point is reached the silver chloride flocculates. Near the end point the solution turns brown, when the end point is reached the color changes sharply to orange. A slight excess of silver nitrate produces a rose or red color.

Volhard Method.—The Volhard thiocyanate method for the determination of chloride in which the silver chloride precipitate is removed by filtration before back-titrating may be improved by eliminating the filtration. Caldwell and Moyer [20] suggest the use of nitrobenzene which inhibits the darkening of silver chloride in the light and improves the end point. This immiscible liquid draws the silver chloride to the interface and thus removes it from the aqueous solution, the nitrobenzene forming an insoluble layer over the precipitate. Other immiscible liquids may also be used.

Procedure.—Titrations may be made in 250-ml. glass-stoppered bottles. Acidify 25 to 50 ml. of the solution containing from 0.048 to 0.26 g. of sodium chloride, free from the usual interfering ions, with 8 to 10 drops of nitric acid, and add 1 ml. of nitrobenzene for each 0.05 g. of chloride. Add standard silver nitrate until an excess of 1 to 4 ml. of 0.1 N solution is present. Stopper the bottle tightly and shake vigorously until the silver chloride settles out in large spongy flakes. Usually 30-40 seconds agitation is required. A perfectly clear supernatant solution is not necessary. Fine droplets of nitrobenzene are left in suspension. However, nearly all the nitrobenzene is so closely attached to the silver chloride that there is little evidence of a separate phase.

Add 1 ml. of ferric alum indicator, prepared by adding concentrated, freshly boiled nitric acid to a saturated solution of ferric alum until the solution becomes greenish yellow, and complete the titration with 0.05 N potassium thiocyanate solution. The ferric alum acts as an effective flocculating agent and coagulates any suspended

[18] Fajans and Wolff, *Z. anorg. allgem. Chem.* 137, 221 (1924).

[19] Kolthoff and Stenger, *Volumetric Analysis . . . Titration Methods.* Interscience, New York, 1947.

[20] Caldwell and Moyer, *Ind. Eng. Chem., Anal. Ed.* 7, 38 (1935).

CHLORIDE 759

matter which is present. Add standard potassium thiocyanate solution slowly with gentle swirling until a pink color is produced. Usually a false end point appears one drop before the true end point. It fades in about 30 seconds and may be due to the desorption of the last traces of silver nitrate from the precipitate. The next drop of thiocyanate produces a decided color change which persists 10 to 15 minutes. Titration should be made at temperatures below 25° C., as is customary in other titrations with thiocyanate. If nitric acid and subsequent boiling were used to put the material into solution, the addition of 5 ml. of a saturated solution of hydrazine sulfate, just prior to the addition of the ferric alum indicator, removes any nitrous acid formed.

Diphenylcarbazide and Diphenylcarbazone Indicator for Chloride.—Diphenylcarbazone was suggested as a specific indicator for the titration of chloride and bromide ions by silver ion by Chirnoaga.[21] Dubsky and Trtilek,[22] on the other hand, suggested the use of both diphenylcarbazide and diphenylcarbazone with mercuric nitrate as the titrating agent. This subject was reinvestigated by Roberts.[23] The method depends upon the formation, from mercuric ion and the indicator, of a deep blue-violet complex, after the chloride ions have combined to form slightly ionized mercuric chloride. Diphenylcarbazide is an acid-base indicator, changing from a light yellow in acid solution to a deep orange in alkaline solution, in the pH range of 6.6 to 7.4. It is probable that the alkaline form of the indicator forms the deep blue-violet complex with the mercuric ion.

Reagents.—Diphenylcarbazide.—A saturated solution of diphenylcarbazide in 95 per cent alcohol. This solution gradually turns red after standing for several days and may be used as the indicator. No apparent difference results if a fresh solution of diphenylcarbazide or diphenylcarbazone solution in alcohol is used.

Mercuric Oxide.—Dissolve mercuric oxide in nitric acid (1:1) and filter. To the filtrate add 8 N sodium hydroxide until precipitation is complete. Filter the precipitate and wash free from alkali. Dry the yellow mercuric oxide over phosphorus pentoxide for 10 days, during which period it should be powdered.

Mercuric Nitrate Solutions.—Weigh out accurately the required amount of mercuric oxide necessary to make 0.1 N, and 0.025 N solutions of mercuric nitrate and suspend in water. Add the calculated equivalent amount of nitric acid. To the well-stirred mixture add nitric acid drop by drop until complete solution takes place. Make

[21] Chirnoaga, Z. anal. Chem. 101, 31 (1935).
[22] Dubsky and Trtilek, Mikrochemie 12, 315 (1933).
[23] Roberts, Ind. Eng. Chem., Anal. Ed. 8, 365 (1937).

up to volume. The solution should be no more than 0.01 N with respect to nitric acid.

Procedure.—With 0.1 N mercuric nitrate solution the following procedure should be followed: Adjust the final volume of the solution to be titrated to about 80 to 100 ml. If the chloride solution to be titrated is acid, it should first be neutralized with 0.1 N sodium hydroxide solution. If the acid titer is also required, add 5 drops of diphenylcarbazide and titrate the solution with the standard sodium hydroxide solution to an orange color. Add 4 ml. of 0.2 N nitric acid and titrate the solution with 0.1 N mercuric nitrate solution. About 5 drops before the end point, a pink-violet color begins to develop. At the end point, one drop changes the color from a light violet to a deep blue violet.

If the chloride solution is dilute and requires 0.025 N mercuric nitrite solution additional precautions need be observed. Adjust the final volume to 65 ± 10 ml. If the chloride solution to be titrated is acid, add 2 drops of 0.2 per cent bromophenol blue, and titrate the solution with standard sodium hydroxide solution to the full blue color. Add 4 ml. of 0.2 N nitric acid, then 5 drops of the diphenyl-carbazide indicator, and titrate the solution with 0.025 N mercuric nitrate solution to a definite pink color which can be reproduced to ±0.02 ml. with the aid of a daylight lamp. The yellow color imparted by the bromophenol blue in no way interferes with the mercuric nitrate end point, and to make conditions uniform for all titrations 2 drops of bromophenol blue should be added whenever 0.025 N mercuric nitrate solution is used.

Determine a blank correction with the 2 drops of bromophenol blue, 4 ml. of 0.2 N nitric acid and with the nitric acid equivalent to the amount of acid in the mercuric nitrate solution used in the titration.

NITRATES AND NITRITES

Phenoldisulfonic Acid Method.—The following method is based on the nitration of phenoldisulfonic acid by any nitrate present with the formation of a colored nitrophenoldisulfonic acid compound. Any nitrite present is oxidized to nitrate and reacts the same way. The method is applicable to meats. The silver sulfate is used to precipitate chlorides.

Reagents.—(a) Phenoldisulfonic Acid Solution.—Heat 6 g. of phenol with 37 ml. of sulfuric acid on a steam bath, cool, and add 3 ml. of water.

(b) Standard Comparison Solution.—Dissolve 1 g. pure, dry sodium nitrate in water and dilute to 1 liter. Evaporate 10 ml. of this solution to dryness on a steam bath, add 2 ml. of the reagent (a), mix quickly and thoroughly by means of a glass rod, heat for about a minute on a steam bath, and dilute to 100 ml. One ml. of the dilute solution is equivalent to 0.1 mg. of sodium nitrate. Prepare a series of standard comparison tubes by introducing quantities ranging from 1 to 20 ml. of the diluted solution (0.1-20 mg.) of sodium nitrate into 50-ml. Nessler tubes, adding 5 ml. of ammonium hydroxide to each and diluting to 50 ml. The standard tubes thus prepared are stable for several weeks if kept tightly stoppered.

Procedure.—Weigh 1 g. of the sample into a 100-ml. flask, add 20-30 ml. of water, and heat on a steam bath for 15 minutes, shaking occasionally. Add 3 ml. of a saturated nitrate-free silver sulfate solution for each per cent of sodium chloride present, then 10 ml. of basic lead acetate solution and 5 ml. of alumina cream, shaking after each addition. Make up to the mark with water, shake, and filter through a folded filter, returning the filtrate to the filter until it runs through clear. Evaporate 25 ml. of the filtrate to dryness, add 1 ml. of the phenoldisulfonic acid solution, mix quickly and thoroughly by means of a glass rod, add 1 ml. of water and 3-4 drops of sulfuric acid, and heat on a steam bath for 2-3 minutes, being careful not to char the material. Then add about 25 ml. of water and an excess of ammonium hydroxide, transfer to a 100-ml. volumetric flask, add 1-2 ml. of alumina cream if not perfectly clear, dilute to volume with water, and filter. Fill a 50-ml. Nessler tube to the mark with the filtrate and determine the quantity of sodium nitrate present in the sample by comparison with the standard comparison tubes. If the solution is too dark for comparison with the standards, dilute with water, and correct the result accordingly.

NITRATES IN FLESH FOODS

Van Voorst [24] suggests the following procedure. Break up the sample and heat 10 g., once with 100 ml. and then with 4-50 ml. portions of water. Dilute the combined extracts to 500 ml. Filter cold and remove the chlorides from a 25-ml. aliquot by the addition of 10 ml. of 0.5 per cent silver sulfate solution. Evaporate 10 ml. of the resulting filtrate to dryness. Dissolve the residue in 5 ml. of water and allow the solution to stand for 10 minutes with 1 ml. of a solution of 3 g. of phenol in 32 g. of 96 per cent sulfuric acid. Make the mixture ammoniacal and match the yellow color produced in a volume of 50 ml. against that produced from a known

[24] Van Voorst, *Chem. Weekblad* **30**, 101 (1933).

volume of a solution of 60 mg. potassium nitrate in 250 ml. of water, treated the same way. The color due to nitrite is negligible. Good results may be obtained in sausage containing from 0.1 to 1 per cent potassium nitrate.

Xylenol Method.—In this method,[25] nitrates and the nitric acid formed from them are used for the simultaneous nitration of *m*-xylenol. The 4-hydroxy-1,3-dimethyl-5-nitrobenzene formed is separated from the reaction mixture by distillation after making it alkaline and the amount formed is estimated colorimetrically, by an evaluation of the deep-yellow color produced, which obeys the Beer-Lambert law.

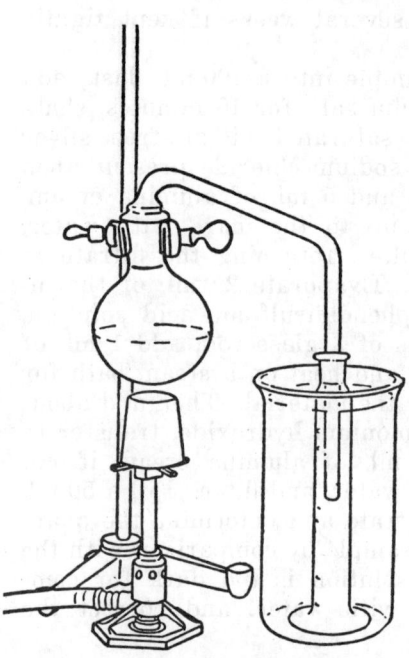

Fig. 77. Distillation Apparatus for Xylenol Method

Procedure.—Transfer a 5-ml. aliquot to a 250-ml. standard taper flask and dilute with an equal volume of water. Add 0.1 ml. of 1 per cent *m*-xylenol dissolved in either triethylene or propylene glycol, which is adequate for 100 micrograms of potassium nitrate, stopper, and cool under running water. Add 17 ml. of concentrated sulfuric acid dropwise with constant cooling so that the temperature does not exceed 35° C. Allow to stand for 10 minutes. Dilute the mixture by the addition of 150 ml. of cold water, add a few chips of porcelain, and connect the flask to the distillation apparatus (Fig. 77).

Bring the solution to the boiling point and reduce the size of the flame to about 2 cm. so that the distillate comes over at a rate of about 1 ml. per minute. Collect the distillate in a water-cooled 25-ml. graduated cylinder containing 1 ml. of 2 per cent sodium hydroxide solution. The bulk of the nitroxylenol should come over in the first few drops of distillate. Quantitative separation is achieved by stopping the distillation when the condensate reaches the 10-ml. mark on the cylinder. Mix the contents of the receiver and set aside for comparison.

---

[25] Yagoda and Goldman, *J. Ind. Hyg. Toxicol.* **25**, 440 (1943).

Prepare a blank by distilling a system consisting of 5 ml. of the glycol used, 5 ml. of water, 0.1 ml. of 1 per cent m-xylenol in the glycol, and 17 ml. of concentrated sulfuric acid, following the identical procedure used for the unknown. Prepare a standard by substituting 5 ml. of 0.002 per cent potassium nitrate, containing 100 micrograms of KNO_3, for the 5 ml. of water used in the blank, add the other reagents, and proceed with the method. When the sample and the standard are of similar intensities, the concentration of the unknown can be evaluated by a comparison of the solutions with the aid of a Duboscq-type colorimeter. When the concentration falls below 50 micrograms of potassium nitrate, compare the unknown with a series of dilute standards prepared by distilling 10, 20, 30, and 40 micrograms of potassium nitrate.

NITRITE

Sulfanilamide Method.—The well-known method of Griess-Ilosvay,[26, 27] in which a-naphthylamine hydrochloride and sulfanilic acid are used for the detection and determination of nitrite, depends upon the diazotization of the sulfanilic acid by the nitrite present with subsequent coupling of a-naphthylamine and the formation of a red dye. The color produced is proportional to the amount of nitrite present. It has been shown by Germuth [28] and others that the alpha-naphthylamine complex is not stable. Other substances such as dimethyl-alpha-naphthylamine, $C_{10}H_7N:(CH_3)_2$, have been suggested as substitutes to overcome this disadvantage.

Marshall and his coworkers [29, 30] used an analogous method for the determination of sulfanilamide, p-aminobenzenesulfonamide, and its analogues. Shinn [31] adopted this method for the determination of nitrite.

The author [32] adapted the sulfanilamide-N-(1-naphthyl)-ethylenediamine dihydrochloride method for the determination of nitrite with but slight modification for estimating nitrite in meat, water, etc., for a range extending from less than 1 p.p.m. to over 1,000 parts of nitrite per million of the substance with the same set of standards. Dyer [33] adapted it for the determination of nitrite in fish.

This method has the marked advantage over the Griess-Ilosvay

26 Griess, *Ber.* 12, 427 (1879).
27 Ilosvay, *Bull. soc. chim.* [3] 2, 317 (1889).
28 Germuth, *Ind. Eng. Chem., Anal. Ed.* 1, 28 (1929).
29 Bratton and Marshall, *J. Biol. Chem.* 128, 537 (1939).
30 Marshall and Litchfield, *Science* 88, 85 (1938).
31 Shinn, *Ind. Eng. Chem., Anal. Ed.* 13, 33 (1941).
32 Jacobs, work performed 1941-1942.
33 Dyer, *J. Fisheries Research Board Can.* 6, 414 (1945).

procedure that maximum color development is obtained in 10 minutes and the color produced remains constant for about 2 hours.

Reagents.—Sulfanilamide Solution.—Prepare a saturated solution of p-aminobenzenesulfonamide in water. This solution contains 0.4 g. per 100 ml.

Coupling Agent.—N-(1-Naphthyl)-ethylenediamine Dihydrochloride.—Dissolve 0.2 g. of this substance in water and dilute to 200 ml. Store in an amber-colored bottle in a refrigerator.

Hydrochloric Acid.—Dilute a volume of concentrated hydrochloric acid with an equal volume of water. This is approximately 6 N.

Standards.—(A) Stock Solution.—Dissolve 1.00 g. of reagent sodium nitrite in water and dilute to 1 liter. This contains 1 mg. $NaNO_2$ per ml.

(B) Dilute 50 ml. of solution A to 500 ml. (0.1 mg. per ml.).

(C) Dilute 5 cc. of solution B to 500 ml. (0.001 mg. per ml.). Add 6-7 drops of chloroform to each solution to act as a preservative. The author [34] found such solutions are stable for years.

Procedure.—Weigh 5 g. of properly ground meat, cured, smoked, etc., into a 50-ml. beaker. Transfer with the aid of hot water, at about 80° C. to a 500-ml. flask. Dilute with hot water to about 300 ml. Allow to stand on a steam bath for 2 hours. Add 5 ml. of saturated mercuric chloride solution while the mixture is still hot. Swirl vigorously, allow to stand for 5 minutes, dilute almost to the mark, cool under running water, make to the mark, shake by inverting, and filter through a dry filter.

To analyze samples of pickle brine transfer 1 ml. to a 500-ml. flask, dilute with water and, after the addition of 5 ml. of saturated mercuric chloride solution, proceed as above. Water may be treated as brine, if badly polluted, or may be tested directly.

Transfer a 1 ml. and a 10-ml. aliquot of the sample filtrate to 100-ml. Nessler tubes. Dilute to 100 ml. with distilled water. Add 1 ml. of 6 N hydrochloric acid, 2 ml. of saturated sulfanilamide solution and 1 ml. of the coupling reagent. Invert each tube 3 times to mix and allow to stand for at least 10 minutes. Compare with standards prepared at the same time.

Standards.—Transfer 0.5, 1.0, 1.5, 2.0, 2.5, and 3.0 ml. of standard solution C to Nessler tubes. Dilute to 100 ml. and treat as directed in the method. Each of these solutions contains 0.0005 mg., 0.001, 0.0015, 0.002, 0.0025, 0.003 mg. of sodium nitrite respectively.

Calculation of Results.—Each milliliter of test solution is equivalent to 10 mg. of meat. If 1 ml. of test solution develops a color that

---

[34] Jacobs, work performed 1941-1946.

is stronger than 0.003 mg. standard, it is necessary to repeat the determination with a smaller aliquot, say, 0.5 ml. It is unnecessary to prepare new standards since they are stable for at least 2 hours. If the 10-ml. aliquot is too weak, use 25 ml., 50 ml., or 100 ml. of the sample solution for comparison.

To calculate place the standard in milligrams over the milligrams of meat in the aliquot used and multiply by an appropriate factor. Suppose 1 ml. of the sample matches 0.5 ml. of the standard solution C, then

$$\frac{0.0005}{10} = \frac{5}{100,000} = \frac{50}{1,000,000} = 50 \text{ p.p.m.}$$

Photoelectric Colorimeter Modification.—Dyer [35] modified the sulfanilamide method for determining the nitrite concentration in fish.

Procedure.—Mix 10 g. of fish with water and dilute to 500 ml. Blend in a Waring blendor for 2 minutes. Transfer 0.5 ml. of the tissue extract or an aliquot of the sample solution containing 0.0002 to 0.005 mg. of sodium nitrite equivalent to 0.00005-0.001 mg. nitrite nitrogen, to 10.0 ml. in a colorimeter tube. The sample solution should be neutral or only slightly acid. Add 1 ml. of a solution of 0.2 g. of sulfanilamide dissolved in 20 ml. of concentrated hydrochloric acid and 80 ml. of water and mix thoroughly. Add 1 ml. of coupling reagent, 0.02 g. of N-(1-naphthyl)-ethylenediamine dihydrochloride dissolved in 100 ml. of water, and mix thoroughly again. Measure the red color in a photoelectric colorimeter using a filter with maximum transmission at 5400 Å. A 5200 Å filter may also be used but the absorption is not at a maximum. Maximum color development is obtained in 5 minutes and the color is stable for 2-3 hours.

Griess-Ilosvay Method.—The nitrite ion present is used to diazotize some added sulfanilic acid which then is coupled with α-naphthylamine hydrochloride.

Reagents.—(a) Sulfanilic Acid Solution.—Dissolve 1 g. of sulfanilic acid in hot water, cool, and dilute to 100 ml.

(b) α-Naphthylamine Hydrochloride Solution.—Boil 0.5 g. of the salt with 100 ml. of water, kept at constant volume, for 10 minutes.

(c) Standard Nitrite Solution.—Dissolve 1.1 g. of silver nitrite in nitrite-free water, precipitate the silver with sodium chloride solution, dilute to 1 liter, mix, allow to settle. Dilute 100 ml. of the supernatant liquid to 1 liter and then 10 ml. of this solution to 1 liter, using in each case nitrite-free water. One ml. of the final dilution is equivalent to 0.0001 mg. of nitrogen as nitrite.

35 Dyer, *J. Fisheries Research Board Can.* 6, 414 (1946).

Procedure.—Weigh 5 g. of the sample into a 50-ml. beaker. Add approximately 40 ml. of nitrite-free water heated to a temperature of 80° C. Mix thoroughly by stirring with a glass rod, taking care to break up all lumps, and transfer to a 500-ml. graduated flask. Wash out the beaker and rod thoroughly with successive portions of the hot water, adding all washings to the flask. Add sufficient hot water to bring the contents of the flask to a volume of approximately 300 ml., transfer the flask to the steam bath, and let stand for 2 hours, shaking occasionally. Add 5 ml. of saturated mercuric chloride solution and mix. Cool to room temperature, make up to the mark with nitrite-free water and mix again. Filter and determine nitrogen as nitrite in a suitable aliquot.

Place 100 ml. of the suitably diluted aliquot in a 100-ml. Nessler tube and treat with 1 or 2 drops of hydrochloric acid. Add 1 ml. of the sulfanilic acid solution, 1 ml. of the α-naphthylamine hydrochloride solution, and mix thoroughly. Set aside for 30 minutes with other Nessler tubes containing known quantities of the standard nitrite solution made up to 100 ml. with the nitrite-free water and treated with hydrochloric acid, sulfanilic acid, and α-naphthylamine hydrochloride solution in the same manner as the sample. Determine the quantity of the nitrite by comparison with the depth of pink color in the known and unknown solutions.

Germuth [36] recommends, as noted above, that, because of the instability of the α-naphthylamine complex, dimethyl-α-naphthylamine $C_{10}H_7N:(CH_3)_2$ be substituted for the α-naphthylamine. A solution of 5.25 g. of dimethyl-α-naphthylamine dissolved in 1 liter of 4 N acetic in 95 per cent methanol insures the most satisfactory results in the colorimetric estimation of nitrites. Such a solution is stable and is used in exactly the same manner as the α-naphthylamine as directed above.

AMMONIA

Aeration Method.—Uncombined ammonia may be estimated by aeration in preference to distillation, for boiling even in dilute alkali, at times, will cause some decomposition with the formation of more ammonia. This procedure is generally applied to flesh products. The ammonia is liberated by sodium or potassium carbonate solution and is then aspirated into a measured quantity of standard acid. The ammonia is then estimated by titrating the excess acid with standard alkali solution or by nesslerizing.

Procedure.—Place in the wash bottle on the right in Fig. 78, 125

[36] Germuth, *Ind. Eng. Chem., Anal. Ed.* 1, 28 (1929).

ml. of sulfuric acid (1:9) which is used to wash the incoming air. In the adjoining test tube, place a weighed quantity of sample of from 2 to 4 g. and 20 ml. of ammonia-free water. In the next test tube introduce a measured quantity of 0.02 to 0.05 N sulfuric or hydrochloric acid, according to the amount of ammonia evolved. The end wash bottle is used as a safety bottle.

Add to the sample 1 ml. of saturated potassium oxalate solution and a few drops of kerosene to minimize frothing. Then add potas-

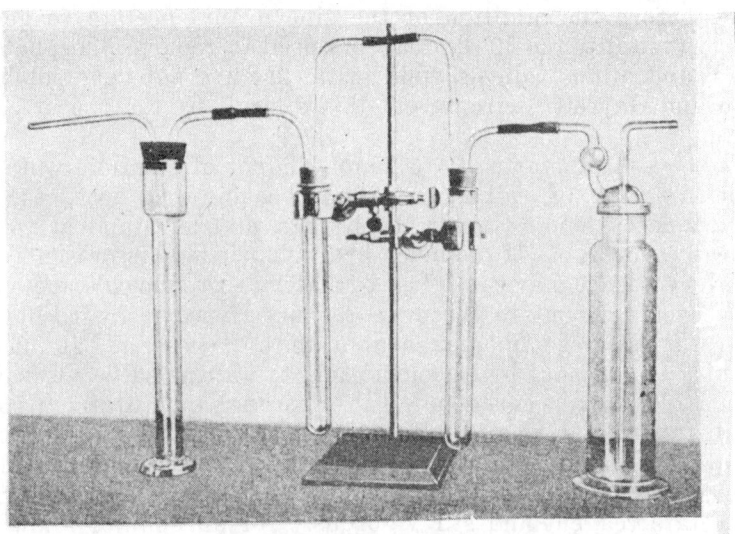

Fig. 78. Ammonia Aeration Apparatus

sium or sodium carbonate solution until the mixture is just alkaline. Replace the tubes in position immediately and pass air through the system by means of an aspirator. Titrate the standard acid at hourly intervals until ammonia ceases to be evolved using methyl red as the indicator or else nesslerize as described in the Koch-McMeekin method in Chapter I.

SODIUM

Zinc Uranyl Acetate Method.—With the introduction of the zinc uranyl acetate reagent for sodium by Barber and Kolthoff,[37] a certain specificity was obtained in analyses for sodium. The *reagent* is made from two solutions. Solution (1) is composed of 100 g. of

37 Barber and Kolthoff, *J. Am. Chem. Soc.* **50**, 1625 (1928).

uranyl acetate, $UO_2(OAc)_2 \cdot 2H_2O$, 60 g. of acetic acid and sufficient water to make 650 g. Solution (2) is composed of 300 g. of zinc acetate, $Zn(OAc)_2 \cdot 3H_2O$, 30 g. of 30 per cent acetic acid, and sufficient water to make 650 g. After the salts in solutions (1) and (2) are dissolved by warming on a steam bath to 70° C., the solutions are mixed, a few milligrams of sodium chloride are added to saturate the solution with sodium, and the reagent is allowed to stand for 24 hours. The solution is filtered immediately before use.

In reducing the volume of sodium solution to 1 or 2 ml., which is necessary before the addition of the zinc uranyl acetate reagent, salts may crystallize out if the reagents used to remove phosphates, arsenates, and other radicals yield salts that are not very soluble. Overman and Garrett [38] circumvent this difficult by the use of zinc carbonate.

Procedure.—Transfer a 10- to 15-ml. aliquot of solution containing from 2 to 8 mg. of sodium in a 50-ml. beaker. Add an excess of powdered zinc carbonate, cover the beaker, and let stand at room temperature overnight. If too much hydrochloric acid is present, the violent effervescence may cause loss of material, or zinc chloride may possibly crystallize out in the evaporation. Hence if hydrochloric acid is present in large amounts, evaporate to dryness, add 10 ml. of water and just sufficient hydrochloric acid to put the salts into solution. Then add the zinc carbonate. Filter through quantitative paper in a small funnel and wash thoroughly with cold water 5 or 6 times with small portions of water. Catch the filtrate and washings in a small beaker and evaporate to 1 or 2 ml. Add 100 ml. of the zinc uranyl acetate reagent and stir vigorously for an hour or allow to stand overnight. Filter through a weighed Gooch crucible, wash well with 95 per cent alcohol saturated with the sodium-zinc uranyl acetate salt and then with ether saturated with the same salt. Dry in air and weigh. The salt has the formula: $(UO_2)_3ZnNa(OAc)_9 \cdot 6H_2O$.

Caley and Foulk Magnesium Uranyl Acetate Method.—Caley and Foulk [39] suggested the use of the complex reagent magnesium uranyl acetate instead of zinc uranyl acetate. This reagent yields a precipitate of $(UO_2)_3MgNa(OAc)_9 \cdot 6\frac{1}{2}H_2O$.

The reagent is also made from two solutions. Solution (1) consists of 85 g. uranyl acetate, $UO_2(OAc)_3 \cdot 2H_2O$, 60 g. of glacial acetic acid and sufficient water to make a liter. Solution (2) consists of 500 g. of magnesium acetate, $Mg(OAc)_2 \cdot 4H_2O$, 60 g. of glacial acetic acid and sufficient water to make a liter. The two solutions are heated separately at approximately 70° C. on a water or steam bath until the

[38] Overman and Garrett, *Ind. Eng. Chem., Anal. Ed.* **9**, 72 (1937).
[39] Caley and Foulk, *J. Am. Chem. Soc.* **51**, 1664 (1929).

salts are dissolved. The solutions are then mixed at that temperature and ths mixture is allowed to cool to 20° C. Add a few milligrams of sodium chloride, and then allow the mixture to stand at 20° C. overnight. Filter and catch the reagent in a dry bottle.

Procedure.—The sodium solution may be prepared as directed above or the following procedure may be used.

Add to the sample of which 1-10 g. is used, according to the sodium content, sufficient sulfuric acid (1:10) to make the sample acid and dry. Ash in a muffle at low red heat. Dissolve the residue in about 2-5 ml. of hydrochloric acid by warming on a water bath, add about 40 ml. of water and heat to boiling. Add sufficient calcium chloride to precipitate all the phosphates and make the precipitation complete by the addition of ammonium hydroxide until the mixture is slightly alkaline. Filter through a small filter. Wash thoroughly and catch the filtrate and washings in a 150-ml. beaker. Evaporate the filtrate to 1 or 2 ml. or, if salts crystallize out, only to 5 ml. Cool and add 100 ml. of the magnesium uranyl acetate reagent. Stir vigorously and allow to stand overnight. Filter through a weighed Gooch crucible and wash the precipitate with 95 per cent alcohol and ether saturated with the magnesium sodium uranyl acetate salt. Dry at 105-110° C. for 30 minutes, cool and weigh. Weight of sodium-magnesium uranyl acetate × 0.0153 = sodium. The presence of small quantities of ammonium and potassium do not interfere but lithium does.

The sodium may be estimated colorimetrically by dissolving the precipitate in hot water. A yellow solution is formed which may be compared in a colorimeter against known amounts of sodium prepared the same way.

Flame Photometer Method.—The flame photometer method will in all probability become a method of choice for the determination of sodium, potassium and calcium in foods.

POTASSIUM

It is at times necessary for the food analyst to detect and estimate potassium in foodstuffs. The flame test, precipitation as the chloroplatinate, and cobaltinitrite are the usual tests. Clark and Willits [40] suggest that naphthol yellow S may be used for the detection of potassium for naphthol yellow S, that is, the disodium salt of 2:4 dinitro-1-naphthol-7-sulfonic acid forms a difficultly soluble lake with potassium.

Naphthol Yellow S Test.—To 10 ml. of the aqueous solution to be tested, containing only the soluble group, add 3 ml. of 2 per cent solu-

[40] Clark and Willits, *Ind. Eng. Chem., Anal. Ed.* 8, 209 (1936).

tion of naphthol yellow S and set aside at room temperature. The appearance of a precipitate within 65 minutes or less indicates the presence of at least 0.79 mg. of potassium per liter. Three ml. of a 5 per cent solution of the dye should produce a precipitate with 0.39 mg. of potassium per liter in 20 minutes or less.

Cobaltinitrite Method.—In order to test for potassium and sodium separately on the same solution Adams, Hall and Bailey [41] suggest the use of zinc cobaltinitrite as a reagent for potassium. The reagent of zinc cobaltinitrite may be prepared by passing the oxides of nitrogen obtained by the action of nitric acid on copper foil through a solution of cobalt acetate and zinc acetate for from 45 to 60 minutes. The resulting brown solution is kept in brown bottles tightly stoppered and is decanted from any precipitate formed.

To the solution containing no barium, arsenate, phosphate, or ammonium, add an equal volume of zinc cobaltinitrite and allow to stand for 15 minutes. A yellow precipitate of potassium cobaltinitrite indicates potassium. The filtrate may be used for the detection of sodium by the uranyl acetate method.

Wilcox [42] uses an aqueous solution of trisodium cobaltinitrite in the presence of nitric acid as the precipitating agent for potassium. He points out the following advantages for the method: The precipitate is crystalline and heavy, comparing favorably with barium sulfate in ease of filtering and washing. The composition appears to be constant and practically independent of the sodium ion concentration. The determination can be completed either gravimetrically by drying and weighing the precipitate or volumetrically by titration with potassium permanganate. The volumetric method depends upon the oxidation of the nitrite of the precipitate with permanganate in acid solution. The cobalt is present as Co^{+++} and under these conditions is a strong oxidizing agent equal to potassium permanganate and that it therefore oxidizes an equivalent amount of nitrite. The reaction:

$$5K_2NaCo(NO_2)_6 + 11KMnO_4 + 14H_2SO_4 \rightarrow 5CoSO_4 + 9MnSO_4 +$$
$$2Mn(NO_3)_2 + 5NaNO_3 + 21KNO_3 + 14H_2O$$

Here 30 NO_2 or 60 reducing equivalents are balanced by 11 $KMnO_4$ or 55 oxidizing equivalents plus 5 oxidizing equivalents from the cobaltic-cobaltous couple. Therefore 11 $KMnO_4$ or 55 equivalents are required for 10 K or

$$\frac{10K}{55} = \frac{390.96}{55} = 7.1084 \text{ g. of } K \text{ per equivalent of } KMnO_4$$

[41] Adams, Hall, and Bailey, *Ind. Eng. Chem., Anal. Ed.* 7, 310 (1935).
[42] Wilcox, *Ind. Eng. Chem., Anal Ed.* 9, 136 (1937).

Gravimetric Procedure.—Prepare an aqueous solution containing 1 g. of the trisodium cobaltinitrite in each of 5 ml., allowing 5 ml. for each determination. Filter before use. Prepare freshly if necessary.

The aliquot for analysis should contain between 2 and 15 mg. of potassium in a neutral aqueous solution of 10 ml. volume. Add 1 ml. of 1 N nitric acid and 5 ml. of the sodium cobaltinitrite solution, mix, and allow to stand for 2 hours. Filter in a porous-bottomed porcelain crucible, the tare weight of which is known, using 0.01 N nitric acid in a wash bottle to make the transfer. Wash 10 times with 2-ml. portions of the dilute nitric acid and 5 times with 2-ml. portions of 95 per cent alcohol. Evacuate until dry. Wipe the outside with a cloth, and dry for 1 hour at 110° C. Cool, in a desiccator, and weigh.

The composition of the precipitate can be represented by the formula $K_2NaCo(NO_3)_6 \cdot H_2O$. $K = 17.216$ per cent.

Titrimetric Procedure. — Follow the gravimetric procedure through the precipitation and washing with nitric acid. Omit washing with alcohol. Wash the precipitate into a 250-ml. beaker, place the crucible in the beaker, and make to about 100 ml. with water. Add 20 ml. of 0.5 N sodium hydroxide solution and boil for 3 minutes. Withdraw into another beaker a slight excess of standard 0.05 N potassium permanganate, make to 50 ml. with water, and add 5 ml. sulfuric acid. Pour the hot potassium cobaltinitrite solution into the cold potassium permanganate solution, transfer the crucible and wash the beaker with a small amount of water. Add an excess of standard 0.05 N sodium oxalate solution, heat to boiling, and complete the titration with potassium permanganate. Ml. of $KMnO_4 \times$ normality of $KMnO_4 \times 7.1084 = $ mg. of K in sample titrated.

There are many colorimetric methods for the determination of potassium based on the estimation of the nitrite content in the potassium cobaltinitrite precipitate.

Potassium forms a slightly soluble, yellow to dark red salt with dipicrylamine. The potassium is precipitated in solutions neutral to methyl red by the addition of 0.5 normal solution of the magnesium salt of dipicrylamine.[43]

IRON

Determinations of iron in foods are performed for various reasons, among which is the need to estimate the iron content for nutri-

[43] Winkel and Mass, *Z. angew. Chem.* **49**, 827 (1936).

tional reasons and to measure the amount of iron contamination of a food for trace amounts of iron promote oxidative and metallic flavors, rancidity, and vitamin decomposition. In addition it becomes necessary at times to distinguish between available and total iron present in a foodstuff.

Thiocyanate Method.—In the variation of this method used by Stugart,[44] amyl alcohol is used to extract the iron thiocyanate formed which is then estimated colorimetrically.

Preparation of Sample.—Evaporate 100 ml. of milk to dryness on a water bath and ash the material as described below. Dissolve the ash in just enough iron-free hydrochloric acid to effect solution, filter, and set the filtrate aside. Ignite the residue, which consists largely of iron, take up in a little dilute hydrochloric acid, filter, and wash, add the filtrate and washings to the original filtrate, and make up to 50 ml. in a volumetric flask.

For other foods grind the desiccated material finely, place in a platinum dish, dry at 110° C. to constant weight, and place in a muffle oven. Allow to carbonize very slowly in the open muffle. As carbonization proceeds, close the door of the oven gradually, allow to remain overnight with the temperature regulated so that it remains just below dull redness and then proceed as above.

Standard Iron Solution.—Weigh accurately 0.5 g. of pure iron wire and dissolve in 20 per cent hydrochloric acid with the aid of 1 ml. of concentrated nitric acid. Carefully evaporate to dryness and dissolve the ferric chloride in hydrochloric acid avoiding a large excess. Dilute to 100 ml. and preserve carefully. Each milliliter of this stock solution is equivalent to 5 mg. of iron. For a working standard, take 1 ml. of this stock solution and dilute to 100 ml.; each milliliter is equivalent to 0.05 mg. of iron. It is convenient to make up two standards containing 0.01 and 0.02 mg. of iron.

Procedure.—Pipette an aliquot part of the unknown solution containing approximately 0.025 to 0.10 mg. of iron into a 150-ml. beaker. Add 5 ml. of concentrated hydrochloric acid and make up to approximately 25 ml. Hydrolyze by boiling 20 minutes, allow to cool, add 1 drop of concentrated nitric acid, and make up to 25 ml. in a volumetric flask. Transfer an aliquot portion of 10 ml. to a 30-ml. or 50-ml. separatory funnel, add 1 ml. of hydrochloric acid and enough 0.1 N potassium permanganate solution to produce a pink color which persists for 20 seconds. Usually 1 or 2 drops are adequate. Add 5 ml. of isoamyl alcohol with the aid of a safety pipette and 5 ml. of 20 per cent potassium thiocyanate solution. Shake vigorously for

44 Stugart, *Ind. Eng. Chem.*, *Anal. Ed.* 3, 390 (1931).

30 seconds, allow to separate, and draw off the water layer. Transfer a portion of the isoamyl alcohol solution to a colorimeter tube, taking precaution to exclude any water droplets and compare with standards.

Transfer with a pipette 1.0 and 2.0 ml. of the working standard equivalent to 0.05 and 0.1 mg. of iron, respectively, into each of two 50-ml. volumetric flasks. Add 1 drop of concentrated nitric acid, 10 ml. of hydrochloric acid, and make up to volume with water. Transfer 10 ml. to a 30-ml. separatory funnel, add 1 ml. of hydrochloric acid, 1 or 2 drops of 0.1 N potassium permanganate solution, and mix. Add 5 ml. of isoamyl alcohol and 5 ml. of 20 per cent potassium thiocyanate solution. Shake vigorously for 30 seconds, draw off the aqueous layer, and compare the isoamyl alcohol layer with the unknown. In the amounts present in biological materials, cobalt, manganese, nickel, and zinc do not interfere with the color development. Copper, if present in appreciable amounts, should be removed.

Thompson Modification.—The thiocyanate method was also used by Thompson [45] who employed a high concentration of hydrochloric acid, a high concentration of potassium thiocyanate, and redistilled isobutyl alcohol as the extractant to avoid interference of calcium and phosphates.

Standard Iron Solutions.—Stock Solution.—Weigh exactly 1.0000 g. of iron wire into a dry, iron-free beaker. Dissolve in 20 per cent hydrochloric acid to which 1 to 2 ml. of concentrated nitric acid have been added. Carefully evaporate to dryness and dissolve in the minimum amount of hydrochloric acid. Transfer quantitatively to a liter volumetric flask and dilute to volume. This stock solution contains 1 mg. of iron per milliliter.

Working Standard.—Dilute 10 ml. of the stock solution to 1000 ml., adding a few drops of bromine water just prior to adjusting to volume. This solution contains 10 micrograms of iron per milliliter.

Sample Preparation.—Transfer an accurately weighed sample (3 to 5 g., depending on the suspected iron content) to a 300-ml. Kjeldahl flask, add 10 ml. of nitric acid, and warm slightly to start oxidation. When the initial oxidation has subsided, add 2 ml. of concentrated sulfuric acid and boil gently until charring commences. Prepare liquid samples by taking appropriate volumes, depending on the iron content, and concentrating to a small volume in the presence of 10 ml. of nitric acid before the sulfuric acid is added.

Add nitric acid, a few milliliters at a time or, preferably, dropwise, until the oxidation is nearly completed as evidenced by only

45 Thompson, *Ind. Eng. Chem., Anal. Ed.* 16, 646 (1944).

slight darkening upon evolution of sulfur trioxide fumes. Remove the flame and allow the flask to cool slightly. Add 1 ml. of perchloric acid (care) and continue heating until the solution has clarified. It may be necessary to add a few more drops of nitric acid at this point to complete the oxidation. Heat to the point where copious white fumes of sulfur trioxide appear and the perchloric acid has been destroyed. The final solution should be colorless, or, at most, a light straw color. Cool, add 40 ml. of redistilled water, and boil until copious white fumes of sulfur trioxide again appear. Continue heating for about 5 minutes to assure complete oxidation and elimination of perchloric and nitric acids. Cool, add about 10 ml. of redistilled water, and quantitatively transfer to a 100-ml. volumetric flask by washing with small portions of redistilled water until the volume is nearly 100 ml. Cool to room temperature and dilute to volume.

Alternate Method of Sample Preparation.—This method is particularly advantageous for foods of high fat content.

Transfer an accurately weighed sample to a 300-ml. Kjeldahl flask, add 5 ml. of 30 per cent hydrogen peroxide and 2 ml. of concentrated sulfuric acid, and heat gently until charring commences. Add 5 ml. more of the 30 per cent hydrogen peroxide and continue heating until charring again occurs. Proceed as in the regular method, beginning with the addition of the nitric acid, a few milliliters at a time.

Procedure.—Transfer a 25-ml. aliquot of the prepared solution to a 125-ml. separatory funnel and add exactly 5 ml. of concentrated hydrochloric acid. Add 1 ml. of 2 per cent potassium peroxydisulfate (potassium persulfate) and swirl the separatory funnel to insure complete mixing. Add exactly 10 ml. of the 20 per cent potassium thiocyanate reagent to develop the color, then add exactly 25 ml. of isobutyl alcohol and shake for 2 minutes. Draw off and discard the aqueous layer. Invert and slowly revolve the funnel to dislodge any water particles clinging to the walls and allow to stand for 10 minutes. Draw off the small amount of water which has separated from the alcohol, and transfer the alcohol layer to a dry 50-ml. Erlenmeyer flask. Immediately prior to reading the per cent transmission add a small amount (about 0.1 g.) of anhydrous sodium sulfate and agitate, to remove suspended particles of water from the alcohol extract. Read at 485 mμ, setting a reagent blank at 100 per cent transmission. Obtain the micrograms of iron from a standard curve and convert to parts per million.

If the color is too intense to read (in excess of 50 micrograms) repeat the determination, using a smaller aliquot of the prepared sample. As it is important that the volume ratio be kept constant, the difference in the aliquot size must be made up by the addition of redistilled water—for example, if a 15-ml. aliquot is used in place of the usual 25-ml., correct the difference in volume by adding 10 ml. of redistilled water.

Preparation of Standard Curve.—Develop the color on increments of the working standard in the range of 0 to 60 micrograms of iron. A convenient formula to follow is:

5 ml. of concentrated hydrochloric acid
x ml. of standard
$(25 - x)$ ml. of redistilled water

From this point proceed exactly as outlined in the method, beginning with the addition of the potassium persulfate. Plot per cent transmission against the concentration on semilogarithmic paper.

o-**Phenanthroline Method.**— The *o*-phenanthroline method for iron has been adapted for its determination in enriched, enriched self-rising, and phosphated flours and bread.

Preparation of Sample.—Bread may be ground in a mill but the ground material should be checked to see if the iron content is increased. Preferably, slice the bread, allow to dry in air, and crush to pass a 20-mesh sieve on a wooden board with a wooden rolling pin.

Reagents.—Acetate Buffer Solution.—Dissolve 8.3 g. of anhydrous sodium acetate, dried at 100° C., in water, add 12 ml. of acetic acid, and dilute to 100 ml.

o-Phenanthroline Solution.—Dissolve 0.1 g. of the reagent in about 80 ml. of water at 80° C., cool, and dilute to 100 ml.

Preparation of Standard Curve.—Dissolve 0.1 g. of iron, analytical grade in a mixture of 20 ml. of hydrochloric acid and 50 ml. of water. Dilute to 1 liter, Transfer 100 ml. of this solution to a liter volumetric flask and make to the mark with water. Each milliliter of the latter solution contains 0.01 mg. of iron.

Place 2.0, 5.0, 10.0, 20.0, 25.0, 30.0, 35.0, 40.0, and 45.0 ml. of the latter solution into 100-ml. flasks, add 2.0 ml. of hydrochloric acid, and dilute to 100 ml. Use 10-ml. aliquots and proceed with the method detailed below, starting at the point—"add 1 ml. of hydroxylamine hydrochloride solution." Plot concentration against scale readings. Use a blank containing only hydrochloric acid.

Procedure.—Ash 10.0 g. of air dried bread or flour or other product in a silica or porcelain dish, about 60 mm. in diameter and having a capacity of 35 ml. as directed in Chapter I, in a muffle furnace at dull red heat. Ash aids such as 50 per cent magnesium nitrate solution, prepared by dissolving 50 g. of $Mg(NO_3)_2 \cdot 6H_2O$ in water and diluting to 100 ml. or redistilled nitric acid may be used. Cool, add 5 ml. of hydrochloric acid using it to rinse the upper part of the dish and evaporate to dryness. Dissolve the residue by adding 2.0 ml. of hydrochloric acid with the aid of a pipette. Cover the dish with a watch glass, heat for 5 minutes on a steam bath, wash off the watch glass, and filter, catching the filtrate in a 100-ml. volumetric flask. Cool and dilute to volume. Transfer a 10-ml. aliquot to a 25-ml. volumetric flask, add 1 ml. of 10 per cent hydroxylamine hydrochloride solution, prepared by dissolving 10 g. of $NH_2OH \cdot HCl$ in water and diluting to 100 ml., after a few minutes 5 ml. of buffer solution and 1 ml. of the o-phenanthroline solution, and make to volume. Transfer an aliquot to a 2-inch cell and read in a neutral wedge photometer using a 510 mμ filter or use another appropriate instrument.

Instead of o-phenanthroline, 2 ml. of α, α-dipyridyl solution, prepared by dissolving 0.1 g. of the reagent in water and diluting to 100 ml., may be used. Both the o-phenanthroline and α, α-dipyridyl solutions should be stored in amber bottles and kept in a refrigerator.

MANGANESE

It is occasionally necessary to determine the manganese content of foods since manganese is considered to be a micronutrient.

Willard-Greathouse Periodate Method.—The periodate method [46-48] is suitable for estimating quantities of manganese of less than 1 mg. After removal of chlorides, manganous salts are oxidized to permanganate by means of periodate and the color produced is compared with that of standards. Fairhall [49] prefers the periodate method to the persulfate method described below.

If necessary, ash the sample or a portion of it in a silica dish. Take up in hydrochloric acid and evaporate to dryness. Evaporate liquid samples to dryness with concentrated hydrochloric acid. Add a few milliliters of sulfuric acid (1:2) and 3 to 4 drops of concentrated nitric acid. Evaporate carefully to dryness on a water bath and sand bath, finishing the evaporation by gentle ignition with a Bunsen flame. Add 2 to 2.5 ml. of sulfuric acid (1:2) and a little water, and evaporate to white fumes of sulfur trioxide, thus removing all traces of chlorides. Cool, dilute, and filter into a 50-ml. flask for oxidation. Add to the solution one or two small pieces of pumice

[46] Willard and Greathouse, *J. Am. Chem. Soc.* 39, 2366 (1917).

[47] Richards, *Analyst* 55, 554 (1930).

[48] *Standard Methods of Water Analysis* (8th ed.), Am. Pub. Health Assoc., New York, 1936.

[49] Fairhall, *U. S. Pub. Health Service, Bull.* 247, 31 (1940).

stone, previously purified by boiling with 5 per cent sulfuric acid, and a little periodate. Evaporate down to about 10 ml., when the concentration of sulfuric acid will be a 5-6 per cent solution.

Add 0.3 g. of sodium periodate or potassium periodate and insert a loosely fitting pear-shaped glass stopper or similar arrangement in the neck of the flask. Heat to boiling, immerse in a boiling water bath, and heat for 30 minutes. Cool and transfer to colorimeter tubes. Depending on the depth of color, use 10-, 22.5-, 50-, or 100-ml. tubes, and dilute to the mark accordingly. If diluted with water before matching, the solution should be boiled in the water bath for 15 minutes longer.

Preparation of Standard.—Standard Manganous Sulfate.—Dissolve 0.1438 g. of potassium permanganate in water containing 2 to 3 ml. of 2 N sulfuric acid. Reduce by the addition of 0.4 g. of sodium bisulfite. Boil off the excess sulfur dioxide, cool, transfer to a 1-liter volumetric flask, and make to volume. One ml. is equivalent to 0.05 mg. of manganese.

Sulfuric Acid Reagent.—Add 120 ml. of concentrated sulfuric acid to 1,500 ml. of water. Dilute to 2 liters. Add 2.4 g. of sodium periodate (sodium paraperiodate, $Na_2H_3IO_6$), heat to boiling, and place in a boiling water bath for 30 minutes. This gives a 6 per cent by volume solution of sulfuric acid.

Color Standard.—Oxidize exactly 20 ml. of the standard manganous sulfate solution to which is added 1.2 ml. of concentrated sulfuric acid and 30 ml. of 6 per cent sulfuric acid, with 0.3 g. of periodate in the usual way. Cool, transfer to a 1-liter volumetric flask, and make up to volume with the 6 per cent sulfuric acid reagent to 1 liter. One ml. of this solution is equivalent to 0.001 mg. of manganese.

Persulfate Method.—In this method [50-52] the manganous sulfate is oxidized to permanganate by means of ammonium persulfate (ammonium peroxydisulfate).

Samples containing much chloride and organic matter are best freed of chloride and the organic matter in the manner described under the periodate method. Samples containing little chloride or organic matter may be treated as follows.

Procedure.—Take an aliquot containing not more than 0.2 mg. of manganese. Add 2 ml. of nitric acid and adjust to 50 ml. volume. Precipitate any chloride by the addition of silver nitrate solution, con-

50 *Standard Methods of Water Analysis* (8th ed.), Am. Pub. Health Assoc., New York, 1936.
51 Marshall, *Chem. News* 83, 76 (1901).
52 Wester, *Rec. trav. Chim.* 39, 414 (1920).

taining 20 g. of silver nitrate, $AgNO_3$, in 1 liter of water, and add at least 1 ml. in excess. Add about 0.5 g. of ammonium persulfate crystals and warm the solution until the maximum permanganate color is developed. This usually takes about 10 minutes. At the same time prepare standards by diluting portions of 0.2, 0.4, 0.6 ml., etc., of the standard manganous sulfate solution to about 50 ml., and treat them exactly as the sample was treated. Transfer the sample and the standards to 50-ml. Nessler tubes and compare the colors immediately.

To prepare the standard manganous sulfate solution for this method, dissolve 0.2873 g. of potassium permanganate in about 100 ml. of distilled water. Acidify the solution with sulfuric acid and heat to boiling. Add slowly a sufficient quantity of dilute solution of oxalic acid to discharge the color. Cool and dilute to 1 liter. One ml. of this solution contains 0.1 mg. of manganese.

NICKEL

Dioxime Method.—Nickel is used as a catalyst in the hydrogenation of oils to make fats. It may at times contaminate the fats so produced. A simple and convenient means of detection and estimation in fats is to isolate the nickel by solution in acid and, subsequently, precipitate it with dimethylglyoxime or α-benzildioxime.

Procedure.—Place a weighed quantity of fat, about 250 g., in a flask, add an equal amount of hydrochloric acid, and heat on a water bath for 1 to 2 hours with frequent swirling. Transfer the mixture to a separatory funnel and draw off the acid layer. Filter and evaporate in a porcelain dish on a water bath.

Dissolve the residue in 50 ml. of hot absolute alcohol, rendered just alkaline with ammonium hydroxide, and add 50 ml. of a hot saturated solution of dimethylglyoxime or α-benzildioxime. Heat the mixture for a few minutes on the bath and then filter through a tared Gooch crucible, wash with hot alcohol, dry at 110° C., cool, and weigh.

If α-benzildioxime was used, then the weight of precipitate multiplied by the factor, 0.1093 equals the weight of nickel. If dimethylglyoxime was used, then the weight of precipitate multiplied by the factor, 0.2032 equals the weight of nickel.

Potassium Dithiooxalate Method.—Very small amounts of nickel may be determined by the use of this method.[53-56] The nickel is sepa-

---

[53] Jones and Tasker, *J. Chem. Soc.* **95**, 1904 (1909).
[54] Fairhall, *J. Ind. Hyg.* **8**, 528 (1926).
[55] Drinker, Fairhall, Ray, and Drinker, *J. Ind. Hyg.* **6**, 346 (1924).
[56] Yoe and Wirsing, *J. Am. Chem. Soc.* **54**, 1866 (1932).

rated from iron and if necessary, from cobalt. Its concentration is then determined by the formation of magenta-colored nickel dithiooxalate.

Procedure.—Evaporate, dry, and char the specimen in a porcelain dish. Ash at low red heat, being careful not to fuse the ash. Cool, add 15 ml. of hydrochloric acid (1:1) and sufficient water, if necessary, to cover the residue. Cover the dish with a watch glass and heat to boiling. Filter, and extract 2 more times with hot water. If a clean ash has not been obtained, return the filter paper and its residue to the original ashing dish, dry, and re-ash in a muffle. Extract as directed above. Combine all the filtrate-extracts and washings, and neutralize the hydrochloric acid with ammonium hydroxide, using methyl orange as indicator. Add a few drops of hydrochloric acid until the solution is just acid. Saturate the cold solution with hydrogen sulfide and allow to stand overnight. Filter and wash the precipitate with hydrogen sulfide water. Combine the filtrate and washings, which contain the nickel, and boil until free of hydrogen sulfide. Add bromine water to oxidize iron to the ferric state.

If cobalt is present, proceed as directed below. If cobalt is absent, it is necessary to free the solution only of iron, which interferes with the determination. To the cold, slightly acid solution, add 10 ml. of 50 per cent ammonium acetate solution and 0.5 ml. of glacial acetic acid. Under these conditions, iron is precipitated in the cold. Warming should be avoided to prevent reduction of iron to the ferrous state. Filter the cold solution through quantitative filter paper into a volumetric flask. Dilute to a known volume depending upon the nickel concentration. Transfer 50 ml. of this solution to a Nessler tube and add a small amount of potassium dithiooxalate. If nickel is present a clear magenta color develops at once. If nickel is absent no color will develop except a slight yellow.

Procedure in the Presence of Cobalt.—If cobalt is present, it must be separated from the nickel. To do this both calcium and magnesium are precipitated as oxalate and phosphate respectively. Then nickel is isolated with α-benzildioxime and subsequently can be estimated as detailed above.

To the nickel solution, add 10 ml. of 20 per cent sodium citrate solution to prevent precipitation of iron. Add saturated ammonium oxalate solution to precipitate calcium. When precipitation is complete, add dilute ammonium hydroxide solution slowly to precipitate ammonium magnesium phosphate in the same solution. Filter. Dissolve the precipitates in hydrochloric acid and then reprecipitate calcium and magnesium as oxalate and phosphate as above. Filter and combine this filtrate with the main filtrate. This step recovers any nickel occluded on the calcium and magnesium precipitates. To the alkaline filtrate, add an excess of α-benzildioxime, filter the nickel precipitate, dissolve in aqua regia, and evaporate the acid solution to dryness in a porcelain dish. Dissolve in a few drops of dilute hydrochloric acid and make to volume in a volumetric flask. Determine nickel colorimetrically with potassium dithiooxalate.

Standards containing from 0.005 to 0.05 mg. of nickel can be prepared by dissolving a weighed portion of nickel dimethylglyoxime, which contains 20.32 per cent of nickel, in aqua regia, evaporating, redissolving in hydrochloric acid, evaporating again, redissolving in hydrochloric acid, and making up to a known volume. To prepare standards make further dilutions. Higher concentrations of nickel can be matched in a colorimeter.

COBALT

Cobalt is a micronutrient. As noted in the chapter on vitamins, it is part of the vitamin B_{12} molecule. The nitroso R salt method [57, 58] can be modified for its determination in foods.

Standard Cobalt Solution.—Dissolve 0.0249 g. of cobalt oxalate, CoC_2O_4, in 10 ml. of 6 N hydrochloric acid and dilute to 1000 ml. with water. Transfer 100 ml. of this solution to a liter volumetric flask, add 10 ml. of 6 N hydrochloric acid, and dilute to volume. One ml. of this standard solution is equivalent to 0.001 mg. of cobalt.

Procedure.—Transfer a suitable aliquot portion of the prepared sample to a beaker. Filter, if necessary, and wash 3 times with water. Evaporate almost to dryness, add 2 ml. of concentrated nitric acid and evaporate to dryness. Dissolve the sample by boiling in 10 ml. of water and 2 ml. of 6 N hydrochloric acid. Neutralize the sample with 20 per cent sodium hydroxide solution using phenolphthalein as indicator. Add 2 ml. of Spekker acid (prepared from 150 ml. of phosphoric acid, specific gravity 1.75, and 150 ml. of sulfuric acid, specific gravity 1.84, diluted to 1 liter with water), 10 ml. of 0.1 per cent aqueous solution of nitroso R salt, and 10 ml. of 50 per cent sodium acetate trihydrate (W/V solution). Bring the mixture to a vigorous boil, add 5 ml. of concentrated nitric acid, and boil the mixture for 1 to 2 minutes. Cool, dilute to 100 ml. with water, and obtain the optical density difference between the test solution and a reagent blank balanced at zero on a Coleman Universal spectrophotometer at 510 mμ. Estimate the amount of cobalt from a standard curved graph of optical density against concentration, prepared

[57] Keenan and Flick, *Anal. Chem.* **20**, 1238 (1948).
[58] Young, Pinkney, and Dick, *Ind. Eng. Chem., Anal. Ed.* **18**, 474 (1946).

from a series of 0- to 0.500-mg. cobalt standards carried through the same procedure.

CALCIUM

Calcium may be estimated in foods by several methods. After preparing an ash in the manner described in Chapter V, the calcium may be precipitated as the oxalate at a pH at which magnesium does not come down. The calcium oxalate precipitate may be collected by centrifugation, ashed to obtain the oxide and estimated titrimetrically in the presence of boric acid.[59]

Reagent.—Ma-Zuzaga Indicator Solution.—Prepare a 0.1 per cent solution of methyl red in 95 per cent alcohol and a 0.1 per cent solution of bromocresol green in 95 per cent alcohol. Mix five volumes of the bromocresol green indicator solution with 1 volume of the methyl red indicator solution.

Procedure.—Place in a centrifuge tube sufficient neutral or slightly acid sample solution to contain 0.1 to 0.2 mg. of calcium in the aliquot. Add 1 drop of thymol blue indicator solution. Add an amount of saturated ammonium oxalate solution equivalent to the volume of the sample taken for analysis. Add sodium hydroxide solution dropwise until the thymol blue turns yellow. Allow to stand for 1-3 hours to permit the calcium oxalate to settle. Centrifuge at 2000 r.p.m. for 10 minutes, aspirate off the supernatant liquid, add 1 ml. of 0.5 per cent ammonium oxalate solution and spin again as above. Aspirate off the supernatant liquid, dry the centrifuge tube in an oven at 100° C., remove from the oven and place in a muffle at 450-500° C. for 1 hour. Remove the tube from the muffle, allow to cool, and place in a boiling water bath. Add, with the aid of a heated pipette, 0.5 ml. of a *hot* 20 per cent solution of boric acid and keep the tube in the water bath with occasional shaking until the precipitate is completely dissolved. Remove the tube from the bath, add 2.5 ml. water and shake immediately. Add 2 drops of Ma-Zuzaga indicator solution and titrate with 0.01 N hydrochloric acid to the first pink of the indicator. Use a comparison tube containing 0.5 ml. of boric acid, 2.5 ml. of water and 2 drops of the indicator solution.

Permanganate Method.—In this method, calcium is precipitated as the oxalate and the amount of calcium is determined by means of a permanganate titration of the equivalent oxalic acid. It is applicable to flour and bread.

Procedure.—Ash 10 g. of air-dried bread or flour as detailed in Chapter I or as detailed in the *o*-phenanthroline method for iron.

---

[59] Sobel and Sobel, *J. Biol. Chem.* **129**, 721 (1939).

Continue with that procedure to the point at which the hydrochloric acid solution is filtered into the flask and filter into a beaker instead or use 50 ml. of the solution from the iron determination, and place this 50 ml. in a beaker. Add 8 to 10 drops of bromocresol green indicator solution. Add sufficient 20 per cent sodium acetate solution, prepared by dissolving 20 g. of $NaC_2H_3O_2$ in water and diluting to 100 ml. to adjust the pH to 4.8-5.0 at which point it is blue. Cover with a watch glass and heat to boiling. Add slowly 3 per cent oxalic acid solution, one drop every 3 to 5 seconds, until the pH is reduced to 4.4-4.6 which is the optimum for the precipitation of calcium oxalate and at which pH the indicator is green. Boil for 1 to 2 minutes and allow to stand until clear or overnight. Filter the supernatant solution through a Gooch crucible, or a sintered-glass filter, or through quantitative filter paper and wash the beaker, the precipitate, and filter with about 50 ml. of dilute ammonium hydroxide solution (1:50) using small portions of wash solution delivered with the aid of a wash bottle with a fine tip. Wash the filters, breaking the tip of the filter paper, if used, with a mixture of 125 ml. of water and 5 ml. of sulfuric acid at 80-90° C. catching the acid washings in the original beaker. Titrate at 70-90° C. with 0.05 N potassium permanganate solution until a slight pink color is obtained, add the filter paper and continue the titration if necessary. Run a blank determination and make a correction. One milliliter of 0.05 N potassium permanganate solution is equivalent to 1 mg. of calcium. The latter is usually expressed as milligrams per pound.

ALUMINUM

Aluminum is not considered a toxic metal. However, it sometimes occurs in foodstuffs in fairly high amounts due to the addition of alum to pickling solutions and as a component of some baking powders. It may be determined simply with a fair degree of accuracy, in the presence of iron, by the Chancel method. The food material is ashed in the dry way and the silica dehydrated as usual. If no heavy metals are present, the residue is now taken up in 5 ml. of hydrochloric acid and 50 ml. of water, filtered and washed. The filtrate and washings are made up to about 200 ml. and dilute ammonium hydroxide is added until the precipitate which forms dissolves with difficulty. An excess of sodium thiosulfate is added and the solution boiled carefully, until all of the sulfur dioxide is expelled. Filter the precipitate of aluminum hydroxide on ashless filter paper and wash with 2 per cent ammonium nitrate solution. Ignite and blast in a tared crucible to constant weight. Cool and weigh as aluminum oxide, Al_2O_3. The reaction is:

$$2AlCl_3 + 3Na_2S_2O_3 + 3H_2O \rightarrow 2Al(OH)_3 + 6NaCl + 3SO_2 + 3S$$

The method, as may be seen, depends on the precipitation of aluminum as the hydroxide by boiling with sodium thiosulfate in neutral solution.

Small amounts of aluminum may be estimated colorimetrically as its lake with aurintricarboxylic acid which is also known as aluminon.

Rapid Titrimetric Method.—A method modified by the author [60] from those of Viebock and Brecher [61] and of Snyder [62] is based on the solution of the aluminum bearing material in acid, the buffering of the mixture with a tartrate buffer, the neutralization of the test mixture with barium hydroxide, sequestration of the aluminum by use of potassium fluoride with consequent liberation of three moles of alkali hydroxide for every mole of aluminum,

$$Al^{+++} + 3OH^- \rightarrow Al(OH)_3$$

$$Al(OH)_3 + 6KF \rightarrow AlF_3 \cdot 3KF + 3KOH$$

and finally estimation of the liberated alkali acidimetrically.

Procedure.—Prepare an acid solution of the ash as in the previous method. Take an appropriate aliquot. Add 3 drops of phenolphthalein indicator solution and 20 ml. of 30 per cent potassium sodium tartrate solution. Mix gently and neutralize with saturated barium hydroxide solution to a definite pink end point. Add 20 ml. of 30 per cent potassium fluoride solution and mix. Allow the test solution to stand 5 minutes. Titrate with standardized 0.03 N hydrochloric acid until the pink coloration is discharged and remains discharged for 30 seconds.

SELECTED REFERENCES

Jacobs, *Analytical Chemistry of Industrial Poisons, Hazards, and Solvents*. Interscience, New York, 1949.
Methods of Analysis, A. O. A. C. Washington, 1945.
Monier-Williams, *Trace Elements in Food*. Wiley, New York, 1949.

[60] Jacobs, *J. Am. Pharm. Assoc., Sci. Section* 39, 523 (1950).
[61] Viebock and Brecher, *Arch Pharm.* 270, 114 (1932).
[62] Snyder, *Ind. Eng. Chem., Anal. Ed.* 17, 37 (1945).

CHAPTER XIX

FLAVOR AND QUALITY MEASUREMENT

THE investment of capital in a food of untried flavor is a risk that can be evaluated better by an adequate appraisal of the acceptability of that flavor. Experience has shown that, in the long run, maintenance and enhancement of flavor pays in increased profits through greater acceptability of a product, for it is a fact that the acceptance of any given food by the public can be correlated with the acceptance of its flavor.

FLAVOR ACCEPTANCE [1]

Food Flavor Problem.—The shortage of food throughout the world has made the problem of food supply of world-wide interest. It is noteworthy that, along with this interest in the general food problem, there has been a marked growth of interest in the food flavor problem as well. One of the stimuli in this direction has been the work of the U. S. Quartermaster Corps in its effort to provide food, during the past war and in the present wide distribution of our armed forces, that could maintain its flavor and its acceptability in all climates and temperatures and under many varied conditions.

Many attempts have been made to place flavor appraisal on an objective basis so that a quantitative estimate could be made of a flavor grade or quality. While some advance has been made in this direction, virtually all flavor appraisal really has a subjective basis since an estimate of the reaction of an individual to a given flavor is being made.

The preference of a person for a given food item is the resultant of many factors, among which we might mention, physiological response, personal idiosyncrasy, habitual usage, geographical and regional factors, age, and taboos and religious codes. Each of these factors plays a role in flavor appraisal.

Personal Idiosyncrasy.—It is needless to dwell on the factor of personal idiosyncrasy, because certain taste preferences not only in flavor and food but in clothing, books, and, for that matter, almost anything are wholly unaccountable.

[1] Jacobs, *Am. Perfumer Essential Oil Review* 52, 46 (1948).

Habitual Usage.—Habitual usage of a given item often makes a flavor acceptable. The old adage of having to eat 9 olives or a certain, number of Brussels sprouts before one gets accustomed to the taste of these items are cases in point. Through continued use we become accustomed to many foods and find a flavor once rejected as undesirable fully acceptable and even desirable after habit formation. Thus as long as the chlorine concentration in potable water is not too high, most people do not object to chlorinated water after a period of indoctrination.

Geographical Factors.—Geographical and regional factors are very important. It is a commonplace that in every country there are "national" dishes and foods. Many illustrations will occur to the reader. It is less well known that there are marked regional differences. In the New York area, for many years, only white shell eggs were deemed acceptable, whereas in the Boston area a proper egg had a brown shell. Butter in the South has to have a deep yellow shade to be deemed adequate, but in the North a lighter yellow color is desirable. Yellow cornmeal is the type preferred in the North, while white cornmeal is the preference of the South. These color preferences are reflected in flavor preferences. Many other instances of such flavor and color preference can be cited.

Age.—Age is a factor which must also be considered in the appraisal of certain flavors. Milk is generally preferred by the youth of our country to many other beverages. This is attributable not only to habitual usage but also to effective promotion. Despite a preference for milk in their youth many persons, as they grow older, generally tend to use less milk. There may be a definite age flavor preference exhibited here.

Codes and Taboos.—The codes of those who observe the Hebrew, Mohammedan, Hindu, and other faiths contain certain prohibitions which have a marked influence on the acceptability of certain foods, and, undoubtedly, the flavor perception of such peoples is modified by a conscious, or perhaps a subconscious attempt to adhere closely to the rules of the code.

Physiological Response.—Physiological response is very likely the most important factor in flavor acceptance. There are, it is commonly agreed, four fundamental tastes—namely, salty, sweet, bitter, and sour (acid). Some authorities postulate, in addition, such fundamental tastes as the taste of pungency attributable to components present in pepper, ginger, and the like and the "meaty" taste which can be enhanced by sodium glutamate and analogous materials. Auxiliary tastes like the metallic and the astringent are also postulated.

Flavor is, however, a much more complex sensation because it is influenced not only by the fundamental tastes and the many odors, but also by sensations such as those of touch or feeling, temperature difference as hot or cold, and kinesthetic sensations, that is, the sense of muscular effort as in the case of chewing candy or gum.

Just as there are marked differences in the taste perception of individuals, so there are even more marked differences in flavor perception because it is a more complex sensation.

Appraisal Methods.—It is customary to consider the work of professional tasters, sometimes called graders, testers, or scorers (although in reality many of them are actually appraising flavor value) as objective as opposed to that of the layman or general consumer as subjective, because the former have been trained to overcome or at least to make allowance for the factors mentioned above in their testing. The layman, on the other hand, is guided solely by his likes and dislikes and the flavor of a food makes a marked difference to him on a subjective basis alone.

Some experts such as tea tasters, and coffee cuppers and whiskey tasters have adjusted their testing so that they are able to interpret in large measure what the public desires. But often even experts under specialized conditions can fail in judgment. A well-known trick of this type used in psychology experiments and on experts is to submit a number of samples of exactly the same material, beverage, or product to the expert and ask him to differentiate between the samples. Almost all tasters will indicate differences, some being totally at variance with one another.

Often, however, there is little need for precise differentiation. The difference in flavor value of various specimens of a product may be so great that no difficulty will be encountered in their differentiation.

How can one, then, in view of such psychological and physiological factors, arrive at an objective appraisal of flavor acceptance? There are probably four methods that one can use. These are: first, chemical and physical evaluation; second, individual expert appraisal; third, panel appraisal; and fourth, consumer preference tests.

There are instances where chemical and physical methods of appraisal are applicable. We can cite a few to make this point clear. In testing alcoholic beverages, it has been shown that there is a good degree of correlation between the values of certain "constants" such as volatile acids, esters, etc., and the flavor of the given beverage. Reduction values obtained by the chemical analysis of vinegars is indicative of their source and often of their flavor. As another example, it has been shown that peas can be tested physically for succulence by a puncturing device. Less succulent peas are more difficult to puncture. Chemical methods are available for testing the maturity of oranges. Physical and chemical testing methods are undoubtedly the most objective methods available.

Individual appraisal is at times adequate. Some instances have been mentioned above. The use of such experts for the organoleptic testing of such products as nuts, dates, figs, olives, butter, tea, coffee, and many other products has been well standardized so that objectivity is obtained. A high degree of correlation with consumer acceptability has been achieved.

The use of panels for estimating the flavor acceptance of products is gaining in favor. Generally from six to ten persons comprise a panel, and their scores on flavor testing provide the means of obtaining averages and deviations from such averages. By this means a certain amount of objectivity is obtained.

Flavor Measurement [2]

The concept of flavor is, as mentioned, the resultant of subjective responses to a number of different sensations, namely, taste, odor, appearance, temperature difference, and cutaneous and kinesthetic sensations. Its measurement is, then, dependent upon an evaluation of such subjective responses.

Despite these subjective overtones, in many instances an objective evaluation, hence measurement, can be obtained for flavor estimation by (1) determining chemical and physical factors for which a high degree of correlation has been established with subjective flavor factors; (2) use of trained expert appraisal, (3) use of trained panels, and (4) consumer preference tests.

Organoleptic Analysis.—The measurement of flavor can be considered to be a division of the more general problem of organoleptic analysis, that is, analysis by means of the senses. Such analyses are accepted methods of evaluation of certain foods. Actually in each of the means enumerated above for the measurement of flavor, organoleptic analysis plays an important role.

Customary organoleptic analysis as used in food and beverage analysis is designed primarily to obtain some estimate of the taste, odor, and appearance of the product and to note certain gross features of adulteration such as whether or not the product is rancid or has a foul odor; if it is wormy, or moldy, or if it contains worm excreta or worm tracks; or if it is contaminated with any extraneous

---

[2] Jacobs, *Am. Perfumer Essential Oil Review* **52**, 133 (1948).

matter such as insects, rodent hair, etc. Indeed such objectionable features can be classified under certain headings as wormy, shriveled or empty, moldy, decomposed or rancid, dirty or filthy, and miscellaneous objectionable features. Methods for the detection and estimation of filth are detailed in the succeeding chapter.

Various food products such as cocoa beans, coffee beans, dates, figs, fruits, legumes, nuts, alimentary pastes, candy, olives, and vegetables, can be evaluated by this method. But these factors, as mentioned, are gross adulterations and thus play a gross role in flavor appraisal.

Quality factors are more difficult to evaluate and require finer and more subtle distinctions.

It is often the function of the food analyst to analyze a food product organoleptically, that is, by the use of his senses. The taste, odor, and appearance of a food product are very often indicative of the quality of that product and very often are part of the specifications of that food product. Thus, for instance, the specifications for powdered egg [3] state that the finished product shall comply with the following requirements:

They shall be free from glass, metal, wood splinters, paper, packaging material, or other foreign matter.

They shall be smooth and free from lumps that do not fall apart under light pressure.

They shall reconstitute readily with cold water to produce a smooth mixture.

They shall be free from cheesy, moldy, musty, sour, kerosene, and any other off-odors and flavors, both in the dried form and when reconstituted.

Naturally, discretion must be observed in interpretation. Thus, although a fruit would be considered bad food if it were slimy or moldy or had a putrid odor, certain types of cheese, as is well known, may have all of these characteristics and still be fit food.

The objectionable features of food materials may be classified and described as follows:

1. Wormy.—Food products may be considered wormy if they show evidence of worms, their excreta, their infestation, or their tracks.

2. Shriveled or Empty.—Foods, such as nuts or fruits, may be considered shriveled or empty if more than one-half of the kernel or

---

[3] *U. S. Quartermaster Corps, Specification C.Q.D.* 117B, **May, 1945.**

contents are entirely withered or if the contents are shrunken to more than one-half of the cavity or interior or if more than one-half of the interior is dry.

3. Moldy.—A foodstuff may be considered moldy if it shows definite evidence of hairy masses of mycelium.

4. Decomposed or Rancid.—Food products may be considered decomposed or rancid if they show definite evidence of decomposition and spoilage. This is characterized by the presence of foul, putrid or rank odor, marked discoloration, slimy feel and appearance, decayed texture, soft spots, rot, rancid taste, and similar characteristics.

5. Dirty or Filthy.—Foodstuffs may be considered dirty or filthy if dirt, filth, or extraneous matter, such as insects, parts of insects, hair and excreta of animals and the like are present.

6. Miscellaneous Objectionable Features.—Any other objectionable feature in a food material such as bitter taste in almonds, salty or sour taste in milk, cheesy, yeasty or fruity taste in butter or any other unnatural concomitant taste, odor or appearance of a food may be placed in this classification.

Systematically, this type of analysis may be made by selecting a given number, say 100, of pieces of a properly sampled material, breaking, cutting open, tasting, smelling, observing with the aid of a hand lens, and otherwise examining the pieces and then noting the number in each group listed above. The total number in each classification type may then be expressed in terms of percentage. If this number examined shows the foodstuff to be substandard, a greater number of pieces should be examined to substantiate the first findings.

Canned goods are frequently examined organoleptically. The analyst should note the external appearance of the containers to detect the presence of "swells," "springers" or "leakers."

Swells are canned goods in which decomposition has taken place so that the formation of gas causes the ends to bulge and be convex instead of the normal conditions, namely, slightly concave.

Springers are canned goods that may appear normal in ordinary weather but bulge in hot weather. There is little doubt this condition results from improper canning.

Leakers are canned goods which have openings or orifices from which the contents of the can are able to escape.

After opening the can, the odor, appearance, color, flavor and size of the food material should be noted. The liquid or brine should also be examined, to see if it be clear or turbid. Any variation from the normal, which the analyst knows from experience, will generally be evident. If it is desired, the gases in the can may be collected, before opening the can, with the aid of a Doremus gas collector.[4] Macroscopic and microscopic examination can then be made according to the classification listed above.

The container, itself, should be examined for evidence of corrosion or blackening. The distance of the height of the food material and liquor from the top seam should be noted to detect slack filling.

The examinations described in the preceding part of this section help to determine if a food be adulterated or deleterious. However, organoleptic examination may be made to ascertain if a food be standard or substandard, although still good food.

The Food, Drug, and Cosmetic Act of 1938, as amended, gave control to the Secretary of the Department of Health, Education, and Welfare over the standard and fill of canned goods except of meat and meat products. Such standards and definitions are promulgated in Food and Drug Administration, Service and Regulatory Announcements, Food, Drug, and Cosmetic No. 2, as amended and extended.

In order to ascertain if the canned food is standard or substandard, although still good food, according to these promulgations and definitions, the following factors are examined and evaluated.

1. Color.—The food material is "normally colored" if it has a naturally developed general effect of color. If the color is designated by a system, it must conform to that system. (See Munsell Color System, Chapter II.)

2. Size.—(a) The food material is "normal sized" if the units are of the designated size or over.

(b) The food material is "uniform sized" if the weight of the piece of largest size in the can be not more than twice the weight of the smallest piece in the can.

3. Tender.—The food material is "tender" when it complies with the definitions referred to above. An apparatus for determining the tenderness of canned fruits and vegetables by crushing and penetration tests is described by Bonney and Lepper.[5] Other tests are detailed in Chapters XII and XIII.

4. Peeled.—The food material is "peeled" if there is present per pound of net contents not more than a prescribed area of peel. In the case of peaches and pears, 1 sq. in. and in the case of tomatoes, sq. in. are the prescribed amounts.

5. Unblemished.—The food material is "unblemished" if 80 per

---

[4] Leach, *Food Inspection and Analysis*. Wiley, New York, 1920.
[5] Bonney and Lepper, *U. S. Dept. Agr., Circ.* 164 (1931).

cent or more of the pieces in the container are free from unsightly scabs, bruises, frostbites, sunburn, hail injury, cracks, checks, raggedness (frayed condition of the edges), unnatural colorations, or other unsightly blemishes.

The standards deal mainly with fruits and vegetables, consequently the aforementioned descriptions deal with canned fruits and vegetables. A number of the tests designed to measure these factors are detailed in Chapters XII and XIII concerning fruits and vegetables.

Physical Methods.—Physical and chemical methods of flavor appraisal are the most objective means available for evaluating flavor. These have been developed in some measure for judging canned goods as standard or substandard. It has been mentioned that appearance often influences flavor evaluation and appearance is one of the factors readily measured in canned goods by physical methods. Such items are *color* and *unblemished appearance*. Often the former is designated by a system, say the Munsell color system, then the color of the product must conform to that system.

Blemished fruits and vegetables have an unattractive appearance which often adversely influences their flavor appeal. Such products are considered unblemished if 80 per cent or more of the pieces in a container are free from unsightly scabs, bruises, cracks, checks, unnatural colorations or other blemishes. If the canned goods are labeled as *peeled,* then the amount of peel per pound of product must not exceed a given value as 1 sq. in. in the case of pears and peaches and 3 sq. in. in the case of tomatoes.

One factor that is capable of physical measurement and that is more closely allied to flavor is tenderness or succulence. The tenderness of fruits and vegetables can be estimated by various crushing and penetration tests.

Still another factor which can be determined by physical methods is that of size. Food is considered *uniform sized* if the weight of the piece of the largest size in a can is not more than twice the weight of the smallest piece.

Methods for the estimation of these factors are given in detail in Chapters XII and XIII on fruits and fruit products and on vegetables and vegetable products.

Chemical Methods.—Chemical methods of flavor evaluation also are objective. Maturity in oranges can be evaluated by determinations of the sugar and acids concentration and their ratio to each other. It has been shown that esters are not produced by ordinary quick-aging processes, hence determination of esters in whiskey presents a reliable index to the age of a whiskey. A wine containing over 0.15 per cent of volatile acid is considered unsound and such high volatile acidity easily shows its effect on wine flavor. High acid concentration and high chloride concentration in milk are closely associated with undesirable flavors. Several types of apparatus have been designed for driving or distilling off volatile, unpleasant odors and measuring their amount by oxidation-reduction methods.

Chemical methods for the evaluation of flavor have been detailed in various chapters throughout the text in connection with the particular food in question. Thus the estimation of rancidity in oils has been detailed in Chapter IX. Other methods are detailed in Chapters XII and XIII on fruits and fruit products and vegetable products. Methods for the appraisal of eggs are detailed in Chapters XVI and XX, for tea in Chapter XIV, and for mustard in Chapter XIV.

Expert Tasters.—It is well known that in some divisions of the food industry expert tasters are employed to evaluate raw materials and finished products. Professional·tasters, who are also known as testers, graders, scorers, and by other titles, have trained their taste so that they become skilled in detecting slight variations in odor and taste. They can often associate such variations with particular raw materials or with definite modifications in a given process. This training gives the results of individual expert tasters a certain amount of objectivity and it can be shown that there is a high degree of correlation between their results and consumer acceptance. Nevertheless the personal or subjective element cannot be entirely eliminated by such methods. This will be clear from a consideration of the flavor testing of butter.

The full score for butter is composed of flavor 45, body 25, color 15, salt, 10, package 5, totaling 100. There are, however, over 30 different flavors which have been identified for listing in making such standard scores. Weighted scores have been given to certain flavors, for example, 93—fine or slightly cooked; 92—pleasing, or definitely cooked; 87—slightly onion or garlic; 86—definitely onion or garlic; 85 pronouncedly onion or garlic. It can be seen then that individual reaction must play an important role in butter flavor appraisal when performed by an individual expert. It should be noted that in 1943 letters were substituted for numerical designations in butter scoring.

Among the food products for which U. S. grades or identity standards have been established and for which expert graders are used are meats, dressed poultry, eggs, butter, cheese, fresh fruits and vegetables, canned fruits and vegetables, flour, tea, coffee, and spices.

Since the subjective element persists in individual expert appraisal of flavors, more objectivity is obtained by the employment of expert panels. These panels generally consist of six to twelve trained persons, although at times only four and often more than twelve are used. Most often the results of their work are reported individually but at other times discussion is permitted before a conclusion is made. Results obtained by use of panels permit one to make average and weighted conclusions. Thus if three persons on a panel of six agree on one sample as the freshest of a group of samples, it is fair to accept that sample as the freshest by weighted order.

One aspect of panel testing, although it is to be admitted that it is seldom assigned that term, should be mentioned. Many small firms do not have or cannot afford the expense of a panel of experts to taste their product or their raw materials. Such firms often use their employees as the taste panel. For instance, several samples prepared either by members of the staff or submitted to the firm will be given to employees in the plant and their reaction will be noted. Generally they are merely asked which sample they prefer although sometimes more specific reactions are requested.

While this method may be scientifically unsound, since the tasters have not been trained, it may have a high degree of correlation with the results obtained by consumer panels. It is to be noted that when plant employees are used as tasters, the firm's own product often is deemed to have the better flavor. This may indeed be true, but the result may be influenced by the fact that the plant employees are familiar with and may like the plant product as a result of habitual usage.

Consumer Testing.—There is little doubt that the most certain method of ascertaining consumer acceptance is by making preference tests with consumers. This is not always feasible; hence the methods mentioned above have a definite place in flavor measurement.

In the last analysis probably the surest test of consumer acceptance is to make a test with consumers. It is beyond the scope of this text to discuss such methods. However, it is essential in making such consumer surveys to bear in mind certain precautions. These have been classified as, first, an accurate description of the consumer group from whom it is desired to obtain an opinion; second, a specific purpose in the survey; third, a proper sample of the population to be surveyed; and fourth, an adequate technique for obtaining the information desired.

SELECTED REFERENCES

Cartwright and Nanz, "Comparative Evaluation of Spices," *Food Technol.* 2, 330 (1948).

Cartwright and Nanz, *Food Industries* 20, 1608, 1710 (1948).

Crocker, *Flavor.* McGraw-Hill, New York, 1945.

Metzner, *Wallerstein Laboratories Communications,* 6, 5 (1943).

Paull and Campbell, *Taste Testing.* Anheuser-Busch, St. Louis, 1948.

Scientific Analysis of Flavor and Odor. Evans Research and Development Corp., New York, 1953.

CHAPTER XX

FILTH AND DECOMPOSITION IN FOODS

OVER the past two decades there has been a marked increase in interest shown in the detection of spoilage and filth in foods. Undoubtedly the lead in this work has been taken by the staff of the Food and Drug Administration and a considerable portion of this chapter is abstracted from circulars [1] published by this Agency and papers published by members of its staff. It is beyond the scope of this book to discuss in detail the topics of microscopy and histology of foods and food products. The discussion in this chapter will be limited to filth recovery methods, chemical methods for the detection and estimation of soluble filth and decomposition products, and miscellaneous methods.

EXAMINATION OF WHOLE FOODS FOR GROSS INSECT AND RODENT CONTAMINATION

Insect Chewing, Tunneling, Webbing.—Filth extraction procedures represent the best available means for separating filth from a food so that it may be examined and counted. If insect, rodent, or mold contamination can be judged adequately by a visual examination of the untreated product such as a consumer would purchase, this means should be employed.

In certain samples insects may be swarming over the product and a brief examination will suffice, but usually the damage will be less obvious and it is necessary to examine each piece with considerable care, although still on a semimacro basis, to determine the presence or absence of insects or insect-damaged pieces.

Insect tunneling and chewing leave characteristic clues; it is often accompanied by an accumulation of chewed bits of the food, insect fragments, and insect excreta commonly called "frass." The presence of frass is one character that may be used to distinguish cracks and blemishes from infested areas. Walnuts may be infested while the nut is immature and still growing; consequently the worm-cut areas turn dark and show a pronounced discoloration. However, when the insect-eaten areas are open, accessible pits, the frass usu-

---

[1] *Microanalysis of Food and Drug Products*, Food, Drug Admin., *Food Drug Circ.* 1 (1948).

ally is removed in cleaning. This is particularly true in such cases as cut green beans where the somewhat rounded edge-darkened "craters" remain, but usually the frass has been washed out. Orchard-damaged walnut meats and storage insect infested almonds show the darkening characteristic of damage to the growing plant and the undercut and rounded edges characteristic of all insect feeding. This feeding is in marked contrast to irregular chipped or mechanically damaged areas.

Tunnels through the material may follow either an irregular or straight course. The tunnels themselves are approximately circular and may have enlargements and dead-end branches where the insect paused to feed. Some insects literally honeycomb the food with their galleries. Others eat into an area and stay in one place eating out one large chamber. Where the feeding is along tunnels in the food, the detection based upon frass, or the presence of the adult or larva insect itself, is rather simple, but surface feeding, in which the frass is less firmly retained, may be less readily separated from mechanical chipping or cutting. Insect-eaten areas usually are rounded crater-like cuts where the insect stood in approximately one place and fanned out in its eating in various directions. At times insect feeding is selective, so that the germ of seeds may be eaten before other areas are attacked. On green beans the insects will cut deep pits but the asparagus beetle eats in a shallow expansive pattern, not going much below the epidermal layers.

The pattern of webbing will depend upon the insect and the food. It may trail endlessly all through the food, matting it together in a loose conglomerate; or it may appear, for instance, on individual nuts as a fine cottony covering; or it may be in the form of silken tubes. In granular or powdery material, such as flour or meal, there will be an accumulation of particles on the web so that in comminuted cereals it is not the web which is readily visible but the material held to it.

The sample should be studied with great care so that the differences between mechanical injuries and insect-eaten areas can be distinguished. Indentations along the edges of nuts should be viewed with caution since insects are most likely to work down in protected areas and a nut may readily be chipped along an edge as in the case of shelled peanuts and almonds. At times a rounded cut, in which dark particles of dirt have become lodged, will appear to be an insect cut filled with frass. More than casual scrutiny is necessary to resolve the difference between bits of dirt and excreta.

A more common source of error is to mistake tannin cells for insect excreta. Tanning occurs as inclusions in the cells of many plants and in certain tissues it may fill entire cells so that it appears as small, brownish-red granules that somewhat resemble pellets. It is this tanning which imparts the bitter astringent flavor to the septa of pecan nuts. Tannin cells are translucent, glistening, homogeneous bodies with smooth outline. Some are round and others more ovate but none are heterogeneous and laminated, nor do they have the irregular outlines of insect excreta. In contrast to frass, tannin cells turn deep blue when treated with ferric chloride.

Certain plant diseases, for example, anthracnose on green beans, form deeply pitted areas which are similar to pits that have been eaten out by insects. These may be differentiated by the presence of mold in the depressions, the dark discoloration, and the general shape of the damage.

Whole Larvae and Adults.—Where the whole insects are present the examination should be carried out so that their presence is noted.

Rodent Contamination—*Rodent ex-creta pellets* may readily be distinguished from the nondescript filth that may be found either around a food-packing establishment or in the food itself. However, care should be taken that *cockroach excreta*, which is somewhat the same size as small mouse pellets, is not mistaken for that of mice and rats. Rodent excreta may be identi-

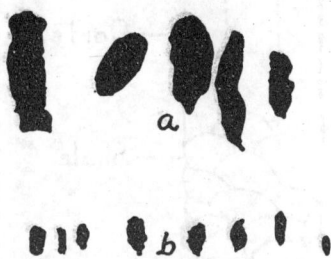

Fig. 79. Shadow Profiles. A, Rat Excreta; B, Mouse Excreta [1]

fied by its size and shape (see Fig. 79), by the fact that it has a mucous-like coating, and by embedded rodent hairs. In infested plants rodent excreta is usually found along the walls, behind sacks, or in any storage space that is not regularly cleaned. Since rodents need both food and water, their pellets are usually found adjacent to such sources. Suspected rodent excreta pellets always should be checked for the presence of rodent hairs.

In addition to feces, urine may be voided by rats, mice, and cats in food-processing establishments. The urine stains appear as somewhat darkened soiled areas on or about the food, or on the sacks, boxes, or other containers. Urine fluoresces under ultraviolet light. When there is any doubt as to the nature of suspected stains, portions of the stained material can be removed to the laboratory for a confirmatory examination for the presence of urea. *Bird excreta* also may occur where food is being stored or processed. However it is readily distinguished by its white, chalky matrix.

In certain more or less hard substances such as cheese, dried fruits, nuts, grains, and spices the *rodent teeth marks* sometimes may be detected. These teeth marks are usually short parallel scratches or indentations spaced by high sharp ridges and are made by the rodent gnawing with the two sharp incisors in the front of the mouth.

Animal hairs may be divided into four general classes: (1) Guard hairs, or over-hairs, which are coarse, heavy hairs that in the living animal act as protection for the softer hairs; (2) soft, under, or fur, hairs which are soft fine hairs used to keep the animal warm; (3) intermediate or curly over hairs which have the characteristics of both guard and fur hairs alternating along their length, and (4) tactile hairs such as the vibrissae of animals. Tactile hairs are rarely found in food products.

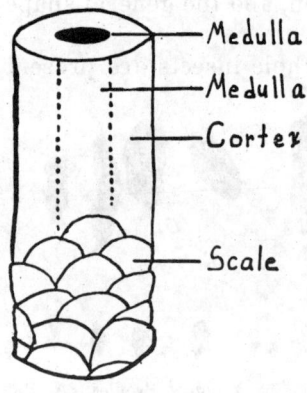

All animal hairs, whether they are fur or guard hairs, exhibit the following struc-tures (Fig. 80). (1) They are covered with a cuticle composed of overlapping keratinous scales of variable shapes. The shape of the scales sometimes is of value in identifying some types of hairs, but in cat and rodent hairs this feature is of limited impor-tance; (2) inside the cuticle the hair contains a cortex. This area may or may

not contain some pigment granules and in some hairs it may appear to be evenly stained with color because of the presence of very fine pigmentation. The size and density of the pigment granules, as well as the size of the cortex and relative sizes of the cortex to the underlying medulla, sometimes are of diagnostic value; (3) inside the cortex the hair contains a medullary shaft. It is composed of various loosely arranged cells, interspersed with air spaces. In some hairs the pigmented areas and the air spaces give a characteristic microscopic pattern to the medulla, and in other hairs the loosely arranged cells are not noticeable and the cortex appears to be a tube with a hollow medulla.

The color effect of hair is based on two different factors, air spaces and pigment granules. The air spaces, when examined microscopically with transmitted light, appear to be black because the light is reflected within them and does not pass on through. If the air spaces are filled with balsam or some other substance, the hair appears almost completely clear; if some true pigment is present, the pattern of distribution of the pigment may be distinctly seen. When pigment is present in the cortex it is present as true pigment granules, but in the medulla it may be present as true pigment or confused with air spaces. In the examination of a hair or hair fragment to determine whether it is cat or rodent, the first character to note is the medulla. If the medulla is multiple-rowed, discontinuous, it is the guard hair of a rodent; if it is smooth or erose (continuous), it is the guard hair of a cat. It is only when the hair has a single-rowed, discontinuous medulla that difficulty in differentiation is encountered. This type of medulla occurs in the fur hairs of both rodents and cats. Fortunately, along the length of the hairs of these animals there are a number of constrictions or internodes.

The internodes of our common house mouse and brown rat usually are very prominent, while those of the cat are much less prominent. Unfortunately, the internodes of some rats are not overly pronounced. For example, approximately 10 per cent of the last internodes toward the proximal end of the hairs of brown rats are as indistinct as the more distinct ones of the cat. The internodes of the black rat similarly are about as indistinct as those of the cat and, therefore, are frequently difficult to differentiate. However, the black rat is common only in a few coastal localities in this country, and this factor is of relatively small importance. If a prominent internode is present, the hair is that of a rodent, but if the hair fragment is short and the internode is either indistinct or lacking, further study may be necessary.

In general *cat hairs are larger* in diameter than either rat or mouse hairs, but when individual measurements are made it is found that considerable overlapping occurs.

The *cortex of cat hairs is thicker generally* than that of rodent hairs, and the relative size of the cortex to the medulla often shows the cortex of cat hairs to be relatively thick whereas that of rodents often is relatively thin. Again actual measurements have shown that some overlapping of the character occurs, and it should be used only in conjunction with other features.

Cortex pigment is of more common occurrence in the hairs of cats than it is in those of rodents. Usually, when large numbers of cortex pigment granules are present, the hair is that of a cat rather than of a rodent. When pigment is present in the cortex of a rodent hair there will invariably be some in the medulla and, *when pigment is present in the cortex only, the hair is definitely that of a cat.* Most cat fur hairs have more densely concentrated pigment granules than most rat or mouse hairs. At 200-400 × this pigment is visible as a blotched darkening of the cortex. At 800-1000 × (oil immersion objective) the pigmentation resolves itself into distinct and separate

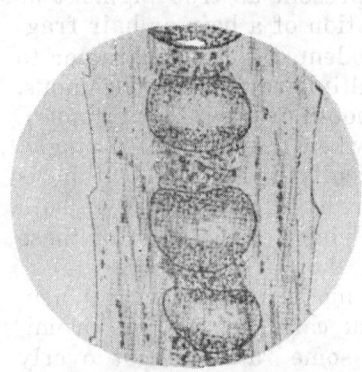

granules. Hairs swollen in sodium hydroxide solution or chlorine (hypochlorite) solution and then examined using the oil immersion objective show the granules even more plainly. In cat hairs (Fig. 81) they are arranged in short broken parallel rows, whereas in rat and mouse hairs the rows are usually widely separate and less obviously parallel. Granules are ovate-elongate to round in both types of hair, although there may be more elongate granules in cat hairs.

Fig. 81. Cat hair showing elongate rows of pigment granules in the cortex, as seen in optical section. Granules below and above focus appear as rows of fuzzy pigment. Medulla consists of air-filled chambers and pigment granules. (Drawn as examined at X 1000)[2]

The flattened cuticular scal s of cat hairs are much smaller and finer than those of rats and mice. In mice the cuticular scales of guard hairs are ovate to crenate and the typically small flattened scales are not found. In rats they are ovate elongate to flattened, the ovate elongate scales being the most pointed of any of these three.

When hairs are treated with 10 per cent sodium hydroxide solution on a microscope slide, the hairs swell, and within about half an hour additional diagnostic characters are visible at 400-500 ×. After several hours the hairs more or less disintegrate and are no longer usable for study. When the hairs are swollen by sodium hydroxide the cuticular scales are enlarged and stand out against the swollen cortex.. Within the cortex the medullary shaft appears as individual segments (also termed cells or chambers) variously shaped and arranged.

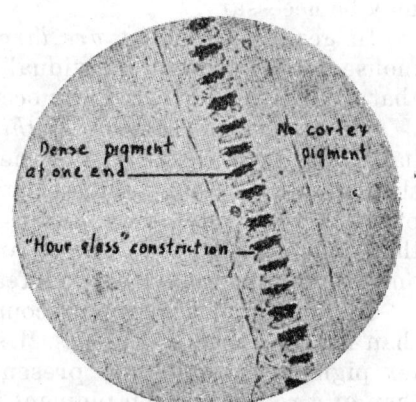

Fig. 82. Rodent hair swollen in NaOH, showing medulla, pigmentation, constricted medulla shape, and clear cortex[2]

---

[2] *Microanalysis of Food and Drug Products*, Food, Drug Admin., *Food Drug Circ.* **1** (1948).

The following characters apply to fur hairs treated with sodium hydroxide solution:

Rodent Fur Hairs.—The medulla segments of rodent fur hairs, if pigmented, are compactly pigmented, usually one end of the segment is clear with the compacted pigment at the other although not infrequently both ends of the segment are clear with the compacted pigment in the middle. (See Fig. 82.)

Individually, the pigment granules are black or extremely dark brown, although occasionally light brown granules occur. Segments often are as long (in the direction of hair elongation) or longer than they are wide and have an hour-glass-like constriction in the middle. The segments are relatively far apart when compared to the segments in cat hairs. The cortex as a rule is not heavily pigmented, but the pigment granules are rather large and appear compact. The cortex seldom or never appears stained with color except at the extreme tip of the hair. The cortex of rodent hairs has a peculiar clearness which cat hairs rarely if ever show.

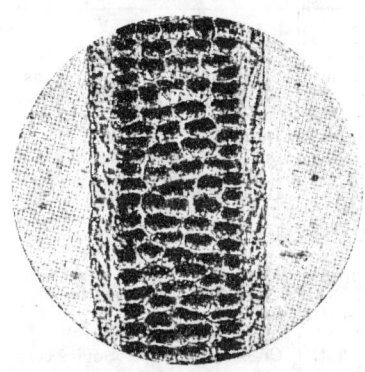

Fig. 83. Rodent guard hair treated with NaOH and showing the swollen medulla segments [2]

Cat Fur Hairs.—In cat fur hairs the segments of the medulla, if colored, are loosely pigmented so that the granules appear scattered.

The pigment granules seldom appear black or dark brown even in black cats. Their usual color is brown or reddish brown. The segments of the medulla, except in Angora cats, whether pigmented or not, usually are wider than they are long. These segments are convex on one end and concave on the other. When examined under the microscope they appear as though, if pushed together, adjacent segments would fit similarly to a ball-and-socket joint. They do not touch one another, although they are relatively close together and a straight line drawn across the tips of the concave end of one of the segments will in many cases touch the convex end of the adjacent one. Pigment granules often are rather abundant in the cortex of cat hairs. Sometimes the cortex appears stained with color because of dispersed very fine pigment granules. In any case, it seldom or never has the clearness of the rodent cortex and this is due in part to faint parallel longitudinal lines that occur in the cortex of the cat hair. *Rodent hairs seldom if ever contain these longitudinal lines.*

Guard Hairs (cat or rodent).—In addition to the compound medulla of the untreated rodent guard hair and the continuous medulla of the cat guard hair some additional observations on the swollen hairs after treatment with sodium hydroxide solution also appear diagnostic. The medulla of the rodent guard hair breaks up into two or more rows of irregularly shaped blocks that appear to have been fitted closely together until forced apart by the chemical action. (See Fig. 83.) These blocks may or may not contain pigment. The medulla of the cat guard hair, on the other hand, breaks up into thin

TABLE 101. CHARACTERISTICS OF RODENT AND CAT HAIRS [a]

Character	Rodent	Cat
Medulla fur hair	Single-rowed, discontinuous	Single-rowed, discontinuous
Medulla guard hair	Multiple-rowed, discontinuous	Smooth or erose, continuous
Internodes	Prominent or a few indistinct	Indistinct
Diameter	Usually smaller	Usually larger
Cortex size in relation to medulla	Usually narrow	Usually wide
Cortex pigment	Usually sparsely scattered or absent	Usually abundant
	None when none in medulla	May be present when medulla is unpigmented
	Pigment granules in scattered rows	Granules often in dense parallel rows
Cuticular scales of guard hairs	Ovate, elongate to flat. Some *very* ovate elongate present	Crenate to flat. Some *very* flat present
In NaOH: Medulla pigment	If present, usually compact, black or dark brown; often localized at one end of segment with other end clear, or at both ends with center clear	If present, composed of loosely scattered granules. Granules brown or reddish; seldom black or dark brown
Shape of medulla segment	Usually long in direction of hair elongation	Usually wider than long
	Usually have hour-glass like constriction	Sometimes have hour-glass like construction
Distance apart of adjacent medulla segments	Usually relatively far	Usually relatively close
Contact of adjacent segments	Often squared across	Usually ball-and-socket effect.
Cortex appearance	Usually has a clear gelatin-like appearance	Usually not clear, and with faint parallel, longitudinal lines.
Cortex——in hypochlorite, NaOH, or untreated	Usually not heavily pigmented	Usually abundantly pigmented, with dense pattern of pigment granules arranged in broken parallel rows

[a] *Microanalysis of Food and Drug Products, Food Drug Admin., Food Drug Circ.* 1 (1948).

plates which appear to have been pressed one upon another along the length of the medulla. The individual plates often do not cover the full circumference of the medulla but, when they do, they appear as a great number of wafer-like discs not unlike the chambered pith of certain plants.

FILTH RECOVERY METHODS

The types of filth most commonly met in food products are usually insoluble pieces of insects, rodent filth, or molds. Certain soluble components of urine may be detected by microchemical methods, and a relatively insoluble substance such as uric acid from bird or insect excreta may be precipitated and its identity confirmed, but the majority of substances that are encountered are pieces of plant or animal material. Occasionally glass and sand or quartz are found and these usually require special examination.

Insect fragments, rodent hairs, and molds can stand rather vigorous treatment and still retain their microscopic characteristics although hairs will dissolve in strong alkali and be lost. The most satisfactory way to recover filth from food is to retain it on a filter paper after dissolving the food and filtering it. Such a process is possible in only a limited number of food products. Sugars, some hard candies, and a few other materials such as dextrose-malt preparations fall into this group. By far the great majority of separations are made by some procedure involving a differential of wetting, specific gravity, size, solubility, and/or appearance of the filth and the food involved. Often a combination of steps is necessary before a satisfactory separation is obtained. As is the case with chemical analyses, each precipitation or extraction is incomplete to some small extent, and the fewer the manipulations involved, other factors being equal, the better the recovery.

Heavy material such as sand can be retained as a sediment by using heavier-than-water organic solvents with such material as peanut butter and ground spices. Canned leafy vegetables can be floated in strong salt solutions while adhering soil settles out. In either case it is necessary to work the soil free from the plant material, for when the bulk of the separation rises to the top there is a tendency for small particles to become entrapped and consequently not settle out.

Sedimentation in Heavier-than-Water Liquids.—This procedure can be used to separate rodent excreta pellet fragments from cereals. In a liquid with a specific gravity near 1.49 the pellet fragments tend to settle out while much of the cereal floats. As the specific gravity is raised, more cereal is floated but some pellet fragments also will rise and be lost. It is necessary to strike a practical balance between the need for floating the plant tissue with the possible loss of excreta. Factors other than density play a part in this separation. The particles must be soaked in the liquid long enough to become fully permeated. The density balance is delicate and the containers should be covered and handled in such a manner as to avoid strong convection currents. The separations cannot always be completed in one operation. It is easier with cereals simply to carry out the separation in a beaker and gradually remove the plant tissue with successive decantations rather than repeatedly transferring from a separatory funnel to a beaker and back to a funnel.

Sieving.—In certain instances, a size separation offers a rapid means of separating filth from food. Adult insects, larvae, insect eggs, and excreta, as well as rodent excreta, bits of metal, etc., may be sifted from foods, or a food-filth separation may be accomplished by washing the food (or filth as the case may be) through a screen leaving the filth (or food) on the screen. A screening operation can be used for a preliminary separation before other operations are carried out. For example, in a candy mixture containing chocolate and nuts, the nuts can be removed immediately; and, once they are out of the way, further operations are simplified. Sifting often is useful for a rough qualitative separation to give some indication of the advisability of performing a more time-consuming quantitative segregation.

Wildman Trap Flask Methods.—Since for most foods neither a plain water solution nor sifting will give adequate separation, other procedures must be used. When gasoline is mixed with an aqueous mixture containing insects or insect fragments, the insects float up with the gasoline layer. This principle has been utilized repeatedly for filth extractions.

Specific Gravity and Oil Wetting.—Part of this effect is brought about when the hydrophobic and oleophilic insect cuticle is wet by oil, which is lighter than the liquid in which the food is soaking, and, as the oil droplets rise, the insects are carried up. Some of the oil wets the smooth insect cuticle, some clings to external processes, and water is repelled. Most plant tissues will settle out. This explanation is only part of the picture, for in some instances the separation appears to be due mainly to the low specific gravity of the insects. This is particularly noticeable when the insects or fragments are dried out and contain air. Both the relative densities of the liquid, plant material, and insects, and the increased differential caused by the oil increment are utilized in the flour method where the insects are floated from the white wheat flour in a saturated-salt medium aided by gasoline.

Alcohol Penetration of Bran.—Bran reacts somewhat similarly to the insect tissues and floats into the upper oil layer. This difficulty can be overcome, in part, by substituting a water-alcohol solution for the plain water. The alcohol reduces the specific gravity of the solution and it also wets the bran better than does plain water so that it soaks into the bran, driving out the air. There is some indication that oils in alcohol-water solutions wet the insect fragments better than oil in water, although any increase in the wetting of the plant material by the oil is offset by the other factors.

Detergents.—Better wetting might be obtained through the use of detergents, although so far in practice it has not been found that they can be used to obtain a sharper plant separation or a more complete recovery.

Boiling.—Troublesome deposits of starch or bran in the oily layer can be prevented by boiling before the oil is added, unless there is so much starch that boiling forms a considerable gel. Boiling drives out air and saturates the bran or chaff tissues with water so that there is less tendency for them to float in the oil layer. When an abundance of starch is giving trouble, boiling usually is to be avoided since it gelatinizes the starch. This gel then forms an emulsion when the oil or gasoline is stirred into the mixture and it is impossible to obtain a clean separation unless, as will be discussed later, additional digestion processes are used. Starch in a mixture which is being extracted in a Wildman trap flask is best handled in cold water, although usually when starch is present, bran is also, and alcohol solutions are used.

Emulsions.—The aim of oil flotations of filth is to float the light filth fragments and still allow the plant material to settle out. In practice this separation is incomplete because of the factors noted above and several additional factors. It is difficult to wet all of the insect material with oil without creating a frothy emulsion. Although emulsions formerly were thought to be troublesome only because they included plant material which obscured the filth in subsequent microscopic examinations, further experience has shown that, in addition to this, fairly stable emulsions can hold filth down in the trap flask and out of the neck of the flask and, once the emulsions form, it is difficult to break them and thus set the insect fragments free. At times emulsions may be broken with a few drops of ethyl or caprylic alcohol. The oil or gasoline should be worked into the food mixture with as little inclusion of air as possible, and intermittent agitation should be provided while the separation is taking place. Persistent emulsions sometimes are caused by the release of dissolved air from the water, and, because of this, de-aerated water is recommended for extractions in a Wildman trap flask.

Oil Types.—Both the formation of an emulsion and the wetting of the filth and/or plant material by the oil are dependent upon surface phenomena which will vary with any of the components of the mixtures or solutions. Light oils, such as gasoline, form more troublesome emulsions as a rule than do light and medium mineral oils, although if an emulsion once forms with the heavier oils it is very difficult to break. The oils most commonly used are gasoline, kerosene, mineral oil, and castor oil. Castor oil is heavier than the other oils, yet it will carry insect fragments up with it as it floats, but because of its density it must be used in water. Also, it is very viscous, and its use so far has been confined to hot mixtures. For some reason, rodent hair recoveries are poor when castor oil is used.

pH.—Some recent work has confirmed the surmise that the acidity of the solution in which a gasoline flotation is being carried out has a significant effect on the insect fragment recovery. At pH of 7 and above the recovery of insect fragments appears to be lower than that from an acid medium.

Cleaning the Flask Sides and Dilution.—In addition to the oil-water interface, the side of the flask presents a very troublesome surface. Oil droplets, some of them holding filth, cling to the sides, and they must be cleaned off repeatedly with the rubber stopper plunger. When oily liquids are used in trapping filth they should be added only to a wet flask since, if added to a dry flask, they will wet the sides and filth may be held on the sides and not float into the upper oil layer. Because the filth must rise up through a mass of descending material the descending material must be dilute, so that in both Wildman trap extraction and in heavy filth sedimentations, crowding too large a sample into a small container is inadvisable, for it will cut down on the filth recovery. It is necessary to stir the bottom (or top) layer to release entrapped filth fragments during extractions.

MAGGOT RECOVERY

Fly eggs and maggots (and at least some nematodes) respond differently to oil-water notations than do other insects. They settle out while other insects float. Two theories have been advanced to explain this; both may play a part. While it is known that certain maggot parts, such as the spiracles, are hydrophobic and oleophilic it may be that the remainder of the maggot body surface does not attract oils. The maggots encountered in foods generally live in a watery medium and perhaps they are too full of water and thus are too heavy to be floated. Because of these peculiarities they are separated by a process opposite to that used for other insects and insect fragments. They are permitted to settle out while the plant material is floated. Consequently, instead of gently stirring gasoline into the mixture, as is carried out in a Wildman trap, the mixing is carried out vigorously in a separatory funnel so that the plant tissue is caught in the rising gasoline and air and carried up and away from the maggots and eggs which settle. Since even small fly eggs and maggots can be retained on a 10 xx bolting cloth, a mixture containing such filth can be filtered on the cloth rather than on a filter paper. By its use much of the extraneous matter is lost through the cloth and the microscopic examination is simplified.

Maggots may be recovered by other sedimentation procedures which are somewhat similar to the "panning" or washing for gold. Pulped fruits, jams, etc., sometimes are handled in this way, one method being to dilute the food with water and stir it into a wide pan. Then after the maggots have settled out, the top water mixture is poured off so that the maggots are retained in the lower corner of the pan. A method for the recovery of larvae from blueberries is detailed in this chapter.

REMOVAL OF INTERFERING COMPONENTS BEFORE EXTRACTION
OF FILTH

The above-discussed sedimentation, sifting, and flotation of filth constitute virtually ideal conditions from the analyst's point of view. In practice many foods present difficulties which make it necessary to examine each by an individual procedure. Some mention has been made of the trouble bran and emulsions cause. At times the troublesome elements become so troublesome that they must be removed before a filth extraction can be attempted.

Starch and Protein Digestions.—Starch can be hydrolyzed by different methods. Boiling, boiling in dilute acid, or enzymatic digestion all have been used. Proteins can be broken down by comparable methods and they, along with starch, can be hydrolyzed with pancreatin. Enzymatic digestions have been most helpful with bakery products where the enzymes are used to release filth from the food and to produce a liquid mixture which can be extracted or filtered. The digestions are accomplished readily if the right conditions are maintained. Pancreatin has been the most useful enzyme. Its pH and temperature requirements are rather easily fulfilled while it acts on carbohydrates, fats, and proteins. It sometimes is necessary to boil or otherwise soften the food to prepare it for digestion, but at other times a dilute pancreatin solution can simply be soaked into the food and digestion will start immediately.

Fat and Oil Removal.—Fats and oils are troublesome under some conditions and of minor importance in others. Chloroform or carbon tetrachloride used for the sedimentation of heavy filth will remove enough of the oil from corn meal, and some of the spices, so that any subsequent flotation for light filth can be carried out without further attention to the oils. In other instances petroleum ether should be used to remove fats and oils. Bakery products with large amounts of shortening can be digested with pancreatin more readily if most of the shortening is first removed, by soaking in an organic solvent and decanting off the oil-bearing solvent, to permit more complete penetration by water and the enzymes. This principle of oil removal is one of the simplest in the filth methods. At times, sodium hydroxide solution can be added to emulsify the oils, and while they are not removed, the soapy emulsion formed can be handled more readily than could the original droplets of fat.

Treatments to Avoid with Hairs and Insect Fragments.—Fortunately insect fragments, hairs, and mold will withstand rather vigorous treatment and yet retain their identity. However, hairs are susceptible to the action of alkaline solutions; and strong alkalies, such as sodium hydroxide or potassium hydroxide when used hot even in 1 per cent solutions, will dissolve rodent hair. Hence, when rodent hairs are to be recovered the use of these and other relatively strong alkaline reacting substances, such as trisodium phosphate, sodium carbonate, and ammonium hydroxide should be avoided.

Hairs are much more resistant to the action of acids, although the use of hot sulfuric acid and nitric acid at even 5 per cent should be avoided. Rodent hairs will hold up for 15-40 minutes in boiling 5 per cent hydrochloric acid, the degree of attack on the hairs depending upon the protection given them by the particular food. Phosphoric acid and the other relatively weak acids are much safer. After treatment with acids the hairs may be soft or brittle and should not be subjected to vigorous mixing action. On a filter paper they cannot be teased or turned readily without breaking. A further difficulty introduced by severe acid treatment is that the characteristic morphological pattern may be altered somewhat; and, in identifying hairs by the sodium hydroxide swelling technique, the treatment they have been subjected to must be taken into account since a characteristic pattern may not be obtained. It is believed that rodent hairs processed in acid foods, such as tomatoes, apples, and grapefruit, give a slightly modified pattern when swollen in 10 per cent sodium hydroxide solution as compared with the unprocessed hairs.

Insect fragments will withstand any chemical action to which it has been found necessary to subject the food. Alkali will clear insect fragments and remove much of the pigment, especially when used in strong concentrations; but, if the treatment is taken into account, even fragments boiled in saturated aqueous potassium hydroxide solution can be identified as insect parts almost as readily as could the original untreated pieces. However, it must be remembered that strong sodium hydroxide or potassium hydroxide solution softens insects and, after boiling in aqueous 10 per cent sodium hydroxide solution for 5-10 minutes, they are more subject to mechanical breakage. For all practical purposes in filth work, insect fragments and insects may be considered to be unaffected by acids, even though they can be carbonized by extended boiling in strong sulfuric acid.

Violent mechanical stirring or grinding of foods should not be used for filth extractions. Tests have demonstrated that rodent hairs agitated in a liquid by a malted milk stirrer sometimes survive unbroken and sometimes are broken. Inasmuch as there is no way of determining what the effect will be in the particular sample being examined, stirrers should not be used for any quantitative analysis. Insects are broken by such stirring. When the insects are dry and large the damage is greater than when they are pliable and small. Dry grinding comminutes the filth even more than does wet agitation.

Solution and Filtration.—Sometimes it is possible to complete a separation by simply dissolving the food and leaving the filth in a beaker or on a filter paper without further treatment. Chewing gum can be handled in this manner. Boiling in dilute acid solution will dissolve the water-soluble portions and also hydrolyze a starchy coating or dissolve any carbonates that may be present in the coating. The acid treatment also serves to soften and disperse the gum so that it is accessible to other reagents. However, the boiling must be stopped before carmelization of the carbohydrates has gone too far. Acetone can be added to the water to facilitate the solution of the chicle component by an organic solvent such as turpentine or chloroform. By attacking each component separately it is possible to filter the gum through a No. 100 or 150 sieve or bolting cloth.

Washing the Filth from the Food.—Washing is advantageous with such products as nut meats and dried fruits where either plain water or water and detergents can be used to remove and concentrate a surface contamination for microscopic observation.

Centrifuging and Other Treatment.—On occasion centrifuging may solve an otherwise difficult separation problem. In general, procedures involving few transfers of material and few manipulations will give better recoveries than more complicated methods. Each manipulation introduces a point at which material may be lost. This does not mean that a separation in a 100-ml. beaker can be more efficiently carried out than in a 400-ml. beaker, for at times it is necessary to dilute the menstruum in order to loosen adhering filth. Whenever a method is under question it should be checked by repeated runs with known amounts of material added.

Oil Clearing and Sedimentation for Excreta [3]

There are two general methods by which insect excreta in flour may be counted. One utilizes sedimentation of the excreta in a heavier-than-water liquid, which floats the flour, and the other consists in treating the flour so that the excreta are made visible in the matrix.

Several difficulties may be encountered in flotation procedures because different particles of flour vary in their specific gravity, depending upon the part of the grain from which they were taken. For example, the bran is heavier than the starch, and even similar parts vary in weight according to the presence or absence of entrapped air or other factors. Insect excreta pellets similarly show a wide density range depending upon the type of food eaten, physiological condition of the insect, and the species or stage of the insect. A liquid with specific gravity 1.52 will support most flour particles, but in it many insect excreta pellets may rise into the flour layer. At a specific gravity of 1.40, the insect pellets settle out adequately, but much floury material settles with the excreta. Because of these difficulties it is impractical to depend solely on specific gravity for separation.

Insect excreta pellets behave differently when moistened with an oil than do wheat flour granules. The latter tend to become clear and transparent when immersed in an oil medium, whereas the excreta pellets tend to retain their white, opaque appearance. The refractive index is important in rendering flour particles transparent; with higher refractive indices the flour is less apparent than it is with lower indices.

Factors other than refractive index must be considered in making the flour transparent and the pellets visible by contrast. The oil must readily penetrate the flour and drive out most of the air, and it must not be too volatile or the prepared amounts will be unstable while an examination is being made. In actual practice clove oil has

---

[3] *J. Assoc. Official Agr. Chem.* **26**, 257 (1943).

been found to satisfy the requirements of such an examination, although with any of the oils certain phosphates and salts found in self-rising flour remain opaque in oil and may bear a superficial resemblance to fragments of excreta pellets while some fragments of cereal fail to clear completely and so resemble excreta. Mineral material will appear crystalline unless its crystals have been destroyed by grinding, but in this case the appearance will be dissimilar to insect excreta in that the excreta will be smooth and rounded while the minerals will be angular and have flat or irregular cleavage sides. Internally the excreta show a heterogeneous laminated appearance while the salts are homogeneous.

In examining a product for whole rat or mouse excreta a simple visual examination may be used, or sifting employed where necessary. Where an oil flotation of a product is carried out in a Wildman trap flask the rodent excreta will be found settled to the bottom with the other heavy matter.

With comminuted products, for example flour, meal, and spices, it is advisable to float off most of the food material so that the rodent excreta is concentrated. Such a separation may be obtained by using an organic heavier-than-water solvent such as chloroform (specific gravity 1.498). In such a solvent most of the plant material will float and the rodent excreta pellet fragments will settle out. Where trouble is encountered because too much of the plant material settles out with the rodent excreta, the specific gravity may be raised by the addition of carbon tetrachloride (specific gravity 1.595) to the chloroform, but the specific gravity should at no time go above 1.546 (chloroform to carbon tetrachloride 1:1) because at this point the rodent excreta fragments may rise and be lost in the floating layer. Such separations may be carried out in a separatory funnel or simply in a beaker, depending upon the nature of the particular substance involved. In tea, for example, where all of the leafy material readily floats and the particles are not finely ground, the separation can be carried out in a crude drug percolator, but when repeated decantations are necessary and the product is not so finely ground that a separatory funnel can be used, sedimentation can be carried out in a beaker and the heavy filth gradually concentrated at the bottom of the beaker.

GENERAL METHOD

In general, methods for the recovery of filth in foods depend upon the digestion of the material being examined, the separation of the extraneous matter, and the identification of the latter.[4] The

---

[4] Wildman, *Science* **75**, 268 (1932).

methods of digestion range from simple solution of the product, say candy in water, or the digestion of cheese in sodium citrate solution, to the more complex methods of acid, alkali, and enzyme hydrolysis of proteinaceous material. The principles behind these separations have been detailed in the preceding sections.

Apparatus.—The apparatus consists of a 2-liter Erlenmeyer flask fitted with a rod and rubber stopper so that the lower part of the neck of the flask can be sealed as shown in the Fig. 84. To prepare the rod and stopper, bore a hole in a suitable rubber stopper that will permit the rod to fit in snugly. Then countersink a slightly larger hole on the bottom side of the stopper. Insert the rod, prefer-

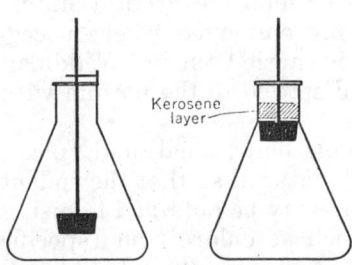

Kerosene layer

FIG. 84. Apparatus for Separation of Filth in Food

ably of copper tubing having a threaded end, with the threaded end through the stopper and then fix it in the stopper by means of a washer and nut. Jam the washer in the countersunk hole so that no metal protrudes from the bottom end, which precaution will minimize breakage. If possible reseal the bottom with rubber. Then jam or screw the rubber stopper into the flask.

An alternative device that can be used for this purpose is the type of milk bottle, Fig. 85, designed to show the cream line and to make it easier for the consumer to pour off the cream, if desired.

Procedure.—Weigh 50 g. of the sample into a beaker. Add sufficient saturated borax solution to cover the material. Transfer to the 2-liter flask fitted with the rod and stopper described above with the aid of more saturated borax solution. By means of a pinch clamp, hold the rod and stopper off the bottom of the flask and boil the contents of the flask for about 15 minutes.

Cool, add 35 ml. of kerosene, and shake well. Add sufficient saturated borax solution so that the kerosene layer will rise to a point just above the rubber seal. Allow the contents of the flask to settle for about half an hour. Agitate gently with the rubber stopper, if necessary to get all of the kerosene to the top. Hold the rubber stopper in the sealing position by means of the rod, and pour off the kerosene layer and the accompanying filth and small quantity of borax solution into a beaker. Wash the neck of the flask thoroughly with hot water and catch the washings in the same beaker. Discard the contents of the flask, wash with water, and discard the washings.

Return the kerosene layer and the hot water washings to the flask and add about 500 ml. of 40 per cent alcohol. Shake well. Make up to the mark in the neck of the flask, previously noted, with 40 per cent alcohol solution. Allow to settle for about 15 minutes. Seal again with the rubber stopper and pour the kerosene layer on a Büchner funnel (which will hold a 7-cm. filter paper), washing the neck and funnel with 95 per cel alcohol.

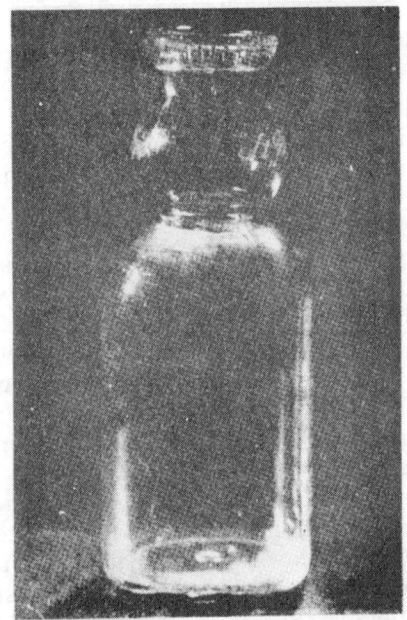

Examine the paper and residue, when dry, under a low power microscope fitted with a scanning device consisting of wires set 5 mm. apart in a circle the size of the filter paper. If desired, pick suspicious material out with a dissecting needle, place on a slide, fix with filtered glycerol, and then examine microscopically.

Omit alcohol wash for hard candies. Use 100 g. of material for coconuts and peanuts. Use saturated salt solution instead of borax for flour and cereals, and omit the alcohol treatment for these materials.

Processed Milk

These methods include dried milks, evaporated milk, and con-

FIG. 85. Apparatus for Separation of Filth in Food—Milk Bottle Type.

densed milk. All types may be contaminated with filth from the barn, utensils, and while en route to the processing plants. The filth encountered may include insects, rodent hairs, and nondescript dirt and manure particles.

Sediment Pads.—To 300 ml. of cold water in a Waring Blendor gradually add 100 g. of the milk powder. Avoid prolonged mixing because of excessive foam formation. The total mixing time need not exceed 15-20 seconds. Transfer the mixture to a liter beaker, rinsing the blendor with about 150 ml. of water. Stir into the milk mixtur 100 ml. of filtered 40 per cent sodium citrate solution. Heat to 60° while stirring and maintain at this temperature with frequent stirring for 10-15 minutes or until the milk powder is dissolved and the solution appears translucent. Filter the hot solution on a milk-sediment pad with a cloth covering. Any foam on the solution will pass through the pad without difficulty. Use a little hot water from a wash bottle to assist in transferring the foam and in overcoming any tendency of the filtration to slow down.

Butter

Mold.—A high mold count in butter indicates the use of decomposed cream in its preparation.

Reagent.—Gum Solution.—To prepare solution, make a thin paste-like mixture of the required amount of the dry powder in alcohol, add cold water, mix, and heat gradually. Allow to boil until gum is dissolved. Adjust volume, if necessary, after cooling. Preserve with 2 per cent formaldehyde solution U.S.P. In the case of carob bean gum or other gum allow cellular elements in the mixture to settle out and use clear supernatant fluid.

Procedure.—Make a careful examination of the surface of the sample and note any visible mold growth. To remove the possibility of contamination by surface mold, scrape off and discard ⅛ in. of surface. Weigh out 1 g. of butter obtained from exposed surface in a tared ¼-teaspoon measure and place spoon in a 50-ml. beaker. Add 7 g. of hot (50-60° C.) gum solution prepared by making either a 0.75 per cent solution of carob bean gum or a 3 per cent solution of pectin or other gum solution of similar viscosity. Stir until mixture is uniform and fat globules are 0.1-0.2 mm. in diameter. Mount a portion of the mixture on the mold-counting slide and estimate mold as described under tomato products below. Consider fields positive when a single filament or combined length of the two longest filaments exceeds ⅙ of diameter of field.

Alternate Procedure (Staining).—Add 1 or 2 drops of 5 per cent crystal violet solution to the gum-butter mixture after butter is melted. Mix preparation thoroughly and prepare slide as directed above.

Microscopic Examination and Extraneous Matter.—Microscopic examination of butter will show the presence of crystals of other fats if these have been added to the butter. Renovated butter will also show crystals and variegated colors with a selenite plate. If the butter has been made from milk or cream taken from unclean farms or dairies, a microscopic examination will show the presence of dirt, rat hairs, and other contaminants.

Procedure.—Weigh into a beaker 100 g. of butter and add 150 to 200 ml. of borax solution, 40 g. of borax in 1 liter of water. Heat to boiling and filter through a Büchner funnel containing a 7-cm. rapid filter paper [5] supported by a 50-mesh copper screen 6 cm. in diameter. Wash well with gasoline or petroleum ether to remove fat. Wash with hot water. Examine the filter paper under a wide field microscope, Fig. 87, for foreign material. The debris may be transferred to a microscope slide having a drop of glycerol and examined under a compound microscope for identification. If the butter is contaminated with much mold, the filter paper will clog very rapidly. In this case, stop the filtration, wash the paper with the aforementioned solvents. Remove the paper, replace with another sheet, and continue the filtration. Examine each sheet as detailed.

Butter churned from decomposed cream generally retains some of the decomposition products. Among these are increased acidity, higher aldehydes that respond to the Kreis test, indole, and mold. Clarke [6] and co-workers discuss this subject fully and give procedures for the detection of these decomposition products.

Cheese Products

Cheese may be contaminated with insect and/or rodent filth during the handling of the milk, during the manufacture of the cheese, or during curing. Where it is essential that an accurate picture of the amounts of such filth be obtained a procedure given under "Quantitative Procedures" should be used. It will be noted that a number of reagents are listed. In general the choice of reagent will depend upon the type of cheese and the past experience of the analyst. Where a number of samples are to be run, preliminary tests by one of the methods listed under "Qualitative procedures" may be found to be advantageous and more useful as a sorting procedure. Where the main purpose is to obtain sediment comparison one of the qualitative procedures may be used. In all cases use a sample of 225 g.

Qualitative Procedures.—Pancreatin Reagent Procedure.—Mix 225 g. of cheese with 300 ml. of water in a Waring blendor and adjust the pH to 8. Add 300 ml. of pancreatin solution which has been centrifuged and filtered through cotton and paper. Keep the mixture at 35 to 45° C. for about 5 minutes, readjust the pH to 8, and then continue the digestion for an additional hour. Acidify to pH 2 with phosphoric acid and boil for 15 minutes. Add 500 ml. of 95 per cent alcohol, heat to boiling again, filter through a sediment pad, and examine.

---

[5] Greene, *Food Industries* 7, 442 (1935).
[6] Clarke, *J. Assoc. Official Agr. Chem.* 20, 475 (1937).

Hydrochloric Acid-Sodium Hydroxide Procedure.—This variation is used for cheese containing gums. Place 225 g. of cubed cheese and 300 ml. of water at 60° C. into a blendor and mix for 15 seconds. Transfer to a 1500-ml. beaker and boil with stirring. Add 50 ml. of hydrochloric acid (1:1) slowly, stirring vigorously and boil for 5 minutes. Cool to about 60° C., add 1 ml. of phenolphthalein indicator solution, and neutralize with 20 per cent sodium hydroxide solution. Adjust the pH to 11.5 but not higher with this alkali, heat to 60° C., hold at this temperature for 10 minutes, dilute with an equal volume of water at the same temperature, filter through a pad, and examine.

Quantitative Method.—*Preparation of Sample.*—Cut cheese into cubes 25 mm. across or smaller and add to the reagent selected in an appropriate-size beaker. Provide means of heating or maintaining proper temperature and means of stirring the mixture such as a Waring blendor or malted milk mixer.

Sodium Citrate Variation.—Use 400 ml. of 15 per cent sodium citrate in 1,500-ml. beaker and heat to 65° C. Add cheese and stir. Maintain temperature close to 60° C. but not above 62° C. After stirring 5 to 15 minutes, add 200 ml. H_2O and 200 ml. sodium citrate solution both heated to 60° C. Continue stirring 5 to 15 minutes longer and filter through paper, keeping a stream of hot water on the paper to facilitate filtration.

Extraneous Matter.—Cheese that is made from dirty milk or cream or that is made under unsanitary conditions is likely to contain extraneous matter. This foreign matter is usually present in very small particles. In process cheese where a grinding process is part of the method of manufacture, particles of foreign matter are especially likely to be small. The examination of cheese is more difficult than that of butter because of

1. The use of rennet for precipitating the casein,
2. The variable methods used for manufacturing cheese, especially process cheese in which organic substances, inorganic substances, and gums are used as emulsifying agents, and
3. The presence of a large number of bacteria.

Procedure.[7]—Weigh 50 g. of cheese, excluding the rind, and place in a mortar. Add 100 ml. of a freshly filtered solution of 950 ml. of 95 per cent alcohol plus 50 ml. of hydrochloric acid, sp. gr. 1.18, slowly while stirring the cheese to a smooth paste. Pour the paste into a beaker rinsing the mortar with another 100-ml. portion of the solution. Heat slowly and allow to boil gently for 3 minutes. Filter at once through a 7-cm. rapid filter paper placed in a Büchner fun-

---

[7] Greene, *Food Industries* 7, 442 (1935).

nel. The filter paper should be supported by a piece of 50-mesh copper screen 6 cm. in diameter. Wash the filter immediately with ½ liter of previously filtered boiling water until the paper shows no cheese material left. Examine the filter paper with a wide field binocular, Fig. 86, to detect the extraneous material. For exact

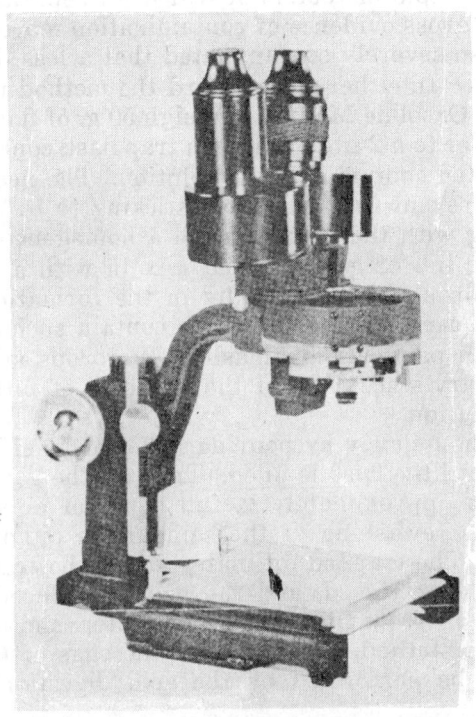

FIG. 86. Wide Field Microscope
(Courtesy of Spencer Lens)

identification, examine the debris by transferring it to a drop of glycerol on a microscope slide and observe with a compound microscope.

Fragments of insects, principally flies or other insects and larvae, rodent and other hairs, paper, wood, and nondescript debris may be found and identified by this method.

Flour

The basic method commonly has been known as the "saturated-salt gasoline-flotation" method and was designed to recover insects and insect fragments from white wheat flour. In this writing there is included an optional procedure for the sedimentation of heavy filth such as rodent pellets and stones. The two procedures may be run in conjunction with each other or each may be carried out separately. A rapid method is provided for separating out whole insects and other gross evidence of contamination more rapidly when the samples are so severely contaminated that a less detailed examination will suffice. In either case, record the method used.

Saturated-salt Gasoline Method.—Weigh 50 g. of flour in a 250-ml. beaker and transfer to a 2-liter Wildman trap flask containing 200 ml. of filtered saturated sodium chloride solution. The neck of the flask must be dry to prevent the flour from sticking to it. Break up the lumps by stirring with the stopper until a homogeneous mixture is obtained. Add 35 ml. of gasoline and mix in with a gentle rotary motion. Too vigorous stirring results in the formation of a dense emulsion which causes the final paper to contain such an amount of flour that the filth particles are mashed. Vigorous stirring at this point is unnecessary since most of the elements of filth will float in saturated salt solution.

Add salt solution slowly by pouring it down the side of the flask and stir gently until the flask is filled. Trap off the gasoline layer to another trap flask approximately ⅓ full of water or saturated salt solution. For this rewashing of the sample use either a 1-liter or 2-liter trap flask. The trapped off material may be transferred to a beaker, the trap flask rinsed, and the material placed back in the same flask. Trap off from the second flask, filter, and examine.

Acid Digestion Method.—Detection of the eggs of *tribolium* species in flour can be performed by the acid digestion method [8] as follows:

(a) Reduction of the bulk of the sample by bolting through a cloth fine enough to retain the eggs. The eggs plus the coarser flour particles which are also retained are designated as "overs."

(b) Digestion of the overs with 5 per cent sulfuric acid to disintegrate the flour clumps and wash the eggs free of adherent flour.

(c) Filtration of the digestion mixture through a Büchner funnel, and staining of the eggs on the filter paper with 1 per cent iodine solution, the excess iodine being removed by washing with 1 per cent sulfuric acid or distilled water.

---

[8] Blumberg and Ballard, *Food Industries* 12, No. 1, 36 (1940)

(d) Inspection of the filter paper for eggs, which appear as yellow-brown or golden-brown ovate structures upon a blue filter paper.
Procedure.—Place the combined corings on a previously weighed 12-cm. watch glass, then weigh and transfer to a 6xx silk bolting cloth. Arrange the cloth on a 7-in. hoop so that there are no crevices or folds in which the eggs can be retained. This set-up bolts 60 to 90 g. of flour. Less than 0.1 g. of sample generally adheres to the watch glass during the transfer.

Continue sifting until no more material from the sample passes through the cloth. Transfer the overs to a 250-ml. beaker and return the throughs to the sample jar.

Wet the overs with 1 to 3 ml. of 95 per cent ethyl alcohol. Then rotate the beaker gently to disintegrate the clumps. Add 30 ml. of 5 per cent sulfuric acid and digest the mixture for 10 minutes on a boiling water bath. Filter the mixture through a 9-cm. Büchner funnel containing two sheets of No. 589 Schleicher and Schull filter paper, or No. 613 Eaton and Dikeman (medium filtering speed) paper. Keep the beaker partially inverted over the funnel and wash thoroughly, using the spray from a wash bottle; pass the washings through the filter. At this point diminish the suction to decrease the speed of filtration and pour 30 ml. of 1 per cent aqueous iodine (1 per cent iodine, 2 per cent potassium iodide, water) over the paper. If this does not maintain contact with the residue on the filter paper for 15 to 30 seconds, add another 25 to 30 ml. When the iodine filters through, wash the paper with 30 to 50 ml. of 1 per cent sulfuric acid or distilled water. Remove the paper from the funnel and examine with a 3x lens for eggs. If higher magnifications are required, observe the paper under a wide-field binocular. Count the number of eggs and, from the weight of sample taken, calculate the number of eggs per kilogram of the flour.

Spices

Rot in Powdered Capsicum (Based on Mold Count).—Occasionally an excessive number of moldy pods are used in preparing ground capsicum products such as chili powder, cayenne pepper, red pepper, paprika, etc. This adulteration may be determined by the following procedure.

Pectin Reagent. –Weigh 4 g. of powdered pectin and add sufficient alcohol to make a thin paste. Add, with stirring, 100 ml. of cold water and heat gradually. Boil until the paste is dissolved and adjust the volume to 100 ml. after cooling. Add 2 ml. of formaldehyde solution U.S.P.

Procedure.—Weigh out 10 g. of the thoroughly mixed sample of ground capsicum and transfer to a Waring blendor. Add 200 ml. of a 1 per cent sodium hydroxide solution in three or four successive portions, stirring the mixture upon each addition, and finally wash down with the final portion any material that may stick to the walls of the blendor. Agitate the mixture in the blendor for 1 minute. Rub down into the mixture any material sticking to the walls of blendor with a rubber policeman and repeat the blending for 2 minutes longer. Add 2 or 3 drops of caprylic alcohol to break the resulting foam. Mix 100 g. of this mixture with 50 g. of a 4 per cent pectin solution and count with the Howard mold counting chamber as for tomato products.

Occasionally a blended mixture will contain particles of seed tissue which make it difficult to obtain "Newton rings" in preparing the slide for mold counting. A clamp for holding the cover slip in place has been devised which removes this difficulty. This consists of a metal plate with a circular opening, 2.5 cm. in diameter, in the center of the plate. Two clips attached to the anterior edge of the plate fasten the cover slip in position when the slide is placed on the plate.

Blackberries and Raspberries

Rot.—Canned and frozen blackberries may contain some rotten fruit which may be detected by a careful macroscopic examination of the berries under water after they have been washed. Frequently canned blackberries will disintegrate to some extent, and the detection of rot in this material presents a special problem.

Procedure.—Drain the contents of a No. 2 can or its equivalent on a No. 20 sieve (5 in. diameter). Immerse the berries in water in a large white pan. Pour off most of the water through a No. 20 sieve and add more water. Repeat the washing if the water is not fairly clear. Examine the berries under water and remove all questionable berries to a black pan containing distilled or deaerated water. Reexamine the suspected berries and note particularly the outline of the berries as they are turned over under a strong beam of light. Class, as rotten, berries and fragments which have at least 3 drupelets containing either external (aerial) or internal mold, or both. Separate into two classes—those with external mold, and those moldy but without external mold. Confirm all questionable rot spots by examining a fragment of the tissue for mold under the compound microscope. Classify the tissue as rotten only when a substantial number of mold filaments is present.

Drain separately the good and the two separations of rotten berries and fragments for 2 minutes on a No. 20 sieve (8 in. diameter). Weigh each rotten portion separately and add these to the good berries.

If the sample contains a large amount of disintegrated berry material, pick out the whole berries and large fragments. Remove the disintegrated material to a No. 20 sieve and allow it to drain for 2 minutes; then weigh. Take an aliquot of approximately one-fourth and separate the rotten from the good material. Calculate the total weight of rot in all the disintegrated material and add this to the weight of whole berries showing internal mold. Mix all the portions together, drain, and determine the total drained weight of the sample. Calculate and record the percentages of external and internal mold. Pulp the berries through a cyclone with openings ca. 0.023 in. in diameter, or through a No. 30 sieve using a stiff brush. Mix the pulp thoroughly, weigh out 50 g., and dilute with an equal weight of pectin solution. Make a mold count on this mixture.

Strawberries

Rot.—Strawberries are most commonly received for examination in the frozen condition. Some of the berries may contain rotten areas. These may be detected by a careful microscopic examination of the berries under water after they have been washed with water. The berries should be thawed either at room temperature or in an electric refrigerator at about 40° F. Do not immerse the sample in hot water, since this makes the berries soft and mushy.

Procedure.—Drain the entire sample, if 1 qt. or less, on a No. 20 sieve (5 in. diameter). Immerse the berries in water in a large white pan. Pour off most of the water through a No. 20 sieve, catching and returning any strawberry tissue, and add more water. Repeat the washing if the water is not fairly clear. Examine the berries under water and remove all questionable berries to another pan containing distilled or deaerated water. Re-examine the suspected berries. Class a berry or fragment as rotten if it has a rot area at least 6 mm. in diameter. Confirm all questionable rot spots by examining a fragment of the berry tissue for mold under the compound microscope. Classify the tissue as rotten only when a substantial number of mold filaments is present. Also separate and count berries and/or fragments with rot areas 12 mm. or more in diameter. Count all rotten berry fragments in the above classes and enter the count in the table. Drain separately the good and rotten separations for 2 minutes on a No. 20 sieve (8 in. in diameter). Weigh the rotten separation, add this to the good, and determine the total drained weight of the sample. Pulp the berries through a cyclone with openings about 0.023 in. in diameter, or through a No. 30 sieve using a stiff brush. Mix the pulp thoroughly, weigh out 50 g. and dilute with an equal weight of pectin solution. Make a mold count on this mixture.

Blueberries

Larvae.—The following method [9] is a rapid simplified method for the estimation of larvae in blueberries. It is advantageous because it reduces the amount and size of glassware needed, it is less tedious and saves a great deal of time and effort. It depends upon the decolorization of the color of the berries thus permitting rapid work. *Procedure.*—Transfer a No. 2 can of berries to a 1-liter beaker cover with a watch glass and boil until the berries have checked and disintegrated. Transfer to a sieve and mash. Transfer the sifted liquid with its sediment to a 2-liter Erlenmeyer flask. Wash the skins in the sieve twice with water, add the washings to the Erlenmeyer flask. Permit the material in the Erlenmeyer flask to stand for about 5 minutes to allow it to settle and decant the supernatant liquid. Decolorize the residual liquid with a minimum of sodium hydrosulfite (sodium hyposulfite $Na_2S_2O_4$) or similar reducing agent. Dilute to 2 liters, permit to settle, and again decant. Count larvae as usual on a black background.

Very likely this procedure can be adopted to other berries.

Fig Paste

Insect Fragments and Other Light Filth.—Place 100 g. of the sample in a beaker with about 300 ml. of water. Boil and stir until the paste is thoroughly softened and mixed with water. Transfer to a 2-liter Wildman trap flask, add 20 ml. of castor oil, and stir thoroughly. Add sufficient warm tap water (50° C.) to fill the flask. Allow to stand for at least half an hour with occasional stirring. Trap off the oil layer. Add a little water to the flask, stir, and trap off again after 10 minutes. Filter the trapped off portions through a rapid paper and examine with a Greenough microscope at 20 to 30 diameters.

Cranberry Sauce

The mold mycelia count method as used for tomato products is also applicable to cranberry products for the purpose of detecting

---

[9] Goldstone and Jacobs, unpublished work 1938.

the use of rotten fruit. The method as used by many laboratories has consisted of making a mold count directly on a portion of the gelled sauce. However, the direct method of counting is not entirely suitable, since the microscopic field is often too dense to permit clear vision and prompt detection of the mold mycelia in the cellular mass. This condition has been observed in making counts at the factory of both the pulp and the finished sauce. The use of a pectin solution for dilution purposes, such as is used in making mold counts of many other fruit products has been shown by Eisenberg and Tillson [10] to be preferable.

Manufacturing plants using this method for control purposes can take samples of the strained pulp or the finished strained sauce from the production line and make their dilutions directly, thus foregoing the need for breaking the gel as directed in the method for strained sauce. It has been found, however, that control is best exercised on the sorting belt in order to eliminate rotten fruit.

Procedure for Strained Sauce.—Immerse the unopened can of sauce in a boiling water bath for 30-45 minutes, in order to facilitate breaking the gel. Remove can from bath and open carefully to avoid loss of sauce through sudden release of pressure. Empty contents of can into a suitable sized beaker (1-liter beaker for No. 2 can). Stir the sauce in order to break the gel. A slow-speed electric mixer (350-450 r.p.m.) may be used for this purpose.

Mix thoroughly 50 g. of the stirred sauce with 50 g. of a 3 per cent pectin solution prepared as directed on page 819. Make a mold count of this mixture using official mold count method as directed in the section on tomato products.

Procedure for Whole Sauce (Seeds and Skins Included).—Pulp contents of container (if considerably greater than 1 lb., such as No. 10 can, remove well-mixed aliquot of 1 lb.) through cyclone with screen openings about 0.027 in. in diameter. This will remove skins and seeds and prepare a homogeneous pulp for mold counting. Mix 50 g. of pulp with 50 g. of 3 per cent pectin solution. Make mold count of this mixture as directed in the section on tomato products.

Tomato Products

Rot Fragments.—*Apparatus.*—Sieve Holder.—Make a suitable holder for sieve by cutting a circular opening in a rectangular piece of tin slightly larger in diameter than the outer diameter of the lower part of the sieve. Bend corners of tin down to fit a 1000-ml. beaker and trim to convenient size.

---

[10] Eisenberg and Tillson, *J. Assoc. Official Agr. Chem.* **31**, 783 (1948).

Counting Plates and Covers.—Plates are of glass such as is used for photographic plates. Dimensions.—55 mm. x 100 mm. Rulings.— Crosswise, parallel lines 3.5 mm. apart, with one 15 mm. space at each end.

One half of a square cover slip about 22 mm. on a side and about 0.25 mm. thick is fastened at each end of the counting plate by balsam. These act to separate counting plate and cover plate. A cover for each plate (50 mm. x 85 mm.) is prepared from the same type of glass as counting plate.

Reagent.—Algin Solution.—To prepare algin solution mix a thin paste of the required amount of dry algin powder and alcohol, and add the required amount of cold water, heating gently until dissolved. When cool add 2 ml. of commercial formaldehyde per 100 ml. of solution. Adjust solution to pH 7.5-8.0 with sodium hydroxide solution.

Procedure.—Weigh 10 g. of juice (5 g. of puree or catsup or 2 g. of paste) into a 50-ml. tared beaker and rinse with 100 ml. of tap water into a 400-ml. beaker. Add about 2 ml. of a saturated aqueous gentian violet solution, stir, and allow to stain for 3 minutes. Add 200 ml. of water, stir, and pour through a No. 60 sieve about 7.5 cm. in diameter. Keep sieve in horizontal position set in the circular opening of holder over a 1-liter beaker. In pouring, distribute material over entire surface of sieve to insure rapid drainage, using a glass rod held against the lip of the pouring beaker with the lower end of the rod about 2 cm. from the screen. If sample weight given above does not drain rapidly, reduce the size of sample and take this reduction into consideration in computing number of fragments. Rinse the beaker with two 100-ml. portions of tap water pouring each portion over the strained tomato debris on the sieve using the glass rod as before. Tilt the sieve in holder to about a 30° angle and wash debris to the lower part with water. This usually requires about 100 ml. Allow the debris to drain and transfer as much as possible to the bottom of a 50-ml. centrifuge tube with a thick, metal, square-ended spatula about 6 mm. wide. Wash the remaining debris again to the lowermost part of screen and add to the centrifuge tube as before. This procedure may be repeated until most of the debris has been transferred to the tube. Transfer the final remaining debris by means of a pipette, about 20 cm. in length, 6 mm. inside bore, fitted with a rubber bulb. To do this wash the debris down with water from pipette and immediately take up the debris in wash water before it has run through screen. Usually 300 ml. of water will be ample for the complete transference of the tomato debris from the sieve to the centrifuge tube. When completely transferred, the vol-

ume of water and debris should be made up with water to 10 ml.
Add sufficient neutralized 3 per cent algin solution to make volume
up to 20 ml. The clear gum solution prepared from carob bean gum
may also be used for this purpose. Mix the stained suspension with
a spatula, measure out two separate 0.5-ml. amounts to each of two
counting plates, and cover with special cover slip. (The pipette is
prepared by cutting off a 1-ml. pipette squarely at the 1-ml. mark.)
In pipetting, draw the material slightly above the 0.5-ml. mark and
then allow it to drop slowly back to the proper volume. Wipe off the
pipette and then allow the material to flow slowly onto the slide
spreading uniformly in the center of the slide covering an area
roughly 6 cm. x 2 cm. Touch the lower end of pipette several times
to slide to insure removal of material. Blow out the last drop if
necessary. Examine each slide with a Greenough-type microscope
using a magnification of about 40-45 diameters with transmitted light.
The microscope should be fitted with a 4x bi-objective and 10x oculars
or a similar combination. For the light source use a substage
box-type light, with a 15-watt bulb, and blue or daylight ground-glass
filter. Place the light so that center of glass filter is directly below
objective and about 2 cm. below glass stage of microscope. Count
and record the number of rot fragments on each of the two slides,
add, and multiply the sum by 2 (10-g. sample) to obtain number of
rot fragments per gram of product. Where 5 g. of material is used
multiply total of the two plates by 4 and by 10 where a 2-g. sample is
used. If it is found necessary to use a different size sample, calculate
the appropriate factor.

Count as rot all fragments showing abundance of mold with tissue
either apparent or present in slight amounts.

Many of the fragments observed will be heavily stained masses of
mold filaments which are obviously rot. Because of the thickness of
the mycelial mass, tissue cells may or may not be apparent in the
magnification used. Other types of rotten tomato pieces are tomato
tissue elements containing only a loosely woven mat of hyphae. In
most of such cases the tissue elements are quite plain although they
may be somewhat transparent, whereas the mold filaments although
well-stained may be relatively few in number.

Precautions.—(1) Do not count loose filaments that have no, or
only an insignificant amount of, tissue attached. (2) Do not count
as rot, pieces of tissue containing an insignificant amount of mold.

For example in certain types of samples there will be observed
an obvious attachment of a few very short filaments of mold to a
piece of tomato tissue. These should not be counted as rot fragments. The important consideration in making a rot-fragment count

is that emphasis should be placed on those rot fragments which consist of rotten tomato tissue with readily identifiable mold attached.

Corn

The determination of the degree of soundness is of major importance in the grading of corn and other cereal grains. Most methods determine soundness principally by odor and by the percentage of damaged kernels. These have proved to be useful, but because of their shortcomings more accurate methods that will be equally—or even more—practicable from a commercial standpoint are desirable.

As deterioration of the corn kernel is necessarily associated with chemical changes in several of its components, the more important of these chemical changes were studied by Zeleny and Coleman [11] in an effort to discover some easily measurable change to serve as a reasonably accurate index of the degree of deterioration. It was shown that the quantity of free fatty acids, amino acids, and acid phosphates in the kernel tend to increase as deterioration progresses.

Of these three types of acidity only the fat acidity increases significantly with incipient deterioration and is therefore the only type of acidity which appears to be useful in differentiating degrees of soundness in corn.

Procedure.—Grind about 50 g. of a representative portion of each sample to such a degree of fineness that at least 90 per cent of the meal will pass through a 40-mesh gauze sieve. (If the moisture contents of the samples are not known they may be determined with the Tag-Heppenstall moisture meter on the samples before grinding.) Weigh out 20-g. portions of the well-mixed meals to an accuracy of ±0.01 g. and transfer to 100-ml. glass-stoppered flasks or bottles. Add exactly 50 ml. of benzene to each flask, insert the stoppers, shake a few seconds to saturate the air in the flasks with benzene vapor, momentarily loosen the stoppers to release pressure, and replace stoppers. Shake flasks for 30 minutes using the mechanical shaking device. Tilt flasks at such an angle that settling will take place in such a way that decantation will be made easy. Allow flasks to rest in this position at least 3 minutes. Carefully decant as much of the liquid as possible into 15-cm. folded filter papers inserted in 8-cm. glass funnels. Cover the funnels with glass disks or Petri dishes to minimize evaporation. Collect the filtrate in 25-ml. volumetric flasks or accurately calibrated 25-ml. graduated cylinders. When exactly 25 ml. of the filtrates have been collected, transfer the filtrates to

---

[11] Zeleny and Coleman, *U. S. Dept. Agr., Tech. Bull.* **644** (1939).

200-ml. Erlenmeyer flasks. Refill each volumetric flask or cylinder to the 25-ml. mark with 95 per cent ethyl alcohol containing 0.04 per cent of phenolphthalein and transfer to the respective flasks containing the benzene extracts. Titrate the extracts with carbonate-free 0.0178 N potassium hydroxide solution to a distinct pink color. In the case of the yellow extracts from yellow corn the final color should be a little more pink than orange. Run a blank titration on a mixture of 25 ml. of the benzene and 25 ml. of the alcohol. Calculate fat acidity by the formula:

$$\text{Fat acidity} = \frac{100\,(T - B)}{100 - M}$$

where

T = titration value of extract in milliliters
B = titration value of blank in milliliters
M = per cent of moisture in sample.

Fat acidity is expressed in terms of milligrams of potassium hydroxide required to neutralize the free fatty acids from 10 g. of corn (dry-matter basis). This result should be multiplied by 10 in case it is desired to express fat acidity on the basis of 100 g. of corn (dry-matter basis). For routine use it is suggested that the 100-g. basis be adopted.

Dried Eggs

Silver Nitrate Paper Method.—Steffen [12] and co-workers point out that, because of the high percentage of fat contained in dried whole eggs, it may be assumed that the development of undesirable flavors with storage is due to changes in the fat. When negative tests are obtained for rancidity in dried whole eggs which have been held for a long period, attention may be turned to those changes in the proteins which are probably responsible for the unpleasant flavor noted. A modified Eber's test was devised on the premise that a progressive protein denaturation or decomposition would increase either the lability of those sulfhydryl groups resulting from the heat of the drying process or the number of sulfhydryl groups.

In making the test, dilute sulfuric acid is stirred into the dried egg, and air is passed through the mixture at a constant rate. As the air passes from the aeration bottle it comes in contact with a filter paper disk moistened with silver nitrate solution, and a deposit of silver sulfide forms on the paper. The color density of the silver sulfide spot is then compared with a set of standards, the darkness of the spot varying from a faint brownish shading when the test

---

[12] Steffen, Hopkins, Kline, and Whetzell, *U. S. Egg Poultry Mag.* **49**, 308 (1943).

is made with fresh dried whole egg of good quality to a sooty black with inedible dried eggs.

The method consists of bubbling 1½ liters of air at a standardized rate through a cylinder of standard size containing 20 g. of dried whole egg and 60 ml. of 5 per cent sulfuric acid by volume which have been carefully mixed. The hydrogen sulfide evolved is collected on a filter paper disk impregnated with 1 drop of 3 per cent silver nitrate solution. After an aeration period of about 15 minutes the filter paper is thoroughly washed with distilled water to free it of excess silver nitrate, and fixed in "hypo" solution. After the filter paper has thoroughly dried, the density of the spot produced by the egg sample is compared with the standards and scored.

Standards.—Four standards are used and are assigned palatability scores of 3, 4, 5, and 6. If the spot density fell between two of the standards it is scored as a half interval.

Apparatus.—Water is siphoned from a 5-gal. bottle through glass tubing having an internal diameter of 5 mm. A glass stopcock is adjusted so that water is delivered from the siphon at a rate of 115 ml. per minute. A screw clamp serves to stop the flow of water at the end of a determination so that the setting of the stopcock need not be changed. Water is allowed to flow into a 2 or 3-liter Erlenmeyer flask until it reaches a mark placed at the 1.5 liter level. The glass tubing connecting the water receiving flask and the gas washing bottle has the same internal diameter as the siphon (5 mm.). A 5-cm. length of capillary tubing with a bore of 0.5 to 1 mm. is inserted in the air train to give a more even flow of air through the egg-sulfuric acid mixture. The end of the tube through which air enters the gas washing bottle should be placed in the rubber stopper so that it is about ¼ in. from the bottom of a cylinder. This cylinder is a glass test tube 25 cm. long made from 32-mm. glass tubing (29-30 mm. inner diameter). Gas bubbles pass through the exit tube on which the filter disk moistened with silver nitrate solution is set. The wooden disk insulated with a thin layer of cork is held in place with two rubber bands of such elasticity as to hold the wooden disk firmly against the top of the outlet tube; two glass hooks on the exit tube and two projecting wire brads on the wooden disk make it possible for the rubber bands to hold the wooden disk in place. An alternative for the above would be a cork held firmly in place by means of a stiff steel wire.

Procedure.—Weigh a 20-g. dried whole egg sample, to 0.1 g., transfer to the reaction cylinder and add approximately 40 ml. of 5 per cent sulfuric acid. Mix the contents with a glass stirring rod having a loop at right angles to the shaft to a somewhat smooth paste and then add the remainder of the acid (20 ml.) and stir the whole to break up any lumps. The addition of the acid and mixing with the sample should be done in the shortest possible time. After the contents are thoroughly mixed, quickly put the stopper containing the aeration tube and the gas exit tube in place. Put a drop of 3 per cent silver nitrate solution on a filter paper disk 2.5 cm. in diameter, allow to spread evenly, and then carefully place the moistened paper between the top of the gas exit tube and the firmly held insulated wooden disc.

Open the shut-off clamp and allow water to pass into the 2- or 3-liter Erlenmeyer flask at the standardized rate. This incoming water displaces its equivalent volume of air, forcing air through the egg-acid mixture and out the gas exit tube where any hydrogen sulfide reacts with the silver nitrate on the paper disk. Shake the cylinder and its contents about 4 times during the aeration period of about 15 minutes. Remove the silver nitrate paper when 1500 ml. of water have passed into the Erlenmeyer flask.

Give the spot test paper an identification mark and immerse in distilled water for at least 15 minutes in order to remove all the excess silver nitrate. Transfer the paper to a 20 per cent "hypo" solution and allow to fix for 10 minutes, after which wash again in distilled water to remove the "hypo" and then set aside to dry. Compare the silver sulfide spot with a set of standards and score.

DETERMINATION OF VOLATILE FATTY ACIDS IN DRIED EGGS

From the tests and experiments made by the U. S. Food and Drug Administration,[13] it has been found that lactic acid concentration in good eggs never exceeds 50 mg. per 100 g. of dried eggs, and that the acetic acid concentration in good eggs never exceeds 65 mg. per 100 g. of dried eggs. Formic acid is not found at all in good eggs.

Apparatus.—A steam distillation assembly consisting of a boiler flask (3-liter) giving steam at a uniform rate so as to produce a constant rate of distillation, a distillation flask, a condenser, and 50- and 200-ml. volumetric flasks as receivers.

The Pregl tube to be used is a glass tube approximately 75 mm. long, the upper half of which is 7 mm. inside diameter and the lower end 3 mm. diameter. A sintered glass disk about 3 mm. thick is fused into the constriction between the upper and lower halves. A convenient holder for the Pregl tube for filtration is a glass tube large enough to permit seating the Pregl tube at the construction by aid

---

13 *U. S. Food Drug Admin. F.N.J., F.D.&C. 7783,* 1946.

of a rubber sleeve at the upper end of the holder. The holder passes through a rubber stopper into a suction flask. By means of a small rubber stopper the upper end of the Pregl tube is connected with a siphon through which the solution and precipitate are transferred to the filter. Wash water and alcohol from the flask are also passed through the siphon tube.

Standardization of the Distillation Apparatus.—Place the apparatus in the laboratory that is free from drafts and sudden changes in temperature. Make a mark on the 3-liter boiler flask at the 1500-ml. level, fill to this mark with boiled distilled water, heat to boiling, and boil for several minutes before starting a distillation. Transfer about 50 ml. of 0.1 N formic acid to the distillation flask, add one drop of sulfuric acid (1:1) and adjust the volume to 150 ml. with water. Connect the condenser, insert the steam inlet tube into distillation flask, and bring contents of this flask to incipient boiling by means of a burner. Connect the steam inlet tube with the steam supply of boiler and steam distill. The rate of evolution of the steam and height of small flame of burner under the distillation flask should be regulated so that the volume of the liquid in the distillation flask is kept constant at 150 ml. and the distillate is collected at the rate of 20 ml. per hour. The period of collection may vary ±2 min. The 150-ml volume in distilling flask should remain constant within ±5 ml. The boiling may be stopped to permit a test of the constancy of the 150-ml. volume by momentarily interrupting the steam supply. A few trials with water in the distillation flask will show conditions necessary to maintain constant volume in distillation flask and rate of distillation. Collect a 50-ml. portion of distillate followed immediately by a 200-ml. portion. Transfer contents of the 50-ml. volumetric receiver to a 125-ml. Erlenmeyer flask, and those of the 200-ml. receiver to a 300 ml. flask, and titrate each to a phenolphthalein end point with 0.1 N alkali. Make a blank determination using 150 ml. of boiled distilled water and one drop sulfuric acid (1:1) in the distillation flask, collecting a 50-ml. and a 200-ml. portion and titrating. Correct the titrations of each distillate for the respective blanks. Calculate, in terms of percentage, the fractions of the total formic acid originally present, which were carried over into the 50-ml. and 200-ml. distillates. Designate them as TF_1 and TF_2, respectively.

Repeat the standardization using 0.1 N acetic acid and also using 0.1 N butyric acid. Other volatile acids such as propionic and isobutyric can be included in a general standardization of the apparatus, but these acids have not been encountered in either sound or decomposed eggs. Calculate distillation fractions for these acids and designate those of the 50-ml. distillates as TA_1 and TB_1, and of the 200-ml. distillates as TA_2 and TB_2 for acetic and butyric acids, respectively. Distillation fractions for standardization made in triplicate should check within a range of 1 per cent. Calculate the ratio of TA_2/TA_1 and designate as "C."

Preparation of Buffer Solutions and Determination of Distillation Blanks.—Prepare a buffer mixture of pH 8.6 by dissolving 12.404 g. of boric acid and 14.912 g. of potassium chloride in water and diluting to 1 liter. The salts should be as pure as possible. Fifty ml. of this mixture plus 12 ml. of 0.2 M sodium hydroxide solution, or its equivalent, made to 200 ml. with water has a pH of 8.6. Transfer 60 ml. of this solution to a 125-ml. Erlenmeyer flask, and 220 ml. of the same solution to a 300-ml. Erlenmeyer flask; add 3 and 5 drops of phenolphthalein indicator solution, respectively, to the flasks and stopper tightly with rubber stoppers.

Determine the titration blank of a 50-ml. and a 200-ml. portion of the distillate collected as in the standardization of the distillation apparatus from 150 ml. of distilled water and 1 drop of sulfuric acid (1:1) in the distilling flask. Transfer the 50· ml. of distillate to a 125-ml. Erlenmeyer flask with about 10 ml. of water, and the 200-ml. distillate to a 300-ml. Erlenmeyer flask with about 20 ml. of water. Titrate the 50-ml. distillate with 0.01 N barium hydroxide solution until the color matches that of the prepared 60-ml. buffer solution. (The alkali solution should be kept in heavily paraffined bottles.)[14] Titrate the 200-ml. distillate to match the 200-ml. buffer solution. The indicator color should persist for about 10 seconds.

Preparation of Sample.—Weigh 25 g. of dried eggs into a 250-ml. beaker and, with a heavy stirring rod, make into a smooth paste with water. Transfer the contents of the beaker to a tared 500-ml. Erlenmeyer flask using a total of about 200 ml. of water. Add 25 ml. of N sulfuric acid and shake for about 1 minute. Add 40 ml. of a 20 per cent phosphotungstic acid solution, make to 350 g. with water, and shake for 1 minute. Filter through a 24-cm. folded filter paper. Weigh 150 g. of the filtrate into a 300-ml. Erlenmeyer flask, add an excess of silver sulfate, approximately 0.5 g., and heat to boiling on a hot plate under a water cooled reflux condenser. Transfer the contents of the flask to a 200-ml. volumetric flask with water, make to mark, shake, and filter through a folded filter paper. Test the filtrate for complete removal of chlorides with a few crystals of silver sulfate. If all the chlorides have not been removed, add more silver sulfate to the solution, shake for several minutes and again filter, pouring back until bright. In those cases when it is not possible to

[14] Clark and Hillig, *J. Assoc. Official Agr. Chem.* **21**, 684 (1938).

get a bright filtrate, the addition of Filter Cel and refiltration will prove helpful.

Distillation and Titration.—Pipette 150 ml. of the filtrate into the distillation flask of the apparatus and, if not already acid against congo paper, make acid with sulfuric acid (1:1). Steam distill under the conditions specified, collecting one 50-ml. portion followed by one 200-ml. portion of distillate. Titrate each portion with 0.01 N barium hydroxide, following strictly the procedure outlined under determination of distillation blank.

Determination of Formic Acid.—Combine the 50-ml. and 200-ml. portions of titrated distillate obtained above, add 2 drops of saturated barium hydroxide solution and evaporate to dryness on the steam bath. Add about 5 ml. of water to the residue and 1 ml. more of N hydrochloric acid than is necessary to liberate the volatile acids. Filter through a small paper into a 125-ml. Erlenmeyer flask with ground joint and wash the paper with water in such a manner that the total filtrate will equal 30 to 40 ml. Add 10 ml. of sodium acetate-sodium chloride mixture (25 g. of sodium acetate plus 12 g. of sodium chloride made to 500 ml.) and 10 ml. of a 5 per cent mercuric chloride solution. Connect the flask with a ground joint air condenser, place on the steam bath, and allow the liquid to react for 2.5 hours. Transfer the precipitate of calomel to a previously weighed Pregl filter tube provided with a mat of asbestos about 2 mm. thick. Wash the precipitate with water, followed by alcohol, and dry for 0.5 hour at 100° C. Cool and weigh. The tube should be weighed with another Pregl tube as a tare, prepared with asbestos and treated in the same manner as the one containing the precipitate.

The weight of calomel precipitated (in milligrams) from the combined distillates multiplied by 0.0975 gives the milligrams of formic acid in the distillates. To calculate the total formic acid originally present in the aliquot of the sample in the distillation flask before distillation, divide this result by TF_1 plus TF_2, as determined above. Divide by 8.04 and multiply by 100 to obtain milligram per cent (milligrams per 100 g.) of formic acid in the dried eggs being analyzed.

Computation of Volatile Acids Other than Formic.—Formic Acid Present.—For the purpose of computing acids other than formic, the titrations of the total acid in the 50- and 200-ml. distillates, obtained under *distillation* and *titration,* are corrected for the formic acid present. Convert the milligrams of formic acid in the aliquot of the sample in the distillation flask before distillation as calculated to milliliters of 0.01 N formic acid by dividing by 0.46. Multiply this result by TF_1 to obtain the milliliters of 0.01 N alkali required to neutralize the formic acid in the 50-ml. distillate, and subtract from the original titration and designate the corrected titration as t_1. Using TF_2 as the multiplier, correct the original titration of 200-ml. distillate in a like manner and designate as t_2. Calculate the ratio t_2/t_1 and compare with "C" as determined under standardization of the distillation apparatus. If comparison shows a close agreement (± 0.1), acetic and formic acids are the only acids present. The sum of the corrected titrations of the 50-ml. and 200-ml. distillates divided by the sum of TA_1 and TA_2 gives the milliliters of 0.01 N acetic acid in the aliquot sample in the distillation flask before distillation. This divided by 8.04 and multiplied by 60 gives the mg. of acetic acid in 100 g. of the dried eggs being analyzed. If "C" is greater than t_2/t_1 by more than 0.1, butyric acid is present.

Formic Acid Absent.—Divide the titration of the 200-ml. distillate by that of the 50-ml. distillate determined as above. If the result is in close agreement with "C" (± 0.1), then acetic acid only is present, but if the result is less than "C" by more than 0.1, butyric acid is present. In the absence of butyric acid, compute the milligrams of acetic acid per 100 g. of dried eggs as directed above.

When butyric acid is present along with acetic acid, the titration of the 50-ml. distillate (either corrected or uncorrected, depending upon whether formic acid is present) equals the sum of two products—(1) TA_1 times the total milliliters 0.01 N acetic acid in the distillation flask at the start of distillation, plus (2) TB_1 times the total ml. 0.01 N ml. 0.01 N butyric acid in the distillation flask at the start. Analogously, the titration of the 200-ml. distillate is the sum of TA_2 times the total acetic acid plus TB_2 times the total butyric acid in the flask. The values TA_1, TA_2, TB_1, and TB_2 are constants for the particular apparatus as standardized above, and the titrations are known. The quantities of acetic and butyric acids in the flask are unknown and are to be calculated. These data can be set up as simultaneous equations and the unknown values found.[15] The calculated results will be in terms of milliliters 0.01 N acid. The milliliters of 0.01 N acetic acid divided by 8.04 times 60 equals milligrams acetic acid per 100 g. of the dried being analyzed. The milliliters 0.01 N butyric acid divided by 8.04 times 86 equals milligrams butyric acid per 100 g. of dried eggs.

Lactic Acid

The determination of the presence of lactic acid in foods has greater importance from the standpoint of detection and evaluation

---

[15] Hillig and Knudsen, *J. Assoc. Official Agr. Chem.* **25**, 180 (1942).

of decomposition than from the point of view of use as a preservative (see Chapter IV).

Extraction-Colorimetric Method.—In the method devised by Hillig,[16] the lactic acid is extracted with ether in a liquid-liquid extractor and is subsequently estimated colorimetrically with ferric chloride. This method is applicable to nearly all foods, the principal variation being in the preparation of the sample. The general principles involved in correcting for the volume of precipitable solids can be illustrated by the directions given for the instance of evaporated milk. This correction becomes important when the solution is made up to a standard volume rather than to a standard weight.

Preparation of Sample.—Weigh 50 g. of *milk* into a 100-ml. volumetric flask; weigh 20 g. of *cream* or *ice cream* into a 100-ml. volumetric flask and add about 50 ml. of water; weigh 5 g. of *dried whole* or *skim milk* into a 100-ml. beaker, make a paste with water, and transfer with 50 ml. of water to a 100-ml. flask; weigh 25 g. of sweetened condensed milk into a 100-ml. beaker and transfer to a 100-ml. volumetric flask with 50 ml. of water. In each instance add 6 ml. of ·*N* sulfuric acid, mix gently, add 5 ml. of 20 per cent phosphotungstic acid, using, however, 1 ml. for cream and 2 ml. for ice cream, and complete to volume with water. Shake and filter through a fluted filter.

Evaporated Milk.—Dilute 25 g. with an equal volume of water; add, with constant stirring, 6 ml. of *N* sulfuric acid and 5 ml. of 20 per cent phosphotungstic acid solution. Transfer with the aid of water to a 100-ml. volumetric flask, breaking any foam by use of a few drops of alcohol and a stirring rod. Fill to the mark, shake, allow to stand for several minutes, and filter through a fluted filter. Use a 50-ml. aliquot and calculate the amount of evaporated milk represented as follows:

$$25 \times \frac{50}{100 - S}$$

in which S is the number of milliliters occupied by the precipitated fat and protein. For practical purposes S may be taken as 3.67 ml. for an evaporated milk just meeting the minimum Food and Drug Administration standards for milk fat and total solids. The figure is based on an assumed fat and protein content of 7.9 per cent and 7.7 per cent, respectively. Taking the specific gravities of fat and protein as 0.9 and 1.3, respectively, the total volume of precipitate is

$$\frac{25}{100}\left(\frac{7.9}{0.9} + \frac{7.7}{1.3}\right) = 3.67 \text{ ml.}$$

---

[16] Hillig, *J. Assoc. Official Agr. Chem.* **20**, 135 (1937); **25**, 255 (1942); **26**, 199 (1943).

Liquid or Frozen Eggs.—Transfer 20 g. to a tared 200-ml. Erlenmeyer flask, add approximately 30 ml. of water, and shake thoroughly. Add 10 ml. of *N* sulfuric acid and 15 ml. of phosphotungstic acid, add sufficient water to bring the weight to 125 g., shake for 1 minute, and filter through a fluted filter. In this instance, 50 ml. of the filtrate will generally weigh 50 g.

Dried Eggs.—Mix 5 g. of dried eggs with 50 ml. of water to form a uniform paste with the aid of a stirring rod and add with constant stirring 10 ml. of *N* sulfuric acid and then 12 ml. of 20 per cent phosphotungstic acid solution. Transfer the mixture with water to a tared 200-ml. Erlenmeyer flask, and make up to 125 g. with water. Shake for 1 minute and filter through a fluted filter paper; 50 ml. of the filtrate will generally weigh 50 g.

Reagents.—Standard Barium Lactate Solutions.—Dissolve in about 10 ml. of water a quantity of a pure lactic acid salt, such as lithium, zinc, or calcium lactate, that will contain the equivalent of about 300 mg. of free lactic acid. Transfer the material to the extractor (Fig. 87, see also Chapter I), add 0.5 ml. of sulfuric acid (1:1) and adjust the volume to 50 ml. Extract with ether for 2 hours. Add 20 ml. of water to the extraction flask, evaporate the ether on the steam bath and carefully titrate the contents of the flask with 0.1 *N* barium hydroxide. Transfer the neutralized material to a 200-ml. volumetric flask, make to mark, and shake. Transfer with a pipette to a 500-ml. volumetric flask such a quantity of this barium lactate solution as will contain the equivalent of exactly 250 mg. of free lactic acid, make to mark, shake, and designate as *standard lactate solution*. Two ml. of this solution will contain the equivalent of 1 mg. of lactic acid. The standard lactate solutions are for use in plotting the standard curve and must be freshly prepared and promptly used. Transfer 20 ml. of the standard lactate solution to a 100-ml. volumetric flask, make to mark, and designate as *dilute standard lactate solution*. Ten ml. of this solution will contain the equivalent of 1 mg. of lactic acid.

Carbon.—To 10 g. of a high-grade carbon (Nuchar W is suitable, and Suchar, Darko G60, and Carbex E can also be used) in a 600-ml. beaker, add 200 ml. of water and 30 ml. of *N* hydrochloric acid, and place on the steam bath for 20 minutes. Agitate continuously with air passed through cotton. Filter on a Büchner funnel and suck as dry as possible, tamping with a flattened rod. Transfer the cake to the beaker, add 200 ml. of water, mix thoroughly, and refilter. Repeat the washing and filtering twice and dry in the water oven.

Ferric Chloride Solution.—Dissolve 2 g. of $FeCl_3$ (analytical reagent) in water, add 5 ml. of *N* hydrochloric acid and dilute to 200 ml.

Plotting Standard Curve.—Transfer from burette the quantities of standard solutions listed in the left-hand column of Table 102 to volumetric flasks graduated at 50 and 55 ml. In the right-hand column of Table 102 are given the quantities of lactic acid (milligrams) that will be contained in the 40 ml. of filtrate obtained from each sample after the carbon treatment below, and that will therefore be read in the photometer. A blank using 40 ml. water in place of designated quantities of lactate solution must be included in the series.

TABLE 102. DILUTE STANDARD LACTATE SOLUTION

Solution To Be Transferred to 50-55-Ml. Volumetric Flask Ml.	Lactic Acid in 40 Ml. Aliquot Out of 55 Ml. Ml.
6.90	0.5
13.80	1.0
27.60	2.0

STANDARD LACTATE SOLUTION

8.25	3.0
11.00	4.0
13.75	5.0
16.50	6.0
19.25	7.0
22.00	8.0
24.75	9.0
27.50	10.0
30.25	11.0
33.00	12.0

To each flask add 6.6 ml. of 0.1 N hydrochloric acid and water until the volume is about 40 ml. Add 200 ml. of the prepared carbon, shake, and place on the steam bath for 10 minutes, mixing at frequent intervals. Cool, make to the 55-ml. mark with water, 'and as soon as possible filter through ashless filter paper, returning the filtrate to the filter until bright.

Transfer 40 ml. of each clear filtrate to a 50-ml. Nessler tube. As the 40 ml. of filtrate used contains only 4.8 ml. of the acid added during the carbon treatment, add an additional 1.2 ml. of 0.1 N hydrochloric acid, for a total of 6 ml. is required. in each tube. Place each tube in a jacket of black paper. With one tube at a time, add 5 ml. of the ferric chloride solution with a pipette, make to the mark, and mix. Pour the solution into a 4-in. photometer cell with plane parallel fused ends, the side walls of which are painted black. Read in a neutral wedge photometer, using filter No. 46. This precaution enables the color to remain stable for a number of hours for it fades in direct light. With the readings prepare a standard curve, plotting milligrams of lactic acid as abscissa against scale readings as ordinates. With large-scale graph paper more accurate interpolations can be made.

It is generally not necessary to prepare a new standard curve when a new batch of carbon or ferric chloride solution is used, but if a blank determination with water does not coincide with the original blank, readings can be brought into conformity with the curve by simply adding to or subtracting from the readings the observed amount of variance of the new blank from the old one.

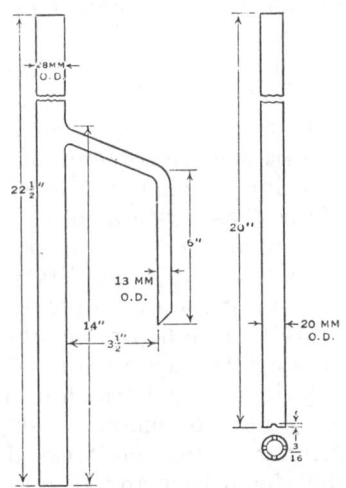

Extraction.—Place 50 ml. of the filtrate obtained from the prepared sample and 0.5 ml. of sulfuric acid (1:1) in the inner tube of the extractor, Fig. 87, and connect to the longest bulb-type condenser available, having an outlet not less than 0.5 in. inside diameter to prevent regurgitation of the ether. Run water through the condenser in sufficient quantity to obtain the maximum condensation efficiency. Connect the extraction flask containing 200 ml. of ether and lower the flask onto a hot plate that has been previously heated in order to prevent super-heat-

FIG. 87.   Liquid-Liquid Extractor [16]

ing of the ether. Protect the extractor from the heat of the hot plate by an upright sheet of asbestos and extract until all the lactic acid is obtained. When the ether in the extraction flask is kept boiling at a rapid rate and the condenser is sufficiently cold to permit the condensed ether to return to the extraction flask in a steady stream, a 3-hour extraction period will deliver all the lactic acid. When this rate of extraction cannot be maintained because of high temperature of the water passing through the condenser, the extraction must be continued until the equivalent of 7500 ml. of ether has passed through the solution being extracted. The time required, T, established for each set of new conditions, is calculated from 2 factors A, the quantity of ether necessary to fill the extractor to overflowing at the side-arm, which is a constant for each apparatus; and B, the time in minutes required for quantity A to pass from the extraction flask and to fill the extractor. To determine A, place 50 ml. of water and 0.5 ml. of sulfuric acid (1:1) in the extractor. With the extractor held upright, carefully pour ether into the inner tube until it just starts passing out of the side-arm. Determine B in the ordinary course of starting each determination. With a stopwatch, record the interval from the time the ether first drops from the condenser and falls into the inner tube to the time the first drops return to the extraction flask from the overflow into the side-arm. The time T necessary for 7500 ml. to pass through the apparatus equals 7500 B/A. The calculated T holds true only if the rate of boiling and condensing is unchanged throughout the extraction period.

Procedure.—To the flask containing the ether extract add 20 ml. of water and expel the ether on the steam bath. Do not allow the flask to remain on the steam bath after the ether has been expelled. Neutralize the contents of the flask with saturated barium hydroxide solution to phenolphthalein indicator. Transfer to a 110-ml. volumetric flask with alcohol until the volume is 90 ml. Heat almost to boiling on the steam bath, cool, make to mark with alcohol, and filter through a quantitative paper. To expel the alcohol evaporate 100 ml. of the filtrate to 10 ml., add 50 ml. of water and again evaporate to 10 ml. (or the 100 ml. of filtrate may be evaporated to dryness on the steam bath). Add from a burette 6.6 ml. of 0.1 N hydrochloric acid and transfer the contents of the beaker with water to a 50-55-ml. volumetric flask until the volume is about 40 ml. Add 200 mg. of the acid-treated carbon, immediately shake, and place on the steam bath for 10 minutes, mixing at frequent intervals. Cool, make to mark with water, and filter through a quantitative paper, pouring back until bright.

Transfer 40 ml. of the filtrate to a Nessler tube. To provide the 6 ml. of 0.1 N hydrochloric acid required, add 1.2 ml. of the acid. Place the tube in a jacket of black paper, add from a burette or pipette 5 ml. of the ferric chloride solution, make to the 50-ml. mark, and mix. After the color has been developed, diluting for the purpose of reducing the color intensity is not permissible. Fill a 4-in. cell, the walls of which are painted black, with the solution and read in the photometer, using filter No. 46. Estimate the quantity of lactic acid present in the 40-ml. portion taken, from the standard curve of the instrument. If the quantity of lactic in the 40-ml. portion of filtrate exceeds the 12 mg. limit of the standard curve, repeat the estimation on a 10-ml. portion of remaining filtrate. The 10-ml. portion will contain 1.2 ml. of 0.1 N hydrochloric acid and 4.2 ml. of the acid will have to be added to complete the 6 ml. required in the Nessler tube. Use the same cell, photometer and color filter as used in obtaining the standard curve. If the photometer is not available comparisons can be made in the conventional manner using Nessler tubes.

Chromatographic Method.—The method [17,18] detailed below consists of five steps: (a) ether extraction, (b) precipitation of the barium succinate from 80 per cent alcohol, (c) isolation of lactic acid from the filtrate of (b), and of succinic acid from the insoluble fraction of (b) by partition chromatography, (d) titration of the acids with barium hydroxide, and (e) identification of the acids by microscopic examination of the barium salt of succinic acid or the zinc salt of lactic acid.

With modifications in the preparation of the sample, and separation of the acids from the sample in a form suitable for passage through the column, these procedures should be applicable to any food product. The procedures below are given for the determination of lactic acid in liquid or dried milks, and for the determination of succinic acid in liquid, frozen, or dried eggs.

Reagents.—Glycerol Indicator Solution.—150 mg. of Alphamine Red-R dissolved in 100 ml. of U.S.P. glycerol. This indicator is the ammonium salt of 3,6-disulfo-β-naphthalene-azo-N-phenyl-α-naphthylamine, also called R-NH$_4$ indicator, and is prepared according to Liddell and Rydon.

Phenol Red Indicator.—Rub 100 mg. of phenol red in a mortar with 5.7 ml. of 0.05 N sodium hydroxide solution until dissolved, then add sufficient water to make the volume 100 ml.

Preparation of Sample.—(a) Liquid, Whole, or Skim Milks.— Weigh 50 g. into a 100-ml. volumetric flask.

(b) Dried Whole or Skim Milks.—Weigh 5 g. into a 100-ml. beaker and make into a smooth paste with small amount of water. Transfer contents of beaker to a 100-ml. volumetric flask with about 50 ml. of water.

To the mixtures add 6 ml. of N sulfuric acid and mix, avoiding vigorous agitation. Add 5 ml. of 20 per cent phosphotungstic acid solution and make to mark with water. Shake and filter through a folded filter paper.

Preparation of Sodium Salt of Lactic Acid.—Place 50 ml. of filtrate in a continuous extractor, add 0.5 ml. of sulfuric acid (1:1) and extract with ether 3-4 hours. To the contents of the extraction flask add 20 ml. of water and evaporate the ether on the steam bath. Add

---

[17] Claborn and Patterson, *J. Assoc. Official Agr. Chem.* **31**, 136 (1948).
[18] Liddell and Rydon, *Biochem. J.* **38**, 68 (1944).

1 drop of phenolphthalein solution and make alkaline by dropwise addition of saturated barium hydroxide solution. Transfer to a 110-ml. volumetric flask, using about 70 ml. of alcohol for the transfer. Heat almost to boiling on the steam bath, cool, and make to volume with alcohol. Filter through a folded filter paper into a 250-ml. beaker. Evaporate the filtrate to dryness on the steam bath. Add 20 ml. of water and 3 or 4 drops of sulfuric acid (1:1), stir well, and filter. Wash the beaker and filter with two 5-ml. portions of water. Make the filtrate alkaline to phenolphthalein by dropwise addition of 40 per cent sodium hydroxide solution. Evaporate to dryness.

   Preparation of the Partition Column.—Place 20 g. of silicic acid [19] in a mortar, add 5 ml. of glycerol indicator solution, 3.7 ml. of water and N ammonium hydroxide solution, dropwise, sufficient to produce an alkaline color with the indicator (1-3 drops). Mix, using a spatula to break up lumps. With a pestle grind into a uniform powder. Make a slurry with 50 ml. of a 20 per cent tertiary butyl alcohol solution in chloroform (V/V) and transfer with aid of a beaker to a glass tube 25 × 200 mm., one end of which has been constricted to an outlet 3 cm. long and 5-6 mm. in diameter, plugged at the constriction with glass wool, and clamped in an upright position. Apply air pressure (3-5 lbs.) to the large end, forcing the excess solvent dropwise out of the small end (about 20 ml.). Release pressure as soon as the liquid disappears at the top of the gel. (Otherwise the gel will dry and crack, becoming useless.) Add 1 ml. of chloroform containing 20 mg. of acetic acid, and 2 ml. of butanol-chloroform solvent containing 20 mg. of lactic acid. Apply pressure until the surface of the solvent just disappears into the gel. Fill the tube with solvent, place a 100-ml. graduated cylinder beneath the outlet and apply pressure. Collect the percolate until the lower edge of the blue band, second from the outlet end, reaches the constriction of the tube. The volume collected is the threshold volume (s) for lactic acid. (The first blue band to leave the column is the acetic acid.)

   Isolation of Lactic Acid.—To the dry residue of sodium lactate in the 50-ml. beaker, add 2 ml. of butanol-chloroform solvent, 5 or 6 drops sulfuric acid (1:1), mix well with glass rod, breaking up all lumps, and stir in 1 g. of anhydrous sodium sulfate.

   Prepare a new partition column from the same batch of silicic acid, proceeding as directed under "Preparation of the Partition Column" above, down to and including the addition of the 1 ml. of chloroform containing 20 mg. of acetic acid. Place a 100-ml. graduated cylinder beneath the outlet of the tube. With a pipette, transfer

---

[19] Ramsey and Patterson, *J. Assoc. Official Agr. Chem.* **31**, 142 (1948).

the solvent containing the lactic acid to the column and apply pressure until the liquid disappears into the gel. Wash the beaker with 3 successive 1-ml. portions of solvent, transferring to the column with the same pipette, and applying pressure each time until the liquid disappears. Fill the tube with solvent and apply pressure. A light placed behind, but not so close as to heat, the tube increases the visibility of the bands because of the translucency of the gel formed by glycerol and silicic acid.

Collect the percolate until the bottom edge of the second band reaches the constriction of the tube, and discard. Place a 50-ml. graduated cylinder beneath the outlet, and collect percolate until the second blue band has passed out of the column (not less than 35 ml.).

With 2 mg. of lactic acid or less, the second blue band may not be distinct enough to be visible. In such case remove and discard the percolate equal in volume to the threshold volume (s) for lactic acid established for the silicic acid being used. Place a 50-ml. graduated cylinder beneath the outlet and collect 35 ml. of percolate.

Transfer the 35 ml. of percolate to a 125-ml. Erlenmeyer flask with 20 ml. of water and titrate with 0.05 N barium hydroxide solution (phenol red indicator). One ml. 0.05 N alkali = 4.5 mg. lactic acid. Recoveries of lactic acid range from 97-99 per cent.

Identification of Lactic Acid.—Separate the aqueous layer in the titration flask. Add 0.05 N zinc sulfate equivalent to the barium hydroxide solution used in the titration. Heat to boiling and boil for 5 minutes, stir in a small amount of Norite and filter. Evaporate to dryness on the steam bath. Add 10 ml. of acetone, heat to boiling, and decant. Repeat heating and decantation once more with acetone, and then once with alcohol. Add a small volume of water, and filter if the solution is not clear. Evaporate on the steam bath to a concentration of 5 mg. of lactic acid per milliliter. Place a few drops of this solution on a microscope slide, allow the water to evaporate at room temperature and observe the crystals of zinc lactate. For amounts of lactic acid less than 5 mg. leave the solution in the beaker and allow the water to evaporate at room temperature. Place the beaker under the microscope and examine the crystals. Compare the crystals with those from a known solution of pure zinc lactate.

SUCCINIC ACID

Succinic acid [20] was not found in shell eggs of acceptable, edible quality or in frozen or dried eggs prepared from such shell eggs.

[20] Lepper and Hillig, *J. Assoc. Official Agr. Chem.* **31**, 734 (1948).

It is formed during some processes of decomposition of eggs, either in the shell or after separation from the shell. This evidence of decomposition is demonstrable in dried eggs.

Chromatographic Method.[21]—*Preparation of Sample.*—(a) Liquid Eggs.—Transfer 200 g. to a tared 1500-ml. beaker. With constant stirring, add 500 ml. of water, 75 ml. of N sulfuric acid, and 125 ml. of 20 per cent phosphotungstic acid solution. Make contents to 1000 g. by addition of water, stir thoroughly, and filter through a folded filter paper.

(b) Dried Eggs.—Weigh 50 g. into a tared 1500-ml. beaker and stir into a uniform paste with 100 ml. of water. With continuous stirring, add 600 ml. of water, 50 ml. of N sulfuric acid and 75 ml. of 20 per cent phosphotungstic acid solution. Make contents to with water, stir thoroughly, and filter through a folded filter paper.

Preparation of Sodium Salt of Succinic Acid.—Evaporate 500 g. of the filtrate in a 1000-ml. beaker on a hot plate to about 75 ml., and concentrate on a steam bath to about 20 ml. Transfer to the continuous extractor by aid of 10 ml. of water. Add 23 g. of ammonium sulfate and 1 ml. sulfuric acid (1:1) to the solution in the extractor. Rinse the beaker with ether into the extractor, and extract with ether for 2.5 hours.

Add 20 ml. of water to the extraction flask and expel the ether on steam bath. Neutralize with saturated barium hydroxide solution (phenolphthalein indicator), transfer to 110-ml. volumetric flask with alcohol until volume is about 90 ml. Heat almost to boiling on steam bath, cool, complete volume with alcohol, and filter through folded 12.5 cm. filter paper.· Place the paper and precipitate in a 50-ml. beaker, add 20 ml. of water, 3 or 4 drops of sulfuric acid (1:1), heat on steam bath about 5 minutes, and filter. Wash the beaker and its paper with 3 successive 8-ml. portions of water, pour through the filter, collecting filtrate and washings in a 50-ml. beaker. Neutralize the filtrate with 40 per cent sodium hydroxide solution (phenolphthalein indicator) and evaporate to dryness on a steam bath.

Isolation of Succinic Acid.—Prepare a partition column as described for lactic acid, using a solution of 20 mg. of succinic acid in 2 ml. of butanol chloroform solvent instead of the lactic acid solution to determine the threshold volume for succinic acid. Prepare another column, acidify the sodium succinate, and put the solution through the partition column, in the same manner as described for "Isolation of Lactic Acid." Transfer the 35-ml. percolate to a 150-ml. Erlenmeyer flask with 20 ml. of water and titrate with 0.05 N barium

---

21 Claborn and Patterson. *J. Assoc. Official Agr. Chem.* **31**, 136 (1948).

hydroxide solution (phenol red indicator). One ml. 0.05 N alkali = 2.95 mg. succinic acid.

Identification of Succinic Acid.—Separate the aqueous solution in the titration flask, transfer to a 50-ml. beaker, evaporate to about 3 ml. on the steam bath, and then allow to evaporate at room temperature until crystals appear. Compare the crystals with known barium succinate prepared by titrating about 25 mg. of pure succinic acid with the barium hydroxide solution, adding 3 or 4 drops of glycerol, and evaporating as directed above.

DETECTION OF DECOMPOSITION IN FISH

Peroxide Number.—Rancidity in fish may be detected organoleptically because it is generally evident by an after taste. Untreated fish when decomposed will have the usual accompanying bad odor and appearance, but smoked fish will not show rancidity as readily. Wheeler [22] determines the peroxide number of oils and this number may be used as an index of rancidity.[23]

Procedure.—Pipette a 20-ml. aliquot of the ether solution of the oil obtained as described on page 675 into a 500-ml. flask. Add 50 ml. of a freshly prepared mixture of 60 per cent glacial acetic acid and 40 per cent chloroform, followed immediately by 1 ml. of a saturated solution of potassium iodide from a pipette. Shake the flask with a rotary motion for exactly 1 minute. Quickly add 100 ml. of a 0.05 per cent starch solution. Immediately titrate the solution with 0.01 N sodium thiosulfate solution to the disappearance of the last purple tinge, shaking vigorously at the end point. Express the peroxide number, M, as the millimoles of peroxide per 1000 g. of oil. The weight of the oil in the 20-ml. aliquot is known from the previous determination. Then,

$$M = 0.5 \frac{(\text{ml. thiosulfate}) (\text{normality})}{\text{weight of oil}}$$

For example, suppose one 20-ml. portion of ether solution contained 0.407 g. of oil and a second 20-ml. aliquot of the ether solution required 3.90 ml. of 0.009 N sodium thiosulfate solution, then the peroxide number,

$$M = \frac{(0.5) (3.90) (0.009)}{0.407} = 0.0431$$

This is the value for 1 g. of oil, hence multiplying by 1000 gives the desired value, namely, 43.1.

---

22 Wheeler, *Oil and Soap* 9, 89 (1932).
23 Stansby, *J. Assoc. Official Agr. Chem.* 18, 618 (1935).

Estimation of free fatty acids serves as a qualitative test for, as decomposition of oils increase, the free fatty acid content also increases.

In mackerel peroxide values and organoleptic examination give the following groups:

```
Not rancid ..............................0.0 to 0.6
Slightly rancid .........................0. to 21.4
Rancid ..................................18.4 to 36.5
Extremely rancid ........................33 to 201
```

Indole.—*Reagents.*—Ehrlich's Reagent.—Dissolve 2 g. *p*-dimethylamino-benzaldehyde in 100 ml. of chloroform.

Preparation of Sample.—(a) If fish is packed in sauce, wash by holding under running tap water a few minutes. Place in a pan and allow to drain a few minutes, pouring off the drained liquid. Pass the fish through an ordinary meat grinder.

(b) If packed in oil, drain off the oil a few minutes before passing through the meat grinder.

Method.—Weigh 50 g. of homogeneous sample in a 250 or 400-ml. beaker. Add 10 ml. of sodium hydroxide (10) and stir about a minute with a stirring rod. Add 100 ml. chloroform. Stir a few minutes to break up large lumps and transfer the whole mixture to a 250-ml. wide-mouthed centrifuge bottle. Close the bottle with a rubber stopper and shake vigorously a few minutes until the mixture becomes a pasty mass. Add about 10 g. ammonium sulfate powder and with a stirring rod incorporate the salt with the pasty mass, beginning at the top and gradually working downward as the emulsion breaks. When sufficient liquid has separated, stopper the bottle and shake a moment to further distribute the ammonium sulfate. (Do not shake too long or the mixture may again become pasty.) Centrifugate a few minutes at about 1,000-1,500 r.p.m. The mixture usually separates into three layers; a small liquid upper layer, a middle solid layer of fish and a lower layer of chloroform.

Pour off the small upper layer and discard it. With a stirring rod pierce the solid fish layer and pour the chloroform layer through a folded filter paper, catching the filtrate in a 50-ml. graduated cylinder.

Place a 25-ml. aliquot (measure in a cylinder) in a 250-ml. separatory funnel. Add 1 ml. Ehrlich's Reagent and 5-ml. phosphoric acid and shake vigorously for half a minute. Allow to stand a few minutes, then add 15 ml. acetic acid, shake about a quarter of a minute and allow the layers to separate. There is usually a clean sharp separation at this point.

Tap off the lower layer into a 50-ml. Nessler tube, being careful not to allow even a drop of the upper layer to pass. Fill to the mark with acetic acid, close with a rubber stopper and mix. Compare the color with the standards by looking through the bottom of the tube. A La Motte colorimeter—Nessler tube comparator box 7 in. long, 4 in. wide, 9 in. high with etched glass and mirror is useful.

Standards.—Stock solution in alcohol 1 ml. = 0.001 g. indole. To make standards dilute the stock solution with chloroform until 1 ml. = 0.000001 g. indole. With a pipette measure into a separatory funnel quantities equal to 2-4-6-8-10 micrograms. Make to 25 ml. with chloroform, 5 ml. phosphoric acid and 1 ml. Ehrlich's reagent, and proceed from this point as described in the method.

Notes.—(1) The lower phosphoric acid layer may be colorless or only slightly pink. This is no indication, however, of the amount of true color which is brought out by dilution in the Nessler tube with acetic acid.

(2) If, after adding the ammonium sulfate and shaking, the mixture again emulsifies, add about 5 g. more of the sulfate and shake again. The ammonium sulfate should be finely divided so as to more quickly dissolve and break the emulsion on first shaking.

(3) The color developed is most comparable with the standards at about 5 micrograms. If the unknown develops a color too strong for comparison, it may be diluted in the Nessler tube by removing 25 ml. and filling to the mark again with acetic acid. Sometimes if the pink color appears to have an off tint, a comparable color in the standard may be obtained by moving the hand over the top of the tube until a proper "shadow" produces a more comparable tint.

(4) If the amount of indole is very small, two 50-ml. aliquots may be used instead of 25 ml. In this case, 25 ml. of acetic acid should be used, but not increasing the amounts of any of the other reagents.

(5) The separatory funnel and Nessler tube should be entirely free from drops of water. If present in the Nessler tube, a turbidity will be formed.

SELECTED REFERENCES

Methods of Analysis, A.O.A.C. 8th ed. Washington, 1955.
Microanalysis of Food and Drug Products, Food, Drug Admin., Food Drug Cir. 1 (1948).

CHAPTER XXI

FIELD TESTS

PURPOSES

The purpose of field testing is to provide a health inspector with an additional tool by which to arrive at a suspicion of adulteration or to indicate the need of more extensive laboratory investigation. Because field tests are of necessity limited they cannot replace the more precise and accurate methods of the chemical laboratory. They are nevertheless very useful not only for the inspector but also for the plant operator as a means of plant control.

GENERAL METHODS

Tests for temperature, specific gravity, pH, and chlorine content are used by nearly all inspectors.

Temperature.—In making a field test to determine the temperature of a given place or material, a standard thermometer is used. It is held in position or immersed in the material to be tested until the thermometer is in equilibrium with its surroundings and then the temperature is read.

Specific Gravity.—In the field specific gravity determinations are usually made with some type of specific gravity spindle such as a hydrometer, lactometer, salometer, or alcoholometer. The liquid to be tested is stirred by pouring from vessel to vessel, taking care to occlude as little air as possible. The spindle is immersed in the fluid and is allowed to rise to its proper level. This is done to overcome surface tension and viscosity effects. The reading is made, the temperature taken, and the reading corrected for temperature, if necessary.

pH Determination.—The simplest method for determination of pH is the use of test papers. The use of litmus paper is an old method. Hydrion test papers can be used for wide range of pH determination. The paper is generally on a spool which has the comparison pH imprinted on the case. A piece may be torn off and the solution to be tested is streaked on it. The paper should not be dipped into the test solution. The test paper is then compared with the standard colors on the spool.

The use of indicators is also well known. Special indicators change color in a given range of H^+-ion concentration. Thus, for instance, is the use of bromothymol blue to detect abnormal pH of milk. This indicator has a slightly greenish yellow color in the normal pH range of milk. It will be yellow if the milk is acid (pH low) and green or blue if the milk is alkaline (pH high).

Simple kits are available for the determination of pH. To use one of the La Motte [1] types, remove the top from the base, fill three of the test tubes to the mark (5 ml.) with the sample to be tested and place them in the holes back of the three slots in the base. To the middle tube add 0.5 ml. of indicator solution by means of the pipette and nipple and mix thoroughly. Place the color standard slide on the base and, holding the instrument toward a window or other source of daylight, move it in front of the test sample until a match is obtained. Then read off the pH directly from the values on the slide.

Determination of Available Chlorine.—The principal methods for the determination of free chlorine in the field are the o-tolidine method, the o-tolidine-arsenite method, and the starch iodide method.

o-Tolidine Method.—This is a colorimetric method [2,3] which depends on the production of a yellow color by the reaction between chlorine and o-tolidine. It is subject to disturbing influences of small amounts of iron, manganic manganese, and nitrite nitrogen.

In a simple test kit the method is to fill the cell with the sample to a given mark, add water if necessary, add 0.5 ml. of o-tolidine reagent, mix, and compare with the standard provided by the kit.

o-Tolidine-Arsenite Test.—This test [4] provides the inspector with a means for differentiating between (a) free chlorine, (b) chloramine, and (c) the errors attributable to nitrite, iron, or manganese. The test is based on the following factors: (1) o-tolidine reacts instantly with active chlorine but only slowly with chloramines; (2) arsenite reacts rapidly with both chlorine and chloramine but not with nitrite, iron, or manganese. By a combination of these reactions the color produced with o-tolidine by (a), (b), and (c) can be evaluated. Commercial kits or a simple one using 2 oz. French bottles may be employed.

Place an equal volume of sample solution in each of two bottles. To the first bottle add 0.5 ml. o-tolidine reagent. Shake and at once

[1] W. A. Taylor Co., *Modern pH and Chlorine Control*. Baltimore, 1943.

[2] American Public Health Association, *Standard Methods of Water Analysis*, 8th Ed. New York, 1936.

[3] Jacobs, *Analytical Chemistry of Industrial Poisons*, 2nd Ed. Interscience, New York, 1949.

[4] Hallinan, *J. Am. Water Works Assoc.*, January (1944).

add 0.5 ml. of 0.5 per cent sodium arsenite solution. The color produced is due to the active chlorine and nitrite, iron, and manganese. To the second bottle add equal volumes first of sodium arsenite solution then o-tolidine reagent. The color is due to the nitrite, iron, and manganese. Back up the first bottle with a bottle containing clear distilled water. Back up the second bottle with a chlorine color standard so that the combined colors of the standard and the second test bottle will equal the color of the water and the first test bottle. Observe the colors. The standard then equals the amount of active chlorine.

Run an additional test on two other portions of the sample. To the first add only o-tolidine reagent. The color produced is due to all components chlorine, chloramine, and nitrite, iron, and manganese. Treat the second portion as if it were the first portion of the first part of this test (add o-tolidine first, arsenite solution second). The color is due, as before, to chlorine, nitrite, iron and manganese. Back up the first bottle with water, the second bottle with a chloramine standard until it equals the color in the first bottle. Observe. Then the chloramine standard equals the amount of chloramine present.

Starch-Iodide Method.—There are many variations of this method. The simplest is to insert starch-iodide test paper in the solution to be tested, note the depth of blue color produced, and compare with standards. This method has poor precision and accuracy.

Another variation used by the Department of Health of Chicago [5] uses solutions. The basis of this test is the neutralization of the test solution with acid, the addition of potassium iodide to liberate free iodine, the reaction between the liberated iodine with a definite amount of thiosulfate, and the reaction of any excess iodine with starch.

In making the test, fill a 4-oz. glass-stoppered bottle up to a 100-ml. mark with the dish water sample, add 3 drops of phenolphthalin indicator solution turning the solution red as a result of the alkalinity. Add 35 per cent acetic acid slowly, shaking after each addition until the red color completely disappears, and then add a few milliliters more of the reagent. Add with the aid of a dip pipette 5 or 7½ ml. of a complex solution starch, potassium iodide, and sodium thiosulfate reagent. For inorganic chlorine, 5 ml. of the solution is used, and for organic 7½ ml. of the solution is used. If the test solution turns blue by using 5 ml. of the complex reagent, 35 p.p.m. or more of chlorine is indicated. When using 7½ ml. of the reagent, 50 p.p.m. or more of chlorine are indicated.

---

[5] Jacobs and Jaffe, work performed in 1939.

TESTS RELATED TO MILK

Field Test for Efficiency of Pasteurization.[6]—This test is based on the distinction produced by the complete or incomplete destruction of the enzyme phosphatase on heating. Phenol is liberated in an improperly pasteurized material by the action of phosphatase on a buffered solution of disodium phenylphosphate and is subsequently estimated by the formation of an indophenol dye with 2,6-dibromoquinonechloroimide. The method was first developed in 1934. Add 0.5 ml. of sample to 5 ml. of buffered substrate prepared by dissolving in water a tablet containing disodium phenylphosphate and a buffer salt such as borax or sodium sesquicarbonate and diluting to 50 ml. Shake briefly. Incubate for 10 minutes in a water bath at 98° F. (If no water bath is available, incubate in pocket for a somewhat longer period.) Remove from bath, add 6 drops of an alcoholic solution of 2,6-dibromoquinonechloroimide, prepared by dissolving a tablet in 5 ml. of alcohol. Shake well immediately. After 5 minutes compare color with standards.

Properly pasteurized milk will be a gray or brown. Properly pasteurized cream will be a gray or white. Raw milk or cream will be an intense blue. The appearance of any blue is indicative of improper pasteurization; the intensity of color is proportional to the amount of raw product or underpasteurization. After development of color as above, add 2 ml. of a neutral solvent [7] such as amyl, isobutyl, or n-butyl alcohol. Invert the test tube slowly at least 10 times and allow to stand. Rapid inversion will result in an emulsion being formed but, if correctly performed, the alcohol will separate clearly and will have extracted the indophenol formed by the test.

The appearance of any blue or blue-green in this alcohol layer is indicative of improper pasteurization. In the absence of a properly pasteurized milk to be used as a control, a boiled milk may be substituted.

Babcock Method.—The field test for determination of fat in milk is the same as the laboratory test (see Chapter VII) except that a hand or field centrifuge is used. This method depends on the solution of all components of milk except fat and lipid bodies in sulfuric acid, and the subsequent estimation of the fat by centrifuging into a graduated narrow neck of a special flask as the supernatant layer over the heavier layer of sulfuric acid.

Measure 18 g. of milk from a properly mixed sample into a standard State branded milk test bottle by using a 17.6-ml. standard State

---

[6] Scharer, J. Milk Technology 1, 35 (1938).
[7] See page 281.

branded pipette, add 17.5 ml. of standard commercial sulfuric acid, specific gravity 1.813, which is best and avoids charring the fat layer, and shake until all the curd has disappeared, then continue the shaking for about one-half minute longer. Before mixing, the milk and acid should have a temperature of about 60° F.; if not, the amount of acid must be adjusted to give the proper rate of color development. Place the test bottles in the Babcock centrifuge and whirl at the proper speed for 5 minutes; then fill the bottles with hot water, having a temperature of at least 200° F., to the bottom of the neck. Whirl for 2 minutes and fill with hot water at 200° F. to the top of the graduations and whirl again for one minute. Read the per cent fat by measuring from the lowest point of the fat column to the highest point of the meniscus at the top of the fat column. The Babcock method makes use of standard pipettes and bottles whose specifications are rigidly drawn.

Causticity of Wash Water.—There are several methods for determining the sodium hydroxide content of wash water. Tablets or standard solutions of known strength are added to a given volume of the alkaline wash water. The number of tablets or cubic centimeters of the solution necessary to change the color of the indicator in the test solution is a measure of the causticity. A key accompanying commercial kits enables one to interpret these results in terms of sodium hydroxide percentage. An estimation is made as follows: [8] Place 10 ml. of the wash solution in a small beaker. Add 2 drops of phenolphthalein indicator solution, then add about 6 ml. of barium chloride solution, stir, and titrate with 2.56 N sulfuric acid until the pink color disappears. The burette reading for the acid used gives the percentage of sodium hydroxide present directly.

Investigations sponsored by the American Bottlers of Carbonated Beverages on bottle-washing compounds [9] have shown that, to sterilize bottles properly under recommended conditions by recognized mechanical methods, washing solutions should comply with the following, or its equivalent; a 3 per cent alkaline solution of which not less than 60 per cent is caustic soda (1.8 per cent sodium hydroxide), with immersion time of 5 minutes and temperature of 130° F.

The concentration and composition of the alkaline washing solution may be determined by the American Bottlers of Carbonated Beverages' alkali and caustic tablet titration methods, or by titration with standard acid solutions and in some cases by electrical

[8] Shrader, ed., *Sanitation Problems in Food Handling and Processing.* Manhattan College School of Engineering, New York, 1943.

[9] Jacobs, ed., *Chemistry and Technology of Food,* Vol. II, p. 725. Interscience, New York, 1944.

methods. The agent used in the alkali test tablet is given in the footnote.[10] An indicator is combined with an acid substance for the purpose of showing when the reaction between the acid substance of the tablet and the caustic of the washing-solution sample has been completed. The amount of this mixture used in each tablet is such that, if 10 ml. of washing solution is used, each tablet will represent 1 per cent alkali as caustic. This is the total alkali strength including caustic and milder alkalis. The indicator of the tablet gives a sharp color change from blue in the presence of an alkali to yellow in the presence of an acid.

Investigations have shown that total alkalinity is not the most important factor to be considered in the composition of washing solutions. It has been proved that caustic alkalinity is very important as a sterilizing agent. The caustic tablet titration process is a two-step procedure differentiating between caustic and total alkalinity. However, the same procedure and equipment are used as in the alkali test. The preliminary tablet is used to add the indicator and remove noncaustic alkalinity from the solution by chemical means for subsequent test with the acid-titration pill.

TESTS RELATED TO FOOD

Organoleptic Analysis.—It is often the function of the health inspector to analyze a food product organoleptically, that is, by the use of the senses. The taste, odor, and appearance of a food product are very often indicative of the quality of that product. Naturally, discretion must be observed in the interpretation. Thus, although a fruit would be considered bad food if it were slimy or moldy or had a putrid odor, certain types of cheese may have all of these characteristics and still be fit food. Such methods have been detailed in Chapter XIX.

Cadmium.[11]—Goldstone devised a method for the detection of cadmium on plated ware which is performed easily in the field. This method is a refinement and adaptation of the laboratory method proposed by Coleman. It depends on the precipitation of yellow cadmium sulfide in the presence of excess cyanide.

Reagents.—(1) Ammonia-Sodium Nitrate Reagent.—Dilute 200 ml. of ammonia water (28 per cent) plus 100 g. of sodium nitrate with water to 1 liter volume.

---

[10] A.B.C.B. Alkali Test Tablets (U. S. Patent 1,721,809) contain bromophenol blue indicator, potassium acid sulfate, and inert binder. A.B.C.B. Caustic Test Tablets No. 1 (U. S. Patents 1,721,809 and 1,912,473) contain tropaeolin O indicator, barium chloride, and inert binder, A.B.C.B. Caustic Test Tablets No. 2 (U. S. Patents 1,721,809 and 1,912,473) contain potassium acid sulfate and inert binder.

[11] Schiftner and Mahler, *Am. J. Pub. Health* 33, 1224 (1943).

(2) Sodium Sulfide Reagent.—Dilute 100 g. of sodium sulfide with water to 1 liter volume.

(3) Potassium Cyanide Reagent.—Dilute 100 g. of potassium cyanide in water and dilute with water to 1 liter volume.

Procedure.—To a small pinch of the metal scrapings in a test tube, add 3 ml. of the ammonia-sodium nitrate reagent; bring the mixture to a boil over a flame and allow to stand for a minute or two. Pour the clear supernatant liquid into another test tube, add 1 ml. of the cyanide reagent and, after shaking, add 1 drop of sodium sulfide reagent. This produces a canary yellow precipitate if cadmium is present. The metals, iron, tin, antimony, arsenic, silver, copper, nickel, chromium, zinc, and aluminum do not interfere. In the case of zinc and aluminum, a whitish gray precipitate is formed which is readily distinguishable from the canary yellow color of cadmium sulfide. If cadmium is present in addition to any of these metals, it is instantly detected. The only metals which do interfere are lead and mercury, but these are rarely, if ever, used as plating metals under these conditions.

Lead.—A simple test for the presence of lead in kitchen utensils and other vessels is based on the relative insolubility of lead iodide in dilute nitric acid.

Procedure.—Add 2 drops of 10 per cent nitric acid to the suspected spot and then 2 drops of 10 per cent potassium iodide solution. The production of a yellow precipitate indicates the presence of lead.

Sulfur Dioxide and Sulfites.—The presence of sulfites as such in a material is not easy to detect qualitatively within the foodstuff because of the possible presence of other reducing agents. However, the following rapid method for the determination of sulfur dioxide, modified by Jacobs from that of Alesi (see page 155), overcomes some of these difficulties. This method is based on the bleaching action of sulfur dioxide on iodine. The sulfur dioxide is carried over the iodine in a foam and current of gas caused by carbon dioxide.

Procedure.—Place 20 to 50 g. of the food material rubbed up to a paste with 20 ml. of water, if solid, in a wide-mouth 8-oz. bottle. Add 20 ml. of a solution containing 2 per cent sodium hydroxide and 3 per cent sodium carbonate. Then acidify with 30 ml. of hydrochloric acid (1:1). When the foam has decreased, suspend in the bottle a piece of starch-potassium iodide paper slightly blued with iodine vapor or by dipping into a dilute solution of chloramine-T. The starch-potassium iodide paper is suspended by inserting it into a slit made in a cork, provided with a vent, that fits the wide-mouth bottle. If sulfur dioxide is present the paper is decolorized. Some measure of the amount of sulfur dioxide present is obtained by the rapidity with which the starch-iodide paper is decolorized. An alternative test has been described by Korff and Kaplan.[12] Transfer 10 g. of ground meat to a wide-mouth bottle containing 30 ml. of 10 per cent sulfuric acid. Stopper the bottle and shake for 10 seconds. Allow several minutes for the foam to subside. Add 3 g. 30-mesh granulated zinc (arsenic and sulfur free). Insert stopper containing a Gutzeit scrubber tube previously fitted with a No. 2 cotton dental roll impregnated with 20 per cent lead acetate solution. The cotton roll will become black in 5 to 10 minutes, and there will be a decided odor of hydrogen sulfide upon opening the bottle if sulfites are present.

White Phosphorus and Cyanides.[14]—Place a few grams and water or 50 ml. of the sample in a bottle and add 10 ml. of 10 per cent tartaric acid solution. Suspend above the surface of the liquid test strips of (1) filter paper moistened with a drop of silver nitrate solution, (2) picric acid paper moistened with a drop of saturated sodium carbonate solution. Warm the mixture if possible to 40-50° C. and allow to stand for 15 minutes. The papers may be suspended by attaching them with the aid of a rubber band to a glass stopper which fits loosely in the mouth of the flask.

The presence of cyanides is indicated by a rose color produced on the picric acid paper. Blackening of the silver nitrate paper may indicate phosphorus.

Fluorides.[13]—The presence of fluorides may be detected by the use of sodium zirconium alizarinate paper.

Dissolve 0.87 g. of zirconium nitrate, $Zr(NO_3)_4 \cdot 5H_2O$, in 100 ml. of water, and 0.17 g. of sodium alizarinate in 100 ml. of water. Place each solution in a large watch glass or evaporating dish. Steep filter paper first in the dye solution and then place the dyed paper in the zirconium nitrate solution. Allow to dry and cut into strips.

Acidify the suspected sample with hydrochloric acid (1:1) and filter. Test with the test paper. If the paper changes from pink to yellow, fluorides are present. Bleaching agents bleach the pink and yellow color and therefore interfere.

Nitrites.[14]—To about 5 g. of sample (a level teaspoonful) add 5 ml. of a 20 per cent solution of potassium alum and dilute to 500 ml. with water. Mix thoroughly. Filter 25 ml. into a Nessler tube. Add 1 ml. of 1 per cent solution of sulfanilic acid solution and 1 ml.

[12] Korff and Kaplan, Am. J. Pub. Health 32, 1110 (1942).
[13] Jacobs and Stebbins, Technical Manual for Gas Reconnaisance Officers, N. Y. C. Dept. Health, 1943.
[14] Shrader, Sanitation Problems in Food Handling and Processing, Manhattan College School of Engineering, New York, 1943.

of 0.5 per cent solution of α-naphthylamine hydrochloride. Make up to 100 ml. in the Nessler tube. Allow to stand 10 minutes for a pink color to develop indicating the presence of nitrites. The depth of color is proportional to the amount of color and must be compared with standards for quantitative results.

Rancidity.[15]—A simple method for the detection of rancidity has been devised by Jacobs.

Procedure.—Measure 5 ml. of the sample of oil or fat, which should be melted into a tube 7¾ x 1 in. Add 5 ml. of concentrated hydrochloric acid and stir for about 30 seconds, taking care not to wet the upper portion of the tube. Insert a piece of phloroglucinol paper, prepared by dipping filter paper into a 0.1 per cent solution of phloroglucinol in ether, in to the cork which has a slit cut in it to hold the paper and a side channel for an air vent. Then drop 5 or 6 chips of marble or an equivalent amount of calcium carbonate lumps into the tube. Stopper immediately and allow the evolution of gas to proceed, taking care that the resultant foam does not touch the paper. If a pink to lavender or orchid coloration is obtained within 10 to 20 minutes on the paper the sample may be considered rancid. Some measure of the degree of rancidity may be obtained from the depth of color produced.

Spoon Test.[16]—A common test for margarine is called the "spoon" test, or "foam" test. Heat 3 or 4 g. of the sample in a large spoon over a flame. Genuine butter will boil quietly and considerable foam will be produced. Renovated butter or margarine will bump and sputter like hot grease containing water, and very little foam will form.

Health inspectors often make use of several other field tests, such as those employed to determine relative humidity, air flow, and exhaust gas flow.

PHYSICAL TESTS

Psychrometer.—A psychrometer is an apparatus for measuring the dew point or the humidity of the atmosphere. The sling psychrometer consists of two thermometers, one the customary dry-bulb thermometer, and one thermometer equipped to be a wet-bulb type. In order to obtain the relative humidity of a given atmosphere the sack around the wet-bulb thermometer is moistened with water and then both thermometers are twirled by means of a sling arrangement at a height at which the determination is to be made. The relative humidity may then be easily obtained from the table, de-

---

[15] Jacobs, work performed in 1938.
[16] Thurston, *Pharmaceutical and Food Analysis.* Van Nostrand, New York, 1922.

pending on the difference between the temperatures of the dry and wet thermometers.

Anemometer.[17]—One of the simplest of the velocity type instruments to use is the anemometer. These instruments are somewhat expensive, if carefully made, and require frequent adjustment. The anemometer is held in the air stream so that the air flow rotates its vanes or the gas is directed so that it hits and rotates the vanes of the instrument. The rotation of the vanes is registered on a calibrated tachometer. The instrument must be used with a stop watch, for the dial readings give the result in linear feet of air travel. From this, feet per minute or equivalent distance over time ratio can be calculated.

Draft Gauge.—The Ellison Draft gauge is a device designed to measure draft in furnaces and flues. It depends on fundamental hydrodynamic theory that, in a stream of moving fluid, the pressure is greatest where the velocity is least and the pressure is least where the velocity is greatest. After adjustment of the gauge to zero, a tube is inserted in the flue or ash pit of a furnace and the drop in pressure is read on the gauge.

SELECTED REFERENCES

American Public Health Association, *Standard Methods of Dairy Products*, 8th ed., New York, 1941.

Burke, *Practical Dairy Tests*. Olsen, Milwaukee, 1935.

Clay and Jameson, *Sanitary Inspector's Handbook*. Lewis, London, 1936.

Geiger, *Health Officers' Manual*. Saunders, Philadelphia, 1939.

Leach and Winton, *Food Inspection and Analysis*. Wiley, New York, 1920.

Methods of Analysis. A.O.A.C. Washington, 1945.

Shrader, *Food Control*. Wiley, New York, 1939.

---

[17] Jacobs, *Analytical Chemistry of Industrial Poisons*, 2nd Ed. Interscience, New York, 1949.

CHAPTER XXII

RADIOCHEMICAL DETERMINATIONS

Foods are analyzed to determine their radioactive content in order to find out (1) if the food is contaminated by radioactive fallout either naturally or accidentally, (2) if the food is contaminated because of processing with radioactive materials or exposure to nuclear and other radiations, (3) if the food contains more than normal quantities of radioactive substances because of the uptake of radionuclides, and (4) the fate of radioisotopes in foods and their effect on foods.

Contamination of Food by Radioactive Substances

Since the explosion of the first "atomic" bomb on July 16, 1945, at Alamogordo, New Mexico, and the subsequent use of plutonium and "hydrogen," or thermonuclear, bombs as weapons and test weapons, there has been intense interest in the possibility of contamination of food by radioactive material not only from the fallout of such test weapons but also by use of radioisotopes for sterilization of food, by use of nuclear radiations and radiations such as gamma rays and hard X-rays, and by electron bombardment and particle bombardment of foods for sterilization and pasteurization purposes.

The accidental contamination of food by radioactive material is no figment of the imagination. Possibly the classic example of such an instance of food contamination is the "atomic mishap" at the Windscale Plutonium Plant in England on October 10, 1957. In this accident an overheated reactor, in a plant presumably shut down for repair, sprayed an area of over 200 square miles with radioactive iodine-131. It was found that the milk produced in this area had become contaminated. While the contamination was not thought to be serious enough to prevent adults from drinking such milk, it was deemed that children, who are more sensitive, might be affected and so a ban was placed on the sale of milk produced in this area. At first it was thought that the milk might be used for the manufacture of cheese since iodine-131 is short-lived, but it was finally decided not to use the milk at all and to bury it in trenches or dump it at sea. In this instance it was fortunate that steps were taken promptly to minimize the danger. It is clear that methods must be available to evaluate such danger.

UPTAKE OF RADIONUCLIDES

A study of the literature reveals that in addition to the possibility of food contamination as discussed in the preceding paragraphs, food may become unwholesome because of the actual concentration of radioactive substances by normal "uptake" processes. For instance, cows grazing on grass contaminated by radioactive fallout or by an accidental spray as in the Windscale atomic mishap will concentrate such radioactive material in the milk they produce.

Setter and Goldin [1] noted that depending on the time of sampling and the kind of sample there was a 200- to 3000-fold increase in the activity of fish and snails over the activity in the water in which the fish and snails lived. Rosenthal [2] also found an uptake of calcium-45 and strontium-90 in tropical fish.

RADIOTOXICITY

It has been stressed by Kahn and Goldin [3] and by others that the mere determination of the amount of radioactivity in a food is not an adequate measure of the possible damage or hazard that may be expected. Such possible damage can only be determined on the basis of the biochemical and physical properties of the radionuclides present in the food. The dose rate to an organ from the ingestion of such food depends on the amount ingested, the fraction reaching the organ in question, the effective energy, the effective half-life, the time of exposure, and the mass of the critical organ, as discussed in detail in *Handbook 52* of the National Bureau of Standards. [4]

For example, the dose rate to bone from continuous ingestion over a long period of three different radionuclides like strontium-90, phosphorus-32, and molybdenum-99 can be calculated. The ingestion of these three isotopes for a 40-year period at a level of 100 $\mu\mu$c. per ml. (10^{-4} μc. per ml.), or 1000 times the maximum permissible concentration for isotopes of unknown composition, would reach the respective bone dose rates of 11 rem per week for strontium-90, 0.033 rem per week for phosphorus-32, and 0.000002 rem per week for molybdenum-99. For comparison, it is to be noted that the occupational permissible dose is 0.3 rem per week. These values indicate that strontium-90 is about 300 times as toxic as phosphorus-32 and

[1] Setter and Goldin, *Ind. Eng. Chem.* **48,** 251 (1956).

[2] Rosenthal, *Science* **124,** 571 (1956); **126,** 699 (1957).

[3] Kahn and Goldin, "Radiochemical Procedures for the Identification of the More Hazardous Nuclides." Presented to the Nuclear Engineering and Science Congress (1957).

[4] "Maximum Permissible Amounts of Radioisotopes in the Human Body and Maximum Permissible Concentrations in Air and Water," *Natl. Bur. Standards, Handbook* **52** (1953).

some 5 million times as toxic as molybdenum-99. Kahn and Goldin point out that while these values would be modified if exposure to the intestinal tract were also considered, they serve to illustrate the difference in dosage resulting from the ingestion of equal quantities of activity of different nuclides.

The measurement of gross activity alone is not adequate for the evaluation of a radioactivity problem. It is necessary to know what particular kinds of radioactivity are present. The determination of strontium-90 in a mixture of the aforementioned three radionuclides would be of far greater value than a determination of the gross activity.

TABLE 103. SOME HAZARDOUS RADIONUCLIDES[a]

Nuclide	Origin	MPC[b] in Water μc./ml.	μμc./l.	pMPC$_W$
Ra-226	Natural	4×10^{-8}	40	7.4
Unknown	....	1×10^{-7}	100	7.0
Sr-90	F. P.[c]	8×10^{-7}	800	6.1
Po-210	Natural	3×10^{-6}(g)	3,000(g)	5.5
U-233	Irradiation	3×10^{-6}(g)	3,000(g)	5.5
Pu-239	Irradiation	3×10^{-6}(g)	3,000(g)	5.5
Am-241	Irradiation	3×10^{-6}(g)	3,000(g)	5.5
I-131	F. P.	6×10^{-5}	60,000	4.2
Sr-89	F. P.	7×10^{-5}	70,000	4.2
Ca-45	Irradiation	1×10^{-4}	100,000	4.0
Ru-106	F. P.	1×10^{-6}(g)	100,000(g)	4.0
Ce-144	F. P.	1×10^{-6}(g)	100,000(g)	4.0
P-32	Irradiation	2×10^{-4}	200,000	3.7
Co-60	Irradiation	4×10^{-4}	400,000	3.4
Nb-95	F. P.	2×10^{-3}	2,000,000	2.7
Cs-137	F. P.	2×10^{-3}	2,000,000	2.7

[a] Kahn and Goldin, "Radiochemical Procedures for the Identification of the more Hazardous Nuclides." Presented to Nuclear Engineering and Science Congress (1957).
[b] MPC values marked (g) are for damage to the intestinal tract and are one-day values from Morgan and Ford (Morgan and Ford, *Nucleonics.* 12, No. 6, 32 (1954).); other values from the Report of Sub-Committee II, ICRP (Recommendations of the Committee on Radiation Protection (ICRP), Supplement 6, British Institute of Radiology, London, 1955, p. 23.
[c] F. P. stands for fission product.

The significance of a radionuclide is due not only to its relative radiotoxicity but also to its concentration. Among the more abundant nuclides and consequently the more likely to be present in appreciable concentrations are the major fission products and heavy elements resulting from the operation of nuclear reactors, the naturally occurring radioactive elements and their compounds, and some radionuclides made in quantity because of their usefulness. Some of the important nuclides are listed in Table 103. The relative importance of such radionuclides depends upon their high toxicity (conversely, their low maximum allowable concentration) or their high abundance or both. Table 103 shows that the toxicity of these radionuclides varies over a wide range for the maximum permissible concentrations range from 40 $\mu\mu$c. per kg. for radium-226 to 2 million $\mu\mu$c. per kg. for cesium-137 and niobium-95, respectively.

Kahn and Goldin suggested that a term, pMPC, representing the comparative toxicity be used in addition to the conventional maximum permissible concentration. This term, pMPC, is analogous to pH for it is defined as the negative logarithm of the concentration, expressed in μc. per g. or μc. per ml., and thus can be employed as a convenient unit, having the advantages of eliminating the exponential notation and of using a larger (rather than a smaller) number for a nuclide of greater toxicity.

Shipman, Simone, and Weiss [5] call attention to the possibility that manganese-54 may be a dangerous radionuclide for it has a relatively long half-life of 291 days.

GROSS RADIOACTIVITY

As a consequence of the work of many investigators, it has been found that there are three series of natural radioactive elements, namely, the actinium series, the thorium series, and the uranium series. There are a number of other natural radioactive elements. As listed by Cowan [6] these include potassium-40, rubidium-87, samarium-147, lutetium-176, and rhenium-187. From the point of view of food chemistry, potassium-40 is the most important. It is to be found everywhere and comprises 0.011 per cent of the natural element. The amount of potassium-40 in the oceans is estimated to be of the order of 5×10^{11} curies or approximately 3×10^{-13} curies per ml.

Sampling.—It is preferable to sample foods using glass or plastic containers and to keep such samples under refrigeration for preservation. If a preservative must be used because of the length of time that elapses before the analysis is made, either alcohol or formaldehyde may be used, but in such instances control determinations must be made on the same amount of alcohol and formaldehyde used to preserve the specimen. If materials such as shellfish are being analyzed, they must be washed free of extraneous matter, preferably

[5] Shipman, Simone, and Weiss, *Science* '**126**, 971 (1957).
[6] Cowan, ''Quantitative Summary of Natural Radiation and Naturally Occurring Isotopes.'' Presented at University of Michigan (Feb. 1951).

before being placed in the sample container, and it is also good practice to weigh the sample at the sampling site.

Preparation of Sample.—The preparation of food samples for radiochemical analysis will be discussed in considerable detail under the methods detailed for the determination of specific radionuclides. The following method is of a more general nature.[7]

Prepare relatively fat-free food samples for counting by determining the total solids by drying at 103° C. and the ash by igniting, at 600° C., sufficiently large representative samples, in silica or stainless steel dishes, to yield from 1 to 20 g. of ash. Because, at times, aluminum counting dishes are used, neutralize the ash with 1 N nitric acid and then knead with a glass rod or pestle to form a smooth thin cream. Transfer suitable aliquots of the moist sample to counting dishes, spreading out the sample uniformly in the tared dishes. Dry in a thermostatically controlled oven, weigh again, and, if the sample shrinks or cracks, wet with a few drops of acetone containing 0.5 per cent of lucite. Evaporate off the acetone, leaving the lucite to bind the sample and hold it dust-free during the counting.

Process small samples that will yield less than 300 mg. of ash directly in stainless steel dishes, 18-8, capable of withstanding heating to 600° C. and nitric acid. If an internal proportional counter is to be used, the dishes should be 2 in. in diameter and ¼ in. in height, with a capacity of about 10 ml.

Fat-bearing foods such as dairy products and muscle tissues may cause some difficulty if dry ashed. It is best in such cases to use wet ashing procedures. Place samples, ranging from 2 to 10 g. in weight, in 100-ml. beakers and digest for periods of from 10 minutes to overnight at room temperature with 5 ml. of nitric acid (sp. gr. 1.42). Place the beakers on a hot plate after the sample is dissolved and raise the heat gradually. Evaporate with care to dryness. Cool to room temperature, add 5 ml. of nitric acid, and evaporate to dryness. Repeat with one more portion of nitric acid. Add 1 ml. of nitric acid and a few drops of 30 per cent hydrogen peroxide solution and again evaporate to dryness. This fourth digestion will be adequate for most samples. Remove the residual acid by twice washing down the sides of the beakers with water and evaporating to dryness.

If the residue comprises less than 50 mg., transfer the entire contents of the beaker with water and the aid of a rubber policeman to a tared counting dish. Evaporate the test solution to dryness,

---
[7] Setter, Hagee, and Straub, "The Analysis of Radioactivity in Surface Waters: Practical Laboratory Methods," *ASTM Bull.* 1958, No. 227, 35.

continue drying in an oven at 103° C., weigh the dish after cooling in a desiccator, and count.

Samples which will yield a large amount of ash, as for instance bone, should be digested in weighed beakers. After digestion, dilute the sample with a weighed quantity of water or transfer to a volumetric flask with water and the aid of a rubber policeman. Make to volume, stopper, and mix. Transfer a known volume with a pipette to a tared counting dish, dry in an oven, cool, weigh, and count. If the sample is dusty, use the acetone-lucite treatment described in a preceding paragraph.

In an alternative step in the wet-ashing procedure, after the digestions have been finished and porcelain or silica dishes were used, the dry ash can be ignited in a muffle furnace at 600° C.

The samples prepared in the manner detailed in this section and in subsequent sections can then be counted as described in the following section.

COUNTING

As will be stressed in the section on identification in this chapter, the characterization of a specific radionuclide may depend on the type of radiation—namely, alpha, beta, or gamma—or the energy of the disintegration. Counting procedures also depend on such characteristics. The disintegration of radioactive nuclides yields alpha or beta particles, which may or may not be accompanied by the emission of gamma rays. These emissions are detected by instruments, whose function is to detect as efficiently as possible.

The actual method of counting will depend in large measure on the instrumentation available. Consequently the analyst must obtain the details of the procedures of counting recommended by the manufacturer of the particular equipment purchased. To give a general picture of such procedures the counting methods suggested by the Sanitary Engineering Center of the United States Public Health Service [8] will be detailed.

Most counters work on the principle of collecting the electrons that are produced by the passage of alpha or beta particles through a counter gas. These electrons are drawn to a thin wire maintained at a high positive voltage. The U. S. P. H. S. suggests the use of an internal proportional counter because it has been found capable of counting either alpha or beta activity with a high efficiency. Such

---

[8] "Measurement of Radioactivity in Water, Bottom Silts, and Biological Materials," Sanitary Engineering Center, U. S. Public Health Service, Cincinnati (Jan. 1957).

counters will detect either alpha or alpha plus beta particles depending upon the operating voltage.

Internal Proportional Counter.—The internal proportional counter comprises a methane-argon flow type of counting chamber, a preamplifier, an electronic scaler, register, timer, counting gas, and a power supply assembly. Setter, Goldin, and Nader [9] describe such an arrangement. A cross section of the counting chamber is given in Fig. 88. The chamber consists of a center wire assembly in a hemisphere-and-piston arrangement.

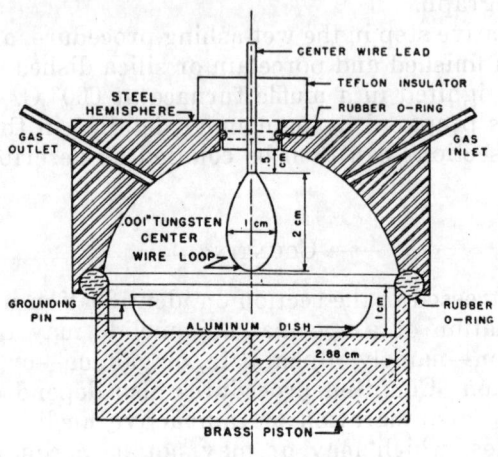

FIG. 88. Internal Counter Chamber [a]

[a] After Setter, Goldin, and Nader, *Anal. Chem.* **26**, 1304 (1954).

By proper adjustment of the operating voltage both alpha and beta emissions can be counted. At the low voltage appropriate for alpha counting, beta particles are not detected because the counting gas is not ionized sufficiently to induce a response in the preamplifier. When the voltage is set at a higher value to count the beta emissions, then both beta and alpha particles are counted for both can activate the preamplifier.

Operating Voltage Plateau.—It is necessary to count at a voltage where the rate does not change substantially for a small change in voltage. To find this plateau, make a series of determinations using an alpha source of counting rates at voltages above and below the approximate alpha operating voltage stated by the manufacturer of the instrument and graph the results. There should be a region ex-

⁹ Setter, Goldin, and Nader, *Anal. Chem.* **26**, 1304 (1954).

tending over a range of 150 volts in which the counting rate is relatively independent of the voltage. Select the alpha counting operating voltage at approximately the midpoint of this range.

Make an analogous series of measurements using a beta particle source. Determine the beta voltage plateau and select the midpoint of this range as the operating voltage.

Background.—To make certain that the instrument is working properly it must be checked 2 or 3 times a day by counting an alpha or a beta source with a known counting rate for a total of at least 10,000 counts. The known counting rate must be determined by counting the source for a total of not less than 40,000 counts. The standard source, actually a secondary standard, for alpha-counting calibration is uranium oxide plated on stainless steel. An analogous standard source is used for the calibration of beta counting, except that the uranium oxide plated on stainless steel is covered with aluminum foil weighing 8 mg. per square centimeter (0.001 in. thick), which completely masks alpha activity.

The counting rate determined in this manner should check the known rate within 3 per cent. If it does not do so, check the operating voltage and make a redetermination of the voltage plateau. If such re-evaluation still does not correct the discrepancy, the instrument must be serviced.

Measure the alpha and beta backgrounds by determining the counting rate at an appropriate voltage using a clean empty similar counting dish in the chamber. Do this two or three times a day, obtain the average background counting rate by calculation. The total background counting time should equal the longest counting time used for any sample.

The alpha radiation background is generally less than 0.5 count per minute if the counting chamber has not become contaminated. With the voltage used for counting beta activity, the background is 50 to 60 counts per minute, attributable to gamma and cosmic radiation. If background counts are made with blank dishes three times a day for two 8- and one 16-minute intervals, following calibration with a standard, the total 32-minute counting period will yield an alpha-background value reliable to ±50 per cent and a beta-background value reliable to ±5 per cent.

A high background count may be attributable to contamination of the counting chamber. Counting that is erratic may be attributable to high voltage discharge from dust or excessive roughness of the electrode. Check the preamplifier tubes, condensers and the resistors, and the scaler tubes and circuitry by trouble shooting in the order mentioned.

Calibration.—No instrument can detect all of the radioactive disintegrations occurring in a sample. For this reason it is necessary to correct the observed counting rate of any sample. The factors involved in this loss of counting efficiency are (1) the geometry of the counting chamber, (2) back-scatter, (3) absorption, and (4) self-absorption. These factors are relatively constant for any given instrument, radionuclide, and method of preparing the sample. If the nature of the radioisotope is not known, the factors cannot be determined accurately. In the U. S. P. H. S. method [8] the corrections used are based on the assumption that alpha activity is due to natural uranium and the beta activity to mixed fission products, that is, to fallout.

Geometry.—Not all of the radiation given off by a sample is sent in the direction of the detecting unit. This factor (G) is a measure of the fraction of the emitted radiation that reaches the detector. For internal proportional counters and for a small sample in the center of a counting dish, this factor is considered to be 0.5. This factor has been shown to be experimentally adequate for a sample spread uniformly over a 2-inch counting dish, hence $G = 0.5$.

Back-scatter.—The reflection of beta and alpha particles into the counter from the sample dish is known as back-scatter (B). For alpha particles there is very little back-scatter. The amount is seldom more than 2 per cent, hence for alpha particles $B = 1.00$ to 1.02. The factor for the back-scatter of beta particles is dependent upon the energy of the radiation and the effective atomic number of the sample dish and its support. Nader, Hagee, and Setter [10] have shown that one-year-old mixed fission products in aluminum dishes mounted in a brass chamber have a back-scatter factor of 1.31.

Absorption.—The stoppage of particles by a window or wall of the detector when the sample is outside the detector is termed absorption. This does not occur in internal proportional counters.

Self-Absorption.—In principle, this factor is the same as that of absorption, but it happens in the sample itself. Thus, some of the particles formed in the disintegrations are absorbed before they can get out of the sample into the counting gas and consequently are not counted. The transmission factor (T) of radiation through a sample is dependent upon the quantity and chemical composition of the ash, the type of radiation, and the energy of the radiation.

---

[10] Nader, Hagee, and Setter, *Nucleonics* 12, No. 6, 29 (1954).

Alpha particles have a very low order of penetrating power and the alpha activity of natural uranium, which has an energy of 4.2 and 4.76 Mev, is markedly reduced by a sample thickness of a few milligrams per square centimeter. Indeed a sheet of paper will stop such alpha particles. Samples of thickness less than 4 mg./cm.^{2} should be used when determining the alpha count of ash of unknown activity.

While the loss by self-absorption of beta particles is less than that of alpha particles, beta particles of low energy are lost by self-absorption.

The factors to be used in calculating the counting efficiency (E) from the equation:

$$E = GBT$$

are given in Table 104.

TABLE 104. COUNTING EFFICIENCY FACTORS[a]

	Alpha	Beta
G	0.50	0.50
B	1.02	1.36 (stainless steel dish)
T	Sample thickness, mg./cm.^{2}	
1.00	0.	0.
0.90	0.8	1.4
0.80	1.6	3.9
0.70	2.4	6.6
0.60	3.2	13.5
0.50	4.0 [b]	

[a] "Measurement of Radioactivity in Water, Bottom Silts, and Biological Materials," Sanitary Engineering Center, U. S. Public Health Service, Cincinnati (January, 1957).
[b] For alpha thicknesses above 4.0 mg./cm.^{2}, divide 2 by the thickness in mg./cm.^{2}; thus, for example, for 10 mg./cm.^{2}, $T = 0.2$; and for 16 mg./cm.^{2}, $T = 0.12$.

Calibrating Standard.—In the U. S. P. H. S. method it is pointed out that the over-all efficiency of counting may be determined, without any knowledge of the efficiency factors discussed above, by adding a known amount of radioactive material to a series of counting dishes and then measuring the counting rate. It is necessary to run a series of determinations in which different weights of ash are used because the self-absorption varies with the thickness of the samples.

Standard solutions of various radioisotopes that can be used for calibrating standards can be obtained from the National Bureau of Standards. Thallium-204 is used most frequently as the standard for mixed fission products and is suggested for the beta determination. Uranium salts are suggested as the standard for alpha deter minations. One milligram of natural uranium emits 1520 alpha particles per minute.

Prepare several nonradioactive samples of different weights up to 300 mg. by processing variable volumes of tap water in tared counting dishes. If the solids content of the water is too low to give this amount of ash, add a sufficient amount of fine calcium carbonate to obtain the amounts desired. Add to each dish an identical aliquot of uranium salt or thallium-204 compound containing about 1000 counts per minute for the alpha or beta calibration, respectively. Add water, disperse the radioactive substances uniformly, dry, weigh, and count. Divide the net counting rate for each sample to obtain the counting efficiency. Plot a curve of counting efficiency against weight of sample ash as a calibration graph.

Counting Procedure.—Count the samples by placing them in the counting chamber at the appropriate operating voltage, flushing the chamber to replace the air with the counting gas, and record the counts over a specified time. Count each sample for two 16-minute periods, preceding each counting period by a 30- to 60-second flushing of the counting chamber with the counting gas. If the two measurements do not agree, it is probable that the counting gas was not properly flushed out and a third count should be made immediately using the same time period.

For this practical time period of counting, the minimum detectable alpha and beta activities [11] are—according to Goldin, Nader, and Setter—0.5 and 4 counts per minute, respectively. When it is desired to obtain greater precision of counting, increase the counting time. When samples of moderate or high activity are being counted, the counting time may be reduced.

Calculations.—The net sample counting rate in counts per minute (cpm.) for alpha activity is given by the expression:

$$(NR)_a = R_a - B_a$$

in which $(NR)_a$ = net sample alpha counting rate, cpm.
 R_a = observed sample alpha counting rate, cpm.
 B_a = background alpha counting rate, cpm.

and the net sample counting rate for beta activity is given by the expression:

$$(NR)_\beta = R_\beta - B_\beta - (NR)_a$$

in which $(NR)_\beta$ = net sample beta counting rate, cpm.
 R_β = observed sample beta counting rate, cpm.
 B_β = background beta counting rate, cpm.

---

[11] Goldin, Nader, and Setter, *J. Am. Water Works Assoc.* **45**, 73 (1953).

The alpha activity can be calculated by use of the expression:

$$A_a = \frac{(NR)_a}{2.2gE_a}$$

in which

A_a = alpha activity in $\mu\mu$c./g.
g = weight of the original sample in grams
E_a = efficiency of alpha counting, counts per minute per disintegrations per minute

The counting error in alpha activity can be calculated, based on the variability as discussed in the succeeding section, by the expression:

$$(CE)_a = \frac{1.96}{2.2gE_a}\sqrt{\frac{R_a}{T_a} + \frac{B_a}{t_a}}$$

in which

T_a = alpha counting time in minutes
t_a = background alpha counting time in minutes

and the other terms have the meanings previously assigned.

The beta activity can be calculated by use of the expression:

$$A_\beta = \frac{(NR)_\beta}{2.2gE_\beta}$$

in which

A_β = beta activity in $\mu\mu$c./g.
E_β = efficiency of beta counting, counts per minute per disintegrations per minute

The counting error in beta activity can be calculated, based on the variability as discussed in the succeeding section, by the expression:

$$(CE)_\beta = \frac{1.96}{2.2gE_\beta}\sqrt{\frac{R_\beta}{T_\beta} + \frac{B_\beta}{t_\beta} + \frac{R_a}{T_a} + \frac{B_a}{t_a}}$$

in which

T_β = beta counting time in minutes
t_β = background beta counting time in minutes

and the other terms have the meanings previously assigned.

The U. S. P. H. S. suggests the following data be reported: the sample identification number; the sample collection station and date; sample type, such as fish, milk, etc.; the weight processed in grams; the method of preparation of sample, that is, whether oven-dried, ashed in a muffle, wet-ashed, etc.; the alpha activity ± the counting error in $\mu\mu$c./g.; the beta activity ± the counting error in $\mu\mu$c./g.; the percentage of total solids; the percentage of ash; and the radioisotopes used as the calibration standards.

Variability.—Radioactive disintegrations occur in a random manner. For this reason, the counting rate of any sample will vary about a mean value. This has been discussed by Goldin, Nader, and Setter.[12] It can be shown that the standard deviation of certain quantities is given by the following:

Quantity	Standard Deviation of Quantity
Number of counts, N	$\sqrt{N}$
Counting rate, R, measured in time, T, that is, $R = \dfrac{N}{T}$	$\sqrt{\dfrac{R}{T}}$
Sum or difference of counting rates: $R_1 \pm R_2$	$\sqrt{\dfrac{R_1}{T_1} + \dfrac{R_2}{T_2}}$

An especially important instance of the last form is that concerned with the counting detailed in the preceding section in which a net counting rate is calculated by subtracting the background counting rate from the observed counting rate when the latter comprises the sample plus the background counting rate.

As noted, determinations will be more accurate as the counting time is increased, consequently the longest counting time practical for the analyses involved should be used. In the method above, a counting time of 16 minutes was suggested. Other suggested minimum counting times are 30 minutes or a total of 2500 counts above the background, whichever is less. By using two equal counting periods as detailed in the counting procedure a check of the observed counting rate is obtained.

Confidence Level.—If the standard deviation is multiplied by a constant, one can obtain limits of error. In the following tabulation, the confidence level is given for each constant.

Confidence level, %	Constant
50	0.67
68	1.00
90	1.64
95	1.96

It is stressed in the U. S. P. H. S. method that the method is not. one of great accuracy. It is questionable that great accuracy is re-

---

12 Goldin, Nader, and Setter, *J. Am. Water Works Assoc.* **45**, 73 (1953).

quired in such instances for there may well be a thousand-fold difference in the significance of the pollution depending upon, as was discussed in the section on radiotoxicity, the particular radioisotopes present.

A major source of inaccuracy in this method is the self-absorption factor, as discussed. The uncertainty in the evaluation of this factor may lead to an error of the order of 20 to 25 per cent in the beta assay of samples containing 100 mg. of ash, with even a larger error in the alpha assay. If there is sufficient activity to warrant making a determination of greater accuracy, then a smaller sample of the order of 25 mg. should be counted. The self-absorption error will in consequence be reduced, but there is a residual error of about 15 per cent in the beta assay attributable largely to the uncertainty in the back-scatter factor and to nonuniform deposition of the sample. This error cannot be materially reduced without additional knowledge of the energy of the radiation.

As mentioned, alpha particles cause an error in the beta counting for the alpha plateau extends into the beta region, consequently more than 1 count is recorded at the beta voltage for each alpha count. The amount of this depends on the sample thickness, the uniformity of deposition, the energy of the radiation, and possibly on other factors. If the alpha count is considerably lower than the beta count, because either the alpha activity itself is low or the alpha self-absorption losses are high, this error will be small. Since alpha particles are of greater significance from a pollution point of view, if the alpha count is of the same order of magnitude as the beta count, the resulting error in the beta count is not too important.

Geiger-Müller Counter Method.—In general, the method detailed for the internal proportional counter is applicable to radioactivity counting with a Geiger-Müller tube and scaler but with some changes depending on the instrument. Thus most end-window Geiger-Müller tubes are designed to be used with samples that are 1 in. in diameter. The standards used for calibration must be of the same dimensions.

The detector of a Geiger-Müller counting unit comprises a tube having a very thin end window with a weight of less than 2 mg./cm.^{2} so that gamma rays and particles can penetrate readily. This tube is encased in a lead shield approximately 1 to 2 in. thick to reduce the background effect. Because the Geiger-Müller tube produces a pulse of sufficient energy, no amplifier is necessary and the tube can be used to actuate a scaler directly. It is, however, an "all or nothing" instrument in that gamma rays and radioactive particles must have sufficient energy, otherwise no pulse is produced.

Instruments of the Geiger-Müller type have the advantage of simplicity since they do not require an amplifier and no difficulties attributable to the electrical or chemical nature of the sample are encountered because the sample is placed externally to the counter. They have certain marked disadvantages. First, they have virtually no sensitivity for alpha particles; second, their sensitivity for beta particles is about 10 to 20 per cent of that of the internal proportional counter; and third, this sensitivity is even less for beta particles of low energy. The latter disadvantages are only slightly offset by the lower background count. Great care must be exercised in establishing the voltage plateau, otherwise the Geiger-Müller tube may be ruined. Such counters have their greatest value in the determination of higher levels of radioactivity.

As representative of such instruments the operation of a Baird-Atomic Model scaler, with an Anton Geiger-Müller tube and lead shield, can be described.

Starting Instrument.—With high voltage switch at "off" position, and the coarse and fine voltage controls turned completely counterclockwise, with the 60-cycle switch at "off" and the "count" switch at "off," turn the power switch on.

60-Cycle Test.—Set the timer and counter at zero, the automatic control at 10 × and the scale selector at 256 (this arrangement sets the scaler to turn off automatically when a count of 2560 is reached) throw the 60-cycle switch to "on" and allow the machine to operate until 10 is reached on the counter, at which time the counting will stop. Repeat for different total counts or run manually for 1 minute.

Thus, with the instrument set as noted, the timer will reach 0.71 minutes when it stops, hence

$$\frac{2560}{0.71} = 3600 \text{ cpm.}$$

This checks the operation of the instrument, for a 60-cycle current should give 3600 impulses or counts per minute. If the instrument is set at 10 ×, a scale-selector setting of 1024 will stop the instrument when a count of 10240 is reached. The time will be 2.84 minutes, hence

$$\frac{10240}{2.84} = 3610 \text{ cpm.}$$

as an alternate check on the instrument.

Voltage Plateau.—The voltage range in which large changes in voltage will give small changes in counts per minute must be determined. With the particular Anton tube used, a voltage of 1000 must not be exceeded.

Set the 60-cycle switch to "off," turn the count indicator to "off," reset the counter and timer, turn the fine and coarse high-voltage knobs completely counterclockwise, and turn the voltage switch to "on." Wait until the indicator light turns bright red and turn coarse knob until the voltage reaches 500.

Place a 0.1 μc. standard 1 in. away from the end window of the Geiger-Müller tube and take counts at 50-volt intervals from 500 to 950 volts. Set the automatic control at 10 $\times$ and the scale selector at 16 so that the instrument will stop automatically at a count of 160 per minute. With such settings assume the following readings were obtained:

Voltage	Time, min.	Cpm.
500	0.0	...
550	0.0	...
600	0.0	...
650	0.0	...
700	5	32
750	0.71	220
800	0.50	320
850	0.46	350
900	0.43	370
950	0.43	370

Thus the plateau is in the 900- to 950-volt range.

Gamma-Ray Method.—The A.O.A.C.[13] describes a gamma-ray method using a Geiger-Müller counter which is applicable to the determination of radium in quantities greater than 10^{-7} g. This is an arbitrary limit and depends on the particular instrumentation used and the accuracy desired. The instrument employed should have a response to gamma rays that is linear to at least 10,000 cpm., and the sensitivity should be such that 0.1 microgram of radium will double the background counts.

Standards.—Obtain a set of 13 gamma-ray standards prepared and evaluated by the National Bureau of Standards. These should range in value from 0.1 to 100 micrograms of radium in the following steps: 0.1, 0.2, 0.2, 0.5, 1.0, 2.0, 2.0, 5.0, 10.0, 20.0, 20.0, 50.0, and 100.

Procedure.—To obtain the total radioactivity, seal the entire sample or one or more subdivisions hermetically in a suitable container or containers and allow to stand 30 days.

---

[13] *Methods of Analysis*, A. O. A. C. 8th ed., Washington, 1955.

Turn the instrument on and allow it to stand for at least 15 minutes before starting the determination. Determine the background count. Place the sample prepared in a convenient manner at such a distance from the detector tube so that the indication on the meter may be read easily. Roughly estimate the value of this reading, remove the sample, and replace it by a radium standard of such strength that about the same reading will be obtained. (If an Anton tube is used, since the position between the sample and the tube is relatively fixed, the size of the sample must be adjusted to accommodate the planchet used.) Remove the sample and standard to such a position that they will not affect the background and wait 5 minutes for the background to become stable again.

Compare the sample against the standard, placing each in the same relative position to the counter. Take check background counts after each pair of observations and allow sufficient time for the instrument to come to equilibrium for each counting of the sample, the standard, and the background. This time of equilibrium will depend on the time constant of the instrument and may vary from 1 to 5 minutes or for even a longer period. Record the individual counts and the total count for each period. Two check determinations should be made.

A series of counts will then comprise: background count begins, after 5 minutes (or other time interval) background count ends; wait 5 minutes; sample run starts, after 5 minutes sample count ends, remove sample; wait 5 minutes; second background run starts, after 5 minutes background count ends; wait 5 minutes; place standard in position, standard count starts, after 5 minutes standard count ends, remove standard; wait 5 minutes; third background count starts, after 5 minutes background count ends; wait 5 minutes; place second sample or check sample in position, etc.

Calculation.—Average each successive pair of background counts and subtract this average count from the sample or standard count. For instance, in the sequence above, average the first and second background counts and subtract this average from the sample count; then average the second and third background counts and subtract this from the standard count; etc.

From the three comparisons three ratios of sample to standard can be obtained. The standard deviation can be calculated as detailed in the internal proportional counter method.

Multiply the value of the standard by the average ratio and express the result as micrograms of radium ± the standard deviation in micrograms of radium. This is the radioactivity of the portion of sample used for the analysis. Express the final result in milli-micrograms or micrograms of radium per gram or per milliliter of sample ± the standard deviation, multiplying or dividing the standard deviation by the same factor used for conversion of the radio-activity of the portion of sample used for the analysis.

Radioactive Contamination—Rapid Method.—An emergency level procedure has been detailed for the determination of radioactive contamination.[14]

Apparatus.—Use a portable count-rate meter comprising a self-quenching Geiger-Müller tube with a side wall not over 32 mg./cm.[2] mounted in a slide opening metal shield with a threshold of approximately 800 volts, operated at about the middle of its voltage plateau, the slope of which does not exceed 10 per cent, and connected with a coaxial cable to a power supply and an electronic amplifier unit with a meter calibrated in milliroentgens per hour, connected through a sensitivity switch to provide three ranges of scale reading, such as 0-20, 0-2, and 0-0.2 mr./hr.; the response should be linear within each range.

Comparison Standard.—The comparison standard should induce a meter response identical to that from the surface of water contaminated with fission products decaying at a rate of 2×10^5 disintegrations per minute per milliliter (the emergency tolerance level for water to be consumed for not longer than a 10-day period). Make the standard in the following manner: suspend about 3 g. of 60-mesh uranium acetate, adjusted as necessary by trial, uniformly in 5 g. of liquid casting plastic, level off, and allow to solidify in a shallow container such as the lid of an ointment can, about 80 mm. in diameter and a side wall 15 mm. deeper than the layer of plastic. Fit the base of the ointment can with an indented ring 15 mm. below its edge to serve as the container for the liquids and the finely divided solids to be tested and to protect the standard when not being used. An additional standard of one-half this activity may be prepared similarly for monitoring supplies to be consumed over a 30-day period.

Procedure.—Set the selectivity switch for the widest measurement range. Open the shield and place the Geiger-Müller tube diametrically across the standard in contact with the edge of the container at 2 points. Adjust the meter pointer to a convenient value about midway in the scale with the calibration screw and record the reading as the average of the fluctuations over 1 to 2 minutes. Dupli-

14 *J. Assoc. Official Agr. Chem.* **38,** 678 (1953).

cate readings should check within ± 5 per cent, making certain to avoid any extraneous radiation such as that that might be derived from a luminous dial of a watch.

Fill the sample container with liquid or finely divided solid to the level of the indented ring and obtain duplicate readings. Sample readings that are within ± 10 per cent of the standard reading are of practical significance for monitoring under emergency conditions.

Scintillation Counter.—There are various types of scintillation counters as there are other radioactivity measuring instruments. Usually this type of instrument comprises a photomultiplier tube equipped with gamma-sensitive phosphor and its circuit enclosed in a light-tight housing. Because the output pulses are too weak to actuate a conventional scaler, it is necessary to use an amplifier. In addition, some method of mounting the samples must be provided so that sample positions can be duplicated.

With this type of instrument it is essential to determine the voltage plateau, in order to avoid placing excessive voltage on the photomultiplier tubes. It is best to operate in a voltage range some 50 to 75 volts above the lower end of the voltage plateau. Care must also be exercised not to expose the photomultiplier tubes to light when the high voltage is on.

Scintillation counters are also used for spectrometers, and such instruments can be used to evaluate a radionuclide like cesium-137, which is a gamma emitter. A Geiger-Müller counter cannot be used for this purpose because it cannot differentiate energy levels; it is, as mentioned, an all or nothing instrument.

By replacing the gamma-sensitive phosphor with anthracene crystals or organic plastic scintillators, scintillation counters can be used for the determination of beta rays.

Scintillation counters have a number of advantages. Thus they can be used at times for direct counting of the sample whether liquid or solid, but their counting efficiency appears to be somewhat lower than that of internal proportional counters.

IDENTIFICATION

The identification of a particular radionuclide may be based on chemical separation techniques, which may include separation of a daughter activity, or on radiation characteristics such as the half-life, type of radiation—namely, alpha, beta, or gamma—or on energy. Usually a combination of chemical and physical procedures is more suitable than either approach alone. In some instances determination of radiation characteristics may not be practical because the radioisotopes being determined or identified are present in extremely low concentrations.

At low concentration levels it is generally necessary to concentrate the active material from relatively large samples. At still smaller concentrations, and in instances in which large samples are not available, specialized types of counting equipment must be used. With these types the background count may be lowered sufficiently by use of heavy shielding and other devices so that it does not interfere in the analysis.

The preliminary sample preparation is a major problem with many kinds of materials as will be clear from the succeeding section on the determination of some radionuclides. The analysis of solid samples requires total dissolution of the sample; thus food is either wet- or dry-ashed, and the ash is then dissolved.

Because radioisotopes are usually present in unweighably small amounts, known quantities of carrier materials are usually added so that the radionuclides can be manipulated. It is not always necessary for the recovery to be quantitative because the results can be corrected on the basis of the recovery of the carrier. For this reason rapid semiquantitative methods can be used. Kahn and Goldin [15] have described a few methods which were developed for the determination of specific radionuclides at low concentration levels. These methods are sensitive for 30-minute counting periods on high-efficiency conventional counting equipment in the order of 10 $\mu\mu$c. for beta emitters and 1 $\mu\mu$c. for alpha emitters. The methods of identification are detailed for analysis of water but, with the modifications given for preparation of samples under the determination of strontium-90, may be applied to various foods.

Radium-226.—Add 200 mg. of lead and 5 to 10 mg. of barium to a liter of water or the equivalent prepared food sample as carriers for the radium and add sufficient ammoniacal citrate to prevent precipitation of the carriers before the interchange with radium has occurred. Precipitate the sulfates with sulfuric acid, allow to settle, and collect the precipitates by decanting the supernatant liquid. Wash the precipitates with concentrated nitric acid and then dissolve them in ammoniacal ethylenediaminetetraacetic acid (EDTA). Reprecipitate the barium sulfate, which carries the radium, by the addition of

15 Kahn and Goldin, "Radiochemical Procedures for the identification of the more Hazardous Nuclides." Presented to the Nuclear Engineering and Science Congress (1957).

excess acetic acid. Wash the precipitate, mount, dry, flame, and weigh.

Determine the radium content by alpha counting. The radium may be checked for radiochemical and isotopic purity by following the growth of radon-222 and its daughters in the precipitate.

Strontium and Barium.—Add 20 mg. each of strontium and barium carriers, and carbonate to 4 liters of water to precipitate the strontium, barium, and calcium carbonates. Allow to settle, discard the supernatant liquid, transfer to a 50-ml. centrifuge tube, redissolve the carbonates in 6 N. hydrochloric acid, and reprecipitate the strontium and barium nitrates (see the determination of strontium-90) in strong nitric acid. The small amount of calcium nitrate that is co-precipitated may be dissolved out by use of anhydrous acetone. After redissolving, remove other contaminating active nuclides by the addition of ferric iron and by precipitating the hydroxide. Separate the barium and strontium as in the Goldin method (page 884) by precipitating barium chromate from a buffered solution. Precipitate strontium oxalate, dry, mount, weigh, and count. Additionally purify the barium by precipitation with hydrochloric acid-ether mixture, dry, mount, weigh, and count.

Strontium-90.—Add 400 mg. of strontium nitrate carrier to a liter of water or equivalent sample preparation. Add sodium carbonate to precipitate the strontium together with the calcium. Collect the precipitate by decanting the supernatant and redissolve in dilute acid. Add barium carrier and precipitate it as chromate from an acetic acid-acetate buffer at pH 5 to remove barium-140, which would subsequently interfere because of its lanthanum-140 daughter. Add ferric ion and precipitate the hydroxides with ammonium hydroxide, which precipitate will include the insoluble rare-earth hydroxides. Add rare-earth and zirconium carriers, and precipitate these with ammonia to assure more complete removal of interfering nuclides. Precipitate the strontium from the supernatant solution with carbonate and allow to stand overnight or longer if possible for the time required for the ingrowth of yttrium-90. Buffer the solution at pH 5 and extract the yttrium with 2-thenoyltrifluoroacetone (TTA) in monochlorobenzene. Re-extract the yttrium-90 from the organic solvent phase with dilute nitric acid and subsequently mount, dry, and count. Determine the yield of strontium by flame photometry.

Determine the strontium-90 by beta counting the yttrium-90 formed, correcting for the fraction of yttrium which has grown into the strontium sample. Yttrium-90 activity reaches 16 per cent of the strontium-90 activity in 16 hours, 40 per cent in 48 hours, and

84 per cent in 168 hours. The purity of the yttrium-90 may be readily checked by decay measurements.

Cesium.—Add 25 mg. cesium chloride and 200 mg. of potassium chloride as carriers to a liter of water or equivalent sample. Precipitate potassium and cesium as the cobaltinitrites (see page 770) from acetic acid solution. Collect the precipitate by decanting the supernatant solution, wash with acetic acid, and dissolve in 6 N hydrochloric acid. Add silicotungstic acid to precipitate cesium silicotungstate and thus separate it from the other alkali metals. Add perchloric acid, heat the solution to precipitate the silicon and tungsten oxides, and centrifuge to pack and remove these oxides. Cool the perchloric acid solution and add absolute alcohol to precipitate cesium perchlorate. Collect, wash with alcohol, mount, dry, weigh, and count. Determine the cesium-137 or cesium-134 content by beta or gamma counting.

Iodine-131.—Add sodium iodide as the carrier to the sample to be analyzed. Oxidize with sodium hypochlorite in alkaline solution. Reduce the iodide to elemental iodine and extract with carbon tetrachloride. Reextract the iodine with sodium bisulfite and precipitate as silver or palladium iodide for beta counting.[16]

As an alternative procedure, if a scintillation counter is available, the carbon tetrachloride or sodium bisulfite extract may be gamma counted. This direct counting will eliminate the troublesome step of handling silver iodide which tends to form colloidal solutions.

Cobalt-60—Add a cobalt carrier to the sample being analyzed and precipitate it as the hydroxide. Dissolve the hydroxide in acetic acid and reprecipitate as potassium cobaltinitrite. Redissolve this precipitate, add a cerium carrier, and precipitate the latter with ammonia to remove contaminating activities. Again precipitate the cobalt as, in this instance, the 1-nitroso-2-naphthol salt, ignite to the oxide, mount, weigh, and beta count.

Manganese-54.—Add a manganese carrier, together with cerium and zirconium hold-back carrier to the dissolved sample [17] and oxidize the mixture with sodium chlorate. Reduce and dissolve the insoluble manganese dioxide with sodium bisulfite and hydrochloric acid, and scavenge the solution with basic ferrous acetate to remove interfering radionuclides. Precipitate the manganese as the ammonium phosphate salt and ignite to form the pyrophosphate which is used to determine the chemical recovery. The resulting precipitate may be

16 Glendenin and Metcalf, in Coryell and Sugarman, eds., *Radiochemical Studies—The Fission Products.* McGraw-Hill, New York, 1951, p. 1625.
17 Shipman, Simone, and Weiss, *Science* **126**, 971 (1957).

gamma counted with a sodium iodide-thallium activated crystal detector and resubmitted to gamma spectral analysis.

RADIOCHEMICAL DETERMINATION OF STRONTIUM-90

AEC Method.—Strontium is separated from calcium, other fission products, and natural radioactive elements by use of fuming nitric acid,[18] which removes the calcium and most of the other interfering ions, for strontium nitrate is insoluble in 80 per cent and higher percentages nitric acid.[19] Radium and lead are removed with barium chromate. Traces of other fission products are scavenged with yttrium hydroxide. After the strontium-90-yttrium-90 equilibrium has been attained, the yttrium-90 is precipitated as the hydroxide and converted to the oxalate for counting.

Preparation of Samples.—(1) *Milk.*—Dry-Ashing Variation.—For each 2 qt. of milk, add 1 ml. of strontium carrier solution containing 20 mg. of strontium per milliliter and evaporate to near dryness with mechanical stirring. As much as 4 qt. can be taken to dryness; however, with this volume some charring will take place but such charring does not affect the final results. With larger quantities the residue will appear charred around the edges but will be pasty in the center. This condition is considered the best state at which to transfer to silica trays for ashing.

Transfer the residue to a fused quartz tray. Place the tray in an electric muffle furnace. It is preferable when sufficient time is available to raise the muffle furnace temperature slowly. Thus, place the dried milk paste in a furnace set at about 200° C. for several hours to reduce the mass to a semicarbonized form. Then raise the temperature slowly to 500° C. over a period of several hours. Hold at 500° C. for 30 hours. The time required to ash large amounts of material can be reduced by mixing the cake periodically in order to break up the lumps and to expose the unashed surfaces to the flow of air.

Grind the ashed milk to a fine powder with a mortar and pestle. Weigh to the nearest tenth of a gram. Transfer the sample, equivalent to 2 qt. of milk, to a 250-ml. beaker and add 50 ml. of water. Heat and add nitric acid dropwise with stirring. The solution should be complete except for traces of carbon and silica. Filter through a 9-cm. glass fiber filter to remove the traces of carbon and insoluble material. Wash the residue with nitric acid (1:9). Discard the residue.

---

[18] Atomic Energy Commission, E-38-01, (1957).

[19] Willard and Goodspeed, *Ind. Eng. Chem., Anal. Ed.* **8**, 414 (1936).

Transfer the solution to an 800-ml. beaker and dilute to 500 ml. with water. Add 2 ml. of phosphoric acid. Add sodium hydroxide solution (240 g. per liter) dropwise with mechanical stirring to obtain a pH of 10. (It is preferable to store the sodium hydroxide solution in polyethylene bottles to prevent the solution of silica that occurs when it is stored in glass bottles). Allow the mixture to cool and the phosphate precipitate to settle. Filter by suction through a 9-cm. glass fiber filter. Test the filtrate for excess phosphate by adding a few drops of a barium chloride solution. If no precipitate is to be seen, repeat the precipitation, using additional phosphoric acid because there is insufficient phosphate present.

Repeat the steps of the preceding paragraph. Transfer the solution to a 600-ml. beaker and evaporate until salting out occurs. Add 60 ml. of water and then add 210 ml. of 90 per cent fuming nitric acid with magnetic stirring, using a Teflon bar. This operation must be performed in a hood. Add the fuming nitric acid slowly in order to dissolve completely all the solid matter before the 75 per cent nitric acid concentration is attained. To achieve this result, add about 100 to 110 ml. of the 90 per cent fuming nitric acid first to dissolve the solid matter and after solution is complete add the remainder of the nitric acid. Stir for 30 minutes. Complete the analysis as detailed under *Procedure.*

Wet-Ashing Variation.—Add 1 ml. of strontium carrier solution containing 20 mg. strontium per milliliter to 2 qt. of milk and evaporate to approximately 700 ml. with mechanical stirring. Add the pasty material in small portions to 500 ml. of hot nitric acid, avoiding excessive foaming. Add additional nitric acid if necessary. Evaporate to approximately 300 ml., then dilute to 1 liter with water. Heat to boiling to dissolve any precipitated salt. Transfer successive portions of the hot solution to a 500-ml. separatory funnel, drawing off the aqueous phase into a 1500-ml. beaker and discarding the fatty layer. Add sufficient pellets of sodium hydroxide to adjust the pH of the solution to 2. Add 2 ml. of phosphoric acid and continue with the preparation of the sample as detailed in the dry-ashing variation from the point of adding sodium hydroxide solution (240 g. per liter).

Powdered Milk.—Weigh out 5 pounds of powdered milk. If a monthly composite is to be analyzed, combine equal aliquots of weekly samples to obtain the total weight of 5 pounds. Place the sample into 2 fused quartz trays. Place the trays in an electric muffle furnace controlled at 500° C. and ash to a white residue. This will require 30 to 35 hours as noted in the milk dry-ashing variation.

Grind the ashed milk to a fine powder with a mortar and pestle, and mix thoroughly. Weigh to the nearest tenth of a gram. Transfer 15 g. of the ash to a 250-ml. beaker. Add 50 ml. of water and 1 ml. of the strontium carrier solution containing 20 mg. of strontium and continue with the preparation of the sample as detailed in the milk dry-ashing variation at the point beginning . . . "Heat and add nitric acid dropwise with stirring. . . ."

Bone.—Place the bone in a fused quartz tray and ash at 500° C. until all the organic matter has been oxidized. This will require overnight ashing if air is fed to the furnace and longer if air is not fed. Grind the ash to a fine powder in an electric grinder or with a mortar and pestle. Weigh out 15 g. of the ashed bone and transfer it to a 1-liter beaker. Add 75 ml. of hydrochloric acid and heat until solution is complete. Add 1 ml. of strontium carrier solution. Filter through a 9-cm. glass fiber filter to remove any carbonaceous material present. Wash the residue on the filter with water. Discard the residue. Dilute to 700 ml. with water. Add 2 ml. of phosphoric acid and continue with the preparation of the sample as detailed in the milk dry-ashing variation from the point of adding sodium hydroxide solution (240 g. per liter).

Plant Material and Tissue.—Dry-Ashing Variation.—Weigh the sample and place it in a fused quartz tray or dish and dry in an oven at 50 to 60° C. for 12 hours. Reweigh the dried material. Ash completely in an electric muffle furnace controlled at 500° C., taking the precautions noted in the preceding preparations. Grind the residue to a fine powder with a mortar and pestle, and weigh. Transfer 0.5 g. of ash to a plastic planchet and beta count for mixed fission products, standardizing with 0.5 g. potassium carbonate. Record the total activity in d./m./g. ash, but if the beta activity is not required, this count may be omitted.

Transfer to a platinum crucible, add four times the amount of sodium carbonate, and mix thoroughly. Fuse to a clear melt in a muffle at 900° C. and cool. Add 25 ml. of water to the crucible and a few milliliters of 60 per cent perchloric acid to dissolve the carbonates partially. Transfer the solution and residue to a 400-ml. beaker and add sufficient 60 per cent perchloric acid to dissolve the carbonates completely. Wash the crucible with 60 per cent perchloric acid and finally with water. Combine the washings with the test solution. Add 1 ml. of strontium carrier solution containing 20 mg. of strontium. Add 25 ml. excess of 60 per cent perchloric acid to the beaker and evaporate until dense white fumes of perchloric acid are produced. Add sufficient water to dissolve the perchlorates completely. Heat the solution to 80° C. and filter hot through Whatman No. 40 filter paper with the aid of suction. Transfer the silica remaining in the beaker to the filter paper with hot hydrochloric acid (1:9). The filtration must be made while the solution is hot in order to keep the salts in solution. Wash the silica on the filter with hot hydrochloric acid (1:9) several times and discard the silica precipitate.

Transfer the filtrate and washings to the original 400-ml. beaker. Add 1 ml. of a calcium carrier solution containing 200 mg. per milliliter, but this step may be omitted if the calcium concentration of the original sample is 100 mg. or more.

Adjust the pH of the solution to 8 by the addition of pellets of sodium hydroxide, using magnetic stirring. Add 2 to 3 grams of sodium carbonate while stirring. Stir for 15 minutes more and allow the precipitate to settle. Filter through a 9-cm. glass fiber filter with suction. Wash the precipitate with a sodium carbonate solution prepared by dissolving 10 grams of sodium carbonate in 90 ml. of water. Dissolve the precipitate on the filter with hot nitric acid (1:9) and collect the solution in a 400-ml. beaker. Evaporate until salting out occurs. Add 60 ml. of water and then a total of 210 ml. of fuming 90 per cent nitric acid. First add from 100 to 110 ml. of the 90 per cent nitric acid to dissolve the solid matter, and after solution is complete, add the remainder of the nitric acid. Add the acid gradually with constant magnetic stirring, using a Teflon coated bar. Perform the nitric acid addition and operations in a hood. Stir for 30 minutes. Complete the analysis as detailed under *Procedure*.

Wet-Ashing Variation.—Add the plant material or tissue to 500 ml. of hot nitric acid in small portions. When all the material has been added and dissolved, add 1 ml. of strontium carrier solution. Evaporate to approximately 50 ml. If the solution is not clear and nitrogen oxides are being formed, repeat the evaporation step with additional nitric acid. Complete the wet ashing by continuous dropwise addition of hydrogen peroxide and nitric acid. Evaporate slowly and with care on a sand bath to a white residue. Redissolve the salt in nitric acid (1:9). Add 1 ml. of a calcium carrier solution containing 200 mg. of calcium per milliliter, but this step may be omitted if the calcium concentration of the sample is 100 mg. or more. Continue with the preparation of sample as in the plant material dry-ashing variation at the point beginning . . . "Adjust the pH of the solution to 8 by the addition of pellets of sodium hydroxide . . ." Complete the analysis as detailed under *Procedure*.

Procedure.—Allow the calcium and strontium nitrates to settle. Filter through a 2.8-cm. glass fiber filter with a Fisher Filtrator and a fluorethene funnel. Be certain to drain the filter thoroughly. Discard the filtrate. With the suction off, transfer the remaining precipitate in the beaker to the fluorethene funnel by means of water and collect the resulting solution of calcium and strontium nitrates in a 250-ml. beaker. Evaporate this slowly to dryness and then dissolve the residue in 23 ml. of water. Add 77 ml. of fuming 90 per cent nitric acid while stirring and continue stirring for 30 minutes with the aid of a magnetic stirrer and a Teflon covered bar. Add the nitric acid gradually in two portions, the first comprising about 35 to 40 ml. and, after solution of the salts is complete, the remainder of the nitric acid is added.

Allow the strontium nitrate precipitate to settle and filter through a 2.8-cm. glass fiber filter with a Fisher Filtrator and a fluorethene funnel. Remove as much of the nitric acid as possible and discard the filtrate. Place a 40-ml. centrifuge tube in position to collect the dissolved precipitate. Dissolve the remaining precipitate in the beaker with water. Transfer to the fluorethene funnel and collect the solution in the 40-ml. centrifuge tube. The volume should be at least 20 ml. Bring the pH up to 8 to check for phosphate ion; if there is no precipitate, adjust the pH to 2; if a precipitate is present, repeat the nitric acid separation.

Add 1 ml. of yttrium carrier solution containing 20 mg. of yttrium per milliliter. Heat in a water bath regulated at 90° C. Adjust the pH to 8 with ammonium hydroxide solution. This adjustment is critical for strontium will precipitate incompletely in more basic solutions. The buffering action using the volumes indicated holds only when the yttrium precipitation is made at pH 8.

Cool to room temperature and allow the precipitate to settle. Centrifuge for 5 minutes. Decant, pouring the supernatant liquid into another 40-ml. centrifuge tube. Dissolve the precipitate with a minimum volume of hydrochloric acid, dilute to 10 ml., heat in a water bath at 90° C., adjust the pH to 8 with ammonium hydroxide solution, cool to room temperature, and allow the precipitate to settle. Centrifuge for 5 minutes. Decant the supernatant solution and combine it with the prior supernatant liquid. Discard the precipitate.

Add 1 ml. of barium carrier solution containing 20 mg. of barium per milliliter to the combined test solution. Add 1 ml. of acetic acid (360 g. per liter) and 2 ml. of ammonium acetate solution (463 g. per liter). The pH should be 5.5 (use narrow range pH paper for testing). This adjustment is critical, for barium chromate will not precipitate completely in more acid solutions.

Heat in a water bath controlled at 90° C. Add dropwise, while stirring, 1 ml. of sodium chromate solution (48.6 g. of anhydrous

Na_2CrO_4 per liter). The supernatant solution should have a chromate color. If it does not, it is necessary to add more sodium chromate solution for all the barium must be removed. Stir vigorously until precipitation is complete. Allow to cool, centrifuge for 5 minutes, and decant the supernatant liquid into a 2-oz. polyethylene bottle. Adjust the pH of the solution to 2 and add 1 ml. of yttrium carrier solution. Store this solution for two weeks.

Transfer the equilibrated solution to a 40-ml. centrifuge tube. Heat in a water bath controlled at 90° C. Adjust the pH to 8 with ammonium hydroxide solution, stirring continously throughout the adjustment. Add 6 drops of 30 per cent hydrogen peroxide solution. Continue heating to remove the hydrogen peroxide. Cool to room temperature before centrifuging. Centrifuge, decant the supernatant liquid into a 150-ml. beaker, and record the hour and date.

Add 25 ml. of water to the precipitate in the centrifuge tube. Dissolve the precipitate by adding hydrochloric acid dropwise with stirring. Adjust the pH to 8 with ammonium hydroxide solution, stirring continuously. Heat in a water bath regulated to 90° C. Check to make certain the pH is at 8. Add 6 drops of 30 per cent hydrogen peroxide and continue heating to remove the peroxide. Cool to room temperature before centrifuging, centrifuge, and decant the supernatant solution into the same 150-ml. beaker holding the first supernatant liquid. Record the date and hour. Reserve the combined supernatant solutions for estimating total radiostrontium.

Add 25 ml. of water to the precipitate. Add hydrochloric acid (1:1) dropwise until the precipitate just dissolves. Heat in a water bath regulated at 90° C. Add 15-20 drops of a saturated oxalic acid solution gradually with stirring. Allow the precipitate to digest and then remove from the bath and cool to room temperature. Filter through a 2.8-cm. glass fiber filter, using suction and a fluorethene funnel. Dry the precipitate in an oven at 110° C. Mount the precipitate on a plastic disc and cover with Mylar. Beta count, recording the hour and date. Standardize with yttrium-90 standard.

Chemical Yield.—Heat the reserved supernatant solutions in the 150-ml. beaker to boiling and add 10 ml. of a saturated sodium carbonate solution with vigorous stirring. Cool and filter through a weighed 5.5-cm. glass fiber filter. Transfer to a tared weighing bottle, dry, and weigh as strontium carbonate to the nearest milligram. The strontium yield is about 85 per cent. The amount of natural strontium present in the original sample must be taken into consideration in calculating the percentage yield.

Goldin Method.—The principle of this method [20] has been detailed in the section on "Identification," this chapter. The preparation of the sample can follow along the lines detailed in the AEC Method.

Concentration.—Add 4 ml. of 1 N strontium nitrate solution to the prepared solution, heat to boiling, and precipitate with 5 ml. of 3 N sodium carbonate solution. Allow the precipitate to settle for 0.5 to 2 hours, decant, and discard the supernatant solution. Transfer the precipitate to a 50-ml. centrifuge tube and centrifuge to collect.

Barium Removal.—Dissolve the precipitate in 2 to 3 ml. of 6 N hydrochloric acid. Add 1 ml. of 0.5 M potassium chromate solution, warm in a water bath, and add slowly 5 ml. of acetic acid-acetate buffer solution (2 N acetic acid and 4 N ammonium acetate). Allow to digest for 5 to 10 minutes, add an additional 1 ml. of 0.1 N barium chloride solution, stir well, centrifuge, and pour off the supernatant solution into another centrifuge tube. Discard the barium chromate precipitate. Add 6 N ammonium hydroxide solution until the test solution is alkaline and reprecipitate the strontium with carbonate. Centrifuge, discard the supernatant solution, wash the precipitate with water containing a drop or two of sodium carbonate solution, and discard the washings. The removal of barium may be omitted at this point if old samples in which it is known that barium-140 is absent are being analyzed.

Removal of Rare Earths.—Dissolve the strontium carbonate in 2 to 3 ml. of 6 N hydrochloric acid, add about 25 ml. of water, 1 ml. of water, 1 ml. of 0.1 N ferric chloride solution, heat, and make alkaline with 6 N ammonium hydroxide solution. Add an additional 1 ml. of 0.1 N ferric chloride solution. Centrifuge, transfer the supernatant to another centrifuge tube, and discard the precipitate. Add to the supernatant solution 1 ml. of 0.02 M zirconyl chloride solution and 1 ml. of a rare-earth carrier such as 1 ml. of 0.01 M cerium nitrate solution or 0.01 M lanthanum nitrate solution. Add 6 N hydrochloric acid to dissolve any precipitate, warm, and again make alkaline with 6 N ammonium hydroxide solution. Add 1 ml. each of the zirconium and rare-earth carrier solution. Note the time of the precipitation (this is the start of the ingrowth time for yttrium-90). Centrifuge, transfer the supernatant solution to another centrifuge tube, and discard the precipitate. Add 3 N sodium carbonate solution to precipitate the strontium, centrifuge, and discard the supernatant. Allow the precipitate to stand overnight. It is preferable to allow the strontium to remain in solid form at this point as strontium carbonate so that

---

[20] Goldin, "Strontium-90 Determination," Sanitary Engineering Center, U. S. Public Health Service, Cincinnati (1957).

the yttrium-90 formed may be trapped in the crystals, thus minimizing absorptive losses to the glassware.

Extraction of Yttrium-90.—Dissolve and transfer the strontium carbonate to a separatory funnel, using 2 to 3 ml. of 6 N hydrochloric acid and 30 ml. of water in this manipulative step. Add 6 N ammonium hydroxide solution dropwise to a methyl orange end point and add 5 ml. of the acetate buffer solution (2 N acetic acid and 4 N ammonium acetate). Extract the test solution with 5 ml. of 5 per cent W/V solution of 2-thenoyltrifluoroacetone in monochlorobenzene and transfer the organic solvent layer to another separatory funnel. Repeat the extraction two more times, adding each additional extract to the first. Note the time again for this is the end of the ingrowth time for yttrium-90 and is the start of the decay time of this radionuclide.

Add sufficient heptane to the organic solvent layer to make the solution lighter than water and wash three times with 5 ml. of acetate buffer, returning the washes to the original separatory funnel. Extract the organic solvent phase three times with 1 N nitric acid. Combine the nitric acid extracts and evaporate to dryness in a stainless steel vessel. Count for yttrium-90. Calculate the strontium-90 content from the growth and decay of yttrium-90.

Chemical Yield.—Sample.—Transfer the combined aqueous layers and washes in the original separatory funnel to a 100-ml. volumetric flask, dilute to the mark with water, and mix well. Transfer 1 ml. with the aid of a pipette to a 50-ml. volumetric flask containing 1 ml. of 6 N hydrochloric acid, dilute to the mark with water, and mix. Determine the strontium content by flame photometry, using the 460.7 mμ line and a hydrogen-oxygen flame. Determine the background reading at 8 mμ above and below 460.7 and subtract from the reading at 460.7 mμ. The exact details of the flame spectrophotometric method depend upon the instrument being used.

Standard.—Add 1 ml. of 1 N strontium nitrate solution with the aid of a pipette to a 1-liter volumetric flask, add 200 ml. of the unknown prepared sample or equivalent solution, add 20 ml. 6 N hydrochloric acid, dilute to the mark with water, stopper the flask, and mix thoroughly. Determine the strontium content as detailed for the "Sample" above.

STRONTIUM-90 AND BARIUM-140

Volchok, Kulp, Eckelmann, and Gaetjen [21] developed a method for the determination of strontium-90 and barium-140 attributable to

---

[21] Volchok, Kulp, Eckelmann, and Gaetjen, *Ann. N. Y. Acad. Sci.* 71, No. 2, 293 (1957).

radioactive fallout material from nuclear tests in bone, cheese, milk, and vegetation. In this method calcium is used as the carrier for the radionuclides, and the sample is prepared in such a manner as to yield a calcium chloride solution carrying the radioisotopes of strontium and barium. In the case of samples that are several months old only the strontium-90 will be detected, for barium-140 has a half-life of 12.8 days. These investigators stress the need for the elimination of phosphate in this method of determination of strontium and barium. The activity measurements are made on the daughters of strontium-90 and barium-140, that is, on yttrium-90 and lanthanum-140, respectively, after these are extracted from the test solution.

The principal objective of the manipulative steps in the preparation of the sample in this method is the formation of a test solution which is suitable for the extraction of the daughters, namely, yttrium-90 and lanthanum-140, of strontium-90 and barium-140. A quantitative separation is difficult in solutions containing phosphate ion, for the phosphates of the two groups precipitate in a narrow pH range. Thus, the alkaline earth phosphates precipitate at a pH of about 2.4, and the rare-earth phosphates precipitate at about pH 1.5. The adjustment of the pH in the range mentioned is particularly difficult, for the solutions are hot and ammonium hydroxide is used to raise the pH. The concentration of the ammonia will vary under such conditions with time and temperature. To avoid this difficulty it is preferable to remove the phosphate ion completely.

The extraction of the mother test solution for the daughter isotopes is essentially a two-step procedure designed first to separate the yttrium-90 and the lanthanum-140 quantitatively from the strontium-90 and the barium-140 and secondly to produce a precipitate of the yttrium-90 and lanthanum-140 that is suitable for measurement of radioactivity.

Because the daughter products have far shorter half-lives than the parent compounds—namely: yttrium-90, 64.24 ± 0.30 hours; lanthanum-140, 40 hours; strontium-90, 19.9 years; and barium-140, 12.8 days—it is far preferable to count the daughters since the decay of the daughter radioisotopes can be followed to detect contamination and additional extractions can be made from the mother radionuclides after a period of only a few days.

Preparation of Sample.—Milk, Cheese, Bone, Vegetation.—Ash the samples of these materials completely at a temperature of about 600° C. The procedure detailed in the AEC Method in this chapter may be followed. Dissolve the ash containing the strontium and barium in calcium phosphate in an excess of concentrated hydro-·

chloric acid and filter the solution to remove the insoluble matter. Add ammonium hydroxide solution to the clear filtrate and adjust the pH to 1.5. It is essential that this adjustment be made carefully for the reasons noted in the preceding paragraph. Add an excess of ammonium oxalate solution. Filter off the calcium oxalate precipitate, which carries the strontium and barium quantitatively, to separate these from the phosphate. Test for completeness of the removal of the phosphate by ashing the calcium oxalate at 600° C. dissolving the calcium oxide formed in hydrochloric acid and adjusting the pH to above 3. The formation of a precipitate is indicative of the presence of appreciable quantities of phosphate. If such a precipitate forms, add additional hydrochloric acid to lower the pH to 1.0 and dissolve all of the phosphate. Readjust the pH to 1.5 and repeat the oxalate precipitation. Once the calcium chloride solution finally formed is free of phosphate, the preparation of the mother solution is complete, and the test solution can be repeatedly extracted to obtain the yttrium-90 and lanthanum-140 without loss of strontium-90 and barium-140.

Extraction.—Add about 10 mg. of nonradioactive yttrium carrier solution to the test solution and mix thoroughly. Adjust the pH to 5 with ammonium hydroxide solution. A white gelatinous precipitate of yttrium hydroxide, with which the yttrium-90, lanthanum-140, and rare earths will coprecipitate, will form. Heat to coagulate. Record the time of precipitation. Filter off the precipitate and reserve the filtrate for additional extraction after the formation of additional daughter nuclides.

Other long-lived rare-earth fission nuclides and natural radioisotopes that were coprecipitated with the calcium oxalate are also coprecipitated in this first extraction.

The gelatinous character of the yttrium hydroxide precipitate makes it unsuitable for counting. It must be treated in the following manner. Dissolve the precipitate from the filter paper with 6 N hydrochloric acid, adjust the pH to a point just below precipitation, and add an excess of oxalate. The yttrium oxalate formed has a coarse granular structure that makes it satisfactory for counting. Filter off the precipitate on a stainless-steel funnel with the aid of suction, dry by drawing air through for about 10 minutes, remove the filter paper from the funnel, mount for counting on a brass disk, cover with Pliofilm or Mylar, and secure with a brass ring.

Chromatographic Separation.—If it is desired to separate the strontium-90 and barium-140, a chromatographic technique may be used. Set up a column 2.2 × 20 cm. containing 200- to 400-mesh

Dowex-50 cation exchange resin (8 per cent cross linkage). Add non-radioactive strontium and barium carrier solutions to the prepared calcium chloride test solution. Transfer the test solution to the column and elute with 4 N hydrochloric acid at a rate of approximately 0.3 ml. per minute. Approximately 5.5 hours are required for the collection of the fractions.

Under these conditions calcium, strontium, and yttrium can be separated completely from barium by collecting the first 480 ml. as one fraction and collecting a second fraction from 480 to 832 ml. Calcium has a peak of 192 ml., strontium has a peak at 256 ml., and yttrium has a peak at 288 ml., whereas barium has a peak at 672 ml. After the chromatographic separation, the radioactive daughters can be extracted as detailed in the extraction step.

Volchok, Kulp, Eckelmann, and Gaetjen made their radiometric determination of the final precipitate in a specially designed beta counter [22] that utilized anticoincidence shielding to obtain a low background count.

It was found that 98 per cent of the calcium present as calcium phosphate is recovered as calcium oxalate and that at least 98 per cent of the strontium-90 is recovered.

RADIUM-226

Goldin Method.—The principle of this method [23] has been explained in the section on identification. It was designed for the determination of radium-226 in water but can be adapted for food by preparation of the sample in question by one of the methods detailed in this chapter.

Procedure.—Add to 1 liter of water or to the equivalent prepared food sample 5 ml. of 1 M citric acid (containing 0.1 per cent of phenol to act as a preservative against microbiological growth), 2.5 ml of concentrated ammonium hydroxide solution, 2 ml. of 1 N lead nitrate solution, and with the aid of a volumetric pipette 1 ml. of 0.1 N barium nitrate solution. Larger volumes of water or prepared sample may be used, but in this case while the volumes of the other reagents are increased, the volume of the barium nitrate should be kept to 1 ml.

Heat to boiling and add, with stirring, 10 drops of methyl orange indicator solution. Add, with stirring, sulfuric acid (1:1) until a

22 Volchok and Kulp, *Nucleonics* 13, No. 8, 49 (1955).

23 Goldin, ''Analysis of Water for Dissolved Radium-226,'' Sanitary Engineering Center, U. S. Public Health Service, Cincinnati (1956).

pink color is obtained and then add 0.25 ml. in excess. Digest for 5 to 10 minutes, allow to settle for 0.5 to 2 hours, and decant and discard the supernatant solution. Transfer the precipitate to a 50-ml centrifuge tube. Wash the precipitate with 10 ml. of concentrated nitric acid, centrifuge, and discard the washing. Repeat the washing with 10 ml. more of concentrated nitric acid.

Dissolve the precipitate in 10 ml. of water, 10 ml of $M/4$ solution of the disodium salt of ethylenediaminetetraacetic acid, and 3 ml. of 6 N ammonium hydroxide solution. Warm and add dropwise to the warm solution 2 ml. of glacial acetic acid. Digest for 5 to 10 minutes, centrifuge, and decant and discard the supernatant solution. The amount of acetic acid added is sufficient to give a 2:1 excess of acetic acid over ammonium hydroxide and to lower the pH to about 4.5. At this pH the barium ethylenediaminetetraacetate is decomposed, but the lead complex is not. Note the time, for from this time on radon and daughters grow into the barium sulfate precipitate.

Wash the reprecipitated barium sulfate with water and transfer to a centrifuge tube adapted for centrifuging precipitates on to planchets. Allow to settle for 5 to 10 minutes, centrifuge, dry, flame, weigh, and count.

Calculate the radium-226 activity from the count and from the time of daughter ingrowth. If there is any question as to the isotopic or chemical purity of the radium-226, retain the precipitate for 24 to 48 hours and recount. Calculate the radium-226 content from the rate of daughter ingrowth. [24]

SELECTED REFERENCES

Coryell and Sugarman, Radiochemical Studies: The Fission Products. McGraw-Hill, New York, 1951.

Friedlander and Kennedy, Nuclear and Radiochemistry. John Wiley, New York, 1955.

Glasstone, Sourcebook of Atomic Energy, 2nd ed. Van Nostrand, Princeton, N.J., 1958.

Heath, Scintillation Spectrometry Gamma-Ray Spectrum Catalogue. Office of Technical Services, U. S. Dept. Commerce, Washington, 1957.

Kahn and Goldin, "Radiochemical Procedures for the Identification of the More Hazardous Nuclides." Presented to the Nuclear Engineering and Science Congress (1957).

"Measurement of Radioactivity in Water, Bottom Silts, and Biological Materials," Sanitary Engineering Center, U. S. Public Health Service, Cincinnati (Jan. 1957).

---

[24] Kirby, Atomic Energy Commission Report MLM-859.

Natl. Bur. Standards, Handbook 52, Washington, 1953.

"Radiochemical Determination of Strontium-90," Atomic Energy Commission, New York Operations Office, E-38-01 (1957).

Setter, Hagee, and Straub, "The Analysis of Radioactivity in Surface Waters: Practical Laboratory Methods," *ASTM Bull.* **1958**, No. 227, 35.

Siegbahn, *Beta- and Gamma-Ray Spectroscopy*. Interscience, New York, 1955.

Taylor, *Measurement of Radio Isotopes*. John Wiley & Sons, New York, 1951.

Chapter XXIII

PESTICIDE RESIDUES

The widespread and increasing use of pesticides, such as insecticides, fungicides, rodenticides, etc., makes it necessary for the food analyst to have available methods that can detect minute quantities of these materials as a check not only on the safety of a particular food being consumed but also on the over-all consumption of such pesticides and on the extent to which they have been removed from the food.

The listing in 1957 by the Food and Drug Administration of tolerances for pesticide residues mentioned dozens of pesticides for some 87 categories of fruits, vegetables, feeds, nuts, etc. It would not be feasible in a text of this nature to give full coverage for all of these. The detection and determination of arsenic and lead in foods attributable to arsenical- and lead-bearing pesticides has been considered in detail in Chapter V. Other pesticides such as fluoride-bearing insecticides and zinc phosphide, a rodenticide, were discussed in Chapter VI.

In this chapter methods for the detection and determination of a number of pesticides will be considered. Among these will be methods for the separation of pesticides, the determination of insecticides containing organic chlorine, DDT, parathion, 1080, and Warfarin, a rodenticide.

Preparation of Sample.—A major factor involved in the detection and determination of pesticides on and in foods is the separation and isolation of the pesticide from the food. There are a number of ways in which this can be done. In "stripping" methods (detailed in the procedure for parathion) applicable to fairly firm and resistant fruits and vegetables—such as apples, pears, lettuce, and cabbage—and somewhat less firm fruits and vegetables—such as peaches, plums, and tomatoes—the fruit or vegetable is placed in a jar of adequate size, is covered with benzene, closed, and then tumbled end-over-end, preferably by a motor mechanism. Anhydrous sodium sulfate is used to dehydrate the extract and some of the coloring matters and wax are removed with Attapulgus clay and activated carbon.

With softer fruits, such as berries, animal tissues, and other types of samples in which the pesticide may have penetrated the product, the food is mixed thoroughly with benzene in a blendor. Anhydrous sodium sulfate is added to the mixture, stirred in thoroughly with the blendor, and the entire mass is centrifuged.

Dry materials—such as grains, flour, cereals, and the like—may be extracted with benzene with the aid of a Soxhlet extractor. Fatty materials are treated to extract the fat and the pesticide isolated from the fat.

Reverse-Phase Partition.—The methods of preparing the sample mentioned have certain disadvantages. Erwin, Schiller, and Hoskins [1] developed a reverse-phase partition column for the removal of interfering extractives prior to the chemical determination of residual insecticides occurring as residues in plant and animal tissues. The column consists of finely ground alumina coated with a 2:1 mixture of petroleum jelly and a low-melting paraffin wax. The eluting liquid is 60 per cent aqueous acetonitrile solution, or a series of solutions containing 40, 60, and 75 per cent of acetonitrile respectively. By use of the series of solutions the pesticides may be separated into groups.

Preparation of Column.—Add 5 g. of paraffin wax, m. p. 160 to 165° F., and 10 g. of white petroleum to 100 ml. of a mixture of benzene and chloroform (1:1) and warm, if necessary, to dissolve all of the paraffin. Add 100 g. of 80- to 200-mesh adsorption alumina and stir while evaporating the solvents on the steam bath. Spread the powder on a flat surface and use air from a small fan or hair blower to remove the last traces of benzene. This removal may be considered adequate if on opening a container in which the powder has been stored, there is no odor of benzene. Pass through cheesecloth to remove and store in a bottle until required.

Use a borosilicate glass tube 20 mm. in outside diameter and 26 cm. long with a 150-ml. reservoir at the top and a medium fine sintered-glass filter plate at the bottom to hold the adsorbent. Arrange the plate to lead into a drip tube surrounded by a 24/40 $\mathbf{\mathcal{S}}$ inner member of a ground-glass joint. Attach the outer half of the ground-glass joint during the loading of the column to a source of vacuum and slowly pour 35 g. of the coated powder into the tube, while tapping it to ensure uniformity of packing. A column about 100 mm. high will be obtained. Cut off the vacuum, place a small plug of cotton above the powder, and rinse the column with 50 ml. of the first eluent to be used. Stop the flow just before all the liquid enters the column.

---
[1] Erwin, Schiller, and Hoskins, *J. Agr. Food Chem.* **3**, 676 (1955).

Preparation of Extract.—Thoroughly subdivide by grinding, chopping, etc. 1 kg. or other convenient weight of sample and transfer the wet mass to a clean gallon paint can or other appropriate container. Add 1 liter of solvent, such as benzene, petroleum ether, or carbon tetrachloride, and 1 liter of 95 per cent ethyl alcohol and shake or roll the mixture, preferably mechanically, for 1 hour. If the sample contains considerable water so that there may be two phases formed, add 100 to 200 g. of anhydrous sodium sulfate before mixing.

Transfer the liquid phase to a large separatory funnel. Dilute with 2 liters of water, shake for 2 to 3 minutes, allow the phases to separate, withdraw the aqueous phase, and repeat the treatment. In the first rinse, the aqueous phase contains about 30 per cent alcohol, in which solvents such as benzene or petroleum ether are soluble only to about 1 per cent. The various insecticides distribute themselves in favor of the water-insoluble phase, which contains only about 3 per cent of alcohol. This small amount of alcohol is removed by the second rinse. Since virtually all of the toxicant is concentrated in the water-insoluble phase, 1 ml. of the remaining volume of solvent will represent 1 g. of sample, or other ratio that was used, assuming that the pesticide is distributed uniformly in both the recovered and unrecovered portions of the solvent.

Place 50 ml. of the extract, corresponding to 50 g. of sample, into a small beaker and evaporate the solvent to approximately 5 ml. at as low a temperature as practicable in a stream of warm air. Add 5 g. of fine white sand to the beaker and continue the evaporation with constant stirring to insure that the plant or animal extractives present are left in a thin coat on the sand. Add 5 ml. of acetonitrile to the beaker and place on a warm plate. Heat and stir the mixture to the boiling point. Add 6 ml. of water to the hot solution, allow to cool, and add the entire volume to the column. Rinse the beaker with two 5-ml. portions of hot 40 per cent acetonitrile and place the washings on the column also. Apply about 2 lbs. of air pressure to the column to hasten the entry of the 40 per cent acetonitrile solution of the extract. When the liquid level just reaches the top of the granular packing, release the pressure, and add 150 ml. of developing eluent. Apply the pressure again and collect the eluate at a rate not exceeding 1 ml. per minute.

Place the eluate into a 500-ml. separatory funnel, add 200 ml. of water, shake, add 100 ml. of petroleum ether, shake thoroughly, and allow to separate. Repeat the extraction with an additional 100 ml. of petroleum ether. Combine the petroleum ether layers and make up to a definite volume. This solution can be used for the chemical test solution and for bio-assay. It is important in diluting the acetonitrile to make certain that the concentration of the acetonitrile is reduced to less than 25 per cent, otherwise the petroleum ether may not extract all of the toxicant.

Acetonitrile-water mixtures hold back nearly all of the extractives from a variety of samples and at the same time elute quantitatively a number of organic pesticides. With the aid of increasing concentrations of acetonitrile the pesticides can be divided into three groups as tabulated in Table 105.

TABLE 105. SEPARATION OF ORGANIC PESTICIDES WITH AQUEOUS ACETONITRILE [a]

40 CH₃CN : 60 H₂O	60 CH₃CN : 40 H₂O	75 CH₃CN : 25 H₂O
Rotenone	Dieldrin [b]	Aldrin
Malathion	DDT	Isodrin
Dilan	Chlordan	
Parathion	Heptachlor	
Lindane	Endrin	
Methoxychlor	Toxaphene	
Dieldrin [b]		

[a] Erwin, Schiller, and Hoskins, *J. Agr. Food Chem.* 3, 676 (1955).
[b] Dieldrin is not completely eluted by 125 ml. of the first eluent and divides between the first and second groups if 40 per cent acetonitrile is used as the first eluent.

CHLORINATED ORGANIC PESTICIDES

Various organic insecticides contain chlorine as part of the molecule. Thus, for instance, DDT contains 50 per cent of chlorine, chlordan from 64 to 66 per cent of chlorine, methoxychlor 30.8 per cent, etc. The chlorine can be separated from the organic compound and can then be determined as detailed in Chapter XVIII on inorganic determinations, pages 757-760. Determination of organic chlorine has the disadvantage that it is nonspecific and that care must be taken to be certain that no inorganic chlorine is included. The A. O. A. C.[2] details in variation of such methods for DDT.

Procedure.—Evaporate aliquots of the "strip" solution or other prepared sample solution from fruits, vegetables, or forage crops, cereals, etc., on a steam bath almost, but not entirely, to dryness to avoid the decomposition of the chlorinated organic insecticide. Add 25 to 50 ml. of 99 per cent isopropyl alcohol, 2.5 g. of sodium in the form of a ribbon or cut into small pieces, and shake the flask to mix the sample thoroughly. Connect the flask to a re-

---

[2] *Methods of Analysis, A. O. A. C.* Washington, 8th Ed., 1955.

flux condenser and boil gently for 1 hour with an excess of sodium present at all times, adding more sodium if necessary and shaking the flask occasionally. Remove the excess sodium by adding cautiously 10 ml. of 50 per cent isopropyl alcohol through the condenser at a rate of 1 to 2 drops per second. Boil for an additional 10 minutes and then add 100 ml. of water. Cool, add 2 or 3 drops of phenolphthalein indicator solution, neutralize by adding nitric acid (1:1) drop by drop, and then add 5 ml. of nitric acid in excess.

If the solution is colored, cool to room temperature, transfer the contents of the flask to a small separatory funnel, wash the original flask, and add the washings to the separatory funnel. Shake out with 15 ml. of a mixture of isoamyl alcohol-ether (1:1). Draw off the aqueous layer into a second separatory funnel and repeat the extraction with the isoamyl alcohol-ether mixture. Draw off the aqueous layer into a 250-ml. beaker and wash the two organic extracts successively with two 10-ml. portions of water. Combine the aqueous wash solutions with the aqueous solution in the beaker. Make it slightly alkaline with sodium hydroxide solution, add 1 ml. of 30 per cent hydrogen peroxide solution, heat to boiling for 10 to 15 minutes, and neutralize with nitric acid (1:1), adding 5 ml. of nitric acid in excess.

Alternatively, after adding 100 ml. of water, transfer the test solution to a 400-ml. beaker, add 10 ml. of 30 per cent hydrogen peroxide solution, and heat to boiling on a steam bath until the peroxide decomposes and most of the alcohol evaporates. Add sufficient water to make the volume about 250 ml., neutralize with nitric acid, and add 3 ml. of the acid in excess. Cool, filter, and determine the chloride as detailed in Chapter XVIII.

DDT

DDT—1,1,1-trichloro-2,2-bis(p-chlorophenyl)ethane—can be isolated from foods by the methods detailed. It can be estimated by the chloride method described in the preceding section or preferably by the colorimetric method which follows.

Colorimetric Method.—A colorimetric method was developed by Schechter, Soloway, Hayes, and Haller [3] for the estimation of small amounts of DDT down to about 10 micrograms. The method depends on intensive nitration to polynitro derivatives and the production of intense colors upon addition of methanolic sodium methylate to a benzene solution of the nitration products. p,p'-DDT and p,p'-DDD give

---

[3] Schechter, Soloway, Hayes, and Haller, *Ind. Eng. Chem., Anal. Ed.* **17**, 704 (1945).

blue colors, and o,p'-DDT gives a violet-red color. Degradation products of DDT, such as dehydrochlorinated p,p'-DDT and p,p'-DDA yield red colors.

The terms used to designate DDT and related compounds are as follows: The generic term "DDT," originally abbreviated from dichlorodiphenyltrichloroethane, refers to the technical product which ordinarily contains 70 to 77 per cent of p,p'-DDT—1-trichloro-2,2-bis (p-chlorophenyl)ethane—and 15 to 25 per cent of o,p'-DDT. One of the minor components is 1,1-dichloro-2,2-bis(p-chlorophenyl)ethane, which has been designated as p,p'-DDD. Dehydrochlorinated p,p'-DDT is a decomposition product, and bis(p-chlorophenyl)acetic acid, which has been called p,p'-DDA, is a metabolite of p,p'-DDT.

Solvents and Reagents.—Nitrating Acid.—A mixture of c.p. fuming nitric acid (sp. gr. 1.49-1.50) and c.p. concentrated sulfuric acid (sp. gr. 1.84), 1 to 1 by volume.

Cotton.—Extracted with acetone in a Soxhlet extractor, dried for several hours at 105° to 110° C., and stored in a tightly stoppered bottle.

Ether.—U. S. P. grade distilled before use. Ether that has been standing long enough to accumulate peroxides and aldehydes or has been recovered after use in this method is unsatisfactory and should be purified before it is used again.

Benzene, c.p., dry.—It is conveniently dried by distilling through a straight condenser until no more water distills over with the benzene and then by replacing the condenser with a dry one and continuing the distillation. Benzene that has been used in this method to dissolve the nitrated residues or to make dilutions may be accumulated and recovered for reuse by distillation.

Sodium Methylate Solution, 10.0 ∓ 0.1 per cent (concentrations are expressed as weight per unit volume) of sodium methylate in dry c.p. methyl alcohol (10.0 g. per 100 ml. of solution).—An excellent method of drying the methyl alcohol is to reflux it with magnesium turnings (5 to 10 g. per liter of methyl alcohol) and a small amount of iodine until the magnesium has completely dissolved, and then to distill with the exclusion of moisture. The solution is prepared by dissolving the requisite amount of perfectly clean sodium or a good grade of powdered sodium methylate (available commercially) in the dried methyl alcohol with cooling, using a stirrer and a reflux condenser protected by a soda-lime tube. An aliquot of a clear portion of this solution should be diluted with water and titrated with standard hydrochloric acid, phenolphthalein being used as the indicator. The concentration of the solution should be adjusted to 10.0

± 0.1 per cent by the addition of sodium or sodium methylate or by dilution with dry methyl alcohol.

The sodium methylate solution that is added to the benzene to develop the color should be colorless and optically clear. If the sediment does not settle completely on standing, the solution should be filtered or centrifuged. Occasionally a turbidity or precipitate of crystalline material (probably sodium carbonate) will form when the centrifuged sodium methylate reagent is added to the benzene solutions. This difficulty can be obviated largely by cooling the standardized solution in a refrigerator for a day or two, centrifuging while cold, and decanting into another container.

Preparation of Sample for Analysis.—Unless the total sample has very little DDT (less than 100 micrograms), it is advantageous to use a portion of the sample which contains a reasonably large amount of DDT (0.5 mg. to several milligrams). It will then be possible to take an aliquot at the end of the procedure for the development of the color. Extract or strip the DDT from the sample with a suitable solvent and evaporate. Using acetone, transfer the residue or an aliquot drawn to a test tube for nitration. In some cases the aliquot may be drawn directly from the extract before its evaporation. Care must be taken not to lose mechanically any of the sample during the evaporation of solvents prior to nitration. The best procedure for evaporating organic solvents is to add a glass bead, immerse the test tube about one third of its length in a steam bath, and shake gently until the glass bead bounces and ebullition starts. When the solvent has been completely boiled out, remove the last traces by inserting a glass tube attached to a source of vacuum one third of the way into the test tube for at least half a minute while the tube is still being heated. Unless the solvent is completely removed, it may react violently with the nitrating mixture in the next step. If benzene or an aromatic solvent has been used, add 5 ml. of ethyl alcohol and evaporate to dryness in the same manner in order to remove the aromatic solvent by azeotropic distillation.

Nitration of Sample.—Cool the test tube in a beaker of cold water and with a safety pipette add 2.0 or 5.0 ml. of nitrating acid. Put the test tube one-third to one-half its length in a steam bath and heat for 1 hour. Since nitrations of even small quantities of materials may sometimes be violent, safety precautions should be observed. If there is much extraneous material, it is advisable to place the test tube in ice-cold water, add cooled nitrating acid, and warm the tube cautiously to prevent a sudden or violent nitration. When the initial reaction has subsided, the tube may be heated at

100° with safety. After the 1-hour nitration, cool the test tube in a beaker of cold water, add 25 ml. of ice-cold water, and mix by gently swirling. This stops the nitration and the test tube may be left overnight if desired.

Extraction of Nitrated Product.—Rinse the contents of the test tube quantitatively through a small funnel into a 125-ml. separatory funnel with about 25 ml. of water from a wash bottle and 50 ml. of ether. A small, irregularly shaped piece of glass placed in the funnel used for the transfer will prevent the glass bead from falling into the separatory funnel. Shake vigorously for at least 1 minute. After the layers have separated clearly, draw off and discard the lower layer. Wash the ether with 10-ml. portions of 2 per cent aqueous sodium hydroxide solution until the washings are alkaline; one washing may be sufficient. Then wash the ether with two 10-ml. portions of saturated salt solution. The final salt wash should be drawn off as completely as possible. Pack a 0.75-inch plug of cotton tightly in a glass Gooch-crucible holder, moisten it with ether, and allow the ether solution from the separatory funnel to filter slowly into a 125-ml. Erlenmeyer flask. Rinse the separatory funnel with 50 ml. of ether in 4 or 5 portions, passing this ether through the cotton in the Gooch funnel. If salt crystallizes in the neck of the separatory funnel, press the stopper of the funnel in place firmly with a rotating motion to prevent leakage of ether. Add a glass bead to the Erlenmeyer flask, warm the flask on a steam bath with a gentle swirling motion until the bead starts bouncing, and evaporate the ether completely. While the flask is still being heated, insert a glass tube connected to a source of vacuum two thirds of the way into the flask for at least half a minute; then remove the flask and stopper it. The analysis may be interrupted at this point if desired.

The whole extraction procedure must be done carefully to avoid any loss, such as ether sprayed from the separatory funnel when the stopcock is opened to release pressure or when the glass stopper is removed. This type of loss can be minimized by allowing time for the ether to drain away from the stopcock or the stopper before performing these operations.

Development of Color.—At this stage there is a choice of procedures, depending on the amount of DDT expected, the amount of solution necessary for use in making the photometric measurements, and whether it is desired to have some solution left to repeat the photometric measurements.

Procedure 1.—Add an accurately measured amount of benzene— for example, 5.00 ml.—to the residue in the Erlenmeyer flask and swirl gently until it is dissolved. Use a volume of benzene at least equal to one third the volume necessary for use in the absorption cell or tube of the photometer. With a pipette add 2 volumes (10.00 ml. for 5.00 ml. of the benzene solution) of the sodium methylate reagent to 1 volume of benzene solution. Swirl gently until the solution is homogeneous, pour into the absorption cell or tube of the photometer, and prepare to make the most important measurements 15 minutes after the sodium methylate reagent has been mixed with the benzene. This procedure should be used only when it is known that the amount of DDT is very low and in the range where the color developed will be suitable for direct measurement in the photometer. If there is a possibility that the color developed will be too dark for direct measurement, it is preferable to use Procedure 2 rather than add more benzene and sodium methylate to the colored solution to dilute it.

Procedure 2.—Add a measured amount of benzene—for example 25.00 ml.—to the Erlenmeyer flask and swirl gently until the residue is dissolved. To an aliquot—for example, ·5.00 ml.—add twice its volume of sodium methylate reagent, mix thoroughly by gently swirling, and pour into the absorption cell or tube. In some cases it is possible to mix the solutions directly in the absorption cell or tube. If the color is too deep, a photometric measurement may be made to obtain a rough estimate. Dilute part or all of the remaining benzene solution to a more suitable volume before withdrawing a new aliquot for development of the color. If the color is too light for good photometric measurement, rinse with benzene the pipette used for the first transfer into the Erlenmeyer flask; evaporate all the solvent on the steam bath, swirling the flask gently to start the bead bouncing; and when all the benzene is evaporated, remove the last traces by inserting a glass tube attached to a source of vacuum. This residue in the Erlenmeyer flask should now be treated as in Procedure 1.

Photometric Measurements.—Spectrophotometric or photometric measurements should be made at the most important wave lengths or with the most important filters as close as possible to 15 minutes after the sodium methylate solution has been mixed with the benzene. Measurements at other wave lengths or with other filters can be made just before or after the most significant readings have been taken.

Absorption cells or tubes should be stoppered tightly. Absorption cells usually have glass covers or stoppers, but if test tubes are used, as in many routine photometric measurements, rubber stoppers washed free of sulfur are preferable to cork stoppers, contact with which will turn the solution yellow. Since the solutions on which optical measurements are made are strongly alkaline, absorption cells constructed with alkali-resistant cement should be used. The solutions should be left in the cells no longer than is necessary to make photometric measurements, after which the cells should be cleaned immediately.

In application of the method it is important to run a blank analysis on a sample of the same type of material being analyzed which has not been treated with DDT. The results, in terms of DDT or extinction values (never in terms of percentage transmission), should be applied as corrections to the values obtained at each wave length or filter used in the analysis of the DDT-treated samples. If appropriate blanks are not run, the results of the analysis may be high. Blank analyses should be made by diluting the blank runs in the same manner as the DDT-treated samples, or else the corrections should be calculated to the same weight of untreated material as used in the analysis of the treated material.

The absorbance for pure DDT is measured at 510 and 580 mμ. Readings at 450 mμ, in the blue region, may be made to test for the presence of extraneous yellow color. For routine work, measurements may be made at 600 mμ, and the results may be interpreted with the aid of a standard curve made in the customary manner, using known quantities of a mixture of 3 parts of p,p'-DDT and 1 part of o,p'-DDT.

CHLORDAN

Technical chlordan is a chlorinated organic insecticide consisting principally of 1,2,4,5,6,7,8,8-octachloro-2,3,3a,4,7,7a-hexahydro-4, 7-methanoindene. The chlorine content ranges between 64 and 66 per cent. The toxicant can be separated from foods by the reverse-phase partition method detailed on page 892, and the total organic chlorine content can be determined as detailed in that section of this chapter, but this is, as mentioned, a nonspecific method.

Ordas, Smith, and Meyer [4] have devised a spectrophotometric method for chlordan based on the Davidow reaction.[5] These investigators also devised chromatographic methods for the separation of the toxicants.

Procedure.—Boil down the pentane or other solvent solution containing 5 to 50 micrograms of the separated chlordan at 40° C. Heat

---

4 Ordas, Smith, and Meyer, *J. Agr. Food Chem.* 4, 444 (1956).

5 Davidow, *J. Assoc. Official Agr. Chem.* 33, 886 (1950).

the residue at 100° C. with 0.2 ml. of the modified Davidow reagent in a reaction bath. The reagent is prepared by mixing 2 volumes of 1.0 N methyl alcohol solution of potassium hydroxide with 1 volume of diethanolamine and diluting with 9 volumes of methyl alcohol.

Cool to room temperature in a beaker of water. Dilute to 0.5 ml. with 90 per cent methyl alcohol and transfer to a microcuvette. Determine the transmittance at 550 mμ. Compare with a standard curve made with known amounts of chlordan treated in the same manner at the same time.

HEPTACHLOR

Heptachlor is a related chlorinated organic insecticide—1,4,5,6,7, 8,8-heptachloro-3a,4,7,7a-tetrahydro-4,7-methanoindene—that is used as a toxicant and is present in commercial chlordan. Technical heptachlor contains about 72 per cent of the compound and 28 per cent of other related compounds.

This toxicant can be separated from foods by the reverse-phase partition method detailed in this chapter and by methods described by Ordas, Smith, and Meyer,[4] who adapted the method of Polen and Silverman [6] for the microdetermination of heptachlor.

Prepare the Polen-Silverman reagent by dissolving 33 g. of 85 per cent potassium hydroxide in 28 g. of water, cooling to room temperature, adding an equal volume of butyl Cellosolve and 30.5 g. of monoethanolamine; and diluting to 1 liter with butyl Cellosolve. Mix, allow to stand for several days, decant from any sediment, dilute with an equal volume of benzene, and mix.

Procedure.—Allow the solvent of the prepared test solution to evaporate to dryness at 40° C. Add 0.2 ml. of the Polen-Silverman reagent to the residue and immerse the tube in a reaction bath, allowing the color to develop at 100° C. for 15 ± 0.5 minutes. It is important to keep the timing in order to be able to reproduce results. Cool the tubes to room temperature in a beaker of cold water. Dilute to 0.5 ml. with benzene-isopropyl alcohol mixture (4:1) and transfer to a microcuvette. Read the transmittance at 567 mμ, using the benzene-isopropyl alcohol as the reference.

BENZENE HEXACHLORIDE

Hexachlorocyclohexane, benzene hexachloride, Lindane, $C_6H_8Cl_6$, is another chlorinated hydrocarbon insecticide. It contains 73 per cent of chlorine. It can be separated and isolated from food products

---

[6] Polen and Silverman, *Anal. Chem.* **24**, 733 (1952).

by the methods detailed. The organic chlorine can be determined in an analogous manner to that described in this chapter.

Schechter and Hornstein [7] detail a method which consists of dechlorinating the compound to benzene by means of zinc in acetic acid. The resultant benzene is absorbed in a nitrating mixture and is converted to m-dinitrobenzene, which is extracted and reacted with ethyl methyl ketone in the presence of strong alkali. The violet-red color formed can be evaluated photometrically. This method, which is known as the butanone method, has been discussed in detail by Jacobs. [8]

PARATHION

A colorimetric procedure for the estimation of small amounts of O,O-diethyl O,p-nitrophenylthiophosphate (parathion, Thiophos 3422 insecticide) depends upon the reduction with zinc to the amino compound, diazotization, and coupling with N-(1-naphthyl)-ethylenediamine dihydrochloride to produce an intense magenta color. [9] Amounts of the insecticide of 20 to 200 micrograms in the final 50-ml. aliquot are readily determined by using a photocolorimeter or spectrophotometer. This method may be applied to the determination of parathion in spray and dust residues on fruit, vegetables, and foliage.

The basic steps in the determination of small amounts of parathion are (1) reduction with zinc dust in acid solution to the amino compound, diethyl p-aminophenylthiophosphate; (2) diazotization of the amino compound with sodium nitrite, removal of the excess nitrite with ammonium sulfamate, and coupling with N-(1-naphthyl)-ethylenediamine dihydrochloride to produce an intense magenta color; (3) evaluation of the developed color, using a photocolorimeter or spectrophotometer and a standard transmittance-concentration curve prepared from data on known amounts of the insecticide.

Parathion is applied as an insecticide to a wide variety of plant materials. It would be difficult, therefore, to detail a concise single procedure for the extraction of residues and preparation of the extract for analysis for such diverse materials as apples, grapes, cabbage, tobacco, etc. The general manipulative steps are outlined along with some mention of vegetable materials treated; the general pro-

---

[7] Schechter and Hornstein, *Anal. Chem.* 24, 544 (1952).

[8] Jacobs, *The Analytical Chemistry of Industrial Poisons, Hazards, and Solvents.* Interscience, New York, 1949.

[9] Averell and Norris, *Anal. Chem.* 20, 753 (1948).

cedure may then be modified as necessary for use on specific plant materials.

Extractant.—Benzene has been found the most satisfactory of the common solvents for use as an extractant of parathion from plant materials.

Apparatus.—The extraction apparatus consists essentially of liquid-tight jars and a mechanical means of rolling or tumbling the jars. The jars may be ordinary preserve jars of 2- or 4-qt. capacity, or larger jars up to 3 gal. capacity. The closures of the jars should have tight gaskets protected from contact with the benzene by heavy tin foil. The agitation machine may be built for the purpose, or ordinary motor-driven rollers, geared down to slow speeds, may be used. For end-over-end tumbling, a large can or fiber container, in which the extraction jars may be placed sideways, may be used.

Extraction Time.—A minimum of 30 minutes extraction time has been employed, and 1 hour is recommended. The extra time does no harm in the extraction of most materials, with the exception of soft-fleshed fruits or vegetables; special effort must be made toward gentler treatment in the determination of surface residues on such material.

Sample Size.—The amount of material taken should be large enough to be representative; in general, the larger the sample, the greater the precision. In some cases, such as pea vines, or wax beans, where there is a high content of natural waxes, too large a sample may mean an excessive amount of wax in the extract, which occasionally appears in the final colored solution as a turbidity. A guiding principle to be kept in mind is that, under the conditions outlined in the procedure, the aliquot of benzene extract taken for analysis should contain 20 to 200 micrograms of parathion, so that the transmittance of the colored solution will be in the right range; for different instruments, filters, or cell depths the proportions may differ.

Decolorization of Extracts.—Benzene extracts of leafy material are usually green or yellow-green with chlorophyll or carotene; other materials as well show a considerable content of carotene. This color may be removed satisfactorily by filtration through Attapulgus clay.

Place a small plug of cotton in the bottom of a 150-ml. separatory funnel and add 10 g. or more of Attapulgus clay adsorbent (2 parts of Attapulgus clay, 1 part of Hyflo-Super-Cel, well mixed); add 50 ml. of benzene and stir the mixture until completely wetted and free of air bubbles; draw off the benzene by suction almost to the surface of the adsorbent. Pour benzene extract of the plant material into the separatory funnel and draw through by suction at about 25 ml. per minute into a beaker or volumetric flask. Wash the container and column with at least 3 separate 25 ml. portions of fresh benzene. Do not permit the column to become dry until the last wash is passed through or some color will pass into the filtrate.

If there is no color in the extract, or only very light tints, the decolorization step may be omitted. Such light tints are usually destroyed during the reduction step.

Reagents.—Parathion.—A sample of Thiophos 3422 insecticide may be obtained from American Cyanamid Company, 30 Rockefeller Plaza, New York 20, N. Y. Determine approximate purity by a total nitrogen assay (theory =4.81%).

Standard Solution 1.—0.1000 g. of parathion (allowing for purity) in benzene to make 500 ml.; 1 ml. is equivalent to 200 micrograms of parathion.

Standard Solution 2.—25 ml. of standard solution 1, diluted to 250 ml. with benzene; 1 ml. is equivalent to 20 micrograms of parathion.

Sodium Nitrite Solution.—Dissolve 0.25 g. of reagent grade sodium nitrite in water and dilute to make 100 ml. Make solution up fresh weekly.

Ammonium Sulfamate Solution.—Dissolve 2.5 g. of technical ammonium sulfamate in water and dilute to make 100 ml. Make solution up fresh weekly.

N-(1-Naphthyl)-ethylenediamine Dihydrochloride Solution.— Dissolve 1 g. in water and dilute to make 100 ml. Make solution up fresh daily or weekly if stored in a refrigerator.

Procedure.—Concentrate the benzene solution of parathion to 10 ml. in a 300-ml. tall form beaker on a steam bath; pass a gentle stream of air over the liquid surface to hasten evaporation. Remove the last 10 ml. of benzene at room temperature by passing air over the surface and stop the evaporation as soon as the residue is just dry.

Take up the residue in 10 ml. of ethyl alcohol, add 10 ml. of water, 2 ml. of 5 N hydrochloric acid, and 0.2 g. of zinc dust. Cover the beaker with a watch glass, heat the solution to boiling on a hot plate, and boil gently for 5 minutes. Wash down the watch glass and sides of the beaker with water, allow the contents to cool, and filter through No. 42 Whatman filter paper into a 50-ml. volumetric flask. Wash the beaker, residue, and paper with small portions of water until the total volume of filtrate is about 40 ml. Add 1 ml. of 0.25 per cent sodium nitrite solution to the flask, mix the solution thoroughly, and let stand 10 minutes. Add 1 ml. of 2.5 per cent ammonium sulfamate solution to the test solution, again mix well, and let stand 10 minutes more. Finally, add 2 ml. of 1 per cent N-(1-naphthyl)-ethylenediamine dihydrochloride solution to the solution, fill to the 50-ml. mark with water, mix the solutions thoroughly, and let stand 10 minutes. Use the blank solution to set the photometer scale at 100 per cent transmittance, and read the transmittance of the test solution.

Run a blank, starting with the same volume of benzene as in the test solution. Use the final blank solution to set the photometer scale at 100 per cent transmittance, or if the same benzene and reagents are used throughout the day, one blank may be run for all the determinations of that day. In the latter case, set the scale at 100 per cent against water and correct the transmittances as follows:

$$\log 100/T \text{ (corr.)} = \log 1/T \text{ (sample)} - \log 1/T \text{ (blank)}$$

or

$$T \text{ (corr.)} = 100 \times T \text{ (sample)}/T \text{ (blank)}$$

where transmittances are expressed as per cent. When the corrected transmittance has been obtained, the number of micrograms of parathion in the final solution is read off the standard curve.

Preparation of Standard Curve.—Run aliquots of standard solution 2 into beakers from a 10-ml. microburette: 1, 2, 3, 4, 5, 7, 9, and 10 ml., equivalent to 20, 40, 60, 80, 100, 140, 180, and 200 micrograms of parathion. Dilute each aliquot to 10 ml. with benzene and place a blank of 10 ml. of benzene in another beaker. Evaporate off completely the benzene at room temperature by passing a gentle stream of air over the liquid surface and stop the evaporation as soon as the residue is just dry. Dissolve the residue in 10 ml. of ethyl alcohol and continue the determination as detailed in the test. Prepare standard curve by plotting transmittance against concentration (micrograms per 50 ml.).

Notes on Procedure.—The loss of parathion from filtration of the benzene extract through clay and by evaporation of the benzene may be as much as 10 per cent of the total present. Although this amount has a negligible effect on the conclusions to be drawn, it should be borne in mind.

The pH of the solution in the coupling step should be in the range 0.6 to 1.0. If the conditions outlined in the procedure are adhered to, the acidity will be in this range.

In the reduction step a 5-minute reaction period is sufficient: Fifteen minutes do not change the recovery in either direction.

The developed color is stable, on standing, to at least 4 hours. In the analysis of plant extracts, however, interfering colors may develop on standing. The transmittance reading should, therefore, be taken 10 minutes after the coupling reagent is added.

The transmittance of the reagent blank will be observed to decrease with the age of the reagents, particularly the naphthylethylenediamine. The reagents should, therefore, be made up fresh weekly.

Control analyses by Averell and Norris [10] of untreated materials showed an apparent parathion content of less than 0.1 p.p.m. A few materials, however, show an appreciable blank. Brussel sprouts and cabbage show a pinkish color in the solution before the coupling agent is added; this color may or may not precipitate out on standing. Some varieties of grapes show an appreciable control blank. It is, therefore, important when working on materials not previously analyzed that control analyses be run, at least at the outset, to determine whether or not interfering colors are introduced by the material itself.

DETECTION OF WARFARIN IN CORN MEAL [11]

The following method depends on the separation of Warfarin, 3-(α-acetonylbenzyl)-4-hydroxycoumarin, in corn meal bait by extraction with ether and subsequent microscopic identification. The method can be applied in poisoning cases. (Federal specifications call for a minimum of 0.025 per cent Warfarin, corresponding to 1.0 mg. per 4.0 g. of corn meal.)

Procedure.—Weigh 8.0 g. of corn meal into a small Erlenmeyer flask, cover with ether to which has been added a drop of hydrochloric acid, stopper, and allow to stand overnight. Filter through paper, washing with a few small portions of ether. Evaporate the filtrate to dryness on a steam bath and keep warm until any residual moisture is removed. Transfer the residue to an 8-ml. conical centrifuge tube with small portions of ether so that the volume does not exceed 3.0 ml. Place the centrifuge tube in a beaker of warm water, and by means of a short glass delivery tube carefully blow a gentle current of air through the solution until the volume is reduced to 0.5 ml. Add 3.0 ml. of Skellysolve B and continue blowing air until the volume is reduced to 0.5 ml., when the ether should be entirely removed. Add 3.0 ml. of Skellysolve B, mix by blowing air through

10 Averell and Norris, *Anal. Chem.* 20, 753 (1948).

11 N. I. Goldstone, personal communication (May, 1955).

the solution for a few seconds, and place in a refrigerator for several hours. A white, gelatinous precipitate of impure Warfarin separates out. Quickly centrifuge and carefully pour off the supernatant liquid.

Recrystallize by dissolving the precipitate in a small volume of ether and repeat the crystallization process with Skellysolve B as described above, again centrifuging and pouring off the supernatant liquid. Place the centrifuge tube containing the delivery tube on a steam bath and evaporate the residual liquid until the residue is completely dry. Cool, until a white opalescence appears. Place in a refrigerator, occasionally scratching the side of centrifuge tube with the delivery tube to promote crystallization, and allow to remain overnight. A crystalline precipitate of purified Warfarin separates out. Add a drop or two of water, mix, scratch, and allow to stand in refrigerator for another hour to crystallize more Warfarin.

Place a drop of the liquid on a slide without cover glass and examine microscopically at 100×. If Warfarin is present, it appears in beautiful clusters of radiating needles. A positive test is obtained with the minimum amount of 2.0 g. of corn meal containing 0.5 mg. of Warfarin.

MICROCHEMICAL DETECTION OF FLUORIDES

Sodium Fluosilicate Crystal Test.—Goldstone [12] modified the composition of the hanging drop solution in the sodium fluosilicate crystal test for the microchemical detection of fluorides in order to render the test considerably more sensitive. The test can be applied to distinguish between inorganic fluorides and monofluoroacetic acid. It is possible to obtain a positive test with as little as 0.2 micrograms of fluoride. With increasing concentrations of fluoride larger numbers of crystals can be observed, thus with 1.0 micrograms of fluoride several thousand crystals of assorted sizes appear in the field.

Along with the increased sensitivity the modified test has the additional advantage of enabling the sodium fluosilicate crystals, which appear in characteristic hexagonal form or as sixpointed stars, to stand out individually and more distinctly from the larger sodium chloride crystals. Furthermore, they are tinted a deeper shade of pink and therefore are more easily recognized.

Apparatus.—Heating Block.—A metal block approximately 2.5 cm. thick and large enough to hold four 10-ml. porcelain crucibles is suitable. A well to hold the bulb of a thermometer is drilled into the

---

12 Goldstone, *Anal. Chem.* **27,** 464 (1955).

block. A few drops of mineral oil are placed in the well to cover the bulb. The block is set on a tripod and preferably heated with a multiple-jet gas burner. A satisfactory block may be constructed by melting sufficient printer's type metal in an aluminum pie plate. A small test tube 1 cm. in diameter is set and held in the molten metal by a clamp on a ring stand, then the metal is allowed to cool slowly and to solidify.

Glass Slides.—Microscope-slide glass is cut into pieces 4 × 4 cm.

Standardized Micropipette.—For convenient delivery of uniform drops of standard fluoride solutions a satisfactory pipette may readily be constructed. A length of thin-walled glass tubing of 5-mm. diameter is drawn out into a fine capillary, which is broken off at a point where its diameter is less than 1 mm. It is standardized by allowing water to flow from it, drop by drop, at a uniform rate into a microburette filled with water exactly to the 1.00-ml. mark. If the zero mark is not reached by the addition of 50 drops, the individual drops are too small, and a short length of capillary is cut off and the trial repeated. The procedure is repeated until a uniform drop of exactly 0.02 ml. is delivered. The pipette is dried and inserted through a rubber stopper fitted to a test tube or small reagent bottle containing standard fluoride solution, from which definite quantities of fluoride may be accurately delivered when required.

Crucibles.—A number of high-form glazed porcelain crucibles of 10-ml. capacity.

Reagents.—Standard Sodium Fluoride Solution.—Dissolve 0.2210 g. of pure sodium fluoride ,in water and dilute to 2000 ml. Each milliliter of this solution contains 50.0 μg. of fluorine.

Standard Sodium Monofluoroacetate Solution.—Dissolve 0.05 g. of sodium monofluoroacetate in water and dilute to 250 ml. Each milliliter of solution contains 0.2 mg. of the salt; 0.05 ml. of solution contains 1.9 micrograms of fluorine.

Standard Sodium Fluosilicate Solution.—Dissolve 0.1650 g. of pure sodium fluosilicate in water and dilute to 200 ml. Each milliliter of solution contains 50 micrograms of fluorine.

Sodium Chloride Hanging Drop Solution.—Dissolve 1.0 g. of pure sodium chloride and 3.0 g. of pure glycerol in water, add 2 drops of 40 per cent formaldehyde to preserve the reagent, dilute to 100 ml., and filter through paper into a glass reagent bottle. Insert a 3-mm.-diameter glass rod, of suitable length, with fire-polished ends through a rubber stopper and keep the bottle well stoppered. This apparatus serves very conveniently for the transfer of a small drop of solution to the surface of the glass slide in the crystal test.

Silver Sulfate.—Pure crystalline silver sulfate stored in a brown bottle.

Saturated Silver Sulfate Solution.—An excess of silver sulfate suspended in water and stored in a brown dropping bottle.

Silica.—Fluorine-free powdered silicon dioxide.

Procedure.—Make a few grams of the material to be tested alkaline with a slight excess of sodium carbonate solution, dry in an oven at 100° C., cautiously burn off the organic matter over a Bunsen flame, then continue heating in a muffle furnace held below 500° C. until a gray or white ash is obtained. Transfer about 20 mg. of the ash to a 15-ml. test tube, add 10 ml. of water, shake until all soluble matter is dissolved, and then transfer half of the solution to another 15-ml. test tube. To the second tube, which serves as a control, add 2.0 micrograms of fluoride and heat both tubes in a beaker of boiling water. Add a pinch of silver sulfate powder to each and shake occasionally until the silver precipitate formed coagulates. Test the clear supernatant liquid by adding a drop of saturated silver sulfate solution, and if additional precipitation occurs, add more powdered silver sulfate; continue to heat, shake, and test until precipitation is complete. Cool the tubes in an ice bath and filter through small paper filters into 10-ml. porcelain crucibles, washing with 2 successive small portions of water. To each crucible add ca. 0.5 mg. of calcium carbonate powder and ca. 2 mg. of powdered silica; then evaporate gently (to avoid spattering) to dryness on a hot plate, allowing the crucibles to bake for a few minutes. Cool to room temperature. To the residue add 2 small drops of sulfuric acid (specific gravity, 1.84) and place the crucible on a metal block, maintained at 170° C. Cover immediately with a glass slide, on the undersurface of which has been placed a small drop (diameter, 0.4 cm.) of modified hanging drop solution. Set a 50-ml. beaker containing an ice cube firmly on top of the slide and allow the distillation to proceed for 20 minutes, after which dry with filter paper and put in a warm place for a few minutes until the hanging drop is dry. If fluoride is present in the sample tested, it is indicated by the presence of the characteristic fluosilicate crystals, the control being, of course, positive.

Microscopic examination (440×) reveals the presence of several thousand decidedly pink crystals of various sizes, either in hexagonal form or as six-pointed stars. These crystals are not uniformly distributed throughout the field but are mainly concentrated along the periphery of the drop. Viewed very slightly out of exact focus, they appear opaquely black. The limit of sensitivity is reached when

0.2 microgram of fluorine is subjected to the test, producing a few tiny crystals, the number increasing to over 100 when 0.3 microgram is used. To perform the test on quantities less than 1.0 microgram the standard solution is diluted to one tenth its fluoride content and the appropriate number of drops taken. When standard sodium fluoride solution is used instead of the fluosilicate, the procedure is not changed except for the addition of ca. 2 mg. of silica powder in the microdistillation; this converts the hydrofluoride into fluosilicic acid. Tests indicate that the recovery in the form of sodium fluosilicate crystals is not quantitative, only part of the fluoride being trapped in the hanging drop.

Interferences.—A number of common negative ions such as chlorides, nitrates, borates, carbonates, and sulfates influence in varying degrees the formation of sodium fluosilicate crystals in the hanging drop test. The presence of these ions tends to inhibit the quantity of fluoride recovered, and in general, the more negative ions present, the fewer crystals appear in the microscopic field. The ions mentioned above are listed in the descending order of their capacity to interfere. When the test is performed on 1.0 microgram of fluoride, to which has been added 1 mg. of sodium chloride or nitrate, interference is complete and no crystals can be observed in the field. Some of these ions influence the shape of the crystals, tending to round off the corners of the hexagon, so that they are more nearly circular. In the distillation the negative ions are volatilized along with the fluosilicic acid and are absorbed in the hanging drop, where they influence the formation of the crystals.

It is essential that conditions of absolute cleanliness be maintained in preparing and handling microscopic slides.

Detection of Fluoroacetate.—The modified crystal test affords a means of distinguishing between inorganic fluorides and sodium monofluoroacetate. The test is based on the stability of the carbon-fluorine linkage in monofluoroacetic acid in contact with hot concentrated sulfuric acid, under which condition no free hydrofluoric acid is released. Therefore, the modified crystal test performed on sodium monofluoroacetate will be negative because no hydrofluoric acid is evolved. If, however, monofluoroacetic acid is first fused with sodium carbonate, the fluorine is converted into sodium fluoride, which, with the addition of silica, will produce a positive test.

Procedure.—Transfer 0.05 ml. (0.01 mg. of the salt or 1.9 micrograms of fluorine) of a solution of sodium monofluoroacetate to each of two 10-ml. porcelain crucibles. To the second crucible add a drop of phenolphthalein solution and a small drop of 0.01 N sodium hydroxide solution, and dry both crucibles on a steam bath. Fuse the contents of the second crucible over a low Bunsen flame or in a muffle furnace below 500° C. for a short time. Allow the crucibles to cool, add about 2 mg. of powdered silica to each, and perform the crystal test as detailed on both.

The unfused sodium monofluoroacetate will give a negative test, while the fused salt will produce large numbers of sodium fluosilicate crystals. If the volume of standard solution is increased to 0.2 ml. (7.6 micrograms of fluorine), great masses of pink crystals are formed in the hanging drop. Commercial sodium monofluoroacetate usually contains traces of free sodium fluoride, and a few crystals are sometimes observed when the test is performed on the unfused salt, but the contrast in numbers between the latter and the fused salt is so sharp that no doubt exists in the interpretation of results. The test is not specific for sodium monofluoroacetate, as other organic fluorine compounds will be converted to sodium fluoride by alkaline fusion. Furthermore, the test is inapplicable to those organic fluorine compounds in which the carbon-fluorine linkage is unstable when in contact with sulfuric acid.

Chapter XXIV

ARTIFICIAL SWEETENING AGENTS

During the decade of the 1950's there was a large increase in the production and sale of artificially sweetened foods and beverages. Some size evaluation of the potential market for such products can be gathered from the statement, "There are approximately 34 million adult Americans who are overweight, plus two million known diabetics. Combined they form a dietetic-diabetic market of more than 20 per cent of our total population." [1] It has been estimated that approximately ¼ billion dollars of dietetic foods was packed in 1956.

There are a variety of foods that are being sweetened with artificial sweeteners. Among the principal products are beverages and dietetic canned fruits. Other products are canned fruit juices and vegetables; cookies and other bakery products; flavoring extracts; gelatin, pudding and frozen desserts; imitation jams, jellies, and marmalades; and salad dressings. In every instance the food product must be properly labeled.

In general, in the United States only two types of artificial sweeteners may be used in foods when any artificial sweetener is permitted. These two types are the saccharins and the cyclamates. Permission for the use of these stems from a series of tests conducted by the Food and Drug Administration [2] on the relative toxicity of saccharin, sodium cyclohexylsulfamate (cyclamate), 2-amino-4-nitro-1-n-propoxybenzene(P-4000), and dulcin. It was found that dulcin produces liver tumors and interferes with the production of red cells in animals, and so it is excluded as an optional ingredient in foods on the basis of harmful effects produced on long-term low-level feeding. P-4000 was found to produce kidney damage and indicated interference with thyroid functioning, consequently it, too, is excluded on the basis of a low margin of safety. Both saccharin and cyclamate produced no apparent effects at high levels of feeding and so were considered safe as food additives.

It must be stressed that no artificial sweetening agent has any

---

[1] "Making the most of the Dietetic Market with Sucaryl," Abbott Laboratories, North Chicago (1956).

[2] Lehman, *Assoc. Food Drug Officials U. S. Quarterly Bull.* **14,** 82 (1950).

food value, consequently their use in foods constitutes an adulteration unless the artificial sweetener is permitted and unless the food product containing the permitted sweetening agent is clearly and properly labeled to show that it contains a non-nutritive synthetic sweetening agent.

The interest of the food analyst in artificial sweeteners lies in (1) the detection of their use as adulterants by unpermitted substitution for nutritive sweeteners, (2) the detection of nonpermitted synthetic sweetening agents, (3) the determination of the amount of permitted non-nutritive sweetening agent, (4) the determination of the amount of each permitted sweetening agent in a mixture, and (5) the control of the concentration of a sweetener during the manufacture of the food product.

SACCHARIN

The main synthetic sweetening agent for over fifty years has been saccharin and its salts, principally sodium and ammonium salts. The name saccharin is commonly used to cover all of these compounds. Saccharin, $C_6H_4SO_2NHCO$, 2,3-dihydro-3-oxobenzisosulfonazole, is also known as benzoysulfonimide, o-sulfonbenzoic imide, and by a number of trade names which are seldom used. It is a white crystalline powder which has no odor but has a very sweet taste, of the order of 300 to 550 times that of sucrose. This wide variation in apparent sweetening power is due to the wide variation in the taste sensation of individuals. Saccharin melts in the range 226 to 230° C. with some decomposition and sublimes on additional heating. It is not very soluble in cold water for about 1 g. dissolves in about 300 ml. of water. One g. dissolves in 25 ml. of boiling water, in 30 ml. of alcohol, and in 50 ml. of glycerol.

The sodium salt of saccharin is about 550 times as sweet as sucrose. It is far more soluble, for 1 g. dissolves in 1.5 ml. of water and in about 5.0 of alcohol. Because of its solubility sodium saccharin is known as soluble saccharin. Crystalline sodium saccharin, $C_7H_4O_3$-SNa.2H$_2$O, effloresces in air.

Identification.—Crystalline or powdered saccharins can be identified by two tests: (1) conversion to salicylic acid and (2) formation of a compound with resorcinol.

Salicylic Acid Test.—Dissolve about 100 mg. of the sample in 5 ml. of sodium hydroxide solution (1:20) and evaporate to dryness. Fuse the residue carefully over a small flame until ammonia is no longer given off. Allow the residue to cool, dissolve it in 20 ml. of water, neutralize the solution with dilute hydrochloric acid, and filter. Add a drop of 1 N ferric chloride solution, prepared by dissolving 9 g. of ferric chloride in sufficient water to make 100 ml. The formation of a violet color in the filtrate is indicative of salicylic acid formed from saccharin. The material being analyzed should be tested for the presence of salicylic acid before making the identification test.

Resorcinol Test.—Mix 20 mg. of the sample being tested with 40 mg. of resorcinol. Add 10 drops of sulfuric acid and heat the mixture for three minutes in a bath capable of reaching 200° C. Allow the test solution to cool. Add 10 ml. of water and an excess of 1 N sodium hydroxide solution. If saccharin is present in the powder, a fluorescent green solution will be obtained.

Detection.—Saccharin can be detected in foods by (1) a taste test, (2) the salicylic acid test, and (3) the phenol-sulfuric acid test.

Taste Test.—The presence of saccharin in a food or beverage can be detected by extracting it from the sample and then tasting the residue. This is known as an organoleptic test.

Procedure.—Prepare the sample as detailed in the section determination. Transfer 50 ml. of an aqueous extract or 50 ml. of the clear liquid sample—as for instance, a carbonated beverage—to a Jacobs-Singer separatory flask or a separatory funnel and acidify with hydrochloric acid. Extract with three 25-ml. portions of ethyl ether. Combine the ether extracts and wash once with 5 ml. of water. Transfer the ether extract to an evaporating dish and allow the ether to evaporate spontaneously. Taste the residue. Concentrations of the order of 1 mg. per 50 ml. of extract or beverage can be detected by taste.

Salicylic Acid Test.—As in the identification test noted above, the saccharin is converted to salicylic acid by fusion with sodium hydroxide, and the latter is detected by means of the ferric chloride reaction.

Procedure.—Transfer the residue obtained as detailed in the taste test to a nickel dish with the aid of ethyl ether and evaporate off the ether. Add 10 ml. of 10 per cent sodium hydroxide solution and evaporate to dryness. Bake for an hour on a hot plate and fuse cautiously over a small flame. Leach with water and transfer to a separatory funnel. Make the solution acid with hydrochloric acid and extract with ether as detailed in the taste method. Transfer the ether layer to an evaporating dish, evaporate off the ether, and add a few drops of ferric chloride solution. The formation of a deep violet color is indicative of salicylic acid.

A.O.A.C. Variation.—In this variation [3] so-called "false-saccharin" and any natural or small amounts of added salicylic acid are destroyed while 5 mg. of saccharin per liter or per kg. of sample can be detected.

Procedure.—Acidify 50 ml. of the aqueous extract or of a beverage with hydrochloric acid and extract successively with 3 portions of ethyl ether as in the taste test. Evaporate off the ether, dissolve the residue in a little hot water and test a small portion by adding 1 drop of 0.5 per cent neutral ferric chloride solution. Dilute the rest of the test solution with water to a volume of 10 ml. Add 2 ml. of sulfuric acid (1:3). Heat to boiling and add a slight excess of 5 per cent potassium permanganate solution dropwise. Cool the solution, add 1 g. of sodium hydroxide, and after solution, filter into a silver dish or crucible lid. Evaporate to dryness and heat for 20 minutes at 210 to 215° C. Dissolve the heated material in water, make acid with hydrochloric acid, and add 1 drop of 0.5 per cent neutral ferric chloride solution.

Phenol-Sulfuric Acid Test.—The following test[3] depends upon the nitration of phenol-sulfuric acid with the nitric acid formed from the imino group of the saccharin and the subsequent production of a color.

Preparation of Sample.—Baked Goods.—Grind 25 g. of the sample and mix it thoroughly with 50 g. of washed and ignited sea sand. Transfer to an extraction thimble and extract for 1 to 2 hours in a Soxhlet extractor with petroleum ether so that it is substantially free of fat. Transfer the extracted sample to a 300-ml. Erlenmeyer flask, add 100 ml. of alcohol, equip with an air condenser, and reflux for 30 minutes on a boiling water bath, shaking the mixture frequently. Filter through a Whatman No. 2 filter paper wet with alcohol in a Büchner funnel and transfer the alcohol filtrate to a 100-ml. beaker. Evaporate to half its volume, add 50 ml. of water and sufficient 10 per cent sodium carbonate solution to make the test solution alkaline, and again evaporate to 50 ml. Transfer the test solution to a Jacobs-Singer separatory flask or a separatory funnel and continue as detailed in the procedure.

The Braverman method, as detailed on page 165, can also be used for the preparation of the sample both for baked goods and other solid foods and semi-solid foods.

Semi-Solid Foods.—Transfer with the aid of a small amount of hot water 25 g. of the sample to a 100-ml. volumetric flask. Add

---

[3] *Methods of Analysis, A. O. A. C.*, 8th ed., Washington, 1955.

enough boiling water to make the volume about 75 ml. Shake occasionally while allowing to stand for 1 hour. Add 3 ml. of acetic acid, mix thoroughly, add 5 ml. of 20 per cent neutral lead acetate solution, dilute to the mark with water, and mix thoroughly again. Allow to stand for 20 minutes and filter. Transfer 60 ml. to a Jacobs-Singer separatory flask or to a separatory funnel and continue as detailed in the procedure.

Procedure.—Place 25 ml. of a carbonated beverage from which the carbon dioxide has been allowed to escape or the volumes of the prepared sample as noted in the preceding paragraphs into a Jacobs-Singer separatory flask or into a separatory funnel. If a Jacobs-Singer separatory flask is used, employ the technique detailed on page 41. Acidify with 3 ml. of hydrochloric acid and if vanillin is present, extract with three 25-ml. portions of petroleum ether. Discard the petroleum ether extract. Extract the test solution successively with 50-, 25-, and 25-ml. portions of an ethyl ether-petroleum ether mixture (1:1). Combine the extracts in a separatory funnel, wash once with 5 ml. of water, and transfer to a 30-ml. beaker. Allow the ethers to evaporate spontaneously and dry at room temperature.

Add 5 ml. of phenol-sulfuric acid reagent—prepared by dissolving pure, colorless, crystalline phenol in an equal weight of sulfuric acid—and heat for 2 hours at 135 to 140° C. Cool, dissolve in a small quantity of hot water, and pour into about 250 ml. of water. Add a small quantity of Filter-Cel, allow to stand for 3 hours or overnight, and filter. Add sufficient 10 per cent sodium hydroxide solution to make the test alkaline and dilute to 500 ml. A magenta or reddish-purple color develops if saccharin was present. Yellow, buff or pale salmon shades are not to be considered significant.

Determination.—The amount of saccharin in a food or a beverage can be estimated by extracting the saccharin, separating it from other components of the food or beverage, and then decomposing it to obtain the SO_2 radical, oxidizing the latter to sulfate and weighing the barium sulfate formed. An alternative method is to separate the saccharin and then sublime it in a sublimation assembly at a temperature of 140 to 160° C. at 1 to 2 mm. Hg, subsequently weighing the sublimate. The amount of saccharin in carbonated beverages can be determined by converting the imino group of saccharin to ammonia and then estimating the ammonia by nesslerization. Saccharin, itself, can be assayed by a titrimetric method.

Sulfate Method.—Preparation of Sample.—Alcoholic Beverages.
—Evaporate 100 to 200 ml. to half the original volume to remove the

alcohol. Repeat by diluting to the original volume with water and again evaporating. Dilute alcoholic syrups with an equal volume of water before evaporating.

Solid or Semi-Solid Foods.—Transfer with the aid of hot water a weighed amount of sample in the range of 50 to 75 g. to a ·250-ml. volumetric flask. Add enough boiling water to bring the volume up to 200 ml. Mix thoroughly and allow to stand for 2 hours, shaking at intervals during this time. Continue with the method as detailed.

Procedure.—To the 200 ml. of prepared sample in the 250-ml. volumetric flask or to 200 ml. of other samples such as fruit juices in such a flask, add 5 ml. of acetic acid and mix thoroughly. Add a slight excess of 20 per cent neutral lead acetate solution, mix well, dilute to volume with water, mix thoroughly, allow to stand for 20 minutes, and filter. Transfer 150-ml. of the filtrate to a separatory funnel or evaporate 150 ml. of the filtrate to 50 ml. and transfer quantitatively to a Jacobs-Singer separatory flask. Add 15 ml. of hydrochloric acid. Extract successively with three 80-ml. portions of ethyl ether and if a Jacobs-Singer separatory flask is being used, employ the technique detailed on page 41. Shake at least 2 minutes for each extraction. Combine the ether extracts in a separatory funnel and wash once with 5 ml. of water. Evaporate off the ether to a small volume and transfer to a nickel crucible. If substances that are only slightly soluble in ether are present, use both water and ether for the transfer. Evaporate off the ether on a steam bath, add 2 to 3 ml. of 10 per cent sodium carbonate solution—or enough to make the test solution distinctly alkaline—rotate the dish so that the carbonate wets all of the saccharin, and evaporate to dryness. Add 4 g. of a mixture of equal weights of sodium and potassium carbonates. Heat cautiously at first, then more vigorously until fusion is complete, and continue baking for 30 minutes. Three to 4 g. of sodium oxide may be used for the fusion, but if this is done, then the time for fusion may be reduced to 5 minutes. Allow to cool, dissolve the fused mass in water, add 5 ml. of bromine water, acidify with hydrochloric acid, and filter. Catch the filtrate in a beaker. Wash the filter paper with a little water and dilute the filtrate and washings to 200 ml. Heat to boiling and add slowly with stirring an excess of 10 per cent barium chloride solution. Digest for a time and allow the test to stand overnight. Filter on quantitative filter paper, wash the precipitate until it is free from chlorides, transfer to a tared crucible, dry, ash, place in a desiccator to cool, and weigh. Multiply the weight of barium sulfate found by the factor 0.7848 to obtain the amount of saccharin. Run a blank on the fusion mixture and correct the result for any sulfur found.

Ammonia Method.—In the case of carbonated beverages, as mentioned, the amount of saccharin can be found by converting the imino group of the saccharin to ammonia and estimating the ammonia by nesslerization.

Procedure.—Place 50 ml. of the beverage or equivalent sample into a Jacobs-Singer separatory flask or into a separatory funnel and add 2 ml. of hydrochloric acid. Extract with two 50-ml. portions of ethyl ether. Filter the ether extracts through a pledget of cotton into a separatory funnel and wash the combined ether extracts with 5 ml. of water acidified with 1 drop of hydrochloric acid. Draw off the water layer and transfer the ether layer through the mouth of the separatory funnel to a beaker. Evaporate to dryness on a steam bath. Add 5 ml. of ammonia-free water and 6 ml. of hydrochloric acid, and evaporate to about 1 ml. while stirring constantly on a hot plate. Repeat the hydrolysis. Transfer the test solution to a 50-ml. volumetric flask and fill to the mark with ammonia-free water. Transfer 2 ml. with a pipette to a 25-ml. Nessler tube, dilute with ammonia-free water to the 25-ml. mark, and add 1 ml. of Nessler reagent. Compare the color produced with standards made with ammonium chloride, equivalent to 200 parts per million of the insoluble form of saccharin. Thus 0.2921 g. of ammonium chloride is equivalent to 1 g. of insoluble saccharin, C_7H_5-NO_3S, and to 1.317 g. of sodium or soluble saccharin, $C_7H_4NNaO_3S$.

Assay.—The method of U. S. P. XV can be followed for the assay of powders, tablets, etc., containing saccharin. Weigh accurately about 500 mg. of saccharin, previously dried at 105° C. for 2 hours, dissolve in 75 ml. of hot water, cool quickly, add 3 drops of phenolphthalein indicator solution, and titrate with 0.1 N sodium hydroxide solution. Each milliliter of 0.1 N sodium hydroxide solution is equivalent to 18.32 mg. of saccharin.

If sodium saccharin or saccharin with some interfering material is being analyzed, a variation has to be used. Weigh accurately about 500 mg. of the sample previously dried at 120° C. for 4 hours. Transfer the sample quantitatively to a separatory funnel with 10 ml. of water. Add 2 ml. of dilute hydrochloric acid and extract the precipitated saccharin first with 20 ml. and then with 4 additional 20-ml. portions of a solvent composed of 9 volumes of chloroform and 1 volume of alcohol, filtering each extract as it is drawn off through a small filter wet with the solvent. Evaporate the combined filtrates to dryness on a steam bath with the aid of a current of air.

Dissolve the residue in 75 ml. of hot water, cool, add 3 drops of phenolphthalein indicator solution, and titrate with 0.1 N sodium hydroxide solution. Each milliliter of 0.1 N sodium hydroxide solution is equivalent to 20.52 mg. of saccharin sodium.

CYCLAMATE

An artificial sweetening agent based on a salt of N-cyclohexylsulfamic acid is known as a cyclamate. Sodium cyclohexylsulfamic acid was discovered accidentally in a study of the antipyretic properties of the sodium salt of N-phenylsulfamic acid, $C_6H_5NHSO_3Na$, and related compounds, for it was noted that the salts of N-cyclohexylsulfamic acid, $C_6H_{14}NHSO_3Na$ etc., were remarkably sweet.[4] Sodium cyclohexylsulfamate is a white, free-flowing, nonhygroscopic, noncorrosive, and practically odorless solid that has no nutritive value. It is compatible with nearly all of the ingredients used in making foods and beverages, has no affect on shelf life, and retains its sweetness indefinitely. It is stable in the pH range of 2 to 10 and is not affected by temperatures up to 260° C. It is very soluble in water and is about 30 times as sweet as sucrose. It is marketed under various trade names such as Sucaryl sodium (Abbott Laboratories) and Cylan sodium cyclamate (DuPont).

Another salt of cyclohexylsulfamic acid that is very sweet is the calcium salt $(C_6H_{11}NHSO_3)_2Ca \cdot 2H_2O$—that is, calcium cyclohexylsulfamate dihydrate—which is a white, crystalline, free-flowing solid. It has no nutritive value and is also about 30 times as sweet as sucrose. It is very soluble in water, but calcium cyclamate reacts with tartaric acid, tartrates, oxalates, and some phosphates to form insoluble calcium salts. It is sold under trademarks such as Sucaryl calcium (Abbott Laboratories) and Cylan calcium cyclamate (Du Pont).

As noted in a previous section of this chapter, the cyclamates produce no apparent effects at high levels of feeding and consequently are considered safe as food additives.

Detection.—Nitrite Test.—Sulfate can be split from either sodium cyclohexylsulfamate or calcium cyclohexylsulfamate by nitrite; consequently if a prepared sample is freed from sulfate ion by use of barium chloride and upon the addition of nitrite additional sulfate is obtained, a positive test for cyclamate is indicated.

Preparation of Sample.—Blend an adequate amount 75 to 200 g. of the food being tested in a Waring or other blendor with 200 ml. of

---

4 Audreith and Sveda, *J. Org. Chem.* 9, 89 (1944).

water, or grind and mix thoroughly about the same amount of sample and transfer to a volumetric flask with enough water to make a volume of about 400 ml. Shake until the mixture becomes uniform; if necessary, add 2 to 5 g. of calcium chloride, shake again thoroughly until the salt is dissolved. Make the solution distinctly alkaline to litmus paper by the addition of 10 per cent sodium hydroxide solution, fill to the mark with water, stopper the flask, mix well, allow to stand for at least 2 hours shaking periodically, and filter.

Procedure.—Transfer 100 ml. of the prepared sample or the beverage being tested to a 100-ml. centrifuge tube. Add 2 g. of barium chloride, allow to stand for 5 minutes, and centrifuge. Filtration can also be performed. Acidify with 10 ml. of hydrochloric acid the supernatant liquid after decanting of the filtrate and add 0.2 g. of sodium nitrite. A white precipitate of barium sulfate is indicative of the presence of cyclamate.

Determination.—The qualitative test described in the preceding section can be adapted for the quantitative determination of cyclamate in foods and carbonated beverages.

Procedure.—Blend a weighed amount of the sample to be analyzed that contains about 100 mg. of the cyclamate in a Waring or other blendor with about 200 ml. of water. Transfer to a centrifuge bottle and centrifuge the suspension. Decant the clear supernatant into another centrifuge bottle. Wash the residue with about 50 ml. of water, centrifuge again, and add this supernatant liquid to the first. Add 10 ml. of hydrochloric acid and 10 ml. of 10 per cent barium chloride solution, mix thoroughly, and if free sulfate is present as evidenced by the formation of a precipitate, centrifuge again at a rate sufficient to pack the barium sulfate formed. Pour off the clear supernatant liquid into a beaker, add 10 ml. of 10 per cent sodium nitrite solution, and digest on a hot plate until the solution clears.

Prepare a Gooch crucible in the customary manner. Ignite, cool in a desiccator, and weigh. Filter off the precipitate on the asbestos pad of the Gooch crucible, wash, and dry in an oven at 100° C. for 15 minutes. Ignite the crucible, place in a desiccator to cool, and reweigh. The difference in the weight of the tared crucible and the crucible plus the precipitate is considered to be barium sulfate. Each milligram of barium sulfate is equivalent to 0.9266 mg. of calcium cyclamate and to 0.8621 mg. of sodium cyclamate.

In the case of beverages, transfer with the aid of a pipette a known volume of the beverage freed from carbon dioxide, containing about 100 mg. of a cyclamate. Dilute to approximately 200 ml. with water. Add 10 ml. of concentrated hydrochloric acid and 10 ml. of 10 per cent barium chloride solution, mix thoroughly, and proceed with the method as detailed above.

DULCIN

Dulcin was the most important synthetic sweetening agent next to saccharin for over 50 years until the Food and Drug Administration ruled that it was too toxic to be used in foods. According to these studies dulcin produces liver tumors and interferes with the production of red cells in animals.

Dulcin—$NH_2CONHC_6H_4OC_2H_5$, p-phenetylurea, p-phenetylcarbamide, and known by various trade names such as Sucrol—is a solid crystallizing in white lustrous crystals. It melts at 173 to 174° C. It is only slightly soluble in water, for only 1 g. dissolves in 800 ml. of cold and 50 ml. of boiling water. It is much more soluble in alcohol, 1 g. dissolving in 25 ml. of alcohol. This synthetic sweetening agent is about 200 times as sweet as sucrose. One of the reasons that dulcin remained a competitor of saccharin for many years was that its sweet taste was more agreeable than that of saccharin. Neither dulcin or its derivatives, some of which are sweet-tasting compounds, have any food value.

Detection.—It was noted on page 168 that dulcin can be separated from other compounds such as saccharin and salicylic acid because after it is extracted from an acidified prepared sample with ether, it is not extracted from the ether solution by an aqueous alkaline extracting solution.

Procedure.—Prepare an aqueous extract of the food as detailed in this chapter in the nitrite test for cyclamate, or if the sample is a beverage, make it alkaline with 10 per cent sodium hydroxide solution. Transfer 60 ml. to a Jacobs-Singer separatory flask or to a separatory funnel and extract successively with three 50-ml. portions of ether. Combine the ether extracts, mix, and divide equally into 3 aliquots. Transfer each aliquot to a porcelain dish, allow the ether to evaporate spontaneously at room temperature, and dry in an oven at 110° C.

Heat the residue in 1 dish for a brief period with 2 drops of phenol and 2 drops of concentrated sulfuric acid. Dilute with a few milliliters of water. Overlay with a small volume of ammonium hydroxide solution. At the liquid interface, a blue or violet-blue color develops.[5]

---

[5] Morpurgo, *Pharm. Zentralhalle* **1893**, 466.

Wet the residue in the second dish with a drop or two of nitric acid. A colored compound is formed which dissolves in an excess of the acid. Add a drop of water at an edge of the nitric acid drop. An orange or brick-red precipitate develops as the water diffuses into the nitric acid. These crystals are soluble in chloroform. The addition of acetic acid to this solution produces crystals of the compounds when evaporated.[6]

Treat the residue in the third dish with hydrogen chloride gas in a hood for 5 minutes. Add a drop of anisaldehyde. If dulcin is present, an orange-red or a blood-red color is formed.[7] This test can be used with 25 mg. of dulcin per kilo.

There are a number of other tests for the detection of dulcin. Bellier[8] suggested dissolving the dulcin in 2 ml. of sulfuric acid, adding a few drops of formaldehyde, and allowing the solution to stand for 15 minutes. The development of a white turbidity on the addition of 5 ml. of water is indicative of dulcin. The test will be given with 1 mg. of dulcin.

Jorissen[9] used a mercury reagent for the detection of dulcin. Dissolve 1 to 2 g. of freshly precipitated mercuric oxide in dilute nitric acid, add sodium hydroxide solution until a precipitate just forms, and add enough water to make 15 ml. Suspend the extracted residue in 5 ml. of water and shake with 2 to 4 drops of the reagent. Heat for 5 to 10 minutes in a boiling water bath. A violet-blue color develops that changes to violet on the addition of lead peroxide.

ALKOXYAMINONITROBENZENES

The 1-alkoxy-2-amino-4-nitrobenzenes were proposed as artificial sweetening agents by Verkade[10] in 1946. Initially they were looked upon with great favor, for they were far more powerful sweetening agents than saccharin, thus very much less had to be used, and in addition they had no bitter aftertaste. Some of the physical properties and the sweetening power of the alkoxyaminonitrobenzenes are given in Table 106. It can be seen from this table that 2-amino-4-nitro-1-n-propoxybenzene, or 5-nitro-2-n-propoxyaniline, as it is also known, or P-4000, its common name, has the greatest sweeten-

6 Deniges and Tourrou, *Compt. rend.* 173, 1184 (1921).

7 La Parola and Mariani, *Ann. chim. appl.* 36, 134 (1946).

8 Bellier, *Ann. chim. anal. chim. appl.* 11, 412 (1900).

9 Jorissen, *J. pharm. Liége* 1896 (Feb.).

10 Verkade, *Chem. Eng. News* 24, No. 21 (Nov. 10, 1946).

ing power of this group of compounds, roughly 4000 times as sweet as sucrose. Because of this, it appeared to have great promise but, as mentioned, tests of the Food and Drug Administration indicated that it might have too high an order of toxicity, even at the low levels of use, to be considered suitable for employment in foods.

TABLE 106. ALKOXYAMINONITROBENZENES (ALKOXYNITROANILINES)

Compound	Crystals [a]	M.p. °C.	Solubility [a] mg./l. 22-23° C.	Sweetening power Verkade[b]	Profft[a]
2-Amino-1-hydroxy-4-nitrobenzene			925	120	
2-Amino-1-methoxy-4-nitrobenzene	Sand-colored	124-125	105	220	2150
2-Amino-1-ethoxy-4-nitrobenzene	yellow	97-98	112	350	830
2-Amino-4-nitro-1-*n*-propoxybenzene	light orange-yellow	49-49.5	91	4100	3100
2-Amino-1-*n*-butoxy-4-nitrobenzene	yellow-brown	58-58.5	81	1000	280
2-Amino-1-isopropoxy-4-nitrobenzene	copper-brown	114-116	432	600	100
1-Allyloxy-2-amino-4-nitrobenzene				2000	

[a] Profft, *Deut. Chem.-Ztg.* **1**, 51 (1949); **2**, 194 (1950).
[b] Verkade, *Chem. Eng. News* **24**, No. 21 (Nov. 10, 1946).

Detection.—P-4000 can be extracted by petroleum ether from an alkaline aqueous solution and, after evaporation of the solvent, can be tested organoleptically or by the formation of a diamine by reduction of the nitro group.

Procedure.—Prepare an aqueous extract of the solid or semisolid samples as detailed in the phenol-sulfuric acid test for saccharin in this chapter or by other convenient means and transfer 200 ml. of this extract or of a liquid sample such as a beverage to a separatory funnel. Adjust the pH to 7.5 to 8.0 with 10 per cent sodium hydroxide solution and extract with three 25-ml. portions of petroleum ether. Combine the petroleum ether extracts and wash once with 5 ml. of water. Draw off the water phase, pour the petroleum ether layer into a small beaker, and allow to evaporate spontaneously at room temperature. Taste the residue: 5 mg. per kilo will yield a residue that gives a distinct sweet taste; 12.5 mg. per kilo will yield a sufficient residue to give an anesthetic effect.[11]

[11] Cox, *J. Assoc. Official Agr. Chem.* **35**, 321 (1952).

Diamine Reaction.—Prepare a petroleum ether extract as detailed above. Transfer to a beaker and allow to evaporate spontaneously to 5 ml. Transfer to a 25-ml. test tube, wash the beaker with small portions of petroleum ether, and add the washings to the test tube. Evaporate cautiously to dryness. Add 2 ml. of hydrochloric acid (1:1) and 2 ml. of a saturated solution of stannous chloride. Immerse the tube in a boiling water bath, heat for 15 minutes, take out of the bath, and cool. If not colorless, dilute with water until nearly colorless. Add saturated bromine water drop by drop until an excess is present. The color of the solution will change from nearly colorless to purplish-red or Burgundy-red if P-4000 is present, and if a large excess of bromine water is added, the solution will become yellow.[12]

2-HALO-5-NITROANILINES

A group of substances relatively closely related to the alkoxynitroanilines of the alkoxyaminonitrobenzenes, of which P-4000 is one example, are the 2-halo-5-nitroanilines, or 2-amino-1-halo-4-nitrobenzenes. A patent was granted to Snelling[13] for the use of such substances for the flavoring and sweetening of chewing gum. Snelling found that these substances have high sweetness and high solubility in chewing gum base with concomitant low solubility in saliva. The 2-iodo-5-nitroaniline had the highest sweetening power, whereas the fluoro derivative had the lowest sweetening power, of this group. The bromo compound appeared to have analgesic action. Such sweetening agents can be detected by the extraction and organoleptic tests detailed in this chapter.

Jacobs[14] has discussed other artificial sweetening agents and their properties, and high-power natural sweetening agents like stevioside.

---

[12] Cox, *J. Assoc. Official Agr. Chem.* **37**, 384 (1954).

[13] Snelling, U. S. Patent 2,484,860 (1949).

[14] Jacobs, *Am. Perfumer Essent. Oil Rev.* **57**, No. 1, 49 (1951); **57**, No. 2 129 (1951); **62**, 295 (1953); **65**, No. 2, 56 (1955); and **66**, No. 6, 44 (1955).

APPENDIX TABLES

TABLE 1. MUNSON AND WALKER SUGAR TABLE [a,b] IN MILLIGRAMS

CUPROUS OXIDE (Cu₂O)	COPPER (Cu)	DEXTROSE (d-GLUCOSE)	INVERT SUGAR	INVERT SUGAR AND SUCROSE 0.4 gram total sugar	INVERT SUGAR AND SUCROSE 2 grams total sugar	LACTOSE $C_{12}H_{22}O_{11}+H_2O$	LACTOSE AND SUCROSE 1 lactose, 4 sucrose	LACTOSE AND SUCROSE 1 lactose, 12 sucrose	MALTOSE $C_{12}H_{22}O_{11}+H_2O$	CUPROUS OXIDE (Cu₂O)
10	8.9	4.0	4.5	1.6	......	6.3	6.1	......	6.2	10
12	10.7	4.9	5.4	2.5	......	7.5	7.3	......	7.9	12
14	12.4	5.7	6.3	3.4	......	8.8	8.5	......	9.5	14
16	14.2	6.6	7.2	4.3	......	10.0	9.7	......	11.2	16
18	16.0	7.5	8.1	5.2	......	11.3	10.9	......	12.9	18
20	17.8	8.3	8.9	6.1	......	12.5	12.1	......	14.6	20
22	19.5	9.2	9.8	7.0	......	13.8	13.3	......	16.2	22
24	21.3	10.0	10.7	7.9	......	15.0	14.5	......	17.9	24
26	23.1	10.9	11.6	8.8	......	16.3	15.8	......	19.6	26
28	24.9	11.8	12.5	9.7	......	17.6	17.0	......	21.2	28
30	26.6	12.6	13.4	10.7	4.3	18.8	18.2	......	22.9	30
32	28.4	13.5	14.3	11.6	5.2	20.1	19.4	......	24.6	32
34	30.2	14.3	15.2	12.5	6.1	21.4	20.7	......	26.2	34
36	32.0	15.2	16.1	13.4	7.0	22.8	22.0	......	27.9	36
38	33.8	16.1	16.9	14.3	7.9	24.2	23.3	......	29.6	38
40	35.5	16.9	17.8	15.2	8.8	25.5	24.7	......	31.3	40
42	37.3	17.8	18.7	16.1	9.7	26.9	26.0	......	32.9	42
44	39.1	18.7	19.6	17.0	10.7	28.3	27.3	......	34.6	44
46	40.9	19.6	20.5	17.9	11.6	29.6	28.6	......	36.3	46
48	42.6	20.4	21.4	18.8	12.5	31.0	30.0	......	37.9	48
50	44.4	21.3	22.3	19.7	13.4	32.3	31.3	......	39.6	50
52	46.2	22.2	23.2	20.7	14.3	33.7	32.6	......	41.3	52
54	48.0	23.0	24.1	21.6	15.2	35.1	34.0	......	42.9	54
56	49.7	23.9	25.0	22.5	16.2	36.4	35.3	......	44.6	56
58	51.5	24.8	25.9	23.4	17.1	37.8	36.6	......	46.3	58
60	53.3	25.6	26.8	24.3	18.0	39.2	37.9	......	48.0	60
62	55.1	26.5	27.7	25.2	18.9	40.5	39.3	......	49.6	62
64	56.8	27.4	28.6	26.2	19.8	41.9	40.6	......	51.3	64
66	58.6	28.3	29.5	27.1	20.8	43.3	41.9	......	53.0	66
68	60.4	29.2	30.4	28.0	21.7	44.7	43.3	40.7	54.6	68
70	62.2	30.0	31.3	28.9	22.6	46.0	44.6	41.9	56.3	70
72	64.0	30.9	32.3	29.8	23.5	47.4	45.9	43.1	58.0	72
74	65.7	31.8	33.2	30.8	24.5	48.8	47.3	44.2	59.6	74
76	67.5	32.7	34.1	31.7	25.4	50.1	48.6	45.4	61.3	76
78	69.3	33.6	35.0	32.6	26.3	51.5	49.9	46.6	63.0	78
80	71.1	34.4	35.9	33.5	27.3	52.9	51.3	47.8	64.6	80
82	72.8	35.3	36.8	34.5	28.2	54.2	52.6	49.0	66.3	82
84	74.6	36.2	37.7	35.4	29.1	55.6	53.9	50.1	68.0	84
86	76.4	37.1	38.6	36.3	30.0	57.0	55.3	51.3	69.7	86
88	78.2	38.0	39.5	37.2	31.0	58.4	56.6	52.5	71.3	88

[a] *Nat. Bur. Standards, U. S. Circ.* 44 (1918).
[b] Given, *Methods of Sugar Analysis and Allied Determinations.* Blakiston, Philadelphia, 1912.

TABLE 1, MUNSON AND WALKER SUGAR TABLE IN MILLIGRAMS—*Continued*

CUPROUS OXIDE (Cu₂O)	COPPER (Cu)	DEXTROSE (d-GLUCOSE)	INVERT SUGAR	INVERT SUGAR AND SUCROSE		LACTOSE	LACTOSE AND SUCROSE		MALTOSE	CUPROUS OXIDE (Cu₂O)
				0.4 gram total sugar	2 grams total sugar	$C_{12}H_{22}O_{11}+H_2O$	1 lactose, 4 sucrose	1 lactose, 12 sucrose	$C_{12}H_{22}O_{11}+H_2O$	
90	79.9	38.9	40.4	38.2	31.9	59.7	57.9	53.7	73.0	90
92	81.7	39.8	41.4	39.1	32.8	61.1	59.3	54.9	74.7	92
94	83.5	40.6	42.3	40.0	33.8	62.5	60.6	56.0	76.3	94
96	85.3	41.5	43.2	41.0	34.7	63.8	61.9	57.2	78.0	96
98	87.1	42.4	44.1	41.9	35.6	65.2	63.3	58.4	79.7	98
100	88.8	43.3	45.0	42.8	36.6	66.6	64.6	59.6	81.3	100
102	90.6	44.2	46.0	43.8	37.5	68.0	66.0	60.8	83.0	102
104	92.4	45.1	46.9	44.7	38.5	69.3	67.3	62.0	84.7	104
106	94.2	46.0	47.8	45.6	39.4	70.7	68.6	63.2	86.3	106
108	95.9	46.9	48.7	46.6	40.3	72.1	70.0	64.4	88.0	108
110	97.7	47.8	49.6	47.5	41.3	73.5	71.3	65.6	89.7	110
112	99.5	48.7	50.6	48.4	42.2	74.8	72.6	66.7	91.3	112
114	101.3	49.6	51.5	49.4	43.2	76.2	74.0	67.9	93.0	114
116	103.0	50.5	52.4	50.3	44.1	77.6	75.3	69.1	94.7	116
118	104.8	51.4	53.3	51.2	45.0	79.0	76.7	70.3	96.4	118
120	106.6	52.3	54.3	52.2	46.0	80.3	78.0	71.5	98.0	120
122	108.4	53.2	55.2	53.1	46.9	81.7	79.3	72.7	99.7	122
124	110.1	54.1	56.1	54.1	47.9	83.1	80.7	73.9	101.4	124
126	111.9	55.0	57.0	55.0	48.8	84.5	82.0	75.1	103.0	126
128	113.7	55.9	58.0	55.9	49.8	85.8	83.4	76.3	104.7	128
130	115.5	56.8	58.9	56.9	50.7	87.2	84.7	77.5	106.4	130
132	117.3	57.7	59.8	57.8	51.7	88.6	86.0	78.7	108.0	132
134	119.0	58.6	60.8	58.8	52.6	90.0	87.4	79.7	109.7	134
136	120.8	59.5	61.7	59.7	53.6	91.3	88.7	81.1	111.4	136
138	122.6	60.4	62.6	60.7	54.5	92.7	90.1	82.3	113.0	138
140	124.4	61.3	63.6	61.6	55.5	94.1	91.4	83.5	114.7	140
142	126.1	62.2	64.5	62.6	56.4	95.5	92.8	84.7	116.4	142
144	127.9	63.1	65.4	63.5	57.4	96.8	94.1	85.9	118.0	144
146	129.7	64.0	66.4	64.5	58.3	98.2	95.4	87.1	119.7	146
148	131.5	65.0	67.3	65.4	59.3	99.6	96.8	88.3	121.4	148
150	133.2	65.9	68.3	66.4	60.2	101.0	98.1	89.5	123.0	150
152	135.0	66.8	69.2	67.3	61.2	102.3	99.5	90.8	124.7	152
154	136.8	67.7	70.1	68.3	62.1	103.7	100.8	92.0	126.4	154
156	138.6	68.6	71.1	69.2	63.1	105.1	102.2	93.2	128.0	156
158	140.3	69.5	72.0	70.2	64.1	106.5	103.5	94.4	129.7	158
160	142.1	70.4	73.0	71.2	65.0	107.9	104.8	95.6	131.4	160
162	143.9	71.4	73.9	72.1	66.0	109.2	106.2	96.8	133.0	162
164	145.7	72.3	74.9	73.1	66.9	110.6	107.5	98.0	134.7	164
166	147.5	73.2	75.8	74.0	67.9	112.0	108.9	99.2	136.4	166
168	149.2	74.1	76.8	75.0	68.9	113.4	110.2	100.4	138.0	168

TABLE 1. MUNSON AND WALKER SUGAR TABLE IN MILLIGRAMS—*Continued*

CUPROUS OXIDE (Cu$_2$O)	COPPER (Cu)	DEXTROSE (d-glucose)	INVERT SUGAR	INVERT SUGAR AND SUCROSE		LACTOSE	LACTOSE AND SUCROSE		MALTOSE	CUPROUS OXIDE (Cu$_2$O)
				0.4 gram total sugar	2 grams total sugar	C$_{12}$H$_{22}$O$_{11}$+H$_2$O	1 lactose, 4 sucrose	1 lactose, 12 sucrose	C$_{12}$H$_{22}$O$_{11}$+H$_2$O	
170	151.0	75.1	77.7	76.0	69.8	114.8	111.6	101.6	139.7	170
172	152.8	76.0	78.7	76.9	70.8	116.1	112.9	102.8	141.4	172
174	154.6	76.9	79.6	77.9	71.7	117.5	114.3	104.1	143.0	174
176	156.3	77.8	80.6	78.8	72.7	118.9	115.6	105.3	144.7	176
178	158.1	78.8	81.5	79.8	73.7	120.3	117.0	106.5	146.4	178
180	159.9	79.7	82.5	80.8	74.6	121.6	118.3	107.7	148.0	180
182	161.7	80.6	83.4	81.7	75.6	123.1	119.7	108.9	149.7	182
184	163.4	81.5	84.4	82.7	76.6	124.3	121.0	110.1	151.4	184
186	165.2	82.5	85.3	83.7	77.6	125.8	122.4	111.3	153.0	186
188	167.0	83.4	86.3	84.6	78.5	127.2	123.7	112.5	154.7	188
190	168.8	84.3	87.2	85.6	79.5	128.5	125.1	113.8	156.4	190
192	170.5	85.3	88.2	86.6	80.5	129.9	126.4	115.0	158.0	192
194	172.3	86.2	89.2	87.6	81.4	131.3	127.8	116.2	159.7	194
196	174.1	87.1	90.1	88.5	82.4	132.7	129.2	117.4	161.4	196
198	175.9	88.1	91.1	89.5	83.4	134.1	130.5	118.6	163.0	198
200	177.7	89.0	92.0	90.5	84.4	135.4	131.9	119.8	164.7	200
202	179.4	89.9	93.0	91.4	85.3	136.8	133.2	121.0	166.4	202
204	181.2	90.9	94.0	92.4	86.3	138.2	134.6	122.3	168.0	204
206	183.0	91.8	94.9	93.4	87.3	139.6	135.9	123.5	169.7	206
208	184.8	92.8	95.9	94.4	88.3	141.0	137.3	124.7	171.4	208
210	186.5	93.7	96.9	95.4	89.2	142.3	138.6	126.0	173.0	210
212	188.3	94.6	97.8	96.3	90.2	143.7	140.0	127.2	174.7	212
214	190.1	95.6	98.8	97.3	91.2	145.1	141.4	128.4	176.4	214
216	191.9	96.5	99.8	98.3	92.2	146.5	142.7	129.6	178.0	216
218	193.6	97.5	100.8	99.3	93.2	147.9	144.1	130.9	179.7	218
220	195.4	98.4	101.7	100.3	94.2	149.3	145.4	132.1	181.4	220
222	197.2	99.4	102.7	101.2	95.1	150.7	146.8	133.3	183.0	222
224	199.0	100.3	103.7	102.2	96.1	152.0	148.1	134.5	184.7	224
226	200.7	101.3	104.6	103.2	97.1	153.4	149.5	135.8	186.4	226
228	202.5	102.2	105.6	104.2	98.1	154.8	150.8	137.0	188.0	228
230	204.3	103.2	106.6	105.2	99.1	156.2	152.2	138.2	189.7	230
232	206.1	104.1	107.6	106.2	100.1	157.6	153.6	139.4	191.3	232
234	207.9	105.1	108.6	107.2	101.1	159.0	154.9	140.7	193.0	234
236	209.6	106.0	109.5	108.2	102.1	160.3	156.3	141.9	194.7	236
238	211.4	107.0	110.5	109.2	103.1	161.7	157.6	143.2	196.3	238
240	213.2	108.0	111.5	110.1	104.0	163.1	159.0	144.4	198.0	240
242	215.0	108.9	112.5	111.1	105.0	164.5	160.3	145.6	199.7	242
244	216.7	109.9	113.5	112.1	106.0	165.9	161.7	146.9	201.3	244
246	218.5	110.8	114.5	113.1	107.0	167.3	163.1	148.1	203.0	246
248	220.3	111.8	115.4	114.1	108.0	168.7	164.4	149.3	204.7	248

TABLE 1. MUNSON AND WALKER SUGAR TABLE IN MILLIGRAMS—*Continued*

CUPROUS OXIDE (Cu_2O)	COPPER (Cu)	DEXTROSE (d-GLUCOSE)	INVERT SUGAR	INVERT SUGAR AND SUCROSE 0.4 gram total sugar	INVERT SUGAR AND SUCROSE 2 grams total sugar	LACTOSE $C_{12}H_{22}O_{11}+H_2O$	LACTOSE AND SUCROSE 1 lactose, 4 sucrose	LACTOSE AND SUCROSE 1 lactose, 12 sucrose	MALTOSE $C_{12}H_{22}O_{11}+H_2O$	CUPROUS OXIDE (Cu_2O)
250	222.1	112.8	116.4	115.1	109.0	170.1	165.8	150.6	206.3	250
252	223.8	113.7	117.4	116.1	110.0	171.5	167.2	151.8	208.0	252
254	225.6	114.7	118.4	117.1	111.0	172.8	168.5	153.1	209.7	254
256	227.4	115.7	119.4	118.1	112.0	174.2	169.9	154.3	211.3	256
258	229.2	116.6	120.4	119.1	113.0	175.6	171.3	155.5	213.0	258
260	231.0	117.6	121.4	120.1	114.0	177.0	172.6	156.8	214.7	260
262	232.7	118.6	122.4	121.1	115.0	178.4	174.0	158.0	216.3	262
264	234.5	119.5	123.4	122.1	116.0	179.8	175.3	159.3	218.0	264
266	236.3	120.5	124.4	123.1	117.0	181.2	176.7	160.5	219.7	266
268	238.1	121.5	125.4	124.1	118.0	182.6	178.1	161.8	221.3	268
270	239.8	122.5	126.4	125.1	119.0	184.0	179.4	163.0	223.0	270
272	241.6	123.4	127.4	126.2	120.0	185.3	180.8	164.3	224.6	272
274	243.4	124.4	128.4	127.2	121.1	186.7	182.2	165.5	226.3	274
276	245.2	125.4	129.4	128.2	122.1	188.1	183.5	166.8	228.0	276
278	246.9	126.4	130.4	129.2	123.1	189.5	184.9	168.0	229.6	278
280	248.7	127.3	131.4	130.2	124.1	190.9	186.3	169.3	231.3	280
282	250.5	128.3	132.4	131.2	125.1	192.3	187.6	170.5	233.0	282
284	252.3	129.3	133.4	132.2	126.1	193.7	189.0	171.8	234.6	284
286	254.0	130.3	134.4	133.2	127.1	195.1	190.4	173.0	236.3	286
288	255.8	131.3	135.4	134.3	128.1	196.5	191.7	174.3	238.0	288
290	257.6	132.3	136.4	135.3	129.2	197.8	193.1	175.5	239.6	290
292	259.4	133.2	137.4	136.3	130.2	199.2	194.4	176.8	241.3	292
294	261.2	134.2	138.4	137.3	131.2	200.6	195.8	178.1	242.9	294
296	262.9	135.2	139.4	138.3	132.2	202.0	197.2	179.3	244.6	296
298	264.7	136.2	140.5	139.4	133.2	203.4	198.6	180.6	246.3	298
300	266.5	137.2	141.5	140.4	134.2	204.8	199.9	181.8	247.9	300
302	268.3	138.2	142.5	141.4	135.3	206.2	201.3	183.1	249.6	302
304	270.0	139.2	143.5	142.4	136.3	207.6	202.7	184.4	251.3	304
306	271.8	140.2	144.5	143.4	137.3	209.0	204.0	185.6	252.9	306
308	273.6	141.2	145.5	144.5	138.3	210.4	205.4	186.9	254.6	308
310	275.4	142.2	146.6	145.5	139.4	211.8	206.8	188.1	256.3	310
312	277.1	143.2	147.6	146.5	140.4	213.2	208.1	189.4	257.9	312
314	278.9	144.2	148.6	147.6	141.4	214.6	209.5	190.7	259.6	314
316	280.7	145.2	149.6	148.6	142.4	216.0	210.9	191.9	261.2	316
318	282.5	146.2	150.7	149.6	143.5	217.3	212.2	193.2	262.9	318
320	284.2	147.2	151.7	150.7	144.5	218.7	213.6	194.4	264.6	320
322	286.0	148.2	152.7	151.7	145.5	220.1	215.5	195.7	266.2	322
324	287.8	149.2	153.7	152.7	146.6	221.5	216.4	197.0	267.9	324
326	289.6	150.2	154.8	153.8	147.6	222.9	217.7	198.2	269.6	326
328	291.4	151.2	155.8	154.8	148.6	224.3	219.1	199.5	271.2	328

Table 1. Munson and Walker Sugar Table in Milligrams—*Continued*

CUPROUS OXIDE (Cu₂O)	COPPER (Cu)	DEXTROSE (d-GLUCOSE)	INVERT SUGAR	INVERT SUGAR AND SUCROSE 0.4 gram total sugar	INVERT SUGAR AND SUCROSE 2 grams total sugar	LACTOSE C₁₂H₂₂O₁₁+H₂O	LACTOSE AND SUCROSE 1 lactose, 4 sucrose	LACTOSE AND SUCROSE 1 lactose, 12 sucrose	MALTOSE C₁₂H₂₂O₁₁+H₂O	CUPROUS OXIDE (Cu₂O)
330	293.1	152.2	156.8	155.8	149.7	225.7	220.5	200.8	272.9	330
332	294.9	153.2	157.9	156.9	150.7	227.1	221.8	202.0	274.6	332
334	296.7	154.2	158.9	157.9	151.7	228.5	223.2	203.3	276.2	334
336	298.5	155.2	159.9	159.0	152.8	229.9	224.6	204.6	277.9	336
338	300.2	156.3	161.0	160.0	153.8	231.3	226.0	205.9	279.5	338
340	302.0	157.3	162.0	161.0	154.8	232.7	227.4	207.1	281.2	340
342	303.8	158.3	163.1	162.1	155.9	234.1	228.7	208.4	282.9	342
344	305.6	159.3	164.1	163.1	156.9	235.5	230.1	209.7	284.5	344
346	307.3	160.3	165.1	164.2	158.0	236.9	231.5	211.0	286.2	346
348	309.1	161.4	166.2	165.2	159.0	238.3	232.9	212.2	287.9	348
350	310.9	162.4	167.2	166.3	160.1	239.7	234.3	213.5	289.5	350
352	312.7	163.4	168.3	167.3	161.1	241.1	235.6	214.8	291.2	352
354	314.4	164.4	169.3	168.4	162.2	242.5	237.0	216.1	292.8	354
356	316.2	165.4	170.4	169.4	163.2	243.9	238.4	217.3	294.5	356
358	318.0	166.5	171.4	170.5	164.3	245.3	239.8	218.6	296.2	358
360	319.8	167.5	172.5	171.5	165.3	246.7	241.2	219.9	297.8	360
362	321.6	168.5	173.5	172.6	166.4	248.1	242.5	221.2	299.5	362
364	323.3	169.6	174.6	173.7	167.4	249.5	243.9	222.5	301.2	364
366	325.1	170.6	175.6	174.7	168.5	250.9	245.3	223.7	302.8	366
368	326.9	171.6	176.7	175.8	169.5	252.3	246.7	225.0	304.5	368
370	328.7	172.7	177.7	176.8	170.6	253.7	248.1	226.3	306.1	370
372	330.4	173.7	178.8	177.9	171.6	255.1	249.5	227.6	307.8	372
374	332.2	174.7	179.8	179.0	172.7	256.5	250.9	228.9	309.5	374
376	334.0	175.8	180.9	180.0	173.7	257.9	252.2	230.2	311.1	376
378	335.8	176.8	182.0	181.1	174.8	259.3	253.6	231.5	312.8	378
380	337.5	177.9	183.0	182.1	175.9	260.7	255.0	232.8	314.5	380
382	339.3	178.9	184.1	183.2	176.9	262.1	256.4	234.1	316.1	382
384	341.1	180.0	185.2	184.3	178.0	263.5	257.8	235.4	317.8	384
386	342.9	181.0	186.2	185.4	179.1	264.9	259.2	236.6	319.4	386
388	344.6	182.0	187.3	186.4	180.1	266.5	260.5	237.9	321.1	388
390	346.4	183.1	188.4	187.5	181.2	267.7	261.9	239.2	322.8	390
392	348.2	184.1	189.4	188.6	182.3	269.1	263.3	240.5	324.4	392
394	350.0	185.2	190.5	189.7	183.3	270.5	264.7	241.8	326.1	394
396	351.8	186.2	191.6	190.7	184.4	271.9	266.1	243.1	327.7	396
398	353.5	187.3	192.7	191.8	185.5	273.3	267.5	244.4	329.4	398
400	355.3	188.4	193.7	192.9	186.5	274.7	268.9	245.7	331.1	400
402	357.1	189.4	194.8	194.0	187.6	276.1	270.3	247.0	332.7	402
404	358.9	190.5	195.9	195.0	188.7	277.5	271.7	248.3	334.4	404
406	360.6	191.5	197.0	196.1	189.8	278.9	273.0	249.6	336.0	406
408	362.4	192.6	198.1	197.2	190.8	280.3	274.4	251.0	337.7	408

TABLE 1. MUNSON AND WALKER SUGAR TABLE IN MILLIGRAMS—*Continued*

CUPROUS OXIDE (Cu₂O)	COPPER (Cu)	DEXTROSE (d-GLUCOSE)	INVERT SUGAR	INVERT SUGAR AND SUCROSE		LACTOSE	LACTOSE AND SUCROSE		MALTOSE	CUPROUS OXIDE (Cu₂O)
				0.4 gram total sugar	2 grams total sugar	C₁₂H₂₂O₁₁+H₂O	1 lactose, 4 sucrose	1 lactose, 12 sucrose	C₁₂H₂₂O₁₁+H₂O	
410	364.2	193.7	199.1	198.3	191.9	281.7	275.8	252.3	339.4	410
412	366.0	194.7	200.2	199.4	193.0	283.2	277.2	253.6	341.0	412
414	367.7	195.8	201.3	200.5	194.1	284.6	278.6	254.9	342.7	414
416	369.5	196.8	202.4	201.6	195.2	286.0	280.0	256.2	344.4	416
418	371.3	197.9	203.5	202.6	196.2	287.4	281.4	257.5	346.0	418
420	373.1	199.0	204.6	203.7	197.3	288.8	282.8	258.8	347.7	420
422	374.8	200.1	205.7	204.8	198.4	290.2	284.2	260.1	349.3	422
424	376.6	201.1	206.7	205.9	199.5	291.6	285.6	261.4	351.0	424
426	378.4	202.2	207.8	207.0	200.6	293.0	287.0	262.7	352.7	426
428	380.2	203.3	208.9	208.1	201.7	294.4	288.4	264.0	354.3	428
430	382.0	204.4	210.0	209.2	202.7	295.8	289.8	265.4	356.0	430
432	383.7	205.5	211.1	210.3	203.8	297.2	291.2	266.6	357.6	432
434	385.5	206.5	212.2	211.4	204.9	298.6	292.6	268.0	359.3	434
436	387.3	207.6	213.3	212.5	206.0	300.0	294.0	269.3	361.0	436
438	389.1	208.7	214.4	213.6	207.1	301.4	295.4	270.6	362.6	438
440	390.8	209.8	215.5	214.7	208.2	302.8	296.8	272.0	364.3	440
442	392.6	210.9	216.6	215.8	209.3	304.2	298.2	273.3	365.9	442
444	394.4	212.0	217.8	216.9	210.4	305.6	299.6	274.6	367.6	444
446	396.2	213.1	218.9	218.0	211.5	307.0	301.0	275.9	369.3	446
448	397.9	214.1	220.0	219.1	212.6	308.4	302.4	277.2	370.9	448
450	399.7	215.2	221.1	220.2	213.7	309.9	303.8	278.6	372.6	450
452	401.5	216.3	222.2	221.4	214.8	311.3	305.2	279.9	374.2	452
454	403.3	217.4	223.3	222.5	215.9	312.7	306.6	281.2	375.9	454
456	405.1	218.5	224.4	223.6	217.0	314.1	308.0	282.5	377.6	456
458	406.8	219.6	225.5	224.7	218.1	315.5	309.4	283.9	379.2	458
460	408.6	220.7	226.7	225.8	219.2	316.9	310.8	285.2	380.9	460
462	410.4	221.8	227.8	226.9	220.3	318.3	312.2	286.5	382.5	462
464	412.2	222.9	228.9	228.1	221.4	319.7	313.6	287.8	384.2	464
466	413.9	224.0	230.0	229.2	222.5	321.1	315.0	289.2	385.9	466
468	415.7	225.1	231.2	230.3	223.7	322.5	316.4	290.5	387.5	468
470	417.5	226.2	232.3	231.4	224.8	323.9	317.7	291.8	389.2	470
472	419.3	227.4	233.4	232.5	225.9	325.3	319.1	293.2	390.8	472
474	421.0	228.3	234.5	233.7	227.0	326.8	320.5	294.5	392.5	474
476	422.8	229.6	235.7	234.8	228.1	328.2	321.9	295.8	394.2	476
478	124.6	230.7	236.8	235.9	229.2	329.6	323.3	297.1	395.8	478
480	126.4	231.8	237.9	237.1	230.3	331.0	324.7	298.5	397.5	480
482	428.1	232.9	239.1	238.2	231.5	332.4	326.1	299.8	399.1	482
484	429.9	234.1	240.2	239.3	232.6	333.8	327.5	301.1	400.8	484
486	431.7	235.2	241.4	240.5	233.7	335.2	328.9	302.5	402.4	486
488	433.5	236.3	242.5	241.6	234.8	336.6	330.3	303.8	404.1	488
490	435.3	237.4	243.6	242.7	236.0	338.0	331.7	305.1	405.8	490

TABLE 2. LANE-EYNON INVERT SUGAR TABLE

(10 ml. of Fehling's solution)

	No sucrose		1 g. sucrose per 100 ml.		5 g. sucrose per 100 ml.	
Ml. of sugar solution required	Invert sugar factor[1]	Mg. invert sugar per 100 ml.	Invert sugar factor[1]	Mg. invert sugar per 100 ml.	Invert sugar factor[1]	Mg. invert sugar per 100 ml.
15	50.5	336	49.9	333	47.6	317
16	50.6	316	50.0	312	47.6	297
17	50.7	298	50.1	295	47.6	280
18	50.8	282	50.1	278	47.6	264
19	50.8	267	50.2	264	47.6	250
20	50.9	254.5	50.2	251.0	47.6	238.0
21	51.0	242.9	50.2	239.0	47.6	226.7
22	51.0	231.8	50.3	228.2	47.6	216.4
23	51.1	222.2	50.3	218.7	47.6	207.0
24	51.2	213.3	50.3	209.8	47.6	198.3
25	51.2	204.8	50.4	201.6	47.6	190.4
26	51.3	197.4	50.4	193.8	47.6	183.1
27	51.4	190.4	50.4	186.7	47.6	176.4
28	51.4	183.7	50.5	180.2	47.7	170.3
29	51.5	177.6	50.5	174.1	47.7	164.5
30	51.5	171.7	50.5	168.3	47.7	159.0
31	51.6	166.3	50.6	163.1	47.7	153.9
32	51.6	161.2	50.6	158.1	47.7	149.1
33	51.7	156.6	50.6	153.3	47.7	144.5
34	51.7	152.2	50.6	148.9	47.7	140.3
35	51.8	147.9	50.7	144.7	47.7	136.3
36	51.8	143.9	50.7	140.7	47.7	132.5
37	51.9	140.2	50.7	137.0	47.7	128.9
38	51.9	136.6	50.7	133.5	47.7	125.5
39	52.0	133.3	50.8	130.2	47.7	122.3
40	52.0	130.1	50.8	127.0	47.7	119.2
41	52.1	127.1	50.8	123.9	47.7	116.3
42	52.1	124.2	50.8	121.0	47.7	113.5
43	52.2	121.4	50.8	118.2	47.7	110.9
44	52.2	118.7	50.9	115.6	47.7	108.4
45	52.3	116.1	50.9	113.1	47.7	106.0
46	52.3	113.7	50.9	110.6	47.7	103.7
47	52.4	111.4	50.9	108.2	47.7	101.5
48	52.4	109.2	50.9	106.0	47.7	99.4
49	52.5	107.1	51.0	104.0	47.7	97.4
50	52.5	105.1	51.0	102.0	47.7	95.4

[1] Lane and Eynon, J. Soc. Chem. Ind. 42, 32T (1923).
[2] Mg. of invert sugar corresponding to 10 ml. of Fehling's solution.

TABLE 2. LANE-EYNON DEXTROSE [1] AND LEVULOSE [2] SUGAR TABLE [3]—*Continued*

(10 ml. of Fehling's solution)

Ml. of sugar solution required	Dextrose [4] factor	Mg. dextrose per 100 ml.	Levulose[5] factor	Mg. levulose per 100 ml.
15	49.1	327	52.2	348
16	49.2	307	52.3	327
17	49.3	289	52.3	308
18	49.3	274	52.4	291
19	49.4	260	52.5	276
20	49.5	247.4	52.5	262.5
21	49.5	235.8	52.6	250.6
22	49.6	225.5	52.7	239.6
23	49.7	216.1	52.7	229.1
24	49.8	207.4	52.8	220.0
25	49.8	199.3	52.8	211.3
26	49.9	191.8	52.9	203.3
27	49.9	184.9	52.9	196.0
28	50.0	178.5	53.0	189.3
29	50.0	172.5	53.1	183.1
30	50.1	167.0	53.2	177.2
31	50.2	161.8	53.2	171.7
32	50.2	156.9	53.3	166.5
33	50.3	152.4	53.3	161.6
34	50.3	148.0	53.4	157.0
35	50.4	143.9	53.4	152.6
36	50.4	140.0	53.5	148.6
37	50.5	136.4	53.5	144.7
38	50.5	132.9	53.6	140.9
39	50.6	129.6	53.6	137.3
40	50.6	126.5	53.6	134.0
41	50.7	123.6	53.7	130.9
42	50.7	120.8	53.7	127.9
43	50.8	118.1	53.8	125.1
44	50.8	115.5	53.8	122.4
45	50.9	113.0	53.9	119.8
46	50.9	110.6	53.9	117.2
47	51.0	108.4	53.9	114.7
48	51.0	106.2	54.0	112.4
49	51.0	104.1	54.0	110.2
50	51.1	102.2	54.0	108.0

[1] All figures relate to anhydrous dextrose.
[2] All figures relate to anhydrous levulose.
[3] Lane and Eynon, *J. Soc. Chem. Ind.* 42, 32T (1923).
[4] Mg. dextrose corresponding to 10 ml. of Fehling's solution.
[5] Mg. levulose corresponding to 10 ml. of Fehling's solution.

TABLE 2. LANE-EYNON LACTOSE SUGAR TABLE [1]—*Continued*

(10 ml. of Fehling's solution)

Ml. of sugar solution required	Hydrated lactose $C_{12}H_{22}O_{11}.H_2O$		Anhydrous lactose $C_{12}H_{22}O_{11}$	
	Factor[2]	Mg. per 100 ml.	Factor[2]	Mg. per 100 ml.
15	68.3	455	64.9	432
16	68.2	426	64.8	405
17	68.2	401	64.8	381
18	68.1	378	64.7	359
19	68.1	358	64.7	340
20	68.0	340.0	64.6	323.0
21	68.0	323.8	64.6	307.6
22	68.0	309.1	64.6	293.6
23	67.9	295.4	64.5	280.6
24	67.9	282.9	64.5	268.8
25	67.9	271.6	64.5	258.0
26	67.9	261.0	64.5	248.0
27	67.8	251.1	64.4	238.5
28	67.8	242.1	64.4	230.0
29	67.8	233.8	64.4	222.2
30	67.8	226.0	64.4	214.7
31	67.8	218.7	64.4	207.8
32	67.8	211.9	64.4	201.3
33	67.8	205.6	64.4	195.3
34	67.9	199.7	64.5	189.7
35	67.9	194.0	64.5	184.3
36	67.9	188.6	64.5	179.2
37	67.9	183.5	64.5	174.3
38	67.9	178.7	64.5	169.8
39	67.9	174.1	64.5	165.4
40	67.9	169.7	64.5	161.2
41	68.0	165.9	64.6	157.6
42	68.0	161.9	64.6	153.8
43	68.0	158.1	64.6	150.2
44	68.0	154.7	64.6	147.0
45	68.1	151.3	64.7	143.7
46	68.1	148.0	64.7	140.6
47	68.2	145.1	64.8	137.8
48	68.2	142.1	64.8	135.0
49	68.2	139.2	64.8	132.2
50	68.3	136.6	64.9	129.8

[1] Lane and Eynon, *J. Soc. Chem. Ind.* 42, 32T (1923).
[2] Mg. lactose corresponding to 10 ml. of Fehling's solution.

TABLE 2. LANE-EYNON MALTOSE SUGAR TABLE [1]—*Continued*

(10 ml. of Fehling's solution)

Ml. of sugar solution required	Hydrated maltose $C_{12}H_{22}O_{11} \cdot H_2O$		Anhydrous maltose $C_{12}H_{22}O_{11}$	
	Factor[2]	Mg. per 100 ml.	Factor[2]	Mg. per 100 ml.
15	81.3	542	77.2	515
16	81.2	507	77.1	482
17	81.1	477	77.0	453
18	81.0	450	77.0	427
19	80.9	426	76.9	405
20	80.8	404.0	76.8	383.8
21	80.7	384.3	76.7	365.1
22	80.6	366.4	76.6	348.1
23	80.5	350.0	76.5	332.5
24	80.4	335.0	76.4	318.3
25	80.4	321.5	76.4	305.4
26	80.3	308.8	76.3	293.4
27	80.2	297.0	76.2	282.2
28	80.1	286.1	76.1	271.8
29	80.0	276.0	76.0	262.2
30	80.0	266.6	76.0	253.3
31	79.9	257.8	75.9	244.9
32	79.9	249.7	75.9	237.2
33	79.8	241.9	75.8	229.8
34	79.8	234.6	75.8	222.9
35	79.7	227.6	75.7	216.2
36	79.6	221.1	75.6	210.0
37	79.6	215.0	75.6	204.3
38	79.5	209.2	75.5	198.7
39	79.5	203.8	75.5	193.6
40	79.4	198.5	75.4	188.6
41	79.4	193.7	75.4	184.3
42	79.3	188.8	75.3	179.4
43	79.3	184.3	75.3	175.1
44	79.2	180.0	75.2	171.0
45	79.2	175.9	75.2	167.1
46	79.1	172.0	75.1	163.4
47	79.1	168.3	75.1	159.9
48	79.1	164.7	75.1	156.5
49	79.0	161.2	75.0	153.1
50	79.0	158.0	75.0	150.1

[1] Lane and Eynon, *J. Soc. Chem. Ind.* 42, 32T (1923).
[2] Mg. of maltose corresponding to 10 ml. of Fehling's solution.

Table 3. Specific Gravity and Refractive Index of Sugar Solutions [a,b]

DEGREES BRIX OR PER CENT BY WEIGHT OF SUCROSE	SPECIFIC GRAVITY AT 20/20°C.	SPECIFIC GRAVITY AT 20/4°C.	DEGREES BAUMÉ (MODULUS 145)	REFRACTIVE INDEX AT 20°	DEGREES BRIX OR PER CENT BY WEIGHT OF SUCROSE	SPECIFIC GRAVITY AT 20/20°C.	SPECIFIC GRAVITY AT 20/4°C.	DEGREES BAUMÉ (MODULUS 145)	REFRACTIVE INDEX AT 20°
0.0	1.00000	0.9982	0.00	1.3330	9.0	1.03586	1.0340	5.02	1.3464
0.2	.00078	0.9990	0.11	.3333	9.2	.03668	.0348	5.13	.3467
0.4	.00155	0.9998	0.22	.3336	9.4	.03750	.0357	5.24	.3470
0.6	.00233	1.0006	0.34	.3338	9.6	.03833	.0365	5.35	.3473
0.8	.00311	.0013	0.45	.3341	9.8	.03915	.0373	5.46	.3476
1.0	.00389	.0021	0.56	.3344	10.0	.03998	.0381	5.57	.3479
1.2	.00467	.0029	0.67	.3347	10.2	.04081	.0390	5.68	.3482
1.4	.00545	.0037	0.79	.3350	10.4	.04164	.0398	5.80	.3485
1.6	.00623	.0045	0.90	.3353	10.6	.04247	.0406	5.91	.3488
1.8	.00701	.0052	1.01	.3356	10.8	.04330	.0415	6.02	.3491
2.0	.00779	.0060	1.12	.3359	11.0	.04413	.0423	6.13	.3494
2.2	.00858	.0068	1.23	.3362	11.2	.04497	.0431	6.24	.3497
2.4	.00936	.0076	1.34	.3365	11.4	.04580	.0440	6.35	.3500
2.6	.01015	.0084	1.46	.3368	11.6	.04664	.0448	6.46	.3504
2.8	.01093	.0091	1.57	.3371	11.8	.04747	.0456	6.57	.3507
3.0	.01172	.0099	1.68	.3373	12.0	.04831	.0465	6.68	.3510
3.2	.01251	.0107	1.79	.3377	12.2	.04915	.0473	6.79	.3513
3.4	.01330	.0115	1.90	.3380	12.4	.04999	.0481	6.90	.3516
3.6	.01409	.0123	2.02	.3382	12.6	.05084	.0490	7.02	.3520
3.8	.01488	.0131	2.13	.3385	12.8	.05168	.0498	7.13	.3523
4.0	.01567	.0139	2.24	.3388	13.0	.05252	.0507	7.24	.3526
4.2	.01647	.0147	2.35	.3391	13.2	.05337	.0515	7.35	.3529
4.4	.01726	.0155	2.46	.3394	13.4	.05422	.0524	7.46	.3532
4.6	.01806	.0163	2.57	.3397	13.6	.05506	.0532	7.57	.3535
4.8	.01886	.0171	2.68	.3400	13.8	.05591	.0540	7.68	.3538
5.0	.01965	.0179	2.79	.3403	14.0	.05677	.0549	7.79	.3541
5.2	.02045	.0187	2.91	.3406	14.2	.05762	.0558	7.90	.3544
5.4	.02125	.0195	3.02	.3409	14.4	.05847	.0566	8.01	.3547
5.6	.02206	.0203	3.13	.3412	14.6	.05933	.0575	8.12	.3551
5.8	.02286	.0211	3.24	.3415	14.8	.06018	.0583	8.23	.3554
6.0	.02366	.0219	3.35	.3418	15.0	.06104	.0592	8.34	.3557
6.2	.02447	.0227	3.46	.3421	15.2	.06190	.0600	8.45	.3560
6.4	.02527	.0235	3.57	.3424	15.4	.06276	.0609	8.56	.3563
6.6	.02608	.0243	3.69	.3427	15.6	.06362	.0617	8.67	.3567
6.8	.02689	.0251	3.80	.3430	15.8	.06448	.0626	8.78	.3570
7.0	.02770	.0259	3.91	.3433	16.0	.06534	.0635	8.89	.3573
7.2	.02851	.0267	4.02	.3436	16.2	.06621	.0643	9.00	.3576
7.4	.02932	.0275	4.13	.3439	16.4	.06707	.0652	9.11	.3580
7.6	.03013	.0283	4.24	.3442	16.6	.06794	.0661	9.22	.3583
7.8	.03095	.0291	4.35	.3445	16.8	.06881	.0669	9.33	.3587
8.0	.03176	.0299	4.46	.3448	17.0	.06968	.0678	9.45	.3590
8.2	.03258	.0308	4.58	.3451	17.2	.07055	.0687	9.56	.3593
8.4	.03340	.0316	4.69	.3454	17.4	.07142	.0695	9.67	.3596
8.6	.03422	.0324	4.80	.3458	17.6	.07229	.0704	9.78	.3600
8.8	.03504	.0332	4.91	.3461	17.8	.07317	.0713	9.89	.3603

[a] *Nat. Bur. Standards, U. S. Circ.* 19 (1924).
[b] Report International Commission Uniform Methods Sugar Analysis, Analyst 62, 97 (1937).

TABLE 3. SPECIFIC GRAVITY AND REFRACTIVE INDEX OF SUGAR SOLUTIONS—Continued

DEGREES BRIX OR PER CENT BY WEIGHT OF SUCROSE	SPECIFIC GRAVITY AT 20/20°C.	SPECIFIC GRAVITY AT 20/4°C.	DEGREES BAUMÉ (MODULUS 145)	REFRACTIVE INDEX AT 20°	DEGREES BRIX OR PER CENT BY WEIGHT OF SUCROSE	SPECIFIC GRAVITY AT 20/20°C.	SPECIFIC GRAVITY AT 20/4°C.	DEGREES BAUMÉ (MODULUS 145)	REFRACTIVE INDEX AT 20°
18.0	1.07404	1.0721	10.00	1.3606	27.0	1.11480	1.1128	14.93	1.3758
18.2	.07492	.0730	10.11	.3609	27.2	.11573	.1138	15.04	.3761
18.4	.07580	.0739	10.22	.3612	27.4	.11667	.1147	15.15	.3765
18.6	.07668	.0748	10.33	.3616	27.6	.11761	.1156	15.26	.3768
18.8	.07756	.0757	10.44	.3619	27.8	.11855	.1166	15.37	.3772
19.0	.07844	.0765	10.55	.3622	28.0	.11949	.1175	15.48	.3775
19.2	.07932	.0774	10.66	.3625	28.2	.12043	.1185	15.59	.3779
19.4	.08021	.0783	10.77	.3629	28.4	.12138	.1194	15.69	.3782
19.6	.08110	.0792	10.88	.3632	28.6	.12232	.1203	15.80	.3886
19.8	.08198	.0801	10.99	.3636	28.8	.12327	.1213	15.91	.3789
20.0	.08287	.0810	11.10	.3639	29.0	.12422	.1222	16.02	.3793
20.2	.08376	.0818	11.21	.3642	29.2	.12517	.1232	16.13	.3797
20.4	.08465	.0827	11.32	.3645	29.4	.12612	.1241	16.24	.3800
20.6	.08554	.0836	11.43	.3649	29.6	.12707	.1251	16.35	.3804
20.8	.08644	.0845	11.54	.3652	29.8	.12802	.1260	16.46	.3807
21.0	.08733	.0854	11.65	.3655	30.0	.12898	.1270	16.57	.3811
21.2	.08823	.0863	11.76	.3658	30.2	.12993	.1279	16.67	.3815
21.4	.08913	.0872	11.87	.3662	30.4	.13089	.1289	16.78	.3818
21.6	.09003	.0881	11.98	.3665	30.6	.13185	.1299	16.89	.3822
21.8	.09093	.0890	12.09	.3669	30.8	.13281	.1308	17.00	.3825
22.0	.09183	.0899	12.20	.3672	31.0	.13378	.1318	17.11	.3829
22.2	.09273	.0908	12.31	.3675	31.2	.13474	.1327	17.22	.3833
22.4	.09364	.0917	12.42	.3679	31.4	.13570	.1337	17.33	.3836
22.6	.09454	.0926	12.52	.3682	31.6	.13667	.1347	17.43	.3840
22.8	.09545	.0935	12.63	.3686	31.8	.13764	.1356	17.54	.3843
23.0	.09636	.0944	12.74	.3689	32.0	.13861	.1366	17.65	.3847
23.2	.09727	.0953	12.85	.3692	32.2	.13958	.1376	17.76	.3851
23.4	.09818	.0962	12.96	.3696	32.4	.14055	.1385	17.87	.3854
23.6	.09909	.0971	13.07	.3699	32.6	.14152	.1395	17.98	.3858
23.8	.10000	.0981	13.18	.3703	32.8	.14250	.1405	18.08	.3861
24.0	.10092	.0990	13.29	.3706	33.0	.14347	.1415	18.19	.3865
24.2	.10183	.0999	13.40	.3709	33.2	.14445	.1424	18.30	.3869
24.4	.10275	.1008	13.51	.3713	33.4	.14543	.1434	18.41	.3872
24.6	.10367	.1017	13.62	.3716	33.6	.14641	.1444	18.52	.3876
24.8	.10459	.1026	13.73	.3720	33.8	.14739	.1454	18.63	.3879
25.0	.10551	.1036	13.84	.3723	34.0	.14837	.1463	18.73	.3883
25.2	.10643	.1045	13.95	.3726	34.2	.14936	.1473	18.84	.3887
25.4	.10736	.1054	14.06	.3730	34.4	.15034	.1483	18.95	.3891
25.6	.10828	.1063	14.17	.3733	34.6	.15133	.1493	19.06	.3894
25.8	.10921	.1072	14.28	.3737	34.8	.15232	.1503	19.17	.3898
26.0	.11014	.1082	14.39	.3740	35.0	.15331	.1513	19.28	.3902
26.2	.11106	.1091	14.49	.3744	35.2	.15430	.1523	19.38	.3906
26.4	.11200	.1100	14.60	.3747	35.4	.15530	.1533	19.49	.3909
26.6	.11293	.1111	14.71	.3751	35.6	.15629	.1542	19.60	.3913
26.8	.11386	.1110	14.82	.3754	35.8	.15729	.1552	19.71	.3916

TABLE 3. SPECIFIC GRAVITY AND REFRACTIVE INDEX OF SUGAR SOLUTIONS
—*Continued*

DEGREES BRIX OR PER CENT BY WEIGHT OF SUCROSE	SPECIFIC GRAVITY AT 20/20°C.	SPECIFIC GRAVITY AT 20/4°C.	DEGREES BAUMÉ (MODULUS 145)	REFRAC-TIVE INDEX AT 20°	DEGREES BRIX OR PER CENT BY WEIGHT OF SUCROSE	SPECIFIC GRAVITY AT 20/20°C.	SPECIFIC GRAVITY AT 20/4°C.	DEGREES BAUMÉ (MODULUS 145)	REFRAC-TIVE INDEX AT 20°
36.0	1.15828	1.1562	19.81	1.3920	45.0	1.20467	1.2025	24.63	1.4096
36.2	.15928	.1572	19.92	.3924	45.2	.20573	.2036	24.74	.4100
36.4	.16028	.1582	20.03	.3928	45.4	.20680	.2047	24.85	.4104
36.6	.16128	.1592	20.14	.3931	45.6	.20787	.2057	24.95	.4109
36.8	.16228	.1602	20.25	.3935	45.8	.20894	.2068	25.06	.4113
37.0	.16329	.1612	20.35	.3939	46.0	.21001	.2079	25.17	.4117
37.2	.16430	.1622	20.46	.3943	46.2	.21108	.2089	25.27	.4121
37.4	.16530	.1632	20.57	.3947	46.4	.21215	.2100	25.38	.4125
37.6	.16631	.1643	20.68	.3950	46.6	.21323	.2111	25.48	.4129
37.8	.16732	.1653	20.78	.3954	46.8	.21431	.2122	25.59	.4133
38.0	.16833	.1663	20.89	.3958	47.0	.21538	.2132	25.70	.4137
38.2	.16934	.1673	21.00	.3962	47.2	.21646	.2143	25.80	.4141
38.4	.17036	.1683	21.11	.3966	47.4	.21755	.2154	25.91	.4145
38.6	.17138	.1693	21.21	.3970	47.6	.21863	.2165	26.01	.4150
38.8	.17239	.1703	21.32	.3974	47.8	.21971	.2176	26.12	.4154
39.0	.17341	.1713	21.43	.3978	48.0	.22080	.2186	26.23	.4158
39.2	.17443	.1724	21.54	.3982	48.2	.22189	.2197	26.33	.4162
39.4	.17545	.1734	21.64	.3986	48.4	.22298	.2208	26.44	.4166
39.6	.17648	.1744	21.75	.3989	48.6	.22406	.2219	26.54	.4171
39.8	.17750	.1754	21.86	.3993	48.8	.22516	.2230	26.65	.4175
40.0	.17853	.1764	21.97	.3997	49.0	.22625	.2241	26.75	.4179
40.2	.17956	.1775	22.07	.4001	49.2	.22735	.2252	26.86	.4183
40.4	.18058	.1785	22.18	.4005	49.4	.22844	.2263	26.96	.4187
40.6	.18162	.1795	22.29	.4008	49.6	.22954	.2274	27.07	.4192
40.8	.18265	.1806	22.39	.4012	49.8	.23064	.2285	27.18	.4196
41.0	.18368	.1816	22.50	.4016	50.0	.23174	.2296	27.28	.4200
41.2	.18472	.1826	22.61	.4020	50.2	.23284	.2307	27.39	.4204
41.4	.18575	.1837	22.72	.4024	50.4	.23395	.2318	27.49	.4208
41.6	.18679	.1847	22.82	.4028	50.6	.23506	.2329	27.60	.4213
41.8	.18783	.1857	22.93	.4032	50.8	.23616	.2340	27.70	.4217
42.0	.18887	.1868	23.04	.4036	51.0	.23727	.2351	27.81	.4221
42.2	.18992	.1878	23.14	.4040	51.2	.23838	.2362	27.91	.4225
42.4	.19096	.1889	23.25	.4044	51.4	.23949	.2373	28.02	.4229
42.6	.19201	.1899	23.36	.4048	51.6	.24060	.2384	28.12	.4234
42.8	.19305	.1909	23.46	.4052	51.8	.24172	.2395	28.23	.4238
43.0	.19410	.1920	23.57	.4056	52.0	.24284	.2406	28.33	.4242
43.2	.19515	.1930	23.68	.4060	52.2	.24395	.2418	28.44	.4246
43.4	.19620	.1941	23.78	.4064	52.4	.24507	.2429	28.54	.4251
43.6	.19726	.1951	23.89	.4068	52.6	.24619	.2440	28.65	.4255
43.8	.19831	.1962	24.00	.4072	52.8	.24731	.2451	28.75	.4260
44.0	.19936	.1972	24.10	.4076	53.0	.24844	.2462	28.86	.4264
44.2	.20042	.1983	24.21	.4080	53.2	.24956	.2474	28.96	.4268
44.4	.20148	.1994	24.32	.4084	53.4	.25069	.2485	29.06	.4272
44.6	.20254	.2004	24.42	.4088	53.6	.25182	.2496	29.17	.4277
44.8	.20360	.2015	24.53	.4092	53.8	.25295	.2507	29.27	.4281

TABLE 3. SPECIFIC GRAVITY AND REFRACTIVE INDEX OF SUGAR SOLUTIONS
—Continued

DEGREES BRIX OR PER CENT BY WEIGHT OF SUCROSE	SPECIFIC GRAVITY AT 20/20°C.	SPECIFIC GRAVITY AT 20/4°C.	DEGREES BAUMÉ (MODULUS 145)	REFRACTIVE INDEX AT 20°	DEGREES BRIX OR PER CENT BY WEIGHT OF SUCROSE	SPECIFIC GRAVITY AT 20/20°C.	SPECIFIC GRAVITY AT 20/4°C.	DEGREES BAUMÉ (MODULUS 145)	REFRACTIVE INDEX AT 20°
54.0	1.25408	1.2519	29.38	1.4285	63.0	1.30657	1.3043	34.02	1.4486
54.2	.25521	.2530	29.48	.4289	63.2	.30778	.3055	34.12	.4491
54.4	.25635	.2541	29.59	.4294	63.4	.30898	.3067	34.23	.4495
54.6	.25748	.2553	29.69	.4298	63.6	.31019	.3079	34.33	.4500
54.8	.25862	.2564	29.80	.4303	63.8	.31139	.3091	34.43	.4504
55.0	.25976	.2575	29.90	.4307	64.0	.31260	.3103	34.53	.4509
55.2	.26090	.2587	30.00	.4311	64.2	.31381	.3115	34.63	.4514
55.4	.26204	.2598	30.11	.4316	64.4	.31502	.3127	34.74	.4518
55.6	.26319	.2610	30.21	.4320	64.6	.31623	.3139	34.84	.4523
55.8	.26433	.2621	30.32	.4325	64.8	.31745	.3151	34.94	.4527
56.0	.26548	.2632	30.42	.4329	65.0	.31866	.3163	35.04	.4532
56.2	.26663	.2644	30.52	.4333	65.2	.31988	.3175	35.14	.4537
56.4	.26778	.2655	30.63	.4338	65.4	.32110	.3188	35.24	.4541
56.6	.26893	.2667	30.73	.4342	65.6	.32232	.3200	35.34	.4546
56.8	.27008	.2678	30.83	.4347	65.8	.32354	.3212	35.45	.4550
57.0	.27123	.2690	30.94	.4351	66.0	.32476	.3224	35.55	.4555
57.2	.27239	.2701	31.04	.4355	66.2	.32599	.3236	35.65	.4560
57.4	.27355	.2713	31.15	.4360	66.4	.32722	.3249	35.75	.4565
57.6	.27471	.2725	31.25	.4364	66.6	.32844	.3261	35.85	.4569
57.8	.27587	.2736	31.35	.4369	66.8	.32967	.3273	35.95	.4574
58.0	.27703	.2748	31.46	.4373	67.0	.33090	.3286	36.05	.4579
58.2	.27819	.2759	31.56	.4378	67.2	.33214	.3298	36.15	.4584
58.4	.27936	.2771	31.66	.4382	67.4	.33337	.3310	36.25	.4589
58.6	.28052	.2783	31.76	.4387	67.6	.33460	.3322	36.35	.4593
58.8	.28169	.2794	31:87	.4391	67.8	.33584	.3335	36.45	.4598
59.0	.28286	.2806	31.97	.4396	68.0	.33708	.3347	36.55	.4603
59.2	.28404	.2818	32.07	.4400	68.2	.33832	.3360	36.66	.4607
59.4	.28520	.2829	32.18	.4405	68.4	.33957	.3372	36.76	.4612
59.6	.28638	.2841	32.28	.4409	68.6	.34081	.3384	36.86	.4617
59.8	.28755	.2853	32.38	.4414	68.8	.34205	.3397	36.96	.4622
60.0	.28873	.2865	32.49	.4418	69.0	.34330	.3409	37.06	.4627
60.2	.28991	.2876	32.59	.4423	69.2	.34455	.3422	37.16	.4631
60.4	.29109	.2888	32.69	.4427	69.4	.34580	.3434	37.26	.4636
60.6	.29227	.2900	32.79	.4432	69.6	.34705	.3447	37.36	.4641
60.8	.29346	.2912	32.90	.4436	69.8	.34830	.3459	37.46	.4646
61.0	.29464	.2924	33.00	.4441	70.0	.34956	.3472	37.56	.4651
61.2	.29583	.2935	33.10	.4446	70.2	.35081	.3484	37.66	.4656
61.4	.29701	.2947	33.20	.4450	70.4	.35207	.3497	37.76	.4661
61.6	.29820	.2959	33.31	.4455	70.6	.35333	.3509	37.86	.4666
61.8	.29940	.2971	33.41	.4459	70.8	.35459	.3522	37.96	.4671
62.0	.30059	.2983	33.51	.4464	71.0	.35585	.3535	38.06	.4676
62.2	.30178	.2995	33.61	.4468	71.2	.35711	.3547	38.16	.4681
62.4	.30298	.3007	33.72	.4473	71.4	.35838	.3560	38.26	.4685
62.6	.30418	.3019	33.82	.4477	71.6	.35964	.3572	38.35	.4690
62.8	.30537	.3031	33.92	.4482	71.8	.36091	.3585	38.45	.4695

TABLE 3. SPECIFIC GRAVITY AND REFRACTIVE INDEX OF SUGAR SOLUTIONS
—*Continued*

DEGREES BRIX OR PER CENT BY WEIGHT OF SUCROSE	SPECIFIC GRAVITY AT 20/20°C.	SPECIFIC GRAVITY AT 20/4°C.	DEGREES BAUMÉ (MODULUS 145)	REFRACTIVE INDEX AT 20°	DEGREES BRIX OR PER CENT BY WEIGHT OF SUCROSE	SPECIFIC GRAVITY AT 20/20°C.	SPECIFIC GRAVITY AT 20/4°C.	DEGREES BAUMÉ (MODULUS 145)	REFRACTIVE INDEX AT 20°
72.0	1.36218	1.3598	38.55	1.4700	81.0	1.42088	1.4184	42.95	1.4927
72.2	.36346	.3610	38.65	.4705	81.2	.42222	.4197	43.05	.4933
72.4	.36473	.3623	38.75	.4710	81.4	.42356	.4210	43.14	.4938
72.6	.36600	.3636	38.85	.4715	81.6	.42490	.4224	43 24	.4943
72.8	.36728	.3649	38.95	.4720	81.8	.42625	.4237	43.33	.4949
73.0	.36856	.3661	39.05	.4725	82.0	.42759	.4251	43.43	.4954
73.2	.36983	.3674	39.15	.4730	82.2	.42894	.4264	43.53	.4959
73.4	.37111	.3687	39.25	.4735	82.4	43029	.4278	43.62	.4964
73.6	.37240	.3700	39.35	.4740	82.6	.43164	.4291	43.72	.4970
73.8	.37368	.3713	39.44	.4744	82.8	.43298	.4305	43.81	.4975
74.0	.37496	.3725	39.54	.4749	83.0	.43434	.4318	43.91	.4980
74.2	.37625	.3738	39.64	.4754	83.2	.43569	.4332	44.00	.4985
74.4	.37754	.3751	39.74	.4759	83.4	.43705	.4345	44.10	.4991
74.6	.37883	.3764	39.84	.4764	83.6	.43841	4359	44.19	.4996
74.8	.38012	.3777	39.94	.4769	83.8	.43976	.4372	44.29	.5001
75.0	.38141	.3790	40.03	.4774	84.0	.44112	.4386	44.38	.5007
75.2	.38270	.3803	40.13	.4779	84.2	.44249	.4399	44.48	.5012
75.4	.38400	.3816	40.23	.4784	84.4	.44385	.4413	44.57	.5017
75.6	.38530	.3829	40.33	.4789	84.6	.44521	.4427	44.67	.5022
75.8	.38660	.3841	40.43	.4794	84.8	.44658	.4440	44.76	.5028
76.0	.38790	.3854	40.53	.4799	85.0	.44794	.4454	44.86	.5033
76.2	.38920	.3867	40.62	.4804	85.2	.44931	4468	44.95	
76.4	.39050	.3880	40.72	.4810	85.4	.45068	.4481	45.05	
76.6	.39180	.3893	40.82	.4815	85.6	.45205	.4495	45.14	
76.8	.39311	.3907	40.92	.4820	85.8	.45343	.4509	45.24	
77.0	.39442	.3920	41.01	.4825	86.0	.45480	4522	45.33	
77.2	.39573	.3933	41.11	.4830	86.2	.45618	4536	45.42	
77.4	.39704	.3946	41.21	.4835	86.4	.45755	4550	45.52	
77.6	.39835	.3959	41.31	.4840	86.6	.45893	4564	45.61	
77.8	.39966	.3972	41.40	.4845	86.8	.46031	4577	45.71	
78.0	.40098	.3985	41.50	.4850	87.0	.46170	.4591	45.80	
78.2	.40230	.3998	41.60	.4855	87.2	.46308	4605	45.89	
78.4	.40361	.4011	41.70	.4860	87.4	.46446	4619	45.99	
78.6	.40493	.4025	41.79	.4865	87.6	46585	4633	46.08	
78.8	.40625	.4038	41 89	.4871	87 8	.46724	4646	46.17	
79.0	.40758	.4051	41.99	.4876	88.0	.46862	4660	46.27	
79.2	.40890	.4064	42.08	.4881	88.2	.47002	4674	46.36	
79.4	.41023	4077	42.18	.4886	88.4	.47141	4688	46.45	
79.6	.41155	.4091	42.28	.4891	88.6	47280	4702	46.55	
79.8	.41288	.4104	42.37	.4896	88.8	47420	4716	46.64	
80.0	.41421	.4117	42.47	.4901	89.0	.47559	4730	46 73	
80.2	.41554	.4130	42.57	.4906	89.2	.47699	4744	46.83	
80.4	.41688	.4144	42.66	.4912	89.4	.47839	.4758	46.92	
80.6	.41821	.4157	42.76	.4917	89.6	.47979	.4772	47.01	
80.8	.41955	.4170	42.85	.4922	89.8	.48119	.4786	47.11	

TABLE 3. SPECIFIC GRAVITY OF SUGAR SOLUTIONS—*Continued*

DEGREES BRIX OR PER CENT BY WEIGHT OF SUCROSE	SPECIFIC GRAVITY AT 20/20°C.	SPECIFIC GRAVITY AT 20/4°C.	DEGREES BAUMÉ (MODULUS 145)	DEGREES BRIX OR PER CENT BY WEIGHT OF SUCROSE	SPECIFIC GRAVITY AT 20/20°C.	SPECIFIC GRAVITY AT 20/4°C.	DEGREES BAUMÉ (MODULUS 145)
90.0	1.48259	1.4800	47.20	95.0	1.51814	1.5155	49.49
90.2	.48400	.4814	47.29	95.2	.51958	.5169	49.58
90.4	.48540	.4828	47.38	95.4	.52102	.5183	49.67
90.6	.48681	.4842	47.48	95.6	.52246	.5198	49.76
90.8	.48822	.4856	47.57	95.8	.52390	.5212	49.85
91.0	.48963	.4870	47.66	96.0	.52535	.5227	49.94
91.2	.49104	.4884	47.75	96.2	.52680	.5241	50.03
91.4	.49246	.4898	47.84	96.4	.52824	.5255	50.12
91.6	.49387	.4912	47.94	96.6	.52969	.5270	50.21
91.8	.49529	.4926	48.03	96.8	.53114	.5284	50.30
92.0	.49671	.4941	48.12	97.0	.53260	.5299	50.39
92.2	.49812	.4955	48.21	97.2	.53405	.5313	50.48
92.4	.49954	.4969	48.30	97.4	.53551	.5328	50.57
92.6	.50097	.4983	48.40	97.6	.53696	.5342	50.66
92.8	.50239	.4997	48.49	97.8	.53842	.5357	50 75
93.0	.50381	.5012	48.58	98.0	.53988	.5372	50.84
93.2	.50524	.5026	48.67	98.2	.54134	.5386	50.93
93.4	.50667	.5040	48.76	98.4	.54280	.5401	51.02
93.6	.50810	.5054	48.85	98.6	.54426	.5415	51.10
93.8	.50952	.5069	48.94	98.8	.54573	.5430	51.19
94.0	.51096	.5083	49.03	99.0	.54719	.5445	51.28
94.2	.51239	.5097	49.12	99.2	.54866	.5459	51.37
94.4	.51382	.5112	49.22	99.4	.55013	.5474	51.46
94.6	.51526	.5126	49.31	99.6	.55160	.5489	51.55
94.8	.51670	.5140	49.40	99.8	.55307	.5503	51.64
				100.0	.55454	.5518	51.73

SUBJECT INDEX

356

Vitamin B₁ (cont.)
  fermentation method, 711
  thiochrome method, 706-711
  B₂, 694, 695, 713-715
  B₆, see pyridoxine
  B₁₂, 724, 780
  C, see ascorbic acid, 695
  D, see vitamins D
    milk, 293, 294
  D₂, 734, 735, 737
  D₃, 734, 735, 737
  H, see biotin
  M, see folic acid
  P, 694
Vitamins, 693-742
  classification, 693-694
  dietary allowances, 695
  fat soluble, 697, 698
  fortification of foods, 705
  D, 244, 694, 695, 734-737
  E, 694, 738-741
  K, 244, 694, 696, 741-742
Vizern and Guillot method, 314-315
Volatile
  acids, 170-171, 382, 829-833
  matter, 3
  oil, 590-595
    boiling point, 544
    extract, 586
Volhard method, 758

Wagner's reagent, 253
Walnuts, 795
Warfarin, 906-907
Wash water, causticity, 850
Water, 694
  capacity, 12
  dietary allowance, 696
  mineral, 1
  poisons, in, 255-260
Waxes, 47
Weight pipette, 269, 270
Westphal balance, 15
Weyl's lutein test, 688
Whale oil, 367, 368, 369
Wheat, 23, 29, 498
Whey, 339-340
Wichmann-Clifford method, 214-217

Wichmann method
  coumarin, 610, 611
  lead number, 613
Wijs iodine value, 375
Wildman trap flask, 804, 806
Willard-Greathouse method, 776-777
Williams method, 747-748
Wintergreen, 651
Winton lead number, 467
  method, 613
Wood smoke, 150
Worm cuts, 797
Wormy, 787, 788
Wurst, see sausage

Xanthophyll, 138
Xerophthalmia, 699
Xylenol method, 762-763
Xylose, 419, 426, 482

Yeasts, 708
Yellow
  AB, 105, 106, 108, 118, 129, 132, 134,
    136, 137, 146
    separation from yellow OB, 133, 134
  OB, 105, 106, 108, 119, 129, 132, 134,
    136, 137, 146
    separation from yellow AB, 133, 134
Yogurt, 340
Yolks, 683
  dried, 683
  frozen, 683
  glycerol, 688-689
  liquid, 683
  vegetable lecithin in, 687-688
Youngberry, 510, 513, 520, 558
Yttrium-90, 878, 883, 885

Zephiran, 244
Zerban and Sattler method, 447-458
Zinc, 182, 184, 200, 226-232, 234, 237, 241,
    246, 252, 258
  chloride, 238
  ferrocyanide method, 227
  nephelometric method, 288
  phosphide, 245, 246, 249, 250
  salts, 244
  turbidometric method, 229
  uranyl acetate method, 767
Zirconium alizarine test, 745